Exercise Endocrinology

Exercise Endocrinology

Katarina T. Borer, PhD

University of Michigan at Ann Arbor

Human Kinetics

Library of Congress Cataloging-in-Publication Data

Borer, Katarina T.
    Exercise endocrinology / Katarina Borer
         p. ; cm.
Includes bibliographical references and index.
   ISBN 0-88011-566-1 (hard cover)
  1. Exercise--Physiological aspects. 2. Hormones--Physiological
effect. 3. Endocrinology.
   [DNLM: 1. Endocrine System--physiology. 2. Exertion--physiology. WK
102 B731e 2003] I. Title.
  QP187.3.E93B67 2003
  612'.044--dc21

                      2003000180

ISBN: 0-88011-566-1

Acquisitions Editor: Michael S. Bahrke, PhD
Developmental Editor: Judy Park
Assistant Editors: Amanda S. Ewing and Lee Alexander
Copyeditor: Ozzievelt Owens
Proofreader: Erin Cler
Permission Manager: Dalene Reeder
Graphic Designer: Andrew Tietz
Graphic Artist: Dawn Sills
Cover Designer: Jack W. Davis
Art Manager: Kelly Hendren
Illustrator: Mic Greenberg
Printer: Sheridan Books

Printed in the United States of America   10 9 8 7 6 5 4 3 2 1

Human Kinetics
Web site: www.HumanKinetics.com

United States: Human Kinetics
P.O. Box 5076
Champaign, IL 61825-5076
800-747-4457
e-mail: humank@hkusa.com

Canada: Human Kinetics
475 Devonshire Road Unit 100
Windsor, ON N8Y 2L5
800-465-7301 (in Canada only)
e-mail: orders@hkcanada.com

Europe: Human Kinetics
107 Bradford Road
Stanningley
Leeds LS28 6AT, United Kingdom
+44 (0) 113 255 5665
e-mail: hk@hkeurope.com

Australia: Human Kinetics
57A Price Avenue
Lower Mitcham, South Australia 5062
08 8277 1555
e-mail: liahka@senet.com.au

New Zealand: Human Kinetics
P.O. Box 105-231, Auckland Central
09-523-3462
e-mail: hkp@ihug.co.nz

Contents

Preface

ABOUT THIS BOOK

This book examines the ways hormones and messengers of the autonomic nervous system affect human biology before, during, and after exercise. It describes the way chemical messengers constantly regulate the body's internal environment, including responses to stress of acute exercise, and facilitate long-term functional and structural adjustments as exercise training programs create mechanical strains and bioenergetic drain. Discussion of these issues is organized by function rather than by a focus on individual messenger systems. The first three chapters deal with general concepts regarding chemical mediators in exercise, their interaction with the autonomic nervous system, and the mechanisms of hormone action. The remaining chapters address specific functional involvement of chemical messengers that affect exercise or are derived from it. These processes are hormonal regulation of temperature and fluid balance in relation to physical exertion (chapter 4); hormonal role in meeting the urgent body needs during acute physical activity (chapter 5); regulation of fuel use, including dietary modulation of chemical messengers in exercise (chapter 6); partitioning of nutrient energy between support of body structure and fuel storage (chapter 7); exercise and reproductive endocrine function, together with other gender-based and age-based effects (chapter 8); and exercise, chemical messengers, and the biological rhythms (chapter 9).

THE ROLE OF CHEMICAL MESSENGERS IN EXERCISE

Exercise affects five major categories of biological function: (1) stress responses, (2) availability and utilization of metabolic energy, (3) maintenance of homeostasis or the constancy of internal environment, (4) growth and maintenance of skeletal and cardiac muscle and other components of the lean body mass, and (5) reproduction. The effects of exercise are at once immediate and chronically pervasive. Acute adjustments in function to meet the challenges of physical work, homeostatic compensations during and after an exercise bout, as well as longer-lasting structural and functional adaptations, including changes in fertility and growth, are all largely controlled by cooperative action of hormones and the chemical messengers of the autonomic nervous system. These chemical messengers coordinate responses of in-dividual organs into an integrated whole-body response, restore the homeostatic balance in the internal environment wherever it is disturbed by exercise, influence the turnover and growth of structural body components, and alter reproductive function in response to some types of chronic exercise.

Several types of signaling molecules control physiological adjustments during exercise, and their designations differ according to the characteristics and organization of cells producing them and the method of their release. Until recently, endocrinology was confined to the study of hormones, signaling molecules produced by cells clustered within specialized endocrine glands that reached their targets through the blood. More recently, it became necessary to broaden the definition of endocrinology to also include actions of chemical messengers produced by nerve cells and cells dispersed within organs with multiple functions and to routes of dispersal other than the circulation. This change is discussed in more detail in chapter 1.

Of particular significance in exercise are chemical messengers produced by the sympathoadrenal system because of their alerting and coordinating role and their control over the secretion of hormones (described more fully in chapter 2). Its components are the sympathetic nerves and chromaffin cells in the adrenal medulla, and these components in turn represent a division of the autonomic nervous system. Sympathetic nerves innervate smooth muscles and endocrine and exocrine glands, release norepinephrine (NE) into the synaptic clefts on target tissues, and trigger the release of epinephrine (E) and NE from the adrenal medulla into systemic circulation. Parasympathetic nerves belong to another division of the autonomic nervous system and release acetylcholine (ACh) into the synaptic clefts on target tissues. Chemical messengers released from the parasympathetic nerve endings and hormones stimulated by these nerves contribute to restorative, biosynthetic, and anabolic processes after acute exercise or prolonged training.

Chemical messengers, whether autonomic or endocrine, exert their biological effect after binding to receptors located on cell surfaces or inside the cells. A variety of factors affect hormone-receptor binding and transduction of this event at the cellular level. Biological action depends on the structural and quantitative characteristics of messenger molecules and temporal characteristics of their release. Alternate molecular species of a given hormone may exist under normal circumstances and be released in response to an environmental challenge such as specific exercise or may result from a genetic accident. Concentration of the messenger molecules affects the number of binding sites and their affinity for the receptor. The temporal pattern of hormone release likewise can affect receptor number, binding affinity, and the nature of the biological response to the messenger. Hormone-receptor transduction entails changes in ion traffic across the cell membrane, in enzyme activity inside the cell, and in transcriptional and translational events. These issues are discussed in chapter 3.

Understanding the role of chemical messengers in exercise is complicated by circumstances that can alter messenger secretion and action. Besides the already mentioned influences on hormone-receptor interactions (chapter 3), more than one hormone may be simultaneously elicited by exercise, and their actions may be complementary, synergistic, or antagonistic. Action of endogenous signaling molecules also can be altered by varying the quantity and quality of ingested nutrients and the timing between food ingestion and exercise. Nutritional manipulations of this kind are often used by athletes to enhance physical performance and are discussed in more detail in chapter 5.

Some chemicals act both as messengers and as nutrients. For instance 1,25-dihydroxycholecalciferol is importantly involved in the maintenance of bone mineral (chapter 7) and can be manufactured in our bodies and released into circulation to act on vitamin D_3 receptors on target cells. Obtained in this way, vitamin D_3 meets the definition of a hormone. However, the same substance can also be obtained from animal tissues and in this context is more correctly classified as a nutrient. In addition, a variety of chemicals produced either by animals, plants, or humans can mimic the action of endogenous messengers. Thus, plant-derived ephedrine and a large number of synthetic compounds activate catecholamine and other receptors and are used either to treat specific health problems or to enhance physical performance (chapter 3).

CHEMICAL MESSENGERS IN ACUTE EXERCISE

Exercise can induce significant deviations in a number of homeostatic functions, including regulation of blood sugar, body temperature, body fluids, and pH. The by-product of anaerobic glycolysis is a rise in plasma concentration of lactic acid (lactate threshold). This rise triggers hyperventilation (ventilatory threshold), a homeostatic compensatory response under autonomic control that maintains blood pH within the normal range. Conversion of metabolic fuels to muscle contraction is accompanied by production of heat that accumulates within the body during exercise. Sweating is the chief method by which this heat load is dissipated. The resulting disturbance of water and plasma mineral

homeostasis triggers autonomic and endocrine compensatory mechanisms as well as thirst. These reactions are discussed in chapter 4.

As emphasized in chapter 5, few circumstances challenge us so powerfully, rapidly, and comprehensively as acute physical activity. Strenuous exercise demands urgent adjustments in mental alertness, perception of pain, mobilization of internal defense mechanisms, delivery of oxygen and metabolic fuels to muscle, and disposal of metabolic wastes. Because physical activities may engage different muscle groups at different intensities for variable lengths of time, circulatory and metabolic functions are adjusted, including the destination and delivery rates of oxygen and metabolic fuels. To meet increased muscle need for oxygen, cardiac output can increase fivefold in minutes, from about 5 to 6 L\min of blood to about 25 to 30 L\min during heavy exercise. Distribution of blood can change more than 20-fold during exercise. Muscles are perfused by 20% of cardiac output, or about 1 L of blood per minute at rest, and can receive up to 90% of cardiac output, or about 22 to 27 L of blood per minute, during peak effort. Autonomic nerves help match increases in cardiorespiratory function to the magnitude of oxygen and fuel need. In anticipation of increased respiratory and cardiovascular demands, increased sympathetic tone produces an anticipatory acceleration of these functions (chapter 2).

In addition to their influence over cardiorespiratory function, sympathetic nerves and adrenal medullary hormones control the rate of fuel mobilization and utilization during acute exercise, in part through direct actions on liver, adipose tissue, and muscle and in part by affecting the secretion of metabolic hormones (chapter 6). With increases in exercise intensity, fuel metabolism shifts from aerobic oxidation of all available fuels (carbohydrates, lipids, and amino acids) to anaerobic extraction of free energy from muscle glycogen or circulating glucose. The shift in the type of fuel used is easily tracked by monitoring the ratio of carbon dioxide produced to oxygen consumed (respiratory quotient, or RQ), which ranges from 0.7 during oxidation of lipids to 1 during anaerobic metabolism (glycolysis) of carbohydrates.

Endocrine and autonomic systems cooperatively control fuel metabolism during exercise. The autonomic nerves suppress secretion of insulin during exercise to ensure increased production and release of glucose from the liver and of fatty acids from the adipose tissue. They also trigger the release of E from the adrenal medulla and the release of pancreatic glucagon, and they contribute to secretion of pituitary growth hormone. All of these hormones facilitate mobilization and utilization of metabolic fuels. Exercise can produce large increases in energy use. Energy expenditure of a sprinter at peak effort is about 3,000 watts, which is about 30 times greater than the energy expenditure of a resting or sleeping person. Muscle adenosine triphosphate (ATP), the free-energy product of metabolism, and its backup metabolite, creatine phosphate (CP), fuel such high-powered activities.

Actions of the autonomic and hormonal systems during acute exercise are interconnected and coordinated. The sympathoadrenal system controls the secretion of a number of hormones during exercise and is influenced, in turn, by several hormones. Sympathoadrenal actions are integrated with those of the hypothalamo-pituitary-adrenal axis and serve to produce a unified endocrine stress response to meet exercise demands (described more fully in chapter 5). CRF, the hypothalamic corticotropin-releasing hormone (or factor) that is released in response to the stress of physical activity, is a stimulant of sympathetic nerve activity. CRF triggers the release of stress hormones ACTH (adrenocorticotropic hormone, or corticotropin), β endorphin, enkephalins, and cortisol. Cortisol in turn stimulates secretion of adrenal E and also suppresses the sympathetic neural outflow. Thus, the coordinated and interconnected sympathoadrenal and endocrine release and action during the stress of exercise ensure the adequacy of oxygen and metabolic fuel supplies through the action of E, NE, and cortisol and provide additional analgesic actions of opiates and immune-defense actions of cortisol.

In addition to being influenced by chemical messengers, muscle metabolism during exercise is mediated by endogenous mechanisms. Sequestration of circulating fuels and mobilization and use of fuels stored in the muscle can proceed in the absence of hormonal stimulation. As the chief user of circulating fuels, muscle metabolism, stimulated by its contractions, maintains insulin sensitivity and can reverse metabolic morbidities associated with obesity and sedentary lifestyle.

LONG-TERM ADAPTATIONS TO HABITUAL EXERCISE

During prolonged intense exercise, muscle and liver glycogen become depleted, lipids are mobilized out of muscle and fat tissue, and some muscle protein is degraded. This process triggers release of anabolic hormones and acute compensatory as well as slower hypertrophic responses. Exercise leads to repartitioning of nutrient energy between body energy stores and body structure and influences the size of energy stores, bone shape and density, body composition, and growth processes (chapter 7). Exercise training reinforces and amplifies these changes and produces structural changes in muscles, bone, and other lean tissues and produces changes in fuel stores to help the organism better meet the challenges of exercise. On the other hand, a sedentary lifestyle or detraining produces structural and functional changes opposite to those seen in response to physical conditioning. Fitness level can affect the quantity and pattern of chemical messenger release, the number of receptors, and the effectiveness of messenger action. Many age-associated changes in endocrine secretion and actions are a consequence of reduced fitness that results from a sedentary lifestyle and can be corrected with physical training.

Finally, some types of habitual exercise can affect fertility and reproductive function in women by influencing the secretion of pituitary, adrenal, and gonadal hormones (chapter 8). Developmental progression and sexual differentiation create differences in structure, function, and physical performance. Therefore, gender, age, and stage of development exert important modifying influences on the type, quantity, and intensity of exercise that can be performed as well as on chemical messenger secretion and action.

Secretion and action of chemical messengers have a temporal structure that can influence both physical performance and the capacity of exercise to elicit hormone secretion. Physical performance is disturbed by transmeridian flight and reverse shift work in large part because of the disturbance in the timing of the secretion and action of chemical messengers and in the timing of some other rhythmic functions. Disturbances in endogenous biological rhythms are associated with a variety of health problems. Through its ability to affect the timing of hormone secretion, exercise can both aggravate and alleviate these problems. These issues are discussed in chapter 9.

This book assumes that the reader will have some familiarity with fields of exercise physiology, physiology, and endocrinology. When necessary, more detailed information should be sought in exercise physiology (McArdle, Katch, & Katch 2001; Powers & Howley 2001), physiology (Guyton 1991; Vander et al. 2001), endocrinology (Baulieu & Kelly 1990; Norman & Litwack 1997; Wilson et al. 1998), endocrine physiology (Griffin & Ojeda 1996), and other exercise endocrinology (Fotherby & Pal 1985; Galbo 1983; Laron & Rogol 1989; Viru 1985; Warren & Constantini 2000) resources.

Acknowledgments

The inspiration for writing this book was my paternal great-grandfather, Julije Domac (MS, 1874; PhD, 1880; University of Graz) who wrote a useful and influential book. He wrote an introduction to pharmacognosy and a section on pharmacology (in Croatian and Latin) in the second edition of the book "Hrvatsko-slavonska farmakopeja." This book, published with Gustav Janeček in 1899, paved the way for the development of the science of pharmacology in the state of Croatia. The practical value of this work was acknowledged for the next half century by featuring the busts of the two authors in all Croatian pharmacies. They still stand in the main pharmacy on Zrinjevac Square in Zagreb. In 1997, the state of Croatia issued a stamp with

Dr. Domac's image commemorating the influence of his book under the caption "Zavod za Farmakognoziju Sveučilišta u Zagrebu" (Institute of Pharmacognosy of the University of Zagreb), the first institute of its kind in the Balkans, which he also founded in 1896.

I wish to thank several individuals who helped bring this book to completion. Drs. Germaine Cornelissen, Franz Halberg, Jeffrey Horowitz, Sarah W. Newman, Steven Segal, Arthur J. Vander, and James B. Young provided useful comments on individual chapters. My husband, Dr. Paul E. Wenger helped make the text more readable and lent support. Human Kinetics publisher, Dr. Rainer Martens, maintained interest in the book and patience with its slow completion; and editors, Ms. Judy Park, Dr. Michael Bahrke, and Ms. Amanda Ewing, and graphic artist Ms. Dawn Sills shaped my manuscript into a finished product.

Credits

Figures

Figure 3.9–From A.J. Vander, J.H. Sherman, and D.S.Luciano, 1990, *Human physiology: The mechanisms of body function*, 5th ed. (New York: McGraw-Hill), 365, 376, 378. Adapted, by permission from The McGraw-Hill Companies.

Figure 3.10–From TEXTBOOK OF ENDOCRINE PHYSIOLOGY, THIRD EDITION, edited by J.E. Griffin and S.R. Ojeda, copyright © 1996 by Oxford University Press, Inc. Used by permission of Oxford University Press, Inc.

Figure 3.11–From TEXTBOOK OF ENDOCRINE PHYSIOLOGY, THIRD EDITION, edited by J.E. Griffin and S.R. Ojeda, copyright © 1996 by Oxford University Press, Inc. Used by permission of Oxford University Press, Inc.

Figure 3.12–Adapted from B. Richelsen, 1997, Action of growth hormone in adipose tissue, *Hormone Research* 48(suppl. 5): 105-110. By permission of S. Karger AG Basel.

Figure 3.13–Adapted, by permission, from E.E. Baulieu, 1990, Hormones: a complex communication network. In *Hormones: From molecules to disease*, edited by E.E. Baulieu & P.A. Kelly (New York: Chapman and Hall), 21.

Figure 4.1–Reprinted from *Neuroscience Letters*, 57, S. Landas et al., "Demonstration of regional blood-brain barrier permeability in human brain," 252, Copyright 1985, with permission from Elsevier Science.

Figure 4.2–From TEXTBOOK OF ENDOCRINE PHYSIOLOGY, THIRD EDITION, edited by James E. Griffin and S.R. Ojeda, copyright © 1996 by Oxford University Press, Inc. Used by permission of Oxford University Press, Inc.

Figure 4.3–Adapted, by permission, from E.E. Baulieu, 1990, Hormones: a complex communication network. In *Hormones: From molecules to disease*, edited by E.E. Baulieu & P.A. Kelly (New York: Chapman and Hall).

Figure 4.4–Adapted, by permission, from F.L. Dunn, et al., 1973, "The role of blood osmolality and volume in regulating vasopressin secretion in the rat," *Journal of Clinical Investigation* 52: 3212-3219.

Figure 4.5–From A.J. Vander, J.H. Sherman, and D.S. Luciano, 2001, *Human physiology: the mechanisms of body function*, 7th ed. (New York: McGraw-Hill), 505, 520. Adapter by permission of The McGraw-Hill Companies.

Figure 4.8–From TEXTBOOK OF ENDOCRINE PHYSIOLOGY, THIRD EDITION, edited by James E. Griffin and S.R. Ojeda, copyright © 1996 by Oxford University Press, Inc. Used by permission of Oxford University Press, Inc, and from E.E. Baulieu, 1990, Hormones: a complex communication network. In *Hormones: From molecules to disease*, edited by E.E. Baulieu & P.A. Kelly (New York: Chapman and Hall).

Figure 4.9–Adapted, by permission, from A. Takamata, et al., 1994, "Sodium appetite, thirst, and body fluid regulation in humans during rehydration without sodium replacement," *American Journal of Physiology* 266: R1493-R1502.

Figure 4.10–Reprinted, by permission, from A. Takamata, et al., 1994, "Sodium appetite, thirst, and body fluid regulation in humans during rehydration without sodium replacement," *American Journal of Physiology* 266: R1493-R1502.

Figure 4.11–From TEXTBOOK OF ENDOCRINE PHYSIOLOGY, THIRD EDITION, edited by James E. Griffin and S. R. Ojeda, copyright © 1996 by Oxford University Press, Inc. Used by permission of Oxford University Press, Inc.

Figure 5.2–Adapted, by permission, from Luger et al., 1988, "Plasma growth hormone and prolactin responses to graded levels of acute exercise and to lactate infusion," *Neuroendocrinology* 56: 112-117, and from P.A. Deuster et al., 1989, "Hormonal and metabolic responses of untrained, moderately trained, and highly trained men to three exercise intensities," *Metabolism* 38:141-148.

Figure 5.3–Adapted, by permission, from W. Vale, et al., "Effects of synthetic ovine corticotropin-releasing factor, glucocorticoids, catecholamines, neurohypophysial peptides, and other substances on cultured corticotropic cells," *Endocrinology* 113:1121-1131, 1983, © The Endocrine Society.

Figure 5.4–Adapted, by permission, from G. Brandenberger, M. Follenius, and B. Hietter, "Feedback from meal-related peaks determines diurnal changes in cortisol response to exercise," *Journal of Clinical Endocrinology and Metabolism* 54:592, 1982, © The Endocrine Society.

Figure 5.5–Adapted, by permission, from B. Tidgren, et al., 1991, "Renal neurohormonal and vascular responses to dynamic exercise in humans," *Journal of Applied Physiology* 70: 2279-2286 and adapted, by permission, from B.J. Freund, et al., 1991, "Hormonal, electrolyte, and renal responses to exercise are intensity dependent," *Journal of Applied Physiology* 70: 900-906.

Figure 5.6–Adapted, by permission, from B.J. Freund, et al., 1991, "Hormonal, electrolyte, and renal responses to exercise are intensity dependent," *Journal of Applied Physiology* 70: 900-906.

Figure 5.7–Reprinted, by permission, from L.B. Rowell, 1974, Circulation to skeletal muscle. In *Physiology and biophysics II*, edited by T.C. Ruch and H.D. Patton (Philadelphia: W.B. Saunders.), 200.

Figure 5.8–Adapted, by permission, from B.J. Freund, et al., 1991, "Hormonal, electrolyte, and renal responses to exercise are intensity dependent," *Journal of Applied Physiology* 70: 900-906.

Figure 5.9–Adapted, by permission, from E.E. Baulieu, 1990, Hormones: a complex communication network. In *Hormones: From molecules to disease*, edited by E.E. Baulieu & P.A. Kelly (New York: Chapman and Hall).

Figure 5.10–From CENTRAL REGULATION OF AUTONOMIC FUNCTIONS, edited by A.D. Loewy and K.M. Spyer, copyright © 1990 by Oxford University Press, Inc. Used by permission of Oxford University Press, Inc.

Figure 6.3–Adapted, by permission, from G.A. Leveille and D.R. Romsos, 1974, "Meal eating and obesity," *Nutrition Today*.

Figure 6.4–Figures a,b,d, e, f, and g adapted, by permission, from G. Ahlborg and P. Felig, 1976, "Influence of glucose ingestion on fuel-hormone response during prolonged exercise," *Journal of Applied Physiology* 41: 683-688. Figure c adapted from B. Saltin, and P. Gollnick, 1988, Fuel for muscular exercise. In *Exercise, nutrition, and energy metabolism* (New York: Macmillan), 51.

Figure 6.6–Figures a,c,d,e,f,g,h, and i adapted, by permission from V. Hodgetts, et al., 1991, "Factors controlling fat mobilization from human subcutaneous adipose tissue during exercise," *Journal of Applied Physiology* 71: 445-451. Figure b adapted from H. Galbo, N.J. Christensen, and J.J. Holst, 1977, "Glucose-induced decrease in glucagon and epinephrine responses to exercise in man," *Journal of Applied Physiology* 42: 525-530.

Figure 6.7–Reprinted from O. Bjorkman, and J. Wahren, 1988, Glucose homeostasis during and after exercise. In *Exercise, nutrition, and energy metabolism*, edited by E.S. Horton and R.L. Terjong (New York: MacMillan Publishing Company), 111. Reproduced by permission of O. Bjorkman.

Figure 6.8–Adapted, by permission, from J.A. Romijn, et al., 1993, "Regulation of endogenous fat and carbohydrate metabolism in relation to exercise intensity and duration," *American Journal of Physiology* 265: E380-E391.

Figure 6.9–Reprinted from B. Saltin and P.D. Gollnick, 1988, Fuel for muscular exercise: Role of carbohydrate. In *Exercise, nutrition, and energy metabolism*, edited by E.S. Horton and R.L. Terjung (New York: MacMillan Publishing Company). Reproduced by permission of B. Saltin.

Figure 6.10–Reprinted from O. Bjorkman and J. Wahren, 1988, Glucose homeostasis during and after exercise. In *Exercise, nutrition, and energy metabolism*, edited by E.S. Horton and R.L. Terjung (New York: MacMillan Publishing Company), 111. Reproduced by permission of O. Bjorkman.

Figure 6.11–Adapted, by permission, from G. Ahlborg and P. Felig, 1976, "Influence of glucose ingestion on fuel-hormone response during prolonged exercise," *Journal of Applied Physiology* 41:683-688, and from V.A. Koivisto, S.L. Karvonen, and E.A. Nikkila, 1981, "Carbohydrate ingestion before exercise: Comparison of glucose, fructose, and sweet placebo," *Journal of Applied Physiology* 51: 783-787.

Figure 6.12–Reprinted, by permission, from K. Acheson, et al., 1984, "Nutritional influences on lipogenesis and thermogenesis after a carbohydrate meal," *American Journal of Physiology* 246: E62-E70.

Figure 7.1–From TEXTBOOK OF ENDOCRINE PHYSIOLOGY, THIRD EDITION, edited by James E. Griffin and R. Ojeda, copyright © 1996 by Oxford University Press, Inc. Used by permission of Oxford University Press, Inc.

Figure 7.2–Adapted from E.A. Newsholme and A.R. Leach, 1983, *Biochemistry for the Medical Sciences* (New York: John Wiley and Sons), 685. © 1983 John Wiley & Sons. Limited. Reproduced with permission.

Figure 7.3–Adapted, by permission, from J. D.Veldhuis, 1998, "Neuroregulatory pathophysiology of impoverished growth hormone (GH) secretion in the aging human," *Journal of Anti-aging Medicine* 1: 173-196. © Springer-Verlage.

Figure 7.4–Figures a and b adapted from G.H. Meyer, 1867, "Die architectur der spongiosa," *Archiv des Anatomisches und Physiologisches Wissenschaft* 34: 615-628. Figure c reproduced, by permission, from T.A. Einhorn, 1992, "Bone strength: The bottom line," *Calcified Tissue International* 51: 333-339. © Springer-Verlag.

Figure 7.5–Adapted, by permission, from A.W. Ham, 1974, *Histology*, 7th ed. (Philadelphia: J.B. Lippincott), 412.

Figure 7.6–Reprinted, by permission, from N. Young, et al., 1994, "Bone density at weight-bearing and nonweight bearing sites in ballet dancers: The effects of exercise, hypogonadism, and body weight," *Journal of Clinical Endocrinology and Metabolism* 78: 449-454.

Figure 7.7–Adapted, by permission, from A.M. Parfitt, 1984, "The cellular basis of bone remodeling: The quantum concept reexamined in light of recent advances in the cell biology of bone," *Calcified Tissue International* 36(Suppl.1): S37-

S45. © Springer-Verlag.

Figure 7.8–Adapted, by permission, from A.W. Ham, 1974, *Histology*, 7th ed. (Philadelphia: J.B. Lippincott), 417.

Figure 7.9–From TEXTBOOK OF ENDOCRINE PHYSIOLOGY, THIRD EDITION, edited by James E. Griffin and S.R. Ojeda, copyright © 1996 by Oxford University Press, Inc. Used by permission of Oxford University Press, Inc.

Figure 8.1–From A.J. Vander, J.H. Sherman, and D.S. Luciano, 1990, *Human Physiology*, 5th ed. (New York: McGraw-Hill), 605. Adapted by permission of The McGraw-Hill Companies.

Figure 8.2–Adapted, by permission, from N. Orentreich, et al., 1984, "Age changes and sex differences in serum dehydroepiandrosterone sulfate concentrations throughout adulthood," *Journal of Clinical Endocrinology and Metabolism* 59: 551-555. © The Endocrine Society.

Figure 8.3–Adapted, by permission, from W.A. Marshall and J.M. Tanner, 1986, Puberty. In *Human growth: A comprehensive treatise*, edited by F. Falkner and J.M. Tanner (New York: Plenum Press), 171.

Figure 8.4–Reprinted, by permission, from J.M. Tanner, 1974, Sequence and tempo in the somatic changes in puberty. In *Control of the onset of puberty*, edited by M.M. Grumbach, G.D. Grave, and F.E. Mayer (Philadelphia: PA: Lippincott, Williams, and Wilkins), 460.

Figure 8.5–From A.Vander, J. Sherman, and D. Luciano, 1998, *Human physiology*, 7th ed. (New York: McGraw-Hill), 646 and 657. Adapted by permission of The McGraw-Hill Companies.

Figure 8.6–Reprinted with permission from P.E. Belchetz, et al., 1978, "Hypophysial responses to continuous and intermittent delivery of hypothalamic gonadotropin-releasing hormone," *Science*, 202: 631. Copyright 1978 American Association for the Advancement of Science.

Figure 8.7–Adapted, by permission, from S.S.C. Yen, 1999, The human menstrual cycle: Neuroendocrine regulation. In *Reproductive endocrinology: Physiology, pathophysiology, and clinical management*, edited by S.S.C. Yen, R.B. Jaffe, and R.L. Barbieri (New York: W.B. Saunders), 194.

Figure 8.8–Figure a from TEXTBOOK OF ENDOCRINE PHYSIOLOGY, THIRD EDITION, edited by James E. Griffin and S.R. Ojeda, copyright © 1996 by Oxford University Press, Inc. Used by permission of Oxford University Press, Inc. Figure b adapted, by permission, from R.M. Boyar et al, 1978, "Anorexia nervosa," *New England Journal of Medicine* 291: 861-865.

Figure 8.9–Reprinted, by permission, from S.S.C. Yen, 1999, The human menstrual cycle: Neuroendocrine regulation. In *Reproductive endocrinology: Physiology, pathophysiology, and clinical management*, edited by S.S.C. Yen, R.B. Jaffe, and R.L. Barbieri (New York: W.B. Saunders), 195.

Figure 8.10–Reprinted, by permission, from S.S.C. Yen, 1999, The human menstrual cycle: Neuroendocrine regulation. In *Reproductive endocrinology: Physiology, pathophysiology, and clinical management*, edited by S.S.C. Yen, R.B. Jaffe, & R.L. Barbieri (New York: W.B. Saunders), 199.

Figure 8.11–Reproduced with permission from *Pediatrics*, Vol. 5, Page 447, Figure 2, Copyright 1972.

Figure 8.12–Reprinted from J.M. Tanner, P.C.R. Hughes, and R.H. Whitehouse, 1981, "Radiographically determined widths of bone, muscle and fat in the upper arm and calf from age 3-18 years," *Annals of Human Biology* 8: 505. By permission from Taylor & Francis Ltd., http://www.tandf.co.uk/journals.

Figure 8.13–Reprinted from J.M. Tanner, P.C.R. Hughes, and R.H. Whitehouse, 1981, "Radiographically determined widths of bone, muscle and fat in the upper arm and calf from age 3-18 years," *Annals of Human Biology* 8: 505. By permission from Taylor & Francis Ltd., http://www.tandf.co.uk/journals.

Figure 8.14–Reprinted, by permission, from J.D. Wilson, 1988, "Androgen abuse by athletes," *Endocrine Reviews* 9: 181-199. © The Endocrine Society.

Figure 8.15–Reprinted, by permission, from J.D. Wilson, 1988, "Androgen abuse by athletes," *Endocrine Reviews* 9: 181-199. © The Endocrine Society.

Figure 8.16–Adapted, by permission, from D.C. Cumming, et al., 1986, "Reproductive hormone increases in response to acute exercise in men," *Medicine*

and *Science in Sports and Exercise* 18: 369-373 and women's data from E.R. Baker et. al., 1982, " Plasma gonadotropins, prolactin, and steroid hormone concentrations in female runners immediately after a long-distance run," *Fertility and Sterility*, 38, 38-41.

Figure 9.1–Figure a adapted, by permission from T. Reilly, 1990, "Human circadian rhythms and exercise," *Critical Reviews in Biomedical Engineering* 18: 165-180. Figure b and c adapted, by permission, from F. Halberg et al., 1967, "Circadian system phase-an aspect of emporal morphology; procedures and illustrative example. In *The cellular aspects of biorhythms. Symposium on biorhythms*, edited by H. Von Mayersback (Springer-Verlag), 46. © Springer-Verlag.

Figure 9.2–Reprinted, by permission, from O. Van Reeth, et al., 1994, "Nocturnal exercise phase delays circadian rhythms of melatonin and thyrotropin secretion in normal men," *American Journal of Physiology* 266: E964-E974.

Figure 9.3–Reprinted, by permission, from O. Van Reeth, et al., 1994, "Nocturnal exercise phase delays circadian rhythms of melatonin and thyrotropin secretion in normal men," *American Journal of Physiology* 266: E964-E974.

Figure 9.4–Reprinted, by permission, from W.B. Webb, 1971, Sleep behavior as a biorhythm. In *Biological rhythms and human performance* edited by W.P. Colquhoun (London: Academic Press), 149.

Figure 9.5–Adapted from *Cardiovascular Research* 27, R. Furlan, Early and late effects of exercise and athletic training on neural mechanisms controlling heart rate, pp. 482-488, Copyright 1993, with permission from Elsevier Science.

Figure 9.6–Adapted from *Life Sciences* 47, P. Monteleone et al., Physical exercise at night blunts the nocturnal increase of plasma melatonin levels in healthy humans, pp. 1989-1995, Copyright 1993, with permission from Elsevier Science.

Figure 9.7–Reprinted, by permission, from O. Van Reeth, et al., 1994, "Nocturnal exercise phase delays circadian rhythms of melatonin and thyrotropin secretion in normal men," *American Journal of Physiology*, 266: E964-E974.

Figure 9.8–Adapted, by permission, from J. Fröberg, et al., 1972, "Circadian variations in performance, psychological ratings, catecholamine excretion, and diuresis during prolonged sleep deprivation," *International Journal of Psychobiology* 2: 23-36.

Figure 9.9–Adapted, by permission, from J. Fröberg, et al., 1972, "Circadian variations in performance, psychological ratings, catecholamine excretion, and diuresis during prolonged sleep deprivation," *International Journal of Psychobiology* 2: 23-36.

Figure 9.10–Figure a adapted, by permission, from J.E. Muller, G.H. Tofler, and P.H. Stone, 1989, "Circadian variation and triggers of onset of acute cardiovascular disease," *Circulation* 9:733-743. Figure b reprinted with permission from the Association for the Advances of Medical Instrumentation. Copyright © 1999.

Figure 9.12–Adapted, by permission, from K.T. Borer, D. Nicoski, and V. Owens, 1986 "Alteration of the pulsatile growth hormone secretion by growth-inducing exercise: Involvement of endogenous opiates and somatostatin," *Endocrinology* 118:844. © The Endocrine Society and adapted from *Brain Research Bulletin*,18, K.T. Borer, Rostromedial septal area controls pulsatile growth hormone release in the golden hamster, 485, Copyright 1987, with permission from Elsevier Science.

Tables

Table 2.1–Reprinted, by permission, from J.B. Young and L. Landsberg, 1998, Catecholamines and the adrenal medulla. In *Williams textbook of endocrinology*, 9th ed., edited by J.D. Wilson, et al. (Philadelphia: W.B. Saunders Company), 695.

Table 5.2–Adapted, by permission, from E.E. Baulieu, 1990, From morphine to opioid peptides. In *Hormones: from molecules to disease*, edited by E.E. Baulieu and P.A. Kelly (New York: Chapman and Hall), 186.

1

Introduction to Endocrinology

Hormones are signaling molecules that regulate and coordinate physiological and metabolic functions by acting on receptors located on or in target cells. Hormones are produced by secretory cells that are localized either in secretory glands (see table 1.1) or in organs that have other primary functions (Baulieu & Kelly 1990) (see table 1.2). These messengers are given distinctive names, depending on cellular origin and the means by which targets are reached. The term "hormone" is derived from the Greek word *hormein* (to excite). The initial use of hormone was restricted to chemical messengers produced within a gland of internal secretion and released into circulation in small quantities to act on distant receptors. The alternative term, "endocrine," was coined from the Greek words *endon* (internal) and *krinein* (to secrete). Examples of endocrine secretions are insulin and glucagon from the pancreatic islets of Langerhans or the thyroid hormones from the thyroid gland.

The definition of hormone has been extended to include chemical messengers produced by nonspecialized cells and signaling molecules that reach their target receptors by routes other than circulation (see figure 1.1). When chemical messengers are released into interstitial fluid to act on receptors of adjacent cells, they are called paracrine secretions (from the Greek word *para*, which means adjacent). An example of a paracrine messenger is somatostatin in the pancreatic islets acting on adjacent insulin and glucagon cells. Sometimes the signaling molecules are released by the cell into the interstitial fluid space to act on receptors of the originating cell. Such messenger action is called "autocrine" (from the Greek word *autos*, which means self). Many growth factors, such as insulinlike growth factor–I (IGF-I), act in autocrine and paracrine fashion. They are elaborated by the same cells or by cells adjacent to those that are the target of hormone actions. Among short-lived chemical messengers that exert autocrine or paracrine and occasionally endocrine action on their receptors are nitric oxide and eicosanoids: prostaglandins, thromboxanes, leukotrienes, prostacyclins, and nitric oxide. (Nitric oxide is the sole inorganic signaling molecule.) These particular messengers control important functions in exercise. Often the same cell distributes its chemical messengers in more than one fashion. For example, some gastrointestinal cells secrete messengers that have endocrine as well as paracrine actions.

The endocrine system shares its signaling and coordinating function with the nervous system. The two systems have evolved to control and integrate vital body functions. When the neuronal messengers are released into the synaptic cleft to activate receptors on adjacent neurons, they are called neurotransmitters. Specificity of the message is ensured not only by spatial and physicochemical matching of the features of particular receptors and their messengers but also by discrete physical connections between axon terminals of a specific neuron type and postsynaptic receptors. Examples of neurotransmission are release of acetylcholine (ACh) from motor nerve endings to act on nicotinic cholinergic receptors of skeletal muscles and from parasympathetic nerves to activate nicotinic receptors on autonomic ganglion cells and muscarinic receptors on cardiac and smooth muscle cells. Thus, neurotransmission is characterized by the speed with which the message is relayed (milliseconds) and the discrete paths of delivery.

By contrast, hormones act over longer periods of time (seconds to hours) and often are distributed diffusely through the circulation to a large number of targets.

Table 1.1 Glandular Sources of Chemical Messengers

Gland	Location	Hormone
Adrenal gland: cortex	Adjacent to the dorsal surface of the kidney	Aldosterone, androstenedione, cortisol, dehydroepiandrosterone (DHEA), estrone
Adrenal gland: medulla	Core of the adrenal gland	Enkephalins, epinephrine (E) or adrenalin (A), norepinephrine (NE) or noradrenalin (NA)
Gonads: ovary	Abdominal cavity	Activin, androstenedione, inhibin, estradiol, FSH-releasing peptide, progesterone, relaxin
Gonads: testis	Scrotum	Activin, androstenedione, estradiol, Müllerian-inhibiting substance, testosterone
Pancreatic islets of Langerhans	Pancreas (abdominal cavity)	Glucagon, insulin, pancreatic polypeptide (PP)
Parathyroid gland	Thyroid gland	Parathyroid hormone (PTH)
Pineal gland	Dorsal midbrain	Biogenic amines, melatonin, various peptides
Pituitary, anterior lobe (adenohypophysis)	Ventral surface of the brain	Adrenocorticotropic hormone (ACTH, or corticotropin), β-lipotropin, β-endorphin, follicle-stimulating hormone (FSH, or gonadotropin), growth hormone (GH), luteinizing hormone (LH, or gonadotropin), prolactin (PRL), thyroid-stimulating hormone (TSH), proopiomelanocortin (POMC)
Pituitary, intermediate lobe	Ventral surface of the brain	β-Endorphin, α melanocyte-stimulating hormone (α-MSH)
Pituitary, posterior lobe (neurohypophysis)	Attached to the ventral surface of brain by infundibular stalk	Antidiuretic hormone (ADH) or arginine vasopressin (AVP), oxytocin
Thyroid gland	Anterior aspect of the neck	Calcitonin, 3,5,3'-triiodothyronine (T_3), thyroxine (T_4)

Delivery of endocrine messages to specific targets is ensured by the specificity of receptors on target cells rather than by discrete physical connections between specific cells, as is the case in neurotransmission. When chemical messengers are produced by nerve cells and released from axonal endings into circulation, they are called neuroendocrine secretions or neurohormones (see figure 1.1). An example of a neurohormone is hypothalamic corticotropin-releasing hormone (CRH), which is discharged into the hypothalamo-hypophyseal portal capillaries during exercise and other forms of stress. Another example is antidiuretic hormone (ADH), or arginine vasopressin (AVP), which is also released during exercise from neuronal terminals in the posterior pituitary gland into systemic circulation (see chapter 4). Some cells disseminate chemical messages by a combination of endocrine and neural transmission. For instance, sympathetic nerves communicate largely by neurotransmission of norepinephrine (NE), but some NE also has neuroendocrine effects when it spills over into general circulation, as is the case during exercise.

HORMONE STRUCTURE AND SYNTHESIS

Most hormones can be categorized by chemical structure as amines, peptides and proteins, and lipid derivatives.

Amine Hormones

Amine hormones are metabolites of aromatic amino acids and include catecholamines, indoleamines, and thyroid hormones. Precursors for the synthesis of amine hormones are absorbed from digested food, and particularly rich sources are milk and meat. Epinephrine (E), NE, and dopamine (DA), collectively called catecholamines, are synthesized in the brain and the adrenal medulla from the aromatic amino acid tyrosine obtained from food or from phenylalanine after its conversion to tyrosine in the liver (see figure 1.2). The rate-limiting step in this biosynthetic pathway is the synthesis or activation of tyrosine hydroxylase (TH), which converts L-tyrosine to L-dihydroxyphenylalanine (L-DOPA). L-aromatic amino acid decarboxylase (L-AADC) next converts L-DOPA to DA, and in the

Table 1.2 Nonglandular Sources of Chemical Messengers

Organ	Location	Hormone
Adipose tissue	Various locations	Adiponectin, leptin, resistin
Bone	Skeletal system	Bone-derived growth factor (Gla), osteocalcin
Brain	Skull	AVP, β-endorphin, corticotropin-releasing hormone (CRH), dynorphin, cholecystokinin (CCK), gastrin, gonadotropin releasing hormone (GnRH), growth hormone–releasing hormone (GHRH), leu-enkephalin, met-enkephalin, neurotensin, oxytocin, somatostatin, substance P, thyrotropin-releasing hormone (TRH), vasoactive intestinal peptide (VIP), others
Gastrointestinal tract	Abdominal cavity	Bombesin, cholecystokinin (CCK), enteroglucagon, enterostatin, gastrin, gastrin-releasing peptide, gastric inhibitory peptide (GIP), ghrelin, secretin, vasoactive intestinal peptide (VIP)
Heart	Atria	Atrial natriuretic peptide (ANP or auriculin or atriopeptin)
Kidney	Abdominal cavity	1,25′-dihydroxycholecalciferol (vitamin D_3), erythropoietin (EPO), renin
Liver	Abdominal cavity	Insulinlike growth factor–I (IGF-I)
Macrophages, monocytes, lymphocytes	Blood	ACTH, cytokines interleukin-1 (IL-1), IL-6, POMC-derived peptides, TGF-β, tumor necrosis factor (TNF)
Muscle	Muscular system	Insulinlike growth factor–I (IGF-I), IGF-II, IL-6, myogenic regulatory factors (MRFs) myogenin, Myo-D, MRF4, Myf-5, others
Placenta	Uterus	Estriol, estrogen, estrone, GH variant, human chorionic gonadotropin (hCG), human placental lactogen (HPL), progesterone
Platelets	Blood	Platelet-derived growth factor (PDGF), transforming growth factor–β (TGF-β)
Skin	Body surface	Epidermal growth factor (EGF), transforming growth factor-α (TGF-α)
Thymus	Upper chest	Thymosin (thymopoietin)
Various locations	Various tissues, organs	Growth factors, including IGF-I, eicosanoids
Vascular endothelium	Blood vessels	Angiotensin II, endothelin-1, endothelium-derived hyperpolarizing factor (EDHF), epidermal growth factor (EGF), nitric oxide (NO), platelet-derived growth factor (PDGF), prostaglandin H_2 (PGH_2), prostacyclin (PGI_2), thromboxane A2, transforming growth factor–β (TGF-β)

remaining biosynthetic steps, L-NE and E are catalyzed, respectively, by dopamine β-hydroxylase (DBH) and phenylethanolamine N-methyltransferase (PNMT). The indoleamine serotonin is synthesized from the aromatic amino acid tryptophan and can serve as a biosynthetic precursor of the pineal hormone melatonin (see figure 1.3). Thyroid hormone thyroxine (T_4) is derived from two tyrosine molecules and four atoms of iodine (see figure 1.4). Increased metabolic demands of exercise mediate its conversion to biologically active triiodothyronine (T_3), while energy deprivation causes its conversion to biologically inactive reverse T_3 (rT_3) and results in suppression of the resting metabolic rate.

Peptide and Protein Hormones

Peptide and protein hormones range in size from three amino acids in hypothalamic thyrotropin-releasing hormone (TRH) to over 400 amino acids in müllerian-inhibiting substance. Of the hormones that have important functions in exercise, AVP and oxytocin, the two posterior pituitary hormones, are nonapeptides (containing nine amino acids); angiotensin II is a deca-peptide (10 amino acids); and growth hormone (GH), prolactin, luteinizing hormone (LH), and follicle-stimulating hormone (FSH) contain about 200 amino acids and have a molecular weight of between 25,000 and 30,000 daltons.

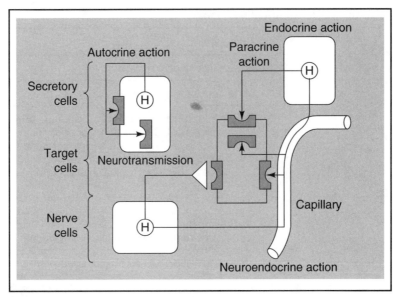

Figure 1.1 *Types of chemical messenger action.*

Chemical messages are transmitted to specific receptors through the extracellular fluids (endocrine, paracrine, autocrine, or neuroendocrine action) or across the synaptic gap (neurotransmission).

H symbolizes chemical messenger.

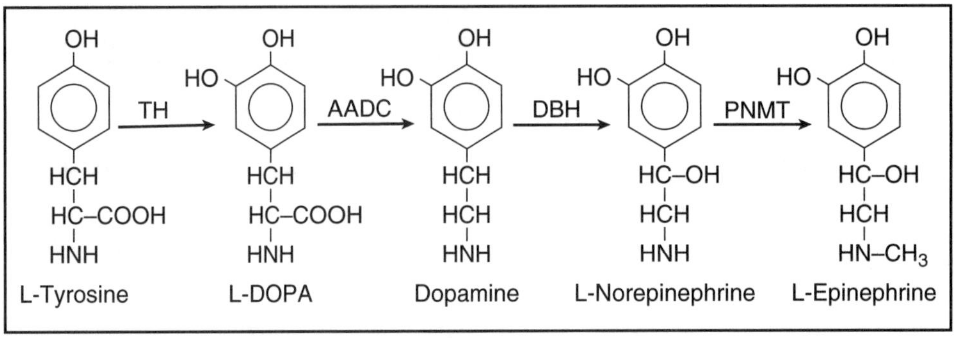

Figure 1.2 *Catecholamine biosynthesis.*

Catecholamines are synthesized from the essential aromatic amino acids tyrosine and phenylalanine. Tyrosine hydroxylase (TH), aromatic-L-amino acid decarboxylase (AADC), and dopamine β-hydroxylase (DBH) mediate formation of norepinephrine from tyrosine. Conversion of norepinephrine to epinephrine in the adrenal medulla and some brain neurons is catalyzed by phenyletha-nolamine-N-methyltransferase (PNMT).

Reprinted from Young and Landsberg 1998.

Amine, peptide, protein, and steroid hormones differ in the details of their synthesis and release. Amine, peptide, and protein hormones are synthesized in the rough endoplasmic reticulum (RER), and steroid hormones are synthesized in the smooth endoplasmic reticulum (SER). Protein hormone genes contain one or more exons or nucleotide sequences encoding the biologically active hormone molecule (see figure 1.5). A gene segment at the 5'-end contains the code for a signal or leader peptide located at the amino end of the hormone precursor molecule. Other nucleotide sequences, or introns, code for molecular fragments with no apparent function. A single-stranded messenger RNA (mRNA) transcript of the protein hormone gene leaves the nucleus through a nuclear membrane pore. Hormone mRNA next attaches to a ribosome to start the translation of the genetic message to a peptide or a protein beginning at its 5'-end. A ribosomal protein called signal recognition particle (SRP) attaches to the mRNA-ribosome complex and facilitates its binding to an SRP receptor on the external membrane of the RER (see figure 1.6). The translation product with a signal or leader peptide at the amino end then penetrates the RER through a pore.

Prohormone is a hormone precursor that contains the signal peptide, the hormone molecule, and inactive fragments. An example of a prohormone is proinsulin, which is made of an A and B chain separated by a C-peptide (see figure 1.7). Cleavage and separation of

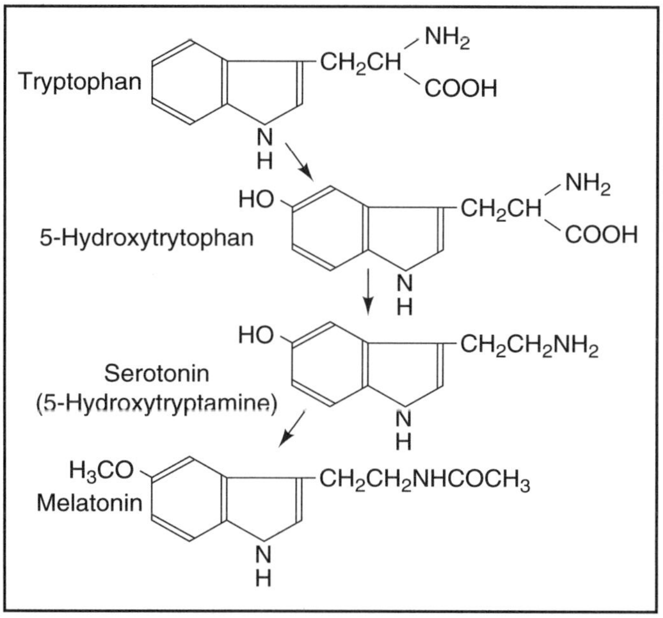

Figure 1.3 *Indoleamine biosynthesis.*

The essential aromatic amino acid tryptophan is the precursor of the hormone and neurotransmitter serotonin, which is synthesized in the brain and the gastrointestinal tract, and of the hormone melatonin, which is made in the pineal gland.

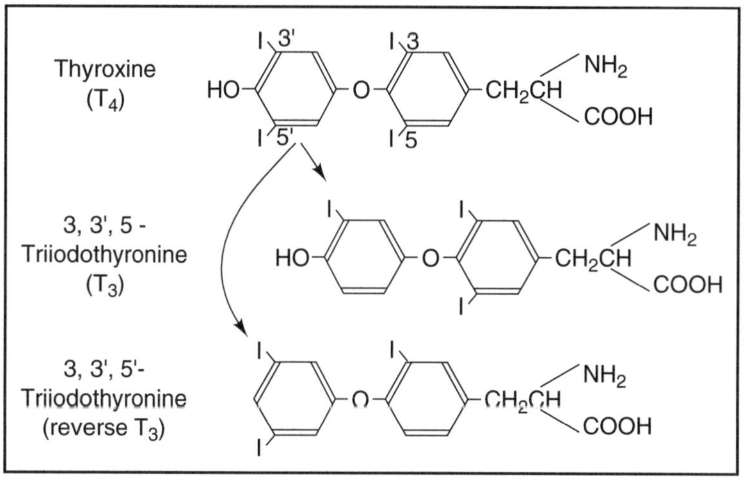

Figure 1.4 *Processing of thyroid hormones.*

In the thyroid gland, tyrosine is iodinated by tyrosine peroxidase to form mostly thyroxine (T_4) and a smaller amount of of triiodothyronine (T_3). In response to thyrotropin-stimulating hormone (TSH), T_4 and T_3 bond to thyroglobulin, a large protein with the molecular weight of 660,000 daltons, and are hydrolyzed by lysosomal proteases. Eighty percent of circulating T_4 is converted to T_3 or reverse T_3 (rT_3) in the liver and the kidney by, respectively, 5'-deiodinase and 5-deiodinase.

Adapted from Griffin and Ojeda 1996.

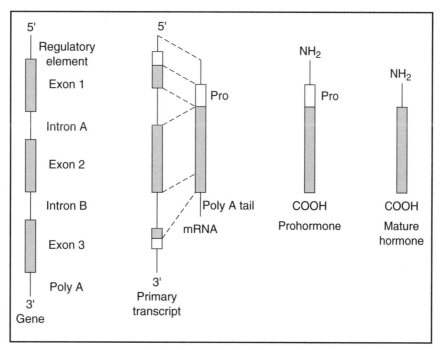

Figure 1.5 *Biosynthesis of a protein hormone.*

During the transcription of a protein hormone gene, the noncoding regions of DNA, introns, are deleted, and the amino acid coding regions, exons, are consolidated. A signal, or leader, peptide at the amino terminal of the prohormone is removed during hormone maturation.

Adapted from Baulieu 1990.

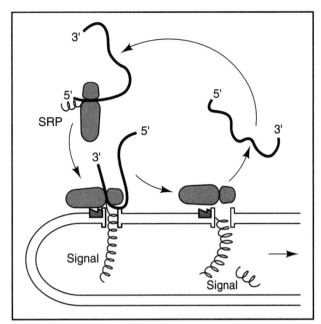

Figure 1.6 *Translation and maturation of a protein hormone.*

The signal recognition particle (SRP) attaches to the newly translated signal peptide and binds to the SRP receptor on the surface of rough endoplasmic reticulum (RER). This process docks the growing prohormone chain and the ribosome to the RER and facilitates the entry of the signal peptide and the growing prohormone chain into the RER. The leader peptide is next cleaved, maturation of the hormone is completed in the Golgi apparatus, and the mature hormone is then stored in secretory vesicles.

Adapted from Baulieu 1990.

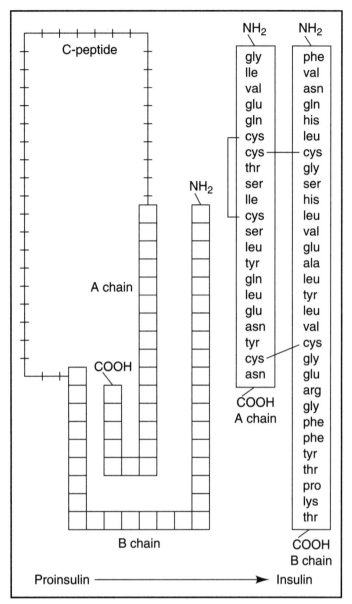

Figure 1.7 *Structure of proinsulin.*

Proinsulin is converted to mature insulin after cleavage of the C-peptide that separates the A and B chains in the prohormone.

Adapted from Baulieu 1990.

the C peptide completes the maturation of circulating proinsulin to insulin. When the initial gene transcript contains more than one active hormone fragment, it is called a preprohormone. Proopiomelanocortin (POMC) (see figure 1.8) is a preprohormone that, in addition to the signal peptide at the amino end of the molecule, can, in stressful situations including exercise, yield corticotropin (ACTH), β-endorphin, α–melanocyte-stimulating hormone (α-MSH), β-MSH, γ-MSH, and met-enkephalin. POMC in the human anterior pituitary gland gives rise to ACTH and β-lipotropin, and the latter is subsequently processed to β-endorphin. Other products such as connecting peptide (CLIP), MSH, and metenkephalin are made in the intermediate lobe of the pituitary in some nonhuman species.

Polyproteins are initial translation products that contain a messenger molecule and its binding protein. The posterior pituitary hormones AVP and oxytocin are synthesized respectively from the polyproteins propressophysin and prooxyphysin (see figure 1.9), which, during packaging into neurosecretory granules, respec-

tively give rise to hormones and the binding proteins neurophysins. The AVP-associated neurophysin II, also known as nicotine stimulated, is associated with a glycoprotein at the C-terminal end. The oxytocin-associated neurophysin I is stimulated by estrogen. Prohormones, preprohormones, and polyproteins lose the signal peptide as they are transported from the RER to the Golgi apparatus, where the processing is generally completed and all but steroid hormones are packaged into storage vesicles or granules.

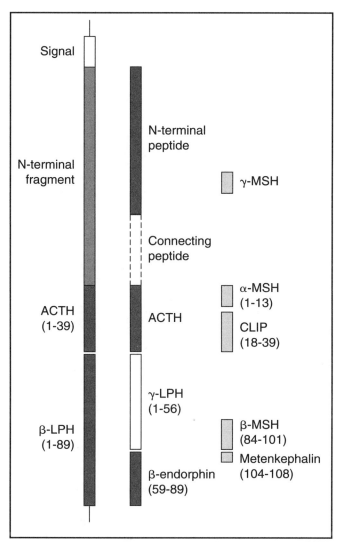

Figure 1.8 *Proopiomelanocortin (POMC) and its translational products.*

After cleavage and removal of signal peptide and the N-terminal fragment, the POMC molecule initially gives rise in the anterior pituitary gland to ACTH and β lipotropin (β-LPH). Subsequent cleavage of β-LPH yields β-endorphin and γ-LPH.

Adapted from Liotta and Krieger 1990.

Steroid Hormones

Cholesterol is a lipid precursor of steroid hormones in the gonads and the adrenal cortex (see figure 1.10) and of 1,25-dihydroxycholecalciferol (vitamin D_3) synthesized in stages in skin, liver, and kidney (see figure 1.11). In the synthesis of adrenal and gonadal steroids, the steroid ring structure of cholesterol is preserved, whereas in the synthesis of vitamin D_3, it is broken. To a limited extent, cholesterol itself is synthesized from acetyl-CoA in the adrenal gland and gonads, but most either is produced by the liver or is of dietary origin and is transported to the adrenal gland and gonads within lipoproteins. Different steroid hormones are synthesized in the three zones of the adrenal cortex (see chapter 5) by enzymes controlling specific biosynthetic pathways (see figure 1.10).

Sex Steroids

Biosynthesis of sex steroids follows the same enzymatic steps in gonads and in adrenal glands. The precursor molecule for sex steroid biosynthesis is cholesterol derived mostly from very-low-density lipoproteins (VLDLs). The key biosynthetic enzymes for sex hormone steroidogenesis are P-450 cytochrome oxidases and hydroxysteroid dehydrogenases (see figure 1.12). Steps 1 through 3 in figure 1.12 occur in glia cells in the brain to produce neurosteroid derivatives of pregnenolone (Mellon 1994). Steps 1 through 5 occur in the innermost zona reticularis layer of the adrenal cortex, and steps 1 through 6 occur in the interstitial cells of the ovary and the testis. The endocrine controllers of steroid biosynthetic process are ACTH in the adrenal gland and LH in the ovary and testis. Both hormones act at step 1 and control the activity of 20,22-desmolase.

The fetal adrenal cortex helps synthesize placental estrogen estriol. It lacks 21-hydroxylase enzymes (CYP21) of adult zona glomerulosa and fasciculata necessary for conversion of progesterone and 17-hydroxyprogesterone into precursors of aldosterone and cortisol. The hypertrophied equivalent of zona reticularis in a fetal adrenal

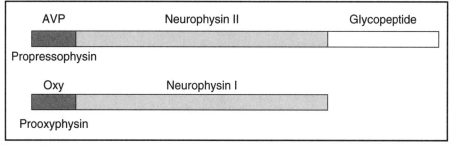

Figure 1.9 *Examples of polyproteins.*

Polyprotein propresophysin gives rise to arginine vasopressin (AVP) and its binding protein neurophysin II, and prooxyphysin is processed to oxytocin and its binding protein neurophysin I.

Adapted from Baulieu 1990.

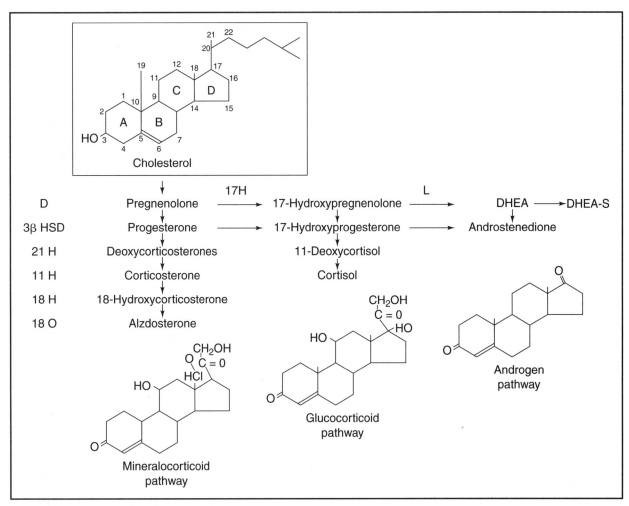

Figure 1.10 *Biosynthetic pathways of adrenal corticosteroid hormones.*

Adrenal corticosteroid biosynthesis from cholesterol is stimulated by the corticotropin or ACTH. ACTH stimulates the activity of 20,22 desmolase (D), the rate-limiting side-chain cleavage enzyme in the P-450 cytochrome. A series of hydroxylases (H), hydroxysteroid dehydrogenases (HSD), or oxidases (O), and chain-cleaving desmolases (D and 17,20 desmolase or lyase [L]) channel the corticosteroid derivatives down either the mineralocorticoid, the glucocorticoid, or the androgen biosynthetic pathway.

DHEA = dehydroepiandosterone; DHEA-S = DHEA sulfate.

Adapted from Griffin and Ojeda 1996.

cortex converts placental pregnenolone into dehydro-epiandrosterone sulfate (DHEA-S). The placenta turns these hormones into estriol, an important estrogen of pregnancy. Shortly after birth, the fetal adrenal cortex regresses and differentiates into three zones with a full complement of enzymes for synthesis of mineralocorticoids in zona granulosa, glucocorticoids in zona fasciculata, and androgens in zona reticularis.

P-450 cytochrome aromatase enzymes (see figure 1.12) in the cells of the adrenal zona reticularis, in ovarian granulosa cells, in Sertoli cells of the testis, in muscle (Longcope et al. 1978; Matsumine et al. 1986), in adipose tissue (Longcope et al. 1978; Nimrod & Ryan 1975), in bone marrow (Frisch et al. 1980), and in the hypothalamus, preoptic area, and limbic forebrain (Balthazart et al. 1990; Naftolin et al. 1971) convert androgens androstenedione and testosterone into, respectively, estrogens estrone and estradiol. About 25% to 30% of aromatization of androstenedione occurs in the adult muscle, and only 10% to 15% occurs in the adipose tissue so that the larger muscle mass and greater capacity for aromatization in the muscle of adult men account for the greater overall aromatization of androgens to estrogens in normal-weight adult men than in normal-weight women (Longcope et al. 1969, 1978; Matsumine et al. 1986). In obesity, however, hypertrophied adipose tissue becomes a predominant site for aromatization of androgens. Other tissues, including the prostate gland and hair follicles on the male scalp, contain 5α-reductase enzymes that convert testosterone into dihydrotestosterone (DHT). Attempts at reduction of androgenic stimulation of prostate cancer and at

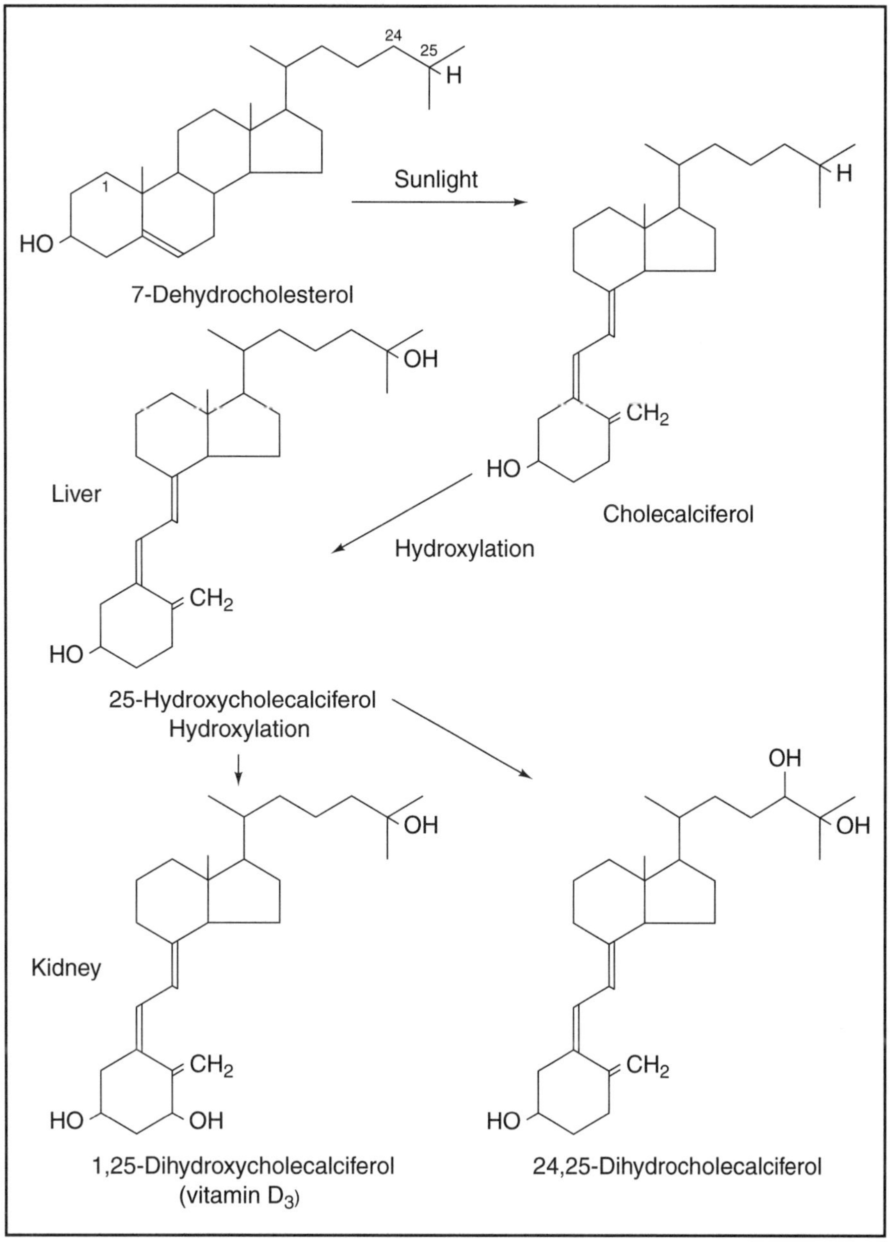

Figure 1.11 *Biosynthesis of vitamin D₃.*

Ultraviolet light breaks the bonds between carbon-9 and carbon-10 in cholesterol derivative 7-dehydrocholesterol in the skin to form the prohormone cholecalciferol. Provitamin D_3 enters circulation and is converted in the liver to 25-hydroxycholecalciferol and in the kidney to the biologically active 1,25 dihydroxycholecalciferol (vitamin D_3). Vitamin D is inactivated by 24-hydroxylation.

Adapted from Griffin and Ojeda 1996.

blocking androgenic alopecia, or male pattern baldness, often include drugs that inhibit activity of 5α-reductase in these tissues.

Eicosanoid Messengers

Eicosanoid messengers are derivatives of polyunsaturated arachidonic fatty acid (AFA). Prostaglandin, prostacyclin, and thromboxane synthesis (see figure 1.13) entails folding by cyclooxygenase of this fatty acid to produce a ring structure. Leukotrienes retain unfolded

linear structure but undergo oxidation of the AFA precursor.

HORMONE MATURATION

Processing of some hormones occurs posttranslationally after the immature molecule has been released into the blood stream (see table 1.3). The vitamin D precursor (see figure 1.11) is modified by sequential hydroxylations in the liver and the kidney before the most

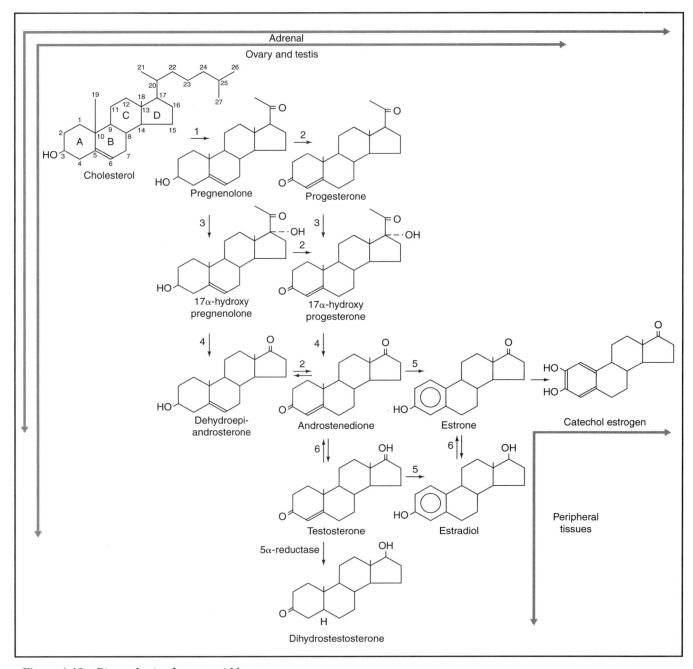

Figure 1.12 *Biosynthesis of sex steroid hormones.*

Arrows identify the range of steroid molecules synthesized in the ovary, testis, adrenal glands, and peripheral tissues.

Adapted from Griffin and Ojeda 1996.

biologically active form of the hormone is produced. Angiotensin II, which is released in response to a decline in plasma volume that often occurs during exercise, likewise is created by serial enzymatic transformations. The precursor angiotensinogen molecule, secreted by the liver into systemic circulation, first is modified by a kidney enzyme renin. The final messenger is produced after further modification by the angiotensin-converting enzyme (ACE) in the vascular endothelium. Yet other biologically active hormones may be converted in some tissues to other messenger molecules. Examples are con-

version of thyroxine (T_4) secreted by thyroid gland to either triiodothyronine (T_3) or to reverse T_3 (rT_3) in target tissues (see figure 1.4), and conversion of testicular testosterone to estrogen in the brain cells and to DHT in the prostate gland. Another example of posttranslational hormone modification is glycosylation of LH, FSH, and TSH. The two gonadotropins and TSH consist of an α chain common to all of them and a structurally distinct β chain. Their side chains, which include sialic acid, fucose, mannose, galactose, and other carbohydrates, are added in the course of posttranslational processing.

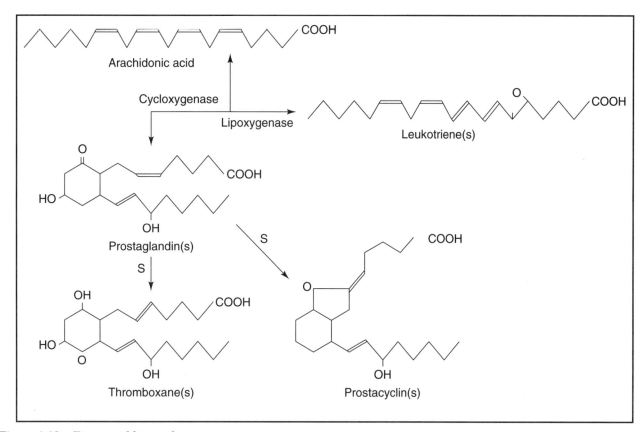

Figure 1.13 *Eicosanoid biosynthesis.*

The polyunsaturated arachidonic fatty acid (AFA) is the precursor of eicosanoids. The cycloxygenase enzyme folds the AFA to form the prostaglandin ring structure. The synthetases (S) convert prostaglandins to thromboxanes and prostacyclins. The lipooxygenase enzyme converts AFA to leukotrienes.

Adapted from Smith and Wood 1992.

Table 1.3   Posttranslational Processing of Prohormones and Hormones

Prohormone or hormone	Secretion site	Processed hormone	Processing site	Target site
Angiotensin I	Kidney	Angiotensin II	Vascular endothelium and lung	Brain, kidney, vasculature
7-Dehydrocholesterol	Skin	25 OH-Cholecalciferol	Liver	Kidney
25 OH-Cholecalciferol	Liver	1,25 Cholecalciferol	Kidney	Intestine, bone
DHEA-S	Adrenal cortex	Estrogen	Placenta	Uterus
Proinsulin	Pancreas	Insulin	Many tissues	Muscle, adipose, other tissues
Testosterone	Testis	DHT	Prostate gland	Prostate gland
Testosterone	Testis	Estrogen	Hypothalamus	Hypothalamus
Thyroxine (T_4)	Thyroid gland	Triiodothyronine (T_3)	Many tissues	Many tissues

HORMONE RELEASE

Packaging of amine and peptide hormones into storage vesicles allows their secretion to be controlled by neurotransmitters, hormones, or metabolites. Extrusion of such stored hormonal messengers into the blood stream or extracellular fluid takes place by exocytosis, a process that entails fusion of vesicles with the cell membrane and rupture of the vesicle wall at its junction with the plasma membrane. A stimulus is required to discharge these hormones from the storage vesicles, and for that reason this control of hormone secretion is called regulated release (see figure 1.14). By contrast, steroid hormones leave the Golgi apparatus in transport vesicles and are released into circulation by exocytosis without being stored in cytoplasm. This latter form of hormone secretion is called constitutive release and permits hormonal response at the level of synthesis rather than release.

Hormones are released into circulation in nannomolar (10^{-9} M) to picomolar (10^{-12} M) concentrations. Because they circulate in the blood at such low concentrations and come into contact indiscriminately with all cells within the body, hormones rely on specificity and affinity of binding to the receptors for successful transduction of their message. Most hormones are secreted in intermittent rather than continuous fashion (see chapter 2). Spontaneous or basal pulsatile hormone secretion supplies hormones in quantities that anticipate usual requirements for their action and influences the number and sensitivity of hormone receptors. In contrast to the anticipatory character of spontaneous hormone secretion, various stimuli such as exercise and secretagogues (e.g., neurotransmitters, metabolites and nutrients) can acutely increase or decrease hormone secretion in compensatory fashion. Compensatory hormone response in this case is proportional to the intensity of eliciting stimulus such as exercise, a change in some variable in the internal environment, or hormonal feedback. Acute changes in hormone secretion are usually caused by changes in the volume of hormone released during a normally occurring spontaneous pulse. Different mechanisms control spontaneous pulsatile and stimulus-elicited hormone secretion.

Frequently the synthesis and release of hormones is under the control of other hormones. As already mentioned, release of anterior pituitary hormones such as GH is controlled by the hypothalamic releasing or inhibiting hormones such as growth hormone–releasing hormone (GHRH) and somatostatin or somatotropin release–inhibiting hormone (SRIF). Pituitary hormones

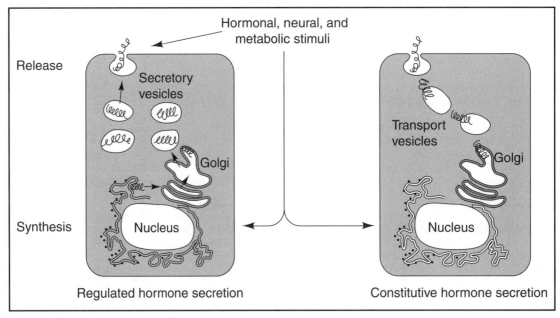

Figure 1.14 *Regulated and constitutive hormone release.*

Amine, peptide, and protein messengers are stored in secretory vesicles, and their regulated release requires transduction of hormonal, neural, and metabolic stimuli. Steroid hormone release is constitutive because stimulated hormone synthesis is immediately followed by hormone release from transport vesicles.

Adapted from Baulieu 1990.

such as ACTH usually control release of other hormones from some other endocrine gland. ACTH elicits secretion of the hormone cortisol from the adrenal cortex (see chapter 5). Another example is the control of adrenal medullary E synthesis by the adrenal cortical hormone cortisol (see figure 1.2). Discussed later are the contributions of the autonomic nervous system (see chapter 2), neural influences that impose rhythmicity (see chapters 3 and 9), hormones (see chapters 4 and 8) and metabolites (see chapters 6 and 7) to this regulated type of hormone secretion (see figure 1.14).

The temporal pattern of hormone secretion is of remarkable biological significance. The same quantity of hormone can have opposite biological functions when it is delivered in intermittent rather than tonic fashion. One example with exercise implications is stimulation of cellular proliferation in muscle and bone growth plates in response to intermittent, but not tonic, GH administration to hypophysectomized rats (Isgaard et al. 1988a, 1988b; Maiter et al. 1988). Another example is the support of normal menstrual function by a circhoral (one pulse per hour) secretion of LH and FSH and suppression of menstrual cyclicity when this pulse frequency is lower or absent (see figure 8.6 on page 154). This phenomenon accounts for exercise-induced amenorrhea and is also used to achieve contraception (see chapter 8).

HORMONE TRANSPORT

Most amine, peptide, and protein hormones are water soluble and circulate dissolved in plasma. GH, some growth factors, and all lipophilic steroid hormones circulate bound to carrier proteins. Albumin and prealbumin nonselectively bind to and transport a variety of small messenger molecules as well as free fatty acids. Globulins, on the other hand, such as sex hormone–binding globulin (SHBG) and others, have single high-affinity binding sites for specific steroid and amine hormones (see table 1.4). Transport of hormones in bound form extends their period of availability and action by preventing their rapid clearance from plasma. By keeping some of the hormone from circulating in free form, binding proteins diminish the magnitude but extend the duration of hormone action. Thus the insulinlike growth factors persist in circulation for several hours because a complex system of binding proteins protects them from rapid proteolytic degradation. Unbound eicosanoids and most protein hormones are inactivated within minutes of being produced. Other binding proteins may facilitate hormone binding to the receptor. GH binding protein has almost identical structure to the GH membrane receptor and facilitates GH binding to its receptors.

HORMONE DEGRADATION

To be effective, hormones must act on their receptors in appropriate intermittent fashion. Whenever receptors are exposed to hormones in unphysiologically high concentrations or for extended periods of time, they become less available for hormone action, or down-regulated. Thus, some hormone degradation is a necessary prerequisite for optimal receptor sensitivity and hormonal action.

Table 1.4   Hormone-Binding Proteins

Binding protein	Abbreviation	Hormone
Albumin		Small hormone molecules
Corticosteroid binding globulin or transcortin	CBG	Glucocorticoids
CRH binding protein	CRH-BP	Corticotropin-releasing hormone
GH binding protein	GH-BP	Growth hormone
IGF binding proteins	IGF-BPs	IGF-I and IGF-II
Neurophysins		AVP and oxytocin
Prealbumin		Small hormone molecules
Sex hormone–binding globulin	SHBG	Testosterone, estradiol
Thyroid hormone–binding globulin	TBG	T_4, T_3
Vitamin D–binding protein	DBP	1,25-Dihydroxycholecalciferol

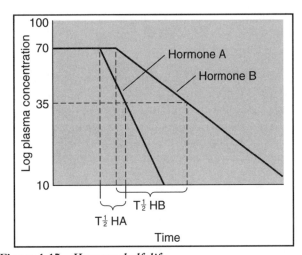

Figure 1.15 *Hormone half-life.*

Hormone A has a shorter half-life than hormone B because of its higher metabolic clearance rate (MCR).

Adapted from Baulieu 1990.

Table 1.5 Hormone Half-Lives

Chemical messenger	Half-life
Eicosanoids	Seconds to minutes
Amines	Minutes
Thyroid hormones:	
T_4	6.7 days
T_3	0.75 day
Polypeptides	4–40 minutes
Proteins	15–170 minutes
Steroid hormones	4–120 minutes

Hormones are changed by enzymes into a metabolically inactive form either in target cells or in the liver or the kidney. The speed with which hormones are metabolized depends on whether they travel free in plasma, on their molecular structure, and on the characteristics and location of their catabolic enzymes. The half-life of hormones (the time it takes to reduce the concentration of circulating hormone by one half when no new hormone is being produced) (see figure 1.15) varies from seconds to hours (see table 1.5). The degradation rate of hormones is usu-ally measured as the metabolic clearance rate (MCR) and may affect plasma hormone concentrations in situations such as exercise where the rate of blood flow through visceral sites of degradation is reduced. MCR represents the volume of plasma cleared per unit of time and defines the relationship between the secretion rate (SR) and the concentration (C) of hormone in circulation:

MCR (volume/time) = SR (mass/time) / C (mass/volume)

2

Role of the Autonomic Nervous System in Exercise

Exercise calls for an increase in oxygen and fuel supply to contracting skeletal muscles and for maintenance of the internal environment in the presence of accumulating body heat and metabolic wastes. These needs generally are met by the sympathoadrenal (SA) system (Loewy & Spyer 1990; Young & Landsberg 1998). The SA system is one of two divisions of the autonomic nervous system (ANS) that comprise the efferent sympathetic (S) nerves and the adrenal medulla, the brain regions that control their actions, and their messengers, the catecholamines. During recovery from exercise, the parasympathetic (PS) division of the ANS and the enteric system of nerve networks and ganglia, coupled with various behavioral responses, correct deviations in the internal environment. They also mediate trophic, growth-promoting functions, as well as physiological adaptations to exercise training.

The SA system has three important general features of particular significance in exercise. First, the release of catecholamines at the S nerve endings and in the adrenal medulla is under direct control of several regions of the brain or central nervous system (CNS). Triggered by the autonomic motor nerves arising in the autonomic control areas in the CNS, catecholamines can transduce mental readiness for physical action into physiological responses. S nerve–mediated actions are rapid. They take place within seconds and usually dissipate quickly, unlike the prolonged effects of many hormones. Second, the brain connection allows mental anticipation of physical performance to activate catecholamine-responsive processes such as heart function or respiration in advance. Third, the SA system integrates multiple physiological processes whenever there is a need for heightened whole-body response.

This coordinated response is achieved through integration of direct and indirect catecholamine actions. Direct effects entail interaction of catecholamines with adrenergic receptors in virtually all tissues of the body (see chapters 3, 5, 6, and 7). Indirect effects are described in this chapter and include alteration in the following:

1. Secretion of other hormones that regulate physiological processes
2. Regulation of blood flow, cardiorespiratory function, hemostasis, and body fluid shifts
3. Mobilization and utilization of metabolic substrates

Vascular, metabolic, and hormonal effects of catecholamines cooperatively provide redundant, backup processes to ensure adequate physiological response even when any one of these systems underperforms. Generalized functional activation of the organism by the SA system in response to danger was recognized by Walter Cannon and labeled a "fight-or-flight" response. The SA system also coordinates physiological reactions to chronic stress as recognized by Hans Selye under the term "general adaptation syndrome." Both of these special roles for the SA system are discussed in more detail in chapter 5.

In addition to global whole-body activation, different components of the SA system, including individual cotransmitters in SA neurons and adrenomedullary cells, can respond selectively to specific physiological stimuli suggesting that they serve discrete functions (Guo & Wakade 1994; Vollmer 1996). For example, during fasting-associated hypoglycemia, ischemic injury,

15

and hemorrhagic hypotension, S activity is suppressed while the adrenomedullary hormone secretion is stimulated (Young et al. 1984; Young & Landsberg, 1997). Special considerations regarding valid approaches for assessment of different components of SA activation are described in chapter 3.

CHARACTERISTICS OF THE AUTONOMIC NERVOUS SYSTEM

The two ANS divisions, S and PS, have predominant, but not exclusive, control over cardiorespiratory function and metabolism, and over biosynthetic and reparative processes, respectively. Traditionally, only the motor component of this complex system has been discussed in relation to exercise, but both divisions engage central nervous centers for integration of autonomic responses. These centers initiate action through motor nerves and their associated ganglia and enteric and other plexuses, and they receive sensory information about the internal environment through the afferent nerves (Furness & Costa 1980). Whereas the visceral organs of the head, thorax, abdomen, and pelvis (smooth muscles, endocrine, and exocrine glands) have dual innervation from S and PS motor nerves, cardiac muscle, the structures of the body wall, including blood vessels, piloerector muscles of the skin, and eccrine sweat glands are innervated only by S nerves.

Autonomic Nerves, Ganglia, and Plexuses

In contrast to somatic motor nerves acting on skeletal muscles by way of a single motor neuron, autonomic motor nerves consist of a preganglionic and a postganglionic neuron that are chemically linked within an autonomic ganglion or plexus (see figure 2.1).

Preganglionic S neurons originate in the intermediolateral (IML) cell column of the gray matter between thoracic T-1 and lumbar L-2 segments of the spinal cord (see figure 2.2). From the IML column, the myelinated white axons of the preganglionic S neurons leave the spinal nerves in the ventral root and reach their ganglion targets in the white communicating ramus (root). Most preganglionic neurons innervate postganglionic neurons in the ganglia of the paired paravertebral S chain. Some preganglionic neurons do not form synapses within the paravertebral chain but instead pass through its ganglia at spinal levels T-5 through L-2 to form the splanchnic nerves (see figure 2.2). These innervate the adrenal medulla directly or synapse with postganglionic S neurons in other preaortic (prevertebral) ganglia (see figure 2.2). The neurons arising from 5th through 12th thoracic segments form greater, lesser, and lowest thoracic splanchnic nerves. They synapse with postganglionic neurons in the celiac ganglion (K in figure 2.2) (Loewy & Spyer 1990). Preganglionic neurons from the lower five thoracic and upper two lumbar segments of the spinal cord

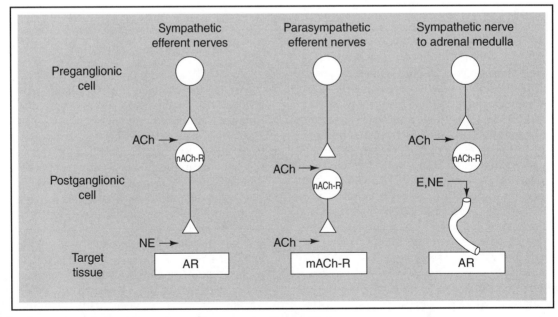

Figure 2.1 *Structure and neurotransmitters of autonomic efferent nerves.*

The autonomic efferent nerves consist of a preganglionic and a postganglionic cell. Acetylcholine (ACh) is the neurotransmitter in preganglionic sympathetic (S) and parasympathetic (PS) efferent neurons, and it activates nicotinic receptors (nACh-R) on postganglionic cells. ACh is the postganglionic PS neurotransmitter that acts on muscarinic receptors (mACh-R) on target tissues. Catecholamines (norepinephrine and epinephrine) activate adrenergic receptors (AR) on target tissues.

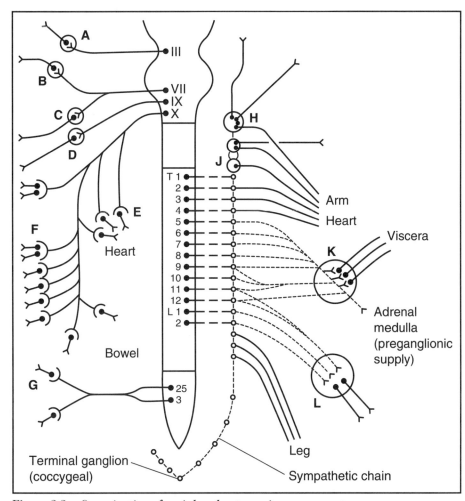

Figure 2.2 *Organization of peripheral autonomic nervous system.*

A schematic diagram of the sympathetic and parasympathetic efferent nerves and ganglia.

Adapted from Young and Landsberg 1998.

below the point of exit of the S nerve out of the S trunk. The preaortic (prevertebral) ganglia include celiac (K) and superior and inferior mesenteric ganglia (L in figure 2.2) that lie, respectively, near the bifurcation of celiac, superior, and inferior mesenteric arteries from the aorta. The first three pairs of paravertebral ganglia, the superior cervical (H), the middle cervical, and the stellate (J), are located in the neck (see figure 2.2). Postganglionic fibers from the first pair innervate the dilator of the pupil in the eye, the glands and the smooth muscles of the head, and the pineal gland. When the SA system is activated during exercise or other stress, contraction of dilator muscles causes pupillary dilation (Loewy & Spyer 1990). Postganglionic fibers from the cervical ganglion stimulate nocturnal melatonin secretion. Of interest in exercise are projections from the other two paravertebral neck ganglia, along with postganglionic neurons from the first five thoracic ganglia to the heart, lungs, bronchi, and trachea via thoracic S cardiac nerves (see figure 2.2). Postganglionic neurons from the celiac ganglion reach the foregut and its associated organs and the kidney.

Adrenal Medulla

Adrenal medulla is a modified sympathetic prevertebral ganglion, innervated by S preganglionic neurons of the greater splanchnic nerve that originate in the IML cell column between T-5 and L-3 (see figures 2.2 and 2.3) and by two celiac branches of the vagus nerve (Berthoud & Powley 1993). Its cells are derived from the neural crest and differentiate in the presence of glucocorticoids into endocrine chromaffin cells that synthesize, store, and secrete catecholamines (Hodel 2001). The term chromaffin denotes the affinity for chromium salts that make these cells darken on exposure to potassium dichromate in a catecholamine oxidation reaction. Within other S ganglia, on the other hand, in the presence of nerve growth factor (NGF) or basic fibroblast growth factor (bFGF), neural crest cells differentiate into postganglionic S neurons (Anderson 1993). As is the case with other postganglionic neurons, the endocrine chromaffin cells of the adrenal medulla are stimulated to release catecholamines

travel in lumbar and sacral splanchnic nerves to the superior and inferior mesenteric ganglia (L in figure 2.2) and plexuses. Their postganglionic neurons innervate, respectively, the midgut and hindgut and the pelvic organs (see figure 2.2).

Autonomic ganglia are simple integrative neuronal units within which interneuron cells modulate neurotransmission between the preganglionic and postganglionic neurons. In S ganglia, this function is carried out by small intensely fluorescent cells that store dopamine (DA) or epinephrine (E). Although there are about 25 pairs of segmental paravertebral ganglia extending from the cranial (see figure 2.2) through the sacral end of the spinal cord, the preganglionic S nerves leave the spinal cord through only the 12 thoracic and the first two lumbar segments and thus form the thoracolumbar S outflow (see figure 2.2). S messages are amplified as each preganglionic neuron synapses with postganglionic neurons in several paravertebral ganglia both above and

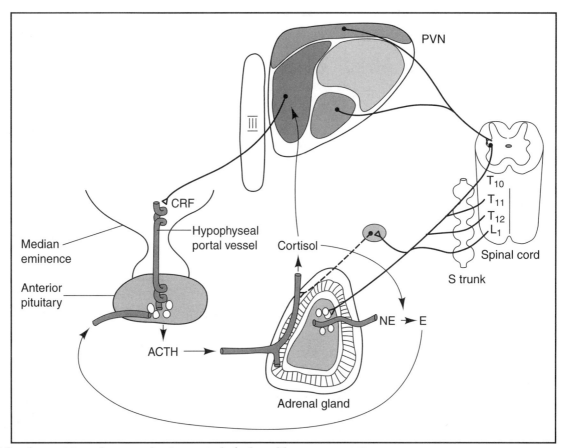

Figure 2.3 *Autonomic and neuroendocrine control of adrenal hormone secretion.*

The dorsal and ventral portions of the paraventricular hypothalamic nucleus (PVN) (lighter gray) are involved in activation of sympathetic (S) nerve outflow and adrenomedullary epinephrine (E) secretion. The medial parvocellular part of the PVN (darker gray) secretes corticotropin-releasing factor (CRF) into the hypophyseal portal vessels in external layer of the median eminence (ME). CRF stimulates release of corticotropin (ACTH) from the anterior pituitary gland, and ACTH, in turn, stimulates cortisol secretion from the adrenal cortex. Cortisol also stimulates biosynthesis of E in the adrenal cortex, and E stimulates ACTH secretion from the anterior pituitary gland. Lateral part of PVN is magnocellular.

ACTH = corticotropin; CRF = corticotropin-releasing factor; E = epinephrine; L = lumbar spine; NE = norepinephrine; PVN = paraventricular hypothalamic nucleus; T = thoracic spine; III = third ventricle.

Adapted from Loewy and Spyer 1990.

into the blood stream in response to S activation. This is yet another way that S messages are amplified in urgent situations that require fast integrated reaction.

The combined weight of the two human adrenal medullae is about 1 g or 10% of the total gland mass. They receive blood from the superior artery (a branch of the inferior phrenic artery), the middle artery that arises from the aorta, and the inferior artery that arises from the renal artery after all three traverse the adrenal cortex. Arteries that supply the adrenal cortex coalesce into venous sinuses in zona reticularis and drain into the medulla, where they again coalesce with medullary capillaries. Adrenocortical blood vessels are innervated by postganglionic S fibers originating in paravertebral or suprarenal prevertebral ganglia. Adrenal medullary chromaffin cells store 80% E and 20% norepinephrine (NE) in addition to enkephalins and other cotransmit-

ters. E is the predominant mammalian circulating catecholamine influencing tissues throughout the body. The adrenal gland contains several milligrams of catecholamines per gram (several millimoles per kilogram) of adrenal tissue. Adrenal enkephalins that are coreleased with catecholamines may mediate stress-associated analgesia (Lewis et al. 1982) and regulate cerebral blood supply (Dora et al. 1992).

Postganglionic S fibers are small and have unmyelinated axons. Postganglionic neurons with axons destined to innervate structures in the body wall, such as sweat glands, blood vessels, and piloerector muscles of the skin, leave the paravertebral ganglia through the gray communicating rami and travel in the segmental spinal nerves. Instead of forming discrete nerve endings, postganglionic S nerves terminate in an extensively ramified plexus (network) that can attain a length of several centimeters. Through this anatomical arrangement, the S message is further amplified as each postganglionic fiber innervates and affects many target cells. NE, the chemical messenger of the S postganglionic cell, is distributed unevenly within this plexus and is concentrated in enlarged areas called varicosities. Each varicosity contains about 1,000 electron–dense small storage or synaptic vesicles containing about 15,000 molecules of NE.

Preganglionic efferent PS cells originate in the motor nuclei of the cranial nerves III (oculomotor), VII (facial), IX (glossopharyngeal), and X (vagus) in the brain stem and in the sacral region of the spinal cord (see figure 2.2). PS nerve fibers within cranial nerves III, VII, and IX innervate smooth muscles and glands associated with the eyes, nose, and mouth. Sacral PS nerves innervate the lower gut and the genital organs. The vagus innervates just about all other visceral organs and provides actions that usually oppose and counterbalance S effects. Preganglionic cells of the efferent vagus originate in two different areas of the brain. Vagal cells that affect gastrointestinal organs, muscles of the upper alimentary canal, and the trachea are located in the dorsal motor nucleus of the vagus (DVN). Vagal cells innervating the heart and the respiratory muscles originate in the nucleus ambiguus (NA) (see figure 2.4). Thus, in contrast to the thoracolumbar S outflow, PS nerves are cranial (III through X) and sacral in origin. Their long preganglionic axons synapse with cells that have very short postganglionic axons. These cells are within PS ganglia that are located near the surface of target organs (see figure 2.2) or, in the intestine, within the walls of the organ.

In addition to the neural plexus formed by the ramified terminals of the S nerves, all hollow organs such as the gut, the heart, and the blood vessels are innervated by autonomic plexuses in different tissue layers. The intestinal wall contains at least six enteric plexuses: periglandular; submucous, or Meissner's; intramuscular (in circular muscles); myenteric, or Auerbach's; intramuscular (in longitudinal muscles); and subserous (see figure 2.5). Neural plexuses are intermingled with ganglia in the walls of the gastrointestinal (GI) organs and are considered by some as an autonomous enteric nervous system. Its sensory and motor cells interact through interneurons within the walls of the GI system to regulate GI function. The heart is innervated by cardiac and coronary plexuses. Intrinsic neurons in the autonomic plexuses often communicate through peptidergic messengers such as somatostatin or the somatotropin release–inhibiting factor (SRIF), dynorphin, neuropeptide Y (NPY), substance P, and serotonin. Frequently more than one peptide is colocalized in the same neuron. Individual plexuses are often differentially chemically coded, with vasoactive intestinal polypeptide (VIP) and en-

kephalin being the most common and cholecystokinin (CCK) the least common messengers in the submucous and myenteric plexuses (Costa et al. 1986).

Chemical Messengers

Acetylcholine (ACh) is the neurotransmitter used by both S and PS preganglionic neurons to activate nicotinic cholinergic receptors on postganglionic cells (see figure 2.1). ACh also is a postganglionic PS neurotransmitter that acts on muscarinic cholinergic receptors (mACh-R) on target cells. Catecholamines E and NE act on adrenergic receptors, which are classified as α or β types, and divided further into subtypes α_1 and α_2 or β_1 through β_3 (see chapter 3). E and NE are potent nonselective agonists of α receptors. The S postganglionic neurotransmitter NE is a more potent agonist than E of β_1 and β_3 adrenergic receptors on target cells. E, to a greater extent than NE, activates β_2 receptors responsible for bronchodilation, vasodilation, and prejunctional stimulation of NE release from S neurons (Young & Lansberg 1998). The exception to S neurotransmission of NE are S fibers to the structures of the body wall, sweat glands, piloerector muscles in the skin, and some blood vessels, which, like the postganglionic PS neurons, are cholinergic. In addition, some autonomic nerves that control smooth muscle relaxation in the GI tract, airways, and pelvic viscera by way of increased nitric oxide production are nonadrenergic and noncholinergic (Moncada et al. 1991).

NE (and also E in chromaffin cells) is stored and released from small dense-core vesicles in the axons of postganglionic S nerves. In addition to NE, S neurons synthesize and release other transmitters, including NPY, SRIF, substance P, galanin, neurotensin, and enkephalins (Morris et al. 1995). These transmitters are stored along with catecholamines in large dense-core vesicles. These colocalized neurotransmitters have specific distributions within the sympathetic nervous system. For instance, NPY is distributed principally in S fibers innervating blood vessels and galanin in pancreatic sympathetic nerves (Dunning & Taborsky 1989). The neuropeptide messengers in S nerves may act as cotransmitters or neuromodulators. Chromostatin, NPY, and enkephalins reduce catecholamine secretion, whereas substance P increases it. Adenosine triphosphate (ATP) also is colocalized with NE and participates in vasomotor control (Burnstock 1995). Cotransmitter NPY can exert a potent pressor response that is not antagonized by adrenergic blockade (Pernow & Lundberg 1988; Potter 1988). Galanin can induce mild hyperglycemia by inhibiting insulin and stimulating glucagon secretion (Ahren et al. 1991). The calcitonin gene–related peptide (CGRP) and VIP are colocalized with ACh in S cholinergic nerves and participate in the control of sweating (Landis & Fredieu 1986).

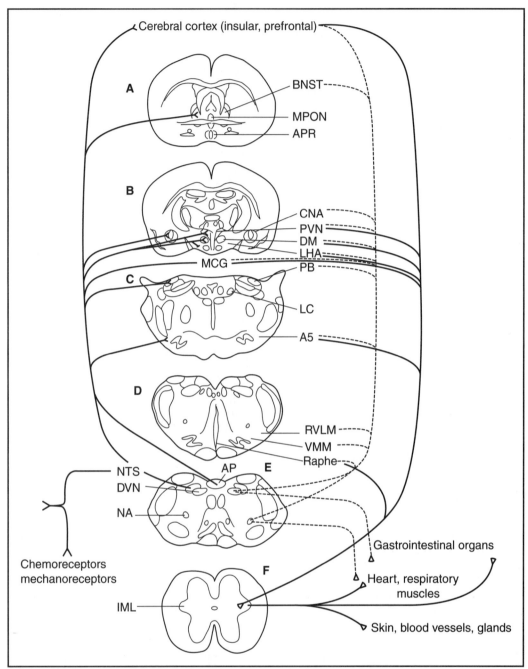

Figure 2.4 *Chemoreceptor and mechanoreceptor connections with their integrative brain centers.*

Chemoreceptor and mechanoreceptor sensory information travels in parasympathetic (PS) nerves to discrete areas of the nucleus of the solitary tract (NTS) as well as to a common integrative commissural area of this nucleus. Ascending neural projections from the NTS are shown on the left. Descending projections from the forebrain to the parasympathetic preganglionic neurons in the dorsal vagal nucleus (DVN) and nucleus ambiguus (NA) (broken line) and to the sympathetic preganglionic neurons in the IML are shown on the right.

AP = area postrema; APR = anterior periventricular region; A5 = noradrenergic cell group; BNST = bed nucleus of stria terminalis; CNA = central nucleus of amygdala; DM = dorsomedial hypothalamic nucleus; IML = intermediolateral cell column; LC = locus coeruleus; LHA = lateral hypothalamic area; MCG = mesencephalic central gray; MPON = medial preoptic nucleus; PB = parabrachial nucleus; PVN = paraventricular hypothalamic nucleus; Raphe = raphe nucleus; RVLM = rostral ventrolateral medulla; VMM = ventromedial medulla. Planes represent, respectively, A = forebrain septum, B = hypothalamus (diencephalon), C = pons, D = rostral medulla, E = caudal medulla, F = thoracic spinal cord.

Adapted from Loewy and Spyer 1990.

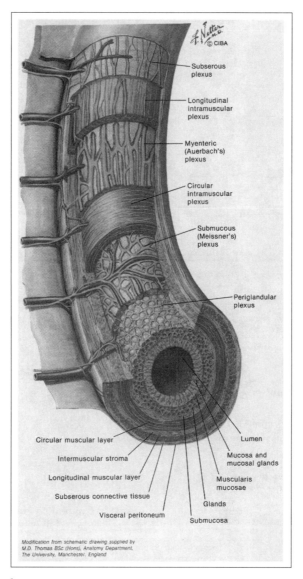

Figure 2.5 *Enteric autonomic nerve plexuses.*

Enteric plexuses in the intestinal wall. Six plexuses innervate tissue layers of the gut. From the inner mucosa to the outer serosa, they are: periglandular (surrounding endocrine cells), submucous or neissuers, circular intramuscular, myenteric or Auerbach's, longitudinal intramuscular, and subserous.

Reprinted from ICON Learning Systems, LLC, 1983.

INTEGRATIVE AUTONOMIC BRAIN CENTERS

Several structures in the forebrain, midbrain, pons, and medulla oblongata have been implicated in the integration of visceral sensory input and facilitation of autonomic function (Loewy & Spyer 1990). They receive both S and PS input, affect both S and PS functions, and have complex reciprocal connections.

In the forebrain, brain structures involved in autonomic integration (see figures 2.6 and 2.7) include the prefrontal and insular cerebral cortex, hypothalamus, and the limbic forebrain. The insular cerebral cortex is involved in emotions of startle, fear, and rage that influence cardiorespiratory function because of its connections with the hypothalamic and brain stem sympatho-excitatory circuits (Loewy & Spyer 1990; Williamson et al. 1997). Even the nonautonomic forebrain areas such as the motor cortex, which is activated during exercise, can stimulate S outflow to the skin (Silber et al. 2000).

Autonomic integrative centers in the hypothalamus are paraventricular (PVN), dorsomedial (DM), and ventromedial (VMH) nuclei and the lateral hypothalamic area (LHA). Limbic telencephalon structures such as the central nucleus of the amygdala (CNA) and the bed nucleus of stria terminalis (BNST) are also autonomic integrative centers (figure 2.7). The DM has been implicated in patterning of respiratory and locomotor rhythms during exercise (Eldridge et al. 1985; Marshall & Timms 1980). The CNA connects with the medial prefrontal cortex that has been shown to inhibit cardiorespiratory function (Loewy & Spyer 1990).

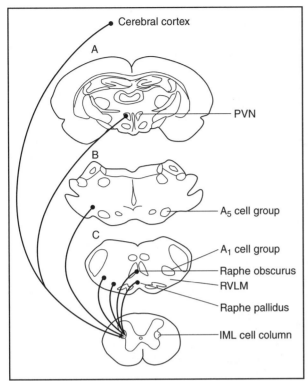

Figure 2.6 *Brain centers that control efferent sympathetic outflow.*

The excitatory brain areas with connections to preganglionic cells in the intermediolateral (IML) column of the spinal cord are insular cerebral cortex, paraventricular nucleus (PVN) of the hypothalamus (plane A), the noradrenergic cell groups in the pons (A5, plane B) and medulla (A1, plane C), and rostral ventrolateral medulla (RVLM) and caudal raphe nuclei in the medulla (plane C).

Raphe obscurus and pallidus = caudal raphe nuclei.

Adatped from Loewy and Spyer 1990.

Paraventricular Hypothalamic Nucleus

The PVN deserves special notice because of its central role in coordination of S responses, regulation of energy and fluid balance, and activation of immune responses. The PVN has three discrete parts: (1) lateral magnocellular, (2) medial parvocellular, and (3) dorsal and ventral (see figure 2.3). The lateral magnocellular part of the PVN (and of the supraoptic [SO] nucleus) controls cardiovascular function and also is an integrative center for water regulation. With respect to cardiovascular function, magnocellular PVN neurons are tonically inhibited by atrial receptors and baroreceptors, and they also receive input from NE and NPY neurons in A1 neuronal cell group, from E neurons in C1 cell group, and from pons and tegmentum (Loewy & Spyer 1990).

Large cells of the magnocellular PVN synthesize hormones arginine vasopressin (AVP) or antidiuretic hormone (ADH) and oxytocin (OXY), transport them in axons along with neurophysins I and II through the

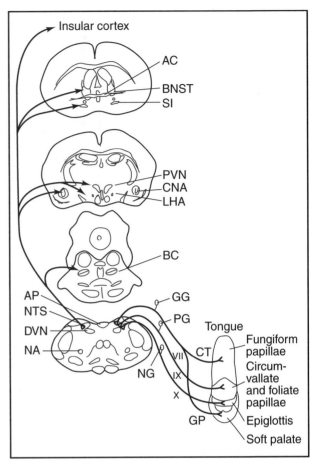

Figure 2.7 *Central projections of taste afferents.*

Projections of taste stimuli to rostral nucleus of the solitary tract (NTS) and regions of the forebrain. Taste stimuli from the front of the tongue, which is covered with fungiform papillae, travel in chorda tympani from the back of the tongue (site of circumvallate and foliate papillae), are transmitted in ninth nerve, and stimuli from the epiglottis and soft palate travel respectively in tenth and a branch of seventh nerve. From NTS taste stimuli ascend to taste areas in the pons (BC), hypothalamus (LHA and PVN), limbic forebrain (CNA, BNST, SI), and insular cortex.

AC = anterior commissure; BC = brachium conjunctivum; BNST = bed nucleus of the stria terminalis; CNA = central nucleus of the amygdala; CT = chorda tympani, branch of VII nerve; DVN = dorsal vagal nucleus; GG = geniculate ganglion; GP = greater petrosal nerve, branch of VII nerve; LHA = lateral hypothalamic area; NA = nucleus ambiguus; NG = nodose ganglion; PG = petrosal ganglion; PVN = paraventricular nucleus of the hypothalamus; SI = substantia innominata.

Adapted from Loewy and Spyer 1990.

hypothalamo-hypophyseal stalk, and store them in the posterior pituitary (Loewy & Spyer 1990). AVP and OXY are released into the fenestrated capillaries of the inferior hypophyseal artery and reach systemic circulation through efferent veins. The magnocellular SO and PVN neurons discharge in a characteristic bursting pattern, and their coordinated discharge is facilitated by cell coupling through tight junctions. These two nuclei receive projections from other brain areas involved in body fluid regulation, among them the circumventricular organs (CVOs). AVP has antidiuretic and vasomotor actions (see chapter 4). OXY, the other hormone secreted by the magnocellular PVN, causes contraction of uterine muscles and myoepithelial cells in the breast.

Small cells in the medial parvocellular PVN synthesize corticotropin-releasing factor (CRF) and secrete this hormone into the fenestrated capillaries of the hypophyseal portal vessels in the external layer of the median eminence (ME) (see figure 2.3). CRF stimulates anterior pituitary corticotrophs to secrete corticotropin (ACTH) from the proopiomelanocortin (POMC) precursor (see figure 1.8). ACTH, in turn, stimulates cortisol (or corticosterone in rodents) secretion from the fascicular zone of the adrenal cortex (see figure 1.10). Parvocellular PVN is sensitive to corticosteroid feedback. Almost all PVN stimulatory actions on food intake and ingestion of carbohydrates in response to NE stimulation of α_2 receptors or to NPY require glucocorticoid presence and feedback. The PVN also is the site of receptors where NE, NPY, and cortisol increase, while serotonin and CRF decrease, carbohydrate intake after glycogen depletion (Bray 1993; Leibowitz et al. 1993; Tempel & Leibowitz 1993). Central control of adrenomedullary hormone release to hypoglycemia is mediated by hypothalamic PVN (see figure 2.4, plane B) and the medullary nuclei (caudal raphe, rostral ventrolateralmedulla [RVLM], ventromedial, and A5) (see figure 2.4, planes C and D) (Loewy & Spyer 1990). Cortisol also stimulates biosynthesis of E in the adrenal medulla, and E stimulates ACTH secretion from the pituitary. This process illustrates the important role of the PVN in integration of endocrine and autonomic actions in response to various forms of stress (see chapter 5). Dorsal and ventral portions of the PVN are involved in activation of S outflow and adrenomedullary E secretion (see figure 2.3), but these processes also depend on the presence and action of CRF (Fisher et al. 1982). In effect, the PVN appears to be one of the few brain centers that regulate the entire S outflow (Brown & Fisher 1985; Luiten, ter Horst, & Steffens 1987; Strack et al. 1989).

Parvocellular PVN also plays a key role in control of immune responses (see chapter 5). Through its control of S outflow, the PVN is involved in neural stimulation of lymphoid organs, spleen, thymus, bone marrow, and lymph nodes to release activated immune cells during stress and exercise (Friedman & Irwin 1997; Hori et al. 1995; Madden et al. 1995). Muscle damage from eccentric exercise triggers release of interleukin-6 (IL-6) from the muscle and IL-1 and tumor necrosis factor–α (TNF-α) from activated macrophages and monocytes (Pedersen et al. 1997). IL-1 triggers CRF release from the parvocellular PVN (Berkenbosch et al. 1987). IL-1

may reach the PVN through circulation, by paracrine action from monocytes and microphages that migrate out of blood vessels into brain tissue, or from neural hypothalamic circuits that use IL-1 as a neurotransmitter. Besides its action on CRF neurons, IL-1 may directly stimulate ACTH production from pituitary corticotrophs (Ruzicka & Akil 1995). Thus in stress, the SNS activates the immune response, and chemical messengers of the immune system in turn stimulate the pituitary stress hormones ACTH and cortisol and terminate their own production, probably through cortisol-negative feedback over parvocellular CRF neurons in the PVN (Uehara et al. 1989).

The brain stem is the "stalk" of the brain including the midbrain and the medulla oblongata underneath the cerebral hemispheres and the cerebellum. It contains catecholamine-releasing and indoleamine-releasing nuclei and the reticular formation, a loose network of neurons essential for arousal and life. A characteristic of the brain stem autonomic centers is that they are tonically active. In the midbrain, the mesencephalic central gray area participates in autonomic control. Several nuclei in this area and in the locus coeruleus (LC), parabrachial (PB) nucleus, and nucleus of the solitary tract (NTS) send projections to the DVN, and from the DVN to gastrointestinal organs and muscles of the upper alimentary canal and trachea. Besides the involvement of the PVN, reflex increases in hepatic glycogenolysis and in E and glucagon release during glucoprivation are also mediated by the lateral PB nucleus, NTS, LC, AP, and DVN (Calingasan & Ritter 1992a; Ritter & Dinh 1994). Projections from the integrative centers destined for the heart and the respiratory muscles are directed to NA and neurons in the IML spinal column (see figure 2.4) (Luiten et al. 1985).

Autonomic Afferent Messages

Afferent pain stimuli from visceral organs such as the heart reach the spinal cord through the S nerves, white rami to the spinal nerves, and spinal nerve dorsal roots. After relay in the spinal cord, these stimuli are transmitted to higher brain centers where they reach consciousness (Loewy & Spyer 1990). The cell bodies of afferent neurons are in segmental dorsal root ganglia projecting to laminae I and V in the spinal gray matter of the thoracic and upper two lumbar segments of the spinal cord. Afferent fibers constitute only 20% of splanchnic neurons, and most of them are unmyelinated. They are joined by the 10 times more numerous sensory fibers from receptors in the muscles and the skin. Because of the quantitatively limited afferent visceral input and its convergence with the more numerous somatic afferents, visceral pain is generally referred to skin areas. The heart pain in angina pectoris is felt in the superficial areas of the arms and upper chest, whereas pain in the esophagus, gall bladder, and duodenum is referred to the overlying superficial areas of the body (Wall & Melzack 1985). Additional receptors are located in the muscles where they monitor changes in the metabolic state and promote cardiorespiratory responses to exercise (Kaufman et al. 1983; Kniffki et al. 1981; MacLean et al. 2000; O'Leary & Augustyniak 1998).

Mechanoreceptors, osmoreceptors, special ion (sodium) receptors, chemoreceptors, and some hormone receptors generate afferent messages that travel in PS nerves. These PS afferents principally convey sensory information from the viscera, the tongue, and the smooth muscles and participate in reflexes controlling lung inflation, heart rate, blood pressure, plasma volume, partial pressures of O_2 and CO_2 in the blood, digestion, and energy regulation. Cell bodies of these afferent neurons are in the ganglia of cranial nerves (geniculate ganglion of the seventh nerve, petrosal [inferior] ganglion of the ninth nerve, and nodose [inferior] ganglion of the vagus nerve) and in the sacral dorsal root ganglia. Most of this sensory information does not reach the cerebral cortex and consciousness. An exception is taste information that is consciously recognized and associated with affective states and specific cravings in situations of energy deficit and specific nutrient deficiencies.

Mechanoreceptors monitor blood pressure in peripheral circulation (Loewy & Spyer 1990). High-pressure arterial baroreceptors in the carotid sinuses and the aortic arch and the renal baroreceptors in the juxtaglomerular apparatus (JGA), the specialized contact area between the ascending limb of the loop of Henle (macula densa) and the afferent arteriole to the glomerulus, sense changes in arterial blood pressure (see chapter 4). Low-pressure atrial receptors at the confluence of great veins with the atria monitor changes in plasma volume and venous return to the heart. Stretch receptors in the lungs and the airways react to alveolar stretching. Arterial baroreceptors relay blood pressure information to the CNS through the sinus nerve, a branch of glossopharyngeal (ninth) nerve, and the other mechanoreceptors through the vagus (tenth) nerve.

Osmoreceptors, sodium receptors, and hormone receptors involved in body fluid homeostasis (angiotensin II and atrial natriuretic factor [ANF] receptors) are located in the CVOs (Loewy & Spyer 1990). CVOs are brain structures at the interface between the brain, systemic circulation, and cerebrovascular fluid that also have direct neuronal connections with brain areas involved in autonomic regulations. They are devoid of blood-brain barrier and have receptors that monitor chemical changes in systemic circulation and in cerebrovascular space. Of the eight CVOs (see chapter 4), two that are involved in regulation of body fluids, the subfornical organ (SFO) and the organum vasculosum of the lamina terminalis

(OVLT), interact with magnocellular PVN and SO nuclei. The OVLT acts as central osmoreceptor and sodium receptor, while SFO has receptors for angiotensin II as well as osmoreceptors and sodium receptors. The area postrema (AP) is involved in the control of both food and fluid homeostasis and receives input from carotid baroreceptors, vagus, and DM and PVN. It sends projections to the commissural NTS and the PB nucleus.

Chemoreceptors monitoring changes in arterial carbon dioxide partial pressure (pCO_2) and arterial oxygen partial pressure (pO_2) are located in the carotid body and aortic body and on the ventral surface of medulla oblongata. Taste receptors, stretch receptors in the stomach wall, and chemoreceptors in the liver, stomach, duodenum (Berthoud & Powley 1992; Prechtl & Powley 1990), and brain, detect changes in concentration and availability of nutrients and associated hormones. Taste chemoreceptors are located on the tongue, epiglottis, and soft palate (see figure 2.7). Chemoreceptors on the fungiform papillae located on the anterior two thirds of the tongue relay sensory information via the chorda tympani nerve, a branch of the facial nerve. Circumvallate and foliate papillae relay taste information from the back of the tongue through the lingual branch of the ninth nerve. The vagus and a branch of the facial nerve carry taste stimuli from the epiglottis and soft palate. Hepatic chemoreceptors include glucoreceptors (Nagase et al. 1993), amino acid receptors (Niijima & Meguid 1995), osmoreceptors (Niijima 1969), and sodium receptors (Contreras & Kosten 1981). The duodenum also has glucoreceptors and sodium receptors (Walls et al. 1995). Glucosensitivity of several hypothalamic areas is discussed in chapter 6. Receptors for CCK, angiotensin II, and galanin also are located on the vagal terminals or cell bodies of vagal afferents (Calingasan et al. 1992; Calingasan & Ritter 1992b; Ritter et al. 1989; Speth et al. 1987).

Sensory information from baroreceptors and chemoreceptors is relayed to the NTS in the posterolateral portion of the medulla and lower pons, to the AP, and from the AP to the RVLM. The lateral PB nucleus, A5 noradrenergic cell group, and reticular formation, and in the medulla, the A1 noradrenergic cell group, caudal raphe nuclei (obscurus and pallidus) and RVLM, all have a role in the integration of this information (see figures 2.6 and 2.7). Mechanoreceptors, stretch receptors in the lungs, taste and other chemoreceptors, gastrointestinal receptors, and the receptors in the CVOs all project to the NTS and to a common integrative area (commissural NTS). Taste afferents project to the most rostral part of the NTS and then to the motor nuclei of cranial nerves that control chewing and swallowing. Afferents from gastrointestinal organs converge in the caudal part of the NTS. The AP, one of the CVOs that receives information from the sodium, osmoreceptors, and glucoreceptors in plasma and in cerebrospinal fluid, relays this information, as does the commissural NTS, to the ascending central autonomic network. Taste afferents ascend to taste areas in the pons (PB nucleus), hypothalamus (LHA and PVN), limbic forebrain (CNA, BNST, and substantia innominata [SI]), and insular cortex (see figure 2.7) (Loewy & Spyer 1990).

Brain Structures in the Control of Heart Function and Vasomotion

The RVLM, the vasomotor center in reticular medulla oblongata and lower pons, and the reticular activating system in the pons, mesencephalon, and diencephalon, are brain centers that control blood pressure and blood flow at rest and during emergencies. As the key integration area of reflexes that control the heart rate and vasoconstriction, the vasomotor center establishes a balance between stimulatory S tone and inhibitory PS tone, both of which determine the level of activation of cardiovascular function (Malliani et al. 1991; Malliani 1997). The C-1 noradrenergic anterolateral region of the upper medulla accelerates heart action and vasoconstricts blood vessels during exercise through its lateral neurons and stimulation of S nerve discharge. NE neurons within the RVLM impart a rhythmic discharge pattern to cardiac neurons and control the timing of S nerve discharges, the pace and intensity of the heart pump, and the periodicities and magnitude of blood pressure changes. The A5 neurons and the RVLM nucleus (see figures 2.4 and 2.6) control regional redistribution of blood from the viscera to the muscles and the skin during exercise (Dampney & McAllen 1988; Stanek et al. 1984). The caudal raphe nuclei provide serotonergic activating influence to S outflow. Raphe nuclei also contain TRH and substance P–releasing neurons responsible for the vasoconstrictor tone (Loewy & Spyer 1990). The vasodilatory A-1 region is located in the anterolateral portion of the lower medulla and sends its commands through its medial neurons and the PS neurons originating in the adjacent DVN. Proximity of the RVLM nucleus to, and interconnections with, the Botzinger complex that initiates and times respiratory movements (Loewy & Spyer 1990) provides a neural basis for respiratory control over cardiovascular function (Daly 1985).

Brain Structures in the Control of Pain Sensation

Pain stimuli reach the integrative brain centers in several distinct pathways (see figure 2.8). Lateral spinothalamic tracts (LSTT) and medial spinothalamic tracts (MSTT) carry the information about tonic aspects of pain and its location from neurons in laminae I and V to the ventroposterolateral and medial thalamus, respectively. Pain also is relayed by the spinoreticular tract to pontine reticular formation and by the spinomesencephalic tract to the brachium conjunctivum (BC), periaqueductal gray (PAG) in the pons, and thalamus, and from the thalamus again to cortical pain areas. Visceral pain such as angina pectoris reaches consciousness in higher brain centers, which then launch autonomic, emotional, and behavioral responses to it. Central autonomic responses to pain stimuli also are mediated by the LHA, the brain stem NTS, and the PB nucleus (Menetrey & Basbaum 1987).

AUTONOMIC CONTROL OF ENDOCRINE SECRETION

ANS influences the secretion of many peptide and steroid hormones (see table 2.1), in addition to generating circulating catecholamines E and NE. SA messengers affect the secretion of hormones in the kidney, adrenal cortex, pancreas, heart, thyroid and parathyroid glands, gonads, gastrointestinal tract, pineal gland, and a number of other organs (Young & Landsberg 1998). PS messengers principally affect the secretion of hormones in the pancreas, heart, and GI organs. S nerves and circulating E can rapidly influence hormone secretion and control circulating hormone mass through positive feedback (feed-forward) mechanisms. This type of control allows graded anticipatory endocrine responses that are proportional to intensity and volume of exercise stress (Luger et al. 1988). Many hormones influenced by the SA system also are responsive to metabolic signals and therefore incorporate a negative feedback–type of endocrine control. Some hormones responsive to SA control in turn influence the SA activity, and this hormonal negative feedback over autonomic activity adds to the complexity of neuroendocrine controls in exercise.

Kidney Hormones Renin and Erythropoietin

Loss of sodium from the extracellular compartment and consequent hypovolemia are sensed by the baroreceptors in the juxtaglomerular apparatus (JGA). This stimulus is sufficient for reflex release of renin from granular cells in the walls of the afferent arteriole in the JGA (renin-angiotensin-aldosterone reflex or RAA; see chapter 4) but is potentiated by S stimulation (Mizelle et al. 1987). Renal S nerve activity or stimulation or infusion of catecholamines triggers renin secretion by acting on prejunctional β_2 or postjunctional β_1 receptors (Kopp & DiBona 1993; Osborn et al. 1982; Saxena 1992). Stimulation of β_1 receptors activates the cyclic adenosine monophosphate (cAMP)-responsive regulatory element of the renin gene and increases its expression (Hack-

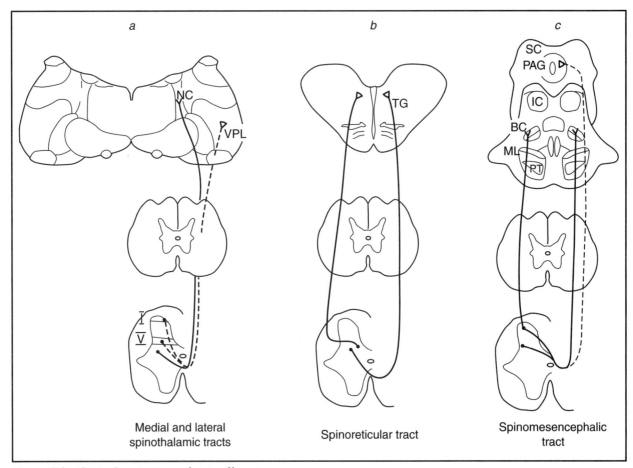

Figure 2.8   *Central projections of pain afferents.*

Sympathetic nerve fibers carry information about visceral pain from laminae I and V of the spinal gray matter to the thalamus *(a)* in the lateral and medial spinothalamic tracts, to the brain stem reticular formation *(b)* in the spinoreticular tract, and to the brachium conjunctivum (BC) and periaqueductal gray *(c)* in the spinomesencephalic tract.

IC = inferior colliculus; ML = medial lemniscus; NC = central nucleus of the thalamus; PT = pyramidal tract; PAG = periaqueductal gray; SC = superior colliculus; TG = tegmental gray in the brain stem reticular formation; VPL = ventroposterolateral nucleus of the thalamus.

Adapted from Loewy and Spyer 1990.

enthal et al. 1990; Holmer et al. 1994). Stimulation by renin of angiotensin I formation and subsequent synthesis of angiotensin II and release of aldosterone (the RAA reflex) are described in chapter 4. Renin potentiates the neurotransmission in the celiac ganglion, causing renal vasoconstriction and reduced glomerular filtration rate. Inhibition of renin secretion by NPY, also released from renal S nerve endings, may be caused by vasoconstriction of tubular arterioles (Persson et al. 1990). Cardiopulmonary low-pressure baroreceptors and, to a lesser extent, carotid baroreceptors exert a tonic inhibition over renin secretion (Kopp & DiBona 1993).

S nerve stimulation potentiates erythropoietin (EPO) secretion in response to hypoxia alone or combined with hemorrhage (Eckard & Kurtz 1992; Eckardt et al. 1992; Fink & Fisher 1976). EPO elevates oxygen-carrying capacity of blood by stimulating synthesis of hemoglobin and red blood cells. EPO secretion is diminished in primary autonomic failure and by weightlessness during space flight (Robertson et al. 1994). EPO

administration in turn stimulates S activity and blood pressure elevation (Hoeldtke & Streeten 1993; Roger et al. 1993).

Adrenocortical Steroid Hormones

ACTH and cortisol secretion is directly proportional to exercise intensity (Luger et al. 1988) partly because of action of the SA system. Catecholamines increase production of glucocorticoids, aldosterone, and adrenal androgens by activating cyclic adenosine monophosphate (cAMP) and expression of CYP genes after acting on β receptors in the adrenal cortex (Ehrhart-Bornstein et al. 1994; Guse-Behling et al. 1992; Lightly et al. 1990; Pratt et al. 1985). They do so directly after hemorrhage or by potentiating corticosteroid secretion in response to other secretagogues (Bernet et al. 1994; Charlton 1990; Edwards & Jones 1993; Hinson et al. 1992). Aldosterone secretion is inhibited by renin when angiotensin II concentration is high (Malchoff et al. 1987), an effect that also may be mediated by S nerves.

Table 2.1 Catecholamine Effects on Hormone Secretion

Endocrine organ	Hormone	Effect	Receptor	Secretagogue/feedback
Kidney				
Juxtaglomerular apparatus	Renin	^	β	Renal baroreceptor, distal tubular Na
Peritubular cells (?)	Erythropoietin	^	β	Arterial PO$_2$
Pancreatic islets				
α cells	Glucagon	^	α, β	Plasma substrate
β cells	Insulin	^	β	Plasma substrate
		v	α	
δ cells	Somatostatin	^	β	?
		v	α	
non-α, non-β, non-δ cells	Pancreatic polypeptide	^	β	?
		v	α	
Thyroid gland				
Follicles	Thyroxine,	^	β	Thyrotropin
	triiodothyronine	v	α	
C cells	Calcitonin	^	β	Plasma ionized Ca
		v	α	
Parathyroid gland	Parathyroid hormone	^	β, DA	Plasma ionized Ca
		v	α	
Gastric antrum and duodenum	Gastrin	^	β	Gastric luminal pH
Adrenal cortex				
Zona glomerulosa	Aldosterone	^	β	Angiotensin II, plasma K
		v	DA	
Zona fasciculata	Cortisol	^	?	Corticotropin (ACTH)
Zona reticularis	Androstenedione	^	β	ACTH
Ovary and placenta				
Granulosa cells or corpus luteum	Progesterone	^	β, DA	LH, human chorionic gonadotropin
	Oxytocin	^	β	?
Theca cells	Androgens	^	β	?
Testis, Leydig cells	Testosterone	^	β	Luteinizing hormone (LH)
Pineal gland	Melatonin	^	β	Light-dark cycle
Heart, atrium	Atrial natriuretic factor	^	α, β	Atrial distension

^ = increase; v = decrease.

Reprinted from Young and Landsberg 1998.

Pancreatic Hormones

Endocrine pancreas is innervated by vagal fibers originating in the DVN, preganglionic S fibers from fifth through ninth thoracic segments, and postganglionic S fibers originating in the celiac ganglion. The autonomic control of pancreatic hormone secretion is in turn directed by the glucoregulatory hypothalamic centers in response to centrally detected changes in glucose availability (Boden et al. 1993; Havel et al. 1988; Young & Landsberg 1979a). In conjunction with feeding, pancreatic hormone secretion is controlled principally by circulating glucose and amino acids, as well as through vagal stimulation. During exercise, or other stresses that increase S outflow, NE from S nerve terminals inhibits insulin secretion by binding to α_2 receptors on β cells of pancreatic islets (Young & Landsberg 1998). S nerves stimulate glucagon and SRIF release (Havel et al. 1991) by acting on α_1 and β receptors located on α and δ cells, respectively. These S effects allow increases in lipolysis and in plasma glucose concentration by blocking the antilipolytic and hypoglycemic actions of insulin, and they thus play an important role in the control of fuel delivery to exercising muscles (see chapter 5). Stimulation of β_2 adrenergic receptors transiently stimulates both insulin and glucagon release, but in humans α receptor–mediated inhibition of insulin secretion usually predominates over β-mediated stimulation (Hirose et al. 1993; Lacey et al. 1993). The SA system is not essential for glucose mobilization during exercise, which can take place without the S nerve involvement and in the absence of the adrenomedullary E (Coker et al. 1999; Howlett et al. 1999a, 1999b). The SA messengers also may inhibit insulin secretion indirectly by paracrine route after first stimulating secretion of SRIF and pancreatic polypeptide (PP) (Ahren et al. 1987). Other messengers released from S nerves, such as galanin and NPY, diminish pancreatic blood flow, suppress insulin and SRIF, and increase glucagon secretion independently of NE (Dunning & Taborsky 1991). In addition to the active inhibitory effect of S nerve activity over insulin secretion, exposure of pancreatic islets to catecholamines potentiates release of insulin and glucagon to their usual secretagogues (Burr et al. 1976).

Heart Hormone

Atrial natriuretic peptide (ANP) is secreted by the heart atria cells in response to β adrenergic stimulation in vivo, and to stimulation of both α and β receptors by catecholamines in vitro (Ambler & Leite 1994; Sanfield et al. 1987; Speake et al. 1993). Stimulation of β adrenergic receptors also increases expression of the ANP gene. Both the ANP and the brain natriuretic peptide (BNP) are secreted during dynamic and, to a lesser extent, during isometric exercise (Barletta et al. 1998).

Thyroid and Parathyroid Hormones

The S nerves originating in cervical ganglia and the vagus nerve innervate thyroid follicles, blood vessels to the thyroid gland, and chief cells in the parathyroid glands (Ahren 1986; Norberg et al. 1975; Tice & Creveling, 1986). Stimulation by S nerves increases thyroid hormone release and is associated with food intake and weight gain, whereas suppressed S activity and secretion of biologically inactive reverse T_3 (rT_3) accompanies fasting and weight loss (Rosenbaum et al. 2000). E promotes iodine uptake by the gland, but catecholamines also suppress thyroxine (T_4) release in response to thyrotropin (TSH).

When cellular calcium levels are low, β-receptor stimulation increases secretion of both parathyroid hormone (PTH) and calcitonin, two hormones that influence bone remodeling and bone mineral content, whereas α-receptor stimulation decreases it (Fischer et al. 1982; Heath 1980). The S nerves also directly innervate the periosteum, bone blood vessels, epiphyseal growth zone, and bone marrow. S stimulation of the β receptors increases osteoblast, and decreases osteoclast, activity and prevents bone demineralization associated with inactivity (Herskovits & Singh 1984; Sherman & Chole 1995; Zeman et al. 1991). Dopamine D_1 receptors also appear to participate in PTH secretion. Overall, the role of the SA system in the control of the parathyroid gland is unclear, but exercise may be influencing bone remodeling in part by stimulating SA function.

Gonadal Hormones

S innervation of the interstitial tissue and vasculature of the ovary (Burden 1985) reaches adult levels at the time of puberty (Schultea et al. 1992). Catecholamines increase progesterone production by follicles and diminish its degradation during the luteal phase and in pregnancy by a β_2 adrenergic mechanism. In contrast, α adrenergic stimulation inhibits gonadotropin-induced progesterone release (Hsueh et al. 1984). Catecholamines also stimulate milk ejection by causing contraction of myoepithelial cells in the milk ducts of lactating women. Stimulation by catecholamines of myoepithelial cells in the ovary facilitates ovulation. In addition, gonadotropins modulate the responsiveness of granulosa cells to catecholamines (Aguado & Ojeda 1986; Rani et al. 1983), progesterone administration activates S nerve discharge (Young & Cohen 1991), and catecholamines preserve ovarian responsiveness to gonadotropins (Aguado & Ojeda 1984). The SA system may be involved in pubertal maturation of the female reproductive axis and the pathophysiology of polycystic ovary syndrome, conditions in which S activity may be elevated (Barria et al. 1993; Lara et al. 1993).

The role of catecholamines in control of testicular androgen release at rest and during exercise remains uncertain. The S nerves innervate Leydig cells (Prince 1992), and S nerve stimulation increases testosterone (T) synthesis and release in vitro (Eik-Nes 1969). It also potentiates androgen production to gonadotropic hormones (Campos et al. 1993). However, catecholamines lower circulating T concentrations in vivo (Eik-Nes 1969).

Gastrointestinal Hormones

The mucosa of the stomach and intestine contains endocrine cells that produce a number of different hormones that complement a parallel series of hormones in the brain (Baulieu & Kelly 1990; Pearse 1969). Most GI hormones communicate by endocrine as well as paracrine routes, and a few (gastrin, SRIF, and secretin) also are released into the GI lumen. Hormones that increase GI motility are motilin, SP, pancreatic polypeptide (PP), and enkephalins, whereas secretin, glucagon, VIP, GIP, NPY, and neurotensin (collectively called enterogastrones) inhibit it. Gastrin and CCK increase motility in some GI regions and decrease it in others (see figure 2.9). Almost all GI hormones stimulate GI blood flow, particularly SP and neurotensin. NPY, however, is a potent vasoconstrictor. Serotonin-secreting and SRIF-secreting cells are ubiquitous throughout the GI tract. Cells producing secretin, CCK, gastrin colocalized with CCK, β-endorphin, neurotensin, and gastric inhibitory peptide (GIP) are mostly found in the duodenum and jejunum, while glucagon-secreting cells are more prevalent in the jejunum, ileum, and colon. Ghrelin is released from the antrum (Kojima et al. 1999). The S fibers containing NE and SRIF predominantly inhibit GI hormone secretion (Costa et al. 1986) and stimulate gastric inhibitory peptide secretion (Salera et al. 1982), whereas the vagus does the reverse.

Pineal Hormone

Pineal melatonin secretion is triggered by S nerves originating in superior cervical ganglion (Ebadi & Govitrapong 1986; Reiter 1991) and is controlled by a neuronal pathway that originates in the retina and responds to diurnal changes in solar illumination. Its effects on performance and biological rhythms are described in chapter 9.

INTEGRATIVE AUTONOMIC FUNCTIONS

An important role of catecholamines during exercise is to produce integrated adjustments in heart function, circulation, and metabolism that anticipate or meet increased energy needs. Through its automatic vasomotor, sudomotor, and endocrine reflexes, the ANS complements behavioral thermoregulation and contributes to regulation of energy supply and body fluids during exercise.

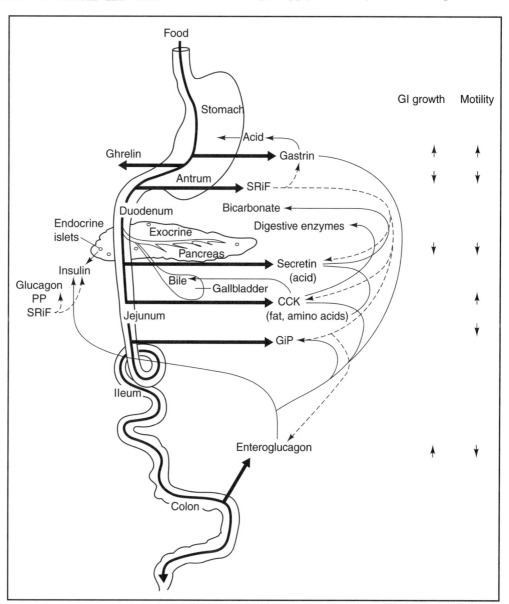

Figure 2.9   *Gastrointestinal endocrine reflexes in control of digestion.*

Food elicits gastrin release, and fat and protein elicit the release of cholecystokinin (CCK). Both hormones stimulate gastric motility, and CCK causes contraction of the gall bladder and release of bile acids necessary for the emulsification of fats. Gastrin releases hydrochloric acid as the preliminary step in digestion of proteins. The acidity of chyme is the stimulus for secretin release from the duodenum, and it triggers secretion of bicarbonate and digestive enzymes from the pancreas.

—— = stimulation; ---- = inhibition; CCK = cholecystokinin; GIP = gastric inhibitory peptide; SRIF = somatotropin release inhibiting hormone or somatostatin.

Autonomic reflexes represent the simplest level of autonomic control. A sensory message is sent in afferent neurons out of muscle or visceral receptors to integrative spinal or brain centers. There it triggers an anticipatory or corrective message that is carried back in S or PS efferent neurons. Autonomic reflexes may be neuronal, coupled to hormonal reflexes, or embedded within complex neuroendocrine and behavioral motivations and drives. This redundancy provides more precise regulatory control and a backup mechanism for defense of internal environment when any one of the feedback or feed-forward circuits malfunctions. Value of such redundancy is illustrated in common congenital deficiency of the adrenocortical enzyme 21-hydroxylase required for synthesis of cortisol and aldosterone (see figure 1.10). In its absence, aldosterone-mediated sodium reabsorption is deficient, and sodium deficiency develops because of the malfunction of RAA reflex. Plasma sodium is maintained however, and death from sodium deficiency is prevented, as higher brain centers take over the malfunctioning RAA reflex and generate increased craving and drive for salt (Wilkins & Richter 1940).

Autonomic Control of Cardiovascular and Respiratory Function

During exercise, catecholamines augment cardiac output and redistribute blood flow from predominantly visceral vascular beds to contracting cardiac and skeletal muscles. Cardiovascular functions are managed either in a regulatory fashion to maintain cardiac output and blood pressure within homeostatic norms at rest, or they are managed in an emergency fashion to provide life-saving adjustments during extended or intense exercise and other types of stress. Activation of the motor cortex during exercise and of lateral regions of the reticular activating system in the pons, mesencephalon, and diencephalon during various forms of stress are believed to stimulate cardioacceleratory and vasoconstrictor areas of the vasomotor center to meet variable energy, oxygen, or waste-disposal needs. Increased heart rate is achieved through withdrawal of vagal influence and through S nerve–induced stimulation of the sinoatrial cardiac pacemaker (Pagani et al. 1997). Increased cardiac contractility is achieved through stronger cardiac muscle contractions independently of a similar effect caused by cardiac distension by greater end-diastolic blood volume. Increased cardiac output is achieved through a direct action of catecholamines on β_1 receptors to increase heart rate, contractility, and conduction velocity. Ventricular contractility also is enhanced by catecholamine-induced venoconstriction. This increases venous return, cardiac distension, and end-diastolic blood volume (Frank-Starling's law). During maximal exercise intensity, these autonomic and other reflex mechanisms can increase heart rate to between threefold and fourfold and increase ejection fraction by about 25%, resulting in up to a fourfold to fivefold increase in cardiac output (see chapter 5). As

a consequence, systolic pressure, mean arterial pressure (MAP = diastolic pressure plus one third of the difference between systolic and diastolic pressure), and pulse pressure (the difference between systolic and diastolic pressure) all increase in proportion to exercise intensity. Systolic blood pressure rises because of simultaneous activation of vasoconstrictor and cardioacceleratory areas of the vasomotor center. Diastolic pressure between heartbeats also depends on the adjustments in the diameter of blood vessels by vasomotion. Catecholamine-induced cardiac acceleration and enhanced contractility during exercise increase myocardial oxygen needs.

Blood flow, which is predominantly directed to visceral organs at rest, is redistributed to muscle during exercise. This redistribution is achieved by vasodilation in the contracting muscle and selective constriction of arteries and arterioles in individual vascular beds in the viscera and tissues other than muscle. Constriction of arteries, arterioles (resistance vessels) in visceral vascular beds (McAllister 1998), and the lymphatic system is achieved through catecholamine action on both α_1 and α_2 receptors to modulate peripheral resistance and tissue perfusion (Ruffolo & Hieble 1994). Constriction of veins, which predominantly contain α_1 receptors, affects venous capacitance and plasma volume (Ping & Faber 1993). At high exercise intensities, catecholamines also restrain blood flow in exercising muscles to prevent muscle blood flow from outstripping cardiac output (Hansen et al. 2000). ATP, NPY, galanin, and opioids that are colocalized with catecholamines in S nerves also cause vasoconstriction. Vasoconstrictive action of chemical mediators other than catecholamines is discussed in chapter 3.

Vasodilation of blood vessels in contracting skeletal muscles is influenced by circulating and locally produced chemical messengers on one hand (see chapter 3) and by central and peripheral actions of catecholamines on the other. In the brain, E mediates vasodilation by stimulating A_{2a} adenosine receptors in NTS (Kitchen et al. 2000). At low concentrations, circulating E causes vasodilation by acting on noninnervated β_2 receptors (Ghaleh et al. 1995; Russel & Moran 1980; Vanhoutte & Miller 1989). Increases in acidity, temperature, and hypoxia that develop in exercising muscles contribute to vasodilation by inhibiting vasoconstriction mediated by S nerve stimulation of α_2 and, to a lesser extent, of α_1 receptors (Freedman et al. 1992; Tateishi & Faber 1995; Thomas et al. 1994). Stimulation of α and β receptors, as well as cholinergic receptors, by S nerves increases synthesis of endothelial vasodilatory messenger nitric oxide (NO) (Lundberg et al. 1989; Morris et al. 1995). Exercise training enhances blood flow to the skeletal muscles in part through increased S cholinergic and decreased S adrenergic nerve activity, and decreased vascular sensitivity to NE (Delp 1998).

The reflexes that control blood volume, flow, and pressure during exercise and at rest are the atrial mechanoreceptor reflex, the baroreflex, and the metaboreflex.

Atrial and Pulmonary Mechanoreceptor Reflexes

Atrial mechanoreceptor reflex is a response to increases in plasma volume detected by the low-pressure atrial baroreceptors (Castellani et al. 1999). Low-pressure cardiopulmonary stretch receptors also detect changes in plasma volume returning to the lungs. Through their tonic inhibition of vasomotor center, these stretch receptors participate in the long-term regulation of blood pressure. They reduce plasma volume after increases in venous return by initiating diuretic and natriuretic endocrine reflex through secretion of ANP and by reducing vasoconstriction in the kidney (Roy et al. 2000). Diuresis is achieved through increased release of ANP as well as increased glomerular filtration rate (GFR).

Baroreflex

Baroreflex, or orthostatic reflex, regulates systemic blood pressure through adjustments in heart rate and car-diac output (Loewy & Spyer 1990). Cardiac output and blood pressure are modulated by baroreceptors and by chemoreceptors (see later) that provide inhibitory parasympathetic feedback or positive sympathetic as well as endocrine stimulation (Persson et al. 1989). High-pressure baroreceptors in the carotid sinus and aortic arch buffer short-term blood pressure fluctuations (see figure 2.10).

The discharge rate of the afferent fibers in carotid sinus branch of the glossopharyngeal and in depressor branch of the vagus that originate in arterial high-pressure baroreceptors is directly proportional to arterial blood pressure (see figure 2.11). When the blood pressure falls, the baroreceptor discharge decreases and reduces PS and increases cardioacceleratory and vasoconstrictive S nerve action. The result is an increase in heart rate and contractility through stimulation of S cardioaccelerator nerves and an increase in the rhythmic cardiac and

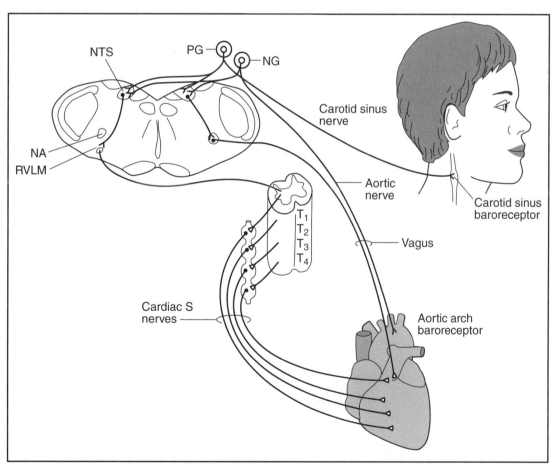

Figure 2.10   *The baroreflex arc.*

High-pressure baroreceptors in the aortic arch and carotic sinus send afferent messages, respectively, in the vagus and carotid sinus branch of the glossopharyngeal nerves to the brain stem circulatory center. The inhibition of the heart rate and contractility and reduction in vasoconstriction of blood vessels is carried out by depressor nerve branch of the vagus with preganglionic neurons in the nucleus ambiguus (NA).

NG = nodose ganglion of the tenth nerve; NTS = nucleus of the tractus solitarius; PG = petrosal ganglion of the ninth nerve; RVLM = rostral ventrolateral medulla; S = sympathetic.

Adapted from Thews and Vaupel 1985.

vasomotor drive from the medullary RVLM and raphe nuclei. In addition, peripheral resistance in the muscle and, to a lesser extent, in the skin also increases. When the blood pressure increases, opposite adjustments occur. Baroreflex redistributes blood to, or away from, the head and upper regions of the body to compensate for gravitational shifts in blood volume to postural changes to upright or to recumbent.

Baroreflex function adapts to long-term changes in blood pressure (Brooks 1997) and is adjusted in some physiological and pathological conditions. Its gain is reduced during exercise, to allow concomitant increases in heart rate and blood pressure (Iellamo et al. 1997; Norton et al. 1999). In postural and postprandial hypotension seen in aging or in acute postexercise hypotension in hypertensive subjects, orthostatic reflex operates sluggishly and causes transient cerebral ischemia and dizziness (Jansen & Lipsitz 1995; Puvi-Rajasingham et al. 1998; Shoemaker et al. 1999). Exercise training in-

creases baroreflex gain by increasing the PS and decreasing the S tone to the cardiovascular system (Kingwell et al. 1992; Somers et al. 1991).

Renin-Angiotensin-Aldosterone Reflex

Baroreflex operates in parallel and interactively with endocrine blood pressure reflexes. Renal baroreceptors contribute to blood pressure elevation by triggering AVP release and the renin-angiotensin-aldosterone (RAA) reflex that is discussed in chapter 4 (Caverson & Ciriello 1987). Potentiating influence of AVP and attenuating influence of angiotensin II on baroreflex sensitivity also are discussed in chapter 4. β Adrenergic stimulation increases aldosterone secretion by the adrenal cortex (Pratt et al. 1985).

Metaboreflex

Metaboreflex, or muscle pressor reflex, is a reflex increase in systemic blood pressure in response to buildup of

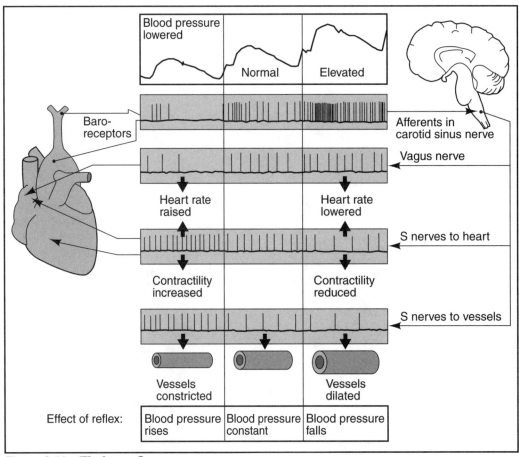

Figure 2.11   The baroreflex.

The baroreflex regulates systemic blood pressure through adjustments in heart rate and peripheral resistance. The discharge rate of high-pressure baroreceptors is proportional to arterial blood pressure, whereas in the sympathetic (S) cardiac and vascular nerves, it is inversely proportional.

Adapted from Thews and Vaupel 1985.

metabolic waste products in the muscle during ischemia or isometric muscle contractions. Blood pressure is raised through muscle arteriole vasoconstriction caused by both increased S discharge (Nishiyasu et al. 1998, 2000; Rowell & O'Leary 1990) and increased AVP release.

Autonomic Control of Pulmonary Function

Pulmonary functions are regulated both by a rich PS and by a sparser S innervation of bronchial muscles. The former causes contraction, and the latter causes vasodilation mediated by β_2 adrenergic receptors (Barnes 1995). In addition, catecholamines stimulate secretion of mucus from the bronchial epithelium. Chemoreceptor and diving reflexes influence cardiovascular and cardiorespiratory functions at rest and during exercise. S nerves increase bronchial ciliary motility by acting on β_3 receptors (Tamaoki et al. 1993).

Chemoreceptor Reflex

Chemoreceptor reflex consists of increases in minute ventilation ($\dot{V}_E$) and cardiac output ($\dot{Q}$) and of adjustments in vasoconstriction in response to reduced pO_2 and increased pCO_2. These changes are detected, respectively, by chemoreceptors in the aortic sinus and carotid body and by cells on the ventral surface of the medulla. Inhibitory vagal influence from the NA and cardiovascular and respiratory nuclei in the medulla (see figure 2.1) modulate this reflex (Loewy & Spyer 1990). During exercise, coordinated patterning of cardiovascular and respiratory responses is apparently more controlled by the posterior hypothalamus or the DM than by the chemoreflex (Eldridge et al. 1985; Saper et al. 1976; Waldrop et al. 1988).

Diving Reflex

Diving reflex consists of bronchoconstriction and suppression of the $\dot{Q}$ and respiratory drive in response to face cooling and breath holding during swimming or exposure to cold air (Kawakami et al. 1967; Smith et al. 1997a). It is initiated by PS afferents in the facial (seventh) nerve and is mediated by the vagus nerve. Chemoreceptor reflex, elicited by the rise in pCO_2 and fall in pO_2 during breath holding, is subordinated to diving reflex through interactions between the neurons in the Botzinger complex and the RVLM nucleus that initiate and time respiratory and cardiovascular functions.

Autonomic Control of Hemostasis and Fibrinolysis

In addition to their circulatory and respiratory functions, catecholamines also influence clotting (thrombosis) and fibrinolysis. E increases platelet count and stimulates platelet adhesion and aggregation through α_2-receptor mechanisms (Kjeldsen et al. 1995; Wallen et al. 1999; Wang & Cheng 1999). E increases plasma concentrations of factor VIII, von Willebrandt's factor, and tissue plasminogen activator (tPA) and decreases concentration of PA inhibitor-1 (Larsson et al. 1990,1992). E also stimulates hepatic synthesis of fibrinogen, a constituent of clots and atherosclerotic plaques (Roy et al. 1985).

Autonomic Control of Body Fluids and Temperature

Catecholamines strongly influence body water balance (discussed in chapter 4) and water and electrolyte shifts among body fluid compartments through a multiplicity of mechanisms (Young & Landsberg 1998). Stimulation of β_2 receptors helps preserve body water and plasma volume by increasing intestinal absorption of water and sodium and by increasing renal sodium reabsorption directly and indirectly. Direct effects are on the JGA through stimulation of RAA axis. Indirect effects are mediated by decreases in GFR and redirection of blood flow from kidney cortex to its medulla (see chapter 4). E causes decreases in plasma concentrations of magnesium and calcium by increasing their urinary excretion. β_2-Receptor stimulation suppresses urinary excretion of phosphate. β Adrenergic stimulation increases potassium uptake by the muscle by acting on the Na^+-K^+ membrane pump. This S action counterbalances K outflow during exercise (Lundborg 1983; Williams et al. 1985) and thus helps prevent hyperkalemia during exercise. In general, all of these fluid and electrolyte effects are antagonized by α adrenergic or DA stimulation.

Thermoregulation during exercise is another ANS function that is integrated with other hormonal and behavioral mechanisms involved in regulation of body fluid balance (Sawka & Coyle 1999). These are discussed in more detail in chapter 4. Two reflexes, vasomotor and sudomotor, play a central role in temperature regulation at rest and during exercise.

Sudomotor Reflex

Sudomotor reflex helps maintain stable core temperature by facilitating heat loss through sweating. Temperature change is detected by somatic temperature-sensitive neurons in the skin or in the anterior hypothalamus, and the eccrine sweat glands are activated by cholinergic S neurons. This reflex is less sensitive to internal or external temperature change than the vasomotor reflex (Stolwijk & Hardy 1977). Apocrine (odor-producing) sweat glands in the axillary and genital areas are stimulated by noradrenergic S fibers.

Vasomotor Reflex

Vasomotor reflex helps maintain stable core temperature at rest or during exercise by controlling vasoconstriction or vasodilation in response to signals from central (brain) or skin temperature receptors. This reflex adjusts radiative heat loss from the body surface. In the cold, constriction of the peripheral veins and shunting of blood to deep limb veins help prevent body heat loss. During exercise and in the heat, peripheral vasodilation facilitates radiative loss of excess body heat (Castellani et al. 1999).

Autonomic Control of Metabolism

Another major role of catecholamines in exercise or other conditions of increased energy need is to rapidly increase utilization of stored fuels. They do so by facilitating breakdown of stored fuels to metabolizable substrates, controlling delivery of these substrates to contracting muscles by affecting blood flow and by influencing uptake and metabolism of substrates in muscle and other tissues. These effects are discussed in more detail in chapter 6. Catecholamine effects are either direct or mediated indirectly through release of other metabolic hormones.

Both the S nerves and the circulating E cause hydrolysis of triglycerides into free fatty acids (FFAs) in white adipose tissue as well as in muscles (Lafontan & Berlan 1993; Oscai et al. 1990). In both tissues, this activity is mediated by β adrenergic mechanism and by phosphorylation of hormone-sensitive lipase (HSL), processes that are strongly assisted by decline in plasma insulin concentration (Kjaer et al. 2000; Seip et al. 1997). All three β adrenergic receptors (β_1, β_2, and β_3) respond to lipolytic actions of catecholamines, and the outcome is influenced by variation in relative receptor numbers and sensitivity in a given tissue as a function of gender, age, and other physiological conditions (Lafontan & Berlan 1993). Reduced S activation of β_3 adrenoceptors and lipolysis in the muscle and the adipose tissue has been implicated in the etiology of obesity (Snitker et al. 1998; Webber & MacDonald 2000). NE released from S nerve endings is the stimulus for lipolysis in brown adipose tissue (Palou et al. 1998). This process is the principal source of body heat production by nonshivering thermogenesis in newborn humans and hibernators. S thermogenesis is augmented in the presence of thyroid hormones, as α_1 receptor stimulation leads to deiodination of T_4 to T_3. Thermogenic adaptation to cold is in part achieved through S facilitation of brown adipose tissue hyperplasia (Geloen et al. 1992). Of interest in the context of cardiovascular health, stimulation by catecholamines of β adrenergic receptors increases the plasma concentration of cholesterol in part through their stimulation of hepatic rate-limiting enzyme 3-hydroxy-3-methyl glutaryl coenzyme A (HMG-CoA) reductase (Devery & Tomkin 1986; Devery et al. 1986; Dimsdale et al. 1983; Edwards 1975; Kunihara & Oshima 1983). Elevation of plasma cholesterol in turn causes sympathetic activation (Smith et al. 1992a). Catecholamines also raise plasma triglyceride concentrations in part by facilitating resynthesis of triglycerides from FFAs in the liver (Brindle & Ontko 1988).

Catecholamines indirectly influence metabolism in favor of lipid utilization in part by affecting cellular uptake of metabolic substrates and visceral blood flow. Catecholamines selectively increase FFA uptake by skeletal muscles and decrease FFA uptake by adipose tissue through a β adrenergic mechanism (Oscai et al. 1988). This process occurs during low-intensity exercise and immediately after intense or prolonged exhausting exercise through activation of lipoprotein lipase (LPL) in the skeletal muscle and suppression of this enzyme's activity in adipose tissue (Deshaies et al. 1993; Miller et al. 1989). At low exercise intensities, when both plasma insulin and catecholamine concentrations are low, NE

and E stimulate glucose uptake into cardiac and skeletal muscle as well as into white and brown adipose tissues by both α and β adrenergic processes (Marette & Bukowiecki 1989; Rattigan et al. 1991; Shirakura et al. 1990). They do so in part by triggering transcription of hexokinase II gene and by stimulating glucose phosphorylation (Osawa et al. 1995). At high exercise intensities or during prolonged exhausting exercise, high E levels inhibit both insulin and non–insulin-mediated cellular glucose uptake by β adrenoceptor and adenosine receptor mechanisms (Diebert & DeFronzo 1980, Joost et al. 1986; Rizza et al. 1980). Blood flow to the adipose tissue is controlled by α-mediated vasoconstriction and β-mediated vasodilation triggered by S nerves (Coppack et al. 1994). As exercise intensity and plasma catecholamine concentrations rise, blood flow from adipose tissue is curtailed in a dose-dependent fashion, and metabolic fuels that are stored within the muscle are preferentially used instead of substrates originating in the liver or adipose tissue (see chapter 6).

Glucose is mobilized from liver and muscle glycogen stores by glycogenolysis, a process that is catalyzed by the active form of enzyme glycogen phosphorylase. In the liver, glycogenolysis is directly activated by stimulation of β_2 receptors via cAMP transduction pathway or of α_1 receptors causing increased cytoplasmic calcium release (Exton 1988; Katz et al. 1993). During exercise, neither S nerves nor E are necessary for glycogenolysis to take place (Coker et al. 1999; Howlett et al. 1999a, 1999b). In the human liver, β adrenergic effects predominate over α adrenergic actions of catecholamines. In the skeletal muscles, catecholamines stimulate glycogen breakdown in the presence of glucocorticoids by acting on β_2 receptors (Clark et al. 1995; Young et al. 1985). Glucose thus produced is metabolized by the muscle and is not released into circulation. An additional source of glucose production and release is gluconeogenesis from lactate, pyruvate, amino acids, and glycerol. This process is mediated by α_1 receptor stimulation in the liver and proximal convoluted tubule of the kidney (Exton 1988; McGuinness et al. 1993; Stumvoll et al. 1995; Wirthenson & Guder 1986). In both the muscle and the liver, glycogenolysis and gluconeogenesis are suppressed by insulin.

Protein also can be utilized as a fuel both during prolonged, intense, and glycogen-depleting exercise and during starvation. Normally, catecholamines inhibit breakdown of muscle proteins into amino acids by a β adrenergic mechanism (Fryburg et al. 1995; Garber et al. 1976; Kraenzlin et al. 1989). This effect is opposite to that of glucocorticoid action and thus balances glucocorticoid proteolysis during exercise and other forms of stress when hormones of both the adrenal cortex and the medulla are simultaneously activated (Del Corral et al. 1998; Straumann et al. 1988). Chronic β_2 agonist administration increases skeletal muscle mass more by decreasing protein degradation than by increasing protein synthesis (Fryburg et al. 1995). This process is achieved through suppression of the ATP-ubiquitin–dependent proteolysis (Costelli et al. 1995; Yang & McElligott 1989). Catecholamine stimulation contributes to adaptive hypertrophy of cardiac, skeletal, and vascular smooth muscles. Hypertrophy is mediated through an α adrenergic mechanism associated with the inhibition of adenylylcyclase (Kim & Sainz 1992; Lee et al. 1995; Terzic et al. 1993).

Catecholamines also indirectly affect carbohydrate utilization by influencing their delivery to, and uptake by, the tissues. E and NE decrease, and DA increases, hepatic blood flow and therefore hepatic glucose output in a dose-dependent fashion (Gardemann et al. 1992; Richardson & Withrington 1982). Catecholamine-induced glucagon secretion however decreases hepatic artery vasoconstriction and facilitates hepatic glucose output, whereas NPY released from S nerve endings has the opposite effect (Ahlborg & Lundberg 1994; Richardson & Withrington 1982). Catecholamines affect hepatic and muscle substrate uptake by stimulating gluconeogenic Cori cycle. This action entails increased hepatic uptake of lactate and release of glucose and increased glucose uptake by the muscle and production of lactate (Cori 1925, 1981). Catecholamines have a greater influence on substrate cycling than on direct stimulation of glucose or lactate metabolism (Forichon et al. 1977; Kusaka & Ui 1977).

Stimulation of β_3, but not of β_1 or β_2 receptors, increases glucose uptake in red and white skeletal muscles (Liu et al. 1996). Stimulation of β adrenoceptors causes short-term (one to two days) reduction in insulin sensitivity. Longer-term (days to weeks) exposure to adrenergic stimulation increases the sensitivities of glucose transport and glycolysis to insulin (Budohoski et al. 1987; Cawthorne et al. 1992). This increased sensitivity of muscle to insulin after acute exercise or exercise training (Prigeon et al. 1995; Scheidigger et al. 1984; Yang & McElligott 1989) may, in part, be a consequence of increased adrenergic stimulation of expression of GLUT4 glucose transporters or of hexokinase II in skeletal muscle and adipose tissue (Torgan et al. 1995; Tsukazaki et al. 1995).

SA catecholamines also directly increase energy metabolism in skeletal muscles and adipose tissue and are therefore important mediators of oxidative calorigenesis (Clark et al. 1995; Landsberg & Young 1983). Catecholamines stimulate lipid oxidation during prolonged submaximal exercise (Fernandez-Pastor et al. 1999). Increased triglyceride–fatty acid cycling after intense exercise is stimulated by high titers of catecholamines and growth hormone (GH) but is not mediated by β or β_1 receptor stimulation (Borsheim et al. 1998; Pritzlaff et al. 2000). Catecholamines cause increased nonshivering thermogenesis by mediating uncoupled oxidation of FFAs in brown-fat mitochondria. In this process, the metabolic heat production is unaccompanied by ATP formation (Dicker et al. 1996; Palou et al. 1998). Both the stimulation of thermogenesis and the transcription of the gene for the mitochondrial uncoupling protein

thermogenin (UCP I and II) are mediated by β_3 adrenergic receptors (Cannon & Nedergaard 1996; Klaus et al. 1991).

Reflexes that contribute to autonomic control of energy balance and fuel mobilization are the GI reflexes and the hyperglycemic reflex to glucoprivation.

Gastrointestinal Reflexes

GI reflexes entail vagal sensory information traveling from the viscera to the spinal cord or the brain and commands from brain integrative centers carried back in splanchnic and vagus nerves to enteric plexuses of the GI tract and their embedded PS ganglia. At rest and after meals, the S tone to the GI plexus is reduced, and the vagus nerve stimulates myenteric and submucous ganglia, gastric motility, and secretion of GI hormones, enzymes, and other products (Berthoud et al. 1991; Holst et al. 1997). Feeding initiation is associated with reflex release of the hormone ghrelin from the stomach and duodenum (Cummings et al. 2001; Dornonville de la Cour et al. 2001; Kojima et al. 1999). Increases in gastric motility and acid secretion induced by ghrelin are mediated by the vagus nerve (Masuda et al. 2000). Intragastric food constituents trigger serial release of gastrin, secretin CCK, enterostatin, motilin, GIP, SRIF, and neurotensin according to the regional distribution of the endocrine cells producing them, type of ingested nutrients, and timing of progression of chyme through the GI tract. GI hormones that are released early in the digestive process facilitate serial release of other GI and pancreatic hormones, and hormones released later in the digestive process terminate the early steps of this endocrine cascade (see figure 2.9). Gastrin is released from parietal cells in the antral stomach wall in response to food entry and PS nerve stimulation. It triggers secretion of hydrochloric acid, which increases acidity of the chyme (to a pH of about 2), aids in initial digestion of proteins, and serves as the stimulus for the release of secretin. Secretin triggers the release of bicarbonate and digestive enzymes from the exocrine pancreas. This release restores neutral pH of the chyme and permits digestive enzymes to continue nutrient breakdown. Dietary fats and peptides trigger the release of CCK and enterostatin, and both hormones bind to their receptors on the vagus nerve (Corp et al. 1993; Erlanson-Albertsson & York 1997). The role of CCK is to release bile acids from the gall bladder (and digestive enzymes from the pancreas) and assist in emulsification and digestion of fats, and both hormones have been implicated in suppression of appetite (Gibbs et al. 1973; Liu et al. 1999). Food and caloric drinks also trigger secretion of SRIF from the antrum, duodenum, and throughout the GI tract and of enteroglucagon from the colon. These two hormones as well as serotonin inhibit secretion of gastrin and gastric acid production. SRIF also inhibits secretion of all other GI and pancreatic hormones and suppresses splanchnic circulation. The presence of SRIF is necessary for the GIF, VIP, and GIP to inhibit gastrin release.

Digestive hormones produce incretin effect, a facilitation of release of insulin, pancreatic polypeptide (PP), and glucagon mediated by the vagus. For that reason, insulin secretion is greater when sugars are eaten than when they are administered intravenously. Upon absorption, increased plasma concentrations of nutrients directly stimulate secretion from the pancreas of insulin, glucagon, and PP and inhibit SRIF release. Insulin secretion is triggered when sugars stimulate intestinal, hepatic, and pancreatic islet glucoreceptors and activate the branches of the vagus (Berthoud & Powley 1990). Glycogen synthesis is concurrently stimulated through the hepatic branch of the vagus (Niijima 1989).

During exercise and other forms of stress, the S nerves inhibit cholinergic and probably NO neurotransmission in enteric ganglia and increase contraction of the sphincters by acting on α and particularly α_2 receptors. The S nerves also inhibit GI motility, causing smooth muscle relaxation by acting on β_2 and sometimes β_3 receptors (Boeckastaens et al. 1993; Bulbring & Tomita 1987; de Boer et al. 1993; MacDonald et al. 1994). NPY in S fibers vasoconstricts splanchnic circulation.

Hyperglycemic Reflex to Glucoprivation

Acute or prolonged hypoglycemia selectively activates adrenomedullary E secretion for reflex increases in hepatic glucose production (Cannon et al. 1924; Houssay et al. 1924; Howlett et al. 1999a). In addition to E, S activation during hypoglycemia triggers secretion of NPY and enkephalins from the adrenal medulla and secretion of galanin from the pancreas (Damase-Michel et al. 1994; Havel et al. 1992). Regulation of E release during exercise is controlled to a greater extent by feed-forward mechanisms related to the level of stress rather than by feedback afferent input from glucoreceptors in the liver and other sites (Latour et al. 1995). S activity in general is decreased in chronic starvation that is accompanied by hypoglycemia and low insulin concentrations (Egawa et al. 1989; Katafuchi et al. 1988; Ritter et al. 1995; Young & Landsberg 1979b, 1997). Only the S outflow to muscle, skin, and pancreas remains active under these conditions (Berne & Fagius 1986; Fagius et al. 1986; Havel et al. 1988).

SUMMARY

Onset of exercise is marked by PS withdrawal and increased activity of S nerves (see figure 2.12) (Nakamura et al. 1993; Yamamoto et al. 1991). Most increases in plasma NE during exercise arise from NE spillover out of S nerves to skeletal muscle and the heart (Johansson et al. 1997; Peronnet et al. 1988; Wallin et al. 1992). At high intensities or during prolonged exhausting exercise, adrenomedullary secretion of E increases (see figure 2.12). Thus, both parts of the SA system control many different physiological functions in exercise to achieve an integrated whole-body response. The SA system responds to intensity and magnitude of overall exercise stress in a

dose-dependent fashion so that physiological, endocrine, and metabolic effects of catecholamines represent graded feed-forward responses that increase in anticipation of actual need. For example, increases in respiration and cardiac output result from S nerve activation and usually anticipate the onset of movement (Delp 1998, Mason et al. 1973). Catecholamine-mediated circulatory changes consisting of increased cardiac output, splanchnic and renal vasoconstriction, and variable dilation in skeletal muscles and the skin ensure that blood pressure is maintained during exercise. ANS also contributes to general arousal and causes pupillary dilation and piloerection. Other defense mechanisms that are activated during exercise include shifts in body fluids and minerals, effects on hemostasis, release of endogenous opiates, and activation of the immune system (Larsson et al. 1994). During exercise, some homeostatic control mechanisms may undergo readjustment to meet increased functional demands. For instance, baroreflex is reset to operate at increased blood pressure ranges encountered during exercise (Rowell & O'Leary 1990).

The S nerves effect redistribution of blood from the viscera to contracting skeletal muscles in response to metabolic signals in contracting skeletal muscles. At higher exercise intensities and during prolonged exhausting exercise, the SA system brings into action thermoregulatory sudomotor and vasomotor reflexes to help dissipate excess heat. These examples illustrate the homeostatic role of the SA system, which activates physiological compensations for the perturbations caused by exercise. Priorities for maintenance of different aspects of the internal environment shift during exercise. Maintenance of blood pressure through vasoconstrictive S and hormonal reflexes and regulation of body temperature through sweating assume higher priority than conservation of plasma volume by antidiuresis.

S suppression of insulin secretion and facilitation of glucagon secretion set the stage for increased mobilization and utilization of metabolic fuels. Predominant activation of S nerves during early or low-intensity exercise stimulates β_1 and α receptors respectively in the adipose tissue and the liver. This process facilitates increased mobilization and utilization of FFAs and activity of the Cori cycle. High FFA levels reduce muscle glucose uptake and sensitivity to α adrenergic stimulation (Burns et al. 1978). High FFA utilization during mild exercise intensities gives way to greater carbohydrate use at intermediate intensities. Glucose uptake by skeletal muscles is facilitated by S activation of α adrenergic receptors at lower exercise intensities and inhibited by β receptor stimulation at high intensities. Decreases in plasma insulin and increases in plasma glucagon, and E concentrations all contribute to increased hepatic glucose production. As exercise intensity increases further, circulatory constraints alter metabolism from a pattern characterized by predominant reliance on circulating substrates to one that relies more heavily on muscle utilization of en-

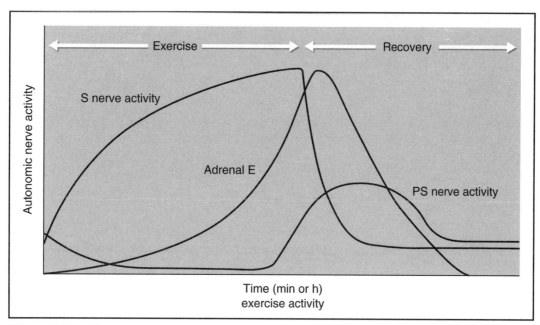

Figure 2.12 *Change in sympathovagal tone during exercise in response to increasing exercise intensity and during recovery from exercise.*

A conceptual illustration of changes in the relative activities of sympathetic (S) and parasympathetic (PS) nerves and in secretion of adrenomedullary epinephrine (E) as a function of intensity or duration of exercise and during recovery from exercise. At the start of exercise, PS tone declines and is transiently increased during recovery when it participates in restorative biosynthetic actions. At the start of exercise, S tone increases in proportion to intensity or stressfulness of exercise. Higher exercise intensities are required to elicit E secretion, which also is secreted in dose-dependent fashion. The decay of S nerve activity is faster than the disappearance of E from circulation.

Adapted from Young and Landsberg 1984.

dogenous fuels. During prolonged submaximal exercise, depletion of muscle glycogen and high plasma titers of catecholamines and other counter-regulatory hormones facilitate lipolysis and lipid utilization and proteolysis and protein utilization.

E secretion from the adrenal medulla is initiated at higher exercise intensities. As plasma catecholamine concentration increases, proportional increases in mobilization of FFAs out of triglycerides are not matched by equivalent increases in their plasma concentration or tissue utilization. This situation occurs because at high exercise intensities, catecholamine-induced vasomotor effects increasingly impair transport of FFAs out of adipose tissue. S elicitation of a number of other metabolic hormones at higher exercise intensities allows for a distinct progression in the pattern of metabolic fuel use (see chapter 5).

The SA system plays an important role in regulation of internal environment during recovery from exercise, when PS tone increases and S tone decreases. This change permits restoration of homeostasis, replenish-

ment of fuel stores, and repair of any structural damage to the musculoskeletal system that may have occurred. A decrease in S tone can lead to transient postexercise hypotension (Halliwill et al. 1996). After intense exercise, the decline in SA action allows plasma insulin and FFA concentrations to increase sharply, and this endocrine milieu contributes to a fall in hepatic glucose output and facilitates increased fat utilization and glucose uptake by the muscle (Pritzlaff et al. 2000; Wahren et al. 1971). During recovery from long-term exercise that has resulted in glycogen depletion and hypoglycemia, plasma insulin remains low and adrenomedullary release of E and secretion of glucagon are sustained. Thus, the SA and pancreatic reflexes here compensate for exercise-induced disturbances in glucose homeostasis and permit a more extended period of hepatic gluconeogenesis (Horton & Terjung 1988).

Habitual physical activity produces adaptations in SA and PS function. For instance, it leads to smaller NE responses and adrenal E secretion and greater PS responses at equivalent absolute exercise loads (Hartley et al. 1972a, 1972b; Goldsmith et al. 1993). The SA system also facilitates structural adaptations to increased functional demands of habitual exercise through stimulation of hypertrophy of cardiac and musculoskeletal systems. The ANS thus controls functional adjustments to multiple needs during acute exercise bouts and participates in physiological adaptations to sustained high levels of physical activity.

3

Receptors and Chemical Message Transduction in Exercise

Receptors (Rs) are composite proteins located on the cell membrane or in the interior of the cell that contain a specific recognition site to which hormone (H) binds. They provide a link to an effector mechanism through which the chemical message is transduced into biological action. Interpretation of studies in exercise physiology that use manipulations to modify H release or action requires an understanding of the characteristics of Rs and the way they interact with Hs. This chapter first outlines basic principles of H-R interactions. After hormones bind with receptors, messages are relayed to various effector processes within the cells through molecular transduction mechanisms. This chapter next describes principal types of receptors and transduction pathways classified first by their molecular characteristics, then organized by the biological effects achieved. As the selection of an appropriate method of H measurement requires an understanding of the limitations of available approaches, methods of H measurement also are described briefly from the perspective of H-R interactions.

HORMONE AND RECEPTOR INTERACTIONS

H-R binding can be distinguished from other types of chemical bonds by its specificity, reversibility, affinity (or sensitivity), and saturability. Specificity refers to the capacity of Rs to selectively bind a particular H molecule often present in circulation at very low concentrations of between 10^{-9} and 10^{-12} M. At the molecular level, specificity of interactions between an H and an R depends on their stereochemical correspondence and capacity to bind. Specificity is seldom absolute, as fami-

lies of structurally similar Hs exist, structurally similar Hs often bind to the same Rs, and one H may bind to several types of Rs. Catecholamines epinephrine (E) and norepinephrine (NE) can both bind to the same receptor types, and each can bind to different subcategories of α and β adrenergic receptors (see tables 3.1 and 3.2). Whenever there is some latitude in H-R specificity, Hs generally bind to individual Rs with differential affinity or binding strength. Insulin and insulinlike growth factor–I (IGF-I) can both bind to either insulin or IGF-I Rs, but each has greater affinity for its own R than for the R of its congener.

Specificity also can be defined by the characteristics of the H transduction mechanism, specific biological effects that an H produces in different tissues, and by nonrandom distribution of Rs throughout the body. Some Hs produce specific effects in different tissues because their transduction machinery triggers tissue-specific biological processes. Through its activation of cyclic adenosine monophosphate (cAMP)–dependent transduction pathway, E stimulates contraction in the heart muscle, conversion of glycogen to glucose in the liver, and breakdown of triglycerides to free fatty acids (FFAs) in the adipose tissue (see tables 3.1 and 3.2). When several different H-R complexes, such as glucocorticoid R (G-R), mineralocorticoid R (M-R), progesterone R (PROG-R), estrogen R (E-R), and androgen R (AN-R) bind to the same region of the DNA, message specificity is achieved through H binding to specific Rs and through activation of specific transduction and effector machinery in different tissues that can mediate only a particular H effect but not others. Tissue specificity of H action on gene expression develops during cellular differentiation when only some

Table 3.1 Families of Membrane-Spanning Receptors and Their Endocrine Ligands

Receptors with a single transmembrane segment	*G protein–coupled receptors with seven transmembrane segments*

Cytokine family:
    Cardiotropin-1
    Ciliary neurotrophic factor (CNTF)
    Erythropoietin (EPO)
    Granulocyte macrophage colony-
        stimulating factor
    Growth hormone (GH)
    Integrins
    Interferons (INFα/β, IFNγ)
    Interleukins (IL-2 to IL-7, IL-9 to IL-13)
    Leptin
    Leukemia inhibitory factor (LIF)
    Oncostatin M
    Prolactin (PRL)
    Thrombopoietin (TPO)
Integrin family
Low-density lipoprotein (LDL)
Nerve growth factor (NGF)
Transferrin
	Adenosine
	Adrenergic family:
	    Epinephrine (E)
	    Norepinephrine (NE)
	Acetylcholine (muscarinic)
	Angiotensin II
	Dopamine
	Endothelin family
	Glycoprotein hormones (LH, FSH, TSH)
	Opioids
	Platelet activating factor
	Prostanoid family
	    Prostacyclins
	    Prostaglandins
	    Thromboxanes
	Serotonin
	Substance P
	Vasopressin family (AVP)

Receptors with a single transmembrane segment and tyrosine kinase domain	
Epidermal growth factor receptor (EGF-R) family Colony-stimulating factor–I receptor (CSF-I R) Insulin receptor (I-R)/insulinlike growth factor–I receptor (IGF-I–R) family Fibroblast growth factor receptor (FGF-R) family Platelet-derived growth factor receptor (PDGF-R) family	

of several genes susceptible to hormonal control and located in DNAse I–sensitive regions of DNA get expressed. Most cellular genes remain tightly packaged with histone proteins in nucleosomes and are inaccessible to DNAse I action during cellular differentiation. Finally, Rs for a particular hormone are unevenly distributed and are represented in different densities in some tissues. This condition allows Hs to exert tissue-specific, age-specific, and gender-specific effects.

Reversibility refers to the property of Hs and Rs to form reversible noncovalent bonds described by the Michaelis-Menten equation for rapid equilibrium kinetics:

$$[H] + [R] \underset{k_d}{\overset{k_a}{\rightleftharpoons}} [HR]$$

[H], [R], and [HR] are, respectively, concentrations of free H (often abbreviated as F), free Rs, and the bound H (often abbreviated as B). The interaction between Hs

and Rs follows the law of mass of action and depends on the concentrations of the two reactants. At equilibrium, one can determine the association constant K_A, and the reciprocal dissociation constant K_D, from the following relationships:

$$K_A = k_a/k_d = [HR]/[H][R]$$
$$K_D = k_d/k_a = [H][R]/[HR]$$

Here K_A is given in M^{-1} and K_D in M at equilibrium, k_a is the association rate constant in $M^{-1} sec^{-1}$, and k_d is dissociation rate constant in sec^{-1} (Baulieu & Kelly 1990).

Saturability characteristic of H-R binding can be illustrated graphically in an H-R saturation plot (see figure 3.1a) or in a Scatchard plot (see figure 3.1c). As the number of Rs in any given system or a cell is finite (N = [R] + [HR]), the point of saturation refers to an H concentration at which all available receptors are bound and maximal binding capacity, or Bmax, has been attained. Maximal R occupancy is seldom needed for a full biological response, but when it occurs, the biological response

Table 3.2 Hormones Acting on G Protein–Coupled Receptors

Hormones with G_s protein–coupled receptors	Hormones with G_i protein–coupled receptors
Corticotropin (ACTH)	Muscarinic acetylcholine (mACh)
Adenosine (A_2-Rs)	Adenosine (A_1-Rs)
α Melanocyte–stimulating hormone (αMSH)	Angiotensin II (AT_1 and AT_2 Rs)
Arginine vasopressin (V_2-Rs in the kidney)	Bradykinin
Calcitonin	Dopamine (D_2, D_3, and D_4 Rs)
Dopamine (D_1 and D_5 Rs)	Endothelin (ET-1)
Epinephrine ($β_1$, $β_2$, and $β_3$ Rs)	Norepinephrine ($α_2$ Rs)
Follicle-stimulating hormone (FSH)	Opioids
Gastrin	Somatostatin (SRIF)
Glucagon	

	Hormones with G_q protein–coupled receptors
Gonadotropin-releasing hormone (GNRH)	Muscarinic acetylcholine (mACh)
Histamine (H_2-Rs)	Adenosine (A_1-R)
Luteinizing hormone (LH)	Angiotensin II
Norepinephrine ($β_1$, $β_2$, and $β_3$ Rs)	Arginine vasopressin (V_1-Rs in the liver and vascular smooth muscle)
Parathyroid hormone (PTH)	
Secretin	Bombesin
Serotonin (5-HT)	Catecholamines ($α_1$ Rs)
Somatostatin (SRIF)	Cholecystokinin (CCK)
Thyrotropin-stimulating hormone (TSH)	Dopamine (D_2-Rs)
Vasoactive intestinal peptide (VIP)	Epidermal growth factor (EGF)

Hormones with G_c protein–coupled Rs	Gonadotropin-releasing hormone (GnRH)
	Histamine (H_1-Rs)
Muscarinic acetylcholine (mACh-Rs)	Nerve growth factor (NGF)
Atrial natriuretic peptide (ANP)	Norepinephrine ($α_1$)
	Parathyroid hormone (PTH)
	Platelet-derived growth factor (PDGF)
	Platelet activating factor (PAF)
	Serotonin (5-HT_2–R)
	Substance P (SP)
	Thromboxanes
	Thyrotropin-releasing hormone (TRH)
	Vasoactive intestinal peptide (VIP)

of the target tissue to the H is directly proportional to the number of R sites that are occupied. Under such a circumstance, for instance during insulin stimulation of nitric oxide (NO) production in vascular endothelia and vasodilation, the binding and biological response curves are superimposable, and maximal biological response occurs when 100% of R sites are occupied. An increase in R sites in this situation causes an increase in biological responsiveness or maximal binding capacity without a change in sensitivity of the response (Zeng et al. 2000).

In the saturation analysis graph, binding saturation is represented by asymptotic leveling-off of H-R binding at high H concentrations (100% binding). With Scatchard analysis (Scatchard 1949), which relies on the Michaelis-Menten equation, the amount of hormone bound at equilibrium can be determined, if [N] and K_D are known, from: $[B] = [N][F]/(K_D + [F])$. Maximal number of re-

ceptors [N] is determined from the equation: $[B]/[F] = K_A ([N]-[B])$. Graphically, [N] is determined as the intercept of the slope with the abscissa after plotting B/F against B (see figure 3.1b).

In contrast to the saturable specific binding of H with Rs, nonspecific binding is irreversible and unsaturable as it occurs between H and non-R entities that have high binding capacity and very low affinity. Nonspecific binding (B_{NS}) is therefore directly proportional to H concentration and needs to be subtracted from the total binding (B_T) to yield specific H-R binding (B_S; see figure 3.1b).

The affinity or sensitivity of H-R binding refers to the strength of H-R association and therefore is proportional to K_A and inversely proportional to K_D. When the Michaelis-Menten equation is displayed in the form of the Scatchard plot, affinity is derived from the formula $[B]/[F] = K_A ([N]-[B])$ or $K_D = [H][R]/[B]$ and graphically

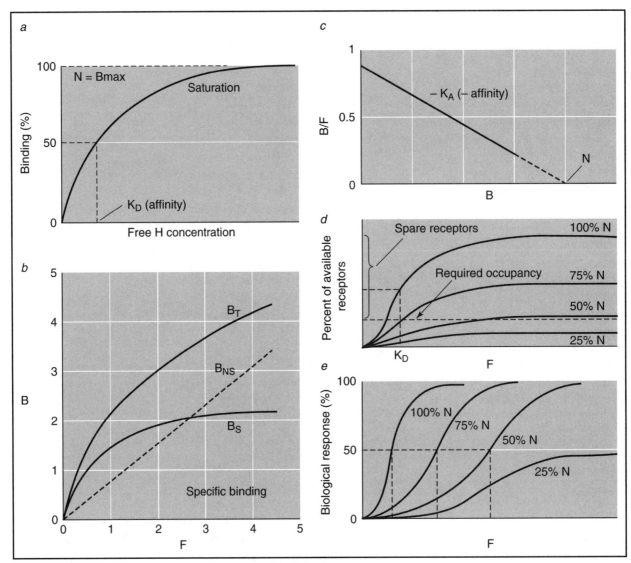

Figure 3.1 *Characteristics of hormone-receptor (H-R) binding.*

(*a*) H-R saturation plot. (*b*) Total (B_T), nonspecific (B_{NS}), and specific (B_S) H-R binding. (*c*) Scatchard plot. (*d*) Binding curves are compared for four situations that differ in the number of available receptors where binding of 50% of the available Rs is sufficient for maximal biological response. Rs in excess of the number needed for maximal biological response are spare Rs. (*e*) As the number of spare Rs declines, a rightward shift of the H-R curves and of K_D illustrates a decline in binding affinity (sensitivity).

Figures *a* and *b* adapted from Baulieu 1990; figures *c* and *d* adapted from Griffin and Ojeda 1996.

from the slope of the plot $-K_A$ or $-1/K_D$ (see figure 3.1b). K_D is operationally defined as H concentration at which H-R binding is half maximal (or 50% of Bmax).

A number of circumstances associated with H-R interactions affect their sensitivity or affinity. The first one is the stereochemical consonance between H and its R. For instance, the two catecholamine Hs E and NE differ in their affinity for subtypes of adrenergic receptors and potency of their biological effects because of such a difference between their stereochemical H-R relationships (see tables 3.1 and 3.2). This principle is the basis for development of pharmacological agents with agonist or antagonist properties that are useful for studying and manipulating biological function. Additional factors that affect H-R sensitivity or affinity include the ratio of R

relative to H concentrations, timing and duration of R stimulation, and influence of competing or modulating chemical messengers.

Maximal biological response is usually achieved at H concentrations that are considerably below maximal R occupancy. Unoccupied Rs in excess of those needed for maximal biological response are referred to as spare Rs. Concentrations of H needed to achieve full biological response are inversely related to the number of spare Rs. In the example shown in figure 3.1d and e, an increase in the number of spare Rs above the minimal necessary R occupancy requires progressively lower H concentrations to elicit a full biological response. This condition occurs because spare Rs confer increased sensitivity and affinity to

H-R interactions (see figure 3.1d). This principle is illustrated by increased cardiac affinity for, and sensitivity to, adrenergic stimulation after exposure to thyroid hormones, which stimulate cardiac synthesis of additional spare β adrenergic receptors. Spare receptors may be reduced in some physiological and pathological conditions and as a result of exposure to a high or invariant H concentration. Fasting causes a reversible reduction in the number of hepatic growth hormone (GH) Rs and an increase in the number of adipocyte insulin Rs. After exposure to high or invariant H concentrations, H-R complexes and additional unbound Rs that cluster in areas of membrane called coated pits are removed from the membrane by endocytosis and released into cytoplasm. This decline in the number of membrane Rs and loss of biological response is referred to as homologous down-regulation, or desensitization, because it is caused by interaction between homologous Hs and Rs. Where H secretion is pulsatile, degradation and removal of the H from circulation after a secretory pulse allows faster recovery of R sensitivity for subsequent H action.

H-R sensitivity also can be affected by interactions between different Hs or other pharmacological agents. In heterologous desensitization, one agonist reduces R responsiveness to other agonists by acting through a different type of R. In competitive binding, a competing ligand binds to a second R site and alters the quality and affinity of H-R binding without a change in N (see figure 3.2. top). In noncompetitive binding, a ligand changes binding capacity reversibly or irreversibly without affecting the affinity of the binding reaction (see figure 3.2, bottom). Finally, alterations in molecular structure of an H, its R, its transduction machinery, and the link between R and the transduction pathway can also alter the sensitivity and responsiveness of H-R interactions. There is a considerable degree of plasticity in H-R interactions. R numbers and types may change with age, exercise training, or as a result of metabolic and endocrine influences. In addition, age, exercise training, or metabolic and endocrine alterations may affect quantity, temporal characteristics of H secretion, and mode of H delivery. These changes will be discussed in the context of exercise throughout this book.

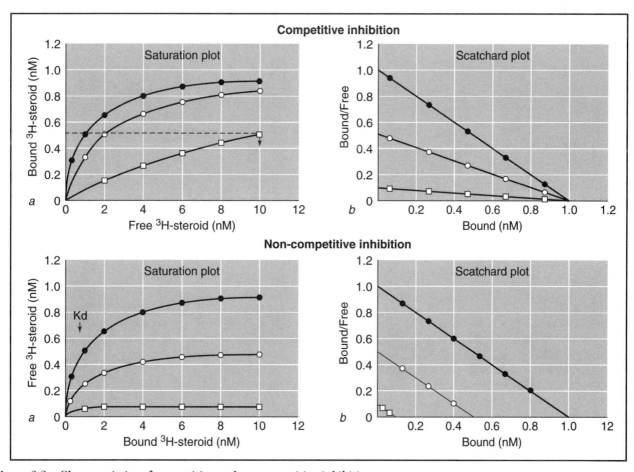

Figure 3.2 *Characteristics of competitive and noncompetitive inhibition.*

Increasing concentrations of competitive hormone (H) alter the K_D for the hormone-receptor (H-R) complex (arrows) but do not change the number of Rs (N) (top). Noncompetitive inhibitors do not alter K_D of the interaction but decrease the number of available Rs (bottom).

Adapted from Clark, Schrader, and O'Malley 1985.

CHEMICAL MESSENGER RECEPTORS

There is a great diversity in the structure and function of chemical messenger receptors, but they can be classified into categories according to their cellular location and structural characteristics (Baulieu & Kelly 1990; Mendelson 1996). Further classification by families is based on similarity of chemical structure and other properties of groups of Hs, Rs, and their transduction pathways. By location, Rs can be found on plasma membrane, in the cytoplasm, and in the nucleus.

Characteristics of Membrane Receptors

Water-soluble messengers such as peptide hormones, catecholamines, and indoleamines have Rs embedded in the plasma membrane. Membrane Rs have (1) an extracellular domain that contains hydrophilic amino acids at the amino terminus, (2) a membrane-spanning hydrophobic and lipophilic domain, and (3) an intracellular domain ending at the carboxy terminus (see figure 3.3, a-d). The extracellular domain is frequently glycosylated in the course of H maturation, and carbohydrates

attached to R proteins may facilitate recognition or hormonal binding processes. By molecular structure, membrane-spanning Rs can be categorized into four groups:

1. Rs with a single transmembrane region (see figure 3.3, a-b)
2. Rs with an intracellular tyrosine kinase (TK) effector attached to the transmembrane region (see figure 3.3b)
3. Rs with seven transmembrane segments (heptahelical Rs) that are associated with guanine nucleotide–binding proteins (G proteins) (see figure 3.3c)
4. Rs with multiple membrane-spanning regions that form ion channels (see figure 3.3d)

Some Rs encompass more than one structural category. Examples of Rs with a single transmembrane region (see figure 3.3a) are those belonging to the large cytokine family and Rs for some other hormones and proteins (see table 3.1) (Kitamura et al. 1994). They can exist as homodimers, heterodimers, heterotrimers, or heterotetramers on the basis of a single transmembrane protein or groupings

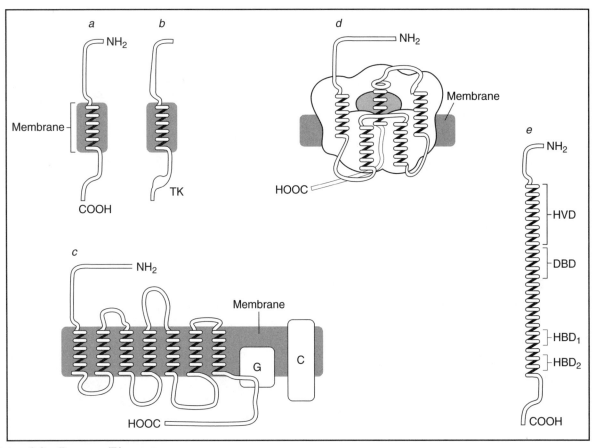

Figure 3.3   *Receptor (R) structure.*

Four types of receptors for water-soluble messengers: *(a, b)* Rs with a single transmembrane region that may have an intracellular tyrosine protein kinase (TK) effector in the intracellular domain; *(c)* Rs with seven transmembrane segments that are associated with G proteins and a catalytic enzyme subunit (C); *(d)* Rs with multiple transmembrane segments that form ion channels; and *(e)* intracellular Rs with a hypervariable hormone-specific domain (HVD), a DNA-binding domain (DBD), and one or more hormone-binding domains (HBDs).

Figures *a-d* adapted from Baulieu 1990; figure *e* adapted from Griffin and Ojeda 1996.

of between two and four such identical or structurally dissimilar proteins, respectively. Some Rs have an intracellular tyrosine kinase effector attached to their single transmembrane region (see figure 3.3b and table 3.1). Of those, insulin R (I-R) and IGF-I–R are tetramers consisting of two membrane-spanning β units that have tyrosine kinase activity and two extracellular α units attached to β units through cysteine cross-bridges (two hetrodimers). Rs in the cytokine family and others with TK activity frequently participate in normal or pathological cellular growth.

Some membrane Rs consisting of proteins with multiple membrane-spanning regions that form ion channels (see figure 3.3d). Among these are hormone-gated (H-gated) channels and ion pumps. Nicotinic ACh-R (nACh-R), γ-aminobutyric acid $_A$ R (GABA$_A$-R), and glycine R are some of the H-gated channels (Stroud & Finer-Moore 1985). The nACh-R contains five transmembrane regions (see figure 3.3d) of which two α subunits constitute the H binding site. Activation of nACh-R opens a channel that permits diffusion of Na$^+$ and K$^+$ across the cell membrane. Na$^+$- K$^+$ ATPase, H$^+$-K$^+$ ATPase, and Ca^{2+} ATPase are ion pumps that also belong to this R family (Bertorello & Katz 1995).

Characteristics of Intracellular Receptors

Steroid and thyroid Hs are hydrophobic lipids that diffuse freely through the lipophilic plasma membrane to bind to intracellular or nuclear Rs and interact with nuclear DNA. Intracellular Rs consist of three parts: (1) a hypervariable hormone-specific domain (HVD) at the amino terminus, (2) a central DNA-binding domain (DBD) consisting of about 50 to 100 amino acids and a characteristic "zinc finger" region that binds to the hormone responsive element (HRE) on DNA—a segment of DNA that acts as a nuclear binding site for a specific H-R complex (see figure 3.4), and (3) a hormone-binding domain (HBD) at the carboxy terminus consisting of one or more clusters of hydrophobic amino acids to which lipophylic steroid and thyroid hormones bind (see figure 3.3e). Examples of cytoplasmic Rs are G-R, M-R, AN-R, E-R, and PROG-R.

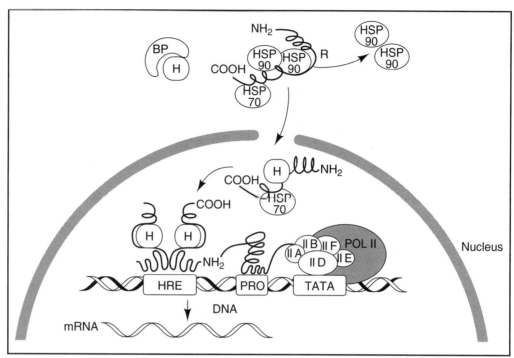

Figure 3.4 *Steroid/thyroid hormone transduction pathways.*

Transduction of the steroid or thyroid hormone message starts when the hormone (H) dissociates from its binding protein and binds to the receptor (R), receptor dimerizes and H-R complex binds to hormone responsive element (HRE) on DNA.

BP = binding protein; HSP = heat-shock proteins 90 and 70; HRE = hormone responsive element on DNA; PRO = promoter region of the gene; TATA = region of the gene where transcription of DNA is limited; IIA, IIB, IID, IIE, IIF, POL II = transcription factors; mRNA = messenger RNA transcript of DNA.

Adapted from Griffin and Ojeda 1996.

Cytoplasmic Rs are bound to two molecules of heat-shock protein 90 (hsp 90) and one molecule of hsp70 (see figure 3.4) (O'Malley 1990; Pratt 1990, 1993; Tsai & O'Malley 1994). After binding with respective Hs, the activated H-R complex releases hsps and translocates to the nucleus. There it dimerizes and binds to an HRE (see figure 3.4) (Glass 1994). Dimerized H-R complex acts as a transcription factor that initiates gene transcription. Other proteins known as IIA through IIF, particularly the transcription factor (TF) IIB, bind to the TATA box on DNA and allow the attachment of RNA polymerase II and initiation of gene transcription. Rs for thyroid Hs, vitamin D$_3$, and vitamin A (retinoic acid) are located in the nucleus and are bound to the HRE on the DNA where they, together with corepressor molecules, silence the expression of respective genes. Some Rs that do not associate with hsps substitute their amino terminus for the carboxy terminus after the release of corepressor molecules to interact with TF IIB. This process then initiates gene transcription.

CHEMICAL MESSAGE TRANSDUCTION

Transduction connects specific Hs with their effector mechanisms. Although distinct and specific molecular transduction mechanisms are associated with different categories of Hs, they cannot be categorized by the nature of the biological response they produce. Instead, biological responses depend on tissue-specific coupling of a particular transduction pathway to a given effector mechanism. Thus, the same transduction mechanism can have pleiotropic effects in different tissues as illustrated by the G_s transduction pathway of catecholamines (see table 3.3). Moreover, most hormones activate more than one transduction pathway. Such cross-talk and transactivation between different signaling pathways adds richness and complexity to intercellular communication. Several transduction pathways that influence biological functions of interest in exercise physiology are explained in some detail, whereas less relevant or incompletely understood pathways are outlined briefly. Presentation of molecular mechanisms of hormone signaling in this chapter reflects available information at the time of writing. Interested readers should seek more recent information, as this field is undergoing rapid growth and transformation.

Anabolic Protein Hormone Transduction Pathways

Rs with a single transmembrane region (see figure 3.5) or with an intracellular TK effector (see figure 3.5) very frequently mediate growth processes and usually utilize

1. the Janus kinase/signal transducers and activators of transcription (JAK/STAT),
2. the mitogen activated protein kinase (MAPK), and
3. the insulin receptor substrate (IRS) transduction pathways (Carter-Su et al. 2000; Herrington et al. 2000; Ullrich & Schlesinger 1990).

GH, a representative of the cytokine family, mostly utilizes the first two but, to a lesser extent, also utilizes the third pathway (Herrington & Carter-Su 2001). In

Table 3.3   Biological Actions of Adrenergic Receptors

Biological function	α_1 A-R	α_2 A-R	β_1 A-R	β_2 A-R	β_3 A-R
Amylase secretion				X	
Bronchodilation				X	
Heart rate			X		
Heart contractility	X				
Glycogenolysis, liver	X		X	X	
Glycogenolysis, muscle	X		X	X	
Glycolysis	X			X	
Gluconeogenesis	X				
Insulin secretion				X	
/ Insulin secretion		X			
Lipolysis, adipose tissue			X	X	X
/ Lipolysis, adipose tissue	X				
Lipolysis, muscle				X	
/ NE release		X			
Nutrient uptake (amino acids, lactate)	X				
Oxidative metabolism				X	
Piloerection	X				
Platelet aggregation		X			
Renin release			X		
Salivation	X				
Sweating	X				
Vasoconstriction	X	X			
Vasodilation				X	

= stimulatory function; / = inhibitory function.

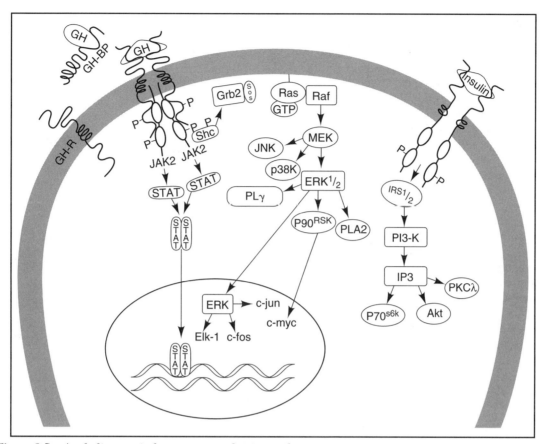

Figure 3.5   *Anabolic protein hormone transduction pathways.*

Growth hormone (GH) signaling is initiated when GH dissociates from its binding protein, binds to a membrane receptor (R), and forms a dimer (top left). Other growth-promoting messengers, such as insulin (right) and platelet-derived growth factor (PDGF), have intrinsic tyrosine kinase activity on their intracellular domain.

JAK2 = Janus kinase 2; STAT = signal transducers and activators of transcription; IRS 1/2 = insulin receptor substrate 1 and 2; Shc, Grb2, and Ras = proteins that are phosphorylated; Raf = serine/threonine kinase; MEK = MAPK kinase; JNK = c-jun NH$_2$ terminal kinase; p38K = p38 kinase; PLγ = phospholipase; c-fos, c-jun = constituents of AP-1 (transcription factor activator protein 1) and proto oncogenes; PLA2 = phospholipase A$_2$; P90^{RSK} = P90^{RS} kinase; PI3-K = phosphatidyl-inositol-3 kinase; Akt = protein kinase; p70^{56k} = ribosomal protein kinase B, PKCγ; PCKγ = protein kinase Cγ; SOS = P13-K, phosphoinositoe triphosphate kinase; GH-BP = growth hormone binding protein; GH-R = growth hormone receptor; GTP = guanosine triphosphate; ERK1/2 = extracellular signal-regulated kinases 1 and 2; Elk-1 = transcription factor I; c-myc = proto-oncogene; IP3 = inositol triphosphate.

Adapted from Griffin and Ojeda 1996.

JAK/STAT transduction, a GH molecule separates from its binding protein to bind to two GH-Rs. This dimerized R next associates with one of several cytoplasmic JAKs. JAK2 is involved in transduction of GH, prolactin (PRL), and erythropoietin (EPO) signals, and JAK1 and JAK3 carry out this function, respectively, for interferon (IFN) Rs and for interleukin-2 (IL-2) Rs. Although the Rs of the cytokine family lack TK activity, their binding with H stimulates JAK to phosphorylate tyrosine residues on the R and on a variety of other proteins. Phosphorylation takes place when proteins containing a sequence of amino acids referred to as src homology 2 (SH2) dock and bind to phosphorylated tyrosines on the R. These recruited proteins Shc and Grb$_2$ then propagate protein phosphorylations and in this way activate enzymes and, in the nucleus, gene transcription. In the JAK/STAT pathway, different members of a family of STATs serve as transcription activators. GH activates STATs 1, 2, 3,

5a. and 5b; IFNγ activates STATs 1 and 5; and leukemia inhibitor factor (LIF) activates STATs 3 and 5. STATs next dimerize, and they either bind to specific HREs within the promoter region of genes where they initiate gene transcription or they act on cytoplasmic translation machinery. The anabolic effects of GH are contingent on the intermittent pattern of GH-R stimulation by GH. STAT5b is responsive to the temporal pattern of GH stimulation in that it requires cycles of activation through phosphorylating kinases and deactivation through dephosphorylating phosphatases (Gebert et al. 1997).

The MAPK transduction pathway diverges downstream of activated JAK. Here as well, transduction of message is propagated by serial protein phosphorylations. Among sequentially activated proteins are Shc, Grb2, and the guanosine triphosphate (GTP)–binding protein Ras. This action results in phosphorylation of the serine/threonine kinase Raf, followed by activation of the enzyme MEK (or MAPK kinase). MEK phosphorylates threonine and tyrosine on several MAPKs: isoforms ERK1 and ERK2 (extracellular signal-regulated kinases), c-jun NH2-terminal kinase (JNK), and the p38 kinase. MAPKs can phosphorylate cellular proteins, including transcription factor activator protein–1 (AP-1) and its constituents c-fos and c-jun and act on gene promoters to cause transcription and anabolic effects (Seger & Krebs 1995). They also can activate phospholipases PLA_2 and $PL\gamma$ in the cytoplasm.

The IRS transduction pathway is more prominently utilized by TK-associated Rs stimulated by Hs such as insulin or IGF-I than by single transmembrane Rs. Here (see figure 3.5), the activated HR-complex triggers phosphorylation of tyrosines on the R and initiates a cascade of other protein phosphorylations. The transcription activators here are IRS1 or IRS2, which are phosphorylated either by JAKs or TKs. As a result of action of activated IRS1/2 on gene promoters, protein synthesis may be initiated. Phosphatidyl-inositol-3 kinase (PI3-K) is the next link in the chain of phosphorylations. Its product, inositol triphosphate (IP3), next activates different protein kinases (PKs) including PKB (or Akt), PKC, and ribosomal $p70^{s6}$ kinase, which produce metabolic effects such as glucose transport or fat synthesis.

Transduction Pathways Mediated by G Protein Coupled Rs

G protein coupled Rs (GPCRs) are a family of proteins that are associated with Rs and mediate several different signaling cascades (see table 3.1). These heptahelical Rs associated with G proteins (see figure 3.3c) (Marinissen & Gutkind 2001) include α adrenergic Rs (α A-Rs), β adrenergic Rs (β A-Rs), calcium Rs on calcitonin and parathyroid hormone (PTH) cells, endothelin (ET_A and ET_B Rs), prostanoids, and many others. These Rs are embedded in plasma membrane in the vicinity of G (GTP-binding regulatory) proteins and catalytic enzymes. G proteins are structurally similar but functionally diverse. They are heterotrimers and consist of identical β and γ subunits and a functionally differentiated α subunit. G proteins are categorized according to their ability to affect enzyme function or ion channels. The stimulatory G protein or G_s stimulates the enzyme adenylylcyclase (AC) to convert the substrate Mg^{2+}-ATP to the second messenger cyclic adenosine monophosphate (cAMP), which in turn activates the enzyme cAMP-dependent protein kinase A (PKA) (Benovic et al. 1988). The G_i protein inhibits the activity of AC by preventing the interaction between the R and its coupled G_s. The G_c protein stimulates guanylylcyclase (GC) to produce cyclic guanosine monophosphate (cGMP) and increase the activity of the enzyme phosphodiesterase (PDE) and also at times close a calcium channel (Garbers & Lowe 1994; Méry et al. 1991). PDE degrades cAMP to the inactive nucleotide 5'-AMP and activates the enzyme PKG. The G_q protein (sometimes also designated as G_p protein) stimulates the enzyme phospholipase C (PL-C)

to increase turnover of membrane phospholipids and produce second messengers inositol triphosphate (IP_3) and diacylglycerol (DAG). DAG in turn activates the enzyme PKC (Berridge 1993; Exton 1990a, 1990b; Nishizuka 1992; Rhee & Choi 1992). Finally, the G_k protein regulates potassium channels.

cAMP (G_s Protein) Transduction Pathway

G_s transduction pathway is triggered by many Hs released during exercise and other forms of stress to produce necessary increases in function and metabolism. Its prominent feature is H-mediated stimulation of AC (see figure 3.6) as a result of interaction between R and G_s protein (Benovic et al. 1988). In the unstimulated state, guanosine diphosphate (GDP) is bound to the α subunit of the G_s protein. After H-R binding, GDP dissociates, and a guanosine triphosphate (GTP) attaches to the α subunit of the G_s protein. This activation step is necessary for the R and G_s protein to aggregate and for the α subunit of the G_s protein to dissociate from the β and γ subunits. G_s-α-GTP then complexes with, and activates, the catalytic enzyme AC to convert Mg^{2+}-ATP to cAMP. GTPase that is associated with the activated G_s-α hydrolyzes GTP to GDP and terminates transduction of the endocrine message.

The cAMP acts as a second messenger that regulates the activity of cAMP-dependent PKA. PKA initiates a cascade of protein phosphorylations and thus activates a number of enzymes or ion-channel proteins necessary for mobilization and utilization of metabolic fuels, muscle contraction, and endocrine and exocrine secretion. Thus, cAMP also influences expression of genes for gluconeogenic enzyme phosphoenolpyruvate carboxykinase (PEPCK), tyrosine aminotransferase, preprosomatostatin, VIP, and cytochrome P-450 that are involved in steroid hydroxylation. Transcription effects are mediated by the cAMP response-element binding protein (CREB). CREB is activated by cAMP and binds to the cAMP responsive element (CRE) on the DNA (Meyer & Habener 1993).

cAMP-Inhibitory (G_i Protein) Transduction Pathway

H-mediated inhibition of AC (see figure 3.6) also entails interaction among three proteins: R for the inhibitory H, G_i protein, and AC. Binding of Hs to Rs that inhibit AC causes dissociation of GDP from the G_i α subunit and substitution of GTP in the place of GDP. The inhibition of AC activity is achieved by interaction of free β-γ subunits from the G_i protein with G_s α subunits. MAPK is next transactivated through a component of the signaling pathway utilized by TK-containing Rs. Toxin of *Bordetella pertussis* interferes with this type of transduction by preventing the dissociation of G_i α and thereby stimulates adenylylcyclase.

IP_3-DAG (G_q Protein) Transduction Pathway

The G_q transduction pathway entails the activation of PL-C after H-R binding (see figure 3.7) (Rhee & Choi 1992). This membrane-associated enzyme then catalyzes hydrolysis of a membrane phospholipid phosphatidylinositol 4,5-biphosphate (PIP_2) to second messengers IP_3 and

DAG (Berridge 1993). IP_3 binds to its R, a membrane-spanning protein on the endoplasmic reticulum (ER) or sarcoplasmic reticulum (SR) that forms a Ca^{2+} channel and triggers the release of Ca^{2+}. Calcium mediates second messenger action by binding to specific Ca^{2+}-binding proteins such as calmodulin and troponin. After binding with Ca^{2+}, activated calmodulin stimulates various enzymes, and activated troponin uncovers myosin binding sites on actin myofilaments and causes skeletal muscle contraction. DAG, the other second messenger in the G_q transduction pathway increases the affinity of PKC for Ca^{2+} (Nishizuka 1992). Activated PKC phosphorylates enzymes and affects the expression of genes. The genetic effect involves activation of a DNA-binding transcription factor AP-1 that contains two protooncogene proteins, c-fos and c-jun, through which DAG stimulates cell proliferation and tumorigenesis. DAG can also be produced from phosphatidylcholine by activation of PL-D in response to muscarinic acetylcholine (mACh) and V1-R binding with ACh and AVP, respectively.

cGMP (G_c Protein) and Nitric Oxide Transduction Pathways

cGMP is a second messenger that is formed from GTP in a reaction associated with G_c protein and catalyzed by GC (Lucas et al. 2000). This enzyme is found both in the plasma membrane and in the cytoplasm. Its main action is activation of PDE, the enzyme that inactivates cAMP (see figures 3.6 and 3.7). After activating GC, cGMP can also phosphorylate and activate hydrolase activity of a bifunctional enzyme (see figure 3.7). Through synthesis of a messenger cyclic adenosine diphosphate ribose (cADPr) from nicotinamide adenine dinucleotide (NAD), this enzyme can open sarcoplasmic Ca^{2+} channels known also as ryanodyne receptors (R-R) (see figure 3.7) (Furukawa & Nakamura 1987). This ribosylcyclase action (see figure 3.7) opens Ca^{2+} channels by causing association of cADPr to its binding site on R-Rs (Lee et al. 1994). Its hydrolase action (see figure 3.7) closes the channel by facilitating cADP-r dissociation from R-R and its degradation. After ACh activates cardiac mACh Rs, the membrane-bound cGMP transduction pathway (see figure 3.7) blocks phosphorylation of the L-type calcium channel through dephosphorylating action of protein kinase–G (PKG) (Petit-Jacques et al. 1994). cGMP also mediates vasodilatory, natriuretic, and diuretic effects of ANP.

GC can also be activated by nitric oxide (NO), the free radical messenger that binds to the enzyme and thus influences a number of physiological processes of impor-

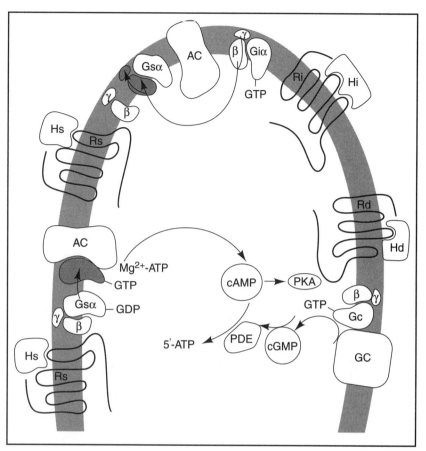

Figure 3.6 *Transduction pathways acting through cyclic adenosine monophosphate (cAMP).*

Hormones (Hs) that stimulate adenylcyclase (AC) bind to their receptors (Rs) and trigger an interaction between the R_s, G_s protein, and AC (left). Hormones (H_i) that inhibit AC bind to their receptors (R_i) and initiate a transduction event (top center) that includes dissociation of the β-γ subunits from the activated G_i protein and their association with the $Gs\alpha_s$, thereby blocking the formation of cAMP. Hormones (H_d) that degrade cAMP engage in analogous transduction process after binding to their receptor (R_d). The resulting enzyme PDE degrades cAMP to 5' AMP.

α, β, γ = components of the G proteins; GTP = guanosine triphosphate; cGMP = cyclic guanosine monophosphate; PDE = phosphodiesterase; PKA = protein kinase A.

Adapted from Griffin and Ojeda 1996.

tance in exercise (Hanafy et al. 2001). NO is produced in both smooth and skeletal muscles, where it suppresses contractility and induces relaxation, as well as in endothelia and nerves (Kobzik et al. 1994; Moncada et al. 1991; Reid 1996b). Nitrergic innervation is found in the heart, blood vessels, airway epithelia, enteric plexuses of the GI tract, and GI sphincters, and VIP is often a cotransmitter (Rand & Li 1995). In addition, NO is produced in macula densa of the JGA where it participates in the regulation of plasma volume (Wilcox et al. 1992).

The highly reactive free radical gas NO represents a new class of signaling molecules that is regulated by GPCRs (Christopoulos & El-Fakahany 1998; Torreilles 2001). It is produced (see figure 3.7) by the enzyme NO synthetase (NOS). NOS is a P-450 cycloxygenase consisting of a reductase and an oxygenase domain (White & Marletta 1992). The reductase contains flavine adenine

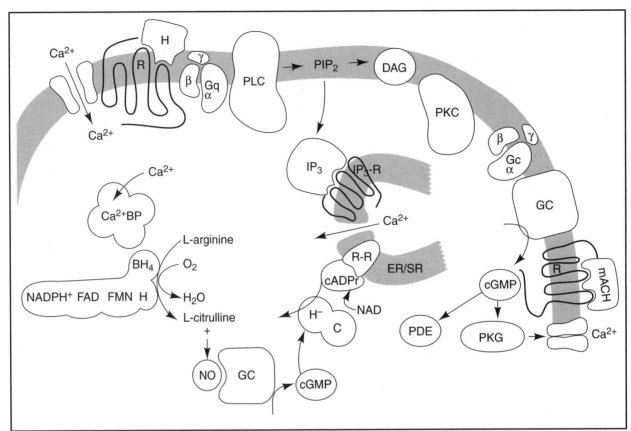

Figure 3.7   *Inositol triphosphate and diacylglycerol (IP$_3$-DAG), cyclic adenosine monophosphate (cGMP), and nitric oxide (NO) transduction pathways.*

IP$_3$ and DAG are second messengers in a transduction pathway (top left) that involves an interaction between hormone (H), receptor (R), and G$_q$ protein. cGMP is a second messenger in a transduction pathway (right) that entails activation of guanylyl cyclase (GC) and stimulation by cGMP of enzymes such as phosphodiesterase (PDE) and protein kinase G (PKG). The NO transduction pathway (left center) entails the activation of NO synthetase (NOS) (center) by calmodulin and the control of cytosolic Ca^{2+} release through H-C.

NADPH = nicotinamide adenine diphosphate reduced from dimoleotide reductase component NOS; FAD = flavine adenine; FMN = flavine mononycleotide, a reductase component of NOS; BH$_4$ = H$_4$biopterin cofactor in NOS; H-C = bifunctional enzyme that can act as either a cyclase or a hydrolase; R-R = ryanodine receptor; NAD = nicotine adenine dinucleotide; CADPr = cyclic adenosive diphosphate ribose; CA^{2+}BP = calnodule or calcium binding protein; ER/SR = endoplasmic/sarcoplasmic reticulum; PKC = protein kinase C; PKG = protein kinase G.

Adapted from Abu-Soud and Stuehr 1993.

dinucleotide (FAD) and flavine mononucleotide (FMN), which reduces nicotinamide adenine dinucleotide phosphate (NADPH$^+$) and passes its electrons on to the oxidase. The oxidase domain has an iron-containing heme group (see figure 3.7) and is associated with an H$_4$ biopterin cofactor (see figure 3.7). It oxidizes L-arginine to L-citrulline and NO and reduces molecular oxygen to water (Abu-Soud & Stuehr 1993). In the absence of the amino acid substrate and the cofactor, NOS generates free radical superoxides and hydrogen peroxide (Mayer et al. 1991).

NOS is expressed either as a constitutive enzyme that is associated with the Ca^{2+} -binding protein calmodulin, as is the case in endothelia (eNOS) and nerve terminals (nNOS), or as an inducible enzyme (iNOS) that requires induction by cytokines, bacterially-derived lipopolysaccharides (LPS), and high cytoplasmic Ca^{2+} for activation of calmodulin (Katusic & Cosentino 1994). In endothelial cells, gene expression of constitutive NOS is up-regulated by shear stress, ADP, and estrogens (Luscher & Barton 1997). Skeletal muscle expresses both forms of constitutive NOS (Kobzik et al. 1995). Endothelium-dependent vasodilation was previously attributed to endothelium-derived relaxation factor (EDRF) before a classic demonstration by Furchgott and Zawadski (1980) identified S cholinergic nerves acting on mACh Rs in the muscle vascular endothelia as a source of NO that caused vasodilation. Since then, NO along with adenosine has been identified as important endothelium-derived vasodilator (Duncker et al. 1995; Luscher & Dohi 1992; Matsunaga et al. 1996).

NO is thought to produce vasodilation by several signaling pathways. Among its actions that prevent smooth and skeletal muscle contraction are closure of L-type Ca^{2+} channels mediated by cGMP and PKG (see figure 3.7; Rapoport et al. 1983) and closure of sarcoplasmic or endoplasmic R-R Ca^{2+} channels in the skeletal muscle of the heart. NO controls R-R closure through hydrolysis of cADP-r. NO can also cause vasodilation by acting on enzymes that dephosphorylate light chain of myosin (Ahlner et al. 1991). Because NO synthesis is linked to IP3-induced activation of calmodulin, and NO acts through the cGMP second messenger, this novel intracellular agent is regulated by two interacting GPCR transduction pathways.

Purinergic R Transduction Pathways

Adenosine and ATP produce distinct biological effects by acting on purinergic Rs, which are usually GPCRs (Burn-stock & Wood 1996). Adenosine is a nucleotide messenger consisting of a purine base adenine bound to ribonucleic acid. It is released through neurotransmission by ATP-containing S fibers (Burnstock 1995) or produced in the cytoplasm by progressive enzymatic degradation of ATP and ADP, nucleotide mediators of cellular metabolism. ADP is converted to adenosine by the enzyme 5'-nucleotidase. Adenosine acts on A_1, A_2, and A_3 receptors (Jacobson et al. 1992). Principal nonselective antagonists are the commonly imbibed xanthine derivatives caffeine and theophylline. Other agonists and antagonists are listed in table 3.4.

Table 3.4   List of Common Hormone Agonists and Antagonists

Hormone	Receptor	Agonist	Antagonist
Adenosine	A		Caffeine, theophylline, 8-phenyltheophylline
	A_1	2-chlorocyclopentadyl adenosine, cyclopentadyladenosine, NECA	DPCPX, 8-cyclopentadyl theophylline
	A_2	CGS21680, CV1808	KF17837, SCH58261, CGS15943, CP66713, ZM241385
	A_3	APNEA, CGS21680	
Adrenergic	α_1	Methoxamine, phenylephrine	BE 2254, corynanthine, phentolamine, prazosin
	α_{1a}		WB4101, 5-m-uradipil, (+/-)-tamsulosin
	α_{1b}		Chloroethylclonidine
	α_{1d}		5-MU, BMY-7378
	α_2	α-Methyl-norepinephrine, clonidine, guanfacin, tramazoline, xylasine	Deriglidole, idazoxan, phentolamine, rauwolscine, RX821002, yohimbine
	α_{2a}	Oxymetazoline, octopamine, synephrine, dexmedetomidine, UK-14304	WB-4101
	α_{2b}		Chlorpromazine, ARC-239, spiroxatrine, SK and F104856, prazosin
	α_{2d}	Oxymetazoline	BRL-44408
	β_1	Dobutamine, isoprenaline, prenaterol, tazolol, tolbutamine	Atenolol, betaxolol, bupranolol, ICI89,407, paraoxyprenolol, practolol, propranolol
	β_2	Albuterol, broxaterol, clenbuterol, epinephrine, hexoprenaline, isoprenaline, procaterol, rimiterol, salbutamol, soterenol, terbutaline, zinterol	Bupranolol, butoxamine, ICI118,551, IPS339, propranolol
	β_3	Bucindolol, BRL37344, CGP12177, CL316243, LY79771, isoprenaline, isopreterenol, pindolol, SR58611A	Bupranolol, CGP20712A, ICI118,551
Angiotensin II	AT_1		Losartan (DUP753)
	AT_2		PD123319
Endothelin	ET		SB209670
	ET_A		BQ-123
	ET_B		Sarafotoxin 6C

A_1-Rs are found in a variety of tissues where they influence different functions. In adipose tissue they inhibit lipolysis and in adipose tissue and muscle they modulate insulin sensitivity (Challis et al. 1992; Londos et al. 1985; Schwabe et al. 1973; Webster et al. 1996). In the skeletal muscle, adenosine causes vasodilation during hypoxia and exercise by increasing endothelial NO synthesis (Marshall 2000; Radegran & Hellsten 2000). In the heart, A_1-Rs have antiadrenergic function, and in respiratory smooth muscles, they cause bronchospasm (Dobson & Fenton 1993; Fozard & Hannon 2000). In the brain, they decrease arousal, physical activity, and NE neurotransmission, and in the anterior pituitary gland, they decrease GH secretion (Dorflinger & Schonbrunn 1985; Fredholm et al. 1997). Transduction of adenosine message after binding to A_1-Rs involves both G_i (see figure 3.6) and G_q proteins (figure 3.7) (Fredholm et al. 1997). Transduction through the G_i pathway produces membrane hyperpolarization through opening of K^+ channels and closing of voltage-dependent Ca^{2+} channels.

A_2-Rs are found in vascular smooth muscles, where they inhibit contraction, and in the brain. Adenosine binding to A_2-Rs is coupled to G_s protein and activates AC to form cAMP (see figure 3.6). Principal biological effects (discussed also in connection with the role of Rs in vasomotor control) are dilatation of vascular smooth muscle, suppression of platelet aggregation and leukocyte adhesion, and stimulation of hepatic glycogenolysis (Anfossi et al. 1996; Feoktisov et al. 1992).

ATP is released in the periphery from S nerve endings in circumstances producing pain, such as traumatic tissue injury, tissue lysis, arthritic tissue abrasion, skeletal and cardiac muscle ischemia, or platelet aggregation in migraine headaches (Hamilton & McMahon 2000). ATP exerts its effects by acting on P2 purinoceptors, which are either ligand-gated ion channels (P2X family) or GPCRs (P2Y family) (Fredholm et al. 1997).

Ceramide Transduction Pathway

The metabolism of another lipid membrane component, sphingomyelin, mediates cytokine tumor necrosis factor–α (TNF-α) message. Binding of TNF-α to its cell membrane receptors activates the enzyme sphingomyelinase, which hydrolyzes sphingomyelin into phosphocholine and ceramide (Cutler & Mattson 2001; Ruvolo 2001). The latter is a second messenger that activates an array of kinases; phosphatases, including the ceramide-activated protein phosphatase (CAPP); and transcription factors. CAPP inhibits cellular proliferation and induces cell differentiation and programmed cell death (apoptosis). Vitamin D_3 induces cell differentiation through this pathway when it acts at membrane sites rather than inside the nucleus.

TRANSDUCTION PATHWAYS THAT INFLUENCE BODY FUNCTION DURING EXERCISE

There is a remarkable diversity in the types of Rs and transduction mechanisms in tissues that support physical activity. This section is an overview of the way autonomic neurotransmitters and endocrine messengers interact with Rs in the heart, skeletal muscle, vascular system, airways, and other organs to affect energy metabolism and tissue repair and growth during, and in response to, exercise.

The Exercising Heart

Two components of heart contraction are controlled by autonomic and chemical messengers: the heart rate (chronotropic function) and heart contractility (inotropic function). The heart syncytium is inherently rhythmic because of the clusters of excitable cells that act as pacemakers. The sinoatrial node (SAN) is one such pacemaker that can spontaneously depolarize about 100 times per minute. The other heart pacemaker, the atrioventricular node (AVN), depolarizes at a slower frequency. SAN and AVN become spontaneously depolarized because of a gradual decrease in the permeability of their membranes to K^+ in the face of a constant permeability to Na^+. The imbalance in the magnitude of inward Na^+ leak and reduced outward K^+ leak triggers membrane depolarization, a pacemaker potential. This depolarization starts in the SAN, spreads across the two atria, and causes the AVN to discharge. The depolarization then rapidly spreads across the two ventricles through the conducting bundle of Hs. The duration of successive cardiac pacemaker depolarization is about 100 times longer than is the skeletal muscle action potential. This slower heart depolarization is imposed by voltage-gated L-type calcium channels that open immediately after initial and transient opening of the sodium channel. For the duration of time that the Ca^{2+} channel is open and that K^+ permeability is reduced below resting level, the heart muscle is depolarized and refractory to catecholamine or pacemaker stimulation. Repolarization of the heart muscle takes place when the permeability of heart membrane to Ca^{2+}, Na^+, and K^+ returns to resting level because of increased activity of Na^+-K^+ pump and Ca^{2+}-ATPase and when the cation channels close.

Heart rate increases during exercise when S nerves or circulating E bind to β_1 and β_2 adrenergic receptors (Leenen et al. 1995). This activates the G_s transduction pathway, generates cAMP, and induces phosphorylation and opening of ligand-gated Na^+ and Ca^{2+}, and inward-flowing K^+ channels (see figure 3.8). A direct consequence of the inward flow of Na^+ and Ca^{2+} is a faster rate of pacemaker depolarization and a shortening of the latency between depolarizations (see figure 3.9a). S nerves and E also act on β_1 and β_2 adrenergic receptors throughout the myocardium to increase heart contractility. An increase in contractility produces more forceful muscle contractions at any given end-diastolic pressure. Thus, a catecholamine-induced increase in inotropic action of the heart is independent of, and additive to, increased force of contraction caused by greater end-diastolic volume (see figure 3.9b). Cardiac contractility increases as a result of threefold action of cAMP:

1. Phosphorylation of proteins in, and opening of, slow L-type of Ca^{2+} channels in the plasma membrane

and of R-R type of Ca^{2+} channels in the SR (see figure 3.8)

2. Phosphorylation of an SR protein phospholamban that controls active transport of calcium via Ca^{2+}-ATPase (see figure 3.7)

3. Phosphorylation of myosin in cardiac myofibrils (Witcher et al. 1991)

Phosphorylation of phospholamban facilitates the Ca^{2+}-ATPase activity and calcium reuptake into SR for repeated release into cytoplasm (see figure 3.8) (Langer 1997). Phosphorylation of myosin accelerates cross-bridge cycling, all of which increases cardiac contractility. Endurance training increases the affinity of the heart to β_2 adrenergic stimulation (Mazzeo et al. 1995). The heart also has α_{1A} and α_{1B} adrenergic receptors, one third as many as β receptors, which modulate its inotropic action (Michel et al. 1994). α_{1B} Rs exert negative inotropic action on R-Rs by reducing Ca^{2+} release out of SR.

The PS nerves reduce the chronotropic action of the heart at rest, in response to baroreceptor signals of increased blood pressure or in response to endurance training (Shi et al. 1995). ACh acts on mACh receptors (of M$_1$ type in intracardiac ganglia and of M$_2$ type in atrial muscle) in the SAN to reduce the heart rate by four different transduction pathways (Buckley & Caulfield 1992; Petit-Jacques et al. 1994) (see figure 3.8):

1. It inhibits the synthesis of cAMP through the G$_i$ pathway.

2. It hastens cAMP degradation through the G$_c$ pathway.

3. It blocks the opening of calcium channel through the G$_q$ pathway.

4. It opens a K$^+$ channel.

The net effect of this action is pacemaker hyperpolarization, reduction in the rate of pacemaker depolarization, and slowing of the heart rate (see figure 3.9a).

In addition to the action of autonomic neurotransmitters and E, purinergic and peptidergic messengers influence heart function during exercise. Purinergic messenger adenosine acts on A$_1$ receptors. Two neurotransmitters that are colocalized with NE in S nerves,

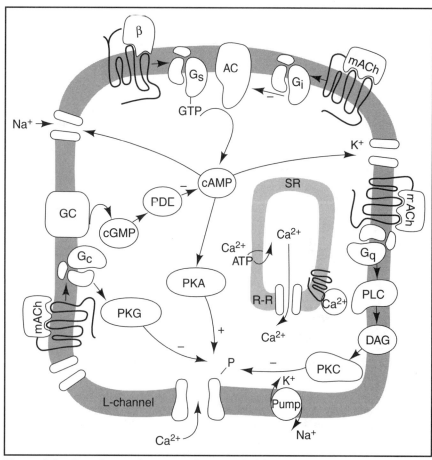

Figure 3.8 *Signal transduction in the control of heartbeat.*

Stimulation of β_1 and β_2 adrenergic receptors (A-Rs) by norepinephrine (NE) or epinephrine (E) triggers the cyclic adenosine monophosphate (cAMP) transduction process, resulting in depolarization and contraction of cardiac muscle. Acetylcholine (ACh) does the reverse through three different transduction pathways.

GC = guanylyl cylcase; G$_c$, G$_i$, G$_g$ = G regulatory proteins; AC = adenylyl cyclase; cGMP = cyclic guanosine monophosphate; PKA, PKG, PKC = protein kinases; PLC = phospholipase C; DAG = diacylglycerol; PDE = phosphodiesterase; mACH = muscarinic acetylcholine receptor; β = β AR receptor; Na$^+$, K$^+$, Ca^{2+} = sodium, potassium, and calcium (L-type) channels; SR = sarcoplasmic reticulum; K$^-$NA$^+$ = sodium-potassium pump.

Adapted from Griffin and Ojeda 1996.

NPY acting on prejunctional Y$_2$ receptors and galanin, prolong catecholamine action by opposing vagal neurotransmission (Potter & Ulman 1994). The duration of catecholamine action is thus increased and may facilitate recovery from exercise.

A number of chemical messengers that are released during exercise such as E, angiotensin II, AVP, IL-1, and Ca^{2+} trigger the release from the endocardial endothelium of another cardioactive peptide, endothelin (ET) (Grossman & Morgan 1997). The ET-1 molecular form of endothelin binds to ET$_A$ and ET$_B$ receptors on cardiac myocytes to increase cardiac contractility and duration of systole by paracrine route (Beyer et al. 1995; Brutsaert 1993). Commonly used ET$_A$ and ET$_B$ antagonists are listed in table 3.4. Angiotensin II also can influence heart function. It acts on AT$_1$ receptors in the central nervous system to reset baroreflex control to higher blood

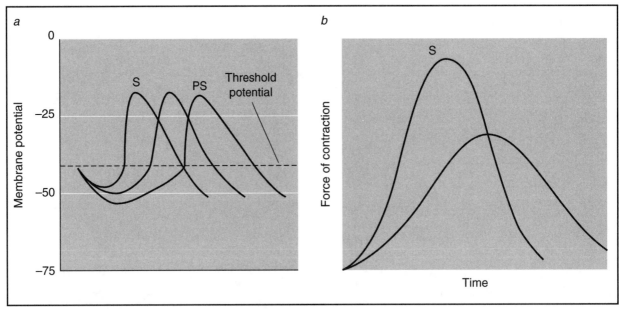

Figure 3.9   *Control of heart function by chemical messengers.*

(a) The effect of sympathetic (S) neurotransmission or adrenomedullary epinephrine (E) and of parasympathetic (PS) neurotransmission on the timing of cardiac pacemaker potentials (the chronotropic effect). *(b)* The effect of S neurotransmission or adrenomedullary E on heart contractility (the inotropic effect).

Adapted from Vander, Sherman, and Luciano 1990.

pressure ranges. By reducing vagal influence over heart function, angiotensin II increases blood pressure without a concomitant decrease in heart rate (Reid 1996a). Finally, nitrergic nerve terminals have been described in the myocardium, but their function has as yet not been elucidated (de Belder et al. 1993; Klimaschewski et al. 1992; Schmidt et al. 1992; Tanaka et al. 1993).

Skeletal Muscle

Neurotransmitters and Hs influence force of contraction and rate of mobilization and storage of metabolic fuels in the skeletal muscle. Skeletal muscle contraction is triggered by action potentials in motoneurons. Coupling of neurogenic excitation and myofilament contraction is achieved in the skeletal muscle and in the heart through an increase in cytosolic Ca^{2+}. In both types of muscle, opening of the L-type Ca^{2+} channels on the plasma membrane causes initial increases in cytoplasmic Ca^{2+}. In the skeletal muscle, the L-type channel (which contains a dihydropyridine, or DHP, receptor) is a voltage sensor (VS) that opens in response to muscle depolarization. It then causes opening of the sarcoplasmic R-R Ca^{2+} channel or receptor through mediation of intracellular messenger, cADPr (Sorrentino & Reggiani 1999; Thorn et al. 1994). R-R on the SR of the heart muscle, on the other hand, opens when Ca^{2+} binds to it. Ca^{2+} is then released through the footlike features of the R-R into the cleft next to T tubules (Fabiato 1985; Innui et al. 1991). This action facilitates Ca^{2+} binding to troponin and formation of actin-myosin cross bridges.

The force of the skeletal muscle contraction can be increased by stimulation of β A-Rs delivered either through S neurotransmitter NE or circulating E and NE.

E is taken up by prejunctional $β_2$ Rs in S nerve endings and, along with NE, also can flow out of S nerve terminals into extracellular fluids when the nerves discharge (Coppes et al. 1995). $β_2$ Adrenergic stimulation increases muscle tension and tetanic contraction by augmenting the amount of Ca^{2+} released from the SR through its R-R Ca^{2+} channels (Cairns et al. 1993). $β_2$ Adrenergic stimulation is more effective in slow-twitch red than in fast-twitch white fibers because of the greater number of A-Rs and greater sensitivity to β adrenergic stimulation in red fibers (Cairns & Dulhunty 1993; Jensen et al. 1995). Stimulation of β A-Rs in skeletal muscle is also necessary for enzymatic and metabolic adaptations to endurance training, increases in oxidative enzyme activity, and muscle sensitivity to insulin (Powers et al. 1995; Torgan et al. 1993). $α_2$ Adrenergic stimulation can also affect the force of muscle contraction, but it does so indirectly. Maximal contraction of the white glycolytic, but not red oxidative, muscles attenuates the action of S stimulation on $α_2$ adrenergic Rs on vascular smooth muscles, and the resultant vasodilation augments the blood flow to the muscles aiding their force of contraction (Thomas et al. 1994).

Vasomotor and Respiratory Control

Smooth muscles that surround blood vessels, respiratory airways, and the GI tract differ from cardiac and skeletal muscles in that their contraction is initiated by diverse stimuli such as autonomic neurotransmission, endocrine messengers, and metabolic and mechanical stimuli rather than all-or-none depolarizations brought about by pacemaker or motoneuron action potentials. These stimuli may increase or decrease smooth muscle contractions, and tension is a result of integrated stimulatory and inhibitory influences.

Innervation of smooth muscles is heterogeneous. Smooth muscles are innervated by both S and PS nerves. Their blood vessels are innervated by cholinergic S fibers (Bulbring & Burn 1935). S nerves form a branching network of fibers and varicosities along the smooth muscle membrane, and NPY and ATP are frequently colocalized with NE. Some sensory nerves that often colocalize SP and 5-HT, or SP and CGRP, also can release transmitters centrifugally and act as effector nerves (Maggi & Meli 1988). Smooth muscles are electrically coupled and allow spreading of membrane depolarization between cells via electrotonic and gap junctions (Burnstock & Hoyle 1992). Depending on the location and strength of the initiating stimulus, only segments of a smooth muscle may contract at times. This section overviews chemical messengers that control smooth muscle contraction and relaxation by neurotransmission and by endocrine and paracrine routes.

Smooth Muscle Contraction

S neurotransmitters NE, NPY, and ACh; circulating messengers adenosine (acting on A_1-Rs), angiotensin II, and AVP (acting on V_1-Rs,); and paracrine messengers eicosanoids, histamine, and endothelin (see figure 3.10) all influence smooth muscle contraction (Ekelund 1996; Marshall et al. 1993). This action plays a major role during exercise in diverting blood from visceral organs to muscle. Constriction of smooth muscles in arterioles by stimulation of α_{1D} A-Rs and of venules by α_{1B} A-Rs (Leech & Faber 1996) and bronchoconstriction by stimulation of adenosine A_1-Rs (counteracted by A_1 blocker aminophylline) are all mediated by G_q protein action in the IP_3-DAG transduction pathway. PLC catalyzes synthesis of IP_3 and DAG and activation of PKC. PKC

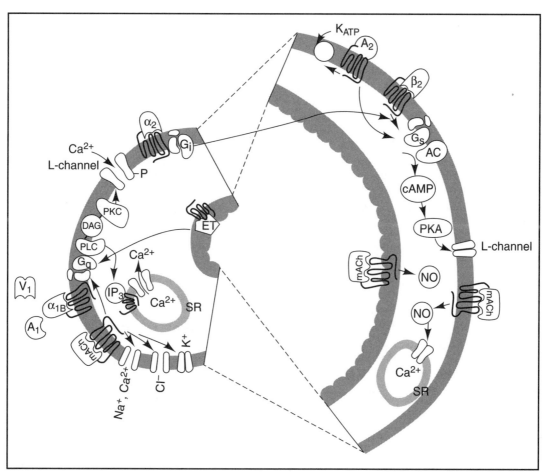

Figure 3.10 *Transduction mechanisms in vasomotor control.*

Constrictive action of the α_1 and α_2 adrenergic receptors (A-Rs), adenosine A_1-R, muscarinic acetylcholine receptors (mACh-R), and endothelin A and B (ET_A and ET_B) Rs is shown on the left. Vasodilating action of β_2 A-Rs, adenosine A_2-R, and nitric oxide (NO) is shown on the right.

α_{1B}, α_2, β_2 = adrenergic receptors; A_1, A_2 = adenosine receptors; cAMP = cyclic adenosine monophosphats; PKA = protein kinase A; AC = adenylyl cyclase; NO = nitric oxide; L-channel = L-type calcium channel; G_s, G_i, G_g = G regulatory proteins; IP_3 = inositol triphosphate; V_1 = AVP_1 receptor; DAG = diacyl glycerol; SR = sarcoplasmic reticulum; ATP = adenosine triphosphate.

Adapted from Griffin and Ojeda 1996.

increases the amplitude of the high-threshold, long-acting L current by opening Ca^{2+} channels on the plasma membrane (Han et al. 1987). This triggers membrane depolarization and muscle contraction (Chick et al. 1996). NPY, which is often coreleased with NE from S nerves, also causes vasoconstriction and potentiates action of NE (Dockray 1992). Vasoconstriction is also triggered by α_2 Rs (Blaak et al. 1993; Frank et al. 1996; Thomas et al. 1994). Constriction of both arterioles and venules in smooth muscles is mediated by α_{2D} Rs, probably through the G_i transduction pathway (see figure 3.9) (Leech & Faber 1996).

Endothelin-1 (ET-1) is the principal endothelium-derived vasoconstrictor messenger (Yanagisawa et al. 1988) that effects smooth-muscle contraction by a paracrine route. ET-1 exerts multiple effects on cardiorespiratory function. It causes ischemia, arrhythmia, and, on occasion, death. It activates the RVLM area of the brain involved in the control of vasomotor function (Kuwaki et al. 1994) and the chemosensitive areas on the ventral surface of the medulla oblongata (Kuwaki et al. 1991). ET-1 stimulates secretion of ANF (Brooks et al. 1994) and inhibits renin and AVP release, and thereby causes increased natriuresis and diuresis. ET-1 release is elicited by TNF-α, IL-1β, E, angiotensin II, AVP, thrombin, transforming growth factor–β (TGF-β), platelet-derived growth factor (PDGF), and shear stress and inhibited by ANP, sodium, heparin, and estrogen (Brooks et al. 1994; Luscher & Barton 1997). ET-1 binds to ET_A and ET_B Rs (see table 3.4), both of which are heptahelical GPCRs and may transduce its message via G_q, G_i, and G_s pathways. ET_A, and to some extent ET_B, Rs mediate contraction of smooth muscle by way of G_q protein, activation of PLC, and release of the second messenger IP_3. IP_3 then acts on a class of Ca^{2+} channels called IP_3 Rs on intracellular Ca^{2+} stores that are also controlled by PKs, Ca^{2+}, and other factors. This IP_3 action results in increased concentration of Ca^{2+} in the cytoplasm (Neylon 1999). ET-1–mediated Ca^{2+} release takes different forms, from single-channel Ca^{2+} "sparks," to oscillations. ET-1 activates R-Rs as well as Ca^{2+}-activated K^+ and Cl^- channels. Both increased cytoplasmic Ca^{2+} and DAG activate myosin light-chain kinase. Resultant phosphorylation of myosin triggers smooth-muscle contraction (Morgan & Suematsu 1990). ET-1 also stimulates matrix synthesis and expression of adhesion molecules and thus contributes to development or atherosclerosis and vascular remodeling. These actions of ET-1 usually involve ET_B Rs, G_q transduction pathway, and the MAPK signaling cascade.

The cycloxygenase pathway also produces endothelium-derived vasoconstrictors thromboxane A_2 or prostaglandin H_2 in response to ACh, histamine, or serotonin (Luscher & Barton 1997). These eicosanoids activate thromboxane Rs in vascular smooth muscle and platelets and counteract actions of NO and prostacyclin. They also neutralize NOS by helping form free-radical superoxide anions.

Angiotensin II (AII) also is a potent vasoconstrictor that is synthesized in the endothelial cell membrane. Besides facilitating synthesis of AII, angiotensin-converting enzyme (ACE) also acts as kininase II, an enzyme that inactivates vasodilatory messenger bradykinin.

Stimulation of the muscarinic m_2ACh and m_3ACh-R of the airway smooth muscle precipitates bronchospasm in sensitive individuals (Buckley & Caulfield 1992; Hargreave et al. 1981). Cholinergic stimulus triggers the opening of several types of ion channels (L-type high-threshold Ca^{2+}, nonselective cation, and a chloride channel) and suppressing of two types of K^+ channels (Janssen & Sims 1992). As is the case with α A-R action on vascular smooth muscles, mACh-R influences bronchial ion channels through the IP_3- DAG transduction pathway. Various irritants can increase endothelial release of histamine and cause tracheal smooth-muscle constriction by a cholinergic vagal reflex (Takahashi et al. 1996). Nocturnal asthma attacks are triggered by an endogenous rhythm that superimposes a decrease in E and cortisol with an increase in ACh (Barnes 1985). In both asthma and aging, hyperreactivity to ACh results from a down-regulation of β_2R number and affinity (Connolly et al. 1994).

Smooth Muscle Relaxation

Relaxation of smooth muscles is mediated by several peptides and amines delivered through neurotransmission, messengers released by contracting skeletal muscle, circulating messengers, and dilatory substances produced in the endothelium. Among the peptidergic vasodilators of neuronal origin are bradykinin, calcitonin gene related peptide (CGRP), SP, and VIP (Brain & Williams 1988; Ekelund 1996; Kubota et al. 1985; Yaoita et al. 1994). As was already discussed in the context of GPCRs, ACh released by S nerves stimulates vasodilation in smooth muscles other than those in bronchi by acting on mACh Rs and activating synthesis of NO in the endothelium. In endothelial cells, gene expression of constitutive NOS is up-regulated by shear stress and estrogens (Luscher & Barton 1997). Adenosine is released by contracting skeletal muscles, oxidative muscles more than glycolytic muscles (Hellsten et al. 1998; Mian et al. 1990). Adenosine maintains coronary artery dilation during exercise through its action on the K_{ATP} channel (Duncker et al. 1995). Adenosine binds to A_2-Rs and causes vasodilation by activating the cAMP transduction pathway, as was the case with E after binding to βA-Rs (see figure 3.10).

Among circulating vasodilating substances, E induces relaxation by acting on β_2 adrenergic Rs (see figure 3.10). When E binds to the smooth muscle β A-Rs, it stimulates the cAMP transduction pathway to activate PKA. The cAMP-dependent PKA closes Ca^{2+} channels on the plasma membrane and as a result reduces the amplitude of the high-threshold, long-acting L current. This condition prevents membrane depolarization and causes vasodilation (Chick et al. 1996). Arteriolar dilation by insulin is mediated by adenosine receptor and membrane hyperpolarization by way of ATP-sensitive potassium channels (McKay & Hester 1996). Endurance training increases vascular sensitivity to dilatory messengers by reducing the action of α_2 A-Rs (Delp 1995).

Among vasodilators of endothelial origin are NO, bradykinin, prostacyclin, and C-type natriuretic peptide (CNP), all released to shear stress. Shear stress activates NOS by opening an endothelial K^+ channel (Cooke et al. 1991). Bradykinin, of either endothelial or endocrine origin, stimulates formation of both NO and endothelium-derived hyperpolarizing factor (EDHF). Prostacyclin has modest vasodilatory action and strongly inhibits platelet aggregation, both through cAMP transduction mechanism. Synergistic action of NO and prostacyclin is needed for maximal inhibition of platelet activation. ET-1, ET-3, and shear stress stimulate release of CNP independently of NO (Zhang et al. 1999b). CNP binds to natriuretic peptide receptor B (NPR_B) and relaxes smooth muscle through activation of particulate GC.

Fuel Metabolism

Release of NE from S nerve terminals and secretion of counter-regulatory hormones E, cortisol, glucagon, and GH help match fuel supply and utilization to the energy needs associated with different types and intensities of exercise. These processes are discussed in more detail in chapter 6. Transduction pathways for these systemic signals meet the changing metabolic fuel needs at rest as well as during exercise and serve to coordinate metabolic, respiratory, and circulatory functions of an exercising individual. They will be briefly outlined below. Contracting muscles also generate signals that reflect changes within the muscle in metabolic fuel and oxygen availability, acidity, temperature, and other types of metabolic stress. These local signals often use transduction pathways, also outlined below, that usually differ from signaling generated by hormones and neurotransmitters.

Hormone-mediated metabolism during exercise is catabolic in character and is controlled in a similar way as the postabsorptive metabolism that takes place 4 to 6 hours after ingestion of a meal. Principal metabolic fuels, glucose and FFAs, are mobilized during exercise from their storage sites and metabolized to provide energy for muscle contraction. This section briefly reviews H-R interactions that mediate

1. hepatic glucose output,
2. FFA release from the adipose tissue,
3. fuel uptake and utilization by muscle during exercise, and
4. messenger-mediated glucose uptake and resynthesis of glycogen and triglyceride during recovery from exercise.

Hepatic Glucose Output
Glucose release by the liver is made possible by two metabolic pathways of glucose formation: (1) glycogenolysis or enzymatic degradation of the inert glucose polymer

glycogen (see figure 3.11) and (2) gluconeogenesis or biosynthesis of glucose from the breakdown products of protein, triglyceride, and carbohydrate metabolism. Both processes are controlled by the S nerves and several endocrine and paracrine messengers. Principal controllers of hepatic glycogenolysis are glucagon and SA catecholamines with a quantitatively less important contribution from adenosine; angiotensin; V_1 receptors of AVP, ATP, EGF; and eicosanoid PG D2 produced by Kuppfer cells in the liver (Ali et al. 1989; Casteleijn et al. 1988; Grau et al. 1997; Keppens 1993; Van Stapel et al 1991).

Hepatic glycogenolysis is increased by stimulation of both α_1 and β adrenergic receptors at rest and during exercise (Coker et al. 1997) (see figure 3.11). E acting on β_1 and β_2 A-Rs, adenosine acting on A_2-R, and glucagon stimulate glycogenolysis in the liver by the cAMP transduction pathway (see figure 3.11). Although both E and glucagon activate AC through the G_s transduction pathway, the glucagon transduction pathway utilizes a smaller type of G_s protein (Yagami 1995). Phosphorylation of phosphorylase kinase is carried out by cAMP-dependent PKA, and the remainder of the enzymatic cascade is the same as in the IP_3-DAG transduction pathway. G_q signaling pathway is activated by NE after it binds to α_1 adrenergic receptors and by AVP after it binds to V_1 receptors in the liver. Stimulation of PLC in this pathway brings about IP_3 synthesis and Ca^{2+} release from SR. Increased Ca^{2+} concentration activates calmodulin to catalyze phosphorylation of phosphorylase kinase. Activated phosphorylase kinase phosphorylates the enzyme phosphorylase, which then detaches a glucose residue from glucogen and phosphorylates to glucose-6 phosphate. There are gender (Moriyama et al. 1997) and age (Van Ermen & Fraeyman 1994) differences in the relative participation of the two pathways in hepatic glucose production. In the female, α adrenergic control of glycogenolysis predominates and is converted to the cAMP-mediated pathway that is prevalent in the males in the absence of adrenal glucocorticosteroids (Moriyama et al. 1997). Gluconeogenic enzymes PEPCK and pyruvate carboxylase (PC) are stimulated by S stimulation of α_1 A-Rs and by glucagon, GH, and cortisol.

Free Fatty Acid Release From the Adipose Tissue
Lipolysis, hydrolysis of a triglyceride to glycerol and free fatty acids in the adipose tissue, is stimulated by catecholamines and by the endocrine messengers β lipotropin, glucagon, glucocorticoids, GH, prolactin, and secretin. Catecholamine stimulation of lipolysis is through binding to β adrenergic receptors (Arner et al. 1990; Galitzky et al. 1993b; Hoffstedt et al. 1995). In adipose tissue this process entails stimulation of β_1, β_2, and β_3 Rs (Emorine et al. 1989). Although there is some disagreement regarding the expression of β_3 Rs in humans, this R type is activated at high stimulus intensities (Atgie et al. 1997) and is functional in human intra-abdominal fat (Hoffstedt et al. 1995). Lipolysis of intracellular lipids in skeletal muscle is achieved through activation of β_2 and β_3 A-Rs (Nagase et al. 1996; Sillence et al. 1993).

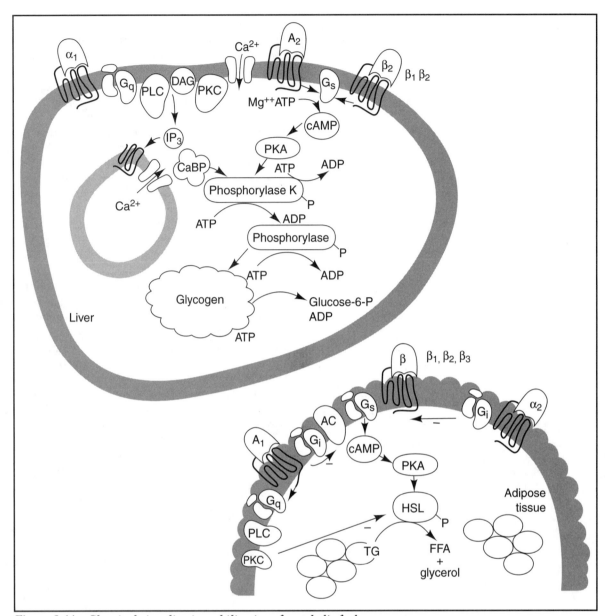

Figure 3.11 *Chemical signaling in mobilization of metabolic fuels.*

Hepatic glycogenolysis is shown above the mobilization of free fatty acids (FFAs) out of a triglyceride (TG) molecule (lipolysis).

α_1, α_2, β_1, β_2 = adrenergic receptors; G_s, G_i, G_q = G regulatory proteins; PLC = phospholipase C; PKA, PKC = protein kinases; IP_3, cAMP = second messenger; DAG = diacylglycerol; phosphorylase K = phosphorylase kinase; A_1, A_2 = adenosine receptors; CaBP = calcium binding proteins or calmodulin.

Adapted from Griffin and Ojeda 1996.

The affinity among the three β A-Rs is different for E (β_2 > β_1 > β_3) and NE (β_1 > β_2 > β_3) (Galitzky et al. 1993; Lafontan et al. 1995). Prolonged exposure to catecholamines results in desensitization of β A-Rs with the relative susceptibility varying by type (β_1 > β_2 > β_3) (Atgie et al. 1997; Granneman 1995; Hoffstedt et al. 1995). After the higher-affinity β_1 and β_2 A-Rs become desensitized by prolonged exposure to high concentrations of catecholamines, β_3 Rs remain active and allow lipolysis to proceed over extended periods of time. Stimulation of β_3 Rs increases energy expenditure by uncoupling oxidation from phosphorylation of ADP in the brown adipose

tissue (Nagase et al. 1996) and β_1 stimulation increases the basal metabolic rate (Lamont et al. 1997; Tremblay et al. 1992).

Action of all of the above chemical messengers except for glucocorticoids and GH is through the cAMP transduction pathway (see figures 3.6 and 3.11). cAMP activates cAMP-dependent PKA, which in turn phosphorylates and activates a triglycerol hydrolase called hormone-sensitive lipase (HSL) (Egan et al. 1992). HSL acts on triglycerides inside a lipid droplet after the protein barrier on its surface, consisting of perilipins, is removed through phosphorylation of these proteins

(Londos et al. 1995; Souza et al. 1998). HSL hydrolyzes storage triglycerides into FFAs and glycerol. Stimulation of lipolysis by GH and glucocorticoids is delayed 1 to 2 hours because these hormones act through transcription of HSL gene and synthesis of HSL. GH also enhances catecholamine lipolytic action by stimulating synthesis of β A-Rs and by inhibiting adipose tissue LPL via the G_i protein pathway (see figure 3.12) (Richelsen 1997).

Chronic GH administration acts by antagonizing the antilipolytic action of adenosine (Doris et al. 1996).

During exercise, lipolysis is also enhanced through suppression of lipogenesis. Hepatic lipogenesis is inhibited by suppression of insulin secretion and by exercise-induced activation of AMP-activated protein kinase (AMPK). AMPK blocks the activity of lipogenic enzyme acetyl CoA carboxylase (Zhou et al. 2001).

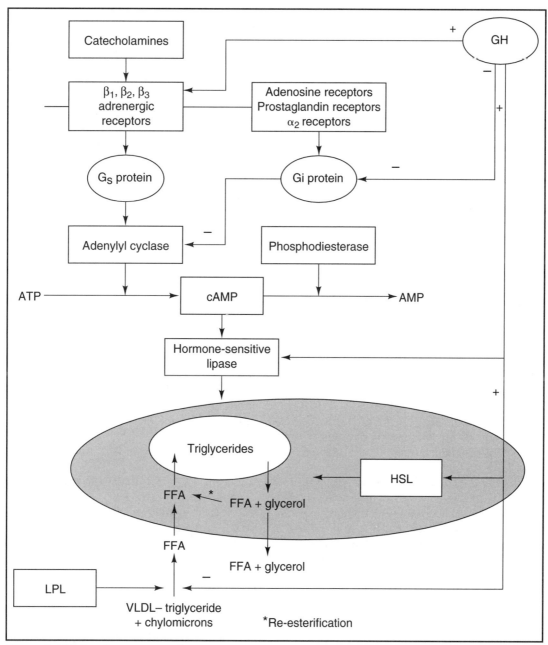

Figure 3.12 *Lipolytic actions of growth hormone.*

Growth hormone (GH) stimulates lipolysis in the adipocytes by (*a*) enhancing catecholamine lipolytic action through increased synthesis of β adrenergic receptors, (*b*) stimulating synthesis of hormone sensitive lipase (HSL), and (*c*) inhibiting action of lipoprotein lipase (LPL). GH stimulates HSL both by increasing its synthesis as well as increasing the activity of adenylate cyclase (AC) as shown in the figure.

FFA = free fatty acids; cAMP = cyclic adenosine monophosphate; ATP = adenosine triphosphate; AMP = adenosine monophosphate; VLDL = very low density lipoproteins.

Adapted from Richelsen 1997.

Lipolysis in the adipose tissue is inhibited by catecholamines acting on α_2 A-Rs through the G_i transduction pathway (see figure 3.11) (Tarkovacs et al. 1994), as α_2 R blockade increases β stimulation by NE and lipolysis (Lafontan et al. 1992). Insulin also is a particularly potent inhibitor of lipolysis (Campbell et al. 1992). The antilipolytic action of insulin is mediated by activation of PDEs, which diminish action of the cAMP transduction pathway (Eriksson et al. 1995, Manganiello et al. 1996). PDE is activated by cGMP, and its activation in turn depends on the IP_3-DAG transduction pathway (Rahn et al. 1994).

Besides insulin, other hormones with antilipolytic action are adenosine, IGFs, oxytocin, GIP, and glucocorticoids, which, along with insulin, can down-regulate β_3 receptors (Langin et al. 1995). Adenosine inhibits lipolysis by acting on A_1-Rs that are coupled with G_i and G_q transduction pathways. Most of these Hs inhibit formation of cAMP, and adenosine blocks HSL phosphorylation by activating PKC (see figure 3.11). A_1-Rs display weaker coupling to the G_i protein than α_2 receptors (Larrouy et al. 1994). Abnormal A_1-R function has been proposed as one of the causes of obesity (La Noue & Martin 1994).

Distribution of A-Rs differs among adipose tissue sites (subcutaneous versus intra-abdominal), between genders, and at different ages. Because of differential R distribution, different A-R affinities for catecholamines, and different susceptibility to inhibition by insulin, lipolytic activity differs among adipose tissue sites (Jensen et al. 1989; Martin & Jensen 1991; Richelsen et al. 1991; Robidoux et al. 1995). Differences in the sensitivity of β Rs and of antilipolytic α_2 Rs in the two genders and in hormonal stimulation were found to cause differential fat loss in exercising or sedentary humans and contribute to regional differences in fat deposition (Hellstrom et al. 1997; Lonnqvist et al. 1997; Robidoux et al. 1995; Wahrenberg et al. 1992). Differential distribution and sensitivities of α_2 and β_1 and β_2 adrenergic receptors in different fat depots may be responsible for regional differences in lipolysis and fat accumulation (Hellmer et al. 1992). Lipolytic β A-R affinities are greater in intra-abdominal than in subcutaneous adipocytes, and antilipolytic α_{2A} A-R affinity is greater in subcutaneous than in intra-abdominal adipocytes (Hellmer et al. 1992; Mauriege et al. 1987; Rebuffe-Scrive et al. 1987; Vikman et al. 1996). Abdominal subcutaneous adipocytes are more sensitive to β R stimulation and less sensitive to α_2 R agonists (Berman et al. 1998; Guo et al. 1997; Lafontan et al. 1979; Rosenbaum et al. 1991; Wahrenberg et al. 1989). Finally, abdominal subcutaneous fat is less sensitive than gluteal adipose tissue to antilipolytic α_2R stimulation and more sensitive than femoral adipose tissue to the lipolytic effect of E (Galitzky et al. 1993a; Horowitz & Klein 2000b).

Fuel Uptake, Utilization, and Storage by Muscle During Exercise

Skeletal and cardiac muscles have enzymatic adaptations and transduction capacities that allow them to take up, utilize, and store metabolic fuels both with and without the help of chemical messengers. Thus, in the muscle, glycogenolysis is either directly triggered by muscle contraction, as in the case of fast-twitch white fibers, or is stimulated by S nerves and endocrine and paracrine messengers (Greenhaff et al. 1991). Stimulation of both β_2 and β_3 adrenergic Rs increases skeletal muscle glycogenolysis, β_2 Rs at lower, and β_3 Rs at high messenger concentrations (Sillence et al. 1993), and the effect is confined to red muscle fibers (Greenhaff et al. 1991).

Metabolic fuel uptake involves transport of glucose and FFAs from plasma into the exercising muscle and other tissues. Its normal operation is important both as a means of regulating plasma concentrations of fuel metabolites within healthful range and as a basis for nutrient storage that also needs to be regulated within healthy range. Analysis of transduction pathways that mediate cellular fuel uptake has been the focus of intense recent research because aberrations in cellular uptake of glucose can cause hypoglycemia and diabetes mellitus, abnormalities in the uptake of lipids, various forms of cardiovascular complications, and defects in lipid storage can cause metabolic syndrome of obesity.

The enzymatic adaptation that provides the muscles with the capacity to take up glucose at exceedingly low concentrations whenever there is a need for that substrate is the low Km of its constitutive enzyme hexokinase (HK) and its near complete inactivation by its metabolic product, glucose-6 phosphate. This condition is in contrast to the adaptive control of a similar enzyme in the liver, the glucokinase (GK), whose higher Km facilitates hepatic glucose uptake at high postprandial glucose concentrations. In addition to this enzymatic adaptation, glucose uptake by the skeletal muscle is mediated by three additional independent mechanisms: insulin action on I-R, insulin-independent processes of muscle contraction, and action of NO (Higaki et al. 2001).

The insulin-independent component is induced by muscle contraction and requires increased cytosolic Ca^{2+} (Hayashi et al. 1997) and the cGMP-NO pathway for signaling (Kapur et al. 1997; Roberts et al. 1997; Young et al. 1997). Stimulation of glucose transport by exercise entails activation of α_2 isomer of AMP-activated protein kinase (AMPK) and its facilitation of translocation of GLUT-4 glucose transporters to the sarcolemma by the p38 MAPK pathway (Goodyear, 2000; Hayashi et al. 1998, 2000; Mu et al., 2001; Ojuka et al. 2002; Sakoda et al. 2002; Thorell et al. 1999; Xi et al. 2001). AMPK is a sensor of reduced cellular energy charge in the muscle (Ai et al. 2002; Hardie & Hawley 2001; Winder 2001). It is activated by an increase in 5'-AMP and suppressed by a rise in intracellular content of PCr. AMPK stimulates glucose uptake by muscles during metabolic stress of exercise but does not contribute to glucose uptake by the adipose tissue (Sakoda et al. 2002). NO and cGMP-dependent PDE stimulate both G uptake and carbohydrate metabolism in the absence of insulin (Musi et al. 2001; Young et al. 1997), and deficient release of

NO was identified as a cause of insulin resistance in the skeletal muscle (Young & Leighton 1998). In type II muscle fibers, glycogen depletion stimulates, and glycogen repletion suppresses, AMPK activity and glucose transport (Aschenbach et al. 2002; Richter et al. 2001; Stephens et al. 2002; Wojtaszewski et al. 2002). In slow, oxidative muscle fibers, exercise stimulates a high rate of glucose uptake that is independent of AMPK activation or glycogen repletion (Derave et al. 2000).

Besides hormonal effects on postexercise glucose uptake and glycogen synthesis, muscle contraction also enhances both processes through activation of AMPK (Aschenbach et al. 2002). Contractions stimulate glycogen synthesis through activation of muscle-specific protein phosphatase type 1, an enzyme that also catalyzes basal glycogen synthesis (Aschenbach et al. 2002).

Messenger-Mediated Glucose Uptake and Resynthesis of Glycogen and Triglyceride During Recovery From Exercise

Insulin-dependent component of glucose uptake by the muscle is mediated by the IP_3-DAG transduction pathway and phosphorylation of IRS, PI3-K, and Akt (see figure 3.4) (Cheatham & Kahn 1995; Coffer & Woodgett 1998; Cortright & Dohm 1997; Czech & Corvera 1999; Hayashi et al. 1997; Houmard et al. 1999; Hickey et al. 1997; Manganiello et al. 1996; Zhou & Dohm 1997). Activation by insulin of dephosphorylating phosphatases is one way that insulin inactivates phosphorylase and HSL. Insulin also acts through the cGMP pathway. The inhibitory effects of insulin on glycogenolysis and lipolysis are mediated through the activation of PDE by the second messenger cGMP. PDE reduces intracellular concentration of cAMP and of cAMP-dependent PKA.

Other messengers that affect skeletal muscle glucose uptake are adenosine, NO, catecholamines, and cortisol. Adenosine and NO are generated by skeletal muscle contraction (Ballard 1995). Adenosine acts on A_1-Rs to reduce insulin-dependent glucose uptake in red muscle fibers (Challis et al. 1992). Insulin resistance in the muscles of genetically obese rats is reversed by administration of adenosine A_1-R antagonist. In the presence of insulin and β A-R stimulation, adenosine increases activity of the glycogen synthetase in red muscle fibers (Vergauwen et al. 1997). Adenosine also mediates glucocorticoid blockade of E-induced glycogenolysis in skeletal muscle (Coderre et al. 1992). α Adrenergic stimulation facilitates glucose uptake by the muscle, particularly in the presence of FFAs (Saitoh et al. 1974) while β adrenergic stimulation by E contributes to insulin resistance in the muscle by decreasing the activity of enzyme hexokinase (Lee et al. 1997). Unlike the effects of insulin and NO, which affect glucose uptake at the level of message transduction, β adrenergic stimulation increases intracellular glucose 6-phosphate formation through glycogenolysis and blocks hexokinase activity by affecting substrate metabolism.

Glycogen synthesis and lipogenesis in the liver and adipose tissue are largely under the control of insulin, which, with the exception of its stimulation of carbohydrate metabolism, exerts mostly anabolic and biosynthetic actions. In addition to stimulating glucose uptake, insulin stimulates glycogen synthesis and inhibits glycogenolysis and gluconeogenesis. It stimulates triglyceride and protein synthesis and inhibits lipolysis and protein degradation.

Biosynthetic actions of insulin include phosphorylation of transcription factors and kinases that belong to the growth-factor signaling cascades (see figure 3.4) (Manganiello et al. 1996). Thus, insulin-stimulated MAPKs phosphorylate and activate ribosomal S6 kinases that are involved in growth and differentiation and protein phosphatases, which dephosphorylate and stimulate glycogen synthetase and inactive HSL. Because insulin transduction mechanisms combine features of IP_3-DAG and cGMP pathways (see figure 3.7), NO could be the connecting link in insulin signaling. At this point however, NO is known to only be involved in the glucose-uptake aspect of insulin action (Kapur et al. 1997; Young & Leighton 1998).

METHODS OF HORMONE MEASUREMENT

Selection of the appropriate method of H measurement (Baulieu & Kelly 1990; Griffin & Ojeda 1996) depends on the type of question asked, secretory and binding characteristics of the H, acceptable level of invasiveness in collection of H sample, and physiological condition of the subject donating the sample. Bioassays, radioimmunoassays, and immunometric assays provide information on H concentration in body fluids and tissue extracts. Receptor-binding assays can assess N and H-R affinity, whereas solution hybridization assay measures H synthesis through assessment of the concentration of specific H mRNA in nucleic acid extracts. For localization of Hs in tissues and cells, immunohistochemical and immunocytochemical approaches are used, while in situ hybridization reveals sites of hormone synthesis.

Bioassays

Bioassays (BAs) utilize some measurable feature of the animal's physiology for determination of hormone concentration. An example of a BA is the hypophysectomized rat tibia test in which the increase in the width of epiphyseal growth plate is used to measure growth-promoting action of GH. A special advantage of BAs is that they directly measure biological action of an H, while their lack of sensitivity is a disadvantage. In vitro BAs are more recent developments. They are performed with cells or cell fragments and provide a more sensitive measure of endocrine biological action.

Radioimmunoassays

Radioimmunoassays (RIAs) utilize the principle of reversibility of H-R interactions described by the Michaelis-Menten equation, to measure H concentration at H and specific antibody (Ab) dilutions that are well below saturation. Because H-R interactions in an RIA take place at low Ab concentration of about 10^{-5} g/L, prolonged incubation (hours to days) is needed to reach equilibrium, and assay sensitivity (measured by half-maximal binding) is increased as Ab concentration is reduced. Amounts of all reactants in an RIA are kept constant with the exception of H in reference preparation (or standard), which is systematically varied, and in unknowns (or samples). Radioactively labeled H allows detection of bound hormone after the removal of unbound radioactive H. Peptide and protein Hs are usually labeled with ^{125}I on tyrosine or histidine residues (which need to be attached to small molecules that lack them) or with ^3H. For best binding, only monoiodinated H is used in an RIA and is separated out of the iodination complex by chromatography. Steroid and thyroid hormones are tagged with ^3H on one or more molecular sites to achieve the necessary level of specific activity.

The specificity in an RIA is imparted by the Ab that consists of IgGs developed to a partially purified (polyclonal Abs) or a highly purified (monoclonal Abs) H preparation. Polyclonal Abs are a mix of heterogeneous IgG molecules that may bind to different parts of H molecule, whereas monoclonal Abs have uniform structure and binding properties. Abs develop to peptides with a molecular weight greater than 1,000 daltons. Smaller Hs need to be complexed to large proteins to increase their antigenicity. Abs will cross-react with more than one H if they bind to parts of a molecule that has primary structure shared by these Hs. Because of such Ab cross-reactivity, it is often impossible to distinguish between some Hs and their biologically inert precursors or derivatives (for example β-LPH from β endorphin) or between Hs that have structural homologies (LH and FSH).

The amount of H can be estimated with an RIA from the comparison of competitive inhibition of binding of radioactively labeled H by a known amount of nonradioactive H in standards. The amount of H in samples is then extrapolated from the binding curve in which the percent of binding suppression (B/B_0) is plotted as a function of free H (or standard) concentration. The curvilinear plot is converted to a straight line by using a log-logit or other type of mathematical transformation, but H measurements are invalid at the low and high ends of the binding curve. Separation of H-R complex from unbound radioactive HR is achieved by centrifugation after incubation with the second Ab against the γ globulins of the animal species that provided the first Ab. Other separation methods are precipitation of H-R complex with polyethylene glycol (PEG), attachment of the complex to a solid phase (assay-tube coating), and precipitation of free H by activated charcoal (steroid RIAs).

The advantage of an RIA method is in its capacity to detect Hs at very low concentrations (10^{-12} M). The disadvantage of the method is that it cannot measure H concentrations outside the range of the standard curve and it may measure H isoforms, precursors or degradation products that have little if any biological activity. The former limitation can be overcome by repeating measurements after H dilution, and the latter by comparison of RIA results with parallel in vitro BA measurements.

Immunometric Assays

Immunometric assays (IMAs) also utilize the specificity of H-R binding for assessment of H concentration. Two Abs with binding sites to different parts of an H molecule are used at saturating concentrations. One Ab is tagged with either a radioisotope, a chemical that imparts to the Ab the capacity for phosphorescence or chemiluminescence, or an enzyme, which permits photometric or colorimetoric measurements, and the second Ab is coupled to a solid phase (test tube, microtiter-plate coating, or synthetic beads). During a very brief incubation (minutes to hours), the H creates a "sandwich" by connecting two Abs.

IMAs can be more sensitive than RIAs, as is the case with chemiluminescent GH assay, and have made it possible to measure some Hs (TSH, PTH) that were difficult to measure with RIAs (Chapman et al. 1994). Additional advantages of IMAs are their capacity to measure Hs throughout the entire range of concentrations, lack of interference by endogenous H Abs, and sensitivity of H-R interaction that is independent of Ab concentration. The disadvantage of IMAs is that they require high Ab concentrations.

Solution Hybridization Assay

Solution hybridization, or nuclease-protection, assay (SHA) is used to assess H synthesis by measurement of the messenger ribonucleic acid (mRNA) for a given hormone (Durnam & Palmiter 1983). This method uses high incubation temperatures to produce hybrids between nonradioactive antisense copies of mRNA (standards) and radioactive sense copies of mRNA of a given H. Only the specific sense and antisense copies of a particular H mRNA with complementary base-pair sequences will hybridize, and the remaining nucleic acids in the tissue extract will remain single stranded and vulnerable to enzyme RNAse that is used to separate radioactive hybrids from unhybridized mRNA.

Receptor-Binding Assays

Receptor-binding assays (RBAs), like RIAs, take advantage of competitive suppression of binding of labeled H with Rs on cells or cell membranes by nonradioactive H. Radioreceptor assays (RRAs) are performed at different concentrations of Rs to determine H concentration at which maximal binding occurs (N) and at different concentrations of free H with R concentration held constant to determine H-R affinity.

Immunohistochemistry and Immunocytochemistry

Immunohistochemical (IHC) and immunocytochemical (ICC) methods utilize the specificity of H-R binding to localize Hs within the body. To that end, Hs are labeled with isotopes, chemicals that convey to them the capacity for fluorescence or chemiluminescence, or colorimetric enzymes, and appropriate procedures are then used for microscopic visualization of the location of labeled hormone in tissue slices.

In Situ Hybridization

In situ hybridization (ISH) is a hybridization procedure that allows cellular and subcellular visualization of H mRNA. The location of newly synthesized H can be determined after incubation of tissue slices with radioactive mRNA for a particular H under the conditions that favor its hybridization with complementary RNA that is present in the tissue.

Assessment of SA Function

Assessment of SA function presents special challenges and is therefore outlined here. SA activity during exercise is usually assessed from catecholamine concentrations, kinetics, and effects or from S nerve activity. Circulating catecholamines can be measured with radioenzymatic, RIA, or high-performance liquid chromatographic (HPLC) methods. Without information about NE appearance and clearance rates, the source of circulating E and NE cannot be identified with certainty. The adrenal medulla secretes both E and NE, and at high exercise intensities, NE, E, and NPY spill over from S nerve endings into plasma (Johansson et al. 1997; Kaijser et al. 1994; Leuenberger et al. 1993). Measurements of arteriovenous NE differences or catecholamine kinetics (E or NE rates of appearance and disappearance) circumvent this limitation. Daily urinary catecholamine output is another valid way to quantify overall SA activity.

Other approaches to assessment of autonomic activation are available. S nerve activity can be directly measured from discharges in the peroneal nerve in the leg (Saito 1995). Pharmacological blockade and stimulation studies also can yield useful information. Finally, analysis of heart rate variability indirectly measures relative contributions of PS and S nerve activity to the control of heart function (Malik & Camm 1995). This method either examines overall HR variability in the temporal domain, or separates S and PS contributions to this variability by power spectral analysis in the spectral domain. Total spectral power (Pt) of HR variability is separated into harmonic and nonharmonic components, and the harmonic component is further differentiated into high frequencies (Ph) and low frequencies (Pl) in the respective ranges of 0.3 to 0.4 and 0 to 0.1 Hz. PS and S influences on the heart are then inferred from respective Ph/Pt and Pl/Pt ratios (Nakamura et al. 1993; Yamamoto & Hughson 1991, 1993; Yamamoto et al. 1991).

Limitations and Strategies

Methods of H measurements often provide limited or misleading information when characteristics of H secretion and binding are not considered. Individual mea-

surements are valid indicators of concentration for those Hs that maintain constant concentration over extended periods of time. Hs that display pulsatile secretory pattern require frequent sequential sampling to adequately define variations in their concentration. H concentration is influenced by the rates of H release and degradation. H degradation can be estimated from the arteriovenous differences in H concentration across organs (e.g., kidney or liver) where degradation occurs and from the rate of decay of H pulses assessed at frequent intervals and subjected to cluster or deconvolution analyses (Veldhuis 1992; Veldhuis & Johnson 1988).

H-binding proteins and endogenous Abs against circulating Hs can confound H measurements and complicate the interpretation of RIA results. Where endogenous Abs are present, exogenous Abs will bind to residual free H, and reduced binding of radioactive H will be misinterpreted as high H concentration. Where endogenous Abs are suspected, RIA should be run without the addition of exogenous Ab and precipitated with ammonium sulfate. Ammonium precipitation is also used for assessment of bioactive steroid Hs, which circulate bound to specific proteins. Although less than 5% of testosterone and cortisol circulate free of binding protein, additional H dissociates from its binding protein within the capillaries and can also be assessed through ammonium precipitation. For assessment of biologically active IGF-I concentration, its binding proteins are usually removed with acid precipitation.

Concentration of H in systemic circulation does not provide information about H concentration in vascular beds into which the H is first released. Yet collection of samples from hypothalamo-hypophyseal or hepatic portal vessels and cerebrospinal fluid is even more invasive than blood collection from systemic circulation. Invasive procedures can be avoided when H concentration in the saliva or urine bears a predictable relationship with concentrations of circulating H. Urinary H measurements are carried out when information about total H secretion during an extended period of time is sought. For example 24-hour urinary hydroxycorticoids or free cortisol are a valid measure of daily glucocorticoid output provided creatinine clearance is also measured to assess glomerular filtration rate.

An additional limitation of individual H measurements is that they yield ambiguous results without the information about their secretagogues. This problem can be avoided by dynamic tests in which H measurements are carried out after administration of eliciting stimuli (stimulation tests) or of Hs that provide negative feedback (suppression tests). Stimulation tests (see figure 3.13) (CRF and ACTH tests) are usually done by administering trophic or releasing Hs (H_{tro}) and by measuring the magnitude of secretory response of the target H (H_{tar}). Four patterns of H concentrations can identify abnormalities in endocrine secretion. Low response of H_{tar} to a high concentration of H_{tro} identifies target gland failure. High response of H_{tar} to a low concentration of

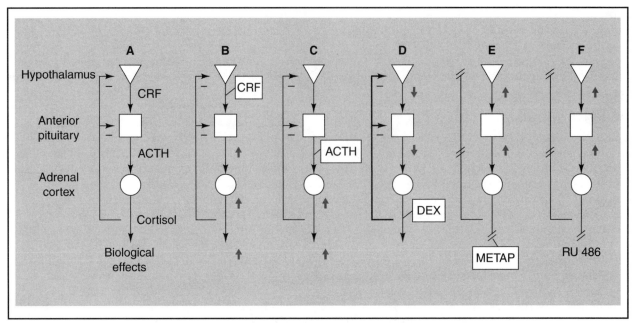

Figure 3.13 *Dynamic tests of hypothalamo-pituitary-adrenal (HPA) function.*

(a) Spontaneous operation of the HPA axis; *(b)* pituitary corticotroph competence test in response to exogenous corticotropin-releasing factor (CRF) administration with increased corticotropin (ACTH) and cortisol secretion (shown as dark arrows on the right); *(c)* adrenocortical response test to ACTH administration; *(d)* dexamethasone (DEX) suppression test (DEX is a synthetic glucocorticosteroid) *(e)* metapyrone (METAP) test (this compound inhibits 11β hydroxylase and production of cortisol, thereby removing negative feedback on ACTH); *(f)* RU486 test (RU486 blocks the effects of cortisol at the receptor level and thereby reduces negative feedback on CRF and ACTH, leading to an increase in plasma cortisol).

Adapted from Baulieu 1990.

H_{tro} identifies autonomous oversecretion of H_{tar}. When both the H_{tro} and H_{tar} are low, and administration of exogenous H_{tro} yields normal response, deficiency is in H_{tro} secretion. An example of this is hyposecretion of GnRH and LH in exercise-induced amenorrhea where normal LH is elicited after administration of exogenous GnRH (De Souza & Metzger 1991). When both the H_{tro} and H_{tar} are high, there may be autonomous oversecretion of H_{tro}, insensitivity of H_{tro} to negative feedback by H_{tar}, or tissue resistance to the actions of H_{tar}. An example of the last possibility is tissue resistance to insulin stimulation of glucose uptake where high insulin concentrations (H_{tar})

accompany high glucose concentrations (equivalent to H_{tro}).

Suppression tests (see figure 3.13) are usually done by administering an excess of H_{tar} negative feedback (dexamethasone test), suppression of H_{tar} synthesis (metapyrone test), or blockade of hormone Rs in target tissues (RU486 test). Supplementary endocrine and metabolic information is often needed for proper interpretation of dynamic H tests as metabolic abnormalities usually affect tissue sensitivity to hormones, and endocrine abnormalities affect secretion and action of other Hs.

4

Regulation of Body Fluids During Exercise

All processes of life occur in a liquid environment, but water, unlike energy nutrients, is not stored by our bodies. Instead, its quantity and distribution among cellular and extracellular compartments are regulated through a complex system of endocrine, autonomic, and behavioral processes. In humans at rest, water intake and endocrine reflexes mediated by hydromineral hormones that control water and sodium reabsorption in the kidney closely match evaporative, urinary, and fecal water losses.

During exercise, a need for water escalates sharply. The increased metabolic rate of contracting muscle requires both a greater delivery of nutrients and oxygen and faster waste and heat removal from the body. To achieve these ends, cardiac output of blood and systolic blood pressure increase, while the vasoconstriction of vascular beds in the kidney and other viscera route a finite volume of plasma to the muscle. Heat dissipation is achieved by diverting part of the cardiac output through skin veins for radiative heat loss and by sacrificing some of the blood volume and body water for evaporative heat losses through sweating. Cardiovascular adjustments and sweating are rapidly initiated during exercise through increased S nerve activity and autonomic reflexes, while several hormones and paracrine-autocrine messengers contribute both to vasoconstriction in the viscera and to vasodilation in the muscle. These emergency functions during exercise and other stressful situations have a higher priority than the fluid regulatory mechanisms that operate at rest. For instance, kidney circulation is suppressed during intense exercise as water-conserving and sodium-conserving actions of hydromineral hormones are curtailed, and their vasoconstrictive actions come into play.

Cardiovascular brain centers that control heart function, circulation, and blood pressure act in conjunction with hydromineral hormones. A blood-brain barrier caused by tight junctions between ependymal (endothelial) cells interferes with such direct hormone-brain interactions throughout most of the brain vascular beds. Exceptions to this general rule are several brain regions where blood vessels are fenestrated and thus permit transit of water-soluble polypeptides from the brain into systemic circulation and from systemic circulation into the brain. During the recovery from exercise, losses of body water and sodium-containing plasma that were incurred during exercise are corrected. Homeostatic signals of water and plasma volume deficit are communicated by autonomic afferent nerve fibers and hormones to brain centers or are detected directly by brain cells. The resulting thirst, sodium hunger, and reinstated reabsorptive actions of hydromineral hormones in the kidney tubule all help to correct body water and plasma volume imbalances. In addition to alteration by acute exercise of the amount and distribution of body water and blood pressure, longer duration exercise training can also affect resting blood pressure. Therefore, exercise is at once an important approach to control hypertension and a tool for the study of mechanisms of hypertension.

This chapter discusses the roles of hormones, autonomic nervous system, and behavior at rest and during exercise in

1. redistribution of body fluids among compartments during dehydration,
2. hormonal regulation of cellular hydration,
3. hormonal regulation of plasma volume, and
4. strategies of fluid management for optimal physical performance.

REDISTRIBUTION OF BODY FLUIDS DURING DEHYDRATION

Approximately 71% of fat-free tissues consists of water. That amount represents about 60% and 52%, respectively, of a 70-kg total body mass of a 25-year-old male and female. An understanding of the distribution of this total body water (TBW) among body compartments (see table 4.1) will facilitate discussion of fluid shifts under different conditions of exercise and dehydration.

Approximately two thirds of TBW is found within the cells (intracellular fluids, or ICF). The remaining third is extracellular fluid (ECF), of which about 79% is in the interstitial space around the cells (ISF), and the remaining 21% constitutes the liquid or plasma portion of blood. Partitioning of water among cellular and extracellular compartments is accomplished primarily through active sequestration of some mineral ions and proteins in separate body compartments. Sodium chloride and albumin are largely confined to extracellular space, and potassium chloride is confined to cells. Movement of water between body fluid compartments depends on the osmotic pressure gradients that these ions create.

Three conditions commonly alter the distribution of water among fluid compartments: (1) loss of body water through dehydration, (2) exercise, and (3) increases in the concentration of plasma sodium. During dehydration, distribution of water losses among fluid compartments changes as TBW losses increase (Costill & Saltin 1974). During exercise lasting less than 1 hour, body fluid losses are primarily drawn from extracellular space, but as sweating increases during prolonged exercise or exercise in the heat, water is increasingly lost from ICF (Maw et al. 1998). After a TBW loss of 1.5 liters (3.6% of TBW), water deficit is predominantly from ECF, but beyond a 2.7 L loss (6% of TBW), water deficit becomes evenly distributed between ECF and ICF compartments (see table 4.2). Plasma volume (PV) losses are restricted to about 10% of TBW loss and grow proportionally with increasing dehydration.

Distribution and movement of fluids among cellular compartments in part depend on osmolality of fluids within them. Osmolarity is a concentration of dissolved solutes confined to a cellular compartment by means of semipermeable membranes and ionic pumps. Osmolality is the concentration of dissociated nonpenetrating ions that control water's rate of passage. Partial permeability of cell membranes of a cell compartment ordinarily precludes transit by nonpenetrating ions but permits free diffusion of water. Osmotic pressure, resulting from the uneven concentrations of nonpenetrating solutes in different fluid compartments, equalizes water concentration by having it diffuse across cell membranes.

Table 4.1   Normal Body Water Distribution

	Male (25 years)	*Female (25 years)*
Body mass (kg)	70.0	70.0
Fat-free mass (kg)	59.5	51.1
Fat mass (kg)	10.5	18.9
Body water (L)	42.0	36.3
Intracellular water (L)	28.0	24.0
Extracellular interstitial water (L)	11.0	9.4
Plasma (L)	3.0	2.6

Table 4.2   Distribution of Water Losses During Dehydration

TBW loss (ml)	As % of TBW	PV loss (ml)	As % of TBW loss	As % of PV	ISF loss (ml)	As % of TBW loss	As % of ISF	ICF loss (ml)	As % of TBW loss	As % of ICF
1,500	3.6	150	10	5	900	60	8.2	450	30	1.6
2,700	6.4	270	10	9	1,026	38	9.3	1,404	52	5
3,800	9	418	11	14	1,482	39	13.5	1,900	50	6.8

TBW = total body water; PV = plasma volume; ISF = interstitial fluids; ICF = intracellular fluids.

Data from Costill & Saltin, 1974.

Table 4.3 illustrates how the distribution of solutes in fluid compartments affects osmolar activity and osmotic pressure (Guyton 1991). Osmolality of the extracellular compartment largely depends on the concentration of sodium and chloride ions. Special autonomic, endocrine, and behavioral mechanisms have evolved to detect increases in extracellular osmotic pressure and to correct resulting cellular dehydration.

Because constituent ions within a fluid compartment are nonpenetrating and draw water by osmotic pressure, the content (absolute amount) of these ions determines the volume of a fluid compartment. The plasma and ISF compartments use sodium to determine their volumes. PV is therefore increased through ingestion and intestinal absorption of dietary sodium, which creates hypervolemia. Hypovolemia occurs when plasma sodium is lost by sweating, hemorrhage, or intercompartmental plasma shifts under conditions generating special hydrostatic or hydrodynamic forces. Evaporation of sweat is the chief method of excess heat dissipation during exercise. Because the concentration of sodium chloride in sweat is lower (about 0.3% NaCl) (Shirreffs & Maughan 1997) than in plasma (0.9% NaCl), sweating results both in hypovolemia and in increases in plasma osmolality.

During brief exercise above 40% of $\dot{V}O_2$max that is too short (6 minutes) to elicit significant water and so-dium losses through sweating, plasma volume declines and its osmolality increases (Convertino et al. 1981). Hypovolemia develops in proportion to exercise intensity (Convertino et al. 1981; Wilkerson et al. 1977). At 40% of $\dot{V}O_2$ max, approximately 110 ml (3.7%) of plasma volume in 22-year-old males shifts to interstitial or cellular compartments. At 90% of $\dot{V}O_2$max, the shift of plasma to interstitial or cellular compartments is approximately 372 ml (12.4%). Increases in osmolar concentration of extracellular fluid rise in proportion to exercise intensity and are a consequence of efflux of water out of the plasma compartment (Convertino et al. 1981; Van Beaumont et al. 1973). Water restriction also increases plasma osmolality, and the effects of exercise and water deprivation are additive (see table 4.4). These fluid shifts from plasma to interstitial compartments, often referred to as exercise-induced hemoconcentration, are caused by an increase in systolic pressure, which raises capillary filtration pressure (Wilkerson et al. 1977). Determination of hematocrit, which is the percentage of total blood volume that is erythrocytes, is used to assess the extent of hemoconcentration. Correcting hormone concentrations for increases in hematocrit during exercise (above 42% in women and 45% in men) helps distinguish changes caused by hormone secretion from those caused by hemoconcentration.

Table 4.3 Osmolar Substances and Osmotic Pressure in Body Fluid Compartments

Substance	Plasma (mOsm/L water)	Interstitial (mOsm/L water)	Intracellular (mOsm/L water)
Sodium (Na⁺)	143	140	14
Potassium (K⁺)	4	4	140
Calcium (Ca⁺)	1	1	0
Chloride (Cl⁻)	108	108	4
Other solutes	46.8	48.3	144.2
Total mosm/L	302.8	301.8	302.2
Corrected osmolar activity	282.5	281.3	281.3
Total osmotic pressure (mm Hg)	5,450	5,430	5,430

Table 4.4 Changes in Plasma Osmolality in Response to Exercise and Dehydration

Exercise intensity (% $\dot{V}O_2$ max)	Rest	25	45	65
Dehydration (% BW loss)				
0	281	284	286	287
3	286	287	291	293
5	294	294	296	299

BW = body weight.
Results from Montain et al., 1997.

HORMONAL REGULATION OF CELLULAR HYDRATION

Preservation of body water is essential for survival and thus even mild dehydration promptly triggers vigorous renal water reabsorption. Loss of body water raises the osmolality of extracellular fluids and draws water out of cells (see table 4.2), including the osmoreceptors in the sensory circumventricular organs (Johnson & Gross 1993). Three circumventricular organs of importance in the regulation of fluid balance are situated at the interface between the circulatory and cerebrospinal compartments. They are the organum vasculosum of the lamina terminalis (OVLT), the subfornical organ (SFO), and the posterior pituitary gland. They provide a breach in the blood-brain barrier because of the fenestrated nature of their capillaries (Simon 2000). Of the three, the OVLT lies on the anteromedial portion of lamina terminalis, the anterior ventral wall of the third ventricle, and the SFO is positioned more dorsally (see figure 4.1) (McKinley et al. 1984). Their osmoreceptors are sensitive to as little as 1% to 2% changes in plasma osmolality above the threshold of about 285 mOsm (Schrier et al. 1979; Share 1996; Thrasher et al. 1980). Although the OVLT and the SFO have direct projections to the osmosensitive magnocellular cells in the paraventricular (PVN) and supraoptic (SO) hypothalamic nuclei, it is likely that nucleus medianus, an integrative structure in the lamina terminalis, serves as a relay of osmotic information to trigger the release of arginine vasopressin (AVP), with angiotensin II serving as a neurotransmitter (Johnson & Thunhorst 1997; McKinley et al. 1984).

Large cells in the lateral, magnocellular part of PVN (see figure 2.3) synthesize hormones AVP (also called antidiuretic hormone [ADH]) and oxytocin (OXY) and their binding proteins, neurophysins, as well as corticotropin-releasing factor (CRF) and some other neurotransmitters (Griffin & Ojeda 1996). These proteins are transported in axons through the hypothalamo-hypophyseal stalk to neurosecretory vesicles in the posterior pituitary (see figure 4.2) (Loewy & Spyer 1990). The posterior pituitary or neurohypophysis is the ventral extension of the hypothalamus and comes to lie adjacent to the endocrine anterior hypophysis during

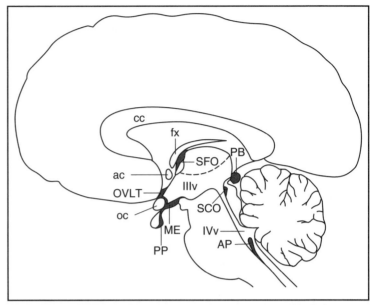

Figure 4.1 *Location of the circumventricular organs (CVOs).*

Midsagittal section of the human brain showing the location of the circumventricular organs depicted in black. These specialized regions surround third and fourth ventricles, lack a blood-brain barrier, and are highly vascularized.

ac = anterior commissure; AP = area postrema; cc = corpus callosum; fx = fornix; ME = median eminence; oc = optic chiasm; OVLT = organum vasculosum of the lamina terminalis; PB = pineal body; PP = posterior pituitary; SCO = subcommissural organ; SFO = subfornical organ.

Reprinted from Landas et al. 1985.

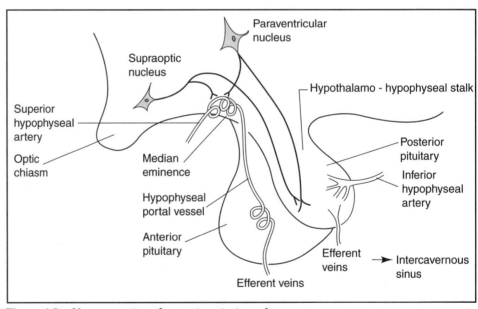

Figure 4.2 *Neurosecretion of posterior pituitary hormones.*

Arginine vasopressin is synthesized in the magnocellular cells of the supraoptic (SO) and paraventricular (PVN) hypothalamic nuclei. It is transported in axons of these cells down the hypothalamo-hypophyseal stalk to the posterior pituitary from where it is released into pituitary efferent veins and general circulation.

Adapted from Griffin and Ojeda 1996.

development. In addition to containing glial cells called pituicytes, the posterior pituitary contains unmyelinated axons and nerve terminals of SO and PV neurons. Because of the fenestrated nature of its capillaries, the posterior pituitary is considered a circumventricular organ with no blood-brain barrier, and for that reason AVP, oxytocin, and associated neurophysins can gain ready access to systemic circulation. Coordinated discharge of SO and PVN neurons is facilitated by cell coupling through tight junctions. SO and PV neurons also are depolarized by neural messages relaying osmotic pressure and other stress information to cause neurosecretion of AVP and OXY. The two hormones are released into the fenestrated capillaries of the inferior hypophyseal artery and reach systemic circulation through efferent veins, whereas arterial blood with hormones from the systemic circulation reaches the posterior pituitary through the internal hypophyseal artery (see figure 4.2). The half-lives of posterior pituitary hormones are shorter than 5 minutes.

The primary action of AVP is antidiuresis, that is, suppression of urine production by increasing urine concentration to restore the osmotic balance of the extracellular fluid compartment. The osmoregulatory role of AVP and brain location of osmoreceptors were discovered by Verney (1947) who showed that hyperosmotic fluid infused into carotid artery supplying the brain reduced urine flow and increased urine osmolality. The response was dependent on the ability to secrete AVP and could be reproduced by infusion of AVP. AVP regulates urine concentration and diuresis in response to either hemoconcentration or hemodilution (see figure 4.3). Hemoconcentration increases ADH secretion and stimulates antidiuresis, whereas a drop in plasma osmolality suppresses ADH secretion and permits increased diuresis

(Schrier et al. 1979; Wade 1984). AVP is secreted when osmolality changes by as little as 1% to 2%. The antidiuretic action of AVP is initiated by its binding to V_2 receptors on the apical membrane of cells in the distal convoluted tubule and in the medullary portion of kidney tubule collecting ducts (see figure 4.4). There, AVP activates a cAMP-dependent transduction pathway, and ensuing phosphorylations lead to exocytotic insertion of a protein aquaporin-2 into the luminal plasma membrane to immediately increase collecting duct water reabsorption (Knepper & Inoue 1997). Aquaporin-2 forms water channels and permits diffusion of water from urine into the hyperosmotic renal interstitium and from there into blood vessels of the cortex and outer renal medulla. The long-term influence of AVP is to increase aquaporin-2 gene expression and thus enhance the water reabsorbing capacity of the kidney (Frokiaer et al. 1998).

Hypertonicity of the renal interstitial fluid is achieved through active sodium chloride transport across segments of the kidney tubule called the loop of Henle (see figure 4.4). Sodium and chloride ions are actively pumped out of the increasingly hypotonic urine flowing in the ascending limb of the loop of Henle. This action makes the surrounding interstitium hypertonic. In contrast to the impermeability of the ascending loop of Henle to water, high water permeability of the descending loop of Henle, mediated by aquaporin-1, leads to the osmosis of water into hypertonic interstitium and makes the urine that flows in the descending loop of Henle progressively more hypertonic (Chou et al. 1999). The interaction between the sodium-pumping action of the ascending loop of Henle, osmotic water flow out of the descending loop of Henle, and unidirectional urine flow through the tubule generates an osmolar gradient in the interstitial fluid that increases from the proximal end of the descending loop to the hairpin turn of the loop. While the ionic concentration difference across the adjacent segments of the loop of Henle is only 200 mOsm/L, osmolar gradient in the renal interstitium and in the urine ranges from 300 mOsm/L at the start to 1,400 mOsm/L at the end of each limb of the loop of Henle. Daily excretion by the human kidney of 600 mOsm/L of various solutes is obligatory. As a consequence, the maximal concentrating ability of a human kidney (1,400 mOsmL) creates an obligatory water loss of 0.430 L/day. Desert animals can reduce this water loss by virtue of their longer loops of Henle, which generate larger osmolar gradients and permit greater urine concentration. The threshold to osmoregulatory secretion of AVP declines and sensitivity of the response increases after reductions in plasma volume (see figure 4.5). In this way both forms of water deficit, cellular dehydration and hypovolemia, stimulate the renal endocrine mechanism of water conservation. With respect to cardiovascular function, magnocellular PVN neurons are tonically inhibited by atrial receptors and baroreceptors, and they also receive input from NE and NPY neurons in A1 neuronal cell group, from E neurons in C1 cell group, and from pons and tegmentum (Harris and Loewy, in Loewy & Spyer 1990).

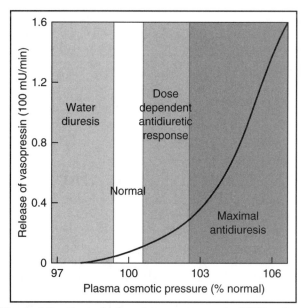

Figure 4.3 *Control of arginine vasopressin (AVP) release by plasma osmotic pressure.*

Relationship between plasma osmotic pressure and concentration of AVP in plasma. Hemodilution suppresses AVP secretion and permits diuresis and hemoconcentration stimulates AVP secretion and antidiuresis.

Adapted from Baulieu 1990.

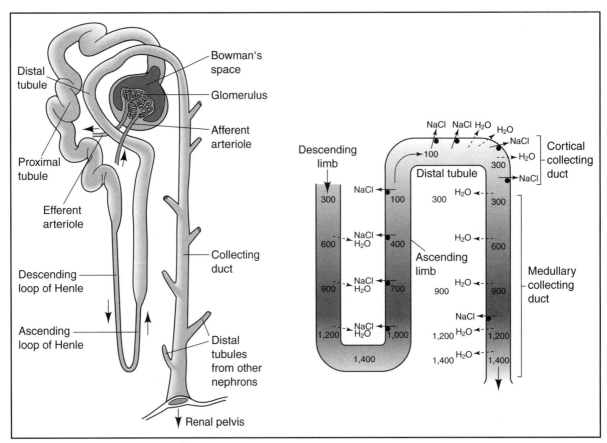

Figure 4.4 *Antidiuretic actions of arginine vasopressin (AVP) in the kidney tubule.*

Kidney tubule creates hyperosmotic renal interstitium through sodium excretion by the ascending loop of Henle. AVP acts on the distal convoluted tubule and medullary collecting ducts to stimulate the reabsorption of water. Reabsorption of water is made possible through the interaction between sodium-pumping action of the ascending loop of Henle, osmotic water flow out of the descending loop of Henle, and unidirectional urine flow through the tubule, which generates an osmolar gradient in the interstitial fluid.

Adapted from Dunn et al. 1973.

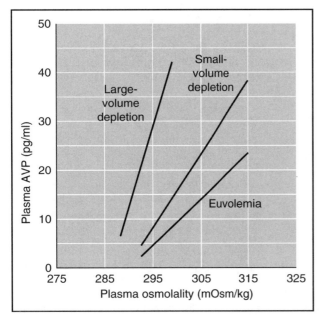

Figure 4.5 *Control of arginine vasopressin (AVP) release by plasma volume.*

Volume depletion lowers the threshold and increases the sensitivity of AVP secretory response to osmotic stimuli.

Adapted from Vander, Sherman, and Luciano 2001.

AVP has no osmoregulatory function during intense exercise because of the reduced renal blood flow. Instead, it acts as a cardioacceleratory and vasopressor hormone that helps accelerate heart action, raise blood pressure, and redistribute blood to the muscle (see chapter 5). Mediation of osmotic thirst by AVP action on the circumventricular organs has not been convincingly demonstrated (Jurzak & Schmidt 1998).

ENDOCRINE REGULATION OF PLASMA VOLUME

Loss of plasma volume compromises the body's ability to maintain adequate perfusion by blood of all tissues that require energy and oxygen. Loss of plasma volume during prolonged exercise by dehydration diminishes performance in part because of the associated reduction in stroke volume and increases in heart rate known as cardiovascular drift (Coyle 1998). High-pressure baroreceptors and low-pressure stretch receptors can initiate vasoconstriction for rapid but temporary normalization of blood pressure after a plasma volume decline. A more complete correction of the lost plasma volume requires that water and salt be recaptured before excretion into urine and that plasma volume deficits be corrected through ingestion of additional water and salt. Angiotensin II plays a key role in each action (see figure 4.6):

1. It increases blood pressure through its vasoconstrictor action.

2. It helps restore plasma volume by stimulating secretion of aldosterone for renal sodium reabsorption and of ADH for renal water reabsorption.

3. It promotes ingestion of additional water and salt by stimulating thirst and sodium hunger.

As water and plasma losses particularly impede physical performance, these topics will be discussed first. When plasma volume exceeds cardiovascular capacity, the excess has to be rapidly reduced. These deviations have less bearing on physical performance but are necessary for a full understanding of fluid regulation and will be discussed last.

A decrease in kidney blood pressure is sensed by intrarenal baroreceptors, granular cells, that respond by secreting renin. Renin secretion also is stimulated by S nerves and by a decline in the concentration of sodium chloride and other plasma electrolytes. Renin-secreting granular cells are in macula densa, a contact area between the distal tubule and the afferent ateriole, jointly called juxta-glomerular apparatus (JGA; see figure 4.7). Chief stimuli for renin release are (1) reduced filtration pressure in kidney arteries in response to a fall in plasma volume, and (2) increased S stimulation of β adrenoceptors in the kidney tubule resulting from increased NE

spillover in the kidney (Peronnet et al. 1988; Saxena 1992; Tidgren et al. 1991; Zambraski et al. 1984).

Renin cleaves the prohormone angiotensinogen to form inactive polypeptide angiotensin I. A conversion to biologically active hormone angiotensin II occurs in vascular endothelia, particularly those in the lung. The endothelial angiotensin-converting enzyme (ACE) stimulates angiotensin II formation and, in its other role as kininase, inhibits synthesis of the vasodilator bradykinin. An intrinsic neuronal angiotensin system also interconnects the CVOs and the hypothalamic and medullary brain areas that are involved in autonomic control of cardiovascular function (Oldfield et al. 1994; Steckelings et al. 1992). These neurons are capable of synthesizing angiotensin II within the blood-brain barrier as well as converting circulating angiotensin I that enters the brain through the sensory circumventricular organs to angiotensin II. Circulating angiotensin II stimulates AT-1 receptors in the adrenal zona glomerulosa to synthesize the mineralocorticoid aldosterone through the IP_3 transduction pathway (Vinson et al. 1995). Angiotensin II also helps restore plasma water by binding to its receptors in the SFO. From there it stimulates PVN and SO nuclei to release ADH for water reabsorption. During exercise, angiotensin II can be formed through an alternative pathway involving a serine protease (Miura et al. 1994).

The chief stimulus for aldosterone secretion is sodium depletion and associated loss of plasma volume (hypovolemia), which triggers RAA reflex and can increase aldosterone secretion twofold to sixfold (Griffin

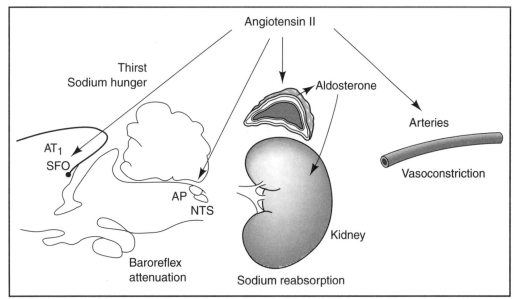

Figure 4.6 *Biological actions of angiotensin II.*

Angiotensin acts on the brain, kidney, and the smooth muscle of the blood vessels.

AT_1 = angiotensin I receptors, NTS = nucleus of the solitary tract; SFO = subfornical organ.
The major pathway of angiotensin II synthesis includes the endocrine cascade of renin-angiotensin-aldosterone (RAA) (see figure 4.7). Renin is secreted by modified myoepithelial cells in the renal afferent arteriole and the cells of macula densa in the distal convoluted tubule, which comprise the juxtaglomerular apparatus (JGA) of the kidney tubule.

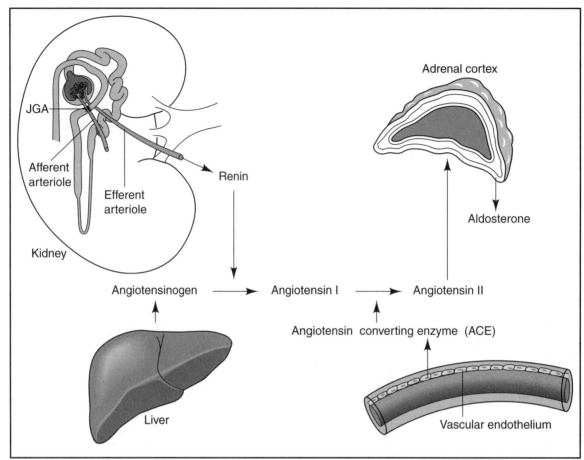

Figure 4.7   *Renin-antiotensin-aldosterone reflex.*

When blood pressure declines, renal sympathetic (S) nerve triggers and potentiates renal release of renin to hypovolemic stimulus. Renin converts circulating angiotensinogen produced in the liver into angiotensin I. The endothelial converting enzyme transforms the angiotensin I into angiotensin II. The angiotensin helps expand plasma volume through three actions. It stimulates release of aldosterone from the adrenal cortex with the consequent increased renal sodium reabsorbtion. It binds to the subfornical organ (SFO), one of the circumventricular organs, and stimulates the magnocellular supraoptic (SO) and paraventricular (PVN) hypothalamic nuclei to release antidiuretic hormone (ADH). Finally it enhances neurotransmission in the S celiac ganglion, which causes renal vasoconstriction and reduced glomerular filtration rate.

JGA = juxta-glomerular apparatus

& Ojeda 1996). Additional stimuli are a sudden increase in plasma potassium ion concentration and increased plasma ACTH concentration at high exercise intensities or during other types of stress (Fallo 1993; Griffin 1982) (see figure 4.8). ACTH triggers aldosterone release via the cAMP transduction pathway, and plasma potassium increases aldosterone release by increasing intracellular calcium through K^+-modulated voltage-gated Ca^{2+} channels.

Aldosterone preferentially binds to high-affinity type I mineralocorticoid receptors in kidney, colon, and salivary and sweat glands. Glucocorticoid hormones bind to type II glucocorticoid receptors in these tissues. An 11-β-hydroxysteroid dehydrogenase is associated with type I receptors in tissue targets of aldosterone. This enzyme prevents high-affinity binding of cortisol with type I receptors by converting cortisol to cortisone.

Glycyrrhizic acid in licorice inactivates this enzyme and allows cortisol to exert massive mineralocorticoid action and increases in blood pressure because cortisol is present in circulation at a significantly greater concentration (10 to 200 ng/ml) than aldosterone (0.05 to 0.2 ng/ml) (Walker & Edwards 1994).

The chief sodium-preserving action of aldosterone is to increase sodium reabsorption and potassium excretion by the kidneys. The sodium reabsorbing action of aldosterone entails synthesis of α and β subunits of sodium permease pump, which is believed to control the rate-limiting step in sodium reabsorption at the mucosal side of the kidney tubule. The Na/K pump then actively transports sodium across the serosal cell membrane into interstitial fluid on its way to the capillaries, convoluted tubule, and cortical portion of the collecting duct in the kidney (see figure 4.7) (O'Neill 1990). Aldosterone also

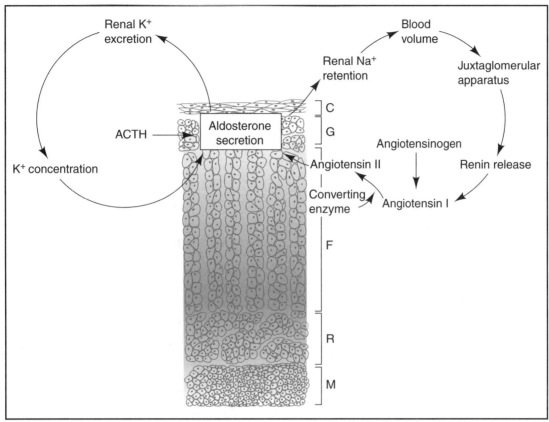

Figure 4.8   *Control of aldosterone secretion.*

Aldosterone secretion from the glomerular zone of the adrenal cortex is stimulated by angiotensin II when there is a deficit in plasma sodium, by corticotropin (ACTH) during stress, and by potassium ions when their plasma concentration increases.

C = capsule; G = zona glomerulosa; F = zona fasciculata; R = zona reticularis; M = medulla.
Adapted from Griffin and Ojeda 1996.

increases sodium reabsorption by cells in the colon, sweat glands, and salivary glands when there is a significant loss of plasma sodium. In addition, osmotic reentry of water into capillaries accompanies active sodium reabsorption, thereby facilitating restoration of plasma volume. Aldosterone-mediated active transport of sodium establishes an electrochemical gradient that then facilitates passive potassium excretion into the urine.

THIRST AND SODIUM HUNGER

Drinking water and ingesting salt are behaviors that provide the ultimate solution to cellular dehydration and plasma volume losses. Common neuroendocrine and endocrine mechanisms coordinate behavioral and physiological mechanisms for body fluid regulation. Neurons in and adjacent to the anteroventral wall of the third ventricle detect osmotic and hypovolemic water deficits, control thirst and sodium hunger, and coordinate reflexes for body fluid conservation and regulation of blood pressure. Sensing of cellular dehydration is localized mostly in the OVLT and partially in the SFO (Simon 2000). Sensing of plasma volume loss is done by high-pressure baroreceptors, by low-pressure cardiopulmonary stretch receptors, and by the JGA.

The key thirst-promoting stimulus (dipsogen) that transmits hypovolemic information to the SFO is angiotensin II, produced as a component of baroreflex that originates in the JGA (Edwards & Ritter 1982). A threshold stimulus of 8% to 10% plasma volume loss is necessary to trigger thirst (Fitzsimmons 1998). This threshold is considerably higher than the 1% to 2% plasma loss necessary to elicit ADH release. The difference in the two thresholds prevents normal variations in plasma volume that are caused by postural changes and movement from triggering thirst before any real fluid deficits are incurred. Angiotensin II elicits thirst by binding to AT1 receptors on the SFO cells. It reaches SFO by blood-borne route or after synthesis within OVLT neurons, within the blood-brain barrier. In the brain, angiotensin II acts on the hypothalamus and other brain structures to elicit drinking (Fitzsimmons 1998).

When cellular dehydration is accompanied by plasma volume loss, correction of increased osmolality takes precedence over correction of plasma volume loss. For instance (Takamata et al. 1994), a combined loss of 4.5% of total body water and of 3% of total sodium balance produced by sweating during exercise in the heat generates a 3 mOsm/L increase in osmolality and a 6.5% plasma volume loss. The immediate behavioral response to such a combined hyperosmotic and hypovolemic stimulus is a fourfold increase in the rating of thirst sensation accompanied by a fivefold increase in plasma AVP concentration (see figure 4.9). A preference for dilute salt solution develops along with a 2.5-fold increase in plasma aldosterone concentration. Water intake is initially rapid but slows down after about 3 minutes of drinking, when plasma AVP concentration starts to decline before the osmotic imbalance has been corrected and while only about 65% of drinking has been carried out (Geelen et al. 1984; Rolls et al. 1980). These changes are initiated by oropharyngeal stimulation and sensations of stomach fullness. Within an hour of spontaneous drinking, urine flow increases sixfold and is accompanied by increased sweating (Greenleaf 1992; Rolls et al. 1980). Sensation of thirst abates and plasma AVP and aldosterone return to baseline within the first 5 hours of drinking (see figure

4.9), which has corrected plasma osmolality imbalance but only a little more than half of plasma volume deficit (Geelen et al. 1984; Takamata et al. 1994). The inability to fully correct for a water deficit within a few hours after the onset of rehydration is called voluntary dehydration, and it impedes restoration of optimal physical performance (Greenleaf 1992).

Ten to 15 hours after the onset of rehydration to a combined plasma volume loss and cellular dehydration, thirst ratings and plasma aldosterone concentrations rise once more in parallel. A concomitant preference for a more concentrated solution of sodium chloride now accompanies a threefold increase in plasma aldosterone concentration (see figure 4.9). The inverse relationship between preference for solutions of variable saltiness and plasma sodium concentration is a manifestation of a specific hunger for salt (see figure 4.10) (Takamata et al. 1994). Thus, initial thirst is accompanied by increases in plasma concentrations of AVP, angiotensin II, and aldosterone and serves to first correct the osmotic imbalance in the plasma fluid compartment. Sodium hunger develops with a delay of 10 to 15 hours, and a joint influence of angiotensin II and aldosterone on the SFO and adjacent brain structures is required for its development and full expression. The delay in its expression and growth in strength with time suggest that a change in either receptor response or in neural connectivity may be required for development of sodium hunger (Fitzsimmons 1998).

Increased plasma fluid loads are the consequence of excessive sodium and water intakes, excessive renal sodium reabsorption, or inadequate diuresis and natriuresis. Hyponatremia or excessive reduction of plasma osmolal-

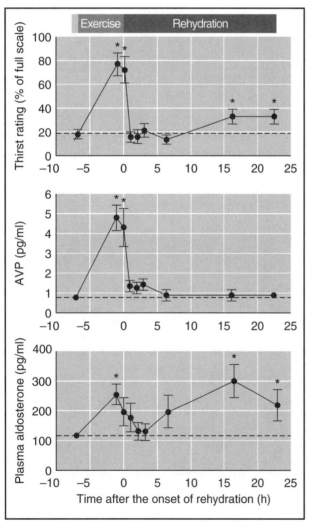

Figure 4.9 *Thirst and hydromineral hormone responses to dehydration.*

Subjective rating of thirst (top panel) and plasma arginine vasopressin (AVP) (middle panel) and aldosterone (lower panel) concentrations before and during rehydration after a hyperosmotic-hypovolemic water loss.

Adapted from Takamata et al. 1994.

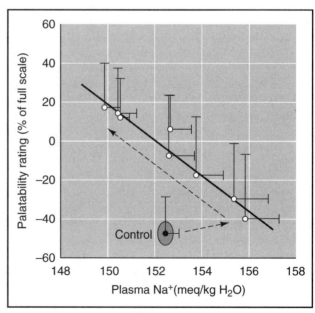

Figure 4.10 *Effect of sodium depletion on salt preference.*

Palatability rating of 1 M (5.5%) NaCl solution as an inverse function of plasma sodium concentration.

Reprinted from Takamata et al. 1994.

ity is occasionally reported when athletes drink in excess of 20L of water during prolonged exercise (Armstrong et al. 1993; Frizzell et al. 1986). The relief of plasma overload is achieved through two mechanisms: the renorenal reflex and the endocrine diuretic and natriuretic reflex.

The renorenal reflex is initiated by increased pressure in the renal pelvis, the central kidney cavity receiving urine from collecting ducts. There, the mechanoreceptors are stretched and electrical discharges increase in substance P and prostaglandin E_2–dependent afferent renal nerves (Kopp & Smith 1991). Renal nerve discharge is relayed to the NTS and SO nuclei, where it inhibits AVP secretion (Stella & Zanchetti 1991). Stretch increases sodium excretion or pressure natriuresis through reflex suppression of sympathetic nerve traffic to the kidney and of renin release.

In addition to this neurally mediated reflex, distension of the right atrium of the heart elicits an endocrine diuretic and natriuretic reflex (see figure 4.11). This reflex can be triggered by net increases in plasma volume as well as by the pooling of blood in the chest cavity during body inversions, immersion in water, or lack of gravitational force. The chief mediator of the reflex is the atrial natriuretic peptide (ANP), which belongs, along with type-B (BNP) and type-C natriuretic peptides (CNP), to the auriculin or atriopeptin family (Chen & Burnett 1988). ANP and BNP are produced in the heart and have similar functions, whereas CNP is a vasodilator of endothelial origin. ANP half-life is shorter than 5 minutes, and its concentration in the aorta is in excess of 150 pg/ml. In systemic circulation, ANP concentrations range between 50 and 150 pg/ml.

ANP exerts biological effects on the kidney, blood vessels, adrenal cortex, and brain (see figure 4.11) (Griffin & Ojeda 1996). In the kidney, ANP increases glomerular filtration rate by constricting postglomerular efferent arterioles to produce increased urine volume (Marin-Grez et al. 1986; Sosa et al. 1986). It also inhibits sodium reabsorption in the distal convoluted tubule to increase natriuresis. It antagonizes the effect of AVP in the distal tubule and the collecting duct to increase diuresis. In the JGA, ANP inhibits renin secretion and subsequent angiotensin II formation (Cuneo et al. 1987). In addition, ANP blocks both basal and angiotensin II–stimulated or ACTH-stimulated aldosterone secretion in the adrenal cortex and thus facilitates natriuresis through several different mechanisms. In the vascular system, ANP exerts several pressure-lowering effects. It facilitates transcapillary fluid shift between the plasma and ISF compartment. It reduces mean arterial pressure in part because of its hypovolemic action and in part because of its vasodilatory effect. In the brain, ANF acts on neural substrates involved in regulation of fluid and electrolyte balance and in control of renal and cardiovascular function. Acting in the central nervous system, ANF inhibits AVP release from magnocellular hypothalamic nuclei (Samson 1985), attenuates baroreflex and blood pressure in NTS (Yang et al. 1992), raises blood pressure in the anterior hypothalamus (Yang et al. 1990), and suppresses thirst (Nakamura et al. 1985) and sodium hunger in the SFO (Antunes-Rodrigues et al. 1985). Thus, both peripheral and most of the central actions of ANP result in reduction of sodium and water overload.

At exercise intensities above 40% of maximal effort, when the GFR and urine flow are reduced, ANP concentration increases exponentially with increases in exercise intensity but does not have diuretic and natriuretic effect. Instead, it participates in plasma fluid shifts and blood pressure regulation, which are discussed in chapter 5.

STRATEGIES OF FLUID MANAGEMENT

An understanding of the neuroendocrine mechanism of thirst and sodium hunger, its time structure, and its interaction with renal sodium and fluid mechanisms is necessary to devise optimal strategies for hydration or rehydration of individuals who exercise in conditions that cause significant fluid losses in the form of sweat. This understanding is necessary because dehydration impairs physical performance in all types of prolonged competition (Sawka & Pandolf 1990), and excess fluid intake usually cannot be stored, as fluid balance is

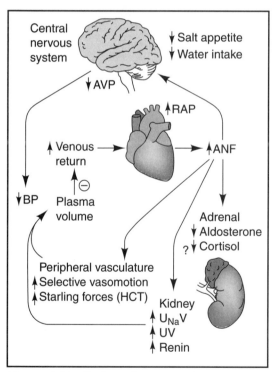

Figure 4.11 *Biological actions of atrial natriuretic peptide (ANP).*

Atrial natriuretic peptide is released in response to plasma volume expansion and associated increase in right atrial pressure. It acts in the brain, adrenal, kidney, and blood vessels to reduce blood pressure and venous return.

AVP = arginine vasopression; RAP = right atrial pressure; ANF = atrial natriuretive factor; BP = blood pressure; HCT = hematocirt; $U_{Na}V$ = urine sodium excretion; UV = urine volume.

Adapted from Griffin and Ojeda 1996.

quickly reestablished through diuresis. Three strategies for avoidance of dehydration during exercise have been developed: prehydration, fluid replenishment during exercise, and rehydration after exercise. Prehydration of 500 ml of fluid on the night before exercise, 500 ml in the morning, 500 to 1,000 ml 1 hour before exercise, and 250 to 500 ml 20 minutes before exercise delays the onset of exercise-induced dehydration if no fluids are given during exercise (Lamb 1999; Latzka et al. 1998). Urinary loss of immediately preingested fluids is prevented by reduced GFR and renal blood flow during exercise. Fluid replacement during exercise, in the amounts that match fluid losses through sweating, prevents performance decrements associated with dehydration but is difficult to achieve under field conditions (Latzka et al. 1997).

To achieve rapid rehydration after total body fluid losses greater than 2% to 3%, deliberate strategies that supersede slow behavioral and renal adjustments to plasma losses and hypertonicity need to be implemented because both volumes ingested and sodium concentration in the fluids influence the rate of body fluid and plasma volume restoration. Initial thirst appears to be responsive to cellular dehydration, and spontaneous water drinking is terminated by correction of plasma hyperosmolality before any significant correction of the hypovolemia. Natural thirst is thus slaked by initial drinking of water in the amount that is a fraction of the incurred plasma volume deficit. Preservation of ingested fluid within the plasma compartment is aided by a simultaneous increase in AVP secretion, which is terminated by reestablishment of isotonicity (see figure 4.11). Deliberate drinking of water in half the amount that was lost through sweating corrects only about half of water deficit, as urinary output remains low for the next 6 hours of hypohydration. Induced urination and sweating limits retention of expanded plasma volume when larger quantities of water are the rehydrating fluid. Thus, urine output triples when water intake matches water deficit and increases 12-fold when intake is twice as large as water deficit (Shirreffs et al. 1996). Preservation of ingested fluids in the ECF compartment is aided by addition of salt or other electrolytes to drinking water in concentrations between 0.3% (50 mM/L) and 0.6% (100 mM/L) or by drinking of water along with an electrolyte-rich solid meal (Maughan et al. 1996). This electrolyte intake suppresses postprandial urine output and brings plasma volume and body water into balance within 6 hours after the onset of drinking (Maughan and Leiper 1995; Maughan et al. 1997; Shirreffs & Maughan 1997, 1998; Shirreffs et al. 1996). Addition of electrolytes and sweeteners to the fluids also improves their palatability and ingestion and thus facilitates spontaneous rehydration (Maughan & Leiper 1993).

5

Exercise As an Emergency and a Stressor

Stress has physical, chemical, or emotional components that cause bodily or mental tension and can affect susceptibility to disease. Through its intensity and duration dimensions, exercise can increase or decrease physical, chemical, and psychological stress. Exercise ranges in intensity from gentle, slow, stretching movements as in yoga, tai-chi, and qi-gong, which, in combination with meditation, have been reported to reduce stress and most stress hormone concentrations (Schmidt et al. 1997), to maximal effort in high-resistance and endurance sports accompanied by high concentrations of stress hormones (Gould et al. 1995; Jin 1992; Ryu et al. 1996; Schell et al. 1994). Outside of recreational activities, intense movement often is part of stressful competition, combat, or escape behaviors. At the high end of the intensity range, when physical effort is maximal or intense and prolonged, exercise is a state of emergency that jeopardizes oxygen supply to tissues, increases core temperature, reduces body fluids and metabolic fuel stores, and causes some tissue damage. In common with other stressors, exercise requires extraordinary physiological adjustments. It engages proactive changes and reactive adjustments to deficiencies and injuries by engaging the SA system, stress hormones, and host defense system. Moderate exercise, performed at intervals that permit full physiological and structural recovery, stimulates growth and increases in function. Although many forms of low-intensity exercise can reduce anxiety and psychological stress and improve psychological mood, when carried out to excess, intense exercise also can augment anxiety and produce changes in HPA function similar to those seen in depression (Bulbulian & Darabos 1986; Gould et al. 1995; Jin 1992; Pierce & Pate 1994; Roth et al. 1990; Roy & Steptoe 1991; Ryu et al. 1996; Schell et al. 1994). To what extent these observations are repeatable and attributable to some aspect of physical activity

rather than to changes in psychological state remains to be elucidated (Harte et al. 1995). Thus, when exercise is carried out at intensities and frequencies beyond the body's capacity for full recovery of structural integrity and function, deleterious outcomes range from minor impairment to major breakdown.

When power, that is the amount of physical work carried out per unit time, increases, expressed either in absolute terms as a rate of oxygen consumption or in relative terms as a percentage of maximal oxygen consumption ($\dot{V}O_2$max), a number of physiological variables rise in parallel. To supply contracting muscles with increasing amounts of oxygen and fuel, and at the same time preserve sufficient perfusion of the brain by blood, the cardiovascular system increases its circulatory rate and shunts most available blood from the viscera to the muscles to effect differential blood pressure gradients. The ventilatory rate also accelerates to provide greater quantities of oxygen and to vent carbon dioxide, the metabolic by-product of oxidation necessary to fuel physical effort. At high exercise intensities, muscles and other tissues may experience hypoxia (shortage of oxygen) and hypercapnia (increased carbon dioxide concentrations). At a certain level of exercise intensity called anaerobic or ventilatory threshold, compensatory hyperventilation is initiated to partially counteract rising plasma acidity (acidosis). The acidity results from accumulation of lactic acid caused by a mandatory shift in metabolism at high exercise intensities from aerobic oxidation to anaerobic glycolysis. In addition to increases in plasma acidity, faster metabolism in contracting muscles generates heat that raises the core body temperature (hyperthermia). Compensatory sweating, initiated to help dissipate some of the heat load, may produce dehydration. Finally, rapid fuel use may cause hypoglycemia, tissue damage may result from forces generated by the musculoskeletal

77

system, and intense physical effort may trigger pain and psychological stress. These consequences of intense exercise elicit endocrine and host-defense reactions that are very similar to those elicited by hypoxia, hypercapnia, acidosis, hyperthermia, dehydration, hypoglycemia, physical injury, and psychological stress, acting alone.

This chapter first describes four groups of autonomic and endocrine messengers that mediate urgent physiological adjustments during intense acute or prolonged exercise. Next, it describes how during strenuous exercise, autonomic and endocrine messengers

1. augment cardiovascular and respiratory functions, redistribute blood flow to increase oxygen and nutrient supply for muscles, and help regulate blood pressure;

2. affect blood clotting in anticipation of possible injury;

3. modulate pain perception;

4. mediate tissue repair after injury; and

5. engage immune defenses.

Finally, the chapter shows how, with habitual exercise, SA and other stress messengers help lower blood pressure in hypertensive individuals.

The additional involvement of autonomic and endocrine messengers in the thermoregulatory reduction of heat overload and metabolic fuel adjustments to exercise of various intensities is discussed in chapters 4 and 6, respectively.

ENDOCRINE SYSTEM SUPPORT OF INTENSE EXERCISE

To provide support for strenuous exercise, four messenger systems are engaged in coordinated fashion (see figure 5.1):

1. The sympathoadrenal (SA) catecholamines

2. The hypothalamo-pituitary-adrenal axis (HPAA) messengers

3. The endogenous opiate messengers

4. The hydromineral hormones (see chapter 4)

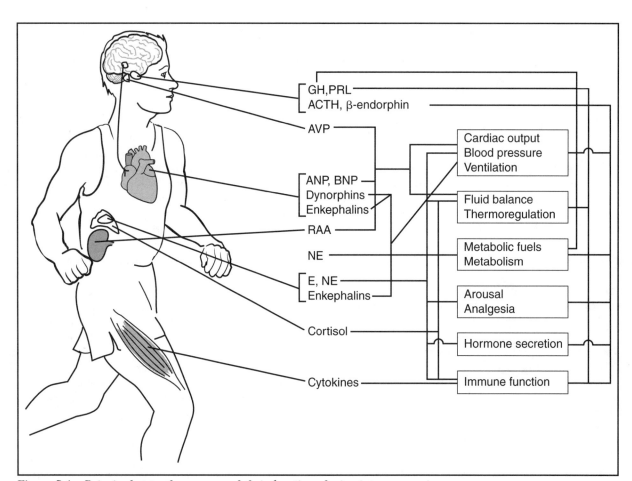

Figure 5.1   *Principal stress hormones and their functions during intense exercise.*

Biological effects of the principal stress hormones released during intense exercise.

GH = growth hormone; PRL = prolactin; ACTH = andrenocorticotropic hormone, or corticotropin; AVP = arginine vasopressin, or antidiuretic hormone; ANP = atrial natriuretic peptide; BNP = brain natriuretic peptide; RAA = renin-angiotensin-aldosterone axis; NE = norepinephrine; E = epinephrine.

All of these messengers are secreted in proportion to exercise intensity, some proactively to induce feed-forward urgent increases in life-saving cardiorespiratory and other functions, and some reactively through negative feedback to correct exercise-induced perturbations in homeostatic physiological balance.

Sympathoadrenal Catecholamines

The sympathoadrenal (SA) catecholamines (see chapter 2) coordinate adjustments to intense exercise and comparable levels of stress. Highlighted here is their proactive feed-forward role in the control of circulation, blood clotting, and immune defense reactions. NE, released by the S nerve endings, activates circulatory adjustments to initial increases in exercise intensity. After PS withdrawal, NE starts to spill out of organs activated by S nerve

discharges at exercise intensities above 70% of $\dot{V}O_2$max, close to the lactate threshold (see figure 5.2) (Esler et al. 1990; Mazzeo et al. 1997; Savard et al. 1989). Thereafter, S tone increases in parallel with the intensity and duration of physical effort (see figure 2.12) (Christensen & Galbo 1983; Rowell & O'Leary 1990; Seals & Victor 1991). The adrenomedullary stress hormones E and NE are secreted at intensities above 80% to 90% of $\dot{V}O_2$max (Deuster et al. 1989) (see figure 5.2).

The emergency functions of E were first noted by Walter Cannon (1922), who used the term "fight or flight response" for defense reactions associated with its release. Cannon showed that E was responsible for redistribution of blood to muscle, increased heart rate (HR), blood pressure, blood clotting, and plasma glucose, together with other physiological challenges to the

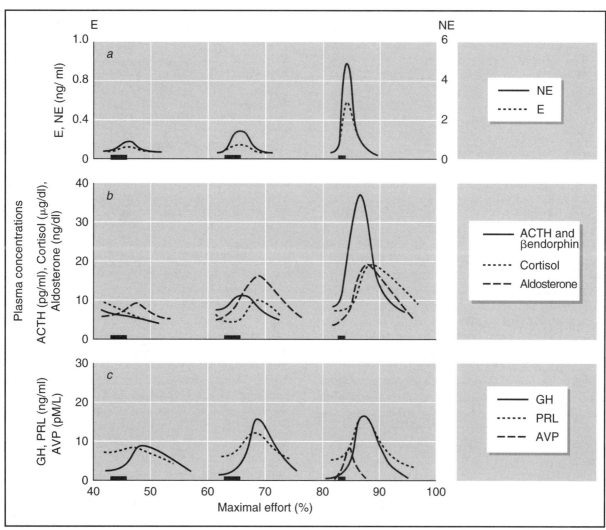

Figure 5.2 *Effect of exercise intensity on stress hormone release.*

Secretory response of (*a*) catecholamines (solid line indicates norepinephrine; broken line indicates epinephrine), (*b*) hormones of the hypothalamo-pituitary-adrenal axis (HPAA) (solid line indicates corticotropin [ACTH] or β-endorphin; broken line indicates cortisol; dashed line indicates aldosterone), (*c*) anterior (GH, PRL) and posterior (AVP) pituitary stress hormones (solid line indicates growth hormone [GH], broken line indicates prolactin [PRL]), and dashed line indicates arginine vasopressin (AVP) in response to 20 minutes of exercise at 50% and 70% of maximal effort and 10 minutes of exercise at 90% of $\dot{V}O_2$max. (Deuster et al. 1989; Luger et al. 1988; Rahkila et al. 1988; Wittert et al. 1991).

Adapted from Luger et al. 1988.

internal environment, reduced muscle fatigue in competitive sport, and emotions of fear, rage, and pain. Catecholamine concentrations in plasma also are affected by their reuptake into prejunctional α_2 receptors in S nerve endings and degradation or clearance in kidneys. At high exercise intensities, decreased renal catecholamine clearance, caused by reduced GFR, contributes to their rise in plasma (Leuenberger et al. 1993).

Hypothalamic-Pituitary-Adrenal Axis Messengers

The second major messenger system that is engaged during intense prolonged exercise, as well as in response to prolonged exposure to other stressors, is the HPAA consisting of hypothalamic paraventricular nucleus (PVN), its neural connections (see chapter 2), and the cascade of stress hormones that culminates with the release of cortisol. Small cells in the medial parvocellular PVN synthesize CRF and secrete it into the fenestrated capillaries of the hypophyseal portal vessels in the external layer of the median eminence (ME) (see figure 2.3). Parvocellular PVN cells are sensitive to corticosteroid feedback. The cells in dorsal and ventral parts of the PVN control the activity of S nerves and release adrenomedullary catecholamines, but these processes depend on the presence and action of CRF (Fisher et al. 1982). PVN appears to be one of the few brain centers that regulates the entire S outflow (Strack et al. 1989).

CRF is the primary stimulus for secretion of ACTH and β-endorphin out of the POMC precursor in the anterior pituitary corticotrophs (see figure 1.10). Other secretagogues of ACTH are AVP, NE, and angiotensin II (Vale et al. 1983). During acute exercise and in response to stressors such as electric shock, bacterial toxins, and surgery, AVP appears to be a more important stimulus for ACTH secretion than CRF (Schmidt et al. 1996; Wittert et al. 1991). Individually, AVP, NE, and angiotensin II can elicit ACTH and β-endorphin release to a limited extent, but each can greatly amplify the capacity of CRF to do so (see figure 5.3). This amplification is achieved through stimulation by CRF of AVP release (Kalogeras et al. 1996). Intense exercise increases the store of AVP in the median eminence and causes oversecretion of AVP into the pituitary portal circulation. AVP then induces oversecretion of ACTH by increasing the responsiveness of CRF and the HPAA to stress (Schmidt et al. 1996). This process is an example of feed-forward amplification of stress hormone release.

ACTH, in turn, stimulates cortisol (or, in rodents, corticosterone) secretion from the fascicular zone of the adrenal cortex (see figure 4.8). Hans Selye (1956, 1974) identified cortisol as the chief hormone secreted in response to chronic intense stress, a component of physiological and pathological responses he called general adaptation syndrome (GAS). Oversecretion of cortisol as well as of SA catecholamines characterizes the early, alarm phase of GAS. A stage of resistance develops after

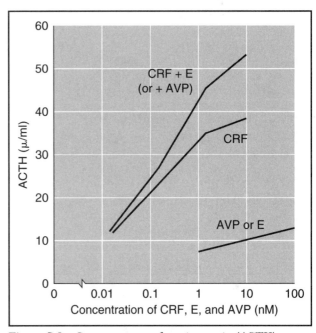

Figure 5.3 *Secretagogues of corticotropin (ACTH).*

Synergistic augmentation of ACTH release by joint actions of either arginine vasopressin (AVP) and corticotropin-releasing factor (CRF) or epinephrine (E) and CRF. Each messenger alone is a less effective secretagogue of ACTH.

Adapted from Vale et al. 1983.

prolonged exposure to intense stress, such as during prolonged and intense exercise training. It is characterized by adrenal glucocorticoid secretion at normal or slightly supranormal levels. Exhaustion occurs when duration and intensity of stress exceed the capacity of adrenal and physiological systems to adjust to the stressor, as in overtraining in sport or prolonged exposure to extreme environmental adversities.

The rise in cortisol secretion follows ACTH release after a 15-minute to 30-minute delay. Secretion of cortisol and adrenomedullary E is elicited at exercise intensities between 80% and 90% of $\dot{V}O_2$max (see figure 5.2) (Brandenberger & Follenius 1975; Deuster et al. 1989). At this exercise intensity, the rising plasma concentrations of cortisol reach the adrenal medulla from the surrounding adrenal cortex and stimulate expression of the PNMT gene and biosynthesis of E (see figure 1.2). E, on the other hand, stimulates ACTH secretion from the anterior pituitary (see figure 2.3). This action is another example of feed-forward amplification of endocrine and autonomic responses during stress.

The magnitude of cortisol secretory response during high-intensity exercise is dependent in part on the time of day and the timing of meals (see figure 5.4) because cortisol release is also controlled by a circadian oscillator, which triggers a major secretory episode in the early morning hours (see figure 5.4) (Linkowski et al. 1985). Cortisol secretion displays 7 to 15 spontaneous or meal-associated secretory bursts as it declines throughout the day (Follenius et al. 1982). To control for circadian and

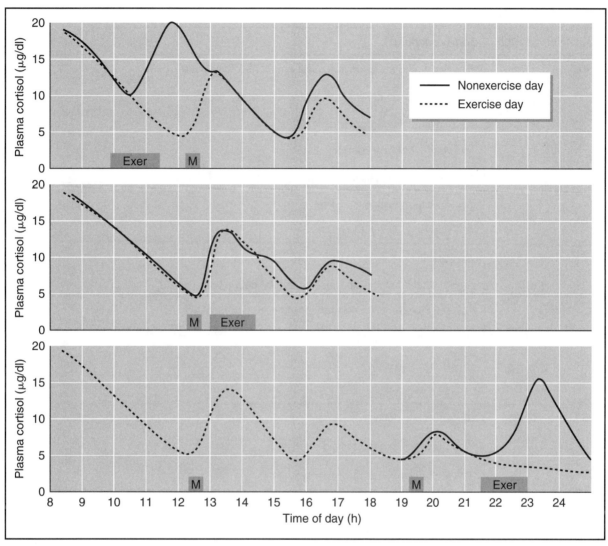

Figure 5.4 *Circadian, prandial, and exercise effects on cortisol secretion.*

The influence of time of day on the magnitude of cortisol elicited by exercise. Ninety minutes of exercise at 25% to 55% of $\dot{V}O_2$max stimulated a large peak of cortisol release when exercise was performed during the time of day when cortisol is not secreted (10 to 11:30 h and 21:30 to 22 h). When exercise was performed during the midday cortisol peak (13 to 14: 30 h), exercise elicited no additional cortisol secretion.

Broken line shows circadian cortisol rhythm on a day when no exercise was performed, and solid line shows cortisol secretion on exercise day. M = meal, EXER = exercise.

Adapted from Brandenberger, Follenius, and Hietter 1982.

meal-associated variation in plasma cortisol concentration, which ranges between 5 and 200 ng/ml, it is necessary to evaluate exercise-induced changes in its secretion with time-control cortisol measurements (Brandenberger et al. 1982; Brandenberger & Follenius 1975; Thuma et al. 1995). Between 10 mg and 30 mg of cortisol is secreted per day, and about 75% of it circulates bound to its binding protein, CBG or transcortin. At concentrations of cortisol above the saturating capacity of CBG (280 ng/ml), another 15% to 20% is loosely bound to albumin, and 5% circulates unbound. Estradiol increases hepatic production of CBG so that circumstances associated with increased

(pregnancy) or decreased (exercise-induced amenorrhea and menopause) plasma estradiol titers affect the amount of circulating free cortisol and its actions.

Cortisol plays a permissive role in a number of vital processes necessary for adaptation to stress and survival. Table 5.1 lists its more important functions. Of relevance for intense exercise are its catabolic proteolytic and anti-inflammatory actions. An additional important circulatory function is maintenance of blood pressure and plasma volume through facilitating GFR and sodium retention (see table 5.1). Cortisol has some residual mineralocorticoid and sodium-retentive action despite

Table 5.1 Biological Actions of Cortisol

Category of function	Cortisol actions
Metabolism	1. ∧ Protein degradation in muscle and connective tissue
	2. ∨ Protein synthesis in muscle and connective tissue
	3. ∨ Amino acid transport into muscle
	4. ∧ Amino acid transport into liver
	5. ∧ Gluconeogenesis in liver (amino acids, glycerol, and alanine)
	6. ∧ Glycogen synthesis in liver
	7. ∨ Glucose uptake and utilization
	8. ∧ Lipolysis
	9. ∧ Truncal and facial lipogenesis and fat deposition
Host defenses	10. ∨ Inflammatory cytokines (IL-1 and IL-6) and eicosanoids
	11. ∨ Capillary permeability
	12. ∨ Phagocytosis by leukocytes
	13. ∨ T-lymphocytes
Circulation and blood	14. ∧ Vasoconstriction
	15. ∧ Blood volume (body fluid retention)
	16. ∧ Erythrocyte and leukocyte synthesis
Kidney function	17. ∧ Glomerular filtration rate
	18. ∧ Sodium retention
Brain functions	19. ∧ Mood (euphoria), ∨ mood (depression)
	20. ∨ Taste, hearing, and smell
	21. ∧ Food intake

∧ = increase or stimulation; ∨ = decrease or suppression.

its enzymatic conversion in the kidney to cortisone, an 11-keto analog with a low affinity for the type-I mineralocorticoid receptor (Griffin & Ojeda 1996). Cortisol also improves the mood but, at high concentrations, can cause depression (Holsboer et al. 1994).

The negative feedback of plasma cortisol over CRF and ACTH release (see figure 2.3) is altered by intense exercise and other types of stress. A 10-minute bout of intermittent exercise at 90% of $\dot{V}O_2$max reduces the steroid negative feedback over ACTH secretion. Whereas the standard DEX suppression test (1 mg) is effective in approximately 85% of normal inactive individuals, a four times higher dose of DEX is needed to suppress ACTH after acute intense exercise, and then it is effective in only 64% of normal individuals (Petrides et al. 1994; Sherman et al. 1984).

Endogenous Opiates

The endogenous opiates (EOPs) are enkephalins, dynorphins, and endorphins, a group of peptide hormones that share the same four-amino-acid sequence (Tyr-Gly-Gly-Phe) at their amino terminal end (see table 5.2). Enkephalins are produced by the adrenal medulla, the posterior pituitary, the brain, and the heart (Barron et al. 1997; Caffrey et al. 1995; de Jong et al. 1983; Hayashi & Nakamura 1984; Laubie & Schmidt

Table 5.2 Families of Endogenous Opiates

Proenkephalin messengers	Molecular structure
Metenkephalin (menk)	4-Met
Leu-enkephalin (L-Enk)	4-Leu
Heptapeptide	4-Met-Arg-Phe
Octapeptide	4-Met-Arg-Gly-Leu

Prodynorphin messengers	Molecular structure
Dynorphin A	4-Leu-Arg-Arg-Ile-Pro-Lys-Leu-Lys-Trp-Asp-Asn-Gln
Dynorphin B	4-Leu-Arg-Arg-Gln-Phe-Lys-Val-Val-Thr
α-Neoendorphin	4-Leu-Arg-Lys-Tyr-Pro-Lys
β-Neoendorphin	4-Leu-Arg-Lys-Tyr-Pro

Proopiomelanocortin (POMC) messenger	Molecular structure
β-Endorphin	4-Met-Thr-Ser-Glu-Lys-Ser-Gln-Thr-Pro-Leu-Val-Thr-Leu-Phe-Lys-Asn-Ala-Ile-Ile-Lys-Asn-Ala-Tyr-Gly-Lys-Lys-Gly-Glu

Messengers from proenkephalin, prodynorphin, and POMC families share the initial Tyr-Gly-Gly-Phe (4) molecular sequence.
Adapted from Baulieu 1990.

1981; Ota et al. 1986). Enkephalins are located in chromaffin cells in the outer rim of the adrenal medulla where they are colocalized with catecholamines, neurotensin, somatostatin, NPY, and chromogranin A (Bloch et al. 1986; Corder et al. 1982; Eiden et al. 1987; Hexum et al. 1987). Enkephalins are released into circulation from the adrenal medulla after stimulation of nicotinic ACh Rs by renal splanchnic nerves and at exercise intensities of 100% of $\dot{V}O_2$max or higher (Boone et al. 1992; Farrell et al. 1987; Grossman et al. 1984; Hexum & Russett 1987, 1989; Hexum et al. 1987; Kraemer et al. 1991b). In the presence of angiotensin II, arachidonic acid, bradykinin, histamine, neurotensin, prostaglandin E2, and VIP, S stimulation increases expression of proenkephalin gene, whereas cortisol is required to initiate expression of the PNMT gene (Bommer & Herz 1989; Livett et al. 1990; Suh et al. 1992a, 1992b). Enkephalin and NE secretion from the adrenal medulla are, therefore, differentially regulated depending on the type of stress and associated stress hormones (Boone et al. 1992; Kraemer et al. 1991b; Livett et al. 1990). The heart also contains dynorphin-secreting cells (Caffrey et al. 1985).

Release of β-endorphin by the anterior pituitary occurs at intensities between 80% and 90% of $\dot{V}O_2$ max (see figure 5.2), in equimolar amounts and concurrently with the release of ACTH, as both messengers are products of the POMC gene (see figure 1.10)

(Rahkila et al. 1988; Schwartz & Kindermann 1992). β-Endorphin affects cardiorespiratory, endocrine, and, to a certain extent, metabolic and psychological functions during intense or prolonged exercise and other emergencies (reviewed previously by Farrell 1985; Grossman 1985; Grossman & Sutton 1985; Harber & Sutton 1984; McArthur 1985; Sforzo 1989) (see figure 5.1).

Hydromineral Hormones

The fourth group of hormones that help increase cardiac output, redistribute blood, and raise blood pressure during intense exercise are the hydromineral hormones AVP, renin, angiotensin II, aldosterone, and ANP (see chapter 4). They, too, are secreted at a rate that is proportional to exercise intensity (see figure 5.5) (Convertino et al. 1981; Freund et al. 1991). At exercise intensities above 60% of $\dot{V}O_2$max, S discharges to the kidney, and NE, DA, and NPY spillover into renal arteries increases (Tidgren et al. 1991). When NE-induced vasoconstriction of afferent renal arteries reduces renal blood flow and GFR by more than 35%, absorptive actions of hydromineral hormones in the kidney are curtailed (see figure 5.6) (Freund et al. 1991; Levinsky et al. 1959; Sosa et al. 1986). Instead, during strenuous exercise, these hormones exert cardioacceleratory and vasopressor effects and contribute to emergency elevation of blood pressure and to redistribution of blood from viscera to the muscles.

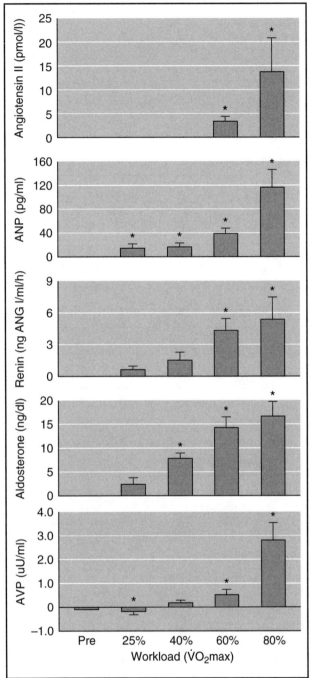

Figure 5.5 *Release of angiotensin II, atrial natriuretic peptide (ANP), renin, aldosterone, and arginine vasopressin (AVP) during exercise of different intensity.*

Adapted from Tidgren et al. 1991.

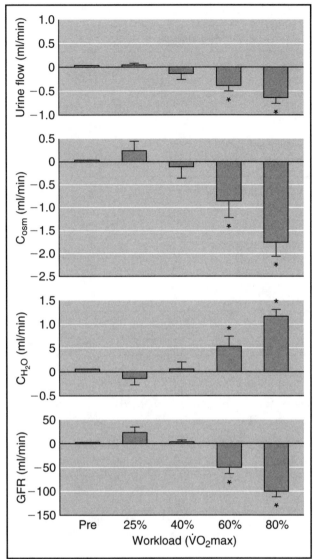

Figure 5.6 *Exercise intensity and renal function.*

Changes in urine flow, osmotic clearance (C_{osm}), free water clearance (C_{H_2O}), and glomerular filtration rate (GFR) as a function of different exercise intensities during 20 minutes of cycle ergometry.

Adapted from Freund et al. 1991.

STRESS MESSENGERS IN SUPPORT OF CIRCULATION DURING INTENSE EXERCISE

Increased cardiac output and vasomotor changes that produce redistribution of blood from visceral organs to exercising muscles are key to meeting oxygen and fuel needs during times of heavy demand and are effected by SA catecholamines, adrenal and endogenous opiates, and hydromineral hormones. During peak exercise intensity, these reflex autonomic and endocrine mechanisms can increase HR to between threefold and fourfold and increase ejection fraction by about 25%, producing a fourfold to fivefold increase in cardiac output (see figure 5.7). Redistribution of blood flow from visceral organs, which receive about 85% of cardiac output at rest, to exercising muscles, which can receive approximately the same amount during maximal effort (see figure 5.7), is achieved by selective constriction of arteries and arterioles in individual vascular beds and by vasodilation in the contracting muscle. This circulatory shift is effected through the vasoconstrictive action of catecholamines, hydromineral hormones, and some other messengers in the viscera and inactive muscles and through vasodilatory action of a number of circulating and local messengers in contracting muscles.

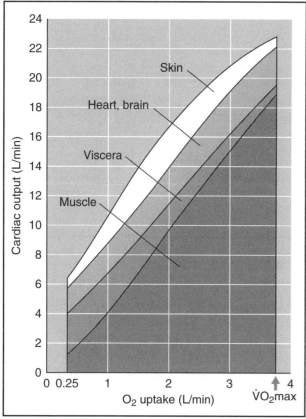

Figure 5.7 *Cardiac output and regional blood distribution during exercise of different intensity.*

With increasing exercise intensity, cardiac output increases from about 5 to 6 L/min to a maximum of about 20 to 25 L/min. Partitioning of cardiac output to the muscle changes from about 1 L/min at rest to about 18 L/min during maximal effort.

Reprinted from Rowell 1974.

SA Messengers

During low-intensity exercise, contractions in a small muscle mass send mechanoreceptor input to cardiac brain control centers to suppress the inhibitory PS tone to the heart and thereby increase heart rate and cardiac output (Mitchell 1990; Rowell & O'Leary 1990; Victor et al. 1987). With increased intensity and duration of exercise, and when a larger muscle mass is contracting, PS withdrawal and S activation occur simultaneously. During even more intense exercise, increased neural activity in the motor cortex and in lateral regions of the reticular activating system in the pons, mesencephalon, and diencephalon, causes simultaneous activation of cardioacceleratory and vasoconstrictor areas of the vasomotor center. This condition then enhances S nerve discharge to the sinoatrial cardiac pacemaker and heart muscle and to visceral arterioles and leads to proportional increases in cardiac output by way of increases in HR and contractility on one hand and vasoconstriction and redistribution of blood on the other (see figure 3.9) (Pagani et al. 1997).

Direct action of catecholamines on β_1 receptors in cardiac muscle stimulates HR and heart contractility.

Ventricular contractility also is enhanced by catecholamine-induced venoconstriction, which increases venous return and cardiac distension by greater end-diastolic blood volume (Frank-Starling's law). As a result, systolic and mean arterial pressures (MAP = diastolic pressure plus one third of the difference between systolic and diastolic pressure) and pulse pressure (difference between systolic and diastolic pressures) all rise in proportion to exercise intensity, but diastolic pressure does not change because of vasodilation and reduced vascular resistance in contracting muscles (see figure 5.8). Catecholamine-

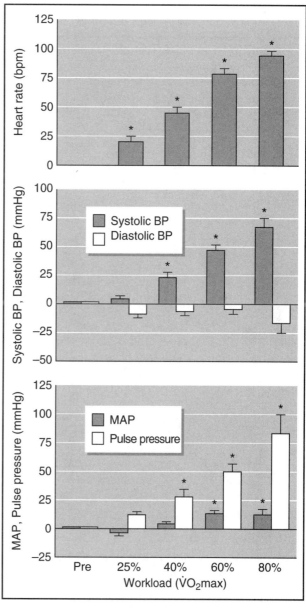

Figure 5.8 *Blood pressure and heart rate changes during exercise of different intensity.*

With increasing exercise intensity, the heart rate, systolic blood pressure, and pulse pressure increase, while diastolic pressure does not change or, at highest exercise intensities, declines.

Adapted from Freund et al. 1991.

induced stimulation of heart contraction during exercise increases myocardial oxygen needs.

Catecholamines help redistribute blood to skeletal and cardiac muscles by causing vasoconstriction in visceral vascular beds and in the lymphatic system. Constriction of both α_1 and α_2 Rs in visceral arterioles reduces blood flow to the viscera (McAllister 1998; Ruffolo & Hieble 1994). At high exercise intensities, catecholamines restrain blood flow in some contracting muscles to prevent it from outstripping cardiac output (Dietz et al. 1997; Hansen et al. 2000). Constriction of veins through catecholamine action on α_1 receptors reduces venous capacitance and plasma volume (Ping & Faber 1993). Cotransmitters ATP, NPY, galanin, and opioids are released with catecholamines from S nerves and also contribute to vasoconstriction.

Catecholamines also can cause vasodilation in contracting skeletal muscles. At low concentrations, circulating E causes vasodilation by acting on noninnervated β_2 receptors (Ghaleh et al. 1995; Russel & Moran 1980) and endothelial α_2 receptors (Vanhoutte & Miller 1989). In the brain, E mediates vasodilation by stimulating A_{2a} adenosine receptors in the NTS (Kitchen et al. 2000). Release of vasodilatory messengers such as adenosine and endothelial NO contributes to vasodilation in isometrically contracting white, fast-twitch muscles and in cardiac muscle in response to blood shear and increases in acidity, temperature, hypoxia, and other by-products of muscle metabolism (Mian et al. 1990; Reid 1998; Rowell & O'Leary 1990; Victor et al. 1987). Metaboreflex contributes to NO release by increasing S stimulation of cholinergic receptors in endothelia of skeletal muscle and of coronary, pulmonary, and cerebral arteries (Bassenge 1994; Hickner et al. 1997; Horvath et al. 1994; Iwamoto et al. 1994; Lundberg et al. 1989; Mankad & Yacoub 1997; Morris et al. 1995; Reid 1998; Sherman et al. 1997; Yasuda et al. 1997; Zhang et al. 1997). At exercise intensities above 65% of $\dot{V}O_2$max, NO is also released from the upper respiratory airways in proportion to exercise intensity (Chirpaz-Oddou et al. 1997). These metabolic stimuli also inhibit vasoconstriction mediated by S stimulation of α_2 and, to a lesser extent, α_1 receptors (Freedman et al. 1992; Tateishi & Faber 1995; Thomas et al. 1994). Exercise training enhances blood flow to the skeletal muscles in part through increased S cholinergic nerve activity, decreased S adrenergic nerve activity, and decreased vascular sensitivity to NE (Delp 1998).

Endogenous Opiates

Endorphins and enkephalins also influence cardiovascular function during exercise. In the brain, enkephalins increase cardiac output, blood redistribution, blood pressure, and ventilation by acting on NA and NTS (de Jong et al. 1983; Hayashi & Nakamura 1984; Laubie & Schmidt 1981). In the periphery, they do so by reducing the PS tone and increasing the S tone (Caffrey et al. 1995; Sander et al. 1989). Met-enkephalin, intrinsic to the heart muscle, also helps accelerate the HR.

By contrast, β-endorphin counteracts the stress-related cardiovascular actions of catecholamines and enkephalins, both directly and indirectly (Caffrey et al. 1985; Musha et al. 1989; de Jong et al. 1983). It slows HR and reduces stroke volume by acting on δ and μ receptors (Haddad et al. 1986; Sandor et al. 1987; Tuggle & Horton 1986). It reduces blood pressure, especially after hemorrhage (Gaumann et al. 1987; Schadt & Gaddis 1985). It does so by stimulating secretion of ANF, by suppressing secretion of AVP, and by resetting the baroreflex (Inoue et al. 1987; Louisy et al. 1989; Rubin et al. 1983; Yamada et al. 1989). β-Endorphin also suppresses ventilatory rate by acting on δ receptors and stimulates shallow breathing by acting on μ receptors (Bonham 1995; Schaeffer & Haddad, 1985). β-Endorphin also indirectly inhibits various emergency functions by suppressing secretion of stress hormones E, glucagon, GH, and PRL, and reproductive function, by suppressing secretion of gonadotropins (Farrell et al. 1986; Grossman et al. 1984; Mannelli et al. 1984; Staessen et al. 1988).

Hydromineral Hormones

The pressor action of AVP comes into play at exercise intensities above 60% of $\dot{V}O_2$max when increases in its plasma concentration are not accompanied by antidiuresis (see figure 5.5) (Freund et al. 1991). Instead, free water clearance in the kidney is increased, and a smaller volume of dilute, rather than concentrated, urine is produced (see figure 5.6) (Refsum & Stromme 1975, 1977). At lower exercise intensities, the antidiuretic and pressor actions of AVP can operate concurrently within the physiological range of 1 to 20 pM/L plasma, the antidiuretic action to correct osmotic imbalances, and the pressor action to redistribute the blood. Stimuli for AVP release during intense exercise are

1. attenuated tonic suppression over AVP secretion,
2. increased plasma osmolarity,
3. loss of plasma volume, and
4. metaboreflexes.

Tonic inhibition of AVP secretion, vasoconstriction, and cardiac function (see figure 5.9) arises from high-pressure carotid and aortic baroreceptors (see figure 2.10) and from left atrial stretch receptors signaling increases in blood pressure to the brain stem centers through vagal and glossopharyngeal afferent nerve discharges (see figure 2.11) (Badoer et al. 1994; Kappagoda et al. 1974; Morita et al. 1986; Share 1996; Share & Levy 1962). This suppression is attenuated at exercise intensities above 90% if plasma volume losses exceed 10% and plasma angiotensin II concentration is increased (Convertino et al. 1981; Kouame et al. 1995; Undesser et al. 1985; Wilkerson et al. 1977). During strenuous exercise, AVP secretion is triggered primarily by increases in plasma osmolarity, as both the osmolarity and AVP concentrations rise in parallel with increases in exercise intensity and magnitude of dehydration (see table 5.4) (Convertino et al. 1981; Wade & Claybaugh 1980). Hypovolemia potentiates AVP release to hyperosmotic stimuli during exercise and hemorrhage that produces plasma volume losses of greater than 10% to 20% (see figure 4.4) (Dunn et al. 1973). AVP secretion and vasoconstrictor and pressor action also are triggered by metaboreflexes during intense isometric exercise that compromises blood flow through the muscle and the kidney and activates group III myelinated and group IV unmyelinated pain receptors in muscles and their arteries (Caverson & Ciriello 1987; O'Leary et al. 1993; Pullan et al. 1980; Yamashita et al. 1984).

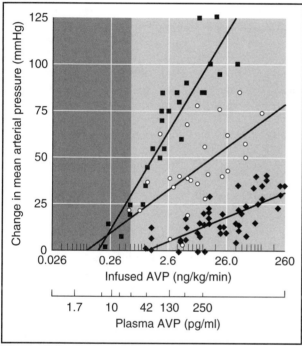

Figure 5.9 *Tonic inhibition of arginine vasopressin (AVP) secretion by high-pressure baroreceptors.*

Tonic inhibition of AVP secretion (solid diamonds) is attenuated by deafferentation of high-pressure baroreceptors (open circles) or removal of their brain connections (squares). Dark shading indicates normal range.

Adapted from Baulieu 1990.

AVP raises blood pressure by three types of action:

1. It enhances neurotransmission in S ganglia (Peters & Kreulen 1985).

2. It causes contraction of vascular smooth muscles by stimulating V1 receptors (Altura & Altura 1984).

3. It potentiates baroreflexes by acting on V1 receptors on the neurons in the AP, the circumventricular organ located near the opening of the fourth ventricle (see figure 4.1) (Cowley et al. 1984; Cox et al. 1990; Hasser et al. 1997).

AVP sensitizes baroreflex control of HR and S tone such that a given rise in arterial pressure results in greater reflex inhibition of HR and S activity (Ebert et al. 1986). This

AVP action prevents increases in systemic mean arterial pressure during exercise that produces increased osmolarity unaccompanied by changes in plasma volume. Thus, AVP-potentiated cardiac baroreflex counterbalances vasoconstriction with reduced cardiac output and S outflow (Bishop & Hay 1993; Jurzak & Schmidt 1998).

Angiotensin II contributes approximately a 10 mm Hg increase to BP elevation and accelerates heart function during maximal exercise and other emergencies such as hemorrhage without affecting secretion of aldosterone or having any sodium-reabsorbing action (see figure 4.6) (Symons & Stebbins 1996; Wade et al. 1987; Watanabe et al. 1998). Its pressor action stems from

- activation of S discharge from the RVLM via AT_1 receptors on the circumventricular organs and the PVN (Brooks 1997; Ito & Sved 1996; Jensen et al. 1992; Noshiro et al. 1994; Saad et al. 1998; Steckelings et al. 1992);

- attenuation, through the brain AT_1 receptors, of the baroreflex control of the heart, which then facilitates AVP secretion and pressor action (Bishop et al. 1995; Bishop & Hay 1993; Brooks et al. 1993; Saxena 1992);

- powerful constriction of small arterioles such that as little as a millionth of a gram of angiotensin II can increase systolic blood pressure by as much as 50 mm Hg; and

- stimulation via AT_1 receptors of release of endothelial vasoconstrictor endothelin-1 (ET-1) (d'Uscio et al. 1998).

By attenuating the reflex inhibition of heart function to increases in blood pressure and shifting the receptor operating range to higher blood pressure values, angiotensin II instigates simultaneous increases in both HR and blood pressure during intense exercise. Whereas AVP effects are transient, angiotensin II and catecholamine pressor actions are strong and sustained (Phillips et al. 1996b; Ponchon & Elghosi 1997).

The third hydromineral hormone, ANP (or ANF), also has cardioacceleratory and vascular actions during intense exercise. Its concentrations increase exponentially with increases in exercise intensity, whereas increases in BNP are less consistent (Freund et al. 1991; Onuoha et al. 1998). During exercise above 40% of maximal effort, ANF causes plasma shifts from vascular to IST space, stimulates blood pressure increases by acting on the anterior hypothalamus, and assists angiotensin II in facilitating concurrent increases of systolic blood pressure and heart rate during periods of great stress (Freund et al. 1988; Weidman et al. 1986).

ENDOCRINE CONTROL OF HEMOSTASIS DURING ACUTE EXERCISE

Exercise triggers formation of both clotting (thrombotic) and anticlotting (fibrinolytic) agents, depending on its intensity, duration, and level of mechanical stress. At low exercise intensities, formation of anticlotting plasmin predominates over production of the clotting protein fibrin (Weiss et al. 1998). Production of both clotting and anticlotting proteins and enzymes increases in a balanced fashion with increases in exercise intensity (El-Sayed 1996). Production of coagulation factors is balanced by increased fibrinolysis during short intense and prolonged strenuous exercise (Bartsch et al. 1990, 1995; Dufaux et al. 1991). During recovery from intense exercise, synthesis of proclotting thrombin and fibrin predominates over that of clot-dissolving plasmin (Van den Burg et al. 1995). Mechanical pounding during running leads to increased production of coagulating factors in the feet (Takashima & Higashi 1994).

The association of increased plasma E with faster clotting during exercise was recognized 80 years ago (Cannon 1922). Both E and, to a lesser extent, NE enhance coagulation during exercise and other stress (Rosenfeld et al. 1994). Platelets facilitate clotting by forming wound plugs and participating in fibrin formation. They are recruited from spleen, bone marrow, and lungs during acute exercise (El-Sayed 1996). E increases their recruitment, potentiates their capacity to aggregate, and stimulates their adhesion and expression of plasmin receptors (Hjemdahl et al. 1994; Kjeldsen et al. 1995; Lanza et al. 1988; Shattil et al. 1989; Wallen et al. 1999; Wang & Cheng 1999). Furthermore, E increases plasma concentrations of factor VIII, von Willebrandt's factor, and tissue plasminogen activator (tPA) and decreases concentration of PA inhibitor-1 (Larsson et al. 1990, 1992). E also stimulates hepatic synthesis of fibrinogen, an agent facilitating the synthesis of fibrin and the formation of clots and atherosclerotic plaques (Roy et al. 1985). Many of these actions of E are mediated through α_2 A-Rs (Kimura & Okuda 1994; Steen et al. 1993; Wang & Cheng 1999). Other stress hormones that affect coagulation during exercise are AVP and ANP (Grant 1990; Jern et al. 1989). Habitual exercise training reduces fibrinogen level and the risk of stroke and heart attacks (Ernst 1993).

ENDOCRINE MEDIATION OF ANALGESIA DURING INTENSE EXERCISE

Exercise and the immediate postexercise recovery period are associated with increased pain threshold and analgesia. Although the role of β-endorphin in analgesia and euphoria during or immediately after exercise has attracted most interest, increased secretion of GH and glucocorticoids also has been linked with increased threshold to dental and other types of pain (Kemppainen et al. 1990; Kiefel et al. 1989; Paulev et al. 1989; Pertovaara et al. 1984; Sforzo 1989; Surbey et al. 1984). Some exercise-associated analgesia is a result of direct activation of pain-inhibiting brain substrates by type III and IV afferent nerves (Droste 1992). Thus, although there is evidence that increased secretion of β-endorphin by the pituitary and by the leu-

kocytes and macrophages in injured tissues contributes to exercise-induced analgesia, it does not account for all observed increases in pain threshold (Droste et al. 1991; Herz 1995; Paulev et al. 1989; Surbey et al. 1984).

ENDOCRINE MEDIATION OF TISSUE REPAIR AFTER INTENSE EXERCISE

Exercise effects are predominantly anabolic but may be preceded by tissue damage and release of catabolic hormones and cytokines. One of the emergency actions of cortisol released during intense exercise is to increase plasma concentrations of amino acids, particularly alanine. Cortisol stimulates muscle and connective tissue proteolysis and facilitates release of alanine by the contracting skeletal muscles (Viru et al. 1994) (see table 5.1). Circulating amino acids are primarily used for hepatic gluconeogenesis during and after prolonged intense exercise but also can serve as substrates for wound healing and protein resynthesis. As high cortisol titers also are associated with muscle wasting, thinning of skin, and loss of cartilage and bone matrix, there has been a long-standing interest in the dynamic interaction between cortisol and its anabolic counterparts, GH, IGF-I, and testosterone, that also are secreted during exercise (Falduto et al. 1990; Gogia et al. 1993; Hickson et al. 1990). This issue is discussed in chapter 7.

Another catabolic hormone released during, or immediately after, acute exercise is the parathyroid hormone (PTH), often in association with low plasma calcium concentrations (Henderson et al. 1989; Nishiyama et al. 1988; Tsai et al. 1997). PTH triggers bone demineralization and turnover by mobilizing mineralized calcium, but the individual vulnerability to this event varies. Calcium intake followed by exercise suppressed PTH release in athletes with normal bone mineral density but increased it in athletes with low mineral density in lumbar spine, who also release more PTH during exercise (Grimston et al. 1993).

SA AND ENDOCRINE MEDIATION OF IMMUNE-DEFENSE RESPONSES DURING INTENSE EXERCISE

Exercise and a number of other stressors elicit a variety of immune responses that are attributable to changes in SA and stress-hormone release and to secretion of cytokines by the immune cells (Eichner & Calabrese 1994; Nieman 1997a, 1997b; Pedersen 1998; Pedersen et al. 1997; Pyne 1994; Shephard et al. 1991). The capacity of exercise to recruit immune cells out of peripheral lymphoid organs, which include lymph nodes, spleen, and linings of respiratory, intestinal, and genitourinary tracts, was documented a century ago (Larrabee 1902). For convenience, types of the diverse immune cells are listed in table 5.3.

B- and T-lymphocytes and monocytes are recruited into circulation out of peripheral lymphoid organs during low-intensity and moderate-intensity and moderate-duration exercise, as well as during very brief (less

Table 5.3 Types of Cells Mediating Immune Responses, Their Actions, and Effects of Exercise

Cell type	Ph	Cyto	Ab	Oxy	Ex-Lo	Ex-Hi	Ex-Ecc
Leukocytes: Lymphocytes: B cells			∧		∧	∨	
T cells Cytotoxic				∧	∧	∨	
Helper	∧				∨		
Suppressor					∧		
NK				∧	∧∧		
Monocytes	∧	∧			∧		∧
Granulocytes: Neutrophils	∧	∧			∧	∧	∧
Eosinophils							
Basophils							
Macrophages (from monocytes)	∧	∧					∧
Plasma cells (from B cells)							
Mast cells (from basophils)							
Platelets (from large bone marrow cells)*							∧

Ph = phagocytosis; Cyto = cytokines, messengers released by lymphocytes (lymphokines), monocytes, and macrophages (monokines); Ab = antibody formation (immunoglubulins IgA from the lining of respiratory, intestinal, and genitourinary tracts and IgG and IgM, which provide the main humoral immunity against pathogens); NK= natural killer cells; Oxy = formation of respiratory burst through an increase in nonmitochondrial oxidative metabolism that produces superoxide anion and other reactive oxygen species by NADH oxidase at plasma membrane; Ex-Lo = low-intensity exercise; Ex-Hi = high-intensity exercise; Ex-Ecc = eccentric exercise; ∧ = increase, ∨ = decrease. * Platelets are cell fragments formed by exocytosis of large bone marrow cells.

than 1 minute) maximal exercise (Field et al. 1991; Gray et al. 1993; Nehlsen-Cannarella et al. 1991; Nieman et al. 1991; Macdonald et al. 1991; Shinkai et al. 1992; Smith 1997). Increased B-cell concentration is associated with IgG, IgA, and IgM immunoglobulin production (Nehlsen-Cannarella et al. 1991). Recruitment of natural killer (NK) cells (CD16+ and CD56+) and of T-suppressor (CD8+) cells also increases, whereas numbers of T-helper lymphocytes (CD4+) decline during exercise (Espersen et al. 1990; Tvede et al. 1993). NK cell cytotoxicity (microbicidal action) is increased, as is monocyte and neutrophil phagocytosis.

Through its control of S outflow, the parvocellular PVN stimulates lymphoid organs to release activated immune cells during stress and exercise (see figure 5.10) (Friedman & Irwin 1997; Hori et al. 1995; Madden et al. 1995). This early recruitment of neutrophils, T lymphocytes, and monocytes is a consequence of the mechanical effect of increased blood flow and of increased S activation of the spleen and other peripheral lymphoid organs (Benschop et al. 1996; Nielsen et al. 1997; Shimizu et al. 1996). Plasma NE, more than E or cortisol, stimulates increases in concentration and proportions of leukocytes,

lymphocytes, and NK cells and increases NK cell activity (Kappel et al. 1998a; Nieman et al. 1991). Density of β A-Rs is higher on NK cells than on T helper and T suppressor cells, and their number is up-regulated by S activation during exercise or after infusion of isoproterenol (Landmann 1992). Thus, during moderate-intensity acute exercise, a difference in adrenergic receptor densities on the cells that mediate immune function and the increase in the concentration and relative proportion of these cell types in the blood are the most likely cause of selective cell mobilization, rather than an increase in the cell-specific activity (Ortega et al. 1993; Tvede et al. 1989). Early leukocytosis is a general stress response that is not affected by physical conditioning (Eichner & Calabrese 1994).

After exercise, neutrophil concentrations continue to increase for several hours, most likely in response to delayed GH surge (Kappel et al. 1993; 1994c). Monocytes and neutrophils contain GH receptors, and GH administration can increase plasma neutrophil concentration and the activity of NK cells in individuals with low endogenous GH secretion (Crist et al. 1987; Kappel et al. 1993; Kiess & Butenand 1985). Postexercise elevations

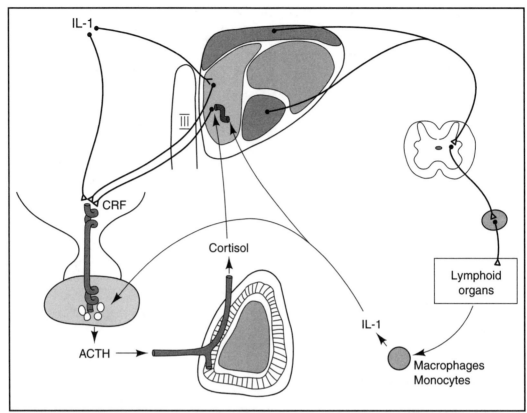

Figure 5.10 *Role of paraventricular hypothalamic nucleus (PVN) in activation of immune response.*

The PVN controls sympathetic (S) outflow to lymphoid organs, spleen, thymus, bone marrow, and lymph nodes and controls release of activated immune cells from lymphoid organs. These actions, in turn, stimulate the release of CRF and activate HPAA, while its ultimate messenger, cortisol, suppresses PVN activation of the immune response.

ACTH = corticotropin, CRF = corticotropin-releasing factor, IL-1 = interleukin-1.

Adapted from Loewy and Spyer 1990.

in plasma concentration of PRL, cortisol, and thyroxine also promote increased phagocytosis by cells mediating immunity (Ortega et al. 1996). Early leukocytosis is followed by a counterregulatory decline in white blood cell number immediately after or within a few hours after exercise (Weinstock et al. 1997).

A different pattern of immune response is seen after prolonged intense exercise. Intense and protracted exercise at greater than 80% of $\dot{V}O_2$max has both an immunostimulatory and an immunosuppressive effect (Fry et al. 1992; Nieman 1997a, 1997b). The stimulatory component entails dose-dependent increases in concentrations of neutrophils and circulating inflammatory cytokines through an intermediate range of exercise intensities (Gabriel & Kindermann 1997; Gabriel et al. 1992; Nieman et al. 1994; Pedersen et al. 1998). The stimulatory immune changes during intense exercise have been attributed to E because the pattern of lymphocyte and neutrophil recruitment obtained after either E infusion or acute exercise is similar (Kappel et al. 1991b, 1998a; Tvede et al. 1994). The immunosuppression after prolonged, intense exercise causes a decline in the number of B- and T-lymphocytes in circulation, salivary IgA secretion and antiviral action

by oral macrophages, resistance to upper-respiratory and other types of infection, and neutrophil bactericidal function (Fukatsu et al. 1996; Gray et al. 1993; Haq et al. 1993; Kohut et al. 1998; MacKinnon et al. 1987; Nieman 1997a; Nieman et al. 1994; Pedersen et al. 1996, 1997, 1998; Smith et al. 1990).

The immunosuppressive effects, have been attributed to the following:

- High plasma cortisol concentrations that prevail after prolonged intense exercise (Fukatsu et al. 1996; Pedersen et al. 1997)

- High catecholamine, particularly E, titers (Kohut et al. 1998)

- High endogenous opiate levels (Carr & Serou 1995; Chiappelli et al. 1991; Fiatarone et al. 1988)

- High concentration of prostaglandins released by monocytes, neutrophils, and B-lymphocytes (Pedersen et al. 1990; Tvede et al. 1989)

Cortisol's anti-inflammatory actions include suppression of proinflammatory cytokines (see table 5.1) (de

Rijk et al. 1996, 1997) and eicosanoids and reduction of capillary permeability to leukocytes (Griffin & Ojeda 1996). Its immunosuppressive actions include inhibition of phagocytosis by leukocytes and of synthesis of thymus-derived lymphocytes. However, cortisol also facilitates leukocyte synthesis, granulocyte release from spleen, delayed neutrophil release, and erythrocyte synthesis (Berk et al. 1990; Gabriel et al. 1992; Haq et al. 1993; McCarthy et al. 1991, 1992; Pedersen et al. 1997; Toft et al. 1994).

Another way that exercise engages immune defenses is through secretion of cytokines by macrophages and monocytes activated by muscle damage (Pedersen et al. 1997). The initial response to injury triggers production of IL-6 by the muscle (which next appears in plasma (Bruunsgaard et al. 1997; Ostrowski et al. 1998). The IL-6 synthesis and increased concentration in plasma coincide with histological evidence of Z-line damage in the muscle and with increased release by the muscle of creatine kinase (Fielding et al. 1993). IL-6 then stimulates production of IL-1 receptor agonist protein from circulating monocytes (Kayashima et al. 1995; Ostrowski et al. 1998). It also promotes muscle proteolysis by stimulating neutrophils, monocytes, and macrophages that have infiltrated the more permeable membranes of the injured muscle to produce proinflammatory cytokines IL-1, TNF-α, and prostaglandin E$_2$ (Cannon et al. 1991; Fielding et al. 1993). This cluster of defense reactions, called the acute phase response (APR), is similar to that elicited by infectious pathogens (Camus et al. 1994; Weight et al. 1991). The overall APR function is to foster proteolysis, superoxide release, and phagocytosis and removal of damaged proteins before new muscle protein synthesis (Cannon et al. 1990, 1991; Fielding et al. 1993). It may represent one of the mechanisms through which high-resistance exercise induces muscle hypertrophy (Cannon and Pierre 1998; Evans & Cannon 1991; Nieman et al. 1995).

IL-1 triggers CRF release from the parvocellular PVN (Berkenbosch et al. 1987). It may reach the PVN through circulation, by paracrine action from monocytes and microphages that migrate out of blood vessels into brain tissue or from neural hypothalamic circuits that use IL-1 as a neurotransmitter. Besides its action on CRF neurons, IL-1 may directly stimulate ACTH production from pituitary corticotrophs, and ACTH stimulates release of cortisol (Ruzicka & Akil 1995). Elevated plasma cortisol may counteract some manifestations of the APR (see table 5.1). It does so by reducing tissue permeability to cellular infiltration, by increasing sequestration of monocytes and lymphocytes in peripheral lymphoid organs, by elimination of some immune cells, and by reducing the adhesion and activity of NK cells (Cupps & Fauci 1982; Pedersen & Beyer 1986; Pedersen et al. 1997). Thus, S messengers initially stimulate immune defenses during most types of stress, chemical messengers of the immune system activate the HPAA and secretion of cortisol, and cortisol suppresses the immune cascade through its negative feedback over parvocellular

CRF neurons in the PVN (see figure 5.10) (Uehara et al. 1989).

Exercise also may enhance immune defenses by producing an abundance of glutamine in contracting muscle (Newsholme & Parry-Billings 1990). Glutamine is needed by lymphocytes and macrophages for their proliferation and synthesis of IL-1 and IL-2 (Newsholme & Calder, 1997). During a very intense and prolonged exercise such as a triathlon, plasma glutamine levels decline and may contribute to observed IgA decreases and immunosuppression (Rohde et al. 1996).

ENDOCRINE RESPONSES TO NONSPECIFIC STRESS OF EXERCISE

In addition to proactive feed-forward hormone secretion that prepares our bodies for physiological needs during intense physical effort, hormones are secreted in a reactive fashion to nonspecific events such as hypoxia, hypercapnia, acidosis, hyperthermia, hypoglycemia (see chapter 6), physical pain, and psychological distress (see tables 5.4 and 5.5). These secretory responses are abolished when these nonspecific stressors are removed.

Hypoxia, Hypercapnia, Hypotension, and Shock

Hypoxia, low ambient oxygen tension or high altitude, high ambient CO$_2$ tension, or a fall in blood pressure resulting from circulatory or endotoxemic shock can each elicit NE, ANP, GH, PRL, and NO release, as do lack of oxygen and pulmonary hypertension during high-intensity exercise (Avontuur et al. 1995; Horvath et al. 1994; Kilbourn et al. 1997; Lordick et al. 1995; Mundinger et al. 1997; Olsen 1995; Skinner & Marshall 1996; VanHelder et al. 1987; Zaccaria et al. 1998). The elicitation threshold for GH and PRL is around the anaerobic threshold or exercise intensity of about 70% $\dot{V}O_2$max (Luger et al. 1992; Van Helder et al. 1984, 1987). Hypoxia elicits differential release of white blood cells and IL-6, and IL-6 may serve here to stimulate erythropoiesis rather than inflammation (Klausen et al. 1997; Pedersen et al. 1994).

Acidosis

Increases in GH and PRL secretion are related to increases in plasma lactic acid (LA) concentration, a metabolic by-product of high-intensity exercise. Infusion of LA causes increases in plasma GH and PRL concentration similar in magnitude to those seen in acute exercise (Luger et al. 1992).

Hyperthermia and Dehydration

Hyperthermia, caused by experimental elevation of core temperature, stimulates GH and PRL secretion; APR, recruitment into circulation of NK cells, neutrophils, and monocytes; and increased cytokine and superoxide production (Christensen et al. 1984; Kappel et al. 1991a, 1994a, 1994b, 1995). The leukocytosis and increased

Table 5.4   Stimuli That Elicit Stress Hormone Release

Stimulus	NE	E	Co	End	Enk	Glu	Re	All	Ald	AVP	ANP	GH	PRL	TSH	Cy
Exercise intensity	∧	∧	∧	∧	∧		∧	∧	∧	∧	∧	∧	∧		
Hypoxia	∧			∧			∨	∨	∨		∧	∧	∧		
Hypoglycemia		∧	∧		∧	∧						∧			
Hypothermia	∧∧		∧∧				∨	∨	∨					∧∧	
Hyperthermia	∧	∧	∧									∧	∧		∧
Dehydration	∧		∧							∧∧					
Hypovolemia	∧	∧	∧	∧	∧		∧	∧	∧		∨				
Hyperosmolarity							∨	∨	∨	∧∧					
Injury	∧	∧	∧	∧								∧	∧		∧
Psychological stress	∧	∧∧	∧									∧	∧	∧	∧

NE = norepinephrine; E = epinephrine; Co = cortisol; End = β-endorphin; Enk = enkephalins; Glu = glucagon; Re = renin; All = angiotensin II; Ald = aldosterone; AVP = arginine vasopressin; ANP = atrial natriuretic peptide; GH = growth hormone; PRL = prolactin; TSH = thyroid-stimulating hormone; Cy = cytokines; ∧ = increased secretion; ∨ = decreased secretion.

cellular defenses during hyperthermia are in part mediated by GH, as they are partially eliminated by blocking GH but not catecholamine or EOP secretion (Kappel et al. 1998b). Increased GH and PRL secretion and associated immune responses are dependent on hyperthermia, as they can be blocked or blunted by preventing the rise in core temperature during exercise or by rehydration (Brisson et al. 1986; Buckler 1971, 1973; Christensen et al. 1984; Cross et al. 1996; Felsing et al. 1992; Frewin et al. 1976; Galbo et al. 1979; Saini et al. 1990).

Fluid losses that result from thermoregulatory sweating, hemorrhage, or other types of dehydration and changes in ambient barometric or water pressure, trigger release of E, NE, GH, PRL, RAA hormones, ANP, and AVP (Boisvert et al. 1993; Brenner et al. 1997; Castellani et al. 1998; Claybaugh et al. 1997; Gaumann et al. 1987; Inoue et al. 1987; Juul et al. 1995; McConnell et al. 1997; Montain et al. 1997; Ota et al. 1986; Vuolteenaho et al. 1992; Zhu & Leadley 1995). GH, secreted in response to hyperthermia and dehydration, helps conserve body fluids and plasma sodium by activiting the renin-angiotensin-aldosterone (RAA) system (Hoffman et al. 1996; Moller et al. 1997).

Physical Injury

Physical injury through surgery or infection is, like exercise, associated with some physical pain, circulatory disruptions, and structural damage (Camus et al. 1994; Rosenfeld et al. 1994). Both sets of circumstances trigger secretion of E, NE, cortisol, β-endorphin, PRL, and GH. Trauma, surgery, burns, and sepsis, like intense

prolonged physical effort, trigger host-defense responses (Lynch & Kirov 1986; Pedersen et al. 1994). IFNα, IL-1β and IL-2, and prostaglandins are the stimulus for the increases in the plasma concentration of lymphocytes and in particular NK cells after these stressors (Klokker et al. 1993a, 1993b).

Psychological Stresses

Psychological stresses that elicit anxiety, fear, or anger can, like exercise, induce secretion of E, cortisol, PRL, GH, and TSH (Cannon 1922; Gerritsen et al. 1996; Richter et al. 1996; Saitoh et al. 1995). Academic examinations, marital discord, and other types of psychological stress also affect cellular and humoral defense mechanisms. Psychological stress can reduce production of salivary IgA, suppress activity of NK cells, and increase the incidence of upper-respiratory infections (Cohen et al. 1991; Graham et al. 1986; Jemmott et al. 1983; Kiecolt-Glaser et al. 1984, 1987). Conversely, relaxation and visualization can increase salivary IgA production (Green et al. 1988; Jasnoski & Kugler 1987; Olness et al. 1989). Psychological stress, like exercise, also affects clotting (Grant 1990; Jern et al. 1989).

In stress, interactions between mind and body are reciprocal. Chemical messengers secreted during moderate-intensity exercise and other types of stress are psychologically and behaviorally arousing. Catecholamines, CRF, and ACTH widen the focus of attention and establish sensory readiness to external cues (Buwalda et al. 1997; Foote et al. 1991; Molle et al. 1997; Opp 1997; Shibasaki et al. 1993; Witte & Marrocco 1997). While

Table 5.5 Biological Actions of Stress Hormones

Action	NE	E	CRF	ACTH	Co	PTH	End	Enk	Ins	Glu	RAA	AVP	ANP	GH	PRL	Cy
Heart rate	∧	∧	∧				∨			∧	∧	∧	∨			
Cardiac output	∧	∧										∧	∨			
Blood pressure	∧	∧			∧				∧		∧	∧	∨			
Ventilation	∧		∧				∨	∧		∧	∧					
Vasomotor	∧	∨									∧	∧				
Glycogenolysis	∧	∧							∨	∧				∧		
Gluconeogenesis	∧	∧			∧				∨	∧				∧		
Lipolysis	∧	∧		∧	∧				∨	∧				∧	∧	
Calorigenesis	∧	∧							∧	∧				∧		
Sudomotor	∧	∧												∧	∧	
Water retention	∧	∧			∧							∧	∨	∧		
Salt retention					∧						∧		∨	∧		
Proteolysis					∧				∨					∨		∧
Bone turnover					∧	∧								∧		
Inflammation	∧	∧			∨									∧		∧
Arousal	∧	∧	∧	∧	∧∨		∧									
Analgesia					∧		∧	∧						∧		

NE = norepinephrine; E = epinephrine; CRF = corticotropin-releasing factor; ACTH = corticotropin; Co = cortisol; PTH = parathyroid hormone; End = β-endorphin; Enk = enkephalins; Ins = insulin; Glu = glucagon; RAA = renin-angiotensin-aldosterone; AVP = arginine vasopressin; ANP = atrial natriuretic peptide; GH = growth hormone; PRL = prolactin; Cy=cytokines; ∧ = increased secretion; ∨ = decreased secretion.

they are positively stimulating and thus beneficial at low stressor intensities, the same messengers have deleterious effects at high intensities. During excessive physical effort or in depression, feedback regulation of CRF and ACTH by cortisol is impaired (Duclos et al. 1997; Petrides et al. 1994; Sherman et al. 1984). In addition, chronically high NE and cortisol titers can produce neuronal damage and memory loss (Galvez et al. 1996; Kitayama et al. 1997; Murphy et al. 1996a; Zahrt et al. 1997).

HYPERTENSION AND ITS CONTROL THROUGH EXERCISE

Hypertension is a state of disordered regulation of blood pressure that manifests as systolic values above 140 mm Hg and diastolic values in excess of 90 mm Hg, when measured at random times during the day. Other forms of hypertension consist of excessive swings in the circadian blood pressure pattern or lack nocturnal blood pressure declines (Otsuka et al. 1997b; Rizzo et al. 1999). All forms of hypertension cause cardiovascular morbidity and damage to kidney circulation and affect approximately 20% of individuals in economically developed societies (Fauvel et al. 1991; Hollenberg et al. 1981; Light & Turner 1992). Hypertension usually develops in genetically susceptible individuals after life patterns characterized by

- inactivity,
- overconsumption of food (particularly salt),
- obesity,
- lack of sleep, and
- acute and chronic stress.

Absence of hypertension in individuals leading unstressed, structured, and disciplined lives, with few time constraints and within supportive social groups, underscores the importance of life-style variables in this disorder (Timio et al. 1997). Because the inappropriate responses by SA and stress messengers interact with life-style variables to precipitate hypertension, they will be briefly reviewed first. The role of physical activity both in the etiology of, and in the protection from, and treatment of hypertension will be discussed throughout.

Food intake acutely stimulates insulin secretion, which is accompanied by S stimulation of blood pressure elevation (Landsberg & Krieger 1989). Acute, meal-associated increases in blood pressure become protracted in obesity, particularly in visceral obesity, which causes stable elevation of blood pressure and increases risk of stroke and coronary heart disease (CHD) (Havlik et al. 1983; Huang et al. 1998).

High concentrations of FFAs in the hepatic portal vein, which characterizes obese individuals with reduced insulin sensitivity, has been suggested as the cause of hypertension in obesity (Grekin et al. 1997). Postprandial rise in plasma lipoproteins after food rich in saturated fat and unaccompanied by intake of anti-

oxidants facilitates their peroxidation by free radicals (Plotnick et al. 1997; Vogel et al. 1997). This process inactivates the endothelial NOS and causes acute endothelial dysfunction, a tonic reduction in vasodilation and compliance in the conductance and resistance arteries (Doi et al. 1999). Impaired arterial compliance attenuates baroreflex sensitivity and leads to chronic increases in S tone and peripheral resistance (Bristow et al. 1969; Eckberg 1979). Protracted endothelial damage involves inflammatory and trophic events mediated by increased platelet adhesion to the endothelia, stimulation of cell migration in response to PDGF, accumulation of lipids in atheromatous aggregations of foam cells, and hypercoagulability of blood (Alexander 1995; Luscher & Barton 1997). Atherosclerotic vascular occlusion coupled to reduced vascular compliance contributes to hypertension and greatly increase the risk of cerebrovascular and coronary infarcts.

Among other behaviors that precipitate hypertension in genetically susceptible individuals, including a relatively large proportion of African Americans, is the high cultural preference for unnecessarily salty diets (Luft et al. 1979). It produces inadequate suppression of RAA axis and a hypersensitivity to pressor actions of angiotensin II (d'Uscio et al. 1998). Lack of sleep is associated with blood pressure elevation in part because it increases fatigue and stress and in part because it curtails blood pressure declines expected to occur during nocturnal rest (Lusardi et al. 1999). Acute stress, which elicits fight or flight reaction, activates the SA system and results in increases in blood pressure (Harshfield et al. 1991; Light & Turner 1992). Finally, the inactivity is associated with increased risk of developing hypertension, but prospective epidemiological studies mostly show no attenuation in development of hypertension despite improving fitness or increasing physical activity (Blair et al. 1984, 1996; Paffenberger et al. 1983; Puddey & Cox 1995). Meta-analyses of exercise intervention studies show that aerobic exercise reduces resting systolic and diastolic pressures, but the effect is relatively small (4–5/3–4 mm Hg) when such analyses use more stringent inclusion criteria (Arroll & Beaglehole 1992; Fagard 1993; Halbert et al. 1997; Kelley 1999; Kelley & Kelley 1999; Kelley & McClellan 1994).

Physiological and endocrine responses that are most frequently identified as functioning abnormally and contributing to development of hypertension are the following:

- Acute endothelial dysfunction and reduced vascular compliance in response to high-fat diets or lack of exercise
- Overactivity of the SA system
- Overactivity of the RAA axis (Freeman et al. 1982)
- Oversecretion of glucocorticoids
- Impaired renal blood flow (Higashi et al. 1997; Freeman et al. 1982; Kohno et al. 1997)

Overactivity of the SA system may be a reflection of a genetic predisposition to respond to life problems with an autonomic defense response, activation of the S system secondary to angiotensin II oversecretion, or reduced baroreflex inhibition of cardiovascular function caused by reduced compliance of the carotid sinus and aortic arch (Blumenfeld et al. 1999; Esler et al. 1977; Grassi et al. 1998). S overactivity has been found in borderline hypertension and in early stages of essential hypertension (Esler et al. 1976; Goldstein 1983; Julius et al. 1991; Rahn et al. 1999). It is often accompanied by higher than normal HRs and development of left ventricular hypertrophy (Julius et al. 1991). Oversecretion of adrenal cortical hormones often occurs during prolonged exposure to stress, in depression, in obesity, or as a result of transcription errors in enzymes regulating biosynthesis of adrenal cortical hormones (Veglio et al. 1999). Chronic stress is also often associated with attenuated negative feedback by glucocorticoids (Ljung et al. 1996; Rosmond et al. 1996; Brown & Fisher 1985; Fisher et al. 1982).

Prevention and treatment of hypertension usually include pharmacological and life-style approaches. Pharmacological approaches are attempts to reduce SA activation of the heart with β-adrenoceptor blockers, to inhibit S vasoconstriction with α-adrenoceptor and calcium channel blockers, and to suppress angiotensin synthesis with ACE inhibitors (Carretero & Oparil 2000a, 2000b). Nonpharmacological approaches are behavior modifications to induce dietary weight loss, reduce sodium intake, reduce stress, and increase physical activity (Appel 1999; Tipton 1999).

Exercise training, particularly of the endurance kind, is attractive as a nonpharmacological approach for control of hypertension because of the multiplicity of ways through which it can contribute to blood pressure decreases (Engstrom et al. 1999). Indirect effects of exercise training are body fat loss and stress reduction. Direct effects are improved arterial compliance resulting from endothelial release of NO, improved baroreflex sensitivity, and reduced SA activation and catecholamine response to stress (Chandler et al. 1998; Clarkson et al. 1999; Hartley et al. 1972a; Higashi et al. 1999; Lehmann et al. 1981; Meredith et al. 1991; Peronnet et al. 1981a). Although the magnitude of resting blood pressure reduction in response to exercise is not very large in normotensive individuals, exercise training attenuates excessive blood pressure elevations to an acute exercise bout in hypertensive subjects (Ketelhut et al. 1997; Peronnet et al. 1981b; Sannerstedt 1966). Acute exercise has a transient hypotensive effect in hypertensive individuals so that lowering of blood pressure by habitual exercise may be a consequence of serial acute and transient blood pressure reductions (Pescatello et al. 1999; Wallace et al. 1999). The inconsistency in the hypotensive effect of exercise reflects incomplete understanding of the mechanisms causing hypertension and of the neuroendocrine modulation of blood pressure by exercise.

6

Hormonal Regulation of Fuel Use in Exercise

Access to metabolic energy is at the heart of human survival and is particularly pressing during exercise, when the rate of energy utilization may increase up to 30-fold. A complex set of metabolic mechanisms has evolved to procure, store, and use this resource. Humans and animals obtain energy by eating and drinking appropriate nutrients for conversion to adenosine triphosphate (ATP) and immediate use. Excess energy is stored by several different organs for future needs as chemically inert molecules of phosphocreatine (PCr), glycogen, and triglycerides (TGs). The location and size of fuel stores differs according to the type of storage molecule (see table 6.1). Small quantities of ATP and PCr are stored within muscle cells to power high-intensity movement for less than 10 seconds. Glycogen is stored by muscle cells and the liver in amounts that are sufficient for 2 to 3 hours of strenuous exercise but inadequate, for instance, for a single marathon run. Modest quantities of TGs are located within some types of muscle cells, and greater quantities are stockpiled in adipose tissue distributed among subcutaneous and intra-abdominal storage sites. The TG store is sufficiently large to fuel between 30 and 40 marathons but can be oxidized only at a slow pace to produce an abundant ATP yield. Glycogen is converted to glucose to produce either a smaller yield of ATP rapidly, by anaerobic glycolysis, or a more abundant yield of ATP aerobically.

Muscle fibers are differentiated for utilization of different fuels so that some are suited for oxidative metabolism, others for anaerobic metabolism, and some for both. Muscle fibers also differ in the extent to which they rely on endogenous fuels stored within their cytoplasm or on glycogen and TGs stored in other organs. Utilization of metabolic fuels therefore depends on activation of appropriate muscle fibers and metabolic pathways for utilization of specific fuels located either within the muscle or in distant fuel depots after fuel mobilization, release into circulation, and uptake by the muscle cells.

Four aspects of the role of SA and other messengers in this complex set of mechanisms that procure, store, and utilize nutrient energy are discussed in this chapter. They include

- metabolic events through which the SA and endocrine messengers influence metabolism at rest and during exercise;

- hormonal and nonhormonal control of metabolism during exercise and recovery from exercise of different intensities and durations;

- the effects of dietary supplementation on hormonal control of metabolism; and

- the modulation by inactivity and exercise of abnormal fuel metabolism in metabolic disease states.

HORMONAL INFLUENCE OF METABOLISM

Activation and metabolism of inert storage molecules during exercise requires that the endocrine milieu shift from one favoring synthesis of storage molecules, primarily mediated by insulin and adenosine, to one that permits their catabolism and release of free energy. Chief mediators of this shift are the catecholamines, as they both exert direct catabolic effects and trigger, or influence, secretion of other metabolic hormones such as insulin, glucagon, cortisol, and GH. The secretion and role of individual chemical messengers during exercise are both influenced by the metabolic role of the messenger and dependent on exercise intensity and nature of the intracellular fuel need.

Table 6.1   Location and Quantities of Metabolic Fuels for Exercise

Tissue	Tissue mass (kg, L)	Total fuel (g)	Fuel conc. (g/kg ww)	Fuel conc. (mM/kg ww)	Total energy (kcal)	Duration of fasting (days)[1]	Duration of walking (days)[2]	Number of mara-thons[3]
Extracel fluids: glucose	13	13			52	0.03	0.01	0.02
Liver: glycogen	2	80			320	0.2	0.04	0.1
Muscle: glycogen	31 (25)	495 (400)			1,980 (1,600)	1.0 (0.8)	0.26 (0.21)	0.8 (0.64)
types I & IIA[4]			16.5					
type IIB[4]			18.5					
Muscle: ATP PCr[5]				6 28	1.9 (1.5) 8.5 (7.0)	0.001 0.004	0.0003 0.001	0.001 0.003
Muscle lipids[9]		945 (765)		7-17	8,500 (6,900)	4 3.5	1 1	3.5 2.75
types I & IIA[6]			45					
type IIB[6]			16					
Adipose tissue: lipids[7]	8.5 (10.5)	8,500 (10,500)			76,500 (94,500)	38 (47)	10 (12.5)	30.5 (38)
Body protein[8]	10 (10)	10,000			40,000 (40,000)	16 (16)	5 (5)	16 (16)

Values are estimates for a 19-24 year old, 70-kg reference man (and woman). [1]A constant basal metabolic rate of 2,000 kcal/day is assumed for a 70-kg person. [2]Caloric expenditure of 7,475 kcal/day for a full day of walking is assumed (Durnin & Passmore 1967). [3]A 2,500 energy cost of running a marathon is assumed (Costill & Fox 1969; Margaria et al. 1963). [4]Information was averaged from Essen 1977 and Greenhaff et al. 1994. A wet to dry ratio of 4 for muscle mass was used. [5]ATP and PCr values are from Green et al. 1995 and Edwards et al. 1982. Formula weight of 507 g/M was assumed for ATP and 208 g/M for PCr. [6]Values from Essen 1977 and Greenhaff et al. 1994. A wet to dry ratio of 4 for muscle mass was used (Chesley et al.1995). Formula weight of 860 g/M was assumed for TG (Frayn 1983). [7]Storage body fat content of 12% for a 70-kg reference male and of 15% for a 70-kg reference female is assumed (Behnke & Wilmore 1974). [8]Boneless lean body mass of 51 kg for a 70-kg reference male and of 50 kg for a 70-kg reference female is assumed (Behnke & Wilmore 1974). Lean body mass water content of 71% and protein content of 20% are assumed. [9]From Gorski, 1992.

Catecholamines

Catecholamines influence metabolism through the following mechanisms:

1. Action on β_1, β_2, and β_3 A-Rs in the adipose tissue and β_2 receptors in the muscle for mobilization and utilization of storage fuels

2. Action on α_2 and β A-Rs in the pancreas for suppression of insulin and stimulation of glucagon release, respectively (see table 2.1 and figure 6.1) (Cleroux et al. 1989; Guillot et al. 1998; Natali et al. 1998)

3. Modulation of GH and cortisol secretion

The SA catecholamines control metabolism throughout the entire range of exercise intensities. Muscle S nerve activity increases during low and moderate intensity exercise in proportion to exercise intensity (Victor & Seals 1989). SA catecholamines are particularly effective stimulators of lipolysis at rest and at low exercise intensities (Arner et al. 1990; Arner 1995a; Coppack et al. 1994; Havel & Goldfien 1959; Kather & Simon 1981; Lafontan et al. 1997; Oro et al. 1965; Rosell 1966).

Adipocytes have neural as well as humoral A-Rs, and these A-Rs include both the antilipolytic α and the

lipolytic β adrenocep-
tor types (Arner et
al. 1990). The latter
type makes adipocytes
accessible to vasodila-
tory action of circulat-
ing E (Belfrage et al.
1977). The vascular
and metabolic actions
of   catecholamines
are particularly well
suited to rapidly in-
crease peripheral lipid
utilization at low ex-
ercise intensities and
to facilitate progres-
sion to greater use of
carbohydrates   and
endogenous   fuels at
intermediate and high
exercise intensities. At
rest, adipose tissue is
under the predominant
antilipolytic influence
of vagus, absorptive
insulin titers, and ad-
enosine and α adren-
ergic stimulation (see
figure 2.12), but rich
vascularization and S
innervation of both ad-
ipocytes and their ar-
terioles permits neural
activation of lipolysis
at rest and during low-
intensity exercise (Arn-
er et al. 1990; Meisner
& Carter 1977). NE
and E content of 5 to
10 μg/mg protein in
the adipose tissue is
similar in magnitude
to the catecholamine

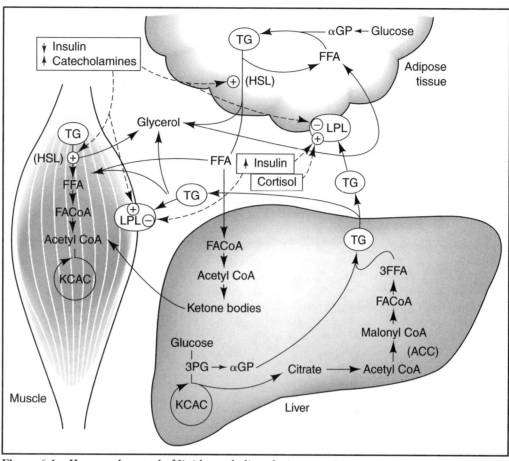

Figure 6.1 *Hormonal control of lipid metabolism during exercise.*

During exercise, sympathetic (S) nerves suppress insulin secretion and increase lipolysis in adipose tissue and muscle through reduction of insulin's antilipolytic effects and stimulation of hormone sensitive lipase (HSL). Lipolysis during and after exercise also is stimulated by growth hormone (GH) and cortisol. The catecholamines increase free fatty acid (FFA) uptake by muscle through stimulation of muscle lipoprotein lipase (LPL) and inhibition of adipose tissue LPL. After carbohydrate feeding, insulin stimulates hepatic fat synthesis and inhibits the activity of HSL in the adipose tissue and the muscle. Insulin facilitates uptake of FFAs by the adipose tissue through stimulation of adipose tissue LPL (an effect that is potentiated by cortisol) and inhibition of muscle LPL.

TG = triglyceride; FFA = free fatty acid; LPL = lipoprotein lipase; FACoA = fatty acyl coenzyme A; Acetyl CoA = acetyl coenzyme A; 3PG = three phospho glycerate; KCAC = Krebs citoric acid cycle; αGP = alpha glycerophosphate; ACC = acetyl coenzyme A carboxylase.

content of brown fat, heart, skeletal muscle, and spleen. Vascular α A-Rs in white adipose tissue are susceptible to rapid S nerve–stimulated vasoconstriction.

Catecholamines stimulate lipolysis by acting on their β_1, β_2, and β_3 A-Rs in the adipose tissue and on β_2 receptors in the muscle, by acting on α A-Rs on the pancreatic α cells to suppress insulin release (see figure 6.1) (Barbe et al. 1996; Cleroux et al. 1989; Guillot et al. 1998; Mathias et al. 1993; Natali et al. 1998; Thirone et al. 1998) and by stimulating secretion of lipolytic hormones glucagon, GH, and cortisol (Gromada et al. 1997).

The lipolytic catecholamine action is mediated by the cAMP transduction pathway and an intracellular cascade of phosphorylations (see chapter 3) that includes activation of the key lipolytic enzyme, hormone sensitive lipase (HSL). HSL catalyzes the hydrolysis of triglycer-

ide (TG) molecule into glycerol and three molecules of free fatty acids (FFAs). This enzyme is found both in adipose tissue and in the skeletal and heart muscle (Langfort et al. 1999; Oscai et al. 1990; Small et al. 1989). Because the HSL is inhibited by α A-Rs (Galitzky et al. 1988; Kather & Simon 1981) through the G_i transduction pathway (Fain 1980; Fain & Garcia-Sainz 1983), lipolysis depends on the variable contribution of SA stimulation and secretion of chemical messengers that differ in their effect on HSL. Lipolysis also depends on relative distribution of α and β adrenergic receptors in the adipose tissue in different body regions and their sensitivity to catecholamines (Galitzky et al. 1993a, 1993b; Hoffstedt et al. 1997; Jensen 1997; Van Harmelen et al. 1997; Wahrenberg et al. 1989, 1991). For instance, lower catecholamine affinity of β_3 than β_1 or β_2 receptors in the adipose tissue keeps β_3 A-Rs responsive to high concentrations of SA messengers after the other two have become desensitized. β_3 A-Rs therefore are essential for lipolysis under conditions of strong and sustained S nervous system activation (Carpene et al. 1993; Granneman 1995; Langin et al. 1995). A difference in the distribution of α and β adrenoceptors in adipose tissue and their differential sensitivity to catecholamines contribute to gender and age differences in distribution and size of fat depots in different body regions.

An additional important role of the catecholamines is to stimulate FFA uptake by the muscle during exercise or fasting. They do so by increasing the activity of lipoprotein lipase (LPL), an endothelial enzyme that hydrolyzes FFAs out of the TG cores of circulating dietary lipoproteins (Broun & Severson 1992; Friedman et al. 1986; Gorski & Stankiewics-Chorouszucha 1982; Miller et al.1989). Muscle LPL is activated by being translocated from its intracellular site of synthesis to the endothelial luminal surface. This translocation is achieved through the cAMP transduction pathway, activated by catecholamine action on β A-Rs and inhibited by insulin (Friedman et al. 1986; Gorski & Stankiewics-Chorouszucha 1982; Miller et al. 1989). Diffusion of FFAs into the muscle is facilitated by transport proteins, and FFAs are then re-esterified into TG droplets or oxidized. Dietary lipoproteins are not considered an important source of energy during exercise (Glatz & van der Vusse 1996; Henriksson 1995; van Nieuwenhoven et al. 1996). Instead, catecholamines activate muscle LPL when intramuscular lipids are depleted after prolonged, moderate-intensity to high-intensity exercise (Lithell et al. 1979; Savard & Greenwood 1988; Seip & Semenkovich 1998). In addition to hormonal stimulation, muscle contractions also can activate muscle LPL (Smol et al. 2001). In contrast to their action in muscle, catecholamines and ACTH inhibit the activity of adipose tissue LPL (Ashby et al. 1978; Chernik 1986). Thus, activation of LPL by contractions and catecholamines diverts nutrient lipids toward muscle and away from adipose tissue after prolonged exercise, and catecholamines alone have this effect during the postabsorptive phase of metabolism.

As physical activity and catechoamine secretion intensify, catecholamines stimulate hepatic glucose production and inhibit insulin secretion. Systemic hypoglycemia raises plasma NE concentration, but a decline in hepatic glucose concentration is also needed to elicit a full E secretory response (Donovan et al. 1991). Stimulation of hepatic glucose output by E is initiated at plasma concentrations (150–200 pg/ml) that are twice as high as threshold concentrations for stimulation of lipolysis (Galster et al. 1981). The catecholamine role in plasma glucose counter-regulation is therefore secondary to their role in lipolysis and auxiliary to pancreatic glucagon but becomes more important during exercise than at rest and dominant at very high exercise intensities (Cryer et al. 1984; Hirsch et al. 1991; Marker et al. 1991; Marliss et al. 1991; Rosen et al. 1984; Sigal et al. 1996). In the liver, glycogenolysis and hepatic glucose release are ini-

tiated primarily by E through activation of both α and β A-Rs (Palmblad et al. 1977; Rosen et al. 1983). β A-Rs activate phosphorylase through the phosphorylating action of cAMP-dependent protein kinase. α A-Rs stimulate glycogenolysis through intracellular Ca^{2+} release and the IP_3-DAG transduction pathway. Under the conditions that favor lipolysis and fat oxidation, S nerve activity and E both stimulate hepatic glucose production through gluconeogenesis (Ruderman et al. 1969). This effect may be mediated by chemical messengers or intracellular metabolic signals that regulate metabolic shifts between lipid and carbohydrate metabolism.

In the muscle, which lacks gluconeogenic enzymes or glucose-6-phosphatase and therefore can not produce and release glucose into circulation, E binds to β_2 A-Rs to stimulate glycogenolysis, production of lactic acid, and glucose oxidation (Cleroux et al. 1989; Greenhaff et al. 1991; Kjaer et al. 2000; Williams et al. 1984). All of the above actions of catecholamines are assisted by their suppression of insulin secretion, which is initiated at E concentrations that are approximately four times higher than those necessary for stimulation of lipolysis (Galster et al. 1981).

Both the lipolytic and hyperglycemic actions of catecholamines during exercise are amplified by their effects on the release of other metabolic hormones (see table 2.1).

At lower and intermediate ranges of exercise intensity, secretion of glucagon, cortisol, E, and GH are predominantly governed by metabolic need and are subject to homeostatic feed-back control, such that decreases in plasma glucose availability, or FFA availability in the case of GH, stimulate their secretion and hyperglycemia or high plasma FFA concentrations suppress their secretion (Casanueva et al. 1987; Pontiroli et al. 1996; Quabbe et al. 1991; Sutton & Lazarus 1976). Because of their collective role in stimulation of hepatic glucose release in opposition to hypoglycemic effects of insulin, this group of hormones is described as counter-regulatory. With insulin, they participate in regulation of plasma glucose and FFA availability. As exercise intensifies, activation of S nerves and secretion of counter-regulatory hormones become exponential and assume feed-forward characteristics of an emergency response (Kjaer et al. 1991) (see chapter 5).

Secretion of adrenomedullary E is under the direct control of S preganglionic nerve fibers (see figure 2.3). S nerves also inhibit insulin secretion by acting on α_2 A-Rs on B cells of pancreatic islets (Porte et al. 1976). Although the stimulation of β_2 A-Rs by NE or circulating E in some circumstances transiently stimulates insulin release, it is overshadowed in humans by the inhibitory α adrenergic action (Edwards, in Loewy and Spyer 1990; Hirose 1993a, 1993b; Lacey et al. 1993). Suppression of insulin secretion during exercise removes its strong facilitation of fuel synthesis and storage and permits unopposed metabolic energy release in the form of lipolysis, glycogenolysis, and gluconeogenesis. In some species, α_1 and β adrenergic stimulation elicits secretion of glucagon, the chief glucose counter-regulatory hormone, but reduced blood sugar concentration, rather than adrenergic stimulation, is necessary for its release in exercising humans (Havel et al. 1991; Galbo et al. 1975, 1976, 1977a; Samols & Weir 1979; Sotsky et al. 1989). Stimulation of β A-Rs is responsible for exercise-induced increase in PP secretion (Berger et al. 1980). Although SA messengers do not directly trigger the secretion of cortisol and GH during exercise, their release is modulated by catecholamines. During intense exercise and other forms of stress, E and cortisol are engaged in a positive feed-forward loop, in which E increases cortisol secretion by stimulating pituitary release of ACTH and cortisol stimulates adrenal synthesis and release of E. A central adrenergic activation may be involved in the secretion of GH during intense exercise as its secretion is temporally coupled to preceding increases in E and NE secretion (Weltman et al. 2000).

The metabolic consequences of this emergency SA and endocrine response to high-intensity exercise is hyperglycemia. FFAs are mobilized at a high rate but are sequestered inside the adipose tissue and released into circulation immediately after the exercise bout. The homeostatic roles of insulin and counter-regulatory hormones are reestablished upon termination of exercise, at which point rapidly declining concentrations of metabolic hormones and metabolic signals from fuel remaining in plasma or from depleted fuel storage organs assume control over metabolism.

Insulin

An important function of catecholamines is to inhibit insulin secretion during exercise because of the many actions of this hormone that interfere with energy needs during exercise. Stimulation of insulin secretion by carbohydrate supplementation before or during exercise elicits insulin actions that interfere with the adaptive pattern of fuel use during exercise.

Insulin and glucagon are produced by the endocrine islets of Langerhans, which constitute only 1% of the 70-g pancreatic mass. Insulin is secreted by the more numerous B (or β) cells, located in the center of the islet. The other three hormones, glucagon, somatostatin (SRIF), and pancreatic polypeptide (PP) are secreted, respectively, by A (or α), D (or δ), and PP cells located at the periphery of the islet. SRIF exerts paracrine inhibitory control over insulin and glucagon secretion (Adrian et al. 1981). As the pancreatico-duodenal vein drains into the portal vein, the liver is exposed to about three times higher basal insulin and glucagon concentrations and about 10 times higher stimulated concentrations than are the other organs encountering the two hormones in systemic circulation (Baulieu & Kelly 1990; Greenway & Stark 1971; Wasserman et al. 1993). If during exercise, systemic plasma glucagon concentrations show modest increases or even decreases, the liver may be exposed to a significantly higher glucagon message (Galbo et al. 1976; Hilsted et al. 1980; Hoelzer et al. 1986; Marliss et al. 1991; Simonson et al. 1984). The biological effects of insulin are mediated through two principal actions: (1) dephosphorylation of enzymes and regulatory proteins to inhibit the activity of cAMP-dependent protein kinase A and (2) activation of cAMP-degrading enzyme phosphodiesterase (PDE) (Enoksson et al. 1998; Gabbay & Lardy 1984; Hagstrom-Toft et al. 1995; Lonroth & Smith 1986; Manganiello et al. 1996) (see chapter 3).

Circulating glucose serves as a principal stimulus of insulin release and also elicits insulin's important hypoglycemic function, stimulation of cellular glucose uptake (Lavoie et al. 1997b). Insulin secretion is linked to a rise in plasma glucose above 80 to 85 mg/dl and to increases in plasma concentration of amino acids during the immediate absorptive or postprandial period after eating. The triggering event is diffusion of glucose into B cells, assisted by the high-Km (15 to 20 mM) GLUT-2 glucose transporter, which, along with two glycolytic enzymes, glucokinase (GK) and phosphofructokinase (PFK), acts as a glucose sensor (Meglasson & Matschinsky 1986). Phosphorylation of glucose by GK to glucose-6-phosphate (G6-P) triggers closure of K^+ channels and the resultant B-cell depolarization opens voltage-gated Ca^{2+} channels, increases intracellular Ca^{2+} concentration, and triggers extrusion of insulin vesicles by exocytosis (Bertram et al. 1995). In addition to being stimulated by plasma glucose concentration, insulin secretion responds to stimulation by the vagus nerve, but such stimulation is not essential for postabsorptive glucose metabolism in man (Corssmit et al. 1995; Tanaka et al. 1990).

The chief functions of insulin (see table 6.2), in order of their decreasing sensitivity to its actions, are inhibition of lipolysis at 13 μU/ml or 96 pM, inhibition of hepatic glucose release at 26 μU/ml, and stimulation of cellular uptake of carbohydrates at several times higher concentrations (Groop et al. 1989; Londos et al. 1985; Lonroth & Smith 1986; Nurjhan et al. 1986). At much higher concentrations (44 μU/ml or 324 pM), insulin inhibits lipid oxidation (Campbell et al. 1992). When insulin concentration declines below 13 μU/ml, lipolysis is powerfully and exponentially stimulated (Campbell et al. 1992; Coppack et al. 1994; Groop et al. 1989). Insulin inhibits HSL through its phosphodiesterase action, which degrades the cAMP (Eriksson et al. 1995). Insulin inhibits hepatic glucose production by dephosphorylating the glycogenolytic enzyme phosphorylase a to its inactive b form and by inhibiting the synthesis of the gluconeogenic enzyme PEPCK (Felig & Wahren 1971). In addition, insulin blocks hepatic glucose release by restraining the activity of glucose 6 phosphatase (G6-Pase), the enzyme necessary for converting G6-P to a more diffusible free glucose (see table 6.3).

Insulin also directs cellular metabolism toward utilization of carbohydrates by stimulating glycolysis and glucose oxidation (see figure 6.2). It promotes glycolysis by activating two enzymes in the Embden-Meyerhoff glycolytic pathway, PFK_1 and pyruvate kinase (PK), and promotes activity of enzymes in the alternate glycolytic pathway, the pentose monophosphate shunt (PMS) (see figure 6.3). Insulin also stimulates synthesis of PK (Cuif

Table 6.2   The Role of Sympathoadrenal and Endocrine Messengers in Fuel Mobilization, Glucose Uptake, and Glycogen Synthesis During Exercise and Recovery

Hormone	Change	Up	Ll	Kg	Gl	Gng	Gs	Td
Insulin (exercise)	∨, =	∧	∨	∨	∨	∧		
Insulin (recovery)	∨∧	∧	∨	∨	∨	∨		∧ cGMP
Glucagon	∧	∨	∧	∧	∨	∧	∨	∧ cAMP
E, NE (β)	∧	∨	∧∨	∧	∧	∧	∨	∧ cAMP
NE (α)	∧				∧			Ca²⁺
GH	∧	∨	∧			∧		mRNA
ACTH	∧		∧					∧ cAMP
Cortisol	∧	∨	∧		∨	∧	∧	mRNA
AVP	∧		∧		∧			Ca²⁺
Secretin	∧		∧					
PRL	∧		∧					mRNA
β-Lipotropin, γ-lipotropin	∧		∧					∧ cAMP

E = epinephrine; NE = norepinephrine; GH = growth hormone; ACTH = corticotropin; AVP = arginine vasopressin; PRL = prolactin; Gl = glycogenolysis; Gng = gluconeogenesis; Gs = glycogen synthesis; Kg = ketogenesis; Ll = lipolysis; Up = glucose uptake; Td = transduction mechanism; cAMP = cyclic adenosine monophosphate; mRNA = messenger RNA; ∧ = increase; ∨ = decrease.

Table 6.3   Metabolic and Anabolic Actions of Insulin

Insulin source	Insulin actions
Nutrient uptake	∧ glucose uptake by way of increased translocation of GLUT-4 glucose transporters in muscle and AT ∧ synthesis of HK and GK ∧ FFA uptake by way of increased synthesis and stimulation of LPL in the AT
Liver glycogen	∨ glycogenolysis ∨ gluconeogenesis by way of increased synthesis of PEPCK ∨ hepatic glucose output by way of increased activity of G-6-Pase ∧ glycogen synthesis through dephosphorylation and increased activity of glycogen synthetase I
Carbohydrate metabolism	∧ glycolysis through increased activity of PFK₁ and PFK₂ ∧ activity of PMS enzymes ∧ carbohydrate oxidation by way of increased activity of PDH

Insulin source	Insulin actions
Lipid metabolism	╱ lipolysis through dephosphorylation and decreased activity of HSL
	lipogenesis by way of dephosphorylation and increased activity of acetyl CoA carboxylase (ACC)
	activity of FAS and fat synthesis
Protein metabolism	╱ protein degradation
	protein synthesis (initiation phase of translation)

= increase; ╱ = decrease; AT = adipose tissue; GLUT-4 = glucose transporter; HK = enzyme hexokinase; GK = glucokinase; FFA = free fatty acid; LPL = lipoprotein lipase; PEPCK = gluconeogenic enzyme; PFK = phosphofructokinase; PMS = pentose monophosphate shunt; PDH = pyruvate dehydrogenase; HSL = hormone sensitive lipase; FAS = fatty acid synthetase.

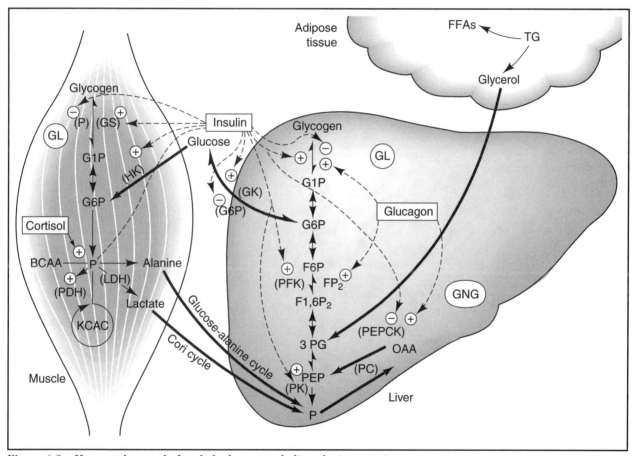

Figure 6.2 *Hormonal control of carbohydrate metabolism during exercise.*

Facilitation by insulin of glucose uptake and phosphorylation via hexokinase (HK) in the muscle is shown on the left and via glucokinase (GK) in the liver is shown on the right. Insulin stimulates glycolysis (anaerobic pathway) and glucose oxidation (aerobic pathway) in the muscle and in the liver. It also blocks hepatic glucose release, glycogenolysis, gluconeogenesis (GNG), and lipolysis. Glucagon, epinephrine, cortisol, and growth hormone stimulate GNG. Glucagon also stimulates glycogenolysis and hepatic uptake of gluconeogenic substrates and inhibits glycolysis. Catecholamines stimulate glycogenolysis, glycolysis, and oxidative metabolism. Cortisol stimulates glycogen synthesis and inhibits glucose uptake and glycolysis. The role of alanine, lactate, and glycerol in stimulation of the gluconeogenic glucose-alanine cycle and the Cori cycle is also shown.

Biochemical reactions in circles, hormones in rectangles, enzymes in parentheses; GL = glycolysis; KCAC = Krebs citoric acid cycle; P = phosphorylase; G1P = glucose 1 phosphate; G6P = glucose 6 phosphate; BCAA = branched-chain amino acids; PDH = pyruvate dehydrogenase; LDH = lactate dehydrogenase; HK = hexokinase; GK = glucokinase; GS = glycogen synthetase; F6P = fructose 6 phosphate; PFK = phosphofructokinase; FP$_2$ = fructose biphosphate; F1,6P$_2$ = fructose 1,6 biphosphate; 3PG = 3 phosphoglycerate; PEPCK = phosphoenolpyruvate carboxykinase; PEP = phosphoenrol pyruvate; P = pyruvate; OAA = oxaloacetate; GNG = gluconeogenesis; FFAs = free fatty acids; TG = triglyceride.

et al. 1997; Laker & Mayes 1984). Insulin activation of the rate-limiting enzyme PFK$_1$ to produce the glycolytic substrate fructose 1,6-biphosphate is achieved indirectly (Pilkis & Claus 1991; Pilkis & Granner 1992). PFK$_1$ substrate fructose 2,6-biphosphate (F2,6P$_2$) is produced from fructose-6 phosphate when insulin stimulates a bifunctional enzyme 6-phosphofructo-2-kinase/fructose-2,6-biphosphatase (PFK$_2$/FBP$_2$ase) to increase its kinase activity. Insulin stimulates carbohydrate oxidation and blocks carbohydrate sparing through dephosphorylation of pyruvate dehydrogenase (PDH) (Mandarino et al. 1987). This enzyme irreversibly commits pyruvate to oxidation in the Krebs citric acid cycle by catalyzing its oxidation and decarboxylation to acetyl-CoA (see figure 6.2).

Insulin stimulates cellular uptake of glucose by the insulin-sensitive muscle, adipose tissue, and a variety of other tissues other than the liver and the brain through translocation of GLUT-4 glucose transporters from an intracellular pool to the plasma membrane (Jones & Dohm 1997; Kahn 1992). In muscle, insulin-medi-

ated glucose flux activates and up-regulates the low-Km (100 μM) enzyme hexokinase (HK), which can phosphorylate glucose at concentrations that can be 50 times lower than fasting plasma glucose (Tsao et al. 1996). In the liver, insulin stimulates the high-Km (10 mM) phosphorylating enzyme GK, which entraps glucose-6 phosphate within hepatic cellular compartment when plasma glucose concentrations rise above the fasting level (DeFronzo et al. 1983; Kietzmann et al. 1998). Increased intracellular glucose availability and activation by insulin of glycogen synthetase by dephosphorylating several of this enzyme's sites, drives carbohydrate carbon flux within hepatocytes, muscle cells, and adipocytes toward glycogen synthesis (see figure 6.2) (Brady et al. 1999; Lawrence & Roach 1997; Miller & Larner 1973; O'Gorman et al. 2000).

Insulin facilitates de novo fat synthesis in the liver in several ways. It increases synthesis of acetyl-CoA carboxylase (ACC), an enzyme catalyzing production of malonyl-CoA in the FFA biosynthetic pathway (see figure 6.3). Through its facilitation of the main glycolytic pathway and the pentose monohosphate shunt (PMS), insulin, respectively, generates α-glycerophosphate and the NADPH reducing equivalents, which are needed for TG and FFA synthesis. Insulin also increases TG synthesis by dephosphorylating and activating as well as by stimulating transcription of fatty acid synthetase (FAS), the enzyme that catalyzes condensation of α glycerol phosphate and FFAs into TG (see figures 6.1 and 6.3) (Stralfors et al. 1984; Sul et al. 2000). Insulin routes plasma lipids into adipose tissue rather than into skeletal or cardiac muscle by stimulating adipose tissue LPL and by inhibiting muscle LPL (see figure 6.1) (Ashby & Robinson 1980; Borensztajn et al. 1972; Kiens et al. 1989; Richelsen et al. 1993). Conversely, a decrease in plasma insulin stimulates the activity of muscle LPL and inhibits the activity of adipose tissue LPL (Taskinen et al. 1980).

The final anabolic insulin action is stimulation of protein synthesis through

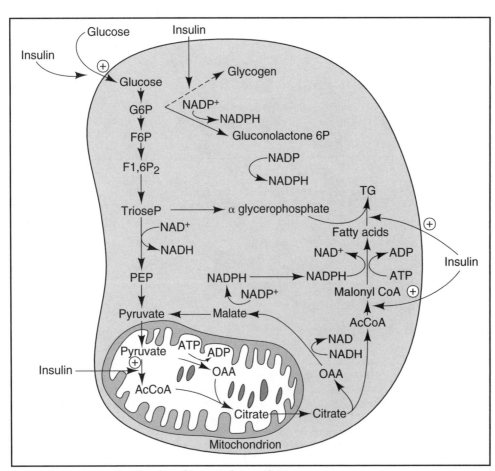

Figure 6.3 *Control by insulin of hepatic fat synthesis.*
The novo synthesis of fat in the liver is facilitated by insulin at several metabolic steps.

G6P = glucose 6 phosphate; F6P = fructose 6 phosphase; F1,6P$_2$ = fructose 1,6 diphosphate; TrioseP = triose phosphate (3 phosphoglucerate); NAD, NADH = nicotinamide adenine dinucleotide, oxidized and reduced forms; PEP = phosphoenol pyruvate; AcCoA = acetyl coenzyme A; OAA = oxalocetate; ADP = adenosine diphosphate; NADP, NADPH = nicotinamide adenine dinucleotide phosphate, oxidized and reduced sources; ATP = adenosine triphosphate.
Adapted from Leveille and Romsos 1974.

increased transport of amino acids into muscle and other cells, initiation of protein translation, and inhibition of protein degradation (Felig et al. 1977).

Glucagon

Glucagon's counter-regulatory role is triggered during exercise when plasma glucose concentration falls below 50 mg/dl and is blocked at glucose concentrations above 150 mg/dl (Cryer et al. 1984; Galbo et al. 1975, 1977a; Sotsky et al. 1989). Additional triggers include stimulation of α_1 adrenergic Rs by S nerves in some species other than humans and increased concentrations of E and cortisol (Samols & Weir 1979). Glucagon's stimulation of hepatic glucose production and hyperglycemic action is achieved through facilitation of glycogenolysis at lower plasma concentrations and of gluconeogenesis at high plasma concentrations (see figure 6.2 and table 5.5) (Chibber et al. 2000; Lavoie et al. 1997a). Glucagon initiates glycogenolysis by activating phosphorylase a, mostly through the cAMP transduction pathway, and to a lesser extent through a calcium-releasing mechanism. Gluconeogenesis becomes a primary source of plasma glucose production after hepatic glycogen is depleted when plasma glucose and insulin decline in parallel. Glucagon promotes gluconeogenesis through the stimulation of the Cori cycle (Cori 1981) and the glucose-alanine cycle (Felig 1973; Felig & Wahren 1971). In the Cori cycle (see figure 6.2), glycolytic substrates pyruvate and lactate are recycled through stimulation by glucagon of the gluconeogenic enzymes PEPCK and F1,6 biphosphatase (Kietzmann et al. 1998; Lange et al. 1992; Pilkis & Granner 1992; Pilkis & Claus 1991). Glucagon participates in the glucose-alanine cycle by increasing hepatic amino acid uptake (Boden et al. 1990). Glucagon stimulates lipolysis indirectly by suppressing activity of hepatic lipogenic enzyme ACC and thus by reducing the availability of malonyl-CoA. Malonyl-CoA is a strong inhibitor of the enzyme carnitine palmitoyl-O-transferase I (CPT I), the enzyme that controls lipolysis through the mitochondrial uptake of FFAs (McGarry 1995; Nathan & Cagliero 2001). Glucagon also facilitates uptake of FFAs into cardiac muscle by stimulating the activity of its LPL (Borensztajn et al. 1973). Glucagon alone cannot achieve adequate glucoregulation without the participation of E (DeFeo et al. 1991a, 1991b).

Growth Hormone and ACTH

Pituitary hormones GH and ACTH are released at relatively high exercise intensities and in response to other forms of stress (see chapter 5). Their contributions to exercise metabolism are auxiliary rather than essential because blockade of catecholamine action and associated energy production during intense exercise fails to stimulate compensatory increases in GH and ACTH secretion (Franz et al. 1983). ACTH has a strong lipolytic action in adipose tissue of some species other than man (Van den Bergh et al. 1992). The slow rise in GH concentration during intense exercise makes its peak secretory response usually occur upon cessation of exercise so that it coincides with the peak FFA outflow from adipose tissue.

GH also is released in response to hypoglycemia, reduction in plasma FFA concentration, and other manifestations of energy need and suppressed by increased plasma concentrations of glucose and FFAs (Casanueva et al. 1987; Pontiroli et al. 1996; Quabbe et al. 1991; Sutton & Lazarus 1976). A single GH pulse, or pulsatile GH delivery, increases basal lipolysis, particularly at night (Bianda et al. 1996; Boyle et al. 1992; Cersosimo et al. 1996; Moller et al. 1990a, 1990b; Vahl et al. 1997b). GH stimulates lipolysis, ketogenesis, and lipid oxidation after a delay ranging from 20 minutes to 3 hours (Fain & Saperstein 1970; Keller et al. 1984; Okuda et al. 2001; Schwartz & Goodman 1976). Increased lipid utilization during recovery from intense exercise has been attributed to the coincidence of high plasma FFA and GH concentrations (Pritzlaff et al. 2000). GH also increases the sensitivity of adipose tissue to catecholamine-stimulated lipolysis by stimulating synthesis of additional β adrenergic receptors (Ottosson et al. 2000; Yang et al. 1996a). GH stimulates the activity of HSL, in part, by interfering with the antilipolytic action of adenosine and its G_i protein–mediated inhibition of adenylcyclase (see figure 3.12) (Doris et al. 1994; Houseknecht & Bauman 1997; Yip & Goodman 1999). It also inhibits fat synthesis by activating protein kinase C and phospholipase C (Vernon 1996). GH and ACTH divert lipids to muscle by stimulating the activity of muscle LPL and away from the adipose tissue by inhibiting the activity of adipose tissue LPL (Oscarsson et al. 1999a, 1999b).

An additional GH metabolic role is to increase hepatic glucose production but suppress glucose uptake and oxidation in muscle. GH stimulates hepatic gluconeogenesis by reducing liver sensitivity to insulin action and by influencing the metabolic shift toward increased lipid oxidation (Moller et al. 1991; Monti et al. 1997; Okuda et al. 1994). Suppression by GH of muscle glucose uptake is more likely caused by interference with activation of insulin receptors than with translocation and synthesis of GLUT-4 glucose transporters (Cartee & Bohn 1995; Napoli et al. 1996; Rizza et al. 1982a; Smith et al. 1997b; Yakar et al. 2001).

Cortisol

Hypoglycemia during exercise and other forms of severe stress also trigger secretion of cortisol, E, and glucagon (Quabbe et al. 1991; Sutton & Lazarus 1976). These stress hormones stimulate muscle protein degradation to a greater extent than muscle protein synthesis and increase the availability of amino acids in circulation (Gore et al. 1993). Whereas suppression of protein synthesis is more prevalent in the muscle and protein catabolism is more prevalent in the liver and other tissues, cortisol selectively degrades type II and spares type I muscle fibers (Clark & Vignos 1979). Cortisol also activates gluconeogenic enzymes fructose 1,6-biphosphatase and PEPCK (Kietzmann et al. 1998; Pilkis & Claus 1991; Pilkis & Granner 1992). Newly formed glucose is either released by liver or serves as a substrate for glycogen synthesis. Cortisol facilitates glycogen synthesis by activating gluconeogenic formation of phosphoenol pyruvate and by stimulating glycogen synthetase (Exton et al. 1976). As is the case for most counter-regulatory hormones, cortisol blocks glucose uptake into adipocytes and muscle. This postreceptor effect requires protein synthesis and is mediated by decreased translocation to the membrane of GLUT-4 glucose transporters (Carter-Su & Okamoto 1985a, 1985b; Rizza et al. 1982b).

The role of cortisol in fat metabolism is more complex (Djurhuss et al. 2002). In the shorter term, cortisol stimulates lipolysis indirectly by inhibiting glucose uptake and glycolytic production of α glycerophosphate and by increasing sensitivity of adipose tissue to catecholamine-stimulated lipolysis (Divertie et al. 1991). However, in the longer term, glucocorticoid excess stimulates FFA uptake and TG synthesis and storage. Cortisol stimulates synthesis of adipose tissue LPL and potentiates the effect of insulin in activating this enzyme (Ashby & Robinson 1980; Cryer 1981). By stimulating appetite and inducing peripheral insulin resistance and insulin oversecretion, glucocorticoids facilitate fat deposition into insulin-sensitive adipose sites and TG wasting in insulin-insensitive sites.

Thyroid Hormones

Thyroid hormones triiodothyronine (T_3) and thyroxine (T_4), at higher concentrations, increase the number of β adrenergic receptors and thus potentiate the lipolytic and glycogenolytic actions of catecholamines. They also increase cellular energy expenditure by stimulating synthesis and activity of Na^+,K^+ ATPase in the plasma membrane (Griffin & Ojeda 1996).

METABOLISM DURING EXERCISE AND RECOVERY FROM EXERCISE

Capacity for rapid or intense movement is of great survival importance to humans and other animals. As a result, muscle can efficiently take up and utilize nutrients with or without the help of chemical messengers. In addition, muscle cells differ in their contractile and metabolic properties so that their differential recruitment during exercise of different intensity and duration allows for switching between fuels with different bioenergetic characteristics. Involvement of hormonal and hormone-independent mechanisms and of specific muscle fiber types in fuel mobilization and metabolism will be described for three levels of exercise intensity and recovery from such exercise.

Low-Intensity Exercise

During exercise at intensities below 50% $\dot{V}O_2$max and during the postabsorptive period at rest, almost all metabolic energy for muscle is supplied by oxidation of peripheral fuels (Andres et al. 1956; Dagenais et al. 1976). These fuels consist most importantly of FFAs mobilized from the adi-

pose tissue and to a lesser extent of glucose released by the liver. At about 25% of $\dot{V}O_2$max, 85% of the energy need is met by FFAs mobilized out of the adipose tissue, and the remaining 15% is supplied in equal measure by intramuscular triglycerides and plasma glucose (Romijn et al. 1993). At 30% to 40% of $\dot{V}O_2$max, circulating FFAs from the peripheral fat depots provide about 40% of energy at the outset and about 60% after 4 hours of exercise, while plasma glucose contributes about 30% of energy (Ahlborg & Felig 1976). Predominance of oxidative metabolism of lipids over oxidation of carbohydrates during low-intensity exercise is a consequence of selective recruitment of oxidative muscle fibers, hormonal and nonhormonal stimulation of metabolism, and activation of enzymes that favor oxidation of fatty acids over oxidation of pyruvate.

Type I and IC red oxidative muscle fibers derive 80% of their metabolic energy from oxidation of endogenous and exogenous TGs (Maggs et al. 1995; Pearce & Connett 1980). The contractile and metabolic functions of muscle fibers are vested in characteristics of slow or fast isoforms of their constituent contractile proteins (Sant'Ana Pereira et al. 1996; Staron & Hikida 1992; Staron & Johnson 1993). Type I and IC muscle fibers differ in that the former expresses only type I slow myosin, whereas the latter also expresses some type IIa fast myosin heavy chains (Staron & Hikida 1992). They attain peak contraction slowly, and their high ATP yield by oxidative phosphorylation of nutrients makes them fatigue resistant (Eberstein & Goodgold 1968). Because they are innervated by the smallest, lowest-threshold α motoneurons, type I fibers are recruited for slow movements at lowest levels of neural stimulation (Horton & Terjung 1988). Their predominant location in the interior of limbs, or attached to the spinal column in rodents (Armstrong & Phelps 1984), underscores their function as extensors in tonic antigravity postural support and in low-intensity to moderate-intensity exercise (Armstrong & Laughlin 1985). In the sedentary human, more deeply located muscles (sartorius, deeper areas of rectus femoris, vastus lateralis, and vastus medialis) and hamstring muscles (biceps femoris) are made up of approximately equal proportions of red and white fibers. A greater proportion of the red fibers in these muscles are used for standing, sitting, or slow movements than is the case in the more superficially located flexors (Edgerton et al. 1975; Johnson et al. 1973).

High oxidative capacity of type I and IC fibers derives from their high mitochondrial density, concentration of oxidative enzymes, and capillarization (Andersen 1975; Baldwin et al. 1972; Essen et al. 1975; Henriksson & Reitman 1976; Winder et al. 1974). Oxidative fibers are richly supplied by intracellular lipid droplets (see table 6.1) in addition to their large store of intracellular glycogen (Essen et al. 1975; Froberg et al. 1971, Froberg & Mossfeldt 1971; Gorski 1990). Capture of circulating FFAs and glucose is ensured by a rich endowment of LPL on vascular endothelia and of HK and GLUT-4 transporters on the membranes of oxidative fibers (Baldwin et al. 1973; Essen et al. 1975). Nutrient uptake also is facilitated by the high sensitivity of these muscle fiber types to metabolic and hemodynamic actions of insulin (Hickey et al. 1995a, 1995b; James et al. 1985a, 1985b, 1986).

Oxidative muscle fibers receive ample exposure to insulin during low-intensity exercise through routing of increased cardiac output through their rich capillary supply. Insulin facilitates diffusion of glucose into muscle by stimulating translocation of GLUT-4 glucose transporters (see chapter 3). Its hemodynamic role is to increase capillary transit time, which is shortened by increased cardiac output during exercise. Insulin vasodilates muscle capillary beds through an NO transduction mechanism (Baron 1994; Baron & Brechtel 1993; Steinberg et al. 1994; Verma et al. 1998; Vicini et al. 1997; Vollenweider et al. 1993). This action increases muscle blood volume (Raitakari et al. 1995) and surface area for optimal exchange of gases, substrates, and metabolites (Saltin et al. 1998). The vasodilatory action is distinct from, and unrelated to, insulin action on glucose uptake, glucose disposal, and stimulation of carbohydrate metabolism and S activity (Anderson et al. 1991; James et al. 1986; Nuutila et al. 1996; Raitakari et al. 1996; Scherrer et al. 1994; Vollenweider et al. 1993).

Glucose uptake by oxidative muscles during low-intensity exercise also is influenced by the exceedingly high affinity of hexokinase (HK) for glucose phosphorylation even in the absence of insulin action. Glucose phosphorylation by HK in the muscle occurs at hypoglycemic concentrations so long as glucose-6 phosphate (G6P) can be used for glycogen synthesis or glycolysis. Enzyme activity is stopped by a near-maximal substrate inhibition when G6P is not used for either process. Oxidative muscles also can take up glucose from circulation in the absence of hormonal influence. Translocation of GLUT-4 glucose transporters from intracellular organelles to the plasma membrane can be activated by muscle contraction without the mediation of insulin (Goodyear 2000; Greiwe et al. 2000; Hayashi et al. 1997; Wojtaszewski et al. 2000) (see chapter 3).

Nutrient mobilization and oxidative metabolism in slow, red muscle fibers are highly sensitive to stimulation by catecholamines (Greenhaff et al. 1991). A rich supply of A-Rs permits lipolysis through noradrenergic stimulation of β A-Rs and by electrical elicitation of muscle contractions (Gorski 1990; Gorski & Stankiewicz-Chorouszucha 1982; Langfort et al. 2000; Liggett et al. 1988). E stimulates LPL activity and FFA uptake in slow oxidative muscle fibers (Miller et al. 1989). Stimulation of β_2 A-Rs also facilitates glycogenolysis, lactic acid production, and glucose oxidation to supplement the activation of these processes by muscle contraction alone (Cleroux et al. 1989; Greenhaff et al. 1991; Kjaer et al. 2000; Williams et al. 1984).

Hormonal environment conducive to high lipid oxidation in slow red fibers during low-intensity exercise consists of increased S tone and a twofold increase in plasma E concentration (see figure 6.4c), which suppress insulin

secretion (see figure 6.4a) (Clutter et al. 1980; Hartley et al. 1972a; Romijn et al. 1993). S nerves discharge at a rate of about one nerve impulse per second at such low exercise intensities, and this discharge stimulates adipose tissue lipolysis without eliciting vasoconstriction (Fredholm & Rosell 1968; Rosell 1966). Noradrenergic stimulation of adipocyte β_1 A-Rs overrides the antilipolytic effect of α A-Rs that prevails at rest (Arner et al. 1990; Wahrenberg et al. 1987). NE action is almost exclusively lipolytic and produces minimal or transient glycogenolysis in liver or muscle (Cleroux et al. 1989; Froberg et al. 1975; Havel & Goldfien 1959). As blood flow through adipose tissue is largely unimpeded, FFA can bind to albumin for export into systemic circulation (Ngai et al. 1966; Rosell 1966).

Withdrawal of insulin action and SA activation to a lesser extent stimulate hepatic glycogenolytic and gluconeogenic enzymes at low exercise intensities (Garceau et al. 1984; Hartmann et al. 1982; Lautt 1979a, 1979b; Lautt & Wong 1978; Shimazu & Ogasawara 1975; Shimazu & Usami 1982). Contractions of oxidative muscle fibers, in the absence of all hormonal stimulation, also can activate HSL as well as phosphorylase (Langfort et al. 1999, 2000). Thus, lipolysis and hepatic glucose production increase during low-intensity contractions in oxidative muscle fibers, either stimulated by hormones or in the absence of direct hormonal mediation.

A shift toward greater oxidation of lipids in oxidative muscle fibers at low exercise intensities is a carbohydrate-sparing mechanism that directs metabolism toward more plentiful lipids whenever the hormonal environment is permissive and carbohydrate utilization is not obligatory. Circulating glucose and liver and muscle glycogen are the only endogenous sources of carbohydrates. Only about 14 g of glucose are available in the extracellular fluid compartment. Plasma glucose concentration is regulated around 90 to 100 mg/dl or about 5 mM/L to fuel brain metabolism and support intense muscle contraction. It is replenished by glycogenolysis of about 80 to 100 g of available liver glycogen, half of which is used up during an overnight fast and 85% of which is depleted in 24 hours of normal activity (Newsholme & Leech 1983; Sugden et al. 1993). A slower and bioenergetically more costly process of hepatic glucose production by gluconeogenesis maintains blood glucose at times of high lipid oxidation (Ruderman et al. 1969; Scrutton & Utter 1967) (see figure 6.5). An additional 300 to 500 g of available muscle glycogen are utilized by muscle in obligatory fashion at higher exercise intensities. The limited size of hepatic and muscle glycogen stores is most likely a consequence of the bioenergetic cost of storing and moving large quantities of a low-energy fuel. The storage form of glycogen yields only about 1 kcal per g as it is bound to 2.7 times its weight of water molecules. By contrast, between 10 and 20 kg of high-energy TGs are stored in subcutaneous and intra-abdominal adipose tissue sites and yield about 9 kcal/g. A shift toward lipid oxidation substitutes the more plentiful and energy-rich lipids for the scarce and energy-poor carbohydrates.

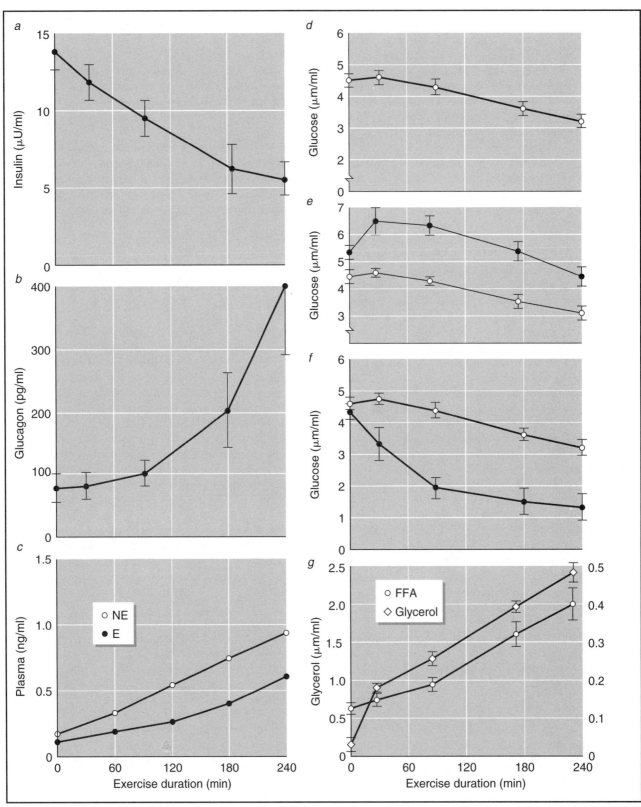

Figure 6.4 *Time course of plasma hormone and metabolite concentrations during prolonged low-intensity exercise.*

During 4 hours of exercise at 30% of V̇O₂max, plasma insulin (*a*) gradually declines and plasma glucagon (*b*) increases exponentially after 2 hours in parallel with decreases in plasma glucose *(d)*. Hepatic glucose production *(e)* and glucose uptake by the leg muscles *(f)* can be seen from the differences between arterial (open symbols) and venous (solid symbols) glucose concentration. Concentrations of plasma free fatty acids (FFAs) (circles in *g*) and glycerol (diamonds in *g*) rise in parallel with those of epinephrine (E) and norepinephrine (NE) *(e)*.

Figures *a*, *b*, *d*, *e*, *f*, and *g* adapted from Ahlborg and Felig 1976; figure *c* adapted from Saltin and Gollnick 1988.

At the cellular level, high lipid oxidation is controlled by an adequate supply of FFAs and activation of metabolic enzymes through changes in concentration of metabolic substrates (see figure 6.5). FFAs are supplied to muscle cytosol through hydrolysis of intramuscular TGs, facilitated diffusion of circulating FFAs mobilized from peripheral fat depots, and hydrolysis by LPL of TGs in circulating lipoproteins. FFAs are converted to metabolically active fatty acyl-CoA (FAcCoA) by the long-chain fatty acyl synthetase (Fas). Their transport into a mitochondrion is controlled by carnitine palmitoyl transferase (CPT I), the rate-limiting enzyme of lipid oxidation (McGarry 1995). The activity of this enzyme increases when its inhibition by malonyl-CoA, an early metabolite in FFA synthesis, declines. Malonyl-CoA is the first step in FFA biosynthesis, catalyzed by acetyl-CoA carboxylase (ACC) under circumstances, including insulin stimulation, that favor pyruvate oxidation and FFA synthesis. For FFA synthesis, acetyl-CoA, the substrate of malonyl-CoA reaction, is exported out of the mitochondrion by being converted to citrate. Citrate inhibits the key glycolytic enzyme PFK1. Acetyl-CoA and NADH, on the other hand, block glucose oxidation by inhibiting the activity of PDH (Randle et al. 1963, 1965). During exercise and fasting, the activity ACC declines (Oakes et al. 1997a; Odland et al. 1996; Winder et al. 1990). As a consequence, increased flux of FAcCoA through the mitochondrial oxidative machinery blocks enzymes controlling glycolysis and carbohydrate oxidation.

Carbohydrate sparing during prolonged low-intensity exercise can occur in the absence of endocrine and SA facilitation. In addition to directly activating of phosphorylase and HSL, muscle contractions can reduce

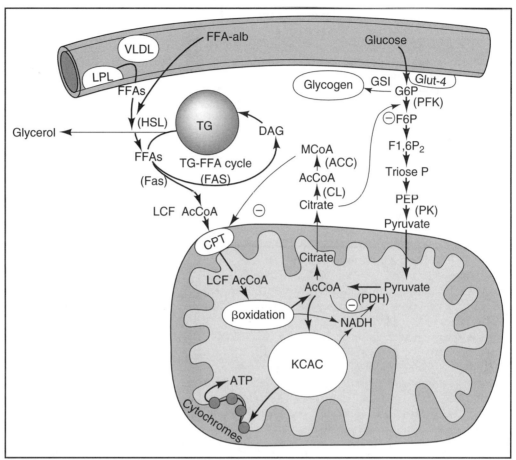

Figure 6.5 *Interactions between lipid and carbohydrate metabolism during exercise.*

Key enzymes of glycolysis (phosphofructokinase [PFK] and pyruvate kinase [PK]) and carbohydrate oxidation (pyruvate dehydrogenase [PDH]) are activated during intense exercise and after intake of carbohydrates. Stimulation of these enzymes is triggered by muscle contraction in the former and by insulin in the latter. During prolonged submaximal exercise or during a fast, hormonal and metabolic signals block carbohydrate and increase fat oxidation. Hormonal signals include stimulation by the catecholamines, cortisol, and growth hormones (GH) of hormone sensitive lipase (HSL) and lipoprotein lipase (LPL) and decline in insulin's antilipolytic actions.

TG = triglyceride droplet; VLDL = very low density lipoprotein; LCF AcCoA = long-chain fatty acid coenzyme A; CPT = carnitine palmitoyl transferase; KCAC = Krebs citric acid cycle; ATP = adenosine triphosphate; MCoA = malonyl coenzyme A; AcCoA = acetyl coenzyme A; ACC = acetyl coenzyme A; carboxylase; CL = citrate lyase; PDH = pyruvate dehydrogenase; NADH = nicotine adenine dinucleotide, reduced form; PEP = phosphoenol pyruvate; G6P = glucose 6 phosphate; F6P = fructose 6 phosphate; F1,6P$_2$ = fructose 1,6 diphosphate; Fas = fatty acyl synthetase.

the activity of ACC and concentrations of malonyl-CoA, more so in type I than in type IIA muscle fibers, even in the absence of SA influence (Langfort et al. 1999, 2000; Winder et al. 1993). Activity of ACC can be reduced in isolated or denervated limb muscles made to contract by electrical stimulation (Duan & Winder 1992, 1993; Saha et al. 1995). These metabolic changes are independent of intracellular concentrations of citrate and acetyl-CoA and preferentially occur at lower rather than higher contraction intensities (Dyck et al. 1996b).

Intermediate Intensity Exercise

Fuels utilized during intermediate exercise intensities of between 50% and 75% $\dot{V}O_2$max are in equal measure derived from endogenous and peripheral energy stores. Because the availability of fuel energy in circulation remains the same at all exercise intensities (approximately 0.075 kcal/kg·min) (Romijn et al. 1993), increases in energy expenditure in support of a higher workload are now derived also from intramuscular lipids and glycogen. Approximately half of expended energy is derived from carbohydrates, and the rest is derived from lipids at the onset of exercise performed at about 65% of $\dot{V}O_2$ max. Muscle glycogen contributes approximately 80% and blood sugar only about 20% to carbohydrate utilization. Hepatic glucose output is about three times higher at this intermediate intensity than at rest, and gluconeogenesis contributes only 13% to 15% of it (Ahlborg & Felig 1982; Ahlborg et al. 1974; Wahren et al. 1971). Adipose tissue FFAs contribute a little more than half of the lipid fuel (70% of what is utilized at low exercise intensity), and the rest is supplied by intramuscular TGs. The upper body or abdominal subcutaneous fat depots contribute FFAs to a greater extent than the gluteal or femoral subcutaneous fat depots (Arner et al. 1981, 1995b; Horowitz et al. 2000a, 2000c). By contrast, intramuscular TG hydrolysis is six times greater than at low exercise intensity (Carlson et al. 1971; Essen 1977, 1978). This pattern of fuel use is a consequence of the metabolic characteristics of the recruited type IIA muscle fibers, SA and metabolic messengers elicited at this exercise intensity, and metabolic events triggered by muscle contractions.

A dual complement of glycolytic as well as oxidative metabolic enzymes, coupled to high hormone sensitivity, permits fast, red, type IIA muscle fibers to switch between anaerobic and oxidative metabolism. This metabolically versatile muscle fiber type can respond to either hormones or contraction-induced signals to accommodate energy demands of exercise of different intensity. It includes fibers designated as IC, IIC, IIAC, and IIA that respectively express type I and type IIa myosin heavy chains as follows, more type I than type IIa, equal proportions of both, more type IIa than type I, and type IIa only (Staron & Hikida 1992). Type I and IIA fibers have greater density of mitochondria, myoglobin, and capillaries than fast, glycolytic, white, type IIX or IIB fibers and receive proportionately more blood when they are recruited (Laughlin & Armstrong 1982, 1985; Krotkiewski 1994). Type IIA fibers comprise between 40% and 45% of fibers in the vastus lateralis of untrained individuals (Hickey et al. 1995a; Jansson & Kaijser 1987). In endurance trained cyclists, type IIA fibers comprise only 26% of fibers, whereas type I fibers are more numerous (70% versus 40%) and type IIX (IIB) fibers less so (4% versus 14%) than in untrained individuals (Jansson & Kaijser 1987). They are innervated by intermediate-size motoneurons and are recruited at intermediate exercise intensities of approximately 50% to 75% of $\dot{V}O_2$ max in accordance with the inverse relationship between motoneuron size and motor unit excitability (Costill et al. 1973; Gollnick et al. 1973a, 1973b, 1974a, 1974b; Henneman & Mendell 1981).

Type IIA muscle fibers contain high concentrations of HK, and their glucose uptake and glycogen synthesis are sensitive to stimulation by insulin (Hickey et al. 1995a, 1995b; James et al. 1985a, 1985b, 1986). Insulin stimulates their uptake of circulating glucose. Their rich store of glycogen (see table 6.1) gives them the capacity to initiate glycogenolysis in response to catecholamine stimulation as well as to direct neural activation (Gorski 1990; Greenhaff et al. 1991; Reitman et al. 1973). Because of their support of prolonged, glycogen-depleting moderate-intensity exercise, type IIA and type I muscle fibers undergo glycogen supercompensation during recovery from exercise.

This muscle fiber type also is sensitive to catecholamines (Greenhaff et al. 1991). It has a large store of intracellular TGs (see table 6.1) and an LPL sensitive to both E and NE stimulation (Gorski 1990; Gorski & Stankiewicz-Chorouszucha 1982; Miller et al. 1989; Stankiewicz-Chorouszucha & Gorski 1978). Its large capacity to store and utilize lipids suggests its probable involvement in increased lipid extraction and lipid oxidation after adaptation to high-fat diets (Jansson & Kaijser 1982, 1987). In type IIA muscle fibers, muscle contractions can directly increase lipolysis, in the absence of hormonal stimulation, by producing faster and larger reduction of malonyl-CoA than in type I muscle fibers (Winder et al. 1990). Thus, type IIA muscle fibers are equipped to both take up lipids from circulation and utilize endogenous TGs.

At moderate exercise intensities of about 50% $\dot{V}O_2$max, E secretion from the adrenal medulla augments the effects of NE released at or spilled out of S nerve terminals in the muscle, the liver, and the adipose tissue (see figure 6.6). Plasma E and NE concentrations are increased fourfold to sixfold and contribute to increased muscle glycogenolysis and lipolysis (Cleroux et al. 1989; Jansson et al. 1986; Langfort et al. 2000; Romijn et al. 1993). E stimulates lipolysis at plasma concentrations as low as three times its basal, unstimulated level (75 to 125 versus about 25 pg/ml), although liver and muscle glycogenolysis are its more prominent actions (Froberg et al. 1975; Galster et al. 1981; Havel 1959). Other gluconeogenic hormones (GH) and cortisol also are released at intermediate exercise intensities (Hodgetts et al. 1991; Galbo et al. 1977c). Catecholamine-induced fall in plasma insulin increases liver sensitivity to glucagon so that any reduction in plasma glucose elicits glucagon release, and small increases in plasma glucagon concentration increase the rate of glucose appearance (Lins et al. 1983). During moderately intense exercise of several hours duration, both the adipose tissue lipolysis and blood flow through the adipose tissue progressively increase about threefold in response to β adrenergic stimulation (Bulow 1982; Bulow & Madsen 1976).

Recovery From Prolonged, Low-Intensity, and Moderate-Intensity Exercise

After 2 hours of low-intensity exercise, liver glycogen and plasma glucose concentrations decline (see figure 6.4d). The liver produces about 75 g of glucose during 4

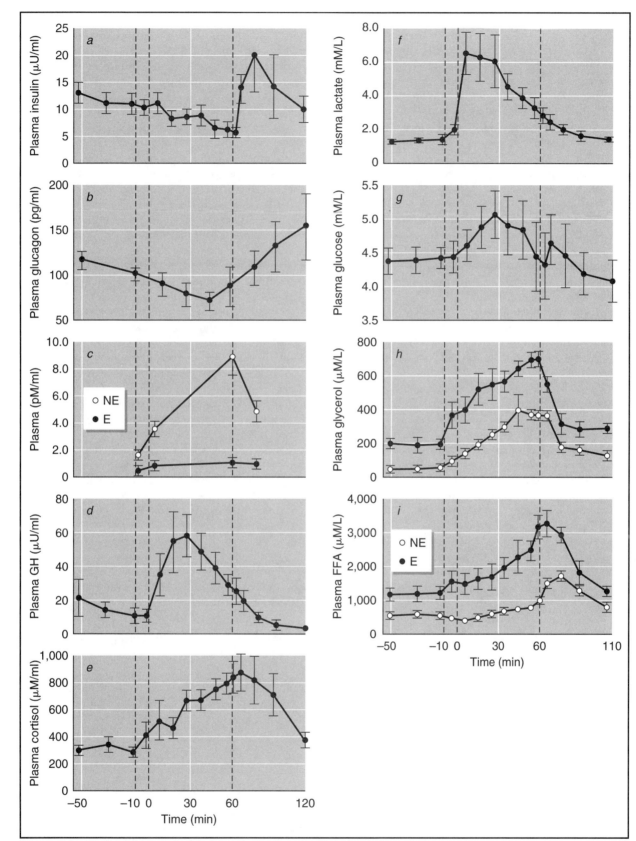

Figure 6.6 *Endocrine control of fuel use during intermediate-intensity exercise.*

Hormonal changes that accompany 1 hour of exercise at 50% to 70% of V̇O₂max are shown on the left (*a* through *e*) and metabolic changes on the right (*f* through *i*). Plasma glucagon (*b*) was measured during exercise at 60% of V̇O₂max to exhaustion. The time course of release by subcutaneous adipose tissue of free fatty acids (FFAs) *(i)* and glycerol (*h*) are shown as the difference between arterial (open circles) and venous (solid circles) concentration of these metabolites.

Figures *a, c, d, e, f, g, h,* and *i* adapted from Hodgetts et al. 1991; figure *b* adapted from Galbo, Christensen, and Holst 1977.

hours of low-intensity exercise, all but 15 to 20 g of it by glycogenolysis (Ahlborg et al. 1974). After four hours of such exercise, contribution of gluconeogenesis to hepatic glucose output increases to 45% (Ahlborg & Felig 1982; Ahlborg et al. 1974). Sustained α_1 adrenergic stimulation and a fall in plasma glucose stimulate glucagon secretion (see figure 6.4b) (Ahlborg & Felig 1977), while β adrenergic stimulation raises FFA concentration (see figure 6.4g) (Romijn et al. 1993). Glucagon facilitates splanchnic uptake of gluconeogenic precursors, but plasma glucose concentration declines, as hepatic gluconeogenesis cannot keep pace with peripheral glucose uptake (Hartley et al. 1972b). Prolonged low-intensity exercise depletes glycogen from type I but not type II muscle fibers (Essen 1977; Gollnick et al. 1974b).

If moderate-intensity exercise is carried out over 2 hours, the relative contribution of muscle-derived and circulating fuels changes. Circulating FFAs now contribute twice as much energy as muscle triglycerides, while a small increase in the utilization of circulating glucose is balanced by a small decrease in muscle glycogen utilization (Romijn et al. 1993). During 3 hours of intermediate-intensity exercise, the liver produces about 75 g of glucose, and both liver and muscle glycogen stores decline (Ahlborg et al. 1974; Ahlborg & Felig 1982; Coyle et al. 1986).

During prolonged moderate-intensity exercise and after carbohydrate stores depletion, low plasma insulin and increased plasma cortisol concentrations facilitate protein breakdown. Disproportionate release of alanine by muscle during exercise is the outcome of branched-chain amino acid transamination and is not related to muscle amino acid composition (Ahlborg et al. 1974;

Felig 1973). Alanine release by muscle is proportional to exercise intensity (Felig & Wahren 1971). High plasma concentrations of glucagon facilitate hepatic uptake of alanine. In the liver, transamination of the amino group from alanine to α-ketoglutarate produces, respectively, pyruvate and glutamate. Pyruvate then is converted to glucose to complete the gluconeogenic glucose-alanine cycle (see figure 6.2) (Felig 1973). High titers of catecholamines and other counter-regulatory hormones also increase the supply of glycerol, lactate, and alanine for gluconeogenesis (Ahlborg & Felig 1982; Ahlborg et al. 1974; Wahren et al. 1971). As is the case with prolonged low-intensity exercise, splanchnic uptake of gluconeogenic precursors and their conversion to glucose account for about 47% of hepatic glucose output, but this process fails to produce enough glucose to meet the needs for peripheral glucose uptake and maintenance of euglycemia (Ahlborg et al. 1974; Ahlborg & Felig 1982).

Hypoglycemia and depletion of liver and muscle glycogen caused by prolonged submaximal exercise trigger the SA and endocrine responses that influence fuel metabolism during recovery from such exercise. Plasma glucose and insulin concentrations remain low and increase slowly during the initial stage of recovery from prolonged, submaximal exercise (see figure 6.7). Glucose uptake and glycogen resynthesis in the previously exercised muscle are increased threefold to fivefold for the next several hours and take precedence over hepatic glycogen resynthesis (Ahlborg & Felig 1982; Ahlborg et al. 1986; Fell et al. 1980). At least five physiological processes favor muscle glycogen resynthesis over hepatic glycogen synthesis after prolonged submaximal exercise:

1. Intestinal absorption of glucose increases (Hamilton et al. 1996; Maehlum et al. 1978).

2. Muscle is more sensitive to insulin.

3. The lower Km of HK relative to hepatic GK favors insulin-dependent and insulin-independent glucose uptake and glycogen synthesis by the muscle (Fell et al. 1980; Hamilton et al. 1996; Maehlum et al. 1978).

4. Gluconeogenesis is less sensitive than glucogenolysis to inhibitory actions of insulin (Chiasson et al. 1976; Felig & Wahren 1971).

5. High plasma concentrations of glucagon and catecholamines are maintained (Ahlborg & Felig

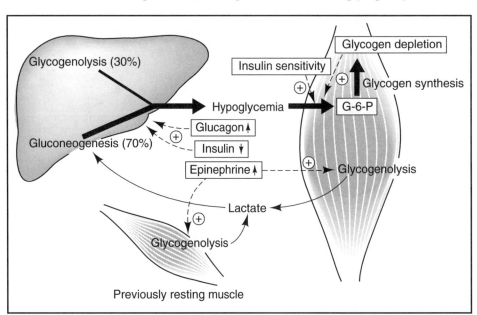

Figure 6.7   *Endocrine control of metabolism during recovery from prolonged exercise.*
High plasma concentrations of counter-regulatory hormones stimulate a high level of hepatic glucose production by gluconeogenesis and lactate from the previously resting muscle. Heightened insulin sensitivity facilitates glucose uptake by the muscle, and increased insulin sensitivity of glycogen-depleted muscle and lower Km of hexokinase relative to glucokinase result in preferential glycogen resynthesis in the muscle.

Reprinted from Bjorkman and Wahren 1988.

1982) and stimulate hepatic gluconeogenesis and glycogenolysis and lactate output from previously resting muscle.

Muscle glycogen synthesis is thus aided by hepatic release of glucose, 65% to 80% of which is derived by gluconeogenesis (Ahlborg et al. 1986; Ahlborg & Felig 1982). An acute postexercise release of ketones is more evident when plasma glucose, alanine, and insulin concentrations are low (Koeslag et al. 1980, 1982).

Muscle sensitivity to insulin is increased during and for about 48 hours after a single bout of prolonged low-intensity to moderate-intensity exercise (Araujo-Vilar et al. 1997; Devlin et al. 1987; Devlin & Horton 1985; Mikines et al. 1988). Adipocyte insulin sensitivity, glucose uptake, and GLUT-4 translocation to their membranes are increased (Hirshman et al. 1989; Vinten & Galbo 1983; Vinten et al. 1985). Increased sensitivity is maintained if individuals engage in regular endurance exercise but is lost within 6 or more days of inactivity (Heath et al. 1983; Pratley et al. 1995). The increase in insulin sensitivity is a consequence of increased insulin binding to its receptors; a change in postreceptor signaling (see chapter 3); insulin-dependent and insulin-independent glucose uptake mechanisms, including GLUT-4 translocation to the membrane; and increased glucose utilization for glycolysis or glycogen synthesis (Devlin 1992; Horton 1992; Houmard et al. 1995; Prigeon et al. 1995). The relative contributions of changes in insulin sensitivity and of hormone-independent muscle contraction to increased muscle glucose uptake are discussed in chapter 3.

When muscle glycogen concentration declines to less than 30 mM, muscle glycogen resynthesis is insulin independent and proceeds at rates that are proportional to the magnitude of glycogen depletion (Casey et al. 1995; Conley et al. 1978). High concentrations of counter-regulatory catecholamines, GH, and a high glucagon/insulin ratio stimulate glycogenolysis in the resting muscle (Conley et al. 1978). This process provides lactate for the Cori cycle and contributes substrates for glycogen resynthesis in previously exercised muscle (Ahlborg 1985; Blomstrand & Saltin 1999). During the early stage of recovery, muscle glucose is taken up rapidly as the substrate inhibition of HK by G-6-P is reduced. With the gluconeogenic supply of glucose, muscle glycogen resynthesis takes precedence over hepatic glycogen resynthesis (Fell et al. 1980; Hamilton et al. 1996; Maehlum et al. 1978). As muscle glycogen content approaches repletion, glycogen synthesis becomes insulin dependent (Price et al. 1994).

Besides its effect on muscle and liver glycogen, prolonged moderate to strenuous exercise also results in muscle TG depletion (Carlson et al. 1971; Kiens et al. 1989). After such exercise, FFA reesterification and TG resynthesis are acutely inhibited probably because of low plasma insulin and increased plasma concentrations of counter-regulatory hormones (Hodgetts et al. 1991; Savard et al. 1987a; Wolfe et al. 1990). A prompt decline in

lipolysis is accompanied by a transient rebound increase in plasma FFA concentrations caused by removal of circulatory restraint over FFA transport out of adipose tissue (see figure 6.8, right center panel) (Jones et al. 1980; Hagenfeldt 1979; Hagenfeldt & Wahren 1972; Hodgetts et al. 1991; Wahren et al. 1975). The circulatory change reflects a decrease in S tone and adrenal catecholamine secretion. Lipid-depleted muscles exhibit an increase of several hours duration in the activity of LPL, which is not immediate but becomes apparent about 3 to 4 hours after termination of exercise (Greiwe et al. 2000; Ladu et al. 1991; Lithell et al. 1979a, 1979b, 1984; Nikkila et al. 1978; Savard et al. 1987a; Savard & Greenwood 1988; Taskinen et al. 1980). Within 1 day of exercise, transcription of muscle LPL is also increased (Seip & Semenkovich 1998). Postexercise increases in muscle LPL activity are accompanied by decreases in plasma concentration of LDL cholesterol and other low-density lipoprotein species (Nikkila et al. 1978; Taskinen et al. 1980). While insulin inhibits LPL activity in inactive muscle in proportion to its stimulation of glucose uptake, this insulin action is blocked in lipid-depleted exercised muscle, revealing again the operation of a hormone-independent mechanism (Taskinen et al. 1980). Contraction-induced increases in LPL are associated with increased plasma E concentrations but do not depend on catecholamine stimulation (Greiwe et al. 2000; Lithell et al. 1981; Miller et al. 1989). Thus, analogous to postexercise glycogen repletion, substrate depletion both activates LPL and facilitate FFA uptake and TG resynthesis in lipid-depleted muscle (Pollare et al. 1991).

Prolonged exercise increases adipose tissue capacity for lipid utilization. After endurance training, an increased lipolytic response to catecholamines is observed in adipocytes at high hormone concentrations of 10^{-6} mol/L but not at lower, more physiological concentrations of 10^{-9} mol/L or in adipose tissue in situ (Bukowiecki et al. 1980; Crampes et al. 1986, 1989; Despres et al. 1984a, 1984b; Stallknecht et al. 1995; Wahrenberg et al. 1987). Increased capacity for whole-body lipid oxidation after endurance training does not appear to be a consequence of increased adipose tissue lipolysis, catecholamine secretion, or adipose tissue sensitivity to catecholamines (Galbo et al. 1977b; Gorski et al. 1990; Henriksson 1977; Holloszy & Coyle 1984; Horowitz et al. 1999a, 1999b, 2000; Horowitz & Klein 2000a; Jansson & Kaijser 1987; Klein et al. 1994b, 1995; Winder et al. 1979). Instead, it most likely results from increased utilization of intramuscular TGs and increased E-stimulated blood flow to the adipose tissue (Horowitz et al. 2000c; Hurley et al. 1986; Martin et al. 1993; Phillips et al. 1996c; Stallknecht et al. 1995).

High-Intensity Exercise

At high exercise intensities above 80% of $\dot{V}O_2$max, only about one quarter of energy for exercise is supplied by lipids, in approximately equal measure by TGs of peripheral adipose tissue origin and muscle (Romijn et al.

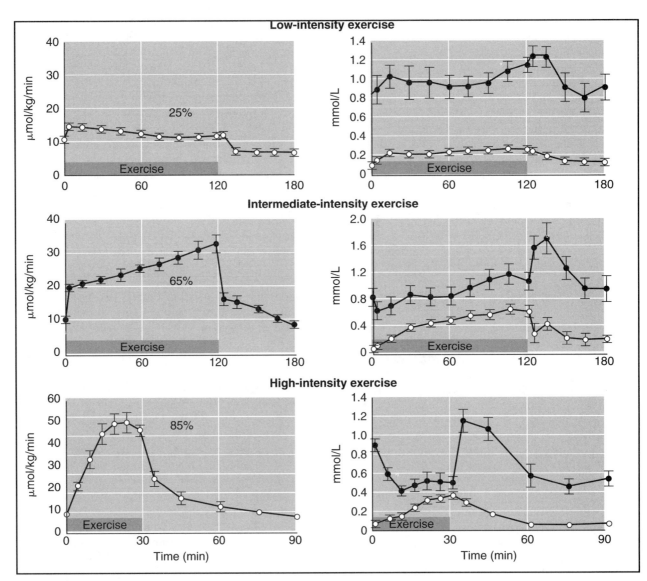

Figure 6.8 *Time course of hepatic glucose production and adipose tissue lipolysis at different exercise intensities.*

The rates of glucose appearance in plasma during exercise at three different intensities are shown on the left, and the rates of glycerol (open symbols) and free fatty acid (FFA) (solid symbols) appearance are shown on the right.

Adapted from Romijn et al. 1993.

1993). Despite high rates of lipolysis at high exercise intensities, FFA entry into circulation is blocked (see figure 6.8, lower right panel) by the vasoconstrictive action of high titers of catecholamines over splanchnic circulation (Jones et al. 1980). Leg uptake of circulating FFAs is thus inversely proportional to exercise intensity (Ahlborg et al. 1974). The rising concentration of lactic acid also exerts a metabolic restraint over whole-body lipolysis (Boyd et al. 1974; Issekutz et al. 1975).

Fast, white, glycolytic, type IIX (previously designated IIB) muscle fibers are preferentially recruited during intense, dynamic exercise at intensities above 85% of V̇O₂ max, with a lesser participation of type IIA fibers (see figure 6.9). The high contractile power of type IIX muscle fibers is vested in their high concentration of glycolytic enzymes, including PDH, and is achieved

through neural activation of ATP hydrolysis and anaerobic breakdown of glucose (Baldwin et al. 1973; Greenhaff et al. 1991; Sugden et al. 1993). These fibers reach peak contraction rapidly, but because of their reliance on anaerobic metabolism with its poor ATP yield, they fatigue rapidly. Their other features are high glycogen and low lipid content, insensitivity to insulin and catecholamines, and low capillary density (Andersen 1975; Essen 1977; Essen et al. 1975; Greenhaff et al. 1994; Saltin et al. 1977). These muscle fibers are prevalent in the superficial flexor muscles recruited for the higher-intensity phasic contractions (Armstrong & Laughlin 1985). The superficial areas of human rectus femoris, vastus medialis, and vastus lateralis have about twice as many white, glycolytic, type IIX muscle fibers (63%) as the more internal regions that have a greater proportion

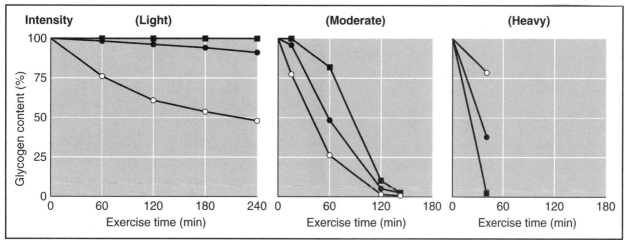

Figure 6.9 *Glycogen utilization by different muscle fiber types at different exercise intensities.*
■ = type IIX fibers; ● = type IIA fibers o = type I fibers.
Reprinted from Saltin and Gollnick 1988.

of slow, oxidative, type I fibers (Edgerton et al. 1975; Johnson et al. 1973).

At exercise intensities above 80% V̇O₂max, catecholamines, GH, and glucagon reach their highest plasma concentrations (see figure 5.2), which in the case of plasma E and NE are 17 to 20 times higher than at rest (Romijn et al. 1993). Despite its high plasma concentration at this exercise intensity, plasma E does not influence muscle glycogenolysis, as can be expected during exercise that predominantly engages catecholamine-insensitive type IIX muscle fibers (Chesley et al. 1995; Greenhaff et al. 1991). The metabolic consequence of high titers of counter-regulatory hormones is hyperglycemia and a high hepatic rate of glucose production that is directly proportional to their plasma concentrations (see figure 6.8) (Wahren et al. 1971). At high exercise intensities, about three quarters of total energy cost of exercise is supplied by glucose and four fifths of it is produced by muscle glycogenolysis. Hepatic glucose output supplies only one fifth of total energy from carbohydrates. Of that, only about 6% is derived from gluconeogenesis, and the rest is derived from glycogenolysis (Ahlborg & Felig 1982; Ahlborg et al. 1974; Wahren et al. 1971). Selective glycogen depletion in type IIX muscle fibers occurs when high-intensity exercise is carried out to exhaustion (see figure 6.9, left panel).

As S nerves increase their discharge rates to greater than three impulses per second, intense vasoconstriction, mediated by α adrenergic action, entraps FFAs within the adipose tissue (Fredholm & Rosell 1967, 1968; Gullestad et al. 1993; Ngai et al. 1966; Oro et al. 1965). Peripheral arterioles and venules in all but white muscle blood vessels constrict in response to activation of α_{1B}, α_{1D}, and α_{2D} adrenoceptors (Leech & Faber 1996). In the fast, white muscles, the vasoconstriction is overridden by declining pH caused by anaerobic production of lactic acid (Thomas et al. 1994). In addition to the catecholamine-induced vasoconstriction at higher exercise intensities, rising FFA/albumin ratios and the plasma

lactic acid concentration in adipose tissue also exert a vasoconstrictive effect (Bulow et al. 1985; Fredholm 1971; Gorski 1977; Hales et al. 1978).

During circulatory restriction, catecholamines stimulate the TG–fatty acid substrate cycle, which consists of increased lipolysis and reesterification. This process permits a more rapid lipolytic response any time exercise intensity declines when the blood flow through the adipose tissue again increases (Wolfe et al. 1990). A threefold increase in lipolysis at moderate exercise intensities produces a sixfold increase in FFA availability for oxidation when half of mobilized FFAs are released into circulation instead of being reesterified. As the need for lipid oxidation declines upon termination of exercise, S vasoconstriction abates, FFAs overflow into systemic circulation, and faster reesterification in adipose tissue is restored (Rosell 1966; Wolfe et al. 1990).

Muscle glycogenolysis provides about 60% of energy for exercise at the intensity of 85% of V̇O₂ max (Romijn et al. 1993). It can be activated during near-maximal exercise in the absence of SA, pituitary, and pancreatic hormones by motoneuron-triggered muscle contractions (Gollnick et al. 1970; Greenhaff et al. 1991), which can directly activate phosphorylase (Langfort et al. 2000). A rapid pace of ATP utilization by contracting fast-twitch muscles increases production of AMP and stimulates the activity of AMP-activated protein kinase (AMPK). These signals stimulate glycolysis through activation of PFK_2, increased production of fructose 2,6 diphosphate, and stimulation of the rate-limiting glycolytic enzyme PFK_1. PFK_1 activation during intense exercise occurs despite high levels of citrate, a metabolite that strongly inhibits its activity at rest (Dyck et al. 1996b). Intense exercise, in the absence of increased plasma insulin, also stimulates the activity of PDH, the rate-limiting enzyme catalyzing irreversible conversion of pyruvate to acetyl-CoA (Constantin-Teodosiu et al. 1991, 1992; Dyck et al. 1993; Spriet et al. 1992). Thus, in contrast to the situation at rest, when activation of PDH activity depends on insulin, intense exercise stimulates this enzyme through a contraction-induced mechanism.

Metabolic by-products of a high glycolytic rate during intense exercise also can stimulate carbohydrate and inhibit lipid utilization in muscle in the absence of hormonal mediation (see figure 6.5). At high exercise intensities, acetyl-CoA of glycolytic origin is both diverted toward lactate and buffered by carnitine to produce acetyl carnitine and reduced coenzyme A (CoASH) (Carter et al. 1981; Spriet et al. 1992). The increased supply of CoASH facilitates glycolytic flux into the Krebs citric acid cycle. A high glycolytic flux of acetyl-CoA at high exercise intensities also activates ACC to produce malonyl-CoA. Malonyl-CoA exerts a strong inhibitory influence over muscle carnitine palmitoyl transferase I (CPT), the enzyme that regulates fat oxidation by allowing FFA entry into the mitochondria. Here again, exercise suppresses lipid and activates carbohydrate metabolism by a mechanism that does not require a high insulin-to-glucagon ratio as is the case at rest. Finally, high concentrations of lactate, a metabolic product of high-intensity exercise, suppress lipolysis and thus indirectly facilitate carbohydrate metabolism (Fredholm 1971; Gorski 1977).

Recovery From High-Intensity Exercise

A rebound increase in plasma insulin concentration that results from decreased S tone occurs within minutes of termination of high-intensity exercise (see figure 6.10) and is accompanied by rapid disappearance of catecholamines from plasma (Hagberg et al. 1979; Wahren et al.

1973). Increased plasma insulin concentration facilitates glucose uptake by recovering types IIX and IIA muscle fibers. Glucose uptake by muscle declines rapidly, but for about an hour remains three to four times higher than resting (Wahren et al. 1973). Splanchnic glucose output falls gradually, reaching basal levels within 40 minutes. During recovery, gluconeogenesis contributes about 40% of hepatic glucose output, which is sevenfold greater than during intense exercise (Ahlborg & Felig 1982; Wahren et al. 1973). This activity is facilitated by a continuing high plasma glucagon concentration, high concentrations of gluconeogeneic precursors, and increased splanchnic blood flow. Oxidative metabolism of lipids, ketone bodies, and lactic acid is gradually replaced by oxidation of mixed fuels as concentrations of insulin and counter-regulatory hormones decline.

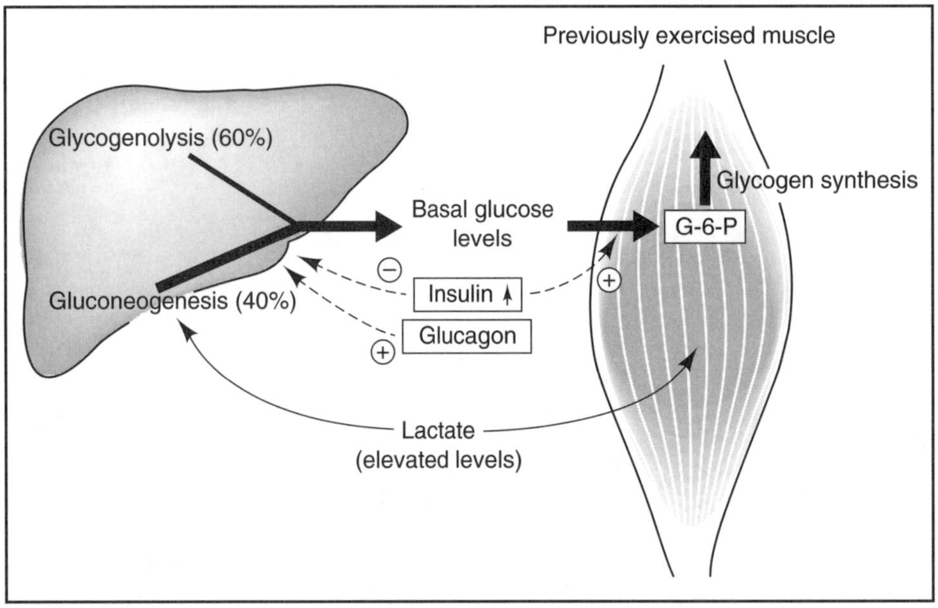

NUTRITIONAL MODULATION OF METABOLISM

Fuel use during exercise depends not only on exercise intensity but also on the level of depletion of endogenous fuels and the recency, type, and duration of particular macronutrient intake. These nutritional variables alter the autonomic and endocrine control of fuel use during exercise.

Carbohydrate Supplementation

Carbohydrate supplementation increases carbohydrate uptake and oxidation and inhibits lipid oxidation in the muscle (Kelley et al. 1990). The metabolic shift toward increased carbohydrate oxidation is mediated by increased post-ingestive insulin secretion, which inhibits the release and metabolic action of catecholamines and other counter-regulatory hormones. Insulin stimulates carbohydrate oxidation by increasing the activity of glycolytic enzymes and PDH (Mandarino et al. 1987). Circulating FFA availability declines because of antipolytic effects of insulin (Kelley et al. 1993). Insulin inhibits fat oxidation by stimulating malonyl-CoA synthesis by ACC (see figures 6.3 and 6.5) (Bavenholm et al. 2000; Duan & Winder 1993; Saha et al. 1995). Malonyl-CoA inhibits CPT I and reduces the concentration of long-chain fatty acyl-CoA (Oakes et al. 1997a).

When between 0.75 and 3 g of glucose per kilogram of body weight is taken at

Figure 6.10 *Endocrine control of metabolism during recovery from short-term moderate-intensity to high-intensity exercise.*

During the first hour of recovery, plasma concentrations of glucagon and gluconeogeneic precursors and splanchnic blood flow are all high. Plasma insulin concentration rapidly rebounds, and catecholamine concentration rapidly declines. This condition facilitates hepatic glucose output (about sevenfold greater than during intense exercise, and about 40% of it is derived from gluconeogenesis) and muscle glucose uptake.

Reprinted from Bjorkman and Wahren 1988.

the onset or during low-intensity to intermediate-intensity exercise, plasma glucose and insulin concentrations increase to about twice the normal level throughout exercise, and circulating glucose utilization by muscle is doubled (Ahlborg & Felig 1976, 1977; Horowitz et al. 1997). As a result of hyperinsulinemia and hyperglycemia (see figure 6.11), the secretion of glucagon and catecholamines is suppressed or blocked, and lipid mobilization and utilization are proportionally reduced (Coyle et al. 1997; De Glisezinski et al. 1998; Horowitz et al. 1997; Kerckhoffs et al. 1998). Muscle uses ingested

glucose for 85% to 90% of its need so that during low-intensity to moderate-intensity exercise of 30 to 120 minute duration, carbohydrate supplementation does not alter muscle glycogen use (Koivisto et al. 1985; Neufer et al. 1987; Pallikarakis et al. 1986).

When about 1 g/kg of glucose is ingested 15 to 45 minutes before intermediate-intensity or high-intensity exercise, activity is initiated at the time of rising plasma glucose and insulin concentrations. Hyperinsulinemia blocks the release of counter-regulatory hormones that ordinarily increase hepatic glucose production at high

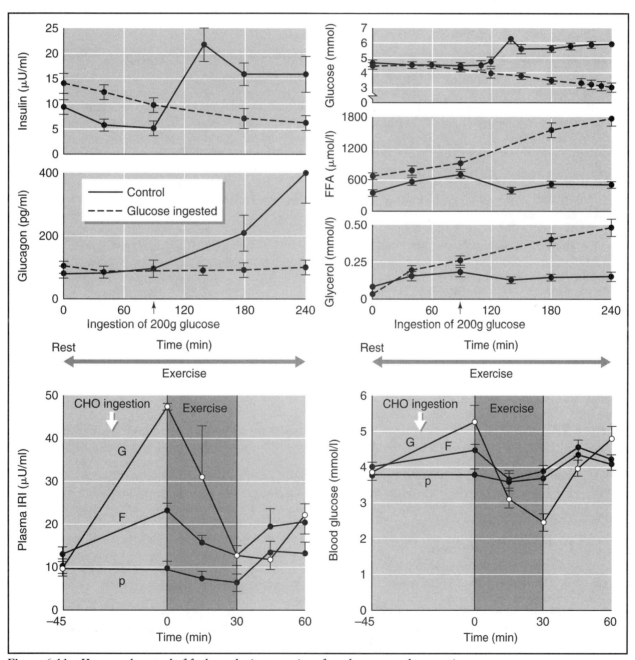

Figure 6.11 *Hormonal control of fuel use during exercise after glucose supplementation.*

Glucose supplementation at 90 minutes of prolonged exercise at 30% of V̇O₂max raises plasma glucose and insulin concentration and suppresses lipolysis and plasma glucagon concentration through subsequent 2.5 hours of exercise (top). Reactive hypoglycemia develops during 30 minutes of exercise at 75% of V̇O₂max when glucose is ingested 45 minutes before the start of exercise.

Adapted from Ahlborg and Felig 1976.

exercise intensities. A reactive hypoglycemia develops (see figure 6.11) from the (1) hypoglycemic action of insulin, (2) a lack of counter-regulatory hyperglycemia, (3) rapid glucose utilization by muscle at high exercise intensities, and (4) diminished contribution of FFAs to the energy cost of exercise (Decombaz et al. 1985; Koivisto et al. 1981). Blood glucose drops during the first 20 minutes of such exercise and is only partially restored during the next 40 minutes (Foster et al. 1979; Horowitz & Coyle, 1993; Koivisto et al. 1981; Levine et al. 1983). Muscle glycogen supplies about 60% to 75% of the energy at intermediate to high exercise intensities (Romijn et al. 1993), and the rate of its depletion is unaffected by glucose supplementation as insulin-insensitive type IIX fibers are significantly involved (Coyle et al. 1986).

When fructose is substituted for glucose and ingested at the start of either brief moderate-intensity exercise or prolonged low-intensity exercise, insulin response is reduced by about 80% (see figure 6.11) and reactive hypoglycemia is prevented (see figure 6.11) (Hasson & Barnes 1987; Jandrain et al. 1993; Koivisto et al. 1981; Massicotte et al. 1990). Fructose is a simple sugar that is absorbed more slowly and utilized less readily by muscle than glucose, as HK has lower affinity for it (Cori 1925; Crane & Sols 1954). Because of lower metabolic availability of fructose during the first hour of moderate-intensity exercise, lipids are utilized more heavily and carbohydrates are spared (Massicotte et al. 1986). However, during the second and third hour of such exercise, carbohydrate utilization is similar after ingestion of either fructose or glucose (Maasicotte et al. 1990).

Onset of fatigue is significantly delayed when sugar intake in the range of 0.8 to 1 g/kg is ingested after about 2 hours of moderate-intensity exercise at a rate in excess of 1 g/min. As muscle and liver glycogen nearly are depleted after about 3 hours of exercise at 70% of $\dot{V}O_2$max, ingestion or intravenous infusion of glucose prolongs such moderate-intensity performance beyond what is possible in the absence of such supplementation (Coggan & Coyle 1987, 1988, 1989; Coyle 1992a, 1992b). At this pace, at which muscle glycogen contributes 80% of carbohydrate utilization, the rate of muscle glycogen depletion is unaffected by carbohydrate supplementation and supplemented glucose substitutes for depleted glycogen and sustains the pace of exercise (Coyle et al. 1986; Hargreaves et al. 1987; Koivisto et al. 1985).

Lipid and PCr Supplementation

Oxidative utilization of lipids during endurance exercise is increased in animals and man after several days of adaptation to a high-fat diet (Costill et al. 1977; Hickson et al. 1977; Jansson & Kaijser, 1982, 1987; Miller et al. 1984). High-fat diets increase muscle FFA uptake as they up-regulate LPL and change the partitioning of muscle fuels toward increased intramuscular TG synthesis and decreased muscle glycogen synthesis (Eaton et al. 1969, Kiens et al. 1987; Kimball et al. 1983; Phillips et al. 1996a). Lipid supplementation at the onset

of or during intermediate-intensity exercise improves endurance performance by sparing intramuscular PCr and glycogen stores (Costill et al. 1977; Dyck et al. 1993, 1996a; Hickson et al. 1977; Miller et al. 1984). High-intensity performance after fat supplementation is impaired because of the reduced muscle glycogen store and reduced carbohydrate intake and utilization (Bergstrom et al. 1967; Jansson & Kaijser 1982; Starling et al. 1997). High-power exercise performance also sometimes is improved by several days of creatine supplementation (Greenhaff et al. 1993). Creatine supplementation is more effective when combined with carbohydrates, suggesting that insulin may mediate this effect (Green et al. 1996a, 1996b).

Exercise in an Energy-Depleted State

Moderate-intensity to high-intensity exercise in an energy-depleted state, whether caused by a prolonged fast or diabetes, hastens the onset of fatigue (Aragon-Vargas 1993; Wahren et al. 1978). The hepatic glycogen store is depleted after 2 to 3 days of food deprivation (Nilsson & Hultman 1973). The hepatic glucose output is only 40% of that after an overnight fast and is generated entirely by gluconeogenesis (Bjorkman & Eriksson 1983). Despite very high concentrations of counter-regulatory hormones and a low concentration of insulin in an energy-depleted state, hepatic glucose production is diminished (Galbo et al. 1981). Because of diminished muscle and liver glycogen stores, a rise in plasma glucose is possible but only by gluconeogenesis (Bjorkman & Eriksson 1983; Hagenfeldt & Wahren 1971a, 1971b). Plasma lactate and pyruvate are also elevated as a result of high plasma catecholamine concentration and increased anaerobic glycolysis or glycogenolysis in exercising as well as in inactive muscle (Ahlborg 1985; Bjorkman & Eriksson 1983; Randle et al. 1964).

ABNORMAL METABOLISM

Inactivity and excessive intake of high-fat food have contributed to the rise of excess weight and obesity in the United States and other developed societies. Obesity usually is accompanied by insulin resistance, dyslipidemias, and hypertension, a cluster of morbidities designated as the metabolic syndrome or syndrome X (Reaven 1993). Its chief complications, cardiovascular and cerebrovascular accidents, are currently the leading cause of death in middle-aged and elderly persons in developed societies. As both the etiology and remission of the metabolic syndrome have connections to exercise, and its central abnormality is a disturbance in the endocrine control of fuel metabolism, inclusion of a brief discussion of the nature of this disorder in this chapter is appropriate.

The metabolic syndrome is a consequence of excessive amassing of storage TGs accompanied by development of abnormal insulin secretion and action. Inactivity combined with excessive food intake and consumption of high-fat foods are the initiating events in the development of obesity (Dreon et al. 1988; Physical Activity and Health 1996). Large carbohydrate meals stimulate insulin secretion and provide hormonal and metabolic support for de novo fat synthesis in the liver (see figure 6.12). These lipids, along with the quantitatively more significant dietary lipids, are transported out of the liver and the small intestine in lipoproteins. With the assistance of the insulin-stimulated adipose tissue LPL, these lipids provide a source of FFAs primarily for storage in the adipose tissue and also for the muscle.

The site of fat accumulation influences the magnitude of the risk of metabolic syndrome. There is a gender difference in regional fat accumulation and in regional distribution of receptors for sex steroid hormones (Rebuffe-Scrive et al. 1990). Accumulation of fat in the abdominal area (upper-body obesity) is a characteristic of obesity in middle-aged men (android obesity) and postmenopausal women (Haarbo et al. 1991; Mauriege et al. 1991). In premenopausal women, fat is predominantly subcutaneous and deposited in the hips and gluteal area (gynoid or lower-body obesity) (Lonnqvist et al. 1997). Accumulation of intra-abdominal (visceral) fat poses a greater risk for development of the metabolic syndrome than deposition of fat in the subcutaneous gluteal area and limbs, designated as lower body (Despres et al. 1985, 1989; Evans et al. 1984a; Fujioka et al. 1987; Ross et al. 1996; Seidell et al. 1988). Deposition of fat in the abdominal area has been attributed to reduced production of sex-hormone binding globulin (SHBG) and resultant increased action of androgens (Pieris et al. 1987b).

Although the visceral component of the upper body fat was initially viewed as the primary risk factor for development of the metabolic syndrome, the subcutaneous abdominal fat was found to be as metabolically labile and potent in producing insulin resistance as the visceral fat (Goodpaster et al. 1997; Horowitz et al. 2000). Most recently, the accumulation of intramuscular fat has been implicated as an independent determinant of insulin resistance, greater than either of the two abdominal sites (Pan et al. 1997; Phillips et al. 1996a).

Inactivity as well postprandial lipemia and hyperinsulinemia caused, respectively, by high-fat and high-carbohydrate diets can alter muscle metabolism and make it resistant to insulin action (Pan et al. 1997; Storlien et al. 1986, 1991). High concentrations of glucose, FFAs, and insulin shift muscle metabolism away from lipid oxidation in the direction of increased TG reesterification and carbohydrate oxidation (see figure 6.5) (Bavenholm et al. 2000; Colberg et al. 1995; Goodpaster & Kelley 1998; Groop et al. 1992; Kelley & Mandarino 1990; Kelley & Simoneau 1994; Mandarino et al. 1996; Sidossis & Wolfe 1996). Increased reesterification of FFAs generates increased amounts of diacylglycerol, or DAG, as well as TGs. DAG acts as an intracellular second messenger that activates PKC and in this manner interferes with the insulin-signaling cascade (Ahmad et al. 1984; Danielsen et al. 1995). Insulin action also is reduced and insulin resistance develops because both the malonyl-CoA and long-chain fatty acyl-Co inhibit glycogen synthetase and insulin-stimulated nonoxidative glucose disposal (Pan et

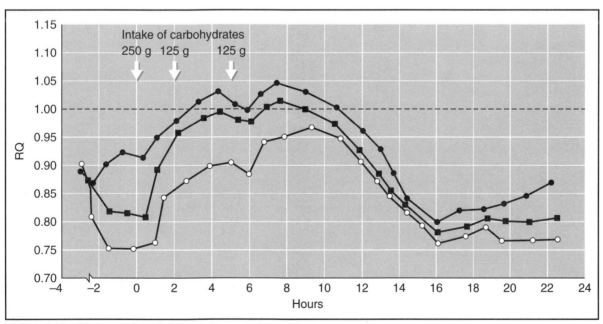

Figure 6.12 *Nutritional modulation of hepatic fat synthesis.*

De novo fat synthesis, evident as non-protein respiratory quotient (RQ) greater than 1, was induced in normal volunteers only after ingestion of extraordinarily large (500 g) carbohydrate meals. The effect of the large carbohydrate meals was enhanced when volunteers ate a high carbohydrate (solid circles) diet during the preceding week. The effect was less readily elicited in volunteers who ate a mixed diet (■), and was not present in those who ate a high-fat diet (o).

Reprinted from Acheson et al. 1984.

al. 1997; Phillips et al. 1996a; Prentki & Corkey 1996; Wititsuwannakul & Kim 1977). Under the conditions of high glucose, FFAs, and insulin, formation of long-chain fatty acyl-Co, the metabolically activated form of FFA needed for initiation of β oxidation, is suppressed (Oakes et al. 1997a). In particular, high titers of insulin and FFAs stimulate the activity of ACC and synthesis of malonyl-CoA (Saha et al. 1995, 1997). These changes inhibit the activity of CPT I and entry of FFAs into the mitochondrion for β oxidation (Colberg et al. 1995; Kelley & Simoneau 1994; Simoneau et al. 1995; Storlien et al. 1991). These processes alter the profile of muscle energy stores in favor of lipids over glycogen.

As muscle is the principal site of glucose utilization, its resistance to insulin action in obese individuals causes the pancreas to oversecrete insulin (Evans et al. 1984b; Pieris et al. 1987a; Shulman et al. 1990). Lipolysis increases both because of reduced sensitivity of the abdominal adipose tissue to the antilipolytic action of insulin and because of disturbances in its control by the S nerves (Horowitz et al. 1999a; Horowitz & Klein 2000b; Rebuffe-Scrive 1991; Reynisdottir et al. 1994). The liver becomes insulin resistant because of exposure to high titers of free fatty acids in the portal vein, which permits increased hepatic glucose production and release into circulation (Bjorntorp 1990; Pieris et al. 1988). Another consequence of hepatic insulin resistance is reduced hepatic insulin extraction and degradation,

which contributes further to increases in plasma insulin concentration (Pieris et al. 1986, 1987a). In addition, excessive release of FFAs in plasma of persons with upper-body obesity impairs the ability of insulin to stimulate glucose uptake, inhibit hepatic glucose production, and stimulate hepatic production of VLDLs, TGs, and cholesterol (Ferrannini et al. 1983; Kelley et al. 1993; Lewis et al. 1995).

Endurance training can normalize fuel metabolism in individuals with metabolic syndrome by increasing the capacity of muscle to oxidize lipids (Holloszy & Coyle 1984; Mole et al. 1971). The aerobic training effects include increases in the volume of mitochondria and of their oxidative enzymes and cytochromes (Holloszy 1967; Holloszy et al. 1970; Kayar et al. 1986). In addition, lipid droplets are relocated to proximity of mitochondria in the trained muscles and away from the plasma membrane location in detrained muscles (Boesch et al. 1997; Vock et al. 1996). Proximity of lipid droplets to the mitochondria facilitates mobilization of FFAs out of intramuscular TGs (see figure 6.5). Reduction in the amount of intramuscular TGs and DAG increases insulin sensitivity by reducing the concentration of long-chain FAcCoA and removing the inhibitory influence of PKC over insulin action (Oakes et al. 1997a, 1997b). With improvement in muscle insulin sensitivity, insulin oversecretion is reduced and multiple complications associated with insulin resistance abate.

7

Exercise and the Nutrient Partitioning Between Structure and Storage

Energy intake and expenditure are processes within a complex neuroendocrine mechanism that ensures our survival by partitioning ingested nutrients among basal metabolism, growth, tissue maintenance and repair, physical activity, and other forms of energy expenditure and nutrient storage. Nutrients are partitioned differently during the developmental stage of incremental growth and in adulthood when growth gives way to body mass maintenance and regulation of energy balance. The neuroendocrine processes responsible for nutrient partitioning are incompletely understood, but chemical messengers play a central role. Exercise affects all partitioning of nutrient energy and thus impacts growth and tissue maintenance and components of energy regulation. The interdependent mechanisms through which these effects occur are slowly being unraveled. This chapter will define how growth, body constituents, and energy regulation are influenced by exercise and how neuroendocrine interactions modulate these processes.

RELATIONSHIP OF GROWTH AND ENERGY REGULATION

Growth is the process through which nutrient energy is assimilated into structural components of the body. Three distinct types of growth differ in their control mechanisms and the extent to which they depend on nutrient abundance and hormones. Whole-body, statural, or incremental growth entails proportional increases in body dimensions and depends on instructions contained in the genome and on guidance by systemic hormones of growth. These hormones have been programmed to confine such growth only to particular, usually early, portions of the life span. How growth hormone (GH) and sex steroids help guide metabolism of ingested nutrients toward synthesis of structural proteins necessary for proportional enlargement of the entire body is described in some detail in chapter 8. Statural growth also is sensitive to nutrient abundance and quality.

After a species-specific adult body mass has been attained, the neuroendocrine stimuli that gave rise to it are suppressed to a level that is sufficient for anabolic tissue turnover and repair but insufficient to produce further body size increments. Incremental growth is reduced to the reparative state through a lower frequency and amplitude of spontaneous GH pulses during the latter portion of human and rodent life span (Corpas et al. 1993; Laartz et al. 1994). In humans, the decline in spontaneous GH secretion is exponential after adolescence with a 7-year half-life (Iranmanesh et al. 1991). It is believed to be caused both by atrophy of some growth hormone–releasing hormone (GHRH) neurons and by increased brain somatostatin action (Coiro et al. 1992; Soule et al. 2001). The largely inactive lifestyle of adult populations in developed countries also contributes to additional age-associated decay in pulsatile GH secretion. In addition to the reduced activity of the neuroendocrine growth stimulatory mechanism that occurs upon attainment of species-specific mature body mass, the skeleton and many other tissues in humans and in most mammals other than rodents (Bourguinon 1988; Donaldson 1919) lose their proliferative responsiveness to hormones of growth.

The transition from rapid, opportunistic whole-body growth to the reparative growth of adulthood is accompanied by two additional changes in control of nutrient intake and disposal. First, the quantity of necessary nutrient intake per unit of body mass declines and is more closely matched to energy expenditure (Kennedy 1957; Young et al. 1975). The consequence of this change is that lean body mass and storage depot mass are maintained at a stable level. Second, as nutrient energy is diverted from predominant support of statural growth toward lean tissue maintenance and maintenance of the storage fat depot, an increased amount of nutrient energy is diverted toward spontaneous physical activity (Kennedy 1953). The redistribution of nutrient energy from incremental growth toward increased energy expenditure of physical activity and approximate matching of energy intake and energy expenditure produce stability of lean body mass and fat mass in mature mammals and mark the onset of energy regulation.

The same neuroendocrine mechanism that was responsible for control of feeding during statural growth is transformed into an energy regulatory mechanism by addition of a neuroendocrine brake over GH secretion. This maturational change can be inferred from neurosurgical experiments that transform mature, nongrowing, spontaneously active, stable-weight, and lean animals into rapidly growing, GH and insulin oversecreting, hypoactive, and obese animals (Borer 1987a, 1987b). This transformation can be produced by surgical separation of limbic structures septum and hippocampus from the hypothalamus in mature hamsters (Borer et al. 1977, 1979a, 1979b, 1979c) and by separation of the paraventricular hypothalamus from catecholamine circuits ascending from the brain stem in mature female rats (Mitchell et al. 1972, 1973; Palka et al 1971). The maturational character of these neural connections is shown by ineffectiveness of these lesions during the developmental stage of rapid growth (Gold & Kapatos 1975).

Hypertrophic growth allows for multiplication of structural proteins, proteins serving specific functions, or protein isoforms that function in such a way as to increase body mass and mechanical strength for physical performance or to meet increased local metabolic demands. This form of growth operates throughout the life span and differs from incremental and reparatory forms in that it is achieved without hyperplasia, or cellular proliferation, and can occur without the guidance of systemic hormones of growth and in the face of scarcity of nutrient energy.

ENERGY PARTITIONING FOR STATURAL GROWTH

Egg fertilization initiates early growth and tissue morphogenesis by activating the nuclear genetic program.

The pace of early statural growth peaks during the third trimester of gestation and subsequently declines after birth and throughout childhood until the onset of the adolescent growth spurt (see figure 7.1). This early growth consists of cellular proliferation that is followed by cessation of cell division and by progression through tissue-specific differentiation. Determinants of statural growth and development are systemic hormones of growth and tissue growth factors, an adequate quantity and quality of nutrient energy, and some form of mechanical stimulation (Fowden 1995).

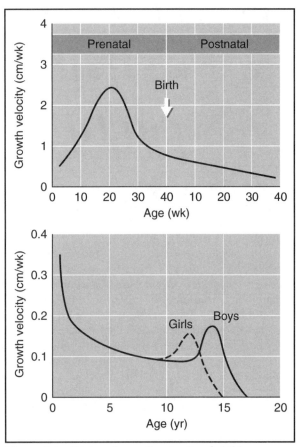

Figure 7.1 *Statural growth velocity of children.*

Growth velocity is highest during the prenatal period and declines throughout childhood. Growth velocity is transiently increased during the adolescent, prepubertal period before it is finally extinguished.

Adapted from Griffin and Ojeda 1996.

Of the systemic hormones of growth, growth hormone (GH) and thyroid hormones play key roles. During pregnancy, the placenta synthesizes and releases several GH variants that increase in maternal circulation through the 35th week of human gestation (Boguszewski et al. 1998; Caufriez et al. 1994; Frankenne et al. 1988). These GH variants stimulate fetal growth directly and indirectly by promoting metabolic utilization of lipids (Chowen et al. 1996). They also suppress maternal GH secretion (Caufriez et al. 1993; Eriksson 1989).

Of the maternal pituitary hormones, only TSH appears to play a role in the control of fetal statural growth by stimulating thyroxine (T_4) and 3,5,3' triiodothyronine (T_3) release (Gluckman 1986). The active hormone is T_3, which is produced in peripheral tissues by deiodination of T_4. T_3 circulates bound to thyroglobulin, and 0.3% of the free unbound hormone is taken up by cells to be bound by cytosolic T_3 binding proteins and to act on nuclear or mitochondrial DNA. T_4 stimulates growth of fetal muscle and long bones to a greater extent than that of the rest of the body (Erenberg et al. 1974). T_3 receptors are expressed both on chondrocytes and on epiphyseal osteoblasts, where the hormone acts to increase activity of alkaline phosphatase, to increase expression of IGF-I receptors, and to decrease cell proliferation (Williams et al. 1998). The action of T_3 on its HRE is usually dependent on dimerization with vitamin D_3 and all-*trans* retinoic acid receptors (Rosen et al. 1993). T_4 (more so than T_3) also affects differentiation and maturation of the fetal lung and development of the nervous system during the perinatal period (Liggins & Schellenberg 1988; Timiras & Nzekwe 1989).

Glucocorticoids derived either from maternal circulation or fetal adrenal glands contribute to fetal tissue differentiation as their concentration rises toward the end of gestation. Among such maturational influences of cortisol are induction of hepatic GH receptors; increases in lung compliance and surfactant release; development of hepatic β adrenergic receptors, glycogen synthetase, and gluconeogenic enzymes; proliferation of intestinal villi; and induction of digestive enzymes (Li et al. 1999; Sangild et al. 1995). Toward the end of gestation, cortisol suppresses fetal growth and affects placental steroidogenesis in ways that precipitate parturition (Fowden et al. 1996; Liggins & Thorburn 1993). Cortisol action in the fetus entails gene transcription as well as suppression of hepatic IGF-II and increase in hepatic IGF-I mRNA synthesis (Challis et al. 1993; Forhead et al. 1998). Systemic hormones usually act cooperatively. Thus, thyroxine enhances and insulin antagonizes cortisol action on pulmonary development and surfactant production, and joint action of cortisol and thyroxine is needed to reduce IGF-II mRNA in late gestation (Forhead et al. 1998; Liggins & Schellenberg 1988).

The importance of abundant nutrition in support of early fetal statural growth manifests itself through nutritional control of insulin, T_3, and IGF-I and IGF-II secretion. Expression and secretion of these important controllers of early growth is suppressed during nutrient shortage, and T_3 production declines in favor of synthesis of the inactive, reverse T_3 (rT_3) hormone variant. In the presence of abundant nutrient availability, insulin promotes growth through its metabolic actions, facilitation of cellular glucose and amino acid uptake, suppression of amino acid catabolism, and preferential use of glucose for fetal metabolism (Fowden 1992, 1997; Hay 1991). In humans, insulin also stimulates adipocyte proliferation during the third trimester of gestation

(Fowden 1993). Both insulin and thyroxine stimulate cellular proliferation directly as well as indirectly by increasing tissue production of IGF-II, IGF-I, and other fetal growth factors and their binding proteins (Ferry et al. 1999; Forhead et al. 1998; Fowden 1993; Thorburn 1974). The IGFs, secretion of which also is dependent on nutrient abundance, stimulate growth through both their metabolic and their anabolic actions (Gluckman 1986; Hill et al. 1987). They increase protein and glycogen synthesis in fetal tissues and, as progression factors, increase DNA synthesis and cell differentiation (Hill et al. 1987). The interdependence of tissue growth factor action and mechanical stimulation during fetal development is illustrated by skeletal muscle growth and development.

Skeletal Myogenesis

Growth and differentiation of muscle cells is under endocrine and mechanical control. The cell growth cycle (see figure 7.2) is either influenced in the direction of cell proliferation and migration (from G_1 to S phase) or directed toward growth arrest and differentiation (from M to G_0 phase). Initial cellular proliferation of myocytes is mediated by IGF-I and several other growth factors. IGF-I promotes a single or a limited number of proliferative cell cycles and a more extended number of cell divisions when it is paired with bFGF (Adi et al. 2000; Florini et al. 1993; Layne & Farmer 1999). Other mediators that stimulate cellular proliferation and inhibit terminal differentiation of myocytes are TNF-α, TGF-β, bone morphogenetic proteins (BMPs), and bFGF (Katagiri et al. 1997; Layne & Farmer 1999; Szalay et al. 1997; Tortorella et al. 2001). The proliferative phase of the cell cycle requires expression of D-type cyclins and their activation for nuclear action of cyclin-dependent kinases (cdks). This activity promotes cell commitment to enter S phase (Bartkova et al. 1998).

These proliferative messengers transduce IGF-I growth stimulus by way of the MAPK/Erk and Raf-MEK-Erk pathways (see figure 3.5). They also inhibit the expression of messengers that promote terminal muscle differentiation when MEK1 forms an inhibitory complex with myogenic regulatory factors (MRFs) (Adi et al. 2000; Florini et al. 1993; Penn et al. 2001; Rommel et al. 1999).

Most of these messengers require mechanical stimulation transmitted by way of proteoglycan molecules located in the matrix surrounding fetal muscle cells to which they bind (Larrain et al. 1998; Yamane et al. 1998). Myocyte adhesion to a firm bone attachment and to each other is a necessary mechanical stimulus for progression of the growth promoting and differentiating action of IGF-I. Hormonal and mechanical stimuli together mediate transition of myocyte from a nonreproductive G_0 phase to mitotic G_1 phase and its subsequent entry into S phase of DNA replication and mitotic M phase (see figure 7.2) (Milasincic et al. 1996; Shall 1981). Mechanical force is transmitted through adhesion molecules

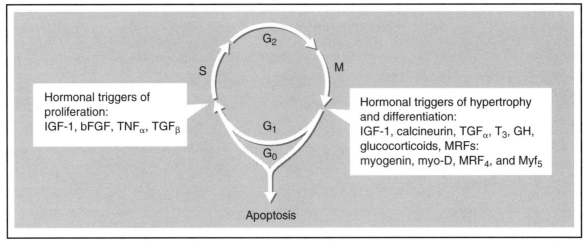

Figure 7.2 *Hormonal facilitation of muscle cell growth cycle.*

The proliferative component of a cell growth cycle (Shall 1981) consists of DNA replication (S phase), preparation for mitosis (G_2 phase), and mitosis (M phase). Initiation factors mediate transition from the quiescent G_0 state to S phase. Progression factors mediate terminal differentiation and transition from M phase to the hypertrophic G_1 phase of the cell cycle. Cells can also enter the quiescent G_0 phase and, under influence of yet other messengers, undergo apoptosis, or cell death.

Adapted from Newsholme and Leach 1983.

N-cadherin and catenins and other proteins in the extracellular matrix, such as collagens, laminin, tenascin, fibronectin, thrombospondin, osteocalcin, osteopontin and proteins in the cytoskeleton to signaling molecules within the cells (Redfield et al. 1997). This mechanotransduction ("tensegrity") is mediated in part by integrin receptors that act as mechanosensors. Integrins are membrane-spanning glycoproteins, usually clustered in dense membrane plaques, that form heterodimeric Rs for various matrix proteins (Davis et al. 2001; Heino 2000). After binding to integrin Rs, matrix proteins transduce the external mechanical stimulus to the cytoskeleton in the form of molecular signaling cascades that include TK, focal adhesion molecule kinases (FAKs), and other proteins (Ingber 1993). Integrin by itself can stimulate myocyte proliferation and block cellular differentiation through the actions of integrin-linked kinase (Huang et al. 2000).

Terminal differentiation of myocytes entails fusion and incorporation of new satellite cells into growing myotubes, fusion of myotubes, and synthesis and assembly of muscle-specific proteins to form myofilaments in mature skeletal muscle fibers. This developmental process is initiated when myocytes exit the proliferative M phase of the cell cycle and enter the G_1 or G_0 phase (see figure 7.2). This switch is mediated by some endocrine-paracrine-autocrine mediators and by intracellular MRFs myogenin, Myo-D, MRF4, and Myf-5, also known as muscle-specific transcription factors.

Among the endocrine-paracrine-autocrine mediators that stimulate MRF expression and terminal muscle differentiation are IGF-I (after its initial proliferative action), calcineurin, AVP, TGF-α, T_3, retinoic acid (RA), and glucocorticoids (Alric et al. 1998; Downes et al. 1995; Florini et al. 1993; Friday et al. 2000; Minotti et al. 1998; Nervi et al. 1995; te Pas et al. 2000; Yamane

et al. 1998). MRF proteins share a basic helix-loop-helix structure and exert their differentiating action by forming heterodimers with E proteins and then by binding to CANNTG elements in the promoters of muscle-specific protein genes (Lu et al. 1999; Zhou & Olson 1994). For instance, all-*trans* RA induces the expression of Myo-D, which then triggers the expression of the mitochondrial uncoupling protein 3 (UCP-3) (Solanes et al. 2000). MRFs in part act collaboratively, sequentially, and substitutively (Mendler et al. 1998; Valdez et al. 2000; Zhu & Miller 1997). Additional facilitators of MRF action are myocyte enhancer factor 2 proteins (MEF 2), which also bind to an adjacent promoter region (Tamir & Bengal 2000; Winter & Arnold 2000). Actions of IGF-I to stimulate skeletal muscle differentiation are mediated through PI3-K/Akt and p70 (S6K) transduction pathways (see figure 3.5) (Xu & Wu 2000). Activation of these transduction pathways and expression of MRFs block the operation of the proliferative MAPK pathway through inhibition of c-fos expression and activation of cdk inhibitors p21 (CIP1) and p57 (KIP2) (Andres & Walsh 1996; Puri et al. 1997; Trouche et al. 1995; Zhang et al. 1999a). These cdk inhibitors prevent proliferative actions of cyclins and their kinases in the nucleus (Guo et al. 1995). After developing muscles become innervated, expression of MRFs is terminated, and mature muscle cells enter the G_0 phase of the cell cycle (Merlie et al. 1994; Weis 1994).

Maternal Exercise and Fetal Statural Growth

Exercise during pregnancy may restrict fetal growth and development by reducing the fetal nutrient supply and by increasing concentrations of stress hormones in maternal and placental circulation. Increased glucose uptake by rat maternal muscles after exercise was associated with reduced glucose uptake by the fetus but no change in fetal liver glycogen (Carlson et al. 1986; Treadway & Young 1989). In energy-restricted exercising pregnant animals, fetal needs were added to, but did not have priority over, maternal energy demands (Chandler et al. 1985). The nutritional drain that large volumes of habitual physical activity generate in pregnant women results in infants with lower birth weights. Thus, at estimated weekly energy expenditure of less than 1,750 kcal, physical activity has either no effect on infant birth weight or is correlated with greater birth weight (Collings et al. 1983; Hall & Kaufmann 1987; Hatch et al. 1993; Klebanoff et al. 1990; Teitelman et al. 1990). However, when energy expenditure reaches 2,000 kcal/week or the maternal body fat level falls to 17%, infant birth weight is reduced by between 0.1 and 0.4 kg, and maternal weight gain during pregnancy is reduced by as much as 3 kg (Briend 1980; Clapp 1990,1998; Clapp & Capeless 1990; Clapp & Dickstein 1984; Kulpa et al. 1987; Tafari et al. 1980). Infant weight shortfall of exercising mothers is attributed to decreased body fat because infant head circumference is unaffected (Clapp & Capeless 1990; Clapp 1998). This condition could be a consequence of increased insulin sensitivity in exercise-trained pregnant women and diminished stimulation by maternal insulin of fetal adipocyte proliferation. Increased corticosterone and androstenedione secretion seen in exercising pregnant animals could affect human fetal sexual and tissue differentiation or hasten onset of labor (Carlberg et al. 1996) (see chapter 8). Earlier onset of labor has been reported in athletic pregnant women (Clapp 1990).

Postnatal Statural Growth

After birth, pituitary GH assumes central control over statural growth, with thyroid hormones, insulin, cortisol, and gonadal hormones playing supporting roles. In a newborn infant, GH is oversecreted relative to its body surface, most likely because of the immaturity of the hypothalamic and pituitary feedback controls over its release (De Zegher et al. 1993).

Postnatally, pulsatile GH secretion is controlled by a number of amine and peptidergic brain neurotransmitter pathways (see figure 7.3) that alter the tandem secretion of two GH regulatory peptides, growth-hormone releasing hormone (GHRH), with cells located in the arcuate hypothalamic nucleus, and somatostatin (SRIF), with cells in the periventricular and adjacent hypothalamic areas (Giustina & Veldhuis 1998; Muller et al. 1999). SRIF inhibits pituitary GH release

by acting on two of five known SRIF receptor subtypes and activating G_i and IP_3 transduction pathways, rather than by affecting GH synthesis (Patel & Srikant 1986). GHRH stimulates both GH synthesis and release through a cAMP-mediated and Ca^{2+}-mediated transduction mechanisms.

Among amine neurotransmitters that control GH secretion, the roles of acetylcholine and catecholamines have been studied most, whereas the involvement of serotonin, histamine, GABA, excitatory amino acids, growth hormone releasing peptides, such as ghrelin, and nitric oxide is less well understood. Acetylcholine acts on muscarinic receptors to increase pulsatile GH secretion, presumably by inhibiting somatostatin release (Friend et al. 1997; Locatelli et al. 1986). Such cholinergic activation plays an important role in GH release in response to exercise and during deep sleep (Casanueva et al. 1984; Peters et al. 1986). Dopamine agonists such as L-dopa, apomorphine, dopamine, and bromocriptine acting on dopamine receptors stimulate GH secretion by an estrogen-sensitive mechanism and also act presumably via somatostatin withdrawal (Chihara et al. 1986). Among adrenergic influences, stimulation of α_2 receptors elicits GH release through stimulation of GHRH release (Miki et al. 1984). Insulin-induced hypoglycemia elicits a GH response through this type of stimulation (Tatar & Vigas 1980). In contrast, β adrenergic stimulation inhibits GH

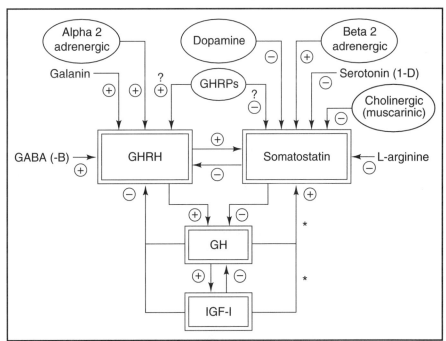

Figure 7.3 *Neuroendocrine control of growth hormone (GH) secretion.*

Brain neurotransmitters that control GH secretion through their action on somatostatin (SRIF) and growth hormone–releasing hormone (GHRH). Exercise and arginine stimulate GH secretion through stimulation of muscarinic cholinergic receptors that inhibit somatotropin release–inhibiting hormone (SRIF) neurons. Hypoglycemia elicits GH secretion by acting on α_2 adrenergic receptors. GH stimulates IGF-1 insulin-like growth factor release, and this messenger exerts negative feedback over GH secretion by stimulating somatistonin and inhibiting GHR secretion. Additional growth hormone releasing peptides (GHRPs), such as ghrelin, stimulate GH secretion.

Adapted from Veldhuis 1998.

secretion, most likely by stimulating somatostatin release (Martha et al. 1988; Mauras et al. 1987). β$_2$ Adrenergic stimulation attenuates GH response to physical exercise in humans (Giustina et al. 1995). Less well understood are the mechanisms of increased basal GH secretion to stimulation of type 1D serotonin receptors, stimulation of both GH secretion and slow-wave sleep by a GABA agonist, and the possible mediation of GH release by nitric oxide in response to L-arginine (Korbonitz et al. 1996; Rolandi et al. 1992; Van Cauter et al. 1997).

Additional factors controlling GH secretion are feedback inhibition by products of its anabolic and metabolic action, nutritional status, various stressors, deep sleep, hormones, and exercise. Feedback inhibition of GH secretion by products of its anabolic action includes stimulation by GH of the hypothalamic SRIF neurons and inhibition by IGF-I of the pituitary somatotrophs (Giustina et al. 1991; Hartman et al. 1993). Feedback inhibition by products of its metabolic action includes inhibition of GH secretion by hyperglycemia and FFAs (Casanueva et al. 1981; Imaki et al. 1985; Roth et al. 1963). Severe undernutrition and body fat loss are associated with increased GH secretion, and this change consists of increases in frequency as well as augmentation of secretory burst and mass (Hartman et al. 1992; Ho et al. 1988; Iranmanesh & Veldhuis 1992). During undernutrition, metabolic action of GH predominates, as GH anabolic action is blocked by down-regulation of hepatic GH receptors and decreased production of IGF-I (Ketelslegers et al. 1996). Obesity on the other hand, particularly in the visceral region, is associated with decreased pulsatile GH secretion and decreased GH release to other provocative stimuli (Iranmanesh et al. 1991; Scacchi et al. 1999; Vahl et al. 1997a). Various stressors, such as exposure to hypoglycemia, cold, heat, surgery, and ether, elicit GH secretion as well (Roth et al. 1963; Shibasaki et al. 1985). Deep sleep elicits GH secretion through cholinergic blockade of SRIF release and possibly also through the influence of GABAergic neurotransmission (Peters et al. 1986; Van Cauter et al. 1997).

Thyroid hormones, glucocorticoids, endogenous opioids, leptin, and sex hormones all influence GH secretion. Thyroid hormone is required for many anabolic actions of GH in statural growth (Nilsson et al. 1994). Thyroid hormones act at the pituitary to increase GH secretion and at the liver to increase IGF-I release (Bonaterra et al. 1998; Weiss & Rebetoff 1996; Wolf et al. 1989). Glucocorticoids inhibit GH secretion by blocking GHRH and stimulating SRIF gene transcription and translation (Fife et al. 1996). Glucocorticoids also can stimulate GHRH receptor synthesis and increase GH response to GHRH (Vale et al. 1983). Endogenous opioids stimulate GH secretion, particularly under some forms of stress such as marathon running and caloric restriction (Delitala et al. 1983; Moretti et al. 1983). Serum leptin concentrations correlate inversely with pulsatile GH secretion in postmenopausal women (Licinio et al. 1998; Roubenoff et al. 1998). During puberty, increases in circulating

testosterone and estradiol induce a 1.5-fold to 3-fold amplification of pulsatile GH secretion and a more disorderly secretory pattern (Martha & Reiter 1991; Martha et al. 1992; Mauras et al. 1996; Veldhuis et al. 1997). The increase is caused by augmentation of GH secretory burst and mass without an effect on burst frequency or GH half-life. Endogenous estradiol concentration also has a major stimulatory influence over GH secretion in adulthood (Ho et al. 1987). Females secrete growth hormone with more process irregularity than males (Pincus et al. 1996). Endogenous as well as exogenous testosterone facilitates spontaneous and GHRH-stimulated GH secretion after being aromatized to estrogen (Corpas et al. 1992; Veldhuis et al. 1997).

Exercise during postnatal period stimulates GH secretion both acutely and chronically. Acute exercise needs to be above the anaerobic threshold to elicit a GH surge, which then rises with a delay and often reaches a peak after the termination of exercise (Sutton & Lazarus 1976; Van Helder 1984, 1987). High-resistance exercise also elicits the release of GH (Kraemer et al 1991a, 1992a, 1992b, 1993b, 1995). Endurance training, on the other hand, increases basal pulsatile GH secretion as well as acute secretory response to a bout of exercise (Bunt et al. 1986; Hurel et al. 1999; Weltman et al. 1992). At peak exercise, the 22 kd GH variant is the predominant secreted isoform. During the recovery, the ratio of non–22 kd GH variants rises because of the longer half-life of the 20 kd variant (Wallace et al., 2001).

GH is expressed and found in circulation in several isoforms, of which the most prevalent ones are 22 kd (21% of circulating GH) (Baumann 1991) and 20 kd variants in addition to a glycosylated 24 kd form and other molecular species (Haro et al. 1996; Lewis et al. 2000). GH circulates bound to a specific binding protein identical in molecular structure to the GH receptor, except for the absence of the transmembrane domain (Leung et al. 1987). GH-BP helps activate the GH receptor by facilitating GH-receptor dimerization with the 20 kd GH variant, the necessary first step in GH message transduction (Behncken & Waters 1999; Finidori 2000; Uchida et al. 1999).

Anabolic GH action consists of stimulation of growth of lean body tissues, including bone mass (Holmes & Shalet 1996; Johannsson & Bengtson 1997). GH has dual growth-promoting action (Green et al. 1985). Its direct action is to induce tissue differentiation, including expression of growth factors such as IGF-I, IGF-II and their receptors in the liver and other target tissues such as the cells in the germinal and proliferative zones of the epiphysial growth zones (EGZs) in the long bones. GH then stimulates clonal cell expansion indirectly by increasing the proliferative action of hepatic IGF-I released into circulation as well as of local growth factors in target tissues (Nilsson et al. 1994). These GH actions cause proliferation of chondrocytes and increased thickness of EGZ. Hepatic production of circulating IGF-I or somatomedin was once considered the main pathway of statural growth stimulation (Salomon & Daughaday

1957). It is now recognized that direct GH stimulation of growth factor gene transcription in target tissues is equally important for statural growth (Isgaard et al. 1988b).

The anabolic action of growth hormone has been attributed to its pulsatile secretory pattern (Isgaard et al. 1988a; Maiter et al. 1988). Pulsatile rather than continuous GH administration induces transcription of IGF-I mRNA in muscle and in epiphyseal growth zones and a faster rate of growth in hypophysectomized rats (Clark et al. 1985; Isgaard et al. 1988a). A more recent finding is that the relative proportions of 22 kd and non–22 kd GH variants may have differential growth-promoting potency and other physiological actions. Thus, a lower proportion of non–22 kd GH variants is responsible for rapid pubertal growth spurt, whereas a higher proportion of this variant circulates at times of reduced growth in normal children or in children with growth failure (Boguszewski et al. 1997a, 1997b). Both GH variants stimulate bone remodeling to an equal extent but may have a differential stimulating potency on PRL secretion, water retention, and lipolysis (Asada et al. 2000; Satozawa et al. 2000; Tsunekawa et al. 1999; Wang et al. 1999). As the 22 kd GH isoform is commercially available, its administration can be detected in circulation as it transiently alters the proportion of naturally occurring variants (Momomura et al. 2000).

IGF-I and IGF-II, the mediators of anabolic GH action, bind to the type I IGF receptor, which triggers their mitogenic effects (Froesch et al. 1985; Spagnoli & Rosenfeld 1997). Both peptides can bind with lower affinity to the insulin receptor. In circulation, IGFs are bound to a family of about 10 binding proteins (IGFBPs), some of which bind to IGFs with high affinity and mediate their binding to membrane receptors, while others act directly on the cells, independently of IGF control. IGFBPs amplify or inhibit the anabolic actions of IGFs (Owens 1991). Among binding proteins, IGFBP-2, -3, -5, and -6 are regulated by GH, and IGFBP-3 facilitates the growth-promoting actions of IGF-I (Peter et al. 1993).

Other systemic hormones augment and modulate postnatal GH anabolic action. Thyroid hormones increase both GH synthesis and action (Bonaterra et al. 1998). T_3 stimulates linear growth by increasing hepatic expression of IGF-I and by enhancing IGF's anabolic effects on cartilage. T_3 also stimulates chondroblast and osteoblast differentiation, which results in skeletal maturation. It also regulates bone and mineral metabolism in the adult (Allain & McGregor 1993; Compston 1993; Williams et al. 1998). Gonadal hormones impose a gender-specific pattern of statural growth that creates different sizes and proportions of pelvic and shoulder girdles, muscles, and fat depots in males and females. Increased secretion of gonadal hormones at puberty is associated with concurrently high concentrations of GH and IGF-I, suggesting that an interaction between GH and sex steroids may be important in the acquisition and maintenance of body and skeletal mass. Simultaneous increases in plasma GH and IGF-I concentration indicate that at puberty, IGF-I

feedback over GH secretion is diminished (Wennink et al. 1991; Wilson et al. 1991).

As was the case with prenatal growth, postnatal statural growth also requires adequate nutrient quantity and quality. Under conditions of nutrient scarcity, the anabolic action of GH, insulin, and thyroid hormones is curtailed, and secretion of growth-inhibiting glucocorticoids is increased (Fichter & Pirke 1986). Insulin secretion is suppressed during negative energy balance at the same time as its receptor sensitivity is increased (Koopmans et al. 1991). Negative energy balance reduces T_4 secretion and redirects thyroid hormone conversion from the pathway producing T_3 to an alternative pathway producing biologically inactive rT_3 (Gardner et al. 1979; Harber et al. 1997a; Kinlaw et al. 1985) (see figure 1.4). Growth-promoting GH action is curtailed during energy restriction through tissue resistance to GH action (Thissen et al. 1999). Resistance to GH action develops principally through down-regulation of GH receptors and manifests itself through reduced transcription and synthesis of IGF-I.

Nutritional deficit has different effects on growth, depending on the time of its operation. In general, the earlier the nutritional deprivation and growth stunting occurs, the more permanent are deficits in ultimate body size and mass (Widdowson & McCance 1960). Permanence of stunting of growth by early deprivation is viewed as evidence of incomplete development of the hypothalamic neuroendocrine controls over growth. As these controls mature, nutritional stunting of growth is followed by accelerated and compensatory catch-up growth. Early nutritional experience may permanently program the neuroendocrine mechanism for different patterns and amounts of GH secretion in adulthood. Thus, different rates of weight gain during infancy differentially affect mean GH and IGF-I concentrations in adulthood, which, in turn, are significantly correlated with bone mineral density during late adulthood (Fall et al. 1998).

Exercise and Statural Growth

Physical activity appears to exert two types of influence on statural growth. The first one takes the form of exercise-induced energy drain, which competes with growth for limited nutrient resources. Exercise does not increase the rate of GH secretion and growth during the developmental phase of peak postnatal statural growth (Borer & Kaplan 1977). Instead, heavy exercise training may slow down the rate of statural growth by competing for limited nutrient resources. Young animals that are forced to exercise grow at a slower pace and achieve smaller adult body size (Oscai et al. 1974). Likewise, children and adolescents who engage in large volumes and intensity of physical training are small for their age, grow more slowly, and frequently have low plasma concentrations of IGF-I (Caldarone et al. 1986a, 1986b; Eliakim et al. 1998; Georgopoulos et al. 1999; Jahreis et al. 1991; Malina 1994; Peltenburg et al. 1984). Energy deficit, whether produced by nutrient deprivation or exercise training delays the onset of puberty, a maturational event that is contingent on the completion of statural growth and attainment of adult body size and bone maturity (Frisch et al. 1981; Warren 1980).

The second effect of exercise on statural growth is a stimulatory one based on its capacity to facilitate pulsatile GH secretion. The clearest demonstration of this stimulatory effect comes from animal studies in which high volumes of voluntary running in slowly growing mature golden hamsters accelerate skeletal and somatic growth (Borer & Kuhns 1977). This growth acceleration is accompanied by increased basal pulsatile GH secretion (Borer et al. 1986). While exercise does not increase GH secretion and accelerate growth while the endogenous rate of GH secretion is high, growth acceleration by endurance exercise is more probable during childhood, when the rate of GH secretion and growth is suppressed (see figure 7.1) but the epiphyseal growth zones remain open. Thus, skeletal maturity is reported to be advanced in children who participate in certain sports, possibly as a consequence of stimulation by exercise of statural growth, unless it is a consequence of selection bias for children who are developmentally advanced (Lariviere & Lafond 1986; Malina et al. 1982; Thompson et al. 1974). In such cross-sectional studies, information on bone age is seldom available, and when it is available, it is seldom used for comparison of growth rates between exercising and less active children. Correction of growth curves of child participants in organized sports for the reported advance in the skeletal age suggests that the growth rates of exercising children are superior to those of their more sedentary controls (Malina et al. 1982; Malina 1994). Although there is little information on the GH-stimulatory effects of exercise during childhood, high-intensity aerobic training in 32-year-old women stimulated basal pulsatile GH secretion (Weltman et al. 1992). This suggests that as in mature hamsters, high-intensity endurance training is capable of removing the maturational inhibition over pulsatile GH secretion.

ENERGY PARTITIONING FOR HYPERTROPHIC GROWTH

Hypertrophic growth is a local increase in mass, density, shape, or function of a tissue, which increases the capacity of the tissue to meet functional or exercise-induced demands. Stimuli for hypertrophic growth originate within the responding tissue after it has been repeatedly subjected to increased functional demands. Hypertrophy is achieved through multiplication of subcellular components such as contractile proteins in skeletal and cardiac muscle, metabolic and biosynthetic enzymes, matrix proteins in cardiac muscle and skeletal tissue, or hydroxyapatite crystals in bone lamellae, rather than through cellular proliferation.

Relative independence of hypertrophic growth from either nutritional support or guidance by systemic hormones of growth is a feature that distinguishes it from incremental and reparative growth. Hypertrophic growth can be supported through reallocation of existing energy supplies and does not depend on abundance of ingested nutrients. Thus, overloaded skeletal muscle will hypertrophy even during complete starvation, although abundance of specific nutrients such as protein or calcium may modulate the magnitude of hypertrophic response (Goldberg et al. 1975).

Hypertrophic growth can occur without the guidance by systemic hormones of growth. Muscle stretch or overload causes increased transcriptional expression of IGF-I and of IGF-I–related mechano growth factor genes within the stimulated muscle (DeVol et al. 1990; Vandenburgh 1987; Yang et al. 1996b, 1997). Transduction of these local growth factor messages takes place via Ca^{2+}/calmodulin-dependent phosphatase, the calcineurin (Dunn et al. 1999; Musaro et al. 1999; Semsarian et al. 1999). These local growth factors then induce protein synthesis by autocrine or paracrine route, independently of pituitary GH or insulin actions (Goldberg et al. 1975; Goldberg 1967, 1968, 1979; Li & Goldberg 1976). Hypertrophy also can occur in the presence of large catabolic doses of glucocorticoids, which cause atrophy in muscles experiencing reduced contractile activity (Goldberg & Goodman 1969). Pharmacological doses of exogenous injected systemic hormones of growth can induce some muscle hypertrophy (see chapter 8), but at physiological concentrations, they only modulate the magnitude of localized hypertrophic response.

Hypertrophic growth is self-limiting in that it depends on the discrepancy between the magnitude of a functional demand and a tissue's ability to meet this demand. It is carried out through changes in transcription of genes for structural proteins, enzymes, and growth factors within the tissue (DeVol et al. 1990; Goldberg et al. 1975). Hypertrophy can also be carried out through translation of the already transcribed genetic message. The translational event consists of a rapid increase in ribosomal RNA and in the initiation of protein synthesis through phosphorylation of initiation factor p70^{S6K} at a time when no changes occur in the rate of gene transcription (Baar & Esser 1999; Wong & Booth 1990a, 1990b). Commensurate atrophy of tissue mass and functional capacity or changes in tissue shape ensue whenever there is a reduction in mechanical or metabolic functional demand within a tissue.

Metabolic Enzyme Hypertrophy

Changes in cellular oxygen tension, reactive oxygen species, ATP energy charge, metabolite concentrations, and metabolic hormones may be the key stimuli that lead to contractile protein and enzyme adaptations (Baar et al. 1999; Vinals et al. 1977). Hypoxia may be sensed by a mitochondrial heme-containing protein, which then activates PKC signaling cascade and affects expression of hypoxia-inducible factor 1 (HIF-1). At present, HIF-1 is known to act as a transcription factor for erythropoietin (EPO), vascular endothelial growth factor (VEGF), and some other genes. A change in oxidation-reduction state affects conformation of proteins that have amino acids with sulfhydryl groups such as cysteine. Generation of NO or reactive oxygen species hydrogen peroxide, superoxide, or hydroxyl radical affects MAPK/Elk-1 signaling through oxidation of its constituent Ras. It also can affect transcriptional activity of proto-oncogenes c-*fos*, c-*jun*, *jun* D, c-*myc* and block the action of protein tyrosine phosphatases, enzymes that dephosphorylate and inactivate protein kinases.

Cytoplasmic and mitochondrial enzymes are selectively genetically expressed and translated to better match oxidative requirements for muscle contractions of different speed, intensity, and duration (Holloszy & Coyle 1984; Holloszy & Booth 1976). Thus, acute exercise as well as chronic exercise up-regulate genes for cytoplasmic glycolytic enzymes, mitochondrial cytochromes, proteins and enzymes mediating intracellular glucose uptake and phosphorylation, intramitochondrial fatty acid uptake, and utilization of metabolic fuels (Jones et al. 1998; Leary et al. 1998; Mascaro et al. 1998; Montessuit & Thorburn 1999; O'Doherty et al. 1994). An acute exercise bout triggers transcription of the hexokinase gene (Hofmann & Pette 1994; O'Doherty et al. 1994).

Both low-intensity muscle contractions and adrenergic agonists trigger transcription of glucose transporter genes (Hofmann & Pette 1994; Jones et al. 1998; Montessuit & Thorburn 1999). Transcription of the succinate dehydrogenase gene is initiated by nuclear respiratory promoters under anaerobic conditions (Au & Scheffler 1998; Compan & Touati 1994). Oxidative metabolism facilitates transcription of the cytochrome oxidase gene (Leary et al. 1998).

Adaptive Shifts in the Skeletal Muscle Fiber Types

Muscle hypertrophy can take the form of changes in the expression of genes for different types of protein. Altered contractile and metabolic demands of exercise can trigger expression of genes for isoforms of myosin, actin, troponin, and tropomyosin proteins that can better meet the needs of changed exercise modality (Swoap et al. 1994; Tsika et al. 1987; Wiedenman et al. 1996).

Skeletal muscle contractile proteins exist as several isoforms differing in the intrinsic velocity of contraction (Goldspink 1998; Goldspink et al. 1992; Pette & Staron 1990). In particular, isoforms of myosin heavy chains are "molecular motors" capable of different rates of force generation (Goldspink 1998). They differ in specific activity of ATPase in their cross-bridge heads, the sites for actin attachment (Reisner et al. 1985). Immobilization or reduced skeletal muscle contractile activity results in the expression of default type IIx (or IIb) myosin heavy chain that has fast and highly fatigable properties (Loughna et al. 1990; Morgan & Loughna 1989; Swynghedauw 1986; Williams & Goldspink 1973). Progressively slower types IIx, IId, IIa, and I isoforms of myosin heavy chain are

expressed in response to progressively less powerful but sustained muscle contractions, while expression of genes for fast myosin isoforms is suppressed (Goldspink et al. 1991; Gregory et al. 1986; Ianuzzo & Chen 1979; Mabuchi et al. 1982; Schiaffino & Reggiani 1994; Sreter et al. 1982; Williams et al. 1986). Changes in heavy-chain myosin molecular motors appear to entail alterations in the configuration and size of electrically charged hypervariable amino acid loops that entrap ATP for different lengths of time (Goldspink 1998). Longer and more flexible loops of type IIx myosin allow faster ADP release at the end of contractile power stroke than do shorter and more tightly closed type I loops.

MRFs participate in muscle fiber differentiation in response to changed functional demand. Their production is triggered by IGF-I via two PI3-K transduction pathways (Adams et al. 1999; Lawlor & Rothwen 2000). Myogenin expression is greater in slow than in fast muscle fibers, and the reverse is true of Myo-D (Delling et al. 2000; Hughes et al. 1997; Sakuma et al. 1999). Increased mitochondrial activation during prolonged low-frequency electrical stimulation causes conversion of fast glycolytic to slow oxidative muscle fibers through increased expression of myogenin mRNA (Hughes et al. 1993, 1999; Kraus & Pette 1997; Muller et al. 1996; Putman et al. 2000; Rochard et al. 2000). In contrast, immobilization is associated with increased expression of Myo-D during fiber conversion from slow to fast (Kraus & Pette 1997). Muscle overloading causes increased Myo-D transcription and translation in rodent muscles (Tamaki et al. 2000). Myo-D then activates the promoter on the myosin heavy-chain IIx (IIB) gene (Wheeler et al. 1999). On the other hand, muscle stretch stimulates expression of myogenin mRNA in type II but not in type I fibers, and it stimulates transcription of MRF-4 in type I but not in type II fibers (Loughna & Brownson 1996). Stretch combined with electrical stimulation increases expression of MRF-4 and Myf-5 mRNA in type II muscle fibers (Jacobs-El et al. 1995). Proliferation of satellite cells is not necessary for expression of MRFs and their hypertrophic effects, although the two processes are usually cooperative (Lowe & Always 1999).

Catecholamines and thyroid hormones aid in such transcriptional shifts in metabolic enzyme and contractile protein adaptations to prolonged contractile activity (Powers et al. 1995). β Adrenergic stimulation promotes expression of myogenin but not Myo-D during immobilization, whereas T_3 increases expression of Myo-D during muscle regeneration and differentiation (Anderson et al. 1998; Delday & Maltin 1997) Similar to skeletal muscle hypertrophy, signaling in the conversion of fast to slow isoforms of contractile proteins entails activation of Ca^{2+}/calmodulin-dependent phosphatase, the calcineurin (Dunn et al. 1999).

Skeletal Muscle Hypertrophy

The stretch of contractile proteins is considered to be the key stimulus for muscle hypertrophy (Dix & Eisenberg

1991; Goldspink et al. 1992; Vandenburgh & Kaufman 1979). It initiates transcriptional and translational growth signals within the muscle, which includes satellite cell proliferation and fusion with existing muscle fibers and expression of MRFs (Adams et al. 1999). Proximal intracellular signaling in response to skeletal muscle stretch involves activation of L-type Ca^{2+} channels and increased calcium entry into the myocyte. Intracellular calcium concentration increases about 100-fold during skeletal muscle contractions as a result of opening of membrane Ca^{2+} channels as well as of intracellular L-type ryanodine Rs and IP_3-gated channels (Suzuki et al. 1998). Depending on the nature of contractile or stretch stimulus, either a fast sarcoplasmic/endoplasmic reticulum Ca^{2+}-transport ATPase (SERCA 1a) or a slow SERCA 1b is transcribed (Zador et al. 1999). Ca^{2+} binds to and activates a series of calcium-sensitive proteins and enzymes that include calmodulin, calcineurin, myosin long-chain kinase, MAPK, and PKC (Hefti et al. 1997) (see figure 3.7). Mechanical stress activates MAPK cascades that can include ERK1/2 c-Jun NH2 terminal kinase (JNK) and p38 kinase, which then activate ERKs (Aronson et al. 1998; Boppart et al. 2001; Wu et al. 2000). Additional signals include activation of protein TK and PLC_γ. Myogenic transcription factors Elk-1, CREB, c-fos, Myf 5, Myo-D, myogenin, and others next affect transcription by acting on gene promoters. While muscle contraction induces some hypertrophy by stimulating transcriptional events, a large portion of hypertrophic growth may be mediated by translational signaling of ribosomal $p70^{s6}$ kinase (Baar & Esser 1999).

Myofilament microdamage caused by stretching of the muscle is a stimulus for release of inflammatory cytokines, leukotrienes, and prostaglandins of the series E by the muscle or by monocytes attracted to the sites of microdamage (Baar et al. 1999; Mundy et al. 1996; Raisz 1996; Somjen et al. 1980). Damaged proteins first are removed after being covalently altered by ubiquitin-dependent signaling (Wilkinson 1999). Through ubiquitination, damaged proteins are targeted for degradation by the 26S proteolytic organelle proteasome. In this process, small proteins containing ubiquitin-like domains are attached to lysine on a target protein. Ubiquitination allows initiation of growth by first removing products of cellular damage caused by mechanical and environmental insults, infection, and mutation.

Loading-induced muscle hypertrophy also is facilitated through down-regulation of myostatin, a 26 kd member of the TGF-β family. Myostatin is a negative regulator of hypertrophic muscle growth that is expressed in the muscle (Gonzalez-Cadavid et al. 1998). It stops the progression of muscle cells from G_1 to S phase, in part through down-regulation of MRF Myo-D and activation of cyclin-dependent kinase inhibitor p21 (Rios et al. 2001, 2002; Thomas et al. 2000). Cattle and mice with a myostatin gene deletion develop pronounced muscle hypertrophy (Bass et al. 1999; Carlson et al. 1999; Grobet et al. 1997, 1998; Kambadur et al. 1997; Oldham et al. 2001; Sakuma et al. 2000; Smith et al.

1997c; Szabo et al. 1998). By contrast, muscle atrophy caused by immobilization, microgravity, HIV virus, and myocardial infarct is associated with increased expression of myostatin mRNA (Gonzalez-Cadavid et al. 1998; Lalani et al. 2000; Reardon et al. 2001; Sharma et al. 1999; Wehling et al. 2000; Zachwieja et al. 1999). As was the case with the four MRFs, myostatin also modulates muscle development and incremental growth (Kocamis et al. 1999).

Skeletal muscle (and skeletal) hypertrophy may be modulated by anabolic actions of several systemic hormones that are released during acute exercise (see chapter 5) such as GH, testosterone, IGF-I, cortisol, β-endorphin, and parathyroid hormone (Kraemer et al. 1989, 1991a, 1992b, 1993a, 1993b, 1995; Rong et al. 1997). GH is released in a dose-dependent fashion at exercise intensities above the anaerobic threshold and with a delay with respect to onset of exercise (VanHelder et al. 1984, 1987; Sutton & Lazarus 1976). The magnitude and pattern of exercise-induced GH release depends on the pattern and intensity of the exercise stimulus and the size and group of muscles employed (Kozlowski et al. 1983). Furthermore, high-resistance exercise performed with smaller loads that allow 5 to 10 repetitions elicits greater GH response than muscle loading done with heavy loads at lower repetition maximum (Kraemer et al. 1991a, 1993b). Exercise-induced increases in plasma GH concentration are usually not accompanied by dose-dependent increases in IGF-I secretion, and, conversely, increases in IGF-I seen during exercise are not always accompanied by concurrent increases in GH (Bang et al. 1990; Cappon et al. 1994; Fry et al. 1993; Hornum et al. 1997; Kraemer et al. 1992a, 1995; Rudman & Mattson 1994; Schwartz et al. 1996). This condition would suggest that acute release of growth-inducing IGF-I during exercise is not triggered by a preceding GH stimulus and that acute GH surges that accompany high-intensity exercise may not have an important anabolic function.

Attempts at selective increases in muscle size and strength through administration of supraphysiological doses of methyonyl GH (8 to 16 mg/week) for up to 18 weeks were uniformly unsuccessful in healthy volunteers and moderately successful in GH-deficient and hypogonadal adults (Crist et al. 1988, 1991; Crist & Kraner 1990; Hayes et al. 2001; Yarasheski 1994; Yarasheski et al. 1995). In GH-deficient men, treatment with GH for longer than 3 months increased thigh muscle mass and strength, whereas discontinuation of treatment reversed both changes (Cuneo et al. 1991a, 1991b; Jorgensen et al. 1989, 1991; Rutherford et al. 1991; Sonksen et al. 1991).

Acute high-resistance exercise and intense aerobic exercise also elicit release of androgens such as testosterone to a greater extent in males than in females (Kraemer et al. 1991a; Kraemer et al. 1992a). This release occurs by a mechanism other than pituitary LH secretion (Cumming et al. 1986). Pharmacological but not physiological doses of testosterone can cause muscle fiber hypertrophy and increases in strength and bone growth by periosteal apposition (Bhasin et al. 1996; Celotti & Negri Cesi 1992). It is probable that acute exercise-associated increases in androgen release may amplify training-induced hypertrophic growth. Acute high-resistance exercise also is associated with increased plasma cortisol concentration (Kraemer et al. 1993a, 1993b). Cortisol causes atrophy in muscle (particularly of the fast type II) and bone (Clark & Vignos 1979; Sambrook et al. 1990).

In addition, high-resistance exercise stimulates the release of a variety of growth factors, including nerve growth factor (Matsuda et al. 1991), epidermal growth factor, FGF, and osteocalcin (Gla protein) (Bell et al. 1988; Konradsen & Nexo 1988; Morrow et al. 1990; Nexo et al. 1988). Actions of the anabolic messengers are mediated by Shc, Grb2-SOS, Ras, Raf, and MEK signaling cascades. GH and growth factors initiate these transductions through JAK/STAT, and insulin initiates them through phosphorylation of TK (Daum et al. 1994). Actions of testosterone, anabolic steroids (see chapter 8), and cortisol are mediated by steroid hormone signaling pathways.

Cardiac Muscle Hypertrophy

Cardiac muscle hypertrophy is a response of cardiac myocytes to active contraction and passive stretch imposed by hemodynamic overload during exercise or pathological elevation of blood pressure (Schluter & Piper 1999; Watson 1996). Effects of contractile force appear to be sensed at the junction of costameres (protein extensions of Z lines in sarcomeres) and cytoskeletal proteins such as integrins (Dankowski et al. 1992; Ross et al. 1998). Stretch within the myocardium initiates intracellular signaling by activating Ca^{2+} ion channels, increasing cytosolic free Ca^{2+}, membrane phospholipases, and adhesion molecules, and it affects a number of different transduction pathways (Sackin 1995; Sadoshima & Izumo 1993a; Sharp et al. 1997; Sigurdson et al. 1992). Hemodynamic myocardial overload can activate TK, $p21^{ras}K$, S6 kinase, and calcium-sensitive PKC and MAPK signaling pathways (Sadoshima & Izumo 1993b; Yamazaki et al. 1996).

Cardiac growth response consists of induction of immediate early oncogene, or proto-oncogenes, c-fos, c-myc, and c-jun followed by transcription of growth factors, hormones, and structural proteins. Thus, increased expression of the c-fos gene is followed by transcription of the angiotensin II gene after myocardial stretch (Sadoshima and Zumo 1993b, 1993c; Sadoshima et al. 1993). Intracardiac angiotensin II stimulates transcription of a series of growth factors (TGF-β1, PDGF, basic FGF, ET-1, and IGF-I) and in this way amplifies its remodeling influence. Cardiac hemodynamic overload together with ET-1, also up-regulates the expression of ANF and brain natriuretic peptide or BNF (Bruneau et al. 1997). Cardiac hypertrophy also can occur by translation via the ribosomal $p70^{s6}$ kinase of already transcribed genes (Boluyt et al. 1997).

The heart muscle also undergoes remodeling in response to a variety of local and systemic chemical messengers that exert anabolic actions, increase blood pressure, or reduce vascular compliance. In addition to mechanical loading, growth factors FGF, TGF-β, IGF-I; cytokines IL-1β and cardiotrophin-1; and hormones NE, E, angiotensin II, endothelin, T_3, and estradiol interact to increase the rate and change the pattern of cardiac structural gene expression (Decker et al. 1997; Schaub et al. 1997; Hefti et al. 1997). FGF is associated with the extracellular matrix proteins glucosaminoglycan heparin and heparan sulfate proteoglycan and stimulates protein synthesis upon injury and activation of matrix degrading enzymes (Pasumarthi et al. 1994). TGF-β is needed for induction of hypertrophic growth by α adrenergic stimulation (Schluter et al. 1995). Hypertrophic growth of the heart can be stimulated by IGF-I arising from the liver in response to increased GH and T_3 secretion as well as by IGF-I expressed in cardiac myocytes (Kupfer & Rubin 1992). Complex interactions of IGF-I with its binding proteins are required for its binding to the IGF-I receptor and transport through cell membrane (Blakesley et al. 1996). IL-1β increases synthesis of β myosin heavy chain and ANF in cardiomyocytes and decreases expression of calcium channel genes (Thaik et al. 1995). It also inhibits hypertrophic effects of α adrenergic stimulation (Patten et al. 1996). Cardiotrophin-1 induces protein synthesis and changes in heart cell shape of the kind seen in volume as opposed to pressure overload (Wollert et al. 1996).

Hypertrophic response of the heart to adrenergic stimulation has been studied most to date (Fuller et al. 1990; Zimmer et al. 1995). Stimulation of α_1 rather than β_1 or β_2 receptors increases the number of ribosomes, increases protein synthesis, and changes the pattern of gene expression toward greater expression of β myosin heavy chain, creatine kinase B, and ANF (Schluter & Piper 1999). The transduction of α adrenergic hypertrophic stimulus entails either G_q, PLC, and PKC pathway and stimulation of MAPKs or activation of PI3 kinase (Haneda & McDermott 1991). PI3 kinase activates protein translational activity in the ribosomes by phosphorylating initiation factors $p70^{s6K}$ and eIF-4E (Schluter et al. 1998). In addition to these effects through α_1 A-Rs, NE can induce hypertrophy by acting through β A-Rs (Iaccarino et al. 2001). This action entails activation of calcium signaling pathways by enhancing L-type calcium channel in the heart.

Two additional vasoconstrictor hormones, angiotensin II and endothelin-1 cause cardiac hypertrophy and changes in gene expression (Paul & Ganten 1992; Rosendorff 1997; Sadoshima et al. 1993c, 1993d; Sugden & Bogoyevitch 1996). Angiotensin II R signaling is particularly complex in that its actions on AT_1- and AT_2-Rs usually are antagonistic. It utilizes a number of different GPCR-linked pathways, and it often transactivates growth processes through Rs that contain TK (Saito & Berk 2001) (see figure 3.5). AT_1-Rs stimulate increases in blood pressure, renal sodium reabsorption, and cardiac and vascular hypertrophy (Geisterfer et al. 1988). Hypertrophic actions of AT_1-R are primarily mediated by intracellular Ca^{2+} release, which engages PKC and activates MAPKs through the JAK/STAT transduction pathway (Dorn & Brown 1999; Marrero et al. 1995; Taubman et al. 1989). Angiotensin II also interacts with the IGF-I system (Delafontaine et al. 1996). Transcriptional effects of AT_1-Rs are mediated by way of proto-oncogenes such as c-*fos* (Babu et al. 2000; Taubman et al. 1989). In contrast, AT_2-Rs mediate reductions in blood pressure, cellular death (apoptosis), and inhibition of growth (Ichiki et al. 1995; Nouet & Nahmias 2000). They do so through G_i protein transduction and activation of protein phosphatases that dephosphorylate and inactivate MAPKs and other anabolic kinases (see figure 3.5) (Nouet & Nahmias 2000).

In developing cardiomyocytes, stretch induces release of angiotensin II, which in turn triggers autocrine release of endothelin, but growth stimulation by these hormones is reduced in adult heart cells (Ito et al. 1993; Wada et al. 1996). NE acting on α_1 and ET1 acting on ET_B Rs also stimulate matrix protein synthesis during cardiac and vascular remodeling (Pham et al. 2000). Stimulation of these Rs increases expression of the matrix adhesion receptor β (1D)–integrin and activates focal adhesion kinase (FAK). Transduction of endothelin-1 messages after ET-1 binding to its receptors is via G_q transduction pathway.

Thyroid hormones stimulate heart rate, contractility, ejection fraction, and coronary blood flow at the same time as they reduce peripheral vascular resistance. Thyroid hormones stimulate heart muscle contractility by increasing the expression of β_1 adrenergic receptor gene, which may lead to cardiac hypertrophy and heart failure (Bahouth 1991). T_3 also controls synthesis of cardiac mitochondria and a number of genes for cardiac contractile proteins and calcium channels (Mutvei et al. 1989a, 1989b; Mutvei & Nelson 1989; Zimmer et al. 1995). T_3 also mediates conversion of cardiac myosin heavy-chain isoforms from slow β variant to fast α variant (Kessler-Icekson 1988).

Estradiol (E2) has cardioprotective properties that are largely mediated by the hormone's capacity to reduce L-type calcium current, induce hemodynamic changes in the cardiac cycle, reduce the activity of visceral adipocyte LPL, increase activity of skeletal muscle LPL, and protect the heart muscle from free radical damage (Kim et al. 1996; Meyer et al. 1998; Pines et al. 1998; Wilson et al. 1976). However, E2 also stimulates growth of cardiac fibroblasts and release of atrial ANF, and it prevents cardiac hypertrophy (Cabral et al. 1988; Deng & Kaufman 1993; Grohe et al. 1996).

Bone Mass and Density

Bone shape and mineral density adjust in response to loading by mechanical forces. Changes in bone shape and mineral density to tensile, compressive, and shearing stresses go on during statural growth and in adulthood,

until the bone can resist the force without undergoing deformation or strain (Carter et al. 1996; Frost 1987). Bones undergo structural remodeling during growth and internal remodeling in adulthood through removal of bone tissue that is not exposed to strain and deposition of additional volumes of bone tissue or more densely packed mineral in bone areas that are experiencing increased strain (van der Meulen et al. 1996). Bone atrophies when there is a reduction in strain (Donaldson et al. 1970).

The relationship of strain to stress is linear within the elastic limit of bone rigidity, whereas above that range, younger bones undergo plastic deformations and older and more brittle bones break. Mechanical deformation of a crystalline bone matrix creates an electrostatic piezoelectric field and electrokinetic currents (streaming potentials) of electrolytes flowing through a bone matrix (Bassett 1971, 1993; McLeod et al. 1993; Yasuda 1954). This, plus local and systemic chemical messengers and local demineralization of the bone matrix caused by increased acidification of plasma and extracellular space during mechanical loading of bones are the stimuli for bone remodeling that are sensed by osteocytes (Ashizawa et al. 1998; Cunningham et al. 1985; Gordeladze et al. 2001; Henderson et al. 1989). Bone structurally remodels either in response to the large-amplitude strains generated by forces in the upper elastic range of stresses or to low-magnitude but high-frequency (10 to 60 Hz) strain energies that characterize actions such as standing and speaking (Rubin & McLeod 1996).

Structural Remodeling

Mineralization of lamellar or mature type of bone is molded by mechanical forces to orient lamellae or bone strands in parallel to prevailing stress trajectories (see figure 7.4), a principle first proposed by J. Wolff (1892) and now recognized as Wolff's law. During statural growth, changes in bone shape are largely accomplished by endosteal resorption and periosteal appositional growth, and as a result, the shafts of long bones also grow in diameter (see figure 7.5) (Carter et al. 1989; Parfitt 1994). As the mineral density of metaphyseal trabeculae increases in response to increased loading, the developmental loading history of adult bone is reflected in its skeletal microstructure and bone mineral density (BMD) (Beaupre et al. 1990; Carter et al. 1976, 1989; Courtney et al. 1994; Dalsky et al. 1988; Donaldson et al. 1970). Significant correlations exist between the size of body mass that is being moved, muscle strength, participation in physical activities or in specific sports, or weight-bearing exercise on the one hand and BMD on the other (Bevier et al. 1989; Block et al. 1989; Huddleston et al. 1980; Kannus et al. 1994; Kral & Dawson-Hughes 1994; Metz et al. 1993; Pirnay et al. 1987; Pruitt et al. 1992; Snow-Harter et al. 1990, 1992; Sowers et al. 1992; Zylstra et al. 1989). These relationships are apparent in cross-sectional and in prospective studies and remain detectable even under conditions that favor bone

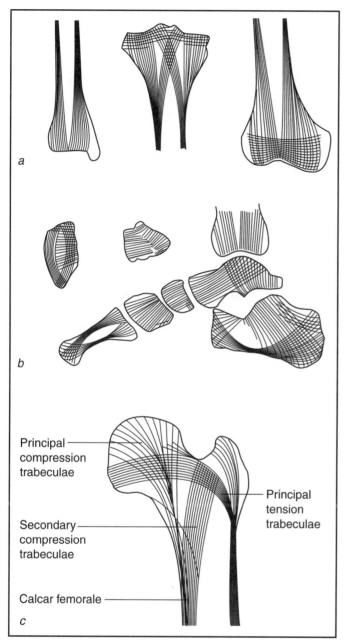

Figure 7.4 *Metaphyseal trabeculae in the long bones reveal stress trajectories.*

Trabecular orientation in the metaphyseal cancellous bone in the appendicular skeleton is aligned with stress trajectories according to Wolff's law and helps transmit mechanical loads from joints to bone diaphyses. (*a*) Long appendicular bones, (*b*) small bones of the feet, and (*c*) proximal femur.

Figures *a* and *b* adapted from Meyer 1867; figure *c* adapted from Einhorn 1992.

resorption such as athletic amenorrhea (see figure 7.6) or loss of estradiol in menopause (Ayalon et al. 1987; Dalsky et al. 1988; Pruitt et al. 1992; Robinson et al. 1995; Snow et al. 1996; Young et al. 1994). In general, increases in BMD are greater in the spine and loaded parts of the appendicular skeleton than in appendicular

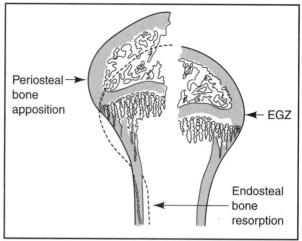

Figure 7.5 *Structural modeling of bone metaphysis during statural growth.*

A growing bone elongates in the epiphyseal growth zone (EGZ) and changes shape through periosteal bone apposition and endosteal bone resorption.

Adapted from Ham 1974.

bones that are not subjected to heavy loading (Bevier et al. 1989; Pocock et al. 1986).

Internal Remodeling

Internal remodeling represents turnover of bone tissue that operates throughout the life span and consists of periodic cycles of demineralization and remineralization of bone segments to assist in regulation of plasma calcium, renewal of structural components of bone, or repair of fatigue fractures. Internal remodeling primarily involves replacement of the resorbed bone with less pronounced change in its volume. It becomes the main process for changes in bone mineral density after the skeletal growth in length and structural modeling of bone shape have ceased. Internal remodeling is sensitive to fluctuations in plasma concentrations of parathyroid hormone (PTH), vitamin D_3, and calcitonin that serve to regulate plasma calcium concentration.

Modeling and remodeling are carried out within a bone modeling unit (BMU) by osteoblasts, bone-building cells that deposit and mineralize bone matrix, and osteoclasts, bone-resorbing cells derived from monocytes and macrophages (see figure 7.7). BMUs differ in size and shape according to bone structure. Immature, woven, or primary bone and bone that is first formed in response to a break consist of coarser and more randomly oriented mineralized strands (Einhorn 1996). Trabecular, spongy, or cancellous variant of mature bone, found at the ends (epiphyses) of long bones and inside cuboid bones such as vertebrae, also has randomly oriented bony lamellae or strands. It constitutes approximately 20% of skeletal mass, 65% to 90% of vertebral bone mass, and about 50% of femoral neck (Einhorn 1996). In these forms of bone, a BMU is a crescent-shaped trabecular segment on the endosteal surface within which osteoclasts resorb a lacuna.

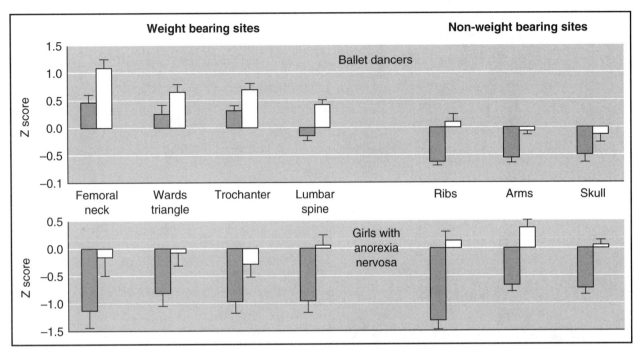

Figure 7.6 *Bone mineral density (BMD) response to mechanical loading and reduced energy availability.*

Protective role of mechanical loading on BMD under the conditions of negative energy balance is seen in loaded predominantly cortical bones in ballet dancers (solid bars, top left) but not in unloaded bones in either dancers (solid bars, top right) or amenorrheic nonathletes (solid bars, bottom). Mineral deficits in non-weight bearing sites containing a large amount of trabecular bone (ribs, lumbar spine) were no longer evident after correction for body fat mass (open bars), but only cortical weight bearing bones of dancers still revealed increases in BMD (open bars, top left).

Reprinted from Young et al. 1994.

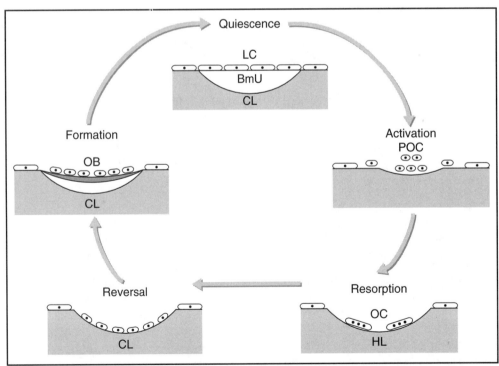

Figure 7.7 *Phases of modeling or remodeling in a bone modeling unit (BMU).*

LC = osteoblast lining cells; OB = osteoblasts; OC = osteoclasts (multinucleated); POC = preosteoclast cells (mononuclear); HL = Howship's lacuna; CL = cement line (reversal line).

Adapted from Parfitt 1984.

Cortical, dense, or compact bone, found in the shafts (diaphyses) of long bones and constituting approximately 80% of the skeletal mass, has tightly packed concentric tubular arrays of mineralized matrix that surround a central canal (a Haversian system) (Foldes et al. 1991). Embedded within the mineralized matrix are osteocytes, a quiescent form of osteoblasts, that are interconnected through threadlike processes (see figure 7.8). In the cortical bone, BMU consists of a Haversian canal within which osteoclasts carve out a "cutting cone" tunnel. Bone remodeling is faster in trabecular than in cortical bone because of the greater trabecular surface area relative to tightly packed Haversian canals and faster in appendicular trabeculae relative to axial trabeculae because of their greater mechanical than metabolic function (Parfitt 1996).

The modeling/remodeling sequence within a BMU (see figure 7.7) (Raisz 1999) begins as previously quiescent osteoblast lining cells on the bone surface activate osteoclasts. The lining cells first express collagenase and plasminogen activator for the synthesis of plasmin, some of the enzymes necessary to resorb organic osteoid from the bone surface (Allan et al. 1986; Heath et al. 1984). These enzymes, along with recruited marrow stromal cells and monocytes, next help release inflammatory cytokines IL-1, IL-6, and TNF-α and prostaglandin E_2 either in response to systemic hormones PTH and vitamin D_3 during internal remodeling or in response to changes in mechanical stimulation such as immobilization or strain (Baron et al. 1986; Bertolini et al. 1986; Manolagas & Jilka 1995; Mundy 1992; Partridge et al. 1987; Pead & Lanyon 1989; Raisz et al. 1972; Suda et al. 1992; Waters et al. 1991).

Released cytokines have both resorptive and anabolic actions. They activate osteoclasts to fuse into a multinucleated unit with a ruffled plasma membrane in contact with the bone (see resorption panel in figure 7.7) (Udagawa et al. 1990). During the next 1 to 2 weeks, activated osteoclasts resorb the mineralized bone matrix by acidifying it through operation of an H⁺-ATPase proton pump (Chatterjee et al. 1992). Coupling of mechanical and hormonal events also is achieved through release of osteogenic growth factors during bone resorption.

The transforming growth factor– β (TGF-β), osteocalcin (or γ-carboxyglutamic acid, Gla protein), osteopontin, integrins, and other proteins that are embedded within bone matrix are released during bone resorption and stimulate subsequent bone formation (Canalis et al. 1989; Hauschka et al. 1986; McKee & Nanci 1995; Mohan & Baylink 1991). Osteopontin facilitates early mineralization, osteocalcin facilitates bone resorption and regulates bone mineralization and geometry, and integrins lead to signal transduction, phosphorylation, and increased gene expression (Hynes 1992).

During the stage of reversal, capillaries invade the resorbed bone surface, and endothelial and mesenchymal cells of mesodermal origin differentiate into bone-forming osteoblasts to express genes for collagen-synthesizing enzymes, alkaline phosphatase, osteocalcin, osteopontin, and IGF-I and IGF-II (Shinar et al. 1993). Cytokines now facilitate bone formation through their stimulation of release of IGF-I and TGF-β (Boyce et al. 1989; Noda & Camilliere 1989). IGF-I dissociates from its binding protein under the influence of newly synthesized plasmin (Campbell et al. 1992). Over the next 3 to 4 months, these growth factors, along with TGF-β, bone morphogenetic proteins, and fibroblast growth factor (FGF) stimulate synthesis of collagen fibers in the bone matrix (Harris et al. 1994). Alkaline phosphatase catalyzes deposition of hydroxyapatite crystals, $Ca_{10}(PO_4)_6(OH)_2$, alongside collagen fibers (Kawaguchi et al. 1994). Conversion of osteoblasts to quiescent lining cells completes the cycle of bone remodeling (Parfitt 1984). With increasing age, the remodeling leads to faster bone resorption, particularly

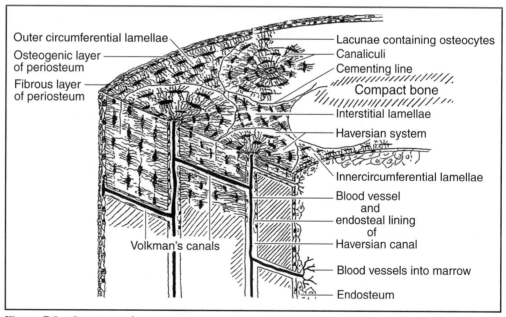

Figure 7.8 *Structure of compact bone.*

Longitudinal and cross-sectional view of Haversian system in the diaphysis of a long bone shows osteocytes embedded within mineralized concentric lamellae surrounding a central canal with a blood vessel. Osteocytes interconnect through tiny canals. Periosteal bone apposition occurs in the osteogenic outer layer, and bone resorption occurs at the inner endosteal surface. Haversian system becomes a bone modeling unit (BMU) during internal remodeling.

Adapted from Ham 1974.

along endosteal surfaces, than bone formation, particularly along periosteal surfaces. As a result, bone diameter grows larger as cortical walls grow thinner and lose mineral with increasing age (Ruff & Hayes 1982).

Systemic hormones such as parathyroid hormone (PTH), vitamin D_3, osteocalcin, GH, sex steroids, and glucocorticoids affect osteoblast and osteoclast behavior and thus influence complex interaction of mechanical and hormonal signals in coupling of bone resorption to bone formation (Farach-Carson & Ridall 1998). Close coupling of bone resorption to bone formation also is aided by actions of inflammatory cytokines and prostaglandin PGE (Jee et al. 1991).

Internal remodeling is influenced by three hormones that cooperate to maintain plasma calcium within the 8.5 to 10.5 mg/dl range (see figure 7.9) (Griffin & Ojeda 1996): parathyroid hormone (PTH), vitamin D_3, and calcitonin. PTH, which is produced by chief cells located in two pairs of parathyroid glands situated adjacent to the thyroid gland, increases plasma calcium concentration through three types of actions. (1) The first type is mobilization of calcium out of the bone. PTH can be released rapidly by stimulation of S nerves or circulating catecholamines and can stimulate calcium mobilization from the bone after acting on osteoclasts for several hours (Abe & Sherwood 1972; Blum et al. 1979; Fischer et al. 1982). (2) Even more rapid mobilization of calcium out of osteoid or recently formed bone is mediated by osteocytes in the presence of vitamin D_3 and without resorption of mineralized bone. (3) In the kidney, PTH increases calcium reabsorption in distal tubule and

stimulates synthesis of vitamin D_3 in proximal tubule (see figure 7.9).

Vitamin D_3 is either obtained from plant nutrients (as precursor vitamin D_2, or ergocalciferol) and milk (D_3, or cholecalciferol) or synthesized from 7-dehydrocholesterol by photostimulation of the skin by the B portion of the UV light spectrum (286 to 310 nm). This precursor undergoes two hydroxylations, the first one in the liver, and the second one, under PTH control, in the kidney (see figure 1.13) (Griffin & Ojeda 1996). The liver variant of vitamin D_3 (25-$(OH)D_3$) is the most abundant form of partially bioactive hormone that is present in circulation at 7 to 42 ng/ml and is stored in adipose tissue and muscle. The fully biologically active kidney variant of vitamin D_3 (1,25-$(OH)_2$ D_3) circulates with a half-life of about 3 h at concentration of between 20 to 50 pg/ml. Vitamin D_3 stimulates intestinal absorption of calcium through the synthesis of several calcium-binding and calcium-pump proteins (Breslau 1996).

Calcitonin helps regulate plasma calcium concentration through inhibition of calcium release by the bone. It is secreted by parafollicular or C cells of the thyroid gland in response to plasma ionized calcium concentration greater than 9 mg/dl and some gastrointestinal hormones such as gastrin. Calcitonin acts via the cAMP transduction mechanism. It inhibits bone mineral and matrix resorption by binding to osteoclasts and causing their dedifferentiation and inactivation. Calcitonin also causes transient increases in renal excretion of calcium, phosphate, sodium, potassium, and magnesium.

In addition to the three hormones that regulate plasma calcium, internal remodeling is influenced by

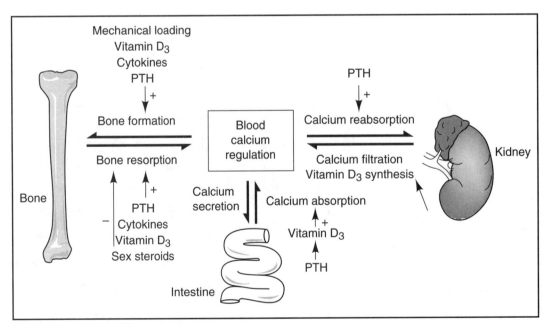

Figure 7.9 *Hormonal regulation of plasma calcium.*

Blood calcium is regulated through hormonal control of calcium flux among bone, intestine, and kidney on one hand and plasma on the other.

PTH = parathyroid hormone.

Adapted from Griffin and Ojeda 1996.

plasma concentrations of sex hormones. Systemic levels of estradiol and, to a lesser extent, androgens that act on the bone by aromatization to estrogens (see chapter 8) determine the overall balance of bone resorption and formation. Estrogen inhibits bone resorption by suppressing osteoclast recruitment and activity. It does so by inhibiting secretion by osteoblasts of the cytokine signals for bone resorption (Jilka et al. 1992; Manolagas & Jilka 1992; Pacifici et al. 1989). Estrogen also may stimulate bone formation by initiating synthesis of IGF-I and TGF-β (Ernst & Rodan 1991; Ernst et al. 1991; Verhaar et al. 1994). After estradiol withdrawal during menopause, with reduced estradiol titers during exercise-induced amenorrhea in young female athletes, and with reduced androgen titers in aging men, the pace of bone resorption becomes rate-limiting, and mechanical loading is effective only in bones of axial and appendicular skeleton that are exposed to stress but not in underloaded appendicular bones (see figure 7.6) (Robinson et al. 1995; Young et al. 1994). For both men and women, the rate of bone mineral loss is only moderately affected by mechanical loading, increased sun exposure, and dietary supplementation with calcium and vitamin D, whereas the estrogen replacement therapy in postmenopausal women is somewhat more effective in reducing this loss (Col et al. 1999; Dawson-Hughes 1996; Falkenbach et al. 1998; Lanyon 1996).

Exercise Influences on Bone Remodeling

Static loading of the bone is generally ineffective in stimulating remodeling, whereas dynamic intermittent mechanical loading, plasma acidification, and hormones elicited by exercise are particularly effective remodeling stimuli (Ashizawa et al. 1998; Cunningham et al. 1985; Henderson et al. 1989; Lanyon & Rubin 1984). Exercise causes a transient decrease in markers of bone formation (Ashizawa et al. 1998). This condition is a consequence of a large increase in plasma ionized and total calcium concentration. Exercise-induced increase in plasma calcium cannot be attributed to a rise in PTH concentration; such increases during exercise are inconsistent and have a time course that does not match that of PTH concentration (Henderson et al. 1989). Instead, it is a result of exercise-induced acidification of plasma (Cunningham et al. 1985). Calcitonin concentrations also do not show a consistent pattern of change with acute exercise (Ashizawa et al. 1998). Exercise training bouts of several weeks duration are associated with decreased plasma 1,25-$(OH)_2D_3$ and calcitonin and increased PTH concentrations (Klausen et al. 1993).

The intermittent nature of exercise may promote bone formation through intermittent exercise-induced increases in PTH and GH secretion. Sustained high titers of PTH and IL-1 promote bone demineralization and cavitation, whereas intermittent stimulation by both messengers promotes bone remodeling and bone formation (Boyce et al. 1989; Tam et al. 1982). By virtue of being episodic, exercise intermittently elicits secretion of chemical messengers such as GH, which stimulates

expression of IGF-I genes in bone and muscle (Isgaard et al. 1988a, 1988b).

Overall bone mineral balance in adults is affected by age-associated changes in bone remodeling as well as by lifestyle practices. With normal dietary intake (though sufficienct intake of calcium and vitamin D are missing even in developed countries) and regular mechanical loading, skeletal mass and mineral density grow in size into the third decade of life (Celotti & Bignamini 1999; Pennington et al. 1989; Tucker 1995). With inadequate diets and inactivity, bone mineral loss of about 1% a year begins in the third decade of life, increases during menopause, and can result in osteoporosis and bone fractures in later life.

ENERGY PARTITIONING FOR ENERGY BALANCE AND REPARATIVE GROWTH

Regulation of energy balance entails matching of energy intake and energy expenditure to produce stability of body mass and composition in nongrowing adult mammals. Application of the term regulation of the constancy of internal environment to either the energy balance or stability of body mass–body composition is at variance with Claude Bernard's (1859) and Walter Cannon's (1932) definition of this concept. Regulation implies operation of compensatory negative-feedback mechanisms that help maintain internal environment variables within a narrow range of values. In the case of regulation of energy balance or of stability of body mass and composition, no unitary regulated variable in the internal environment has been identified as being responsible for the short-term and long-term maintenance of energy equilibria. Neither the individual circulating metabolic fuels, core temperature as the ultimate product of metabolism, nor the amount of storage glycogen and lipids satisfy the criterion of a regulated internal-environment variable that could fully account for the long-term stability of adult body mass and composition.

Likewise, there is incomplete understanding of the short-term and long-term mechanisms that produce stability of body mass and body composition. Meal size is often more strongly influenced by circadian and sensory factors than by preceding energy expenditure, resulting in remarkable misalignment of energy intake and energy expenditure in the short term of hours or a day (Edholm 1977; Edholm et al. 1955). On the other hand, over longer periods of days and weeks, energy intake and expenditure are well matched, resulting in stable body mass and body composition plateaus that are defended against upward or downward displacements. After a caloric restriction and a weight loss of about 10% to 15%, animals display compensatory increases in food intake and decreased metabolic rate until they return to their predeprivation weight (Borer & Kooi 1975; Borer & Kelch 1978; Mitchel & Keesey 1977). Likewise, after being induced to gain weight and overeat through stimulation of the LHA, rats return to their normal body mass and fatness through a reduction in food consumption after the stimulation has been discontinued (Steffens 1975).

Despite this evidence that humans and animals actively maintain body mass and composition over long periods of time, energy regulatory concept cannot account for the prevalence of a wide range of plateaus or "settling points" around which human and animal weight and fat reserves can equilibrate, from anorexic leanness to morbid obesity (Melby et al. 1998). Different energy plateaus reflect variable genetic predisposition for energy conservation, the ease of access to palatable and calorically dense food, and the volume of habitual physical activity (Bouchard 1994). So long as these predisposing conditions remain unchanged, the different energy plateaus are actively maintained in response to homeostatic challenges and do not reflect regulation of any single physiological variable. This discussion of energy balance and exercise will outline the current, but still incomplete, understanding of short-term and long-term neuroendocrine, behavioral, and physiological mechanisms responsible for the stability of body mass and body composition. It will include both facts and hypotheses regarding the relationship of the energy balance mechanism to physical activity.

Short-Term Regulation of Energy Balance

In the time frame of hours, we control energy intake by eating meals and expend energy by varying either our basal metabolic rate (BMR), diet-induced thermogenesis (DIT), or levels of voluntary activity.

Hunger and Meal Initiation

Feeding behavior is under nonhomeostatic as well as homeostatic control. Circadian, sensory, and environmental variables represent nonhomeostatic influences over feeding. Endogenous circadian and ultradian oscillators, of which at least one resides in the suprachiasmatic hypothalamic nucleus (SCN), impart an intermittent and a conspicuously temporal organization to feeding (Zucker & Stephan 1973). The anatomical connections between hypothalamic centers controlling feeding and the hypothalamic SCN nucleus help ensure regular meal taking in anticipation of actual energy needs (Buijs et al. 2001; Watts & Swanson 1987; Watts et al. 1987). The timing of meals every 3 to 6 hours operates in conjunction with short-term signals arising from gastric contractions associated with gut emptying after digestion and absorption of food.

Feeding also has a strong opportunistic character in that meal initiation and meal size are strongly influenced by sensory properties of the food as well as by social and situational variables. Humans and animals overeat palatable, sweet, and energy-rich food and undereat unpalatable food, and these hedonic responses to food may be mediated by endogenous opiates (Drewnowski et al. 1992; Giraudo et al. 1993; Le Magnen 1999a; Le Magnen & Julien 1999). As a result, under nondepriva-tion conditions, there is little correlation between energy expenditure before the onset of a meal and the ensuing meal size (Bernstein et al. 1981; Le Magnen 1999b). This imbalance contributes to development of obesity in an environment of overabundance of commercially designed foods of maximal palatability (Klesges et al. 1992; Pearcey & de Castro 2002).

The endogenous rhythm of spontaneous meal initiation is however amplified by signals of energy shortage. A small drop in plasma glucose concentration precedes meal onset (Campfield 1997; Louis-Sylvestre & Le Magnen 1980). Declines in cellular fuel availability result in compensatory increases in food intake commensurate to energy deficits. Thus, although the meal size is highly variable, sensitive to sensory properties of food or situational variables, and unrelated to preceding energy expenditure, the number of calories eaten directly influences the latency to the next meal (Bernstein et al. 1981). Extended intermeal intervals that result in energy shortage stimulate both meal size and meal frequencies. Homeostatic eating, manifested as increased meal frequency and meal sizes, occurs in response to either greater energy expenditure or extended intermeal intervals (Schwartz et al. 2000).

The neuroendocrine mechanism that transduces sensory encounters with food at appropriate intervals into a feeding response entails activation of lateral hypothalamic (LHA) structures by neurons releasing several appetite-stimulating (orexigenic) peptides and neurotransmitters, which in turn make connections with cerebral cortical, autonomic, and brain stem premotor neurons responsible for motivated and stereotypic feeding behaviors (Bittencourt et al. 1992). The concept of an LHA feeding center was posited more than half a century ago by Hetherington and Ranson (1940). They identified two areas of the hypothalamus as centers regulating feeding and body fatness, which subsequently gave rise to a dual center hypothesis of hunger (Stellar 1954). The perifornical lateral hypothalamic area (LHA) was singled out as the area associated with hunger and feeding initiation, and the medial basal hypothalamus, including arcuate, ventromedial hypothalamic (VMH) neurons, dorsomedial nucleus (DM), and paraventricular nucleus (PVN), as areas associated with satiety and feeding cessation. Subsequent associations were made among LHA, secretion of insulin, and suppression of S activity and between the ventromedial basal hypothalamic centers and activation of hormonal and S mechanisms of energy expenditure. The association of feeding initiation and suppression of energy expenditure with the LHA is now supported by the recent discovery of a variety of neurons in the LHA that release appetite-stimulating peptides, which are the targets of multiple neuronal projections from the PVN and arcuate hypothalamic nuclei that control feeding and suppress S tone (see figure 7.10) (Cowley et al. 1999; Elmquist et al. 1999; Kalra et al. 1999).

Among appetite stimulating neurons in the perifornical LHA are overlapping populations of neurons that contain melanin-concentrating hormone (MCH) and orexins or hypocretins (Bittencourt et al. 1992; Date et al. 1999; de Lecea et al. 1998; Edwards et al. 1999; Elias et al. 1998a, 1998b; Hakansson et al. 1999; Lubkin & Stricker-Krongrad 1998; Sakurai et al. 1998). Their peptide messengers, MCH and orexins, elicit feeding, and their neuronal mRNA content rises with starvation and declines with refeeding (Qu et al. 1996). Feeding stimulation and suppression of metabolic energy expenditure also are mediated by neurons in the medial arcuate hypothalamus that synthesize and release neuropeptide Y (NPY) (Billington et al. 1994; Clark et al. 1984; Kotz et al. 1998a). These neurons overexpress NPY during starvation and stimulate selective intake of carbohydrates in response to reduced glucose utilization (Ahima et al. 1996; Akabayashi et al. 1993; Beck et al. 1994; Stanley et al. 1985; Wang et al. 1998a). Such feeding stimulation at appropriate circadian times (Akabayashi et al. 1994a) is achieved through action of these NPY neurons on the MCH and orexin cells in the perifornical LHA (Stanley et al. 1993). NPY neurons also increase insulin and cortisol secretion, reduce activity of adipose tissue LPL, and decrease S activation of brown adipose tissue and energy expenditure during starvation (Akabayashi et al. 1994a; Cowley et al. 1999; Kotz et al. 1998b; Wang et al. 1998b). NPY upregulation in the arcuate nucleus in turn depends on cortisol that is released in response to eating, particularly of high-carbohydrate diets (Akabayashi et al. 1994b; Wang et al. 1998a). The arcuate NPY neurons coexpress a second hormone, the agouti-related peptide (AgRP), that is also implicated in stimulation of feeding. AgRP binds to melanocortin 3 (MCN3) and melanocortin 4 (MCN4) receptors in PVN and DM, where it blocks their anorexic action. Therefore NPY/AgRP neurons stimulate feeding by acting on orexigenic neurons in the LHA and anorexigenic neurons in the PVN through release of two appetite-stimulating messengers (see figure 7.10).

In contrast to the chemical coding for carbohydrate inges-

tion mediated by stimulation of PVN neurons by NPY and by NE neurons acting on α_2 A-Rs, another peptide, galanin, has been implicated in elicitation of fat intake (Kyrkouli et al. 1990; Leibowitz 1998; Tempel & Leibowitz 1993; Wellman 2000). Galanin is overexpressed in the anterior PVN during food deprivation when fat utilization is increased. Its expression is unaffected by

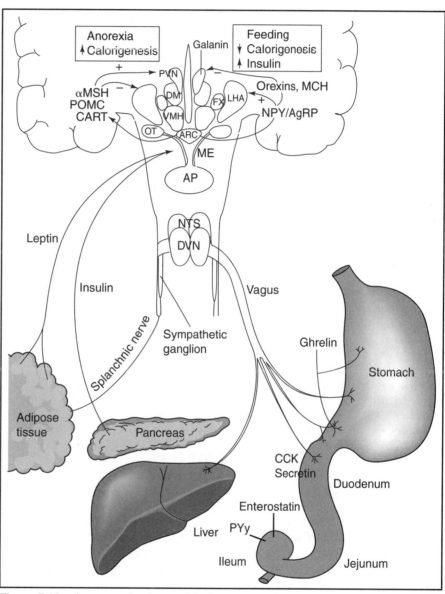

Figure 7.10 *A neuroendocrine model of energy regulation.*

Feeding is stimulated by ghrelin and peripheral sensory stimuli from the mouth and the gastrointestinal tract acting through the ventromedial hypothalamic (VMH) and arcuate nucleus neurons that secrete NPY/AgRP orexins and melanin-concentrating hormone (MCH) cells in the lateral hypothalamic area (LHA). Satiation at the end of the meal is mediated by cholecystokinin (CCK) acting via the vagus nerve on the nucleus of the solitary tract (NTS). Intermeal satiety is mediated by leptin and insulin action on the anorexia-mediating neurons in the arcuate nucleus (secreting αMSH and CART peptide) VMH, dorsomedial nucleus (DM), and paraventricular nucleus (PVN).

FX = fornix; MCH = melanin-concentrating hormone; NPY = neuropeptide Y; AgRP = agouti-related peptide; ARC = arcuate nucleus; ME = median eminence; AP = anterior pituitary; POMC = proopiomelanocortin; CART = cocaine and amphetamine regulated transcript; DVN = dorsal motor nucleus of the vagus; pYY = peptide YY; CCK = cholecystokinin.

reduced glucose utilization but is reduced when lipid utilization is prevented (Wang et al. 1998b). Some of its pleiotropic actions stem from its colocalization with other neurotransmitters in the anterior pituitary, adrenal medulla, and S as well as sensory nerves. In this capacity, it modulates gastrointestinal motility, potentiates NE-mediated stimulation of hepatic glucose production, and modulates GH, PRL, and ACTH release (Taborsky et al. 1999).

The most recently discovered appetite-stimulating peptide, synthesized both in the arcuate hypothalamic nucleus, duodenum, stomach antrum, and the acid-producing parietal (oxyntic) cells of the stomach, is ghrelin (Kamegai et al. 2000; Kojima et al. 1999). Ghrelin can initiate feeding, apparently by activating the hypothalamic NPY, increase gastric motility and acid secretion, and potentiate GHRH-induced GH secretion (Cummings et al. 2001; Hataya et al. 2001; Kojima et al. 1999, Masuda et al. 2000; Nakazato et al. 2001). Ghrelin expression is increased during fasting and after weight loss, suggesting that it may be responsible for compensatory hyperphagia and body weight regain (Cummings et al. 2002; Toshinai et al. 2001). Ghrelin secretion peaks before a meal, whereas feeding and hyperglycemia suppress its release (Cummings et al. 2001; Dornonville de la Cour et al. 2001).

Satiety and Meal Termination

Meal termination is associated with sensations of oropharyngeal satisfaction and stomach fullness and their accompanying chemical, hormonal, and mechanical stimuli (Covasa & Ritter 1999; Rolls 1999). Oral sensation of sweetness may elicit a cephalic phase of insulin secretion, and insulin is one of the hormonal mediators of satiety (Bellisle et al. 1985; Woods et al. 1996). When a similar diet is offered with a covert variation in fat content, the quantity of food eaten is more closely related to dietary bulk than to its caloric density. Compensation for caloric variation of a diet is very limited, even over a period of several weeks (Kendall et al. 1991; Lissner et al. 1987; Stubbs et al. 1995). Several gastrointestinal hormones associated with meal termination, including cholecystokinin (CCK), enterostatin, and gut peptides collectively called enterogastrones, suppress appetite and secretion of gastric acid, hormones, and enzymes associated with meal initiation. Most also suppress gastrointestinal motility and increase metabolism (diet-induced thermogenesis or DIT) through activation of S nerves and up-regulation of thermogenic uncoupling proteins UPC1 and UPC2 in brown and white adipose tissues (Bray 2000b; Rippe et al. 2000).

CCK causes early postprandial satiety and behavioral somnolence (Degen et al. 2001; Gibbs et al. 1973; Moran 2000; Pi-Sunyer et al. 1982; Woltman et al. 1995; Woltman & Reidelberger 1996). It is secreted from the duodenum in response to fat in the diet, as its primary function is ejection of bile from the gall bladder and fat emulsification in addition to digestive enzyme secretion from the pancreas. CCK also mediates estradiol-induced suppression of appetite during estrous and menstrual cy-

cles (Geary et al. 1994). After stimulating CCK A and B receptors of the gastrointestinal vagus nerve, CCK signal is transmitted to the NTS and the brain regions involved in the control of feeding (figure 7.10) (Corp et al. 1993; Glatzle et al. 2001; Schwartz et al. 2000). A pentapeptide enterostatin shortens the meal duration and delays its onset by acting both in the gut and in the brain (Lin et al. 2000; Lin & York 1998a). It selectively reduces fat intake in animals adapted to high-fat diet and decreases insulin secretion to a glucose stimulus (Bray 2000a; Ookuma & York 1998). Enterostatin is located in endocrine cells in stomach antrum, duodenum, and intestine, and it is formed in the intestine from pancreatic colipase (Erlanson-Albertsson & York 1997; Lin & York 1998b; Liu et al. 1999; Sorhede et al. 1996). Enterogastrones are peptides that contribute to satiety by decreasing gastric acid secretion and gut motility. They include peptide YY from the ileum, recently shown to mediate satiety in mice (Batterham et al. 2002), secretin and CCK from the duodenum, gastric inhibitory peptide (GIP), and glucagonlike peptides GLP-1 and GLP-2 (Burckhard et al. 1994; Chen et al. 1997; Li et al. 1998; Wojdemann et al. 1998).

In addition to the satiety-inducing gastrointestinal hormones acting either on the afferent PS fibers arising in gastrointestinal organs or through the circulation on hypothalamic neurons, satiating properties of nutrients also are signaled by several peptides produced by neurons in the hypothalamus (see figure 7.10) (Walls et al. 1995; Yox et al. 1991). Chief appetite suppressing neurons are located in the arcuate nucleus. They coexpress POMC and CART (cocaine and amphetamine regulated transcript) peptide, and their axons project to both the PVN and the LHA (Elmquist et al. 1999; Kristensen et al. 1998). Their α-MSH messenger acts on melanocortin MCN3 and MCN4 receptors on TRH and GABA-ergic cells in the PVN. Their projections to S preganglionic neurons cause S activation (Huszar et al. 1997; Koylu et al. 1997, 1998; Ludwig et al. 1998; Satoh et al. 1998). These neurons also project to the LHA area where they inhibit feeding by acting on MCN3 and MCN4 receptors on MCH cells (Fan et al. 1997). Parvocellular PVN, neurons that inhibit feeding, synthesize TRH, oxytocin, and CRF (Kelly & Watts 1996, 1998; Schwartz et al. 2000; Wilson et al. 1991).

Finally, satiety also is mediated by the brain serotonin (5HT) circuits acting on 1B and 2C 5HT receptors and by noradrenergic pathways acting on β_2 and β_3 receptors in the medial basal and PVN regions of the brain (Bray 2000b; Halford & Blundell 2000). These circuits appear to serve the episodic, short-term, meal-associated satiety, sometimes exerting an influence on the NPY neurons but in general acting independently of the suppression of long-term hunger drive by leptin (Halford & Blundell 2000).

Effects of Exercise on Feeding

As a means of energy expenditure, exercise often is used to generate body fat loss. Its effectiveness in mediating fat loss is based on the expectation that compensatory increases in food consumption will not occur or that metabolic energy expenditure will increase. It is therefore important to establish whether these two assumptions are justified by examining the relationship of exercise to short-term mechanisms of energy regulation.

The impact of exercise on feeding and satiety depends on its volume and intensity, energy density, and possibly palatability of food, and the gender and body fat content of exercising humans or animals. No increase in food intake to 1,200 kcal exercise energy expenditure occurred for 2 days, and feeding compensation was incomplete at higher, compared with lower, exercise volumes (King et al. 1997a; Woo & Pi-Sunyer 1985). In comparison with a missed meal, the energy expenditure of exercise did not produce a sensation of increased hunger (Hubert et al. 1998), and a negative energy balance was incurred when a low-fat meal was fed after exercise to lean men and women (King et al. 1996; King & Blundell 1995). When a high-fat meal was offered after exercise, both genders ate more calories than were expended during exercise.

An anorexic effect of exercise is seen in males more than in females, particularly if exercise is sufficiently intense. Thus in male rats, intense exercise suppresses food consumption and causes body fat loss to a greater extent than less intense exercise (Katch et al. 1979; Nance et al. 1977). In lean male humans, the feeling of hunger is reduced by intense exercise either briefly or over 24 hours, and the onset of eating usually is delayed (King et al. 1994, 1997b; King & Blundell 1995; Thompson et al. 1988). In female rats, exercise produces compensatory increases in food intake and defense of stable weight plateau (Nance et al. 1977). In lean women, an intense exercise session leads to increased palatability rating of foods and no suppression of hunger (King et al. 1996), and several days of exercise at different intensities produces accurate compensation for energy expenditure (Woo & Pi-Sunyer 1985). Thus, in the short term, coupling of energy expenditure to the mechanisms that control feeding is looser in men than in women and in the obese than in the lean and thus produces often inaccurate matching of energy intake to energy expenditure (Blundell & King 1999; Hubert et al. 1998; King 1999; King et al. 1997a). Intense or exhaustive exercise can therefore be exploited for weight loss provided the fat content and palatability of food are not high. This strategy is more likely to be effective in men than in women and in the obese than in lean individuals.

The mechanism through which intense exercise temporarily blunts compensatory increase in food intake is not understood. Stress hormones such as CRF, ACTH, cortisol, and catecholamines that are secreted in proportion to exercise intensity (see chapter 5) could exert an anorexic effect (Lobo et al. 1993; Bagdy et al. 1989). However, other endocrine changes are likely to stimulate feeding. Thus, GH also is secreted at increased exercise intensities, and its releasing peptides, GHRH and ghrelin, are associated with circadian stimulation of appetite and in particular with ingestion of protein (Dickson & Vaccarino 1994; Feifel & Vaccarino 1994; Vaccarino et

al. 1995). In addition, suppression of insulin secretion during intense exercise should diminish its satiating action (Schwartz et al. 2000; Woods et al. 1996).

CCK is a more likely endocrine mediator of temporary postexercise anorexia, as its plasma concentration rises during intense exercise in parallel with the ratio of plasma tryptophan to branched-chain amino acids (BCAAs) (Bailey et al. 2001). Both changes could contribute to satiety through increased brain uptake of tryptophan and a consequent increased synthesis of satiety-inducing 5HT (Wurtman & Wurtman 1996). Female athletes also display lower release of CCK and higher subjective hunger ratings during eating compared with sedentary young women. This condition could reflect a deliberately achieved or unintended state of negative energy balance in the athletes (Hirschberg et al. 1994). Nothing is known about the possible satiating role of enterostatin or enterogastrones after exercise.

Another variable that modifies the effects of exercise on appetite and food consumption is degree of body fatness. A recurrent observation in studies of how exercise affects feeding is that obese men and women fail to compensate for energy costs of exercise either by showing an anorectic response or by inadequately increasing food intake. Incomplete dietary compensation is sometimes inferred from the discrepancy between expected and actual fat loss relative to energy expenditure (Boileau et al. 1971; Dempsey 1964; Duddleston & Benniou 1970; Kissileff et al. 1990; Leon et al. 1979; Oscai & Williams 1968; Woo et al. 1982a, 1982b). A similar phenomenon was reported in female rodents, who lost some of their initially higher body fat and displayed anorexia in response to a moderate volume of exercise energy expenditure. After their body fat settled at a lower level, these animals appropriately increased food consumption to maintain weight stability in response to higher volumes and intensities of exercise (Mayer et al. 1954). The hormonal or physiological basis for this interaction of body fatness and exercise is not fully understood, but it could be related to resistance to insulin and to leptin action that develops in obesity.

Metabolic Adjustments to Energy Deviations
Homeostatic linkage of metabolism to the energy regulatory mechanism would require a decline in basal metabolic rate (BMR) after energy expenditure and an increase in metabolism after food intake. This condition occurs only when energy balance is altered through variation in nutrient availability. A transient postmeal increase in metabolic heat production, DIT, or specific dynamic action of food is homeostatic. An increase in the metabolism corresponding to 10% to 25% of energy content of ingested carbohydrates and proteins is caused in part by increased insulin secretion and in part by insulin-stimulated S activation (Acheson et al. 1983, 1984a, 1984b). In contrast, little if any DIT is elicited by fat ingestion. In addition, peptides that stimulate and inhibit feeding have reciprocal short-term effects on S control of metabolic rate (Bray 2000b).

Changes in BMR in response to energy deficit also occur in a homeostatic fashion. BMR, which accounts for about 65% of total daily oxidative metabolism, decreases in starvation because of suppressed S tone and reduced thyroid hormone secretion (Kinlaw et al. 1985; Young & Landsberg 1997). This decrease in BMR helps prevent loss of body mass and death through emaciation. BMR increases in obesity because of an increase in S tone generated by high plasma insulin concentration (Landsberg 1996; Prentice et al. 1996). This increase in BMR helps contain and reduce obesity. However, obese individuals have a blunted DIT compared with lean individuals (De Jonge & Bray 1997). Reduced DIT in the obese is a consequence of their reduced sensitivity to insulin and reduced insulin-induced S stimulation (Segal et al. 1992).

In contrast to homeostatic metabolic adjustments to nutritional disturbance of energy balance, exercise causes nonhomeostatic changes in metabolic rate. Energy expenditure of exercise is followed by EPOC (excess postexercise oxygen consumption) during the immediate postexercise period (Gaesser & Brooks 1984; Short & Sedlock 1997). With exercise of moderate intensity and duration, increases in metabolic rate are modest and wane within an hour (Freedman-Akabas et al. 1985). When exercise of high intensity is performed over a longer period of time, EPOC may be significant and may contribute substantially to daily energy expenditure (Bahr & Sejersted 1991). In addition, high-resistance exercise may elevate postexercise metabolic rate for up to 48 to 56 hours (Herring et al. 1992; Melby et al. 1993). Furthermore, reduced DIT in the obese can be alleviated if exercise takes place before a meal. Exercise can produce an acute increase in muscle sensitivity to insulin, which then triggers a larger postprandial DIT response (Segal et al. 1992).

Long-Term Regulation of Energy Balance

Mechanisms of long-term regulation of energy balance entail influence of the state of the body mass and body energy reserves on moment-to-moment urges to eat and exercise and on controls of metabolic rate. A half a century ago, Kennedy (1953) postulated that a circulating messenger out of adipose tissue participated in the long-term stability of body composition in adulthood. The anatomical basis for this lipostatic hypothesis was provided by Hetherington and Ranson (1940), who identified the medial basal hypothalamus, corresponding to arcuate, VMH, DM, and PVN nuclei, as areas associated with satiety and feeding cessation, the ablation of which caused overeating, hypoactivity, and obesity. The identification of the LHA as the feeding area and the medial basal hypothalamus as the satiety area led to the concept of homeostatic energy regulation by two antagonistic brain centers (Stellar 1954). The identification of the lipostat with the medial basal hypothalamus has now merged with a new understanding of the energy regulatory actions of leptin and insulin on these brain substrates (Ahima & Flier 2000; Elmquist et al. 1998a,

1998b, 1999; Flier 1998; Schwartz et al. 2000; Spiegelman & Flier 2001; Woods et al. 1996).

Leptin is predominantly synthesized and released by the subcutaneous adipocytes in a stable pulsatile fashion with a secretory peak around midnight (Campfield et al. 1995; Halaas et al. 1995; Licinio et al. 1998; Pelleymounter et al. 1995; Zhang et al. 1994). As its plasma concentrations rise in proportion to the size of adipocyte fat content, whether the level of fatness is influenced by dietary interventions or habitual exercise, leptin may communicate long-term state of body adiposity to the brain (Considine et al. 1996; Elmquist et al. 1999; Friedman & Halaas 1998; Kohrt et al. 1996; Lonnqvist et al. 1995). Obesity in ob/ob mice and in some humans is caused by a mutation in leptin gene that results in synthesis of defective leptin protein (Flier 1998; Flier & Maratos-Flier 1998; Montague et al. 1997; Spiegelman & Flier 1996). Obesity in db/db mice results from incorrect transcriptional splicing of the leptin receptor mRNA that produces a truncated leptin receptor (Flier 1998; Flier & Maratos-Flier 1998; Spiegelman & Flier 1996). The parabiosis experiments, in which ob/ob mice were surgically joined to share systemic circulation with db/db mice, convincingly illustrated the important role of leptin and its receptors in the suppression of appetite and body fatness and lent support to Kennedy's lipostatic hypothesis (Coleman 1973; Coleman & Hummel 1969).

Leptin's ability to link the long-term status of body fat reserves with short-term shifts in energy balance is revealed in disproportionate decreases in its secretion during acute energy shortages, in its potentiation in the brain of the satiating effects of CCK, and in its antagonistic and reciprocal interactions with the orexigenic gut peptide ghrelin (Boden et al. 1996; Emond et al. 1999; Shintani et al. 2001). Although exercise does not acutely influence leptin secretion, short-term exercise-induced changes in energy balance affect leptin secretion during the following night (Perusse et al. 1997). Negative energy balance, caused by either exercise or dietary restriction, suppresses nocturnal leptin secretion. Positive energy balance triggers increased nocturnal leptin secretion whether it is generated through increased diurnal energy intake or decreased diurnal energy expenditure (van Aggel-Leijssen et al. 1999). Acute increases in plasma leptin precede the onset of puberty, and leptin administration advances this maturational event (Ahima et al. 1997; Garcia-Mayor et al. 1997; Mantzoros et al. 1997). In exercise-induced amenorrhea, nocturnal leptin secretion is blunted (Hilton & Loucks 2000). It is thus probable that leptin also links body fat signals with brain mechanisms controlling female fertility.

Leptin as well as insulin gain access to the hypothalamus through the median eminence (ME), a CVO that lacks a blood-brain barrier. Leptin binds to the long form of its receptor on the cells in arcuate hypothalamic nucleus and on parts of VMH, DM, and ventral premammillary nuclei that are located adjacent to the ME (Elmquist et al. 1998b). At these sites, where lesions induce overeating, hypoactivity, and obesity, leptin stimulates anorexigenic neurons, inhibits orexigenic neurons, and stimulates S outflow to increase energy expenditure (figure 7.10). In the arcuate nucleus, leptin stimulates the POMC/CART neurons, which project to the medial parvocellular PVN, LHA, and DVN, to secrete their anorexigenic peptides, α-MSH (Cowley et al. 1999), and CART (Elias et al. 1998a; Kristensen et al. 1998). α-MSH neurons then act on MCN3 and MCN4 receptors on TRH (Flier et al. 2000), oxytocin (Schwartz et al. 2000), CRF (Kelly & Watts 1996, 1998), and GABA-ergic neurons in the PVN to cause anorexia and increase thyroid hormone and cortisol secretion. Conversely, increased secretion of rT_3 during fasting therefore results from withdrawal of leptin action on arcuate AgRP and α-MSH neurons. These leptin-stimulated arcuate neurons also suppress feeding by acting on MCN3 and MCN4 receptors on MCH cells in the LHA. Through these neurons, leptin also activates the S nervous system (Huszar et al. 1997; Koylu et al. 1997, 1998; Satoh et al. 1998). In the arcuate nucleus, leptin also blocks NPY-AgRP cells from stimulating feeding and inhibiting autonomic mechanisms of energy expenditure during starvation (Ahima et al. 1996). In the same way and in the same areas of the brain, insulin produces central satiety by inhibiting NPY/AgRP and stimulating α-MSH/CART neurons. Thus, insulin's overall effect is to decrease feeding behavior (Schwartz et al. 2000).

Other endocrine interactions, through which leptin also affects energy regulation, involve glucocorticoid and sex hormones. During hypoglycemia, leptin suppresses ACTH and cortisol secretion, while in the obese subjects, high cortisol titers stimulate leptin secretion (Giovambattista et al. 2000; Lerario et al. 2001). Male and female gonadal hormones exert a differential effect on leptin expression, which results in greater circulating leptin titers in women than in men (Kennedy et al. 1997; Mystkowski & Schwartz 2000).

Exercise Effects

Two issues relating energy cost of exercise and energy regulation are of particular interest: (1) the relationship of spontaneous physical activity to the energy regulatory mechanism and (2) the effectiveness of exercise in long-term regulation of body fat. If it were designed to operate in a homeostatic fashion, spontaneous physical activity should increase whenever our energy stores and overall energy balance increase and decline during negative energy balance. In effect, the opposite occurs, as obese individuals are hypoactive and underweight individuals behave hyperactively (Obarzanek et al. 1994; Schulz & Schoeller 1994; Yates et al. 1983). This relationship is seen experimentally as occurrence of death from inanition caused by excessive wheel running in rats that are food restricted (Routtenberg & Kuznesof 1967). Conversely, when obesity is generated through dietary means or ablation of hypothalamic substrates of satiety, animals reduce their levels of spontaneous wheel running

(Borer et al. 1983b, 1989). This inactivity is a result of reduced motivation to run and increased septal serotonin content, rather than some physical limitation (Borer et al. 1983b, 1989; Potter et al. 1983). The inverse relationship between levels of spontaneous physical activity and body fat, as well as the stimulatory influence of gonadal hormones and of seasonal increases in day length on physical activity levels, suggests that physical activity may have evolved as a means to search for food, mates, and other environmental opportunities rather than to regulate the level of body fat stores (Borer 1982). To effectively apply exercise for body fat loss, it is necessary to either raise the psychological incentives for physical activity or combine exercise with dieting to increase the spontaneous activation and incentive value of exercise.

Similar to the loose coupling of exercise and feeding, there appears to be a loose coupling between exercise and long-term regulation of energy balance. Endurance exercise of between 10 and 30 weeks duration, at frequencies of between 3 and 5 days a week, at a cost of about 200 to 300 kcal per session without an accompanying dietary restriction, can cause fat losses of between 1% and 2% (Ballor & Kesey 1991; Bergman & Boyungs 1991; Cowan & Gregory 1985; DiPietro 1995; Epstein & Wing 1980; Gaesser & Rich 1984; Girandola 1976; Grediagin et al. 1995; Kohrt et al. 1992; Leon et al. 1979; Moody et al. 1969; Pollock et al. 1972, 1975; Ready et al. 1996; Swenson & Conlee 1979). This magnitude of weight loss is expected in the absence of compensatory increases in food intake. On the other hand, a number of studies report no weight loss with exercise (Hardman et al. 1992; Hinkleman & Nieman 1993; Stensel et al. 1994). The unavoidable selection of more motivated participants for longitudinal studies and difficulty in excluding the confounding influence of deliberate dietary modification for enhanced exercise-induced fat loss make it difficult to assess whether exercise, unaccompanied by dietary changes, can consistently produce fat loss. From the available information, it is equally difficult to assess the relative contribution of exercise modality, intensity, and patterning and the contribution of the type of metabolic fuel used during exercise to any exercise-induced fat loss.

Lifestyle Variables and Obesity

Different dietary practices and levels of physical activity influence body-fat equilibria by affecting secretion and action of several hormones that control nutrient uptake and storage, fuel mobilization and utilization, or reparative and hypertrophic growth. During the past half century, energy has been increasingly partitioned in favor of increased body-fat settling points and has caused obesity in approximately 50% of adult Americans (Flegal et al. 1998). This rate of obesity suggests that mechanisms of energy regulation operate poorly under lifestyle conditions prevailing in contemporary America. The mechanism of human energy regulation may have been designed to opportunistically maximize energy intake and nutrient storage when palatable and energy-rich nutrients are abundant and to increase search for food and homeostatic defense of a lower level of body fat and body mass against inanition, when food supplies become scarce. Genetic variants have been detected in segments of human population that maximize energy accretion and storage. When a lifestyle of inactivity and abundance of rich food replaces nutrient scarcity and physically exacting subsistence, energy plateaus in genetically predisposed ethnic groups change from lean to obese (Ravussin & Gautier 1999). Two features of the long-term energy regulatory mechanism are the most likely reasons for the growing incidence of obesity in contemporary society. The first one is stimulation by large palatable meals of fat synthesis and storage. The second one is the nonhomeostatic association of obesity with inactivity and with reduced sensitivity to hormonal mobilization of storage fuels.

Infrequent intake of large energy-dense and palatable meals triggers hormonal changes conducive to de novo hepatic fat synthesis (see figure 6.12), increased FFA uptake by the adipose tissue, and body fat gain (Oscarsson et al. 1999a; Pearcey & de Castro 2002). Ingestion of about 500 g of carbohydrates in a single meal and adaptation to high carbohydrate diet are needed to induce hepatic fat synthesis and RQs greater than 1 (Acheson et al. 1984b). Such an exceptionally large quantity of carbohydrates exceeds by twofold to threefold amounts customarily eaten during more frequent smaller meals. Increased insulin action after large carbohydrate meals stimulates the activities of hepatic lipogenic enzymes pentose monophosphate shunt and of malic enzyme (see figure 6.3) (Leveille 1970). Although the DIT is increased in proportion to the size of the meal, this increase in energy expenditure is insufficient to overcome the lipogenic action of insulin (Tai et al. 1991). Insulin secreted to high carbohydrate meals suppresses the activity of muscle LPL and utilization of lipids by muscle (Jacobs et al. 1982; Kiens et al. 1987). Appropriately, indices of body fatness are inversely related to customary frequency of meal eating and directly related to customary meal sizes (Drummond et al. 1998; Metzner et al. 1977).

The cycle of fat synthesis initiated by such dietary practices leads to progressively greater levels of obesity and is accompanied by reduced tissue sensitivity to a number of hormones including insulin and leptin. Obesity is accompanied by development of insulin resistance in the liver, muscle, and the adipose tissue and compensatory insulin oversecretion (Evans et al. 1983, 1984b; Kissebah & Pieris 1989). Insulin resistance consists of receptor and postreceptor defects (Patti & Kahn 1998). Hepatic insulin resistance results in inadequate suppression by insulin of hepatic glucose production, increased hepatic glucose production caused by lower insulin suppression of glycogenolysis, and hyperglycemia (Boden et al. 2002). Hepatic insulin resistance develops in response to high FFA titers in the portal vein, which are a consequence of reduced insulin suppression of intra-abdominal adipose tissue lipolysis (Bergman & Mittelman

1998; Bjorntorp 1991; Howard 1999; Kissebah 1987; Kissebah & Pieris 1989). High plasma FFA titers also blunt skeletal muscle insulin sensitivity and LPL activity (Boden 1997). Concurrently, hepatic insulin degradation is reduced (Pieris et al. 1987a).

Insulin resistance is one of several hormonal abnormalities associated with accumulation of intra-abdominal, visceral body fat (Pieris et al. 1986). Others include decreases in plasma growth hormone, increases in plasma cortisol concentration and in androgenic action of gonadal and adrenal steroids in postmenopausal women, and decreases in testosterone in men and estrogen in women (Bjorntorp 1996b, 1997a, 1997b; Evans et al. 1983; Hew et al. 1998; Larsson & Ahren 1996; Marin & Arver 1998; Ottosson et al. 1994; Scacchi et al. 1999; Tchernof et al. 1998; Wilson et al. 1976). Increased cortisol secretion promotes fat deposition in the visceral depot that has a high density of glucocorticoid receptors (Ottosson et al. 1994). Chronic psychological stress also may facilitate cortisol oversecretion and is conducive to development of visceral obesity (Bjorntorp 1991; Rosmond et al. 1996). Once obesity develops, the large adipose tissue mass secretes chemical mediators such as TNF-α and resistin, which further contribute to insulin resistance through paracrine or autocrine action (Halle et al. 1998; Kern 1997; Skolnik & Marcusohn 1996; Steppan et al. 2001).

Prevalence of sustained hyperleptinemia in obese individuals is attributed to development of peripheral resistance to leptin (Mantzoros 1999). Insulin stimulates leptin release, whereas β adrenergic stimulation inhibits it (Mantzoros 1999). Because leptin inhibits insulin secretion and action, hyperleptinemia also contributes to development of insulin resistance in obesity (Girard 1997; Poitout et al. 1998).

Changes in insulin and leptin secretion and action with increasing obesity contribute to stabilization of body fat reserves at different fatness plateaus through compensatory alterations in adipose tissue sensitivity to their action. Body fat stabilizes at a particular level of obesity when insulin and leptin resistance produce an equilibrium between reduced rate of insulin-stimulated lipogenesis and increased rate of lipolysis and fat oxidation caused by blunted insulin antilipolytic action (Pollare et al. 1991). Conversely, when food restriction drives body fat to a lower equilibrium, lipogenesis is increased because of heightened adipose tissue sensitivity to insulin and is kept in check by increases in lipolysis driven by the energy deficit (Halle et al. 1998). Increased insulin sensitivity and activity of adipose LPL increases the probability of increased lipogenesis and a restoration of fat gain in individuals who have reduced their body fat level through dieting (Vessby et al. 1985).

Partitioning of nutrients toward visceral obesity that is accompanied by hypertension, insulin resistance, and dyslipidemia as a result of inactivity and overeating is compounded by preponderance of type IIX muscle fibers in inactive obese individuals and by sarcopenia, a reduction in muscle mass (Carey 1998; De Fronzo 1997; Despres et al. 1985; Grundy 1999; Howard 1999; Krotkiewski & Bjorntorp 1986). Sarcopenia is in part induced by reduced GH secretion and action in obesity and in part by a parallel reduction in sex hormone release (Veldhuis et al. 1994). Sarcopenia contributes to lower BMR, a major contributor to lower daily energy expenditure. Endocrine changes that contribute to sarcopenia and obesity become prominent during aging as a cumulative consequence of inactive lifestyle and overeating (Horber et al. 1996). Inactivity also leads to conversion of muscle fibers to the type IIX that has a limited capacity for fat uptake, storage, and oxidation.

Role of Exercise in Regulation of Lean Body Mass and Reparative Growth

Just as inactivity and consumption of large, energy-rich meals lead to a cycle of energy storage and obesity, hypertrophic actions of high-resistance exercise and endocrine and metabolic effects of acute and habitual endurance exercise partition nutrient energy toward maintenance of lean body mass and reparative growth. Acute exercise increases fat oxidation at an intensity of between 55% and 70% $\dot{V}O_2$max, and habitual endurance exercise increases fat oxidation as a result of endocrine and metabolic adaptations to increased volumes and intensity of exercise (Achten et al. 2002; Melby et al. 1998). When daily fat oxidation exceeds daily fat intake and the RQ declines below the food quotient (FQ, the ratio of oxidative carbon dioxide produced and oxygen consumed when a representative sample of diet is combusted in a bomb calorimeter), fat utilization increases and body fat stores decline (Flatt 1988, 1995).

Postexercise increases in oxygen consumption reflect, in part, increases in fat oxidation, which are proportional to the intensity of the preceding exercise (Bahr & Sejersted 1991; Bielinski et al. 1985). Increased postexercise lipid oxidation is driven by increased S tone, high titers of plasma E acting on β adrenergic receptors (Lowell & Flier 1997), high plasma GH and glucagon concentration, up-regulation of muscle LPL, and expression in the muscle of uncoupling proteins (Lithell et al. 1981, 1984; Vidal-Puig et al. 1997). The role of GH in increased lipid oxidation is underscored by parallel postexercise increases in plasma GH, activity of muscle LPL, and lipid oxidation (Kiens et al. 1989).

In adulthood, lipolysis becomes a more prominent GH action than the stimulation of growth, which now is confined to maintenance and repair of lean body mass. GH action is initiated when GH receptor dimerizes and triggers a cascade of intracellular phosphorylations (Richelsen 1997). Activation by such phosphorylation of insulin-sensitive IRS1 and IRS2 enzymes may be responsible for the short-term insulinlike effects of GH. GH's long-term lipolytic effects are mediated by tyrosine phosphorylation and subsequent dimerization of STAT proteins and by activation of JAK-2 and MAP kinases. Increased lipolysis and enhanced

catecholamine-induced lipolysis are obtained after GH stimulation of adipose tissue of greater than 2-hour duration (Moller et al. 1990a, 1990b). The long latency is apparently caused by induction by GH of β adrenergic receptors, increased expression of hormone-sensitive lipase (HSL), and antagonism of antilipolytic action of other hormones via the G_i transduction pathway (Bjorntorp 1996a; Doris et al. 1994; Oscarsson et al. 1999a, 1999b; Watt et al. 1991). GH inhibits LPL more strongly in visceral than in subcutaneous fat depots and thus has a predominant effect on the size of visceral fat depots (Dietz & Schwartz 1991; Richelsen et al. 1994; Rosenbaum et al. 1989).

Other hormones released during exercise also influence the magnitude and regional aspect of lipolysis. Changes in regional distribution and sensitivity of α_2, β adrenergic, and adenosine receptors contribute to partitioning of dietary fat among different fat stores. Cortisol and GH jointly increase lipolysis and fat oxidation through suppression of adipose tissue LPL and increases in the activity of muscle LPL (Bjorntorp 1996a; Oscarsson et al. 1999b). Estradiol increases FFA and TG content of muscle and adipose tissue, inhibits adipose tissue LPL, and increases muscle LPL activity during exercise (Ellis et al. 1994). GH action is amplified by thyroid hormones to increase skeletal muscle and heart LPL activity and decrease (as does IGF-I) adipose tissue LPL activity (Lithell et al. 1985b; Oscarsson et al. 1999a).

Increases in fat oxidation in endurance-trained individuals are to a large extent a consequence of adaptive changes that result from habitual training activity (Poehlman et al. 1990, 1994; Turcotte et al. 1992). These changes include increases in

- proportions of oxidative muscle type;
- expression of genes and proteins for mitochondrial and metabolic enzymes that facilitate lipid metabolism in these fibers (Holloszy & Coyle 1984; Tunstall et al. 2002);
- muscle LPL activity (Lithell et al. 1979a, 1979b; Seip et al. 1995; Svedenhag et al. 1983);
- muscle triglyceride stores (Lithell et al. 1979a, 1979b);
- capillary supply to these muscles (Lithell et al. 1985a); and
- muscle sensitivity to β adrenergic stimulation (Spina et al. 1998).

In the presence of negative energy balance, exercise-induced and training-induced increases in metabolic rate usually, but not always, counteract fasting-induced suppression of metabolism (Poehlman et al. 1991). Finally, thyroid hormone also increases muscle LPL activity as well as muscle sensitivity to β adrenergic lipolysis (Lithell et al. 1985b).

Sarcopenia is counteracted by exercise-induced increases in muscle mass achieved either through hypertrophic growth or mediated by previously discussed local or systemic messengers released during exercise. Greater lean body mass contributes to greater oxidative energy expenditure. Resulting increases in RMR and exercise-induced EPOC contribute further to partitioning of nutrient energy away from fat storage. Therefore, to achieve healthy partitioning of nutrient energy between lean body mass and fat depots and avoid obesity and sarcopenia, adult humans need to adjust their lifestyle in a way that compensates for deficiencies in the design of the energy regulatory mechanism.

8

Reproductive Hormones and Exercise

Exercise appears to have no obvious connection with reproduction, yet the reproductive hormones interact with exercise in at least two important ways. (1) Hormones controlling growth and reproductive maturation interact to create gender differences in body structure and function that affect physical performance, and for that reason these hormones are manipulated by some individuals in attempts to enhance performance. (2) Exercise can influence secretion of reproductive hormones directly and indirectly and therefore can affect the rate of sexual maturation and gender-specific growth, fertility, reproduction, and gender-associated health risks. This chapter will address the first theme by outlining endocrine mechanisms that mediate sexual maturation and differentiation of growth and adult reproductive function and metabolism. These issues are of interest in sport or equivalent physical activity where gender is a significant variable and where gender differences in strength and fuel use may affect performance. Discussion of the second topic will examine how physical exertion affects sexual maturation and growth, fertility, and reproduction by influencing secretion of reproductive hormones and availability of nutrient energy. Finally, the way exercise affects health issues that are associated with reproduction will be briefly considered.

DEVELOPMENT OF GENDER DIFFERENCES AND THEIR ROLE IN PHYSICAL PERFORMANCE

Human gender is determined by genetic as well as hormonal processes. A single Y sex chromosome is necessary for development of a male gonad and synthesis and secretion of testosterone during fetal development even when

the karyotype includes multiple X chromosomes (Griffin & Ojeda 1996). In the absence of the Y chromosome, a basic female phenotype, or body build, develops, while two X sex chromosomes are needed for female gonad development and maturation. For normal differentiation of either gender, expression of genes on X chromosome and of some autosomal genes also is needed.

Gametogenesis

During gametogenesis, undifferentiated germ cells called, respectively, spermatogonia and oogonia, differentiate into sperm and ova. Initially, they divide and develop into primary spermatocytes and oocytes. This process is repeated episodically throughout the adult male's life, whereas in the female, about one million primary oocytes are formed before birth and are then recruited for further differentiation during approximately the next 40 years. Next, secondary spermatocytes and oocytes are formed by division of their chromosomes from 46 to 23. Secondary spermatocytes that result from reduction division contain either a Y or an X sex chromosome. Of the two secondary oocytes containing an X sex chromosome, only one is functional. The other, called first polar body, is devoid of cytoplasm and appears as an adhering remnant nucleus. A final cell division of the two secondary spermatocytes gives rise to four spermatids, which then differentiate into sperm. In the female, this last division occurs after fertilization and results in a single ovum, which retains all the cytoplasm, while the second replicated nucleus forms another nonfunctional polar body. The polar body can be visualized microscopically as a small mass just inside the nucleus in cell scrapings from inside a woman's cheek. This so-called Barr body has been used as a proof of female chromosomal gender by the International Olympic Committee Medical Commission, which, since 1968, has required athletes to undergo a buccal smear sex test. The test was prompted by gender misidentification of an athlete as female in both the 1932 and 1936 Olympics, and it now detects and allows for exclusion of one athlete with misclassified gender per 400 competitors (Newsholme et al. 1994).

Gender misidentification in sport sometimes is a product of errors in reduction division during gametogenesis, which can produce embryos with an inappropriate number of sex chromosomes. In addition to a single Y and 44 somatic chromosomes, individuals with Klinefelter's syndrome have multiple (two or more) X sex chromosomes (47,XXY; 48,XXXY). This condition results in phenotypic males with low levels of plasma testosterone, small testes, and reduced number of spermatogonia. Their plasma estrogen concentrations are higher than normal, and as a result, such individuals undergo insufficient virilization and enhanced feminization during development and in adulthood. Gonadal dysgenesis or Turner's syndrome is the second abnormal developmental pattern resulting from a karyotype containing a single X chromosome (45,X). Such individuals develop a female body build with immature external genitalia and nonfunctional gonads. As a result, no sex hormone secretion or secondary sex characteristics develop, and statural growth is subnormal. In pure gonadal dysgenesis, the chromosome number is correct for either a male (46,XY) or female (46,XX) pattern of development, but gonadal formation is arrested at the initial stages of embryonic differentiation. Such individuals develop into a female phenotype but do not secrete either sex hormone. Additional forms of sex misidentification can occur because of fetal exposure to inappropriate hormonal environment.

Gonadal Development

Normal development of gonads begins during the 4th week of life for human embryos (Griffin & Ojeda 1996). At this time, coelomic epithelium and underlying mesenchyme form gonadal ridges, which become populated during the 5th week by germ cells that migrate from the primordial gut. By week 6, gonadal ridges have given rise to androgen-producing stromal or interstitial cells called Leydig cells in the testis and theca cells in the ovary (see figure 8.1). Concurrently, there is differentiation of Sertoli cells in the testis, which line seminiferous tubules and facilitate spermatogenesis, and of granulosa cells in the ovary, which line the Graafian follicles. Both of these cells have androgen receptors or binding proteins and the capacity to aromatize androgens into estrogens. They provide a route by which nutrients reach the gametes, and they secrete fluids within which gametes develop. During week 7, a discrete fetal testis with primordial seminiferous tubules is formed. Sertoli cells in these tubules release a müllerian-inhibiting hormone (MIH), which causes disappearance of one of the two systems of male embryonic kidney ducts, the müllerian ducts. At 8 weeks, the androgen-producing fetal Leydig cells differentiate and begin to secrete testosterone, which virilizes the brain, stimulates development of wolffian ducts, and guides differentiation of male external genitalia. Throughout this period, female ovaries grow, but differentiation of follicles and the capacity to secrete estrogens first appear during the 15th week of gestation.

Phenotypic Gender Dimorphism

Sex steroids influence differentiation of dimorphic phenotypes, including secondary sexual characteristics and sexual differentiation of nervous, musculoskeletal, and adipose tissues. In this way, they influence physical and

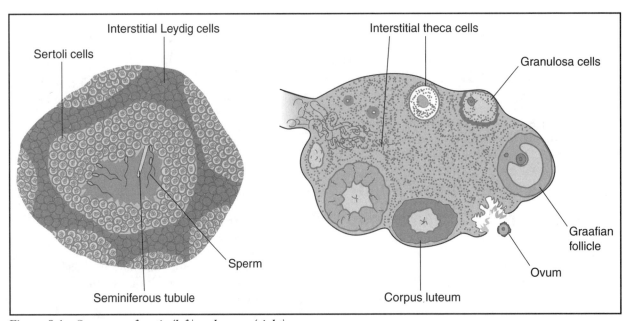

Figure 8.1 *Structure of testis (left) and ovary (right).*
Adapted from Vander, Sherman, and Luciano 1990.

behavioral characteristics that have a bearing on adult physical performance. The onset of secretion of sex Hs initiates development of phenotypic sex and differentiation of external genitalia. Male genitalia begins developing after gestational week 8, when testosterone is first secreted. Epididymis and vas deferens are formed from sections of the wolffian duct that previously served kidney functions. By week 13, seminal vesicles, ejaculatory ducts, and urethra develop from the remaining portions of these ducts. The prostate and membranous portions of the urethra develop from the urogenital sinus at about week 10 of gestation, and the urogenital sinus becomes virilized into a penis and scrotum between weeks 9 and 12. Penis and associated ducts increase in size during the second two trimesters of gestation, and descent of testes into the scrotum is completed around the time of birth.

In the absence of testosterone, development of female external genitalia follows a somewhat different pattern and timetable. Fallopian tubes and the upper one third of the vagina are formed from the müllerian ducts, and the folds of the external genitalia are formed from the urogenital sinus surrounding the openings of the urethra and the vagina. A female clitoris develops from the genital tubercle. The cervix and uterovaginal rudiment appear during the 9th week. The uterine myometrium develops from mesenchyme by week 17 of gestation. The vaginal lumen is first seen during week 11 and completed by week 20. Breast tissue first develops as mammary lines in fetuses of either gender at 5 weeks. They disappear between 6 and 8 weeks of gestation except for a single pair of mammary buds. These buds remain dormant until the 5th month of development when nipples start to form from secondary epithelial buds.

Gender misidentification in sport also can result from inappropriate exposure to androgens during development. Exposure of a female fetus to excess androgens results in development of female pseudohermaphroditism (PHA), and exposure of a male fetus to insufficient androgens results in development of male PHA (Griffin & Ojeda 1996). Female PHA usually results from congenital adrenal hyperplasia associated with inherited deficiency of adrenal cortical enzyme 21-hydrolase (see figure 1.12), which impairs synthesis of cortisol. In the absence of cortisol feedback, ACTH is oversecreted leading to oversecretion of 17-hydroxyprogesterone and increased synthesis of androstenedione and testosterone. Excessive adrenal androgen secretion causes variable degrees of virilization of external genitalia, from clitoral enlargement to partial fusion of labioscrotal folds and formation of penis and penile urethra. In male PHA, insufficient androgen action on the male fetus may be a consequence of congenital defects in any of several enzymes in the androgen synthetic pathway (see figures 1.12 and 1.14), the androgen receptor, or the conversion of testosterone to dihydrotestosterone. Individuals with male PHA display female external genitalia, blind-ending vaginal pouch, or a small phallus and a urethral opening at its base. Because pseudohermaphrodites of

either gender possess gonads appropriate to their genetic sex but inappropriate external genitalia, their gender is often misidentified at birth and the error detected at puberty when they fail to develop appropriate secondary sex characteristics and behaviors.

Exposure to androgens during intrauterine development also has permanent effects on tissues other than the gonads. Of particular interest is virilizing organizational influence of testosterone on the central nervous system and peripheral motor nerves. A number of human brain regions such as the temporoparietal cortex, sexually dimorphic nucleus of the preoptic area (Swaab & Fliers 1985), and hippocampus show sexual dimorphism in size or synaptic connectivity that are dependent on developmental exposure to androgens (Gould et al. 1990; Witelson 1991). Perinatal sex steroid exposure causes gender differences in some spatial and quantitative cognitive tendencies and gross motor ability. Early developmental exposure to androgens in fetuses of either gender that have congenital adrenal hyperplasia is associated with greater spatial orienting abilities and gross motor skill and lower verbal intelligence than is the case after exposure to estrogen (Hampson 1990; Nass & Baker 1991). Some of these effects of androgen are also activational, as shifts in these tendencies and abilities are seen during changes in hormonal environment in transsexual individuals undergoing gender shift and in postmenopausal women receiving either androgen or estrogen hormone replacement therapy (Gooren 1984; Sherwin 1988).

Androgen exposure during early development also determines the number and size of motoneurons innervating skeletal muscles. Number, size, and physiological characteristics of motoneurons, in turn, determine the size and physiological characteristics of muscle fibers (Hauser & Toran-Allerand 1989). Motoneurons express androgen receptors during development (Sar & Stumpf 1977). Testosterone affects motoneuron number by preventing their developmentally programmed death in contrast to motoneuron loss that occurs in female fetuses (Forger & Breedlove 1986; Jones 1994; Yu 1989). Postnatal muscle size and the capacity of muscle to hypertrophy in adulthood may thus be determined in part by intrauterine androgen exposure.

Immediately upon birth, secretion of both sex hormones is high and plasma testosterone reaches levels close to those seen in adulthood. Plasma testosterone declines in the course of the next 18 months to very low levels, which are sustained until the onset of puberty, approximately 10 to 15 years later (Forest et al. 1974). Umbilical cord estradiol in girls reaches levels that are approximately 100 times higher than in the adult woman, despite considerable metabolism of the hormone by the fetus (Griffin & Ojeda 1996). Although the estradiol is the principal sex steroid produced by the neonatal ovary, other estrogens, such as estrone, and androgens, such as androstenedione and testosterone, also are being produced. Estradiol levels decline rapidly during the 1st week after birth and remain at about 8 to 10 pg/ml in girls as well as in boys until the onset of puberty (Klein et al. 1994a).

Pregnancy

Little is known about the possible effects of exercise on the embryonic and early postnatal development. This issue is important, as prenatal exposure of pregnant rats to stress reduces virilization of the brain in male offspring (Anderson et al. 1986). This effect appears to be mediated through increased fetal exposure to CRF (Karaviti et al. 1982) and antagonism between the hypothalamic-pituitary-adrenal (HPA) and the hypothalamic-pituitary-gonadal axes (HPGA). β-Endorphin secretion is greater and more prolonged in response to moderate-intensity exercise during late-stage pregnancy than in a nonpregnant condition (McMurray et al. 1990a; Rauramo et al. 1986). Prenatal exposure to high concentrations of endogenous opiates reduces the ability of rodents to express masculine sexual behavior in adulthood (Johnston et al. 1992; Ward et al. 1983). Although intense exercise can significantly increase plasma androgen and endogenous opiate concentrations in humans of both genders, androgen secretion in adult females is less pronounced than in the males, and pregnant women usually do not reach necessary exercise intensities despite recent more permissive exercise recommendations for pregnant women by the American College of Obstetrics and Gynecology (Anonymous 1994). At moderate intensity requiring about 1 L of oxygen per minute, exercise stimulates progesterone and estriol secretion, the more so the later the stage of pregnancy, but has no impact on secretion of other placental hormones (Bonen et al. 1992).

Additional factors that may influence exercise in pregnancy are hormonal and autonomic modulation of the high-flow, low-resistance circulation of pregnant women and endocrine mediation of body fat gain. Increases in plasma volume and cardiac output and reduced heart-rate variability occur early in pregnancy before any measurable changes in metabolism and thus appear to be of endocrine or autonomic origin (Duvekot & Peeters 1994; Geva et al. 1997; Spaanderman et al. 2000; Stein et al. 1999). Reduced vascular resistance of pregnant women is in part mediated by endothelial release of prostacyclins and by blunted sensitivity to vasoconstrictive and pressor actions of angiotensin II (Ramsay et al. 1992; Sorensen et al. 1992). These vasodilatory changes favor heat loss in hot environments and during exercise, which is also assisted by lowering of the sweating threshold (Clapp 1991b; McMurray et al. 1990b).

As pregnancy progresses, there is increased PS withdrawal and a concomitant increase in S tone, but there is also a blunted responsiveness to various stressors that call for S activation (Blake et al. 2000). Thus in pregnancy, HR and BP responses and NE secretion are blunted in response to orthostatic challenges, isotonic and isometric exercise, the Valsalva maneuver, and psychological stress (Avery et al. 2001; Barron et al. 1986; Bonen et al. 1992; Ekholm & Erkkola 1996; Lucini et al. 1999; Matthews & Rodin 1992; Ramsay et al. 1993; Tur et al. 1992; Wolfe et al. 1999), while phenylephrine, NE, and angiotensin II injections produce blunted cardiovascular responses (Leduc et al. 1991; Broughton Pipkin et al. 1989; Ramsay et al. 1992, 1993). In addition to these systemic changes, supine position can trigger hypotension through compression of aorta and inferior venae cavae by the pregnant uterus (Chen et al. 1999; Leduc et al. 1991; Speranza et al. 1998). These changes of pregnancy are associated with reduced maximal anaerobic power but with unimpaired maximal aerobic capacity (Sady et al. 1990; Spinnewijn et al. 1996; Wolfe & Mottola 1993).

Metabolic fuel use during exercise does not appear to be greatly affected by pregnancy, except possibly for greater utilization of lipids at high intensities (Bonen et al. 1992; Ohtake & Wolfe 1998). On the other hand, other energy expending processes that affect energy regulation, such as resting metabolic rate, diet-induced thermogenesis, and spontaneous physical activity, all decline in pregnancy (Poppitt et al. 1993; Robinson et al. 1993; van Raaij 1995). Although restraint over food intake is reduced during pregnancy, increases in food intake appear too small to account for the magnitude of weight gain (Clapp 1989; Durnin 1991; Heini et al. 1991; Rossner 1999). This discrepancy suggests that work efficiency and metabolic efficiency are increased during pregnancy (Clapp 1989; van Raaij et al. 1987; Durnin et al. 1987), but how hormones and the autonomic nervous system modulate these changes is not known. Later stages of pregnancy are often associated with increases in plasma TNF and with insulin and leptin resistance (Clapp & Kiess 2000; Highman et al. 1998; Young & Treadway 1992). Habitual exercise helps reduce these changes as well as the rate of fat gain (Clapp & Little 1995; Rossner 1999).

Benefits of exercise during pregnancy include better cardiorespiratory function, improved insulin sensitivity, lower body fat gain, and easier labor, which appear to outweigh any costs such as transient fetal distress and reduced infant birth weight (Clapp 1990, 1991a; Clapp & Little 1995; Katz 1991; Rossner 1999; Soultanakis et al. 1996; Stevenson 1997; Wang & Apgar 1998; Winn et al. 1994). Conditions that induce the least fetal distress or uterine contractions are exercise in water, arm rather than leg exercise, movement that does not include heavy lifting, and steady movement of large muscle groups rather than intermittent higher-intensity or isometric exercise (Armstrong et al. 1989; Durak et al. 1990; Katz 1996; Katz et al. 1988, 1990; McMurray et al. 1995; Van Hook et al. 1993; Watson et al. 1991).

Puberty

In addition to their role in differentiation of external genitalia, sex steroid hormones derived from the adrenal zona reticularis and the gonads play a central role in puberty. Puberty is the stage of human development that leads to growth of gonads, maturation of genitalia and reproductive capacity, and gender-specific patterns of statural growth.

During puberty, the onset and timing of sexual maturation can be documented through progressive change of secondary sexual characteristics. One such change is growth of pubic hair, called pubarche. Pubarche and the growth of axillary hair and of axillary apocrine sweat glands are initiated during adrenarche, which is a developmental stage that occurs when increased synthesis of adrenocortical androgens is reinstated. Increased postnatal synthesis of adrenal androgen DHEA-S and secretion of urinary 17-ketosteroids begin around 8 years of age in both girls and boys (see figure 8.2) and precedes the increase in gonadal hormone secretion by about 2 years (Orentreich et al. 1984; Reiter et al. 1977; Sizonenko et al. 1976). The onset of adrenarche and subsequent initiation of puberty are controlled independently because adrenarche occurs normally in adolescents with gonadal dysgenesis and individuals with gonadotropin insufficiency, and gonadarche develops even in adrenal insufficiency (Sklar et al. 1980).

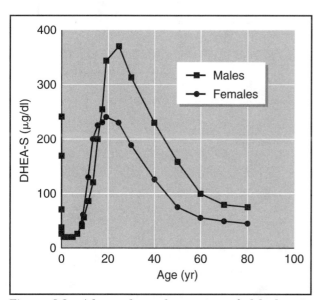

Figure 8.2 *Adrenarche and secretion of dehydroepi-androsterone sulfate (DHEA-S).*

Lifetime pattern of secretion of adrenal androgen DHEA-S, the onset of which is called adrenarche, the first event in the progression of normal puberty.

Adapted from Orentreich et al. 1984.

Pubarche proceeds through five stages (see figure 8.3) (Tanner 1962; Tanner et al. 1966). In stage 1, there is only fine velus hair over pubic area similar to that over other parts of the abdomen. In stage 2, straight or slightly curled and lightly pigmented hair starts growing on either side of the base of the penis or along the labia. In stage 3, the hair is coarser, darker, and curlier and spreads over the pubic symphysis and the junction of the labia. In stage 4, the hair is now adult in type, but the

area covered is considerably smaller than in adults and does not extend to the medial surface of the thighs. In stage 5, the hair is adult in quantity, shaped as an inverse triangle, and spread to the medial surface of the thighs. In men, pubic hair distribution continues to extend up the abdominal wall throughout the second decade of life. In general, axillary hair starts appearing approximately 1.5 years after the onset of male pubarche. Facial hair starts growing at the corners of the upper lip 1 year later and extends over the whole of the upper lip and some on the chin and cheeks after 2 more years (Marshall & Tanner 1969, 1986). These readily perceived markers can be compared easily to other pubertal physiological and hormonal changes.

Thelarche, or breast development, also can be classified as a five-stage progression for females (see figure 8.3). Stage 1 is infantile and consists of a raised papilla that does not substantively change from the postnatal period. Stage 2 is often called the breast-bud stage because of the elevation of the breast underlying the papilla as a small mound with an enlargement of areolar diameter. In stage 3, the breast and areola are bigger and appear as a small adult breast with no separation of their contours. In stage 4, areola and papilla are still bigger and form a secondary mound that projects above the body of the breast. In stage 5, mature and larger breast forms a continuous contour of breast and areola with the projection of the papilla only.

Gonadarche represents pubertal growth of male genitalia caused by a virilizing influence of rising titers of testosterone that again can be divided in five stages (see figure 8.3). Growth of the testis from the infantile size of 1 to 4 ml to 25 ml at sexual maturity can be quantified by comparison of testis dimensions to a Prader orchidometer consisting of a series of elongated spherical models of different sizes (Marshall & Tanner 1969, 1986). Usually the appearance of external genitalia as a whole helps establish stages of gonadarche. In stage 1, genitalia increase slightly in size but remain infantile in appearance. In stage 2, the scrotum has begun to enlarge, and its skin has started to turn red and change in texture. In stage 3, the penis has increased in length and, to a lesser extent, in diameter, and the scrotum also has grown. In stage 4, the length and diameter of the penis have increased further, and the glans has developed. The scrotum has enlarged and its skin has darkened. In stage 5, genitalia reach adult size and shape.

The tempo and timing of pubertal progression differs in girls and boys (see figure 8.4). Although adrenarche begins at approximately the same age in both genders (see figure 8.2), girls enter stage 2 of pubarche, when plasma estradiol titers start to rapidly rise, approximately 1 year ahead of boys (Jenner et al. 1972). Their stage 2 of thelarche is about 1.5 to 2 years ahead of stage 2 of gonadarche (Herman-Giddens et al. 1997; Kaplowitz & Oberfield 1999). Girls of European descent first attain menarche, or initiation of menstrual cycles, at the age of about 12.8 years and African-American girls about 3 months earlier, but the early

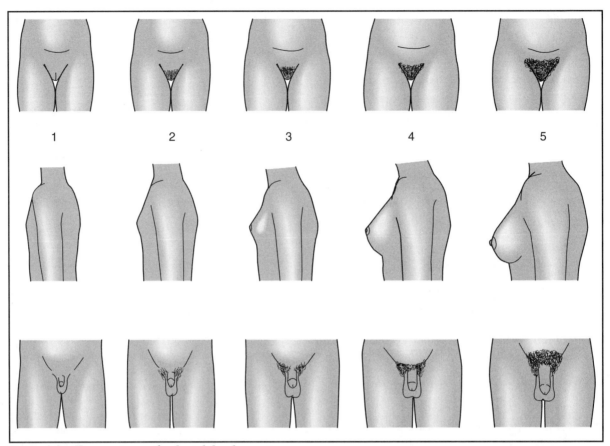

Figure 8.3 *Tanner stages of pubertal development.*

Five stages of development of external genitalia and some secondary sexual characteristics during puberty.

Adapted from Marshall and Tanner 1986.

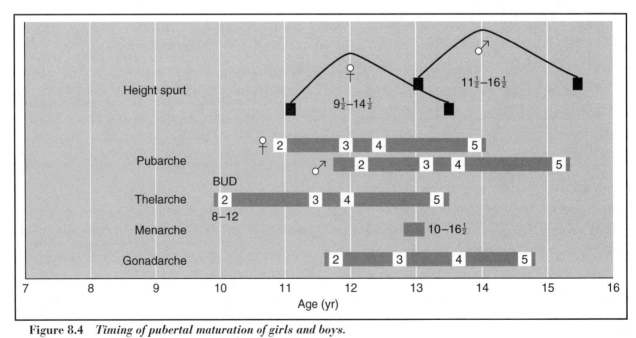

Figure 8.4 *Timing of pubertal maturation of girls and boys.*

Timing of maturational events during pubertal progression in girls (♀) and boys (♂).

Reprinted from Tanner 1974.

cycles are usually anovulatory and infertile for a period of another 6 to 9 months (Apter et al. 1978; Herman-Giddens et al. 1997; MacMahon 1973). Boys on the average do not attain stage 5 of pubarche until 14.5 to 15 years, and thus girls maintain their 1-year to 2-year maturational advance throughout puberty (Marshall & Tanner 1986).

THE HYPOTHALAMO-PITUITARY-GONADAL AXIS IN THE CONTROL OF FERTILITY

Hormonal control of puberty and of adult fertility and reproductive functions depends on the actions of the hypothalamo-pituitary-gonadal axis (HPGA) (see figure 8.5) that includes gonadotropin-releasing hormone (GnRH) or gonadorelin cells in the brain, gonadotroph cells in the anterior pituitary, and cells surrounding the germ cells in the gonads. A diffuse cluster of 1,000 to 3,000 GnRH neurons is distributed in humans between the preoptic area and mediobasal hypothalamus and is concentrated in the arcuate hypothalamic nucleus (Yen 1999a, 1999b). These cells are developmentally derived from the nasal epithelium. An isolated hypogonadotropic form of infertility (Kallman syndrome) occurs when their developmental migration to the hypothalamus fails. GnRH cells interact with a variety of neural circuits and provide a neuroendocrine interface between the brain and the anterior pituitary. GnRH neurons make synaptic contact with structures in the limbic forebrain, CVOs, and the posterior pituitary in addition to their endocrine connection with the anterior pituitary. GnRH axon terminals on the fenestrated capillaries of the hypothalamic hypophysial portal circulation allow for delivery of their endocrine message, the decapeptide GnRH, to gonadotrophic cells in the anterior pituitary. PreproGnRH, which is synthesized in GnRH neurons, contains a 23–amino acid signal peptide, a GnRH decapeptide, an 11– to 13–amino acid fragment, and GAP, a GnRH-associated peptide that may act as a prolactin inhibiting factor.

Brain cells are accessible to steroid hormone influences because of the capacity of most steroid hormones to cross the lipid blood-brain barrier and because the brain can synthesize its own neurosteroids (Yen 1999a). Gonadal steroids exert their influence on the nervous system through genomic and nongenomic effects (Baulieu & Robel 1995). Nongenomic effects are mediated by steroid action on membrane ion permeability. Genomic effects are mediated by two types of estrogen receptors: ERα, which activates gene transcription, and ERβ, which inhibits it. ERs have been located in the hypothalamic areas where GnRH neurons are located but not on GnRH neurons proper. Androgen receptors have also been located in some of the same brain areas but in lower concentrations than estradiol. The major fraction of testosterone that binds to androgen receptors in the brain is aromatized to estradiol. Progesterone and glucocorticoid receptors are also plentiful in the brain, the

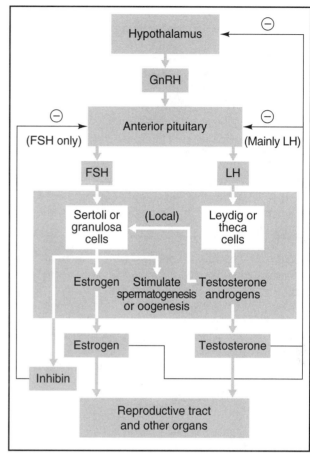

Figure 8.5 *Control of human fertility by hypothalamo-pituitary-gonadal neuroendocrine axis.*

Interactions among hypothalamic gonadotropin-releasing hormone (GnRH), pituitary gonadotropins, luteinizing hormone (LH), and follicle-stimulating hormone (FSH), and gonadal steroids regulate reproductive function in humans.

Adapted from Vander, Sherman, and Luciano 1998.

latter being very abundant in the hippocampus, where they participate in the feedback control of CRF synthesis. It is assumed that sex steroids affect brain control of reproductive and other functions by binding to neurons other than GnRH, which then influence the behavior of GnRH neurons.

Brain control of GnRH neurons includes stimulatory actions of amino acids aspartate and glutamate and of catecholamines (α₁ and β adrenergic) and inhibitory influences of dopamine, endogenous opiates, and CRF (Rivier et al. 1986). It is hypothesized that messages regarding bone maturity, level of body fat, energy availability, and stress are relayed to GnRH neurons through chemical messengers acting on some of these neural circuits. Thus, IGF-I may signal to the brain the information about bone maturity because increases in its secretion in conjunction with increases in GH secretion have been observed during the adolescent growth spurt and the early stages of puberty (Mauras et al. 1996). Plasma leptin concentrations may provide the information to the brain about the size of body fat mass and short-term energy availability because increases in plasma leptin are observed around the time of initiation of puberty, when accumulation of body fat takes place (Cunningham et al. 1999). Acute energy shortages precipitate drops in plasma leptin concentration that overshadow its quantitative relationship to body fat mass and may play a role in delayed onset of puberty (Considine et al. 1996). Initiation of puberty may also be delayed by increased titers of CRF and endogenous opiates released in response to exercise or other types of stress. The titers of both CRF and β-endorphin are chronically increased in amenorrheic athletes, and in anorexia nervosa. Administration of CRF antagonists and opiate receptor blockers usually increase LH pulsatility and stimulate secretion of female reproductive Hs (Hohtari et al. 1991; Judd et al. 1995; Khoury et al. 1987; Laatikainen et al. 1986; Loucks et al. 1989; McArthur et al. 1980; Yen 1998).

Central to the control of reproductive function is pulsatile secretion of GnRH. The intermittent nature of this stimulus is essential for the maintenance of reproductive function because tonic delivery of GnRH abolishes hormonal cycles in women (see figure 8.6) (Belchetz et al. 1978). Therefore, a bolus dose or continuous administration of GnRH can be used as a contraceptive (Yen 1983). Rhythmic neural and hormonal discharges are inherent to the GnRH neurons, but catecholamines and other brain circuits may also contribute to the control of their pulsatility (Kaufman et al. 1985). Fertility in the male is maintained by GnRH pulses at approximately 2-hour intervals. In the female, monthly follicular development is stimulated by circhoral GnRH pulses occurring at 70- to 100-minute intervals. After ovulation, GnRH amplitude increases but the frequency of its pulses (and of LH and FSH pulses) is suppressed to one every 4 hours.

Pulsatile GnRH secretion is first observed during the last trimester of intrauterine life. After a postnatal period of increased GnRH pulsatility, the amplitude of GnRH pulses, and hence the overall activity of GnRH pulse generator, is suppressed (see figure 8.7). Reduced GnRH pulsatility during childhood ensures that sexual maturation does not proceed until a sufficient amount of statural growth, tissue differentiation, and skeletal maturation has taken place. Quiescence of the GnRH pulse generator is believed to be induced either by increased sensitivity of GnRH neurons (or of neural circuits controlling them) to the negative feedback of gonadal or adrenal sex steroids or through influence of other inhibitory brain circuits (Judd et al. 1989; Kulin et al. 1969; Yen et al 1999a). Puberty is initiated when GnRH neurons exhibit renewed pulsatility as a result of reduced steroid negative feedback or of increased stimulation of GnRH neurons by different neural circuits or endocrine messengers (Marshall & Kelch 1986).

Pituitary Gonadotrophins

GnRH stimulates secretion of two anterior pituitary hormones, luteinizing hormone (LH) and follicle-stimulating hormone (FSH), through the cAMP transduction pathway (Labrie 1990; Griffin & Ojeda 1996). A difference in halflives of FSH (170 minutes) and LH (60 minutes) influences their plasma concentrations, which vary at different stages of pubertal development and at different stages of the female menstrual cycle. Increased FSH secretion in excess of LH secretion characterizes the earliest stages of postnatal development (see figure 8.7), and pubertal pattern of gonadotropin secretion, which is predominantly nocturnal (see figure 8.8). FSH concentrations are higher

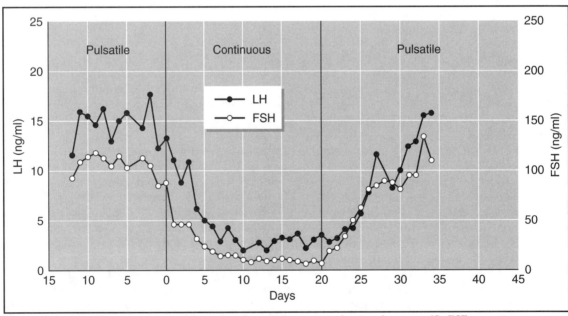

Figure 8.6 *Fertility is maintained by pulsatile gonadotropin-releasing hormone (GnRH) secretion.*

Dependence of human fertility on pulsatile pattern of GnRH secretion. Gonadotropin secretion declines whenever GnRH is delivered in continuous fashion and is maintained within normal concentration range when GnRH secretion is pulsatile.

Reprinted from Belchetz et al. 1978.

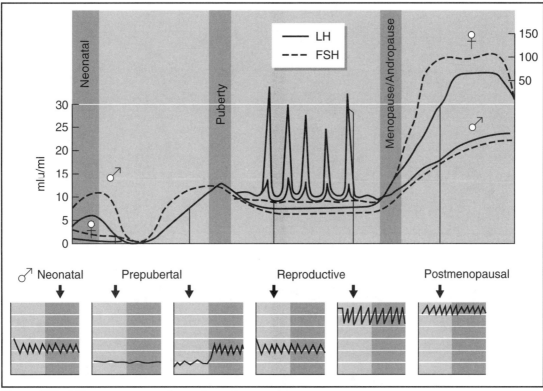

Figure 8.7 *Life span pattern of gonadotropin secretion.*

Changes in the gonadotropin (and by inference in the gonadotropin-releasing hormone [GnRH]) secretory pattern throughout the human life span. In adult males, gonadotropin secretion is under steady negative feedback control. In adult females, gonadotropin secretion shows monthly alternation of negative and positive feedback control. Lower boxes represent changes in pulsatile luteinizing hormone (LH) secretion in a female at different times during her life span as well as during her menstrual cycle. Shading indicates nocturnal period.

Adapted from Yen 1999.

(although not as high as LH) during the follicular phase of the menstrual cycle when development of follicles is first initiated (see figure 8.9). LH acts on the interstitial Leydig and theca cells to stimulate synthesis of androgens (see figure 8.5). FSH, on the other hand, acts on the granulosa cells to stimulate aromatization of androgens to estrogen, proliferation of granulosa cells, and development of the antrum (fluid-filled follicular space enclosing the ovum). FSH also stimulates spermatogenesis within Sertoli cells that line seminiferous tubules through elaboration of local growth factors. In both Sertoli and granulosa cells, FSH initiates the synthesis of LH and FSH receptors. As sex steroid production increases, it augments the stimulatory effect of FSH on spermatogenesis in the male and growth and differentiation of granulosa cells in the female. In the course of puberty, and toward the end of the 14-day follicular phase of the ovarian cycle, FSH secretion declines relative to the pulsatile LH secretion, which increases (Veldhuis 1996).

By making sex steroid synthesis in both male and female possible, pulsatile LH secretion plays a pivotal role in the control of fertility. During puberty, increases in LH pulsatility are first seen during the nocturnal period in association with sleep (figure 8.8), and in later stages of puberty, they become more evenly distributed throughout the day (Yen et al. 1993). Increases in pul-

satile LH release cause increases in concentration of sex hormones during pubertal development that allow for growth of external genitalia, maturation of gonads, and stimulation of statural growth. Similarly, in the course of follicular development, hourly LH pulses stimulate early follicular development and estrogen secretion. LH-stimulated release of sex steroid hormones by Sertoli and granulosa cells within seminal tubules and follicle, respectively, provides the primary stimulus for the maturation of the spermatids into spermatozoa and of secondary oocyte into mature ovum.

Whereas secretion of testosterone is not distinctly episodic, cyclical temporal changes in estrogen secretion by the egg follicle contribute in important ways to orderly progression of the monthly egg maturation cycle. Rising estrogen concentration that results from LH stimulation of follicular development increases GnRH pulse frequency, which in turn causes up-regulation of GnRH receptors. High estrogen concentration and increased concentration of GnRH receptors increase the sensitivity of anterior pituitary gonadotrophs to GnRH stimulation (see figure 8.10). This action constitutes the positive-feedback effect of estrogen over neuroendocrine control of LH secretion. Estradiol positive feedback precipitates the midcycle LH and FSH surges and ovulation within 2 days of peak estradiol secretion (see figure 8.9).

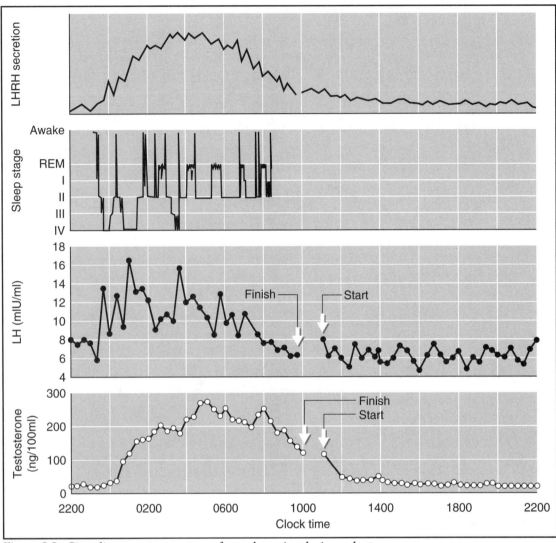

Figure 8.8 *Circadian secretory pattern of gonadotropins during puberty.*

The upper panel shows hypothetical changes in gonadotropin-releasing hormone (GnRH) secretion, which is predominantly nocturnal during the early stages of puberty. GnRH secretion stimulates nocturnal pulsatile secretion of luteinizing hormone (LH), which, in turn, stimulates testosterone secretion in boys and follicular development and estradiol secretion in girls.

Adapted from Griffin and Ojeda 1996 and Boyar et al. 1978.

Programmed releases of more pulsatile LH increase sex hormone secretion and induce sexual maturation during adolescence as well as monthly maturation of ovum and its ejection after menarche. After sexual maturity is achieved in the male, and during phases of menstrual cycle other than the preovulatory period in the female, sex steroid production is regulated to produce a narrow range of concentrations through the sex steroid–negative and inhibin-negative feedbacks. Sex steroids exert an inhibitory influence over their production at two levels. In the hypothalamus, sex steroids inhibit the frequency and increase the amplitude of GnRH bursts and consequently of pulsatile LH and FSH secretion. This action of sex steroid hormones on the hypothalamic GnRH pulse generator is mediated by endogenous opioids during all but the periovulatory period (Ferin et al. 1984). Capacity of EOPs to inhibit GnRH pulse generator dur-

ing the periovulatory period is curtailed by temporary loss of connection between the inhibitory GABA$_A$ neurons, which mediate opiatergic influence, and the GnRH neurons. In the anterior pituitary, sex steroids reduce the amount of LH (but not of FSH) secreted in response to any given amount of GnRH. Secretion of FSH is held in check by the negative feedback exerted by inhibin, the protein hormone secreted by Sertoli and granulosa cells (see figure 8.5).

After discharge of the ovum into the peritoneal cavity, granulosa and theca cells of the ruptured follicle are transformed into corpus luteum, an endocrine organ that synthesizes large amounts of progesterone and substantial quantities of estradiol. This process marks the onset of a 14-day luteal phase of the female menstrual cycle, during which high titers of progesterone and estrogen exert a strong inhibitory influence on the GnRH pulse

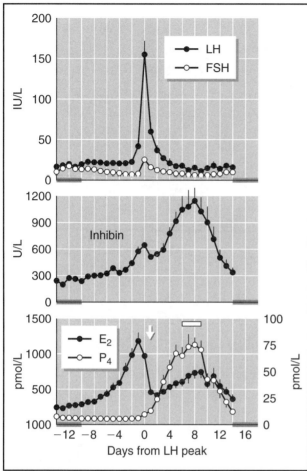

Figure 8.9 *Endocrine basis of human menstrual cycle.*

The arrow shows estimated time of ovulation. The open box shows the estimated time of implantation of the fertilized egg.

E_2 = estradiol; P_4 = progesterone.

Reprinted from Yen 1999.

generator and reduce secretion of gonadotropins, particularly FSH. The steroid negative feedback during luteal phase of the menstrual cycle reduces LH pulse frequency to one secretory episode about every 4 hours accompanied by a doubling of LH pulse amplitude. If pregnancy does not occur, the corpus luteum degenerates causing a sudden decrease in plasma progesterone and estrogen concentrations (see figure 8.9). Sex hormone withdrawal reduces steroid negative feedback over pulsatile FSH and LH secretion and facilitates development of new follicles. The endometrial wall, which has increased in thickness during the period of rising sex steroid hormone titers, now degenerates and is expelled. Its ruptured blood vessels produce menstrual bleeding. New follicles develop in response to increased pulsatile FSH and LH secretion as the cycle continues.

Negative feedback also controls secretion of GH during suppression of statural growth in children and during maintenance of body mass in adulthood. Negative feedback over GH secretion is exerted by IGF-I, the sec-

ond-order endocrine mediator of GH action (see chapter 6). As is the case with sex hormone secretion during puberty, feedback inhibition of GH secretion by IGF-I is attenuated or suspended during the pubertal growth spurt (Mauras et al. 1996). The attenuated negative feedback of IGF-I over GH secretion allows simultaneous coexistence of high plasma concentrations of GH and of IGF-I, which then stimulate the adolescent growth spurt.

Adolescent Growth Spurt

Sexual maturation is coordinated with the control of statural growth. Puberty and adolescent growth spurt do not begin until some critical amount of growth and level of bone maturity have been attained. When the rate of growth is compromised by isolated growth hormone deficiency, onset of puberty is delayed until proper bone age is attained (Tanner & Whitehouse 1975). Initiation of increased growth hormone secretion necessary for the adolescent growth spurt is more closely associated with bone age than with other parameters of statural growth, and likewise, the onset of menarche and of increased secretion of pituitary gonadotropins in gonadarche are linked most closely with bone age (Burr et al. 1970; Ellison 1982; Roemmich et al. 1998). Menarche occurs on average 1.3 years after peak height velocity. Critical body mass levels have been identified as thresholds of attained statural growth necessary for the initiation of landmarks of pubertal progression (see figure 8.11). Thus, on the average, adolescent growth spurt begins only after girls have accumulated a body mass of at least 31 kg, whereas peak growth velocity and menarche are reached after attaining respective threshold weights of 39 and 47 kg (Frisch & Revelle 1969, 1970, 1971a, 1971b, 1972). A corresponding dependence of sexual maturation on attainment of a critical body mass and whole-body growth has previously been described in rodents (Kennedy & Mitra 1963).

The second type of interaction between puberty and growth derives from the stimulatory action of sex steroid hormones on secretion of GH and acceleration of growth. Sex steroids and GH contribute equally to stimulation of adolescent statural growth, because children with either gonadal dysgenesis or isolated GH deficiency attain only about half of the expected amount of statural growth (Brook & Hindmarsh 1992; Zachman 1992). Sex steroid secretion begins in stage 2 of thelarche or gonadarche. In girls, plasma concentrations of 17β estradiol rise from about 10 pg/ml in stage 2 to about 60 pg in stage 5 of puberty, whereas for boys, plasma testosterone increases about 10-fold from about 1 ng/ml between stages 2 and 5 of puberty (Jenner et al. 1972; Kelch et al. 1985; Wu et al. 1993). A low concentration of estradiol powerfully stimulates GH secretion and growth in both girls and boys (Caruso-Nicoletti et al. 1985; Clark & Rogol 1996; Mauras et al. 1989). Testosterone also stimulates GH secretion (Chalew et al. 1988; Martin et al. 1968), but it does so after being aromatized to estradiol (Eakman et al. 1996; Metzger & Kerrigan 1993, 1994; Weissberger & Ho 1993). Testosterone may stimulate statural growth

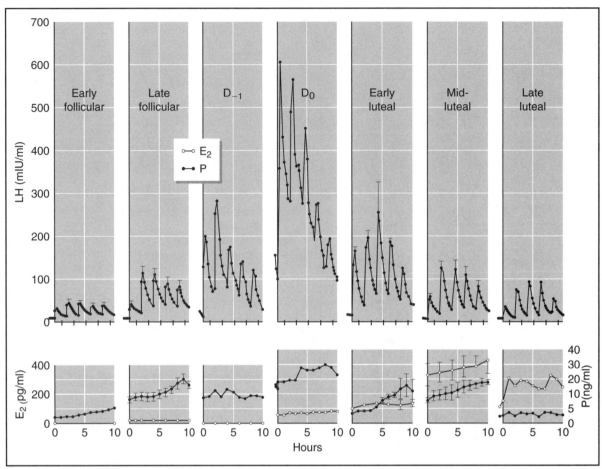

Figure 8.10 *Estradiol positive-feedback control of gonadotropin secretion and ovulation.*

During the periovulatory period in the female menstrual cycle, increasing sensitivity of pituitary luteinizing hormone (LH) secretion is induced by rising plasma estradiol titers (here demonstrated by administration of five consecutive gonadotropin-releasing hormone [GnRH] pulses [solid circles, upper panel]) at intervals shown by the timelines underneath each set of LH measurements.

E_2 = estradiol (solid circles, bottom panel); P = progesterone (open circles, bottom panel); D_{-1} = day before ovulation; D_0 = day of ovulation.

Reprinted from Yen 1999.

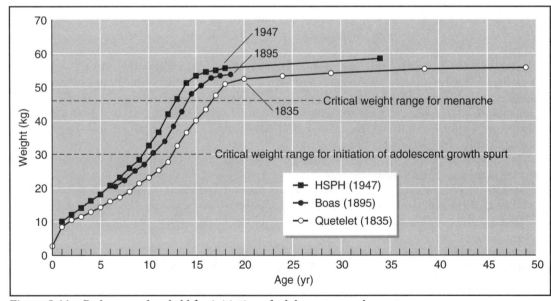

Figure 8.11 *Body mass threshold for initiation of adolescent growth spurt.*

Body mass thresholds of statural growth necessary for initiation of adolescent growth spurt and sexual maturation in girls.

Reprinted from Frisch 1972.

through a dual mechanism of stimulation of GH secretion and direct androgenic stimulation of skeletal and muscle growth (Keenan et al. 1993; Wimalawansa et al. 1999).

In both males and females, stimulation of GH secretion by gonadal steroids appears to be mediated through increased synthesis of the hypothalamic growth hormone–releasing hormone (Shirasu et al. 1990; Zeitler et al. 1990). The pattern of GH secretion consists of low-amplitude secretory episodes that are similar in both genders during childhood and after menopause (Albertsson-Wikland 1994; Ho et al. 1987). The striking sexually dimorphic pattern of GH secretion becomes apparent during puberty as increased GH secretion triggers the adolescent growth spurt (Clark & Rogol 1996; Frantz & Rabkin 1965; Kerrigan & Rogol 1992; Martha et al. 1989, 1992; Rose et al. 1991; Wennink et al. 1991). Females with normal reproductive cycles secrete more GH than males because of their increased GH pulse amplitude (which is the mass of GH released during a secretory episode), a more modest increase in pulse frequency, and an increased GH interpulse nadir (Ho et al. 1987; Mauras et al. 1989; Roemmich et al. 1998). In the rat, and probably in humans, the gender-specific pattern of GH secretion is a consequence of activational effects of sex steroids in adulthood rather than of neonatal virilization of the brain (Jansson et al. 1984; Painson et al. 1992; Tannenbaum 1998). Despite the greater absolute amount of GH released over a 24-hour period in the female, the male pattern of GH secretion consisting of large GH bursts interspersed with low interpulse nadirs is associated with greater stimulation of somatic growth than the more irregular and chronically elevated pattern of female GH concentrations (Jansson et al. 1985).

In addition to gender-specific difference in the pattern of GH secretion and in the effectiveness of this stimulus on growth, a gender difference also develops in responsiveness of peripheral tissues to somatotropic and sex hormone stimulation. Although low concentrations of estrogen stimulate linear growth in girls, estrogen does not stimulate whole-body protein synthesis or muscle growth (Kahlert et al. 1997; Levine-Ross et al. 1986; Mauras 1995). By contrast, estrone, which is derived from the androgen androstenedione, displays anabolic actions on myoblasts in vitro (Kahlert et al. 1997). Androgen stimulates whole-body protein synthesis in pubertal boys and in adults and muscle protein synthesis in children (Mauras et al. 1994; Stanhope et al. 1985, 1988).

The growth-promoting action of androgens depends on density of androgen Rs in myonuclei, which varies by muscle type and location (Janssen et al. 1994; Syms et al. 1985; Takeda et al. 1990). Regional differences in numbers of androgen Rs in specific muscles are either genetically determined or induced neonatally because removal of sex hormones in adulthood does not equalize the rate of muscle growth (Sillence et al. 1995). Fast glycolytic muscles have a higher density of androgen Rs than do slow oxidative muscle fibers, and the Rs are differentially affected by exercise training. Resistance training increases the number of androgen Rs in the former, whereas endurance training does so in the latter (Deschenes et al. 1994). Regional differences in the number of androgen Rs affect the responsiveness of muscle to growth stimulation by androgens. In the human male, limb and neck muscles contain a greater number of androgen Rs than do quadriceps muscles or leg muscles in general (Kadi et al. 2000). Differential increases in androgen R numbers and responsiveness are associated in animals with muscles that play a role in reproductive signaling and sexual behavior. For instance, only those sections of the forelimb flexor muscle in the male frog that are used in the amplexus or grasping of the female during copulation show increased number of androgen Rs and functional changes to androgen stimulation (Dorlochter et al. 1994; Regnier & Herrera 1993a, 1993b).

As a result of these gender-specific differences in the pattern of GH secretion, in anabolic effectiveness of GH and sex steroids, and in regional differences in androgen R number the somatic growth spurt during adolescence follows a different pattern in boys and girls. Three gender differences in the adolescent growth spurt stand out. First, the increase in the rate of growth is almost two times greater in boys than in girls. Height growth velocity, arrested at about 5 cm/year between the ages of 3 and 10 years (see figure 7.1), increases to a peak of about 8 cm/year in girls and a little over 10 cm/year in boys (Marshall & Tanner 1986). This difference in part contributes to greater height (6%), weight (23%), and lean body mass (50%) in average adult young males compared with females (Forbes 1972; Malina 1969; Behnke 1961, 1969).

Second, onset of adolescent growth and peak growth velocity is generally 2 years earlier in girls compared with boys (see figures 7.1 and 8.4). Because boys have about 2 extra years of preadolescent growth, they begin their adolescent spurt roughly 9 cm taller than girls had been before their growth surge, contributing further to gender differences in adult stature. In boys, tempo of weight increases coincide with growth in height, while in girls, linear growth is 1 year advanced relative to changes in weight (see figure 8.12).

Gender dissimilarities in onset of the adolescent growth spurt also contribute to differences in the proportion of appendicular relative to axial skeleton in men and women. During the adolescent growth spurt, increase in length of the trunk is greater than in the length of legs. Therefore, children entering puberty and the adolescent growth surge early will tend to have shorter legs relative to trunk length than late maturers (Marshall & Tanner 1970, 1986). Linear growth of leg bones reaches its peak velocity 0.6 year before trunk length. The longer period of preadolescent growth in boys is therefore largely responsible for greater length of men's legs relative to the length of the trunk.

Third, growth dimorphism is more striking in the muscles of upper extremities than of the lower extremities (see figure 8.12) and in the bones in shoulder and

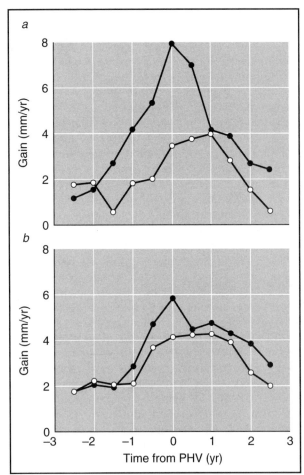

Figure 8.12 *Velocities of muscle growth in boys and girls.*

Mean velocities of arm muscle widths *(a)* and calf muscle widths *(b)* in boys (solid circles) and girls (open circles) classified by years before and after their own peak height velocity (PHV).

Reprinted from Tanner, Hughes, and Whitehouse 1981.

pelvic girdles. This condition suggests the operation of gender differences in tissue distribution of androgen and estrogen receptors. Bone marrow is another site where aromatization of androgens takes place (Frisch et al. 1980a). Increased titers of circulating estrogens in pubertal girls as well as production of estrogens in the bone marrow of both genders end linear growth by closure of the epiphyseal growth zones (Baker 1985).

EFFECTS OF GENDER DIFFERENCES ON PHYSICAL PERFORMANCE

Throughout most of early history, women have been discouraged from participation in sport. They have not been allowed to participate in Olympic Games until 1928 or to run in the Boston Marathon until 1972, but in the past 30 years, their participation in competitive sports has risen exponentially. Significant differences in sport performance have necessitated segregation of the

two genders for competition purpose, but women's times in most running events have been improving much more rapidly than men's times so that the gender difference in these events may disappear by the year 2035 (Drinkwater 1984; Lewis et al. 1986; Pate 1985; Sparling & Cureton 1983; Shephard 2000a; Whipp & Ward 1992). In general, gender differences are less pronounced in endurance physical activities that rely on leg musculature and more strongly expressed in events emphasizing strength that rely on dimorphically divergent arm muscles (Chatterjee & Laudato 1995; Shephard 2000a). While some gender differences in physical performance reflect the already described developmental dimorphism, others may be a function of gender differences in training history and physical fitness. Most frequently studied gender differences that impact physical performance are those in the proportion of musculoskeletal system and body fat, in cardiovascular and respiratory structure and function, thermoregulatory ability, and type of metabolic fuel used during exercise.

Mass, Strength, and Power

The young adult woman has about 20% smaller total muscle mass than the young adult male of equal body mass. Gender differences in skeletal growth produce a broader pelvis, shorter legs, and consequently a lower center of gravity in women compared with men. These physical differences increase stability in the female, change her gait dynamics, and decrease her mechanical efficiency by increasing the angle of the thigh bone and bringing the knees closer together (Horton & Hall 1989; Kerrigan et al. 1998). These biomechanical differences as well as increased female ligament laxity make exercising women more prone than men to anterior cruciate ligament rupture (Hart et al. 1998; Hutchinson & Ireland 1995). Shorter limbs in women increase stride frequency and limit peak running speeds (Hoffman 1972). Achilles tendons, which are important in elastic recoil of running, are shorter in women than in men and thus affect gait as well as risk of injury (Newsholme et al. 1994). Smaller thorax, narrower and more sloping shoulders, and smaller muscle fiber diameter, particularly in the upper limbs, reduce the lever arm and limit throwing performance of women. In addition, smaller and lighter bones in women make them more vulnerable to fracture (Kelly & Bradway 1997).

The young adult woman has more intramuscular fat and connective tissue, and 30% smaller muscle fiber cross-sectional area (Costill et al. 1976; Maughan et al. 1984; Miller et al. 1993). For the same body mass, a young woman has about 80% more fat than a young man (Behnke 1961, 1969), which increases the load to be carried with a smaller muscle mass, and an absolute strength that is only about 60% of values seen in a young man (Wilmore et al. 1974). Gender difference in muscle strength is smaller for eccentric and larger for concentric contractions (Seger & Thorstensson 1994).

Muscular strength increases linearly until the onset of adolescent growth spurt. No gender-specific differences in strength are apparent during the prepubertal period, although generally after 8 years of age, marked acceleration of strength and muscle development occurs in boys but not in girls. Maximum strength in each gender is reached after peak velocity in height and weight (Malina 1974). Gender differences in muscle strength are large in the muscles of the chest, shoulders, arms, and forearms, as brachial muscles almost double in size during male adolescence (Komi 1980; Malina 1986; Miller et al. 1993). Gender differences are smallest in hip flexors and extensors and almost disappear when normalized for body mass (Komi 1980; Wilmore 1974).

The average woman's peak muscle power is only about 70% of that attained by the average man (Karlsson & Jacobs 1980). Because gender difference in body mass is of approximately the same magnitude, performance in anaerobic events where power is used to move the body mass is about equal in the two genders (Wells 1985). In part because of lower absolute strength, the time required to develop peak force is greater in women than in men (Behm & Sale 1994). Finally, women store elastic energy in the stretched muscles more readily than men (Aura & Komi 1986). Increases in absolute and relative strength in response to high-resistance training are the same in the two genders, but increases in lean body mass and muscle hypertrophy are smaller in women compared with men (Brown & Wilmore 1974; Wilmore 1974; Wilmore et al. 1978).

Body Fat and Physical Performance

A gender difference in body fat also develops during the prepubertal period. Subcutaneous fat content becomes significantly greater in girls than in boys about a year and a half before the adolescent growth spurt (see figure 8.13), at the time when a gender difference in body density becomes evident (Parizkova 1961; Tanner et al. 1981). As they enter the adolescent growth spurt, both girls and boys lose subcutaneous fat from the upper arms, with little change occurring in the lower legs. After adolescents reach peak growth velocity, subcutaneous fat accumulation resumes in both boys and girls, with the gender difference becoming even more pronounced. In addition, a distinct gender difference develops in distribution of fat, as the deposition of fat becomes greater in the thigh and gluteal region in females as compared with the abdominal region where fat accumulation predominates in males.

A functional significance to increasing body fat in adolescent girls was postulated by Frisch (1984) who observed a relationship between body fatness and the onset of menarche as well as maintenance of menstrual cyclicity. In girls, body fat increases linearly with increasing lean body mass, but this rate is slower in late maturers than in early maturers. Thus, late maturing girls have less body fat than the early maturing girls at same body weights, but in both, body fat increases from

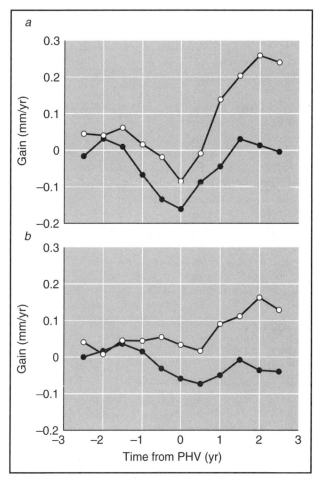

Figure 8.13 *Velocities of subcutaneous fat growth in boys and girls.*

Mean velocities of subcutaneous fat growth of boys (solid circles) and girls (open circles) presented relative to the timing of peak height velocity (PHV). Arm fat folds are shown in *(a)*, and calf fat folds are shown in *(b)*.

Reprinted from Tanner, Hughes, and Whitehouse 1981.

5 to 11 kg during the adolescent growth spurt. Both early and late maturers develop menarche after their body fat level exceeds 17%. The reproductively unstable and infertile period usually ends when female adolescents have gained another 4.5 kg of body fat and achieved a body composition of 28% fat (Frisch & McArthur 1974). Seventeen percent fat is thus the threshold fatness necessary for the onset of menarche, and 22% fat is the threshold level of fatness necessary for the maintenance of fertile menstrual cycles. This correlation has been interpreted to suggest that female fertility is contingent on the availability of 99,000 calories contained in 11 kg of body fat, enough to provide the 80,000 calories needed for a full-term pregnancy (FAO 1957).

Cardiovascular and Respiratory Organs and Aerobic Physical Performance

Growth of cardiorespiratory organs is also dimorphic and contributes to approximately a 10% disadvantage in adult women compared with same-size adult men with respect to stroke volume, cardiac output, vital capacity, hemoglobin concentration, and oxygen-carrying capacity of blood (Wells 1985; Wiebe et al. 1998). Growth of the heart and lung accelerates in synchrony with the timing of the adolescent growth spurt (Simon et al. 1972). Growth in length of the lung reaches peak velocity about 6 months after peak growth of the lung in width. Mean width of the lung in girls tends to be smaller than in boys (Simon et al. 1972). The heart shape changes from elliptical to spherical, though its size remains proportional to body mass (Rakusan 1984). In contrast to uniform growth in overall heart size and shape, the female heart, but not the male heart, develops lower R-wave voltages as girls progress from stage 2 through stage 5 of puberty (Stafford et al. 1989). In adulthood, difference in heart size and greater peak left ventricular ejection fraction and end-systolic volume in men contribute to gender difference in cardiac output (Adams et al. 1987; Hanley et al. 1989; Legget et al. 1996). The size advantage in the left ventricle accounts for about 70% of gender difference in maximal respiratory capacity (Hutchinson et al. 1991).

Gender difference in aerobic performance is caused in part by differences in proportion of muscle and body fat and in part by level of fitness, but this difference is diminished by endurance exercise training (Zwiren et al. 1983). Thus, gender difference in maximal respiratory capacity of untrained individuals is 56% in absolute terms, 28% when normalized by body mass, and 15% when normalized by lean body mass. In a comparison of aerobically trained individuals of equivalent fitness level, these differences shrink, respectively, to 52%, 19%, and 9% (Sparling 1980). In the comparison of men's and women's 12-minute runs, males outperform women by about 20% to 30% (Sparling & Cureton 1983). About 65% of this difference is the result of factors associated with gender, and 95% of gender-associated variance is the result of differences in body fat rather than a difference in running economy. Aerobic training increases mechanical efficiency of running in either gender so that some of the difference in running economy may stem from different training histories or fitness levels in the two genders (Pate et al. 1997). When men and women are matched for percentage body fat and fitness, their maximal aerobic performance is equal (Pate et al. 1985). The male advantage in oxygen-carrying capacity of blood is compensated by higher 2,3-diphosphoglycerate concentrations in women, which allows greater oxygen unloading by hemoglobin in the contracting muscles of women. Finally, central and peripheral cardiovascular adaptations to aerobic training are equal in men and women and unaffected by the mode of training (Eddy et al. 1977; Lewis et al. 1986).

Thermoregulation During Exercise

Women are less tolerant of dry heat stress than men because of larger relative body surface area for transfer of environmental heat load, lower sweat gland density, and lower sensitivity of sweating response to increases in core temperature (Buono & Sjoholm 1988; Cunningham et al. 1978; Kenney 1985). Higher thermoregulatory set point reported for sweating and vasodilation in women may be caused by differences in the level of fitness and heat acclimatization rather than in gender (Shapiro et al. 1980). Gender differences in sweat rate during heat exposure or exercise in heat are attenuated after controlling for level of fitness and acclimatization to heat (Frye & Kamon 1981; Kenney 1985). Women perform better than men in hot, humid weather in that they can complete the same work with less fluid loss, smaller increase in core temperature, and greater sweating efficiency (Avellini et al. 1980; Frye & Kamon 1983). In addition, when the level of physical fitness is controlled, men and women respond physiologically to hypohydration during exercise in a similar way (Sawka et al. 1983).

Exposure to cold water impairs physical performance of women less than it impairs performance of men (Pugh et al. 1960; Wyndham et al. 1964). This dissimilarity is most likely caused by gender difference in the mass and distribution of body fat and, consequently, its insulating and bioenergetic properties. However, women also are at a disadvantage in the cold environment because of their greater body surface for heat loss, higher shivering threshold, and lower capacity for nonshivering thermogenesis compared with men (Graham 1983; Shephard 1993). In contrast to men, women do not show a reduction in heart rate and increased stroke volume during exercise in the cold (Stevens et al. 1987).

Metabolism During Exercise

At submaximal exercise intensities, women release and oxidize more adipose tissue lipids and rely less on muscle lipids and glycogen than do men (Mittendorfer et al. 2002; Tarnopolsky et al. 1997; Tate & Holtz 1998). This difference is apparent at 40% (Horton et al. 1998), 50% (Mittendorfer et al. 2002), 65% (Tarnopolsky et al. 1990), and 75% to 80% of maximal aerobic performance (Friedman & Kinderman 1989; Tarnopolsky et al. 1995, 1997), but may disappear at near-maximal exercise intensities (Froberg & Pedersen 1984). This effect appears to be caused by greater sensitivity of female abdominal subcutaneous adipose tissue to lipolytic beta adrenergic action of catecholamines, higher plasma concentrations of GH, and increased muscle but lower adipose tissue LPL activation by estrogenic stimulation (Arner et al. 1990; Hardman 1998, 1999; Hellstrom et al. 1996; Horton et al. 1998; Jensen et al. 1996; Millet et al. 1998; Tsetsonis et al. 1997; Ruby & Robergs 1994; Schaefer et al. 1995). Endurance training was reported by some, but not the others (Tarnopolsky et al. 1997), to attenuate or abolish gender difference in the type of metabolic fuel used during exercise (Friedman & Kinderman 1989; Ruby & Robergs 1994).

PHASES OF MENSTRUAL CYCLE AND PHYSICAL PERFORMANCE

Variation in concentration of estrogen and progesterone during the menstrual cycle and in response to contraceptive use causes changes in physical performance that are similar to but less pronounced than those caused by gender differences (De Souza et al. 1990; Lebrun 1994; Shephard 2000b). Menstrual cycle has little effect on secretion of stress hormones such as GH during exercise (Hornum et al. 1997; Kanaley et al. 1992a; Keizer et al. 1987). During exercise under the conditions of high estradiol titers that obtain during the luteal phase or contraceptive use, cortisol secretion is either unaffected or suppressed, and aldosterone secretion is significantly higher and may contribute to increased fluid retention (Bonen et al. 1991; De Souza et al. 1989; Kanaley et al. 1992b). Sex hormone variations during the menstrual cycle have a subtle effect on fuel utilization. In some experiments, lipid utilization was found to be higher during the luteal phase, whereas no such differences were seen in others (Galliven et al. 1997; Hackney et al. 1994; Wentz et al. 1997). High estradiol titers are associated with decreased activities of adipose tissue LPL and hepatic lipase (Schaefer et al. 1995). Hepatic lipase helps convert chylomicrons and VLDLs to LDLs. It also participates in degradation of HDL_2, thus contributing to gender difference in plasma HDL concentration. Finally, variations in progesterone concentration affect thermoregulation during exercise (Stephenson et al. 1982). Core temperature is lowest during ovulation and highest during the luteal phase of the cycle at rest when progesterone titers are high, as well as during exercise (Carpenter & Nunneley 1988; Pivarnik et al. 1992; Stephenson & Kolka 1988; Tenaglia et al. 1999). During the luteal phase of the menstrual cycle, thresholds for vasodilation and sweating increase with no change in sensitivity of the thermoregulatory responses (Hirata et al. 1986; Kolka & Stephenson 1989; Stephenson & Kolka 1985). Finally, ventilatory rate is higher under hypoxic conditions during the luteal than during the follicular phase (Schoene et al. 1881; White et al. 1983). Overall, the phase of menstrual cycle and contraceptive use have a modest influence over physical performance of women.

ADMINISTRATION OF SEX STEROIDS TO INCREASE MUSCLE STRENGTH AND POWER

Gender difference in strength and muscle size has not gone unnoticed by individuals interested in developing muscular physique and by athletes eager to improve muscle power. The inference that the male sex hormones were responsible for greater muscle size and strength in men was made shortly after the chemical characterization of androgens in the mid-1930s. Opportunity to test it was made easier by pharmacological modification of androgen structure in anabolic steroids to increase their anabolic relative to their androgenic action. Anabolic steroids were developed through one of three types of structural modification of the testosterone molecule (Wilson 1988) (see figure 8.14): esterification of the 17β-hydroxyl group (type A), alkylation at the 17α position (type B), and modification of the steroid ring structure (type C). Structural properties of anabolic steroids determine their effective route of administration. Type B steroids shown in the top five rows of figure 8.15 are used by oral route because of their slow hepatic degradation, which keeps them in circulation in effective concentrations for an extended period of time. The length of the carbon chain in the ester is inversely related to the speed of its degradation and thus determines the necessary frequency of steroid application for sufficient stimulation. Type C steroids are usually alkylated at positions 1 or 19 of the steroid ring and can be used by oral or parenteral route. Methyl groups in position 19 make anabolic steroids more potent androgenically than testosterone. Only one type A steroid, testosterone undecanoate, is used by oral route because of its nonpolar nature and uptake into the lymph rather than portal venous circulation. This steroid and micronized testosterone is degraded rapidly and need to be administered several times a day. Other type A steroids, shown in the bottom two rows of figure 8.15, are nonpolar and soluble in lipid, which makes them suitable for injection into muscles for slow release into circulation.

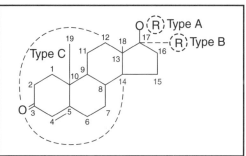

Figure 8.14 *Molecular structure of anabolic steroids.*

Regions of testosterone molecule that are modified in anabolic steroids. Type A, B, and C modifications entail, respectively, esterification of the 17β-hydroxyl group, alkylation at the 17α position, and modifications of the steroid ring structure.

Reprinted from Wilson 1988.

The capacity of anabolic steroids to increase muscle mass and strength was empirically established by the lay community at least a half a century before sufficient scientific evidence could justify these attitudes and steroid usage. Testosterone was used as an ergogenic aid by athletes in the former Soviet Union in the mid-1950s, and the unofficial use of anabolic steroids has spread worldwide since the Olympic Games in Tokyo in 1964 (Strauss & Yesalis 1991; VanHelder et al. 1991). Various surveys estimate that the use of anabolic steroids has spread beyond the competitive arena to nonathlete adolescents who have not yet completed statural growth.

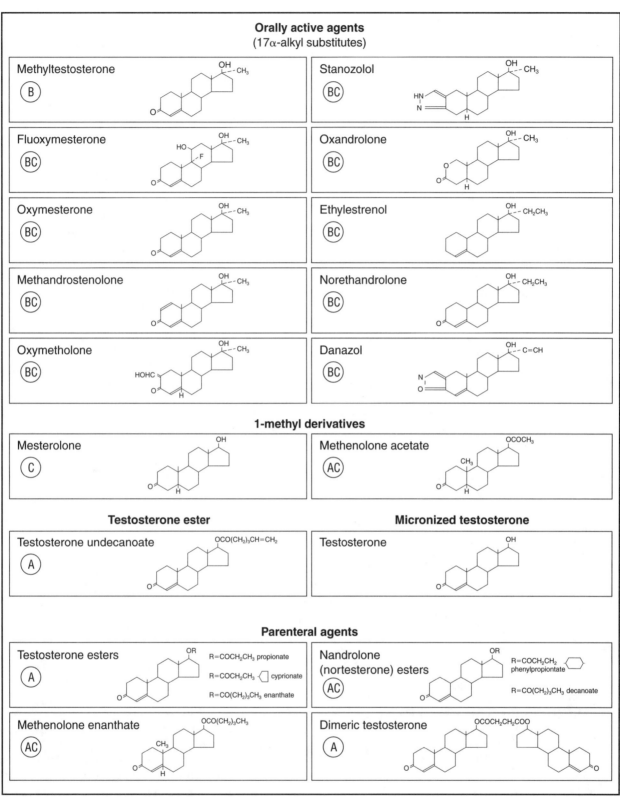

Figure 8.15 *Structure of representative anabolic steroids.*

Orally active agents are 17a-alkyl derivatives (methyltestosterone, fluoxymesterone, oxymesterone, methandrostenolone, oxymetholone, stanozolol, oxandrolone, ethylestrenol, norethandrolone, and danazol), 1-methyl derivatives (mesterolone and methenolone acetate), a testosterone ester (testosterone undecanoate), and micronized testosterone. Parenteral agents are other testosterone esters (propionate, cypronate, and enanthate), methenolone enanthate, nandrolone (nortestosterone) esters, and dimeric testosterone.

Reprinted from Wilson 1988.

Because of the ethical problems associated with the use of hormones or drugs to improve performance and the side effects associated with anabolic steroid use, they have been banned since the 1976 Olympic Games in Montreal, Canada, and sanctioned by the scientific community (American College of Sports Medicine 1977, 1984; Bahrke et al. 1994).

The efficacy of anabolic steroids in increasing muscle mass and strength was not confirmed scientifically until very recently (Bhasin et al. 1996; Sinha-Hikim et al. 2002). Conviction in the weightlifting community of their effectiveness and the lack of scientific confirmation of this effectiveness hinged on four differences in the way their action was tested: steroid dosage, prior exposure of steroid user to weightlifting, adequate protein intake, and method of strength testing. Anabolic steroid users administer doses equivalent to between 20 and 80 mg of testosterone per day, which raises plasma testosterone concentration to between twofold and sixfold above that of male control subjects, and users also add other physiologically active androgen derivatives to circulation (Alen et al. 1985; Bhasin et al. 1996). By contrast, doses used in the laboratory setting, mandated by ethical concerns and institutional review boards, typically ranged between 5 and 20 mg/day (Haupt & Rovere 1984). Plotting of the logarithm of total dose of anabolic steroids administered against increases in lean body mass reveals a clear dose-response curve with approximately half of early studies utilizing ineffective total doses of less than 3 g, whereas total doses of between 4 and 10 g administered in other studies yielded proportional growth of lean body mass (Forbes 1985). In addition, steroid users typically administer a number of different steroid preparations ("stacking") to maximize androgen receptor stimulation and progressively increase their amounts ("pyramiding") to overcome any androgen R down-regulation. Steroid use is progressively reduced before competition ("tapering") to avoid detection in blood or urine. Thus, periods of steroid discontinuation usually alternate with periods of steroid use ("cycling").

Anabolic steroid administration at least at moderate steroid concentrations was ineffective in increasing strength in individuals who had no prior or concurrent involvement in weight training (Haupt & Rovere 1984). A catabolic state generated by resistance training apparently allows the anticatabolic action of anabolic steroids to induce muscle growth and strength gains. Steroids counteract protein degradation and nitrogen loss induced by muscle overloading. They oppose the catabolic state caused by increased secretion of cortisol and increased net protein synthesis (Hickson et al. 1990; Konagaya & Max 1986). In addition, strength and muscle mass increases were predominantly reported in studies where steroid administration was accompanied by dietary protein supplementation (Haupt & Rovere 1984). Muscle overloading that is necessary for strength gains is associated with a negative nitrogen balance when protein intake falls below 1.5 g/kg of body weight per day. Muscle mass and strength during high-resistance training increase in proportion to dietary protein intake of between 1.5 and 2.3 g/kg per day (Lemon & Proctor 1991; Lemon et al. 1992; Tarnopolsky et al. 1988, 1992). Finally, significant increases in strength were reported in studies using maximal weight lifts in a single repetition of weightlifting exercise of the same kind as that used in training. This method is more sensitive in detecting strength gains than the single-joint isolation techniques with dynamometers employing movements that were not used in strength training (Haupt & Rovere 1984).

A recent study (Bhasin et al. 1996) that has avoided all four of the mentioned pitfalls clearly demonstrates that a supraphysiological dose of testosterone (about 85 mg/day) administered for a period of 10 weeks to normal men with prior weightlifting experience and adequate protein intake, but without concurrent resistance training, can significantly increase muscle mass and strength. During this treatment, plasma testosterone concentration increased by about fivefold to sixfold. At the end of 10 weeks, increases in the cross-sectional area of the triceps muscle and in its capacity to lift weight were 14% and 9%, respectively. Corresponding increases in the cross-sectional area of the quadriceps muscle and in the capacity to lift weight from a squatting position were 8% and 10%, respectively. Increases in muscle cross-sectional area in response to anabolic steroid treatment are a consequence of hypertrophy of both type I and II muscle fibers as well as of increased proliferation of satellite cell myonuclei. Activated satellite cells are incorporated into hypertrophied muscle fibers to produce new myonuclei and preserve normal nuclear to cytoplasmic ratio in the enlarged muscle (Kadi et al. 1999; Sinha-Hikim et al. 2002). Resistance training alone led to more moderate increases in muscle growth than testosterone treatment alone, but increases in strength were of equal magnitude in the upper arms and twice as large as testosterone effect alone in the legs. Combined exercise training and testosterone appeared to have additive rather than substitutive effects on increases in muscle size and strength.

The long history of androgen use by athletes has thus found scientific justification complementing successful use of androgens by the medical community to increase muscle protein synthesis for suppression of muscle wasting in patients with AIDS, in elderly men, and in patients with other sarcopenic states (Bross et al. 1998; Rabkin et al. 2000; Urban et al. 1995). Capacity of testosterone to stimulate erythropoiesis has also had medical application in individuals with some forms of anemia (Hamdy et al. 1998; Navarro et al. 1998).

The side effects of anabolic steroid use are related to the inefficiency of hepatic metabolic machinery in excreting them and to the elevation of their plasma concentrations above adult regulated concentration ranges. Because of the structural modifications in testosterone molecules, enzymatic degradation of anabolic steroids in the liver is slow and inefficient. As a consequence, unmetabolized anabolic steroids pool within liver canaliculi, causing jaundice (cholestasis hepatis), blood-filled cysts (peliosis hepatis), and various forms of liver carcinoma (Bernstein et al. 1971; Guy & Auslander 1973; Holder et al. 1975). In contrast to hepatocarcinoma of other etiologies, cancerous pathologies that are associated with anabolic steroid use regress when steroid use is discontinued.

Pathologies associated with elevation of plasma androgen concentrations above the normal adult male range include consequences of increased negative feedback over testosterone production and increased actions of androgens on hair growth, sebum secretion, masculinization of genitalia and secondary sex characteristics in women, and mood. Increased negative feedback from higher circulating androgen concentrations suppresses plasma LH and FSH concentrations below ranges seen in normal men and blunts LH response to a GnRH challenge (Alen et al. 1985; Jarow & Lipshultz 1990). When steroid administration is discontinued, plasma FSH concentrations rebound rapidly, but it takes between several months and a year for the plasma LH and testosterone concentrations to return to normal. During, and for some time after, anabolic steroid use, spermatogenesis is suppressed (Holma 1977; Johnson et al. 1972; Jones et al. 1977).

Exogenous androgens cause alopecia (thinning of hair) and pattern baldness in both genders and growth of excess body hair in women (Davis 1999). There is an increase in secretion of sebum and acne. Deepening of the voice tone and enlargement of clitoris are irreversible effects of anabolic androgens in women (Baker 1999; Davis 1999; Malarkey et al. 1991; Strauss et al. 1985). Significant increases in circulating estrogen resulting from peripheral aromatization of testosterone cause permanent breast enlargement (gynecomastia) in steroid-using men (Alen et al. 1985). In addition, closure of epiphyseal growth zones in steroid-using adolescents occurs prematurely, before they have completed statural growth (Dyment 1982).

Steroid use is also associated with cardiovascular and metabolic pathologies. Androgen use lowers HDL-cholesterol and raises LDL-cholesterol, thus facilitating atherosclerotic vascular changes. Steroid use also is associated with abnormalities in the 24-hour pattern of blood pressure, left ventricular wall thickening, and sudden death linked to atrial fibrillation (Dickerman et al. 1997, 1998; Hausman et al. 1998; Palatini et al. 1996; Sullivan et al. 1998, 1999). Anabolic steroid and contraceptive users display insulin resistance, hypercoagulability of blood, and increased risk of thrombosis (Cohen & Hickman 1987; Diamond et al. 1998; Haffner 1996; Srivastava et al. 1975).

Use of anabolic steroids affects mood and personality traits. At physiological doses, androgens improve the mood, are antidepressant, and do not lead to objectively measurable manifestations of aggressiveness (Bahrke et al. 1992, 1996; Ehrenreich et al. 1999). At pharmacological doses, they have been associated with increased aggressiveness, episodes of uncontrolled aggressive behavior sometimes called "roid rage," and abnormal personality manifestations such as mania, hypomania, paranoid delusions, and excessive preoccupation with physiques and food (Cooper et al. 1996; Gruber & Pope 2000; Kouri et al. 1995; Pope & Katz 1990, 1994; Pope et al. 2000; Porcerelli & Sandler 1998). Steroid use appears to be addictive, and its discontinuation can lead to depression and suicide (Brower et al. 1990; Thiblin et al. 1999). The hypothesis that increased muscle hypertrophy and strength gains are secondary to more intense and aggressive weightlifting in steroid users (VanHelder et al. 1991) is untenable as it cannot account for muscle hypertrophy and strength gains seen after testosterone administration in the absence of weight training (Bhasin et al. 1996).

EFFECT OF EXERCISE ON SEX HORMONE SECRETION

Acute exercise of moderate to high intensity is associated with increases in plasma concentrations of sex hormones in a gender-specific fashion (see figure 8.16). In the male, there is an increase in plasma concentration of androstenedione and total and free testosterone that is dose dependent and ranges between about 30% and 185% at maximal exercise intensities (Dessypris et al. 1976; Ježova et al. 1985; also see Cumming et al. 1989 for review). Males display little change in plasma concentrations of estradiol and estrone in response to exercise (Kuoppasalmi et al. 1976). Plasma androgen increases are seen in men after endurance exercise such as running and cycling as well as after high-resistance exercise, whether it is of high-loading, low-volume type or low-loading, high-volume type (Jensen et al. 1991; Ježova & Vigaš 1981; Kraemer et al. 1991a, 1998; Kuoppasalmi et al. 1976, 1980; Sutton et al. 1973; Webb et al. 1984).

In women, increases in plasma concentrations of estrogen are proportional to exercise intensity during acute exercise and more prominent during the luteal than during the follicular phase of the menstrual cycle. Increases in plasma progesterone occur during the luteal phase of the cycle only (Bonen et al. 1979; Cumming et al. 1987a, 1987b; Jurkowski et al. 1978). Increases in plasma testosterone in response to acute exercise in women are either not seen, are very small, or are delayed compared with men (Cumming et al. 1987a, 1987b; Kraemer et al. 1991a, 1993b; Shangold et al. 1981; Webb et al. 1984; Weiss et al. 1983). Any increases in plasma testosterone and 10-fold greater increases in androstenedione during acute intense exercise in women are most likely of adrenal cortical origin (Weiss et al. 1983). In addition, increased concentrations of catecholestrogens (2-hydroxy estrone) are also observed during strenuous swim training in women (Russell et al. 1984a, 1984b). Catecholestrogens are chemical variants of estradiol or estrone produced in the peripheral tissues and brain, which have two adjacent hydroxyl bonds on either or both the 2 and

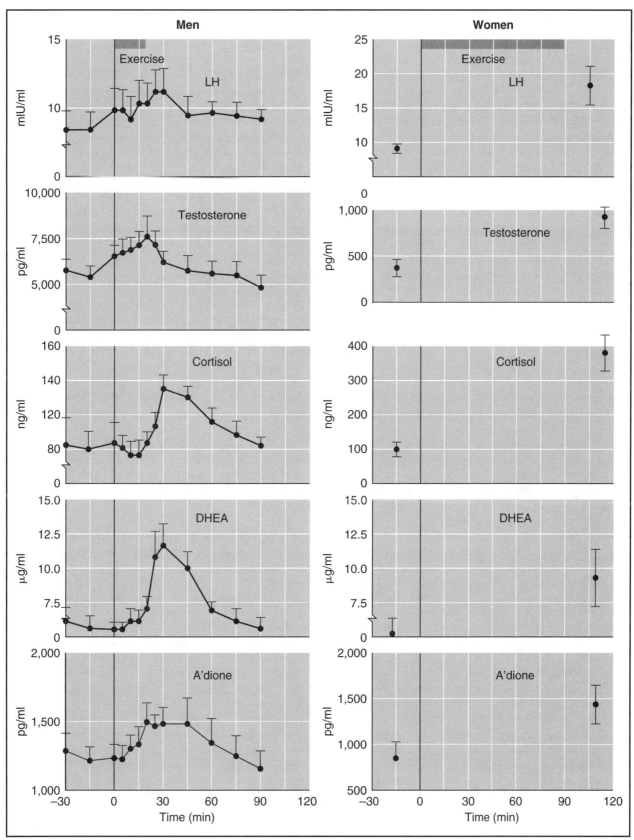

Figure 8.16 *Exercise effects on plasma concentrations of sex hormones in men and women.*

Changes in plasma concentrations of sex hormones in men (left) and women (right) during acute exercise.

DHEA = dihydroepiandosterone; A'dione = androstenedione.

Data for men adapted from Cumming et al. 1986; data for women from Baker et al. 1982.

3 or the 3 and 4 positions of the steroid ring (see figure 1.12). Thus, they derive their name from the structural similarity between the adjacent double hydroxyls on the A portion of their steroid ring and the benzene ring of the catecholamines (see figure 1.2). Because of this structural similarity, catecholestrogens compete with catecholamines for the COMT enzyme responsible for their degradation and, as a result, cause increases in plasma catecholamine concentrations (De Cree 1990).

There is little understanding of the mechanism by which plasma concentrations of sex hormones increase during acute exercise. Exercise-associated increases in hemoconcentration, rate of sex hormone degradation, sex hormone secretion from adrenal cortex, and sex hormone secretion from the gonads in response to stimulation by catecholamines are some of the explanations that have been proposed to account for exercise-associated plasma sex hormone increases. Exercise-associated fluid shifts between vascular compartment and the muscle cannot account in entirety for the changes in the plasma concentrations of sex hormones because of discrepancy in the magnitude and timing of the two sets of changes, as the former account generally for less than 10% of changes and the latter usually exceed this level of change by several-fold (Cumming et al. 1987b). Increases in plasma concentration of sex hormones could also result from their slower degradation during acute exercise because of slower hepatic blood flow. In women, in whom 90% of sex hormones are degraded in the liver, the metabolic clearance rate of estradiol is reduced by about 36% at the end of 10 minutes of exercise at 70% of $\dot{V}O_2$max (Baird et al. 1969; Keizer et al. 1980). In men, the metabolic clearance rate during exercise also declines, but the interpretation of this effect is more complicated, as only 50% of sex hormones are degraded by the liver and the rest by the muscle, the two organs that experience opposite changes in blood flow during exercise (Baird et al. 1969; Cadoux-Hudson et al. 1985; Stenstad & Eik-Nes 1981). If changes in the rate of sex hormone degradation were to completely account for changes in plasma concentrations of sex hormones during acute exercise, all sex steroid concentrations should be changing during exercise in a similar way, which is not the case.

It is probable that increases in plasma androstenedione in men and women and of testosterone in women during acute intense exercise are of adrenal origin (Cumming et al. 1987b; Dessypris et al. 1976; Kuoppasalmi et al. 1976, 1980; Weiss et al. 1983). The intensity-dependent nature of increases in both the concentrations of these sex steroids and correlation between concentrations of androgens and cortisol under some exercise conditions suggest that the adrenal cortex releases androstenedione and cortisol during acute intense exercise (Dessypris et al. 1976; Cumming et al. 1986). It is less certain that increases in plasma concentration of testosterone in exercising men are of adrenal origin despite the argument that the timing of its release makes ACTH a more probable stimulus for its release than LH (figure 8.16) (Cumming et al. 1986). Administration of dexamethasone, which suppresses pituitary secretion of ACTH during exercise, does not prevent exercise-associated increase in plasma testosterone concentration (Sutton et al. 1974).

The possibility that increases in plasma concentration of androgens is triggered by the S nerves or circulating catecholamines is compatible on several grounds but has not been extensively studied or universally accepted, in part because there is not convincing evidence of S innervation of the testes (Cosentino et al. 1984; Kuntz 1919; Wakade et al. 1975). Stimulation of β adrenergic receptors stimulates testosterone synthesis and release in a dose-dependent fashion (Eik-Nes 1969; Ježova & Vigaš 1981). In addition to having adrenoceptors, the testes contain the catecholamine degrading enzyme monoamine oxidase (Penttila & Kormano 1968). Increases in plasma concentrations of testosterone and catecholamines are both directly proportional to exercise intensity, which supports the hypothesis that S nerve activity and catecholamine secretion stimulate testosterone secretion during acute exercise (McMurray et al. 1987). S involvement in testosterone secretion during exercise could also account for anticipatory increases in plasma testosterone concentration occurring in advance of subjects actually engaging in exercise of increasing intensity (Wilkerson et al. 1980). The rate of testicular secretion of testosterone also depends on the rate of blood supply to the testes through spermatic arteries (Eik-Nes 1964). Thus, the support for the S or catecholamine hypothesis of increased testosterone secretion during exercise requires additional information about S innervation of and its effect on the blood supply to the testes during exercise.

EFFECTS OF EXERCISE ON FERTILITY AND ONSET OF MENARCHE

In character with its central role in the control of fertility, LH pulsatility is suppressed in response to extended strenuous exercise or nutritionally adverse conditions to reduce secretion of gonadal steroids and thus either curtail fertility or delay its onset. Conditions that inhibit or delay fertility are exposure to physical and psychological stress and acute or prolonged energy shortage, often producing inadequate bone maturation, statural growth, and fat depot size in adolescents.

In general, neuroendocrine controls of fertility are more resistant to these stimuli in the male than in the female. Intense exercise carried to exhaustion or very heavy routines of prolonged training are associated with progressive and often delayed decreases in plasma testosterone and reduced spermatogenesis in men in a similar way that very heavy and prolonged exercising causes decreases in estradiol concentration and reproductive suppression in women (Brisson et al. 1984; Cumming et al. 1987b, 1989; De Souza & Miller 1997; Dessypris et al. 1976; de Lignieres et al. 1976; Galbo et al. 1977b; Kujala et al. 1990; Kuoppasalmi et al. 1980;

Morville et al. 1979; Schmid et al. 1982; Schurmeyer et al. 1984; Tanaka et al. 1986; Urhausen & Kindermann 1987; Wheeler et al. 1984). Such exhausting exercise is also associated with increased plasma levels of cortisol, androstenedione, and GH, indicating both high levels of stress and depletion of metabolic substrates (Dessypris et al. 1976; Kuoppasalmi et al. 1980). Significant correlations between postexercise increases in androstenedione and cortisol as well as between decreases in testosterone and cortisol, suggest that plasma testosterone changes under the conditions of prolonged and strenuous exercise or high volume of endurance training are influenced by secretion of adrenal cortical and other stress hormones (Dessypris et al. 1976; Frey et al. 1983; Hackney et al. 1988; Wheeler et al. 1986). Decreased LH secretion and blunted LH response to GnRH challenge often but not always accompany postexercise decreases in plasma testosterone, suggesting that such decreases are most likely of hypothalamic origin (de Lignieres et al. 1976; MacConnie et al. 1986). Acute fasting of several days duration or prolonged starvation results in decreased plasma testosterone concentration in normal healthy men (Kyung et al. 1985; Zubiran Gomez-Mont 1952). It is therefore probable that an acute shortage of energy caused by a large volume of acute exercise or by a prolonged regime of strenuous exercise suppresses hypothalamo-pituitary-gonadal axis and activity of GnRH neurons in men.

In women, function of HPGA can be disrupted by signals associated with skeletal immaturity, deficient energy stores, acute reduction in energy availability, and stress. Disruption takes the form of delayed onset of menarche (primary amenorrhea), irregular menstrual cycles (oligomenorrhea), or loss of already established menstrual cycles for at least 6 months (secondary amenorrhea). Disruption is mediated through suppression of GnRH pulsatility (Loucks et al. 1989; Veldhuis et al. 1985). In response to reduced frequency of GnRH pulses, the number of GnRH receptors on gonadotrophs is down-regulated and amplitude of gonadotropin pulses is suppressed. Reinstatement of increased GnRH pulsatility produces gonadotropin priming through up-regulation of pituitary GnRH Rs.

Instead of starting at about 12.8 years, menarche in athletes is often delayed by 1 to several years. Delay in the onset of puberty of about 1 to 2 years has been observed in female ballet dancers, runners, and gymnasts more than in adolescent participants in other sports (Claessens et al. 1992; Frisch et al. 1980b, 1981; Malina 1983; Malina et al. 1973, 1979; Warren 1980). Each year of training before menarche has delayed menarche by 5 months, suggesting that suppressive effects of exercise over factors initiating puberty were cumulative (Frisch et al. 1981). By being associated with delayed bone age and reduced body fat levels, primary amenorrhea clearly reflects protracted inadequate availability of nutrient energy. It also may be mediated by physical and psychological stress associated with competition and intense training schedules in such sports that inhibit secretion of

sex hormones necessary for normal pubertal progression (Georgopoulos et al. 1999). Exercise-induced primary amenorrhea resembles anorexia-induced amenorrhea in that both are characterized by an immature prepubertal pattern of low-frequency LH pulses, increased FSH to LH secretory ratio, and absence of increased secretion of estradiol that is necessary for normal progression through puberty (Boyar et al. 1974a; Yen 1998).

Gonadotropin secretion and pubertal progression revert to normal in both adolescent amenorrheic athletes and in adolescent anorexics, when their demanding training regime is interrupted or when their energy intake increases even in the absence of measurable changes in body weight (Boyar et al. 1974a; Warren 1980). This effect suggests that energy shortage may alter the hormonal events necessary for the initiation of puberty either directly through reduced availability of circulating fuels or indirectly through effects on growth, bone maturation, and accumulation of body fat.

Female hormone cycles in postpubertal women also can be disrupted by habitual exercise that involves large volumes of bioenergetically costly activities. Disruptions may include reduced length of luteal phase, oligomenorrhea, anovulation, and secondary amenorrhea. Incidence of secondary amenorrhea is more frequent in younger women and women who have not borne children, suggesting that the level of reproductive maturation may affect the vulnerability of the female reproductive system to disruption by exercise (Baker et al. 1981; Sanborn et al. 1982). Exercise-induced secondary amenorrhea is associated with sports and physical activities that lower body fat levels, involve energy intakes that do not match levels of energy expenditure, and include demanding training routines that could be causing physical and psychological stress (Dale et al. 1979).

The proximate cause of impaired fertility in exercising females is suppression of the amplitude of LH pulses caused by a decline in the activity of hypothalamic GnRH pulse generator (Cumming et al. 1985; Loucks et al. 1989; Marshall & Griffin 1993; Reame et al. 1985; Veldhuis et al. 1985; Yen 1998). This condition causes either luteal phase defects or anovulation and amenorrhea. Luteal phase defects include either a shortening of the luteal and lengthening of the follicular phase of the menstrual cycle or an inadequate amount and duration of progesterone secretion by the corpus luteum (Beitins et al. 1991). Inadequate secretion of progesterone is a result of deficient corpus luteum function, which is related to reduced LH pulsatility during the follicular phase and reduced estradiol secretion. Ovulatory failure and amenorrhea are the consequences of inadequate follicular development and lack of estradiol positive feedback to gonadotropin secretion (Bullen et al. 1985; Veldhuis et al. 1985). In exercise-induced impairments of fertility, usually no change in pituitary sensitivity to GnRH is observed (Veldhuis et al. 1985).

At least four different mechanisms may contribute to development of exercise-induced amenorrhea. They include the effects of acute energy drain, level of body fat stores, increased levels of inappropriate sex steroids, and increased levels of stress hormones. It is likely that exercise-induced amenorrhea arises from different combinations of these and possibly other exercise-induced endocrine, metabolic, and neural changes. Acute reduction in energy availability to less than 25 kcal/kg of lean body mass, generated by 4 days of increased energy expenditure through exercise, precipitates reductions in LH pulsatility and reductions in T_3 and free T_3 concentrations (Loucks & Heath 1994a, 1994b; Loucks et al. 1998). Below a threshold level of energy availability, adaptation in thyroid hormone secretion indicates a metabolic adaptation to energy shortage that may lead to decreased LH pulsatility. Similar reduction in energy availability through dietary restriction also curtails fertility by inhibiting pulsatile LH secretion (Hoffer et al. 1986; Loucks & Heath 1994a; Pirke et al. 1985). A combination of dietary restriction and exercise has the same effect, and prolonged and severe starvation returns LH secretion to a prepubertal pattern (Boyar et al. 1974a; Williams et al. 1995). Reduction of LH pulsatility is more severe after dietary restriction (23%) than after equivalent reduction in energy availability through increased energy expenditure of exercise (10%). Changes in LH pulsatility in exercising women are prevented when increased food intake adequately compensates for the energy cost of exercise, demonstrating that energy shortage rather than any other aspect of short-term exercise is the cause of suppressed fertility (Loucks et al. 1998). Adequate dietary compensation for the energy cost of exercise is frequently absent in studies documenting feeding behavior of amenorrheic athletes, and selection of macronutrients is often altered, suggesting that energy shortage plays a significant role in triggering exercise-induced amenorrhea (Deuster et al. 1986; Mulligan & Butterfield 1990; Myerson et al. 1991; Schwartz et al. 1981; Wilmore et al. 1982).

Normal LH secretion in chronically undernourished individuals is restored after several days of realimentation before measurable changes can occur in body composition (Boyar et al. 1974a; Loucks & Verdun 1998; Parfitt et al. 1991; Schreihofer et al. 1993). This effect may account for inconsistent association between low levels of body fat with reproductive disturbances and mask the possible independent role of low body fat levels in exercise-induced amenorrhea (see Loucks & Horvath 1985 for a review). Despite the lack of irrefutable evidence that reduction in body fat below some critical threshold causes exercise-induced amenorrhea through a reduced aromatization of androgens to estrogens in the adipose tissue, there is sufficient circumstantial evidence to keep some form of this hypothesis alive (Frisch & McArthur 1974; Frisch & Revelle 1970; Nimrod & Ryan 1975). Demographic data on the correlations between threshold body fat levels on one hand and the initiation of menarche and maintenance of female menstrual cycles on the other that were collated and interpreted by Frisch

lend credence to the notion that the quantity of body fat plays a biologically significant role in the initiation and maintenance of fertility. Recent discovery of leptin, the endocrine product of the adipose tissue, and its several properties make this hypothesis even more compelling. Plasma leptin concentrations bear a quantitative relationship to body fat mass and have the capacity to influence food intake and metabolic energy expenditure in animals and humans (Campfield et al. 1995; Considine et al. 1996). As such, this messenger could affect neural and behavioral controls of energy balance in response to changes in energy content of body fat stores. Fat loss and, in particular, low body fat levels in exercise-induced amenorrhea cause dramatic reductions in plasma leptin concentrations and abolish its circadian pattern of release (Laughlin & Yen 1997). This change in leptin concentration and secretory pattern could be signaling changes in energy and adipose tissue status to the hypothalamic GnRH pulse generator. Leptin has been implicated as a possible factor responsible for initiation of puberty, and its role in the control of the reproductive axis is attracting a lot of interest (Cunningham et al. 1999; Thong & Graham 1999). The relationship of body fat level, plasma leptin changes, and exercise-induced amenorrhea merits further study.

Oversecretion of adrenal and gonadal androgens and other steroid derivatives during exercise has also been proposed as a possible cause of exercise-induced amenorrhea. Increases in plasma concentration of androgens of adrenal origin are associated with acute intense exercise in women. When exercise is of several hours duration, such endocrine environment could exert a detrimental influence over follicular development. Increased secretion and activity of catecholestrogens is found during and after training in eumenorrheic women (De Cree et al. 1997a, 1997b, 1997c; Russell et al. 1984a, 1984b). These changes are even greater in amenorrheic runners (De Cree 1998). Catecholestrogen synthesis occurs in part in the hypothalamus, where it results in increased NE concentration (De Cree et al. 1997c). Catecholestrogens could affect menstrual cyclicity by altering the noradrenergic influence over GnRH pulsatility. Catecholestrogens could also interfere with corpus luteum function by stimulating synthesis of luteolytic prostaglandin F2α (Behrman et al. 1982; De Cree 1990; Demers et al. 1981). The role of exercise-induced androgen and catecholestrogen secretion in the etiology of amenorrhea needs further study.

Habitual intense endurance exercise acts as a stimulus for secretion of several stress hormones (Harber et al. 1997b; Rivier & Rivest 1991; Suh et al. 1988), in particular, CRF, ACTH, cortisol, and β-endorphin, which have been shown to inhibit gonadotropin secretion (Rivier et al. 1986; Ellingboe et al. 1982). Cortisol oversecretion at rest and during exercise is accompanied by blunted ACTH and cortisol responses to a CRF stimulus in amenorrheic athletes, demonstrating a change in the hypothalamo-pituitary-adrenal feedback relationships (Biller et al. 1990; De Souza et al. 1994; Loucks et al. 1989; Suh et al. 1988). Melatonin is also released at night in greater than usual amounts in amenorrheic athletes (Berga et al. 1988; Laughlin et al. 1991). It too has antigonadal actions and could contribute to exercise-associated amenorrhea.

It should be pointed out that large volumes of habitual exercise do not always inhibit, but can under some circumstances stimulate, mammalian reproduction. In some seasonal breeders, a large volume of voluntary running reinstates the estrous cyclicity after the reproductive cycles have been abolished by a short photoperiod (Borer et al. 1983a). This effect is expressed only when nutrient energy is abundant and changes to suppression of reproductive cyclicity when voluntary running is superimposed on severe energy restriction (Powers et al. 1994). This stimulatory influence of exercise over reproductive function may not be expressed in humans because of the nonseasonal character of human reproduction.

SEX HORMONES, PHYSICAL ACTIVITY, AND HEALTH

Although there are significant gender differences in the risk of developing cardiovascular disease that have to do with the sex hormone control of lipid metabolism, this large topic is beyond the scope of this review. It is appropriate, however, to mention that habitual endurance exercise significantly reduces the risk of breast cancer (Albanes et al. 1989; Apter & Vikho 1983; Bernstein et al. 1987; Frisch et al. 1985; Henderson et al. 1985, 1988; Kelsey et al. 1993; MacMahon et al. 1982; Oliveria & Christos 1997; Sternfeld 1992; Vikho et al. 1992). This effect is thought to be mediated through exercise-associated suppression of plasma estradiol secretion and reduced exposure of breast tissue to its potential carcinogenic influence.

9

Endocrinology of Biological Rhythms and Exercise

Biological rhythms are endogenously generated periodicities in body function and in physical activities that reflect a universal design feature of all living organisms (Pittendrigh 1993). Rhythmic design has allowed humans and animals to exploit different physical features of the solar day, lunar month, and annual seasonal environment to increase the success of foraging for food, of avoiding predators, and of finding shelter for rest. Rhythmic design permits bioenergetically efficient alternation between energy-expending and waste-producing functions of the body or behaviors on one hand and energy-conserving and repair functions on the other. As such, rhythmic design serves predictive as well as reactive homeostasis as it both anticipates physiological needs and rapidly adjusts magnitude of response to such needs (Moore-Ede 1986). Rhythms also govern the non-homeostatic functions that are important for survival of the individual or the species, such as wakefulness, sleep, and levels of spontaneous activity or fertility. The identity and location of the clock that initiates rhythmic phenomena and the mechanism through which it times the biological functions (the hands of the clock) have been the focus of intense interest and scrutiny in the new field of chronobiology during the past 4 decades.

There also is a growing interest in the relationship between exercise, hormones, and biological rhythms for at least four reasons. First, exercise can affect the hands of the biological clock. As exercise activates a resupply of energy and disposal of waste to achieve acute homeostatic adjustments and longer-term adaptations, it influences all rhythmic endocrine and neurochemical mechanisms that control these homeostatic functions. Second, optimal physical performance depends on synchronization of many rhythmic functions with periodicities in the environment. Therefore, shifts in day-night work schedules or travel across time zones produces a misalignment between endogenously generated physiological rhythmic functions and the cyclical environmental conditions compatible with their expression. When this misalignment happens, mental and physical performance suffers, and the need to understand the nature of the problem and find a practical solution to it arises. Third, a great many pathological conditions that can be affected by acute exercise and exercise training are associated with changes or alterations in rhythmic processes. Therefore, modulation by exercise of a number of human morbidities cannot be fully understood unless the effects of exercise on the rhythmic characteristics of affected processes are examined. Finally, exercise appears to have the capacity to affect not only the rhythmic processes (hands of the clock) but also the operation of the biological clock itself. For that added reason, the mechanism through which exercise affects biological periodicities is of fundamental scientific interest.

BIOLOGICAL RHYTHMS AND THE INTERNAL BODY CLOCK

Periodicities in physiological processes or activities that originate within human or other living organisms are classified as endogenous, whereas others that reflect a rhythmic change in the external environment are exogenous. When biological rhythms are sinusoidal, each repeating unit represents a cycle taking place within a time frame called a period or tau. The central value in a sinusoidal function is called a MESOR, or midline-estimating statistic of rhythm, and the change in value to either side of the MESOR is the amplitude (see figure 9.1a). In nonlinear oscillations, the amplitude is the distance from the trough (nadir, or the lowest value) of the changing variable to its peak value. Periodicities or repeating cycles can also be represented as a 360-degree circle (see figure 9.1c). The acrophase is the peak of the cosine curve or an angle on the circular representation of the rhythm where the variable displays its greatest value. The phase of a

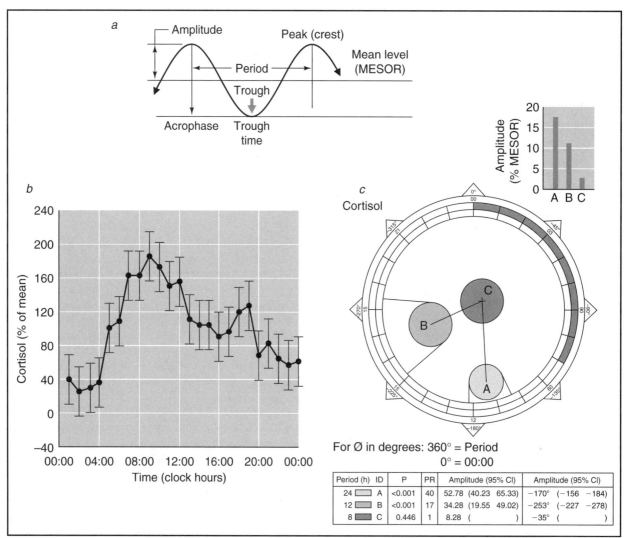

Period (h)	ID	P	PR	Amplitude (95% CI)			Amplitude (95% CI)		
24	A	<0.001	40	52.78	(40.23	65.33)	−170°	(−156	−184)
12	B	<0.001	17	34.28	(19.55	49.02)	−253°	(−227	−278)
8	C	0.446	1	8.28	(	)	−35°	(	)

Figure 9.1 *Features of a sinusoidal biological rhythm.*

Linear (*a* and *b*) and circular (*c*) representation of a sinusoidal rhythm illustrating its quantifiable features: a central value (MESOR, or midline-estimating statistic of rhythm), the amplitude (the extent of deviation from the MESOR), and the acrophase (timing of the crest of the cycle).

(*a*) Linear representation of a feature of a sinusoidal biological rhythm. (*b*) Linear representation of circadian and ultradian cortisol rhythms. (*c*) In a circular representation of a circadian cortisol rhythm, 24 hours represents a full circle with the reference time of midnight being in the top position. The acrophases and amplitudes of the circadian (small circle marked A) and 12-hour ultradian (small circle marked B) cortisol rhythms are represented by two vectors. They point to acrophase times, and their length represents the relative prominence of each rhythm (the 11:00 A.M. circadian rhythm being more prominent than the 5:00 P.M. 12-hour rhythm). The elipse or circle around the tip of the vector shows the 95% confidence region for the joint estimate of the amplitude and the acrophase.

Figure *a* adapted from Reilly 1990; figures *b* and *c* adapted from Halberg et al. 1967.

periodicity refers to any particular point in the periodicity or any particular angle in its circular representation. A phase shift refers to an advance or a delay in time of the appearance of a phase in a periodicity.

Several characteristics have to be present before a rhythm can be classified as endogenous. The principal one is that the rhythm is generated by an oscillator or pacemaker within the body, which keeps time in accordance with its genetically programmed period.

For example, in both young and old humans under constant conditions, the period of the endogenous temperature, cortisol, and melatonin rhythms is approximately 24.2 hours (Czeisler et al. 1999). Under nyctohemeral conditions (meaning within the regular alternation of solar day and night), these rhythms have a period of 24 hours because the orderly onset of solar daylight and sunset serve as synchronizers (or zeitgebers) that daily adjust the period of the clock (Rea 1998). By convention, circadian time is measured either with reference to the onset of solar day, which is designated circadian time 0 (CT0), or to the onset of spontaneous running activity in nocturnal mammals at CT12. When the light and associated clues are absent, the endogenous rhythms free-run in accordance with the period of their controlling internal oscillator. To discern the period of a free-running biological rhythm, its values are usually double-plotted on a chart over successive days. Free-running rhythms present a diagonal pattern, as their period is usually different from 24 hours. A rhythm entrained to a solar or other 24-hour zeitgeber presents a vertical pattern. Additional criteria of an endogenous rhythm are the following:

1. It can be entrained or phase-shifted to a new phase angle by a recognized zeitgeber over a period of several days.

2. If allowed to free-run, the rhythm maintains the new phase angle to which it was most recently entrained rather than the phase it had before it was exposed to the zeitgeber.

3. If reexposed to the original zeitgeber, it again phase shifts after a lag of several days.

Exogenous rhythms instantaneously adjust their phase to a new phase angle imposed by an external stimulus and maintain it only as long as the entraining stimulus operates. Although endogenous rhythms also can be influenced (or masked) by an external rhythmical stimulus, they resume a free-run according to their endogenous phase angle as soon as the external masking stimulus is removed.

In humans, the inferences about the nature of biological rhythms are often limited to observation and description of behavioral or physiological periodicities. A more rigorous approach entails looking for persistence of periodicity after removal of confounding zeitgebers such as solar illumination, intermittent sleep, changing posture from recumbent to upright, becoming physically active, and eating a meal. To that end, humans or animals, over periods of varying duration, are

- deprived of sleep or forced to replace one consolidated sleep period with regularly-spaced naps (Ahnve et al. 1981; Friedman et al. 1979; Fröberg et al. 1972),

- prevented from changing posture from recumbent to upright (Van Reeth et al. 1994), or

- maintained under constant conditions in laboratories, underground caves, or bunkers (Aschoff & Wever 1962; Aschoff et al. 1967a, 1967b; Halberg et al. 1965, 1970; Mills 1964).

Constant routine includes constant temperature and dim light, recumbent posture, individual isolation, and frequent, evenly-spaced meals or constant infusion of a simple nutrient such as glucose (Van Reeth et al. 1994). None of these approaches are entirely satisfactory, as sleep deprivation causes disturbances in the homeostatic component of the sleep rhythm, forced recumbency affects many aspects of human physiology, and parenteral nutrition in constant-routine experiments precludes these studies from being carried out long enough to establish the free-running feature of endogenous rhythms (Ben Sasson et al. 1994; Cugini et al. 1984; Friedman et al. 1979). An approach that avoids many of these pitfalls is the forced desynchrony protocol. Here, the free-running conditions can be created by subjecting humans to cycles of illumination to which their clock mechanism cannot easily adjust, as is, for instance, a 28-hour day (Czeisler et al. 1999).

A common classification of biological rhythms is based on the length of their periods. The term circadian, which was coined by Franz Halberg, refers to a great number of rhythms with a period of approximately 24 hours or the length of a solar day (Halberg 1959, 1963; Halberg et al. 1959, 1967). At least three endogenous oscillators have evolved to govern the circadian organization of human and mammalian biology (Rosenwasser & Adler 1986):

1. The best studied is the light-entrainable oscillator, which governs the circadian rhythm of sleep-wakefulness and usually subordinates all others (Pittendrigh 1993). It most frequently times the rhythms of sleep-wakefulness; GH secretion and slow-wave, or deep, sleep; urinary calcium and sodium excretion; and skin temperature (Aschoff et al. 1967b; Czeisler et al. 1980b; Weitzman et al. 1979; Wever 1979).

2. A food-entrainable circadian oscillator can synchronize mammalian activities with the access to a limited supply of food (Borer & Clover 1994; Stephan 1986a, 1986b, 1986c).

3. The third oscillator is responsive to the state of arousal generated by bouts of physical activity and thus probably serves to help humans and animals capitalize on regular environmental opportunities that require vigorous locomotor action (Buxton et al. 1997a; Eastman et al. 1995b; Lax et al. 1998; Mrosovsky & Salmon 1987; Reebs & Mrosovsky

1989; Van Reeth et al. 1994). This oscillator also may entrain core body temperature; rapid eye movement sleep; cortisol, melatonin, and TSH secretion; and urinary potassium excretion (Aschoff et al. 1967b; Czeisler et al. 1980b; Van Reeth et al. 1994; Weitzman et al. 1979; Wever 1979).

Circadian rhythms have natural periods that usually deviate from 24 hours by up to 1 or 2 hours. When they are allowed to free-run in the absence of external time-keeping cues, zeitgebers correct any deviations in the period of the endogenous rhythm by acting on the oscillators during their sensitive times at either end of the 24-hour period. Thus, in nocturnally active mammals, diurnal humans, and diurnal mammals, the light-entrainable oscillator is phase-advanced when the light is presented during the light-sensitive zone of the rhythm at the dark-light transition and is phase-delayed when it is presented during the light-dark transition but is unresponsive at midpoint of the subjective day (light period) (Honma et al. 1995; Moore-Ede et al. 1982; Pittendrigh & Daan 1976; Van Cauter et al. 1994). Conversely, the light-entrainable oscillator is phase-delayed when the darkness in the form of naps is presented during early morning and is phase-advanced by evening naps (Buxton et al. 2000). Phase-response curves reveal the relationship between an entraining stimulus and the time zones in an oscillator that are sensitive to entrainment (see figure 9.2). Entrainable phases of a rhythm appear to be different for light and for exercise zeitgebers. Thus

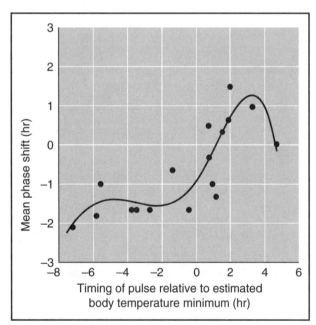

Figure 9.2 Human phase-response curve.

Phase shifts (negative values signify delays and positive values, advances) of nocturnal melatonin and thyroid-stimulating hormone (TSH) peaks in response to a 3-hour pulse of light administered at different phases of the circadian temperature rhythm.

Reprinted from Van Reeth et al. 1994.

physical activity phase-delays human and rodent biological rhythms in middle to late subjective night (dark period) and phase-advances rodent circadian rhythms in the middle of subjective day (Buxton et al. 1997a, 1997b; Eastman et al. 1995b; Leproult et al. 1997b; Mrosovsky et al. 1992; Reebs & Mrosovsky 1989; Smith et al. 1992b).

Rhythms with periods that are shorter than 1 day are called ultradian. Examples are pulsatile GH and gonadotropin secretion, heart rate, and breathing rate. Rhythms with periods that are longer than 1 day are called infradian. Circaseptan rhythms include 7-day periodicities in animals and humans (Haus & Touitou 1994; Herold et al. 2000; Rawson et al. 2000). Circatrigintan or menstrual rhythms have a period length of a 30-day lunar month, and circannual rhythms are approximately 1 year long. Some examples are infradian and about-monthly variation in growth rate and menstrual cycle of female reproductive hormones (Garcia Alonzo et al. 2000; Lampl et al. 1990). Others are annual periodicities in neonatal weight and height, food intake, metabolic rate, sleep, spontaneous activity (Stephens & Casperson 1994), and body fat plateaus in animals and humans (Castex & Sutter 1979; Debry et al. 1975; Garcia Alonzo et al. 2000; Mejean et al. 1994; Walker & Berger 1980; Ward & Armitrage 1981).

In addition to the classification scheme that uses time-measuring features relevant to the environment (hours, day, week, month, or year), rhythms also can be classified with respect to the scale of human physiology and life span. These biological rhythms are timed by a neuroendocrine master clock to appear with a single period or multiple periods within the time frame of human or animal life (Timiras 1978, 1983). Examples are a single life span cycle of statural growth (sometimes also displaying infradian, circaseptan, and annual harmonics), and the multiple menstrual oscillations between menarche and menopause (Garcia Alonzo et al. 2000).

CHARACTERISTICS OF THE BIOLOGICAL CLOCK

The concept of a unitary master clock within the paired suprachiasmatic hypothalamic nuclei (SCN) that generates and synchronizes endogenous physiological rhythms in humans and other mammals to coincide with the geophysical cycles of day and night (the circadian rhythms) and with the seasonal (circannual) rhythms has been inferred from different lines of evidence (Weaver 1998). Anatomical integrity and connectivity of the SCN is necessary for circadian rhythmicity of sleep-wake cycles; feeding; drinking; locomotion; autonomic stimulation of heart, kidney, pancreas, spleen, brown adipose tissue, thyroid gland, and adrenal medulla and cortex; and secretion of many peptides and other hormones, including pineal hormone melatonin (Bartness et al. 2001; Buijs et al. 1998; Sage et al. 2002). This concept of a master clock has taken hold despite suggestions that multiple and dissociable oscillators may be widely expressed throughout animal tissues and that timed functions represent a fundamental "chronome" property of all living tissues (Besharse & Iuvone 1983; Cornelissen et al. 1999; Klein et al. 1991; Plautz et al. 1997; Refinetti et al. 1994; Shirakawa et al. 2001; Tosini & Menaker 1996). The contradictory views concerning operation of a single master clock despite the presence of multiple tissue oscillators are reconciled by the propensity of dispersed oscillators to become coupled. Ordinarily, the SCN appears to impose its period over other oscillators, which become desynchronized and operate independently only under special circumstances.

Synchronization of physiological rhythms with geophysical cycles is possible because of the following:

- the SCN responsiveness to environmental light and some key nonphotic stimuli such as substantial pulses of locomotion or restricted access to food;

- anatomical SCN connections to photoreceptors in the retina on one hand and PVN and autonomic ganglia on the other (Moore 1996);

- temporal structure of changes in SCN sensitivity to photic and nonphotic entraining stimuli (Gillette & Tischkau 1999); and

- changes in the duration and timing of melatonin secretion and in mesencephalic serotonin messages to SCN transduce diurnal and seasonal photic stimuli that help reset the SCN clock (Pickard & Rea 1997).

The biological clock has recently been characterized as a form of molecular feedback mechanism. The clock displays periodicities in transcription of several different genes for proteins that gradually build up during a fixed period of time and reenter the nucleus. When protein concentration exceeds a critical threshold, it exerts negative feedback over the transcriptional activity of the clock genes, which allows the cycle to repeat itself (Yagita et al. 2002). *Clock*, *per* (for period), and *tim* (for timeless) are the recently discovered time-keeping genes in mice (Darlington et al. 1998; Gekakis et al. 1998). Cryptochromes, vitamin B–related flavoproteins that can act as blue-light photoreceptors in bacteria, in plants, and in retinas and possibly other organs of animals and man, are an integral feature of the light-entrainable oscillator (Cashmore et al. 1999; Thresher et al. 1998). These molecules respond to the UV-A–blue range of the light spectrum by engaging an FAD reducing equivalent in transfer of electrons to other catalytic molecules. In mammals, cryptochromes, rather than the chief visual pigment rhodopsin, in retinal rods and cones are responsible for activation of pineal melatonin secretion and entrainment of biological rhythms (Freedman et al. 1999; Lucas et al. 1999).

RECIPROCAL INTERACTIONS BETWEEN EXERCISE AND BIOLOGICAL RHYTHMS

Exercise can influence a number of physiological, hormonal, and autonomic rhythmic functions that are coordinated by endogenous oscillators. These effects of exercise on the hands of the biological clock will be summarized first. In addition, optimal physical performance depends on many rhythmic functions that are impaired when endogenous rhythms become desynchronized from periodicities in the external environment. These interactions between hands of the clock and physical activity will be described next.

Effects of Exercise on the Biological Rhythms

Humans and animals are not continuously physically active, for to be so would be bioenergetically wasteful. Instead, a bioenergetic economy is achieved through ultradian, circadian, and circannual organization of physical activities. Food intake, gastrointestinal function, and physical activity show clear ultradian and circadian periodicities. A solar day permits alternation of a period characterized by rhythmic episodes of energy intake and energy expenditure and by elevated body temperature, with a period when all three functions are reduced and when some small animals exhibit daily torpor (Berger & Phillips 1988; Walker & Berger 1980). A circadian sleep-wake rhythm orchestrates the alternation between bioenergetically active, diurnal or nocturnal, and quiescent periods. At the circannual level, energy economy is achieved through cycles of increased energy accumulation, growth, and energy expenditure in the form of physical activity, which alternate with cycles of energy conservation and reduced physical activity that in some mammals may include hibernation. Vestiges of circannual energy cycles in humans may be represented by reduced physical activity and increased appetite and weight gain during the short-day season that are accompanied by seasonal affective disorder (Krauchi et al. 1999; Lam et al. 1996; Madden et al. 1996; Stephens & Caspersen 1994; Wehr 1992). The effects of exercise on only some of the many periodicities are discussed here in some detail:

- The circadian body temperature rhythm, because of its strong influence over physical performance

- The sleep-wake rhythm, because of recurrent evidence that exercise influences rhythmic properties of sleep and that sleep deprivation can affect performance

- The ultradian and circadian cardiorespiratory rhythms

- Hormonal rhythms, because of their important roles in supporting physical activity

Effects of exercise on several other rhythmic physiological processes will be mentioned briefly.

Body Temperature

The endogenous nature of core body temperature that persists during bed rest, in disease, and in the presence and absence of normal illumination has been known for almost a century and a half (Ogle 1866). In addition, the afternoon acrophase and late night nadir of the temperature rhythm are preserved in fasting subjects, so this rhythm does not simply reflect circadian changes in levels of nutrient intake and physical energy expenditure (Bornstein & Volker 1926). Temperature rhythm free-runs in the absence of external zeitgebers. It is inverted during shift work and gradually entrains to the new phase angle (Czeisler et al. 1999; Eastman et al. 1995a, 1995b; Mosso 1887). Although it becomes dampened during prolonged bed rest, its persistence and robustness are the reason for its frequent use as a prototype endogenous rhythm (Minors & Waterhouse 1981; Winget et al. 1972). Exercise causes acute elevations in core body temperature, in proportion to its intensity (see figure 9.3), and thus may mask its circadian pattern (Buxton et al. 1997a; Van Reeth et al. 1994).

Body temperature is regulated by a hypothalamic thermostat through compensatory vasomotor and sudomotor processes and through shivering and nonshivering thermogenesis. Circadian rhythm in body temperature may be only partly mediated by the SCN because SCN lesions do not completely abolish it. This rhythm also is coordinated by the sleep-wake cycle as well as by the circadian rhythm in heat loss processes, which increase at night as body temperature falls and decrease during daylight as body temperature rises (Van Someren 2000). The key change that drives circadian rhythm of heat loss is variation in threshold body temperature for initiation of sweating and vasodilation. This threshold reaches its nadir when the body temperature is at its lowest, and its acrophase coincides with body temperature acrophase (Torii et al. 1995; Wenger et al. 1976). Circadian change in the threshold of heat loss may be more influential in driving circadian fluctuations in body core temperature than are changes in the gain of heat-loss processes or in diurnal heat production. Nocturnal rise in melatonin secretion may play a role in the lowering of this heat-loss set point, as the set point shift and melatonin secretion can be induced by midday exposure to very bright light (Aizawa & Tokura 1998). It is not clear whether a circadian periodicity exists in the elevation of body temperature produced by exercise. It is reported by some to be absent (Torii et al. 1995). Others claim that circadian

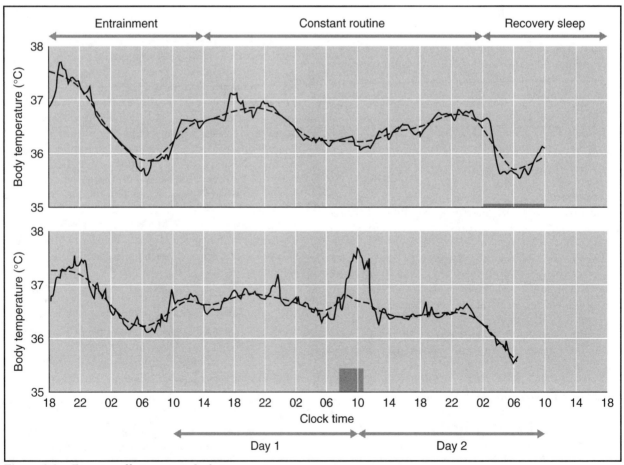

Figure 9.3　Exercise effects on core body temperature.

Profiles of body temperature during baseline period show damping of circadian rhythms during inactivity of constant routine protocol (top) and masking of its nadir resulting from exercise-induced temperature elevation (bottom).

Reprinted from Van Reeth et al. 1994.

periodicity in exercise-induced body temperature elevation is present but report divergent acrophase in the morning or in the evening (Aldemir et al. 2000; Scheen et al. 1998).

Sleep-Wakefulness Rhythm

Sleep-wakefulness is a robust endogenous circadian rhythm that provides a time structure to human and animal activities and physiology. It allows for alternation of activity and rest as well as synchronization of a large number of periodic processes that are increased during the diurnal or nocturnal active phase and suppressed during quiescent phase of this rhythm. Accompanied and enhanced by increases in physical activity, wakefulness also allows for expression of feeding and other behaviors. Sleep on the other hand provides for cessation of all conscious physical activity, secures energy savings, and allows restorative processes to take place (Pierce et al. 1993).

Sleep has attracted a lot of interest both because its biological function remains puzzling and because of its rhythmic substructure (Garcia-Garcia & Drucker-Colin 1999; Turek & Zee 1999). An unresolved controversy pits the view of sleep as a homeostatic restorative process against a view of sleep as a circadian period of enforced inactivity (Adam & Oswald 1977; Horne 1981, 1983; Webb 1971). The restorative view is supported by mounting pressure for sleep as the length of sleep deprivation increases and by compensatory increase in depth or intensity of sleep after sleep deprivation (Friedman et al. 1979; Mistlberger et al. 1987; Webb 1971). Other arguments supporting the restorative function of sleep are its association with increased vegetative PS tone and with secretion of hormones such as GH and melatonin that have anabolic and restorative effects (Dawson & van den Heuvel 1998; Malliani 1999; Malliani et al. 1991).

The contrary view of sleep as a period of enforced inactivity directed by a circadian oscillator and producing energy conservation also has experimental support that seems incompatible with the restorative function (Horne 1981; Webb 1971). Association of sleep with fasting and energy conservation and the absence of serious pathological changes as a consequence of lengthy sleep deprivation would argue against its role as a restorative process and in favor of its role as enforcer of behavioral quiescence (Horne 1981, 1983; Webb 1971). A strong association between sleep and body temperature rhythm provides further support for the circadian rather than homeostatic nature of the sleep (Czeisler et al. 1980a, 1980b; Van Someren 2000). Sleep onset is associated with decreases in body temperature and awakening with increases in body temperature. In addition, duration of sleep is more closely associated with these circadian, body temperature–linked rather than homeostatic variables. Thus, duration of sleep is determined by the timing of sleep onset with respect to body-temperature rises rather than by duration of preceding sleep deprivation (Czeisler et al. 1980a, 1980b; Montgomery et al. 1982). On the other hand, the circadian rhythm of sleepiness is strongly associated with circadian rhythms of hormone secretion that modulate the metabolic fuel availability. Thus, the acrophase of sleepiness coincides with the circadian nadir of plasma glucose availability (Leproult et al. 1997b). This condition suggests that the circadian and homeostatic functions of sleep are integrally interconnected.

Rhythmic ultradian microstructure of sleep has been defined on the basis of its electroencephalographic (EEG) characteristics (Kleitman 1963; Webb 1971). Sleep has five discrete stages, four of which (1 through 4) represent progressively deeper sleep and a fifth one, or dreaming stage, is usually called paradoxical or rapid eye movement sleep (REMS) (see figure 9.4). REMS is characterized by high-frequency EEG activity similar to that of the awake state, an inhibition of muscle tone, and periodic spastic movements of eyes and limb muscles. Stages 1 and 2 represent light sleep, with EEG waves of the former in the range of 2 to 7 Hz and with the appearance of EEG sleep spindles in the range of 12 to 14 Hz in the latter. Stages 3 and 4 represent progressively deeper slow-wave sleep (SWS) with, respectively, 20% to 50% and greater than 50% of EEG delta waves that are of relatively high voltages and slow frequencies of 2 Hz or less. Human sleep progresses from stage 1 through stage 4 and then oscillates between these stages, including occasional awakenings, with REMS usually occurring before or after stage 2. Incidence and duration of SWS is greater at the onset of sleep, whereas incidence and duration of REMS increases as sleep progresses (see figure 9.4). Sleep deprivation increases the pressure for stages 3 and 4 of SWS and for REMS so that rebound from sleep deprivation causes increased duration and frequency of SWS and REMS at the expense of sleep stages 1 and 2, rather than increased duration of normally structured sleep (Spiegel et al. 2000). Besides its ultradian pattern, individual sleep phases also exhibit circadian periodicity with the acrophase for REMS occurring shortly after the nadir of body core temperature and the acrophase of the SWS spindles close to the onset of sleep (Dijk & Czeisler 1995).

Stage 4 of SWS is functionally connected to GH secretion, and brain hormones controlling GH secretion are now viewed as somnogens or sleep factors (Adamson et al. 1974; Obal & Krueger 2001; Sassin et al. 1969; Toppila et al. 1996; Van Cauter et al. 1997, 2000; Zir et al. 1971). Sleep-associated GH secretion has been linked to nocturnal rise in FFA release (Boyle et al. 1992). Sleep deprivation results in parallel rebound increases in GH secretion and SWS (Spiegel et al. 2000). Secretion of prolactin, necessary for growth of mammary tissue and synthesis of milk proteins, is also associated with several phases of sleep (Beck & Marquetand 1976; Linkowski et al. 1998). Aging brings about fragmentation and additional negative changes in the quality of sleep. Duration of sleep is shortened in old individuals, the percent of stage 4 of the SWS selectively declines from about 20 to about 3 in parallel with exponential reduction in nocturnal GH secretion, and the phase relationship between timing of sleep-wake cycle and melatonin secretion is altered (Duffy et al. 2002; Edinger et al. 1993; Van Cauter

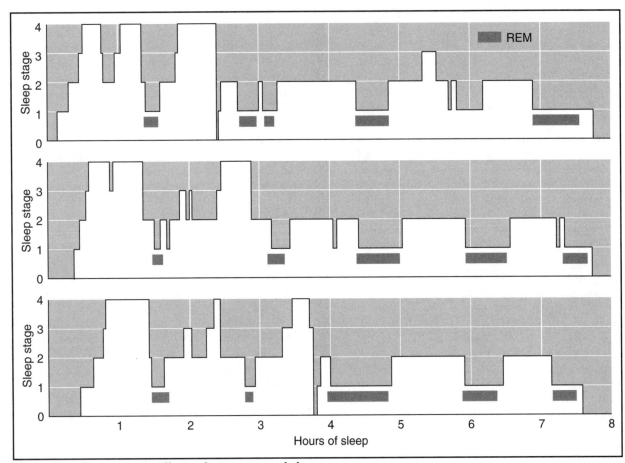

Figure 9.4 *Sleep-stage oscillation during nocturnal sleep.*

Sleep-stage progression is shown in one person on three successive nights. Sleep stages oscillate between wakefulness (stage 0) and deepest slow-wave (SW) sleep (stage 4) through the night. Dark bars indicate incidence of rapid eye movement sleep (REMS). Deepest stages of sleep occur during early sleep and progressively drop out.

Reprinted from Webb 1971.

et al. 2000; Webb 1971). In addition to the association of sleep stages with secretion of hormones, they also differ in relative activities of divisions of the autonomic nervous system. S tone is reduced during SWS, leading to reduced heart rate, respiration, and hypoxia and is increased during REMS, leading to increased heart rate and hyperventilation (Rosen et al. 1991; Schafer et al. 1997; Snyder et al. 1964).

Exercise affects sleep in several ways: it reduces the latency to sleep onset, increases duration of stage 4 and sometimes stage 3 of SWS, and reduces the duration of REMS in young as well as in old humans (Edinger et al. 1993; Kubitz et al. 1996; Naylor et al. 2000; Shapiro et al. 1981). These effects are not large, and they are not always present (O'Connor et al. 1998; Youngstedt et al. 1997). They are more often absent in untrained individuals engaging in exhausting exercise, which suggests an inhibitory influence of stress (Driver et al. 1994). Facilitation of SWS by exercise occurs when substantial increases in body temperature last more than 1 hour, whether they are produced by exercise or by passive external body heating (Horne & Reid 1985; Horne & Staff 1983). Changes in SWS are most readily observed in trained individuals at rest or after acute exercise, possibly because they can sustain intense exercise

without stress for sufficiently long to substantially increase body temperature (Bunnell et al. 1983, 1984; Driver et al. 1988; Taylor et al. 1997; Walsh et al. 1984; Walker et al. 1978). Increased temperature may also be the cause of decreased REMS (Buguet et al. 1998; Bunnell et al. 1984). Marathon running in cold ambient temperatures has no effect on sleep, but it is followed by an increase in SWS if performed at moderate ambient temperatures (Shapiro et al. 1981; Torsvall et al. 1984). In addition, the nyctohemeral timing of exercise affects sleep in the following ways:

1. Exercise during the early part of the day has little effect on later sleep.

2. Exercise late in the day increases stage 3 of SWS.

3. Sleep disruption increases with increased intensity of exercise (Horne & Porter 1975, 1976; Montgomery et al. 1988; Vuori et al. 1988).

Hormonal balance during sleep shifts from high GH and low cortisol secretion during the first half of sleep periods and the reverse endocrine pattern during the second half of sleep periods in a sedentary person to the reverse timing of secretion of those two hormones after prolonged moderately-intense exercise during daytime (Kern et al. 1995).

Cardiorespiratory Rhythms

Human and animal movement patterns, particularly those serving locomotion, exhibit ultradian as well as circadian rhythms that are supported by similarly timed cardiorespiratory and other physiological processes. The resting ultradian heart contraction rhythm permits basic life support and removal of waste products, and changes in its basal frequency and amplitude (force of contraction) are modulated by oxygen and metabolic demands. HR periodicity originates within two clusters of rhythmically depolarizing pacemaker cells, the SAN, with a period of about 600 milliseconds, and the AVN, with a period of about 1 second. These cardiac oscillators are subordinate to rhythmic commands from medullary autonomic nuclei (described in chapter two) transmitted through the autonomic nerves, of which the S nerves exert a stimulatory and PS vagus exerts an inhibitory influence over the heart rate period and amplitude. The autonomic nerves also determine the variability of ultradian HR periodicity with S tone decreasing it and PS vagus nerve increasing it (Malliani 1999; Montano et al. 1994). Circadian increases in heart function also are mediated by upper thoracic S nerves and initiated during the dark-light transition of the solar day, with a peak around 17 to 18 hours and decreases in function during the light-dark transition (Bevier et al. 1987; Imai et al. 1990; Miller & Hellander 1979; Munakata et al. 1997; Reilly & Robinson 1984; Reilly et al. 1984; Richards et al. 1986; Sayer et al. 1998). Circadian heart rhythm is strongly coupled to the circadian rhythm of body core temperature but is not affected by nocturnal melatonin secretion (Crockford & Davies 1969; Davies & Sargeant 1975; Witte et al. 1998).

Exercise at its peak shortens the period of the ultradian heart contraction rhythm from between approximately 750 milliseconds and 1.5 seconds at rest to between as little as 270 to 450 milliseconds. Assuming upright posture and exercise lead to a withdrawal of PS tone and increased S activation of the heart, and these changes are proportional to intensity of physical effort (Malliani 1999; Malliani et al. 1991; Pagani et al. 1986). Increased S activation of the heart after maximal intensity exercise decays slowly over a 48-hour period (see figure 9.5) (Furlan et al. 1993). Daytime exercise causes acute increases in nocturnal HR that are proportional to its duration and intensity (Bunnell et al. 1985; Bevier et al. 1987; Torsvall et al. 1984; Roussel & Buguet 1982; Walker et al. 1978). These effects of exercise on the autonomic control of the heart function can be assessed noninvasively through electrocardiographic (ECG) monitoring of ultradian HR periodicity, in particular of intervals between successive R waves. Contribution of S tone to HR control is reflected in the proportion of the 0 to 0.05 Hz R-R frequencies and of PS vagal influence (also known as respiratory sinus arrythmia frequencies) in the proportion of 0.15 to 0.4 Hz R-R frequencies. These frequencies can be separated and quantified by power spectral and wavelet analyses (Lotric et al. 2000; Malik & Camm 1995; Yamamoto et al. 1991; Yamamoto & Hughson 1991, 1993).

Respiratory and blood pressure rhythms also show ultradian and circadian periodicities (Conway et al. 1984; Pickering 1988; Spengler et al. 2000) (see table 9.1). The circadian acrophases of systolic and diastolic blood pressure (and HR) rhythms occur between 11 and

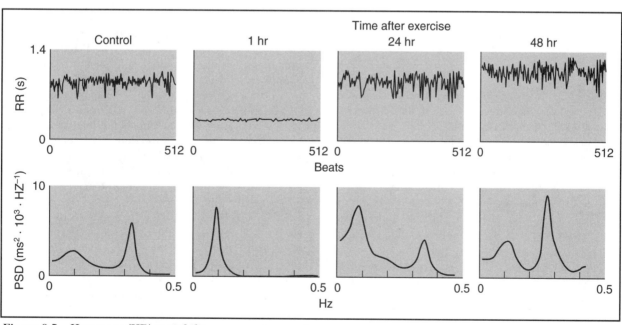

Figure 9.5 *Heart rate (HR) variability response to exercise.*

HR variability represented as a time series of R-R intervals (top left) (tachogram) and as power spectral densities (bottom) at rest (left) and 1 hour, 24 hours, and 48 hours after exercise (successive panels to the right). Predominant sympathetic control is evident from the presence of the low-frequency component of the power spectrum shortly after exercise, while the high-frequency component, centered around 0.3 to 0.4 Hz, reappears 48 hours after exercise and reflects parasympathetic control at rest.

Adapted from Furlan 1993.

16 hours, and nadir occurs around 2 to 4 hours (see table 9.1) (Chiang et al. 1994; Trasforini et al. 1991). Blood pressure and HR are not reflexly linked during the nocturnal nadir of the two rhythms, in contrast to orthostatic control of HR by blood pressure during wakefulness and assumption of upright posture (Munakata et al. 1997). During inactivity and sleep, cardiovascular and respiratory rhythms are dampened, revealing the important contribution of postural changes and movement in the control of their amplitude (Gauquelin et al. 1996; Montelpare et al. 1992). Exercise at its peak increases the frequency of quiet respiration from a breath about every 5 seconds at rest to a breath every 1 to 2 seconds.

Endurance training increases

1. Inhibitory influence over nocturnal as well as diurnal HR that may, in part, be of PS origin (Goldsmith et al. 1992; Hatfield et al. 1998; Shi et al. 1995),

2. speed of sleep-associated decline in HR (O'Connor et al. 1993),

3. amplitude and regularity of circadian blood pressure rhythm, and

4. ultradian R-R interval variability (Gregoire et al. 1996; Seals & Chase 1989; Schuit et al. 1999; Stein et al. 1999).

In contrast, immobilization or prolonged bed rest shifts the sympathovagal balance in favor of S control of the heart, which is then evident through decreased amplitude and regularity of circadian blood pressure rhythm and increased resting HR and decreased interbeat interval variability (Tsuchihashi et al. 1990; Winget et al. 1972).

Hormonal and Associated Physiological Rhythms

Secretion of most hormones is endogenously rhythmic and under the control of a circadian clock, ultradian oscillators, or sleep oscillator (Czeisler & Klerman 1999; Luboshitzky 2000; Van Cauter et al. 1998; Veldhuis et al. 1994). The effects of exercise are usually stimulatory to the amplitude and sometimes also to the frequency of these hormone secretory rhythms.

Synthesis and secretion of melatonin are triggered by light-dark transition and inhibited in daytime by the light, which acts both as a masking and an entraining stimulus. Light enters the body through the eyes, where it acts on retinal cryptochrome pigment. The message is next relayed via the retino-hypothalamic pathway to the circadian pacemaker, the SCN. The circadian signal is transmit-ted to the superior cervical ganglion via the parvocellular PVN and from there through the postganglionic S fibers to the pineal gland (Moore 1996). S nerves stimulate pineal biosynthetic enzymes serotonin n-acetyltransferase and hydroxyindole-o-methyltransferase to first convert serotonin to N-acetyl-serotonin and then complete the reaction to melatonin (Foulkes et al. 1997; Klein & Moore 1979). Thus, the pineal gland transduces the light signal into an endocrine signal that, along with the neural output from the SCN, entrains some circadian rhythms and controls seasonal fertility cycles in seasonal breeders (Pevet 2000; Wurtman 1985). Entrainment by melatonin of the temperature and some other rhythms is vested in responsiveness of the pineal gland to changes in duration and intensity of daily and seasonal illumination and the capacity of melatonin to entrain the SCN (Sharkey & Eastman 2002). Melatonin does not entrain HR, blood pressure, spontaneous activity, cortisol, or TSH rhythms (Witte et al. 1998; Zeitzer et al. 2000). The capacity of melatonin to entrain other circadian rhythms can be simulated by injecting it during the dark-light transition, when it causes phase delays, or during the light-dark transition, when it causes phase advances (Arendt et al. 1987; Lewy et al. 1992; Van Reeth et al. 1997, 1998).

Exercise triggers melatonin secretion when it is carried out during daylight or after 6 hours of exposure to darkness but inhibits it when it is carried out during the initial nocturnal rise in melatonin synthesis and secretion (see figure 9.6) (Buxton et al. 1997a; Carr et al. 1981; Monteleone et al. 1990, 1993; Reiter & Richardson 1992; Theron et al. 1984). When undertaken close to the dark-light transition, exercise can phase-delay the onset of nocturnal melatonin rhythm, and both exercise and melatonin can phase-delay several other circadian and hormonal rhythms (see figure 9.7) (Arendt et al. 1987; Buxton et al. 1997a, 1997b; Van Reeth et al. 1994). Extended nocturnal melatonin surge in long-night/short-day photoperiods suppresses fertility in some seasonal breeders, but habitual endurance exercise can block this antigonadal effect of melatonin (Pieper et al. 1988; Wurtman 1985).

Secretion of adrenal catecholamines and their excretion in the urine have a circadian rhythmicity with an acrophase during the wakeful phase and a nadir during the quiescent phase (see figure 9.8) (Fujiwara et al. 1992). Most anterior pituitary hormones exhibit circadian as well as ultradian rhythms (Waldstreicher et al. 1996). CRF, ACTH, and cortisol circadian secretory rhythms are closely coupled to the sleep-wake rhythm. Acrophase

Table 9.1   Rhythms of Resting Cardiorespiratory Functions and Body Temperature

Variable	Amplitude (% of mean)	Acrophase (24-hour clock)
Heart rate (bpm)	6.0–6.1	13:50–15:31
Minute ventilation (L/min)	7.0–9.7	16:39–17:01
Oxygen consumption (L/min)	6.4–6.5	17:20
Rectal temperature (°C)	0.6–0.8	17:44–19:26

Data from Reilly et al. 1997a.

of the cortisol rhythm occurs between 7 and 9 in the morning, the time of dark-light transition, and its nadir occurs in the late afternoon and evening (McCance et al. 1989; Trasforini et al. 1991). β-Endorphin secretion is coupled to ACTH secretion and therefore also shows a circadian periodicity with an acrophase at about 6:30 hours. Exercise and meal eating can amplify the 15 or so smaller diurnal ultradian cortisol pulses, and the magni-

tude of the effect depends on the time of day when these stimuli are applied (see figure 5.4) (Brandenberger et al. 1982; Brandenberger and Follenius 1975).

In anterior pituitary hormones such as GH, TSH, and PRL that display ultradian periodicities, acute exercise leads to increases in the amplitude of the ultradian pulse during or immediately after exercise. Increase in the mass of secreted hormone is proportional to exercise intensity (Sutton & Lazarus 1976) (see chapter 5). The endogenous circadian TSH rhythm has a nocturnal acrophase (Allan & Czeisler 1994; Van Reeth et al. 1994). Nocturnal exercise increases the amplitude of the nocturnal TSH pulse (Buxton et al. 1997a; Van Reeth et al. 1994). In addition to these acute effects, endurance training can increase the basal ultradian GH pulse frequency and amplitude in mature hamsters and baseline GH secretion and GH pulse amplitude in premenopausal women (Borer et al. 1986; Weltman et al. 1992).

At the onset of puberty, but not after sexual maturation, gonadotropin secretion displays a circadian rhythm with a nocturnal acrophase (see figure 8.8). In mature women, secretion of androgens and progesterone has circadian rhythmicity only during the early follicular phase. In contrast, LH surges exhibit a seasonal and circannual pattern (Blomquist & Holt 1994). Of the estrogens, only estradiol secretion displays a circadian rhythm during late luteal phase with an acrophase in the early afternoon (Carandente et al. 1989; Rebar & Yen 1979). In men, concentrations of androgens fluctuate in ultradian and circadian fashion, with a nadir in the evening and acrophase in early morning (Krieger 1979; Southern et al. 1965; Veldhuis et al. 1987). Endurance exercise that is accompanied with reduced energy availability can suppress ultradian LH pulse frequency in premenopausal women (see chapter 8).

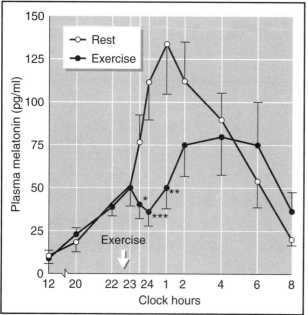

Figure 9.6 *Exercise effect on nocturnal melatonin secretion.*

Suppression by exercise of early nocturnal melatonin secretion.
Adapted from Monteleone et al. 1993.

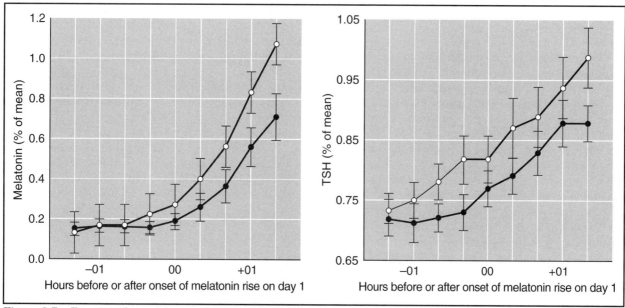

Figure 9.7 *Exercise phase-delays endogenous rhythms.*

A 3-hour bout of moderate-intensity exercise, performed at the time of body temperature nadir, phase delayed melatonin and thyroid-stimulating hormone (TSH) rhythms by about 1 hour (dark circles).
Reprinted from Van Reeth et al. 1994.

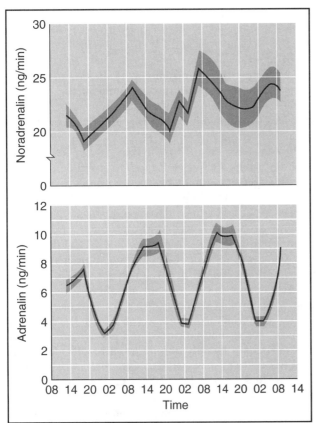

Figure 9.8 *Circadian rhythm of urinary catecholamine excretion.*

Circadian excretory rhythm of both catecholamines persists during a 3-day sleep deprivation but is also amplified in this experiment by moderate environmental stress.

Adapted from Fröberg et al. 1972.

Among hormones that control energy balance, insulin is secreted in an ultradian oscillatory manner that is responsible for maintaining tissue sensitivity to this hormone's action (Goodner et al. 1977; Porksen 2002; Sturis et al. 1995). Low nocturnal and high diurnal plasma concentrations of insulin and increased afternoon insulin resistance are to a large extent of exogenous origin, as they reflect nyctohemeral variations in nutrient intake (Owens et al. 1996; Scheen et al. 1998). These fluctuations also account for the greater hypoglycemic effects of exercise at night than during daytime and in the morning than in the afternoon (Cardosso et al. 1990; Scheen et al. 1998). Plasma leptin displays a circadian rhythm with an acrophase at night around 1 hour. It reflects not only the size of body fat stores but also acute changes in energy balance. The amplitude of this circadian rhythm is proportional to the state of energy availability during the preceding diurnal period and can be altered by both dietary manipulation and exercise (Hilton & Loucks 2000; Van Aggel-Leijssen et al. 1999).

Finally, kidney functions such as urine flow, glomerular filtration rate (GFR), and regulation of hydromineral balance also display circadian periodicities (Martel et al. 1962). Renal excretion of sodium, potassium, calcium, magnesium, chloride, and bicarbonate is maximal at midday and suppressed at night (Doe et al. 1960; Fiorica et

al. 1968; Simpson 1926; Solomon et al. 1991; Stanbury & Thompson 1951). That urine formation is largely diurnal has been recognized for almost a century and a half (Roberts 1860). Urine production rhythm becomes inverted when the nyctohemeral sequence is reversed and is preserved under constant conditions of bed rest (Hart and Verney 1934; Volker 1927). S reflexes that are activated by postural shifts and movement control renal circulation (Wade and Morey-Holton 1998). They increase GFR during the day and inhibit it at night (Buijsen et al. 1994; Gordon and Lavie 1985; Okuguchi et al. 1999; Rosansky et al. 1995; Toor et al. 1965; Wade and Morey-Holton 1998). Diurnal exercise inhibits and phase shifts peak urinary excretion of some minerals from morning to the afternoon (Lobban 1960; Toor et al. 1965).

Some of these periodicities are mediated by the rhythms of hydromineral hormones renin, angiotensin II, aldosterone, AVP, and ANP and rhythmic changes in sympathovagal control. In general, secretion of hydromineral hormones in diurnal humans is high at night when it leads to increased reabsorption of water and salt and production of reduced volume of more concentrated urine. Secretion wanes during daytime to allow excretion of a greater volume of more dilute urine. Renin, angiotensin, and aldosterone rhythms have an acrophase between 4 and 5:30 hours (Cugini et al. 1992; McCance 1989; Trasforini et al. 1991). The ultradian 100-minute renin oscillations are coupled to delta oscillations of the SWS and are inhibited by REMS (Brandenberger et al. 1990, 1996, 1998; Charloux et al. 1998). Under dietary stimulation of a low-sodium diet, plasma aldosterone concentration oscillates together with that of renin (Krauth et al. 1990). Secretion of renin, angiotensin, and aldosterone also is affected by postural changes and exercise (Fyhrquist et al. 1983; Matsui et al. 1997; Rosansky et al. 1995; Stephenson et al. 1989). Daytime exercise shifts the morning aldosterone acrophase to the afternoon (Cugini et al. 1984). Magnitude of renin, angiotensin, and aldosterone secretory response is smaller after endurance training (Hespel et al. 1988).

Rhythms of AVP (or ADH) have been described both in the brain and in the plasma. In the hypothalamus, AVP is the neurotransmitter through which the SCN entrains cortisol and melatonin rhythms to light and to restricted access to food (Buijs et al. 1998; Kalsbeek et al. 1993, 1996, 1998). Peptide content, expression, and activity of AVP-containing neurons in SCN and PVN nuclei of human and animal brains display circadian as well as seasonal fluctuations (Hofman 2000; Hofman & Swab 1993; Ingram et al. 1998; Murphy et al. 1996b). AVP neurons are more active during the day than at night (Forsling 2000). The acrophase of the seasonal neuronal AVP rhythm occurs in the fall and its nadir in the spring and summer (Hofman & Swaab 1995).

The periodicities of circulating AVP are out of phase with those in the hypothalamus and are not always distinct (Brandenberger et al. 1998). Although the AVP rhythm becomes inverted when the nyctohemeral schedule is shifted, circulating AVP rhythm may be exogenously driven by postural changes and the pattern of physical activity, which in turn are influenced by periodic increases in S tone (Nadal 1996; Szollar et al. 1997; Windle et al. 1992). Nocturnal melatonin secretion may act as a zeitgeber for the circulating AVP rhythm (Forsling et al. 1993). Increased diuresis in the elderly may be a consequence of reduced nocturnal amplitude of AVP secretion (Asplund & Aberg 1991). A seasonal rhythm in the concentration of circulating AVP has also been described with higher concentrations in the winter than in the spring (Asplund et al. 1995).

ANP secretion is predominantly nocturnal with an acrophase reported at 17 and at 1:30 hours (Cugini et al. 1992; McCance 1989; Trasforini et al. 1991). This apparent periodicity may not represent an endogenous rhythm but may be responsive to the nocturnal supine posture (Bell et al. 1990; Brandenberger et al. 1998; Chiang et al. 1994). ANP fluctuations are reciprocal and compensatory to circadian changes in blood pressure and plasma aldosterone (Ivarsen et al. 1995; Janssen et al. 1992; Portaluppi et al. 1993; Trasforini et al. 1991).

The endogenous secretion of PTH also is largely nocturnal with an acrophase between 3 and 4 hours and a secondary peak between 17 and 18 hours (el-Hajj Fuleihan et al. 1997). This rhythm, assisted by the stimulatory effect of dynamic loading of rhythmic movement, is the most likely cause of enzyme rhythms that influence bone remodeling (see chapter 6). In osteoporotic patients however, increased nocturnal bone resorption initiated in part by nocturnal PTH secretion is not adequately matched by nocturnal suppression of calcium excretion (Eastell et al. 1992).

Other Rhythmic Functions

A diurnal fluctuation in liver and muscle glycogen significantly influences duration of intense physical activities (Clark & Conlee 1979; Garetto & Armstrong 1983). These fluctuations are most likely exogenous in nature and reflect circadian variation in the timing of feeding and physical activities, both of which can influence muscle and liver glycogen kinetics and content (Suzuki et al. 1983).

Influence of Biological Rhythms on Human Performance

Whether human performance fluctuates in response to circadian or seasonal periodicities has attracted considerable attention, as it has important implications for success in sport (Reilly 1994; Reilly & Brooks 1982; Reilly & Garrett 1998; Reilly et al. 1997a, 1997b; Shephard 1984; Shephard & Shek 1997; Winget et al. 1985, 1992). Additional issues related to physical performance are the effects of nyctohemeral phase reversal through shifted work schedules or through rapid transmeridian travel, use of exercise and other strategies to alleviate these disturbances, and the effects of extended periods of sleeplessness.

A nonscientific concept of "biorhythms" posits that at the moment of birth, three rhythms are activated. A 23-day rhythm governs physical performance, a 28-day rhythm controls emotional well-being, and a 33-day rhythm controls intellectual effectiveness (Reilly et al. 1997b). This deterministic notion, of course, has no empirical support and does not allow for entrainment to environmental zeitgebers. There is however interindividual variability in

the phase of endogenous rhythms and their entrainment to the environment that influences whether an individual will function more effectively in the morning (as a lark) or in the evening (as an owl) (see table 9.2) (Horne & Ostberg 1976). Self-reported preference for morning or evening activity is correlated with corresponding difference in the phase of circadian melatonin and temperature rhythms with both rhythms being phase advanced in larks compared with owls (Duffy et al. 1999).

Detection of periodicities in human performance is complicated by the following:

- Usually small amplitude of these rhythms (see table 9.1)
- Interaction of some physiological rhythms with psychological variables such as arousal, motivation, and learning
- Influence of confounding exogenous variables

Overall, circadian periodicities in physical performance usually have an afternoon or evening acrophase between 15 and 20 hours (see table 9.3). This feature applies to different measures of anaerobic performance including peak performance during 5 seconds, maximal performance sustained during 30 seconds, exercise at 95% of $\dot{V}O_2max$, level of lactic acid accumulation, maximal accumulated oxygen deficit, and maximal swim performance (Baxter & Reilly 1983; Hill & Smith 1991; Javierre et al. 1997; Marth et al. 1998; Melhim 1993; Reilly & Baxter 1983).

Strength performance also has a circadian rhythm with an afternoon acrophase that is revealed in tests using hand-grip strength, isokinetic leg strength, and torque during rapid leg extension (Deschenes et al. 1998; Pearson et al. 1982; Wyse et al. 1994). Measures of maximal aerobic performance show little circadian periodicity when they are collected using graded workload tests but exhibit an afternoon acrophase when an exhaustive constant-power test is used (Hill 1996; Hill et al. 1992; Hill & Smith 1991; Reilly & Brooks 1990). Thus, time to exhaustion is 9% greater, peak $\dot{V}O_2$ is 7% higher, and the time constant describing $\dot{V}O_2$ kinetics is 2 seconds greater in the afternoon than in the morning (Hill 1996). Submaximal oxygen consumption is higher in the afternoon than in the morning (Reilly & Brooks 1990). The psychological perception of effort expressed relative to HR is higher but perception of effort expressed relative to oxygen consumption is lower in the morning than in the afternoon (Faria & Drummond 1982; Hill et al. 1989a). Sensation of pain as well the cardiorespiratory cost of submaximal physical activity is greater at night and during early morning than in the afternoon (Banaszkiewicz & Wojctak-Jaroszowa 1980; Winget et al. 1985). The intensity of self-selected work rate is higher at 17 hours than at 7 hours during the first 50 minutes of exercise (Atkinson & Reilly 1995). Maximal work tolerance at 95% on $\dot{V}O_2max$ is greater in the evening than in the morning (Reilly & Baxter 1983). Maximal aerobic performance is greater at the time of day when the training was carried out. This nyctohemeral specificity of the training effect is superimposed on the circadian influence over maximal performance (Hill et al. 1989b, 1998). In all of

Table 9.2   Questionnaire for Characterization of a Person As a Morning Type or Evening Type

Number	Question
1	What time would you choose to get up if you were free to plan your day? A 5–6 B 6–7:30 C 7:30–10 D 10–11 E 11–12
2	You have some important business to attend to, for which you want to feel at the peak of your mental powers. When would you prefer this meeting to take place? A 8–10 B 11–13 C 15–17 D 19–21
3	What time would you choose to go to bed if you were entirely free to plan your evening? A 20–21 B 21–22:15 C 22:15–0:30 D 0:30–1:45 E 1:45–3
4	A friend wishes to go jogging with you and suggests starting at 7–8 h. How would you feel at this time? A Good B Reasonably good C It would be difficult D It would be very difficult
5	You now have some physical work to do. At what time would you feel able to do it best? A 8–10 B 11–13 C 15–17 D 19–21
6	You have to go to bed at 23 h. How would you feel? A Not tired enough B Slightly tired but awake C Fairly tired and sleepy D Very tired and very likely to fall asleep quickly
7	When you have been up for half an hour on a normal working day, how do you feel? A Very tired B Fairly tired C Fairly refreshed D Very refreshed
8	At what time of day do you feel best? A 8–10 B 11–13 C 15–17 D 19–21
9	Another friend suggests jogging at 22–23 h. How would you now feel? A Good B Reasonably good C It would be difficult D It would be very difficult
Score	1–5 and 8: A = 1, B = 2, C = 3, D = 4, E = 5. 6,7, and 9: A = 4, B = 3, C = 2, D = 1
Rate	9–15 = Definite lark, 16–20 = Moderate lark, 21–26 = Intermediate, 27–31 = Moderate owl, 32–38 = Definite owl

Adapted from Horne and Ostberg, 1976.

Table 9.3   Rhythms of Several Measures of Physical Performance

Variable	Amplitude (% of mean)	Acrophase (24-hour clock)
Peak torque (leg extension)[a]	3	20
Peak power (30 sec)[b]	9	15 and 21 h
Leg strength[c]	7	17
HR max[d]	1.5	16
$\dot{V}O_2$max[d]	3	16
Endurance (max test)[d]	9	16
Body core temperature[c]	1–1.3	15:30–17:30

In most experiments, comparisons were done between performance during afternoon and morning without sufficient number of time points to determine a true acrophase. HR = heart rate.

[a]Deschenes et al. 1998; [b]Melhim 1993; [c]Reilly et al. 1997a; [d]Hill 1996.

these situations, the consonance in timing of the rhythms of physical performance and the circadian rhythm of core body temperature suggest that circadian variation in performance may be governed exogenously by the temperature-dependent facilitation of biochemical reactions and of nerve-conduction velocity (Winget et al. 1985).

Physical performance that requires accuracy of movement, practice, learning, and recall of psychomotor skills in general displays a circadian periodicity with an acrophase in the morning (Reilly et al. 1997b). In part, this effect is a consequence of higher levels of subjective alertness and arousal and more positive mood in the morning than in the afternoon or evening (Folkard 1990; Reilly et al. 1997b). Intensity of arousal may confound these results in that cognitive performance is proportional to low and moderate levels of arousal, but declines at high and stressful levels of arousal (Blake 1971; Colquhoun 1971). Finally, immediate retention of learning and speed of working memory are usually greatest in the morning (Folkard et al. 1990). Thus tasks that require memory and accuracy such as target shooting or tennis serves are performed more successfully in the morning than in the afternoon, whereas higher power of afternoon tennis serves is correlated with higher afternoon body temperature (Atkinson & Speirs 1998).

Jetlag

Rapid time-zone transitions caused by transmeridian travel or shifting of nyctohemeral working schedules produce jetlag or misalignment between the phases of endogenous circadian rhythms and of external rhythms to which they are entrained. This condition results in inappropriate timing of sleep-wakefulness and urine production, in feeding and digestive disturbances, and in suboptimal mood states and mental performance. Crossing several time zones or an inversion of day-night schedule produces a several-day period of decreased physical performance that includes longer reaction times; longer sprint times; and reduced arm, elbow flexor, leg, and back strength (Klein et al. 1977; Reilly et al. 1997a; Shephard 1984; Wright et al. 1983). Performance can sometimes improve after rapid time zone transition, when timing of scheduled competition coincides with the acrophase of endogenous performance rhythms, as for instance after an eastward flight (Jehue et al. 1993).

Westward phase-delays are easier to make than the eastward phase-advances because of the tendency of the human clock to run slow. A variety of strategies have been tried to accelerate phase shifting of the endogenous clock to the new zeitgeber phase (Van Reeth 1998; Van Reeth et al. 1998). Light has been successfully used at the light-dark transition for phase-delays and at dark-light transition for phase-advances of the clock (Boivin et al. 1996; Boulos et al. 1995; Czeisler et al. 1989; Eastman & Martin 1999; Eastman et al. 1995a, 1995b; Minors et al. 1996; Shanahan et al. 1997). Naps can also be used at dusk for phase-advances and at dawn for phase-delays of the clock (Buxton et al. 2000). More recently, attempts have been made to use exercise for entrainment to the new time zone (Harma 1993). Entrainment to a 9-hour phase delay was accelerated when hourly exercise bouts were performed during the dark phase of the first 3 days after the shift (Eastman et al. 1995b). However, this exercise routine was ineffective when combined with bright light (Baehr et al. 1999). Use of sleep-inducing hypnotics such as benzodiazepines and melatonin has also been attempted, but the carryover effects of drugs to the next day had a deleterious

effect on physical performance (Arendt et al. 1987; Lewy et al. 1992; Van Reeth et al. 1998; Copinschi et al. 1995; Zinzen et al. 1994). By shifting physical activity, sleep-wake cycle, and feeding schedules to the new time zone, and by strategic timing of naps, entrainment of biological rhythms is accelerated through phase-shifting influence of these zeitgebers on the endogenous light-entrainable and exercise-entrainable oscillators.

Sleep Deprivation

Prolonged periods of wakefulness, similar to prolonged periods of exercise, are associated with increased sensations of fatigue and sleepiness. Sleep deprivation beyond 1 day, when only mood and not performance shows decrements, progressively reduces physical and mental performance (Ahnve et al. 1981; Bugge et al. 1979; Lubin et al. 1976; Meney et al. 1998). Against this backdrop, circadian oscillations in performance are preserved and coincide with circadian increases in plasma epinephrine and body core temperature (see figure 9.9) (Fröberg et

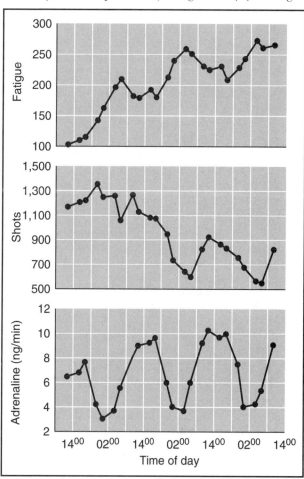

Figure 9.9 *Effects of sleep deprivation on arousal and performance.*

Arousal levels and shooting performance are synchronized with the circadian rhythm of epinephrine excretion during 3 successive days of sleep deprivation. Circadian decreases in fatigue ratings on days 2 and 3 (top panel) correspond to circadian increases in urinary E (bottom panel) and improvements in marksmanship (middle panel).

Adapted from Fröberg et al. 1972.

al. 1972). Naps help reduce deterioration in mental and physical performance (Lubin et al. 1976). On the other hand, naps do not fully compensate for sleep loss, and performance is impaired during 10 or so minutes of "sleep inertia" immediately upon awakening (Jewett et al. 1999).

HEALTH AND BIOLOGICAL RHYTHMS

Health and biological rhythms are interconnected in many ways, but only three of these will be described briefly. (1) A number of psychological and physical disabilities arise from disturbances in endogenous rhythms. Exercise usually ameliorates these disturbances by improving some property of the aberrant biological rhythm. (2) Some health risks are increased during specific phases of a circadian rhythm. Exercise can either ameliorate or exacerbate these health risks, depending on its phase relationship to the variables in question. (3) Effectiveness of many drugs and medical interventions varies with a circadian periodicity, and exercise might alter the processes that underlie this rhythm.

Exercise and Biorhythmic Disturbances

Although a number of different psychiatric disabilities are associated with sleep disturbances, a more comprehensive pattern of biorhythmic disturbances characterizes seasonal affective disorder (SAD) and the metabolic syndrome, or syndrome X (Gierz et al. 1987). Seasonal changes in the length of the photoperiod have been singled out as the most probable cause of SAD (Pinchasov et al. 2000; Young et al. 1997). SAD symptoms include psychological depression, increased sleepiness, feeding abnormalities, and weight gain, and these symptoms increase in frequency and severity during climatic winter in some individuals who live at latitudes away from the equator (Birtwistle & Martin 1999). Disturbances of endogenous rhythms in SAD include a greater sensitivity of unstable circadian rhythms to phase shifting by light and dampened nocturnal temperature decline in climatic winter. These disorders are corrected, along with SAD symptoms, with light treatment (Eastman et al. 1998; Schwartz et al. 1997; Thompson et al. 1997). Other biorhythmic problems in SAD are blunted circadian sleep-wakefulness rhythm and increased secretion of melatonin during daylight, which when blocked also alleviates SAD symptoms (Danilenko et al. 1994; Glod et al. 1997). In addition to light exposure, engaging in midday exercise also reduces SAD symptoms (Pinchasov et al. 2000). However, beneficial effects of exercise on mood do not appear to depend on a particular circadian phase (O'Connor & Davis 1992).

The metabolic syndrome often develops during middle age in individuals leading sedentary, stressful, and energy-replete lifestyles. Its symptoms include obesity, insulin resistance (or type II diabetes), hypertension, hyperlipidemia, and hypercoagulability of blood, all of which contribute individually and collectively to increased risk of coronary heart disease and stroke. In addition to the obvious role of stress and the energy-replete and sedentary lifestyle in development of metabolic syndrome, metabolic abnormalities also may be mediated by alterations in amplitude or pattern of several endogenous rhythms. Evening elevation of plasma cortisol is linked to increases in plasma glucose and insulin concentration and to reduced insulin sensitivity and clearance (Plat et al. 1999). These changes are not an inevitable consequence of aging because they can be induced by sleep deprivation in young subjects (Leproult et al. 1997a; Spiegel et al. 1999). Deterioration of sleep rhythm with age is accompanied by reduced nocturnal GH secretion and to increases in visceral obesity (Dijk et al. 2000; Iranmanesh et al. 1991; Veldhuis et al. 1994). However, chronic sleep deprivation in young individuals also reduces nocturnal GH secretion and may contribute to premature development of the metabolic syndrome (Van Cauter et al. 1998). Exercise can prevent or attenuate development of the metabolic syndrome by increasing the duration of SWS and by stimulating acute GH pulses as well as sleep-associated GH secretion. Whether exercise is equally effective at any circadian time in preventing these metabolic disturbances is not yet known.

Timing of Exercise and Health Risk Factors

Since clots in coronary and cerebral arteries constitute the leading cause of death in middle-aged and older Americans, conditions that precipitate these thrombotic events have attracted much attention. At least three aspects of cardiovascular chronobiology affect the risk of cardiovascular disease and stroke:

1. Circadian and other periodicities in the incidence of thrombotic accidents
2. The influence of the amplitude of circadian blood pressure swings on the incidence of these accidents
3. The role of heart-rate variability in development of cardiovascular disease

Since 1985, it has been known that there is a distinct morning peak in the incidence of sudden cardiac deaths (see figure 9.10), subarachnoid hemorrhage, and strokes (Kleinpeter et al. 1995; Moser et al. 1994; Muller et al. 1985, 1989; Tsementzis et al. 1985; Willich et al. 1987). Most of these events occur in the morning between 8 and 12 hours, although this rhythm is not truly endogenous but is instead contingent on the assumption of upright posture (Kleinpeter et al. 1995; Muller et al. 1987, 1989; Willich et al. 1991, 1993). There is a threefold increase in the incidence of thrombosis during the initial 2 to 4 hours upon arising than at other times of day (Rocco et al. 1987; Willich et al. 1992). When the period of wakefulness is shifted, peak incidence of thrombotic events shifts to early hours after arising from bed and is abolished after confinement to bed rest (Brzezinski et al. 1988b).

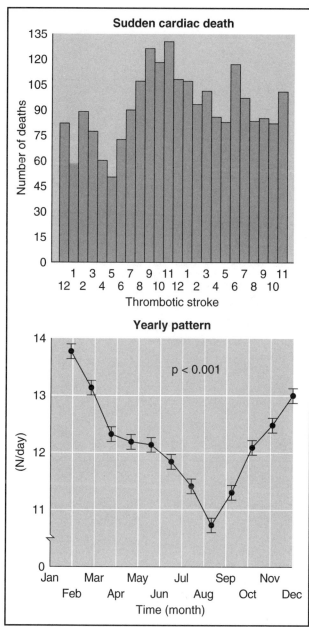

Figure 9.10 *Sudden cardiac deaths occur rhythmically with a circadian peak in the morning (a) and a circannual peak in winter (b).*

Figure *a* adapted from Muller, Tofler, and Stone 1989; figure *b* adapted from Cornelissen et al. 1999.

Circadian pattern of cardiovascular morbidities could be facilitated by some predisposing health conditions (Hjalmarson et al. 1988). Morning acrophase of sudden cardiac deaths is more clearly expressed in hypertensive than in normotensive individuals (Kleinpeter et al. 1995). In chronic heart failure, phases of circadian S and PS rhythms are reversed relative to healthy individuals (Adamopoulos et al. 1995). However, the predominant underlying cause of circadian peak of strokes and fatal heart attacks is a circadian increase in S drive, which is the basis of morning increases in HR and blood pressure

(Dimsdale et al. 1984; Millar-Craig et al. 1978; Turton & Deegan 1974). This increased S drive also is responsible for morning increase in the incidence of chest pain, heart arrhythmias, coronary ischemias accompanied by ST segment depression, stroke, and coronary artery constriction and spasms (Barry et al. 1988; Burkhart & Oswald 1992; Cohn 1988; Gage et al. 1986; Kupari et al. 1990; Panza et al. 1991; Quyyumi et al. 1988, 1992; Rocco et al. 1987; Selwyn et al. 1986; Sloan et al. 1992; Stern & Tzivoni 1974; Thompson et al. 1985; Twidale et al. 1989; Zehender et al. 1992).

Platelet aggregability also displays a diurnal rhythm with a morning peak that depends on postural shift and S activation (Brzezinski et al. 1988; Jovicic et al. 1991; Tofler et al. 1987; Willich et al. 1991, 1992, 1993; Winther et al. 1992). A similar morning peak is evident in blood viscosity and concentrations of the clotting factors PAI-1 (plasminogen activator inhibitor–1) and fibrinogen (Eber & Schumacher 1993; Kubova et al. 1987; Toni et al. 1991). On the other hand, fibrinolysis and activity of the anticlotting agent tissue plasminogen activator (tPA) exhibit a morning nadir (Andreotti et al. 1988; Angleton et al. 1989; Kluft et al. 1988; Rosing et al. 1970). Morning peak of thrombotic events and cardiovascular accidents disappears after application of β adrenergic blockade demonstrating convincingly their dependence on increased morning S drive (Beta-Blocker Heart Attack Trial Research Group 1982; Peters et al. 1989; Willich et al. 1980). The morning acrophase of the cortisol rhythm may interact with circadian increase in S drive to augment the morning risk of coronary or cerebral thrombosis (Reis 1960). There also is a circannual rhythm in cardiac deaths that peaks in winter (see figure 9.10) (Cornelissen et al. 1999; Ornato et al. 1990).

In addition to the effects of mean or MESOR hypertension in precipitating heart attacks and stroke, the circadian amplitude of blood pressure swings was found to constitute an independent risk of stroke. Excessive circadian swings of blood pressure amplitudes beyond the 95% confidence limits for age-appropriate and gender-appropriate norms (also known as CHAT, Circadian Hyper-Amplitude Tension) are associated with sixfold to eightfold greater risk of stroke (see figure 9.11) (Halberg et al. 1998; Otsuka et al. 1996, 1997b). CHAT may be a manifestation of increased S stimulation, as β adrenergic blockade attenuates it (Borer et al. 2002). Hypertensive patients whose blood pressure rhythm exhibits excessive swings also experience greater frequency of silent cerebrovascular damage (Kario et al. 1996). Finally, reduced ultradian heart rate variability has been implicated in development of cardiovascular disease. Advancing age or heart muscle damage caused by a heart attack is associated with reduced R to R interval variability and with greater risk of sudden cardiac death resulting from greater S drive and reduced vagal influence (Kleiger et al. 1987; Otsuka et al. 1997a).

The possibility that intensity or timing of acute and of habitual exercise may diminish or augment the circadian

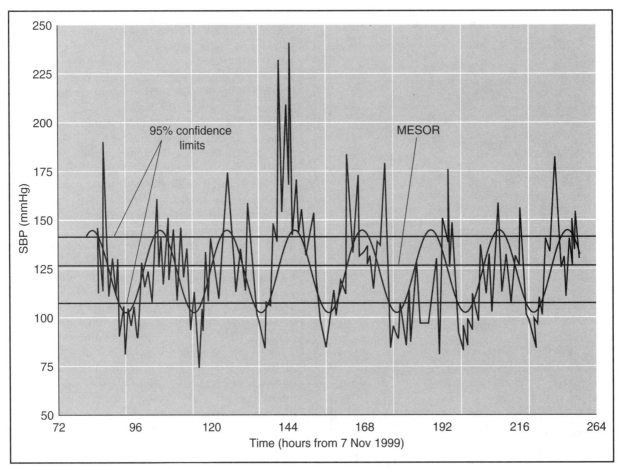

Figure 9.11 *Circadian hyperamplitude tension (CHAT).*

CHAT signals increased risk of stroke. Circadian double amplitudes swing outside the 95% confidence limits (horizontal lines) for age and gender in a 53-year-old African-American woman with normal blood pressure MESOR (midline-estimating statistic of rhythm). Both her parents and her three siblings have died prematurely as a consequence of high blood pressure complications.

risk of fatal cerebral and cardiovascular accidents and coronary heart disease has not been systematically explored. With respect to intensity, vigorous exercise stimulates fibrinolysis but also stimulates platelet activation and aggregation (Naesh et al. 1990; Speiser et al. 1988; Winther et al. 1992). It also precipitates sudden cardiac death more often than moderate exercise (Friedewald & Spence 1990; Gibbons et al. 1980; Moritz & Zamcheck 1946; Romo 1972; Siskovick et al. 1985; Thompson et al. 1982; Vuori et al. 1978; Wroe et al. 1992). The risk of these adverse events ranges from 0 per 375,000 to 1 per 50,000 man-hours (Marti et al. 1989; Siskovick et al. 1984). Such thrombotic accidents are more likely in, but are not confined to, unfit individuals (Siskovick et al. 1984).

With respect to timing of exercise, isometric exercise increases blood pressure in the afternoon more than in the morning (Hickey et al. 1993). Diurnal physical activities increase diurnal amplitude of HR rhythm and decrease nocturnal nadir of blood pressure rhythm (Mann et al. 1979). Morning exercise increases angina pectoris

symptoms and occurrence of coronary artery spasms more than afternoon exercise (Joy et al. 1982; Yasue et al. 1979). Intense morning exercise may lower early morning PAI-1 concentration, but the fibrinolytic effect of exercise is larger in the afternoon than in the morning (Rosing et al. 1970; Szymanski & Pate 1994). Morning exercise did not raise the frequency of thrombotic events above that seen after afternoon exercise in patients who were treated with β blockers and the anticoagulant drug aspirin (Murray et al. 1993).

With respect to habitual exercise, endurance training lowers early morning PAI-1 concentration, appears to increase vagal control over heart function and heart rate variability, and reduces overall morbidity and mortality from cardiovascular disease (Chandler et al. 1996; Gregoire et al. 1996; LaCroix et al. 1996; Schuit et al. 1999; Stein et al. 1999). Greater sympathovagal control over heart function may account for reduced risk of cardiovascular disease and greater longevity of the centenarians (Piccirillo et al. 1998). However, endurance training does not appear to attenuate CHAT (Borer et al. 2002).

Relationship of the intensity and timing of acute as well as habitual exercise on the risk of thrombotic events still awaits additional evaluation.

Exercise and the Circadian Rhythms of Drug Action

The interaction of exercise with chronopharmacology is possibly the least studied area of chronobiology (Halberg et al. 1980a). Drug action may be affected by the timing of drug application because of variable susceptibility of infecting organisms or because of the variable sensitivity of the host to the drug. Infective hosts often display a periodicity in their proliferative cycle as has for many years been known to be the case for malaria (Sanchez de la Pena et al. 1984). Similarly, several forms of cancer have rhythms of cellular activity mirrored in temperature rhythms that can be distinguished from the rhythm of host skin temperature (Wilson et al. 1984). Drug action may be altered by inherent physiological or proliferative rhythmicity of specific tissues (Scheving et al. 1983; Tsai et al. 1987). Rhythms of receptor sensitivity and tissue reactivity should affect binding with drugs, and rhythms of kidney excretory functions affect metabolic neutralization of drugs. It is therefore not surprising that periodicities in the effectiveness of drug action have been described for aspirin and for a number of anticancer drugs (Cornelissen et al. 1991; Halberg et al. 1980b; Sothern et al. 1989). Because exercise can acutely redistribute blood among different organs, increase blood flow through the muscle, and affect immune function as well as renal excretory rhythms, it is remarkable that it has attracted so little attention as a means of modulating pharmacological effectiveness of drugs. The dose and timing of insulin administration is customarily adjusted in accordance with the timing of meals and planned exercise, but we are not aware of any studies in which circadian timing of exercise has been used to alter the action of drugs (Sane et al. 1988).

EXERCISE INFLUENCE ON THE BIOLOGICAL CLOCK MECHANISM

Exercise affects hands of the biological clock mostly by modulating the amplitude of oscillating homeostatic functions. It is less well appreciated that exercise can influence the clock mechanism itself. Exercise effects on

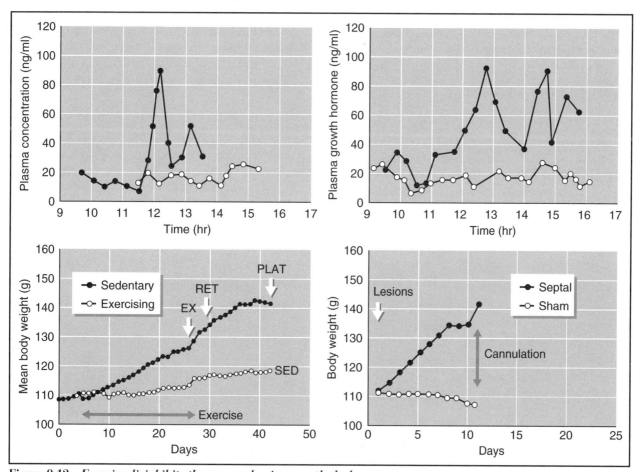

Figure 9.12 *Exercise disinhibits the neuroendocrine growth clock.*

Voluntary running stimulates increased growth hormone (GH) secretion (top left) and skeletal and somatic growth (bottom left) in mature hamsters after the cessation of these functions under the influence of a neuroendocrine clock. Similar effects can be produced by lesions of limbic structure septum, indicating central neuroendocrine location of the clock affected by exercise.

Adapted from Borer, Nicoski, and Owens 1986.

the clock mechanism include its ability to consolidate the sleep-wakefulness rhythm; alter the period, or tau, of the rhythm; entrain endogenous rhythms; and perturb the neuroendocrine clock that controls growth and fertility. Exercise counteracts fragmentation of sleep by increasing the proportion of SWS episodes at the expense of lighter stages of sleep and awakenings. Exercise also facilitates emergence of circadian rhythmicity in arrhythmic animals (Edgar & Dement 1991; Lax et al. 1998). Although spontaneous physical activity has primarily been exploited by chronobiologists as a prototype of a stable endogenous rhythm, spontaneous running in many animals alters the period of their own and other endogenous rhythms (Mrosovsky 1993; Weisberger et al. 1997).

An even more striking example of exercise influencing the clock mechanism includes capacity of exercise to phase-shift the clock. When it is carried out during nocturnal nadir of body core temperature, exercise phase-delays temperature, cortisol, melatonin, and TSH rhythms in humans (see figure 9.7) (Van Reeth et al. 1994; Buxton et al. 1997a). A bout of vigorous running during the subjective day also phase-shifts a number of endogenous rhythms in rodents. This property of exercise to act as an entraining zeitgeber has been successfully utilized for faster reentrainment of endogenous rhythms to jetlag nyctohemeral activity shifts.

Even more remarkably, exercise can alter the operation of the neuroendocrine mechanisms that time lifetime cycles of growth and fertility. In mature golden hamsters, voluntary running reinstates skeletal growth and increases pulsatile GH secretion at an age when growth has been terminated by the neuroendocrine clock (Borer & Kuhns 1977; Borer et al. 1986). A similar effect is produced through interruption of connections between the limbic forebrain and the anterior pituitary, prompting the inference that voluntary running reversibly removes a growth-inhibitory influence of the neuroendocrine clock (Borer 1987b; Borer et al. 1979c) (see figure 9.12). Exercise also can remove the inhibitory neuroendocrine control over hamster reproductive function after it has been suppressed by environmental circumstances of climatic winter (Borer et al. 1983a; Gibbs & Peterborg 1986; Pieper et al. 1995). As was discussed in chapter 8, energy-depleting exercise also can suppress menstrual cycles in women. Mechanisms through which exercise affects the neuroendocrine restraint over growth and reproductive cycles are not understood. It is possible that in some instances, exercise produces these effects by altering melatonin secretion or action, since exercise in hamsters suppresses melatonin action, and melatonin affects hamster reproduction as well as GH secretion and growth (Pieper et al. 1988; Vriend et al. 1987, 1990; Wurtman 1985).

References

Abe, M., & Sherwood, L.M. (1972.) Regulation of parathyroid hormone secretion by adenyl cyclase. *Biochemical and Biophysical Research Communications, 48,* 396-401.

Abu-Soud, H.M., & Stuehr, D.J. (1993.) Nitric oxide synthases reveal a role for calmodulin in controlling electron transfer. *Proceedings of the National Academy of Sciences, 90,* 10769-10772.

Acheson, K., Jequier, E., & Wahren, J. (1983.) Influence of beta-adrenergic blockade on glucose induced thermogenesis in man. *Journal of Clinical Investigation, 72,* 981-986.

Acheson, K., Ravussin, E., Wahren, J., & Jequier, E. (1984a.) Thermic effect of glucose in man. Obligatory and facultative thermogenesis. *Journal of Clinical Investigation, 74,* 1572-1580.

Acheson, K., Schutz, Y., Bessard, T., Ravussin, E., Jequier, E., & Flatt, J.P. (1984b.) Nutritional influences on lipogenesis and thermogenesis after a carbohydrate meal. *American Journal of Physiology, 246,* E62-E70.

Achten, J., Gleeson, M., & Jeukendrup, A.E. (2002.) Determination of the exercise intensity that elicits maximal fat oxidation. *Medicine and Science in Sports and Exercise, 34,* 92-97.

Adam, K., & Oswald, I. (1977.) Sleep is for tissue restitution. *Journal of the Royal College of Physicians,* 376-388.

Adamopoulos, S., Ponikowski, P., Cerquetani, E., Piepoli, M., Rosano, G., Sleight, P., & Coats, A.J.S. (1995.) Circadian pattern of heart rate variability in chronic heart failure patients. *European Heart Journal, 16,* 1380-1386.

Adams, G.R., Haddad, F., & Baldwin, K.M. (1999.) Time course of changes in markers of myogenesis in overloaded rat skeletal muscles. *Journal of Applied Physiology, 87,* 1705-1712.

Adams, K.F., Vincent, L.M., McAllister, S.M., el-Ashmawy, H., & Sheps, D.S. (1987.) The influence of age and gender on left ventricular response to supine exercise in asymptomatic normal subjects. *American Heart Journal, 113,* 732-742.

Adamson, L., Hunter, W.M., Ogunremi, O.O., Oswald, I., & Percy-Robb, I.W. (1974.) Growth hormone increase during sleep after daytime exercise. *Journal of Endocrinology, 62,* 473-478.

Adi, S., Cheng, Z.Q., Zhang, P.L., Wu, N.Y., Mellon, S.H., & Rosenthal, S.M. (2000.) Opposing early inhibitory and late stimulatory effects of insulin-like growth factor-I on myogenin gene transcription. *Journal of Cellular Biochemistry, 78,* 617-626.

Adrian, T.E., Barnes, A.J., Long, R.G., O'Shaughnessy, D.J., Brown, M.R., Rivier, J., Vale, W., Blackburn, A.M., & Bloom, S.R. (1981.) The effect of somatostatin analogs on secretion of growth, pancreatic, and gastrointestinal hormones in man. *Journal of Clinical Endocrinology and Metabolism, 53,* 675-681.

Aguado, L.I., & Ojeda, S.R. (1984.) Ovarian adrenergic nerves play a role in maintaining preovulatory steroid secretion. *Endocrinology, 114,* 1944-1946.

Aguado, L.I., & Ojeda, S.R. (1986.) Prepubertal rat ovary: hormonal modulation of β- adrenergic receptors and of progesterone response to adrenergic stimulation. *Biology of Reproduction, 34,* 45-50.

Ahima, R.S., Dushay, J., Flier, S.N. Prabakaran, D., & Flier, J.S. (1997.) Leptin accelerates the onset of puberty in normal female mice. *Journal of Clinical Investigation, 99,* 391-395.

Ahima, R.S., & Flier, J.S. (2000.) Leptin. *Annual Reviews of Physiology, 62,* 413-437.

Ahima, R.S., Prabakaran, D., Mantzoros, C., Qu, D., Lowell, B., Maratos-Flier, E., & Flier, J.S. (1996.) Role of leptin in the neuroendocrine response to fasting. *Nature, 382,* 250-252.

Ahlborg, G. (1985.) Mechanism of glycogenolysis in nonexercising human muscle during and after exercise. *American Journal of Physiology, 248,* E540-E545.

Ahlborg, G., & Felig, P. (1976.) Influence of glucose ingestion on fuel-hormone response during prolonged exercise. *Journal of Applied Physiology, 41,* 683-688.

Ahlborg, G., & Felig, P. (1977.) Substrate utilization during prolonged exercise preceded by ingestion of glucose. *American Journal of Physiology, 233,* E188-E194.

Ahlborg, G., & Felig, P. (1982.) Lactate and glucose exchange across the forearm, legs, and splanchnic bed during and after prolonged leg exercise. *Journal of Clinical Investigation, 69,* 45-54.

Ahlborg, G., & Lundberg, J.M. (1994.) Inhibitory effects of neuropeptide Y on splanchnic glycogenolysis and renin release in humans. *Clinical Physiology, 14,* 187-196.

Ahlborg, G., Felig, P., Hagenfeldt, L., Hendler, R., & Wahren, J. (1974.) Substrate turnover during prolonged exercise in man: Splanchnic and leg metabolism of glucose, free fatty acid and amino acids. *Journal of Clinical Investigation, 53,* 1080-1090.

Ahlborg, G., Wahren, J., & Felig, P. (1986.) Splanchnic and peripheral glucose and lactate metabolism during and after prolonged arm exercise. *Journal of Clinical Investigation, 77,* 690-699.

Ahlner, J., Andersson, R.G.G., Torfgard, K., & Axelsson, K.L. (1991.) Organic nitrate esters: clinical use and mechanisms of actions. *Physiological Reviews, 43,* 351-423.

Ahmad, Z., Lee, F.T., De Paoli-Roach, A., & Roach, P.J. (1984.) Phosphorylation of glycogen synthase by the Ca^{2+} and phospholipid-activate protein kinase (protein kinase C). *Journal of Biological Chemistry, 267,* 12673-12681.

Ahnve, S., Theorell, T., Akerstedt, T., Fröberg, J.E., & Halberg, F. (1981.) Circadian variations in cardiovascular parameters during sleep deprivation. *European Journal of Applied Physiology, 46,* 9-19.

Ahren, B. (1986.) Thyroid nueroendocrinology: neural regulation of thyroid hormone secretion. *Endocrine Reviews, 7,* 149-155.

Ahren, B., Ar'Rajab, A., Bottcher, G., Sundler, F., & Dunning, B.E. (1991.) Presence of galanin in human pancreatic nerves and inhibition of insulin secretion from isolated human islets. *Cell and Tissue Research, 264,* 263-267.

Ahren, B., Veith, R.C., Paquette, T.L., & Taborsky, G.J., Jr. (1987.) Sympathetic nerve stimulation versus pancreatic norepinephrine infusion in the dog. II: Effects on basal release of somatostatin and pancreatic polypeptide. *Endocrinology, 121,* 332-339.

Ai, H., Ihlemann, J., Hellsten, Y., Lauritzen, H.P., Hardie, D.G., Galbo, H., & Ploug, T. (2002.) Effect of fiber type and nutritional state on AICAR- and contraction-stimulated glucose transport in rat muscle. *American Journal of Physiology, 282,* E1291-E1300.

Aizawa, S., & Tokura, H. (1998.) Influence of bright light exposure for several hours during the daytime on cutaneous vasodilation and local sweating induced by an exercise heat load. *European Journal of Applied Pysiology and Occupational Physiology, 78,* 303-307.

Akabayashi, A., Levin, N., Paez, X., Alexander, J.T., & Leibowitz, S.F. (1994a.) Hypothalamic neuropeptide Y and its gene expression: relation to light/dark cycle and circulating corticosterone. *Molecular and Cellular Neurosciences, 5,* 210-218.

Akabayashi, A., Watanabe,Y., Wahlestedt, C., McEwen, B.S., Paez, X., & Leibowitz, S.F. (1994b.) Hypothalamic neuropeptide, its gene expression and receptor activity: relation to circulating corticosterone in adrenalectomized rats. *Brain Research, 665,* 201-212.

Akabayashi, A., Zaia, C.T., Silva, I., Chae, H.J., & Leibowitz, S.F. (1993.) Neuropeptide Y in the arcuate nucleus is modulated by alterations in glucose utilization. *Brain Research, 621,* 343-348.

Albanes, D., Blair, A., & Taylor, P.R. (1989.) Physical activity and risk of cancer in the NHANES I population. *American Journal of Public Health, 79,* 744-750.

Albertsson-Wikland, K., Rosberg, S., Karlberg, J., & Groth, T. (1994.) Analysis of 24-hour growth hormone profiles in healthy boys and girls of normal stature: relation to puberty. *Journal of Clinical Endocrinology and Metabolism, 78,* 1195-1201.

Aldemir, H., Atkinson, G., Cable, T., Edwards, B., Waterhouse, J., & Reilly, T. (2000.) A comparison of the immediate effects of moderate exercise in the late morning and late afternoon on core temperature and cutaneous thermoregulatory mechanisms. *Chronobiology International, 17,* 197-207.

Alen, M., Reinila, M., & Vihko, R. (1985.) Response of serum hormones to androgen administration in power athletes. *Medicine and Science in Sports and Exercise, 17,* 354-359.

Alexander, R.W. (1995.) Hypertension and the pathogenesis of atherosclerosis. Oxidative stress and the mediation of arterial inflammatory response: a new perspective. *Hypertension, 25*, 155-161.

Ali, M., Cantau, B., & Clos, S. (1989.) Glycogenolytic responsiveness to glucagon, epinephrine, vasopression and angiotensin in the liver of developing hypothyroid rats: a comparative study of in vitro hormonal binding and in vivo biological response. *Journal of Developmental Physiology, 11*, 360-367.

Allain, T.J., & McGregor, A.M. (1993.) Thyroid hormones and bone. *Journal of Endocrinology, 139*, 9-18.

Allan, E.H., Hamilton, J.A., Medcalf, R.L., Kubota, M., & Martin, T.J. (1986.) Cyclic-AMP-dependent and -independent effects of tissue-type plasminogen activator in osteogenic sarcoma cells: evidence from phosphodiesterase inhibition and parathyroid hormone antagonists. *Biochemica et Biophysica Acta, 888*, 199-207.

Allan, J.S., & Czeisler, C.A. (1994.) Persistence of the circadian thyrotropin rhythm under constant conditions and after light-induced shifts of circadian phase. *Journal of Clinical Endocrinology and Metabolism, 79*, 508-512.

Alric, S., Froeschle, A., Piquemal, D., Carnac, G., & Bonnieu, A. (1998.) Functional specificity of the two retinoic acid receptor RAR and RXR families in myogenesis. *Oncogene, 16*, 273-282.

Altura, B.M., & Altura, B.T. (1984.) Actions of vasopressin, oxytocin, and synthetic analogs on vascular smooth muscle. *Federation Proceedings, 43*, 80-86.

Ambler, S.K., & Leite, M.F. (1994.) Regulation of atrial natriuretic peptide secretion by α1- adrenergic receptors: The role of different second messenger pathways. *Journal of Molecular and Cellular Cardiology, 26*, 391-402.

American College of Sports Medicine position statement on the use and abuse of anabolic-androgenic steroids in sport.(1977.) *Medicine and Science in Sports, 9*, xi-xiii.

American College of Sports Medicine position statement on the use of anabolic-androgenic steroids in sport.(1984.) *Medicine and Science in Sports and Exercise, 19*, 534-539.

Andersen, P. (1975.) Capillary density in skeletal muscle of man. *Acta Physiologica Scandinavica, 95*, 203-205.

Anderson, D.J. (1993.) Cell fate determination in the peripheral nervous system: the sympathoadrenal progenitor. *Journal of Neurobiology 24*, 185-198.

Anderson, D.K., Fleming, D.E., Rhees, R.W., & Kinghorn, E. (1986.) Relationships between sexual activity, plasma testosterone, and the volume of the sexually dimorphic nucleus of the proptic area in prenatally stressed and non-stressed rats. *Brain Research, 370*, 1-10.

Anderson, E.A., Hoffman, R.P., Balon, T.W., Sinkey, C.A., & Mark, A.L. (1991.) Hyperinsulinemia produces both sympathetic neural activation and vasodilation in normal humans. *Journal of Clinical Investigation, 87*, 2246-2252.

Anderson, J.E., McIntosh, L.M., Moor, A.N., & Yablonka-Reuveni, Z. (1998.) Levels of MyoD protein expression following injury of mdx and normal limb muscle are modified by thyroid hormone. *Journal of Histochemistry and Cytochemistry, 46*, 59-67.

Andreotti, F., Davies, G.J., Hackett, D.R., Khan, M.I., DeBart, A.C.W., Aber, V.R., Maseri, A., & Kluft, C. (1988.) Major circadian fluctuations in fibrinolytic factors and possible relevance to time of onset of myocardial infarction, sudden cardiac death, and stroke. *American Journal of Cardiology, 62*, 635-637.

Andres, R., Cader, G., & Zierler, K.L. (1956.) The quantitatively minor role of carbohydrate in oxidative metabolism by skeletal muscle in intact man in the basal state: measurements of oxygen and glucose uptake and carbon dioxide and lactate production. *Journal of Clinical Investigation, 35*: 671-682.

Andres, V., & Walsh, K. (1996.) Myogenin expression, cell cycle withdrawal, and phenotypic differentiation are temporally separable events that precede cell fusion upon myogenesis. *Journal of Cell Biology, 132*, 657-666.

Anfossi, G., Massucco, P., Piretto, V., Mularoni, E., Cavalot, F., Mattiello, L., & Trovati, M. (1996.) Interplay between milrinone and adenosine in the inhibition of human platelet response. *General Pharmacology, 27*, 11149-11154.

Angleton, P., Chandler, W.L., & Schmer, G. (1989.) Diurnal variation of tissue-type plasminogen activator and its rapid inhibitor (PA-1). *Circulation, 79*, 101-106.

Anonymous. (1994.) ACOG issues recommendations on exercise during pregnancy and the postpartum period. *American Family Physician, 49*, 1258-1259.

Antunes-Rodrigues, J., McCann, S.M., Rogers, L.C., & Samson, W.K. (1985.) Atrial natriuretic factor inhibits dehydration- and angiotensin-induced water intake in the conscious unrestrained rat. *Proceedings of the National Academy of Sciences, 82*, 8720-8723.

Appel, L.J. (1999.) Nonpharmacological therapies that reduce blood pressure: a fresh perspective. *Clinical Cardiology, 22* (Suppl. 7), III1-IIII5.

Apter, D., & Vihko, R. (1983.) Early menarche, a risk factor for breast cancer, indicates early onset of ovulatory cycles. *Journal of Clinical Endocrinology and Metabolism, 57*, 82-86.

Apter, D., Viinikka, L., & Vihko, R. (1978.) Hormonal pattern of adolescent menstrual cycles. *Journal of Clinical Endocrinology and Metabolism, 47*, 944-954.

Aragon-Vargas, L.F. (1993.) Effects of fasting on endurance exercise. *Sports Medicine, 16*, 255-265.

Araujo-Vilar, D., Osifo, E., Kirk, M., Garcia-Estevez, D.A., Cabezas-Cerrato, J., & Hockaday, T.D.R. (1997.) Influence of moderate physical exercise on insulin-mediated and non-insulin-mediated glucose uptake in healthy subjects. *Metabolism, 46*, 203-209.

Arendt, J., Aldous, M., English, J., Marks, V., & Folkard, S. (1987.) Some effects of jet-lag and their alleviation by melatonin. *Ergonomics, 30*, 1379-1393.

Armstrong, B.G., Nolin, A.D., & McDonald, A.D. (1989.) Work in pregnancy and birth weight for gestational age. *British Journal of Industrial Medicine, 46*, 196-199.

Armstrong, L.E., Curtis, W.C., Hubbard, R.W., Francesconi, R.P., Moore, R., & Askew, E.W. (1993.) Symptomatic hyponatremia during prolonged exercise in heat. *Medicine and Science in Sports and Exercise, 25*, 543-549.

Armstrong, R.B., & Laughlin, M.H. (1985.) Muscle function during locomotion in mammals. In R. Gilles (Ed.), *Circulation, Respiration, and Metabolism. Current Comparative Approaches* (pp.56-63). Berlin: Springer-Verlag.

Armstrong, R.B., & Phelps, R.O. (1984.) Muscle fiber type composition of the rat hindlimb. *American Journal of Anatomy, 171*, 259-272.

Arner, P.(1995a.) Impact of exercise on adipose tissue metabolism in humans. *International Journal of Obesity and Related Metabolic Disorders, 19* (Suppl. 4), S18-S21.

Arner, P. (1995b.) Differences in lipolysis between human subcutaneous and omental adipose tissues. *Annals of Medicine, 27*, 435-438.

Arner, P., Engfeldt, P., & Lithell, H. (1981.) Site differences in the basal metabolism of subcutaneous fat in obese women. *Journal of Clinical Endocrinology and Metabolism, 53*, 948-952.

Arner, P., Kriegholm, E., Engfeldt, P., & Bolinder J. (1990.) Adrenergic regulation of lipolysis in situ at rest and during exercise. *Journal of Clinical Investigation, 85*, 893-898.

Aronson, D., Wojtaszewski, J.F., Thorell, A., Nygren, J., Zangen, D., Richter, E.A., Ljungqvist, O., Fielding, R.A., & Goodyear, L.J. (1998.) Extracellular-regulated protein kinase cascades are activated in response to injury in human skeletal muscle. *American Journal of Physiology, 275*, C555-C561.

Arroll, B., & Beaglehole, R. (1992.) Does physical activity lower blood pressure? A critical review of clinical trials. *Journal of Clinical Epidemiology, 45*, 439-447.

Asada, N., Takahashi,Y., Wada, M., Naito, N., Uchida, H., Ikeda, M., & Honjo, M. (2000.) GH induced lipolysis stimulation in 3T3-L1 adipocytes stably expressing hGHR: analysis on signaling pathway and activity of 20K hGH. *Molecular and Cellular Endocrinology, 162*, 121-129.

Aschenbach, W.G., Hirshman, M.F., Fujii, N., Sakamoto, K., Howlett, K.F., & Goodyear, L.J. (2002.) Effect of AICAR treatment on glycogen metabolism in skeletal muscle. *Diabetes, 51*, 567-573.

Aschoff, J., Gerecke, U., & Wever, R. (1967a.) Phasenbeziehungen zwichen den circadianen Perioden der Aktivitat und der Kerntemperatur beim Menschen. *Pflugers Archiv die Gesamte Physiologie des Menschen und der Tiere, 295*, 173-183.

Aschoff, J., Gerecke, U., & Wever, R. (1967b.) Desynchronization of human circadian rhythms. *Japanese Journal of Physiology, 17*, 450-457.

Aschoff, J., & Wever, R. (1962.) Spontanperiodik des Menschen bei Ausschluss aller Zeitgeber. *Naturwissenschaften, 49*, 337-342.

Ashby, P., Bennett, D.P., Spencer, I.M., & Robinson, D.S. (1978.) Post-translational regulation of lipoprotein lipase activity in adipose tissue. *Biochemical Journal, 176*, 865-872.

Ashby, P., & Robinson, D.S. (1980.) Effects of insulin, glucocorticoids and adrenaline on the activity of rat adipose tissue lipoprotein lipids. *Biochemical Journal, 188*, 185-192.

Ashizawa, N., Ouchi, G., Fujimura, R., Yoshida, Y., Tokuyama, K., & Suzuki, M. (1998.) Effects of a single bout of resistance exercise on calcium and bone metabolism in untrained young males. *Calcified Tissue International, 62,* 104-108.

Asplund, R., & Aberg, H. (1991.) Diurnal variation in the levels of antidiuretic hormones in the elderly. *Journal of Internal Medicine, 229,* 131-134.

Asplund, R., Aberg, H., & Wetterberg, L. (1995.) Seasonal changes in the levels of antidiuretic hormone and melatonin in the elderly. *Journal of Pineal Research, 18,* 154-158.

Atgie, C., D'Allaire, F., & Bukowiecki, L.J. (1997.) Role of beta$_1$ and beta$_3$-adrenoceptors in the regulation of lipolysis and thermogenesis in rat brown adipocytes. *American Journal of Physiology, 273,* C1136-C1142.

Atkinson, G., & Reilly, T. (1995.) Effects of age and time of day on preferred work rates during prolonged exercise. *Chronobiology International, 12,* 121-134.

Atkinson, G., & Speirs, L. (1998.) Diurnal variation in tennis service. *Perceptual and Motor Skills, 86,* 1335-1338.

Au, H.C., & Scheffler, I.E. (1998.) Promoter analysis of the human succinate dehydrogenase iron-protein gene both nuclear repiratory factors NRF-1 and NRF-2 are required. *European Journal of Biochemistry, 251,*164-174.

Aura, O., & Komi, P.V. (1986.) The mechanical efficiency of locomotion in men and women with special emphasis on stretch-shortening cycle exercises. *European Journal of Applied Physiology and Occupational Physiology, 55,* 37-43.

Avellini, E.A., Kamon, E., & Krajewski, J.T. (1980.) Physiological responses of physically fit men and women to acclimation to humid heat. *Journal of Applied Physiology, 44,* 254-261.

Avery, N.D., Wolfe, L.A., Amara, C.E., Davies, G.A., & McGrath, M.J. (2001.) Effects of human pregnancy on cardiac autonomic function above and below the ventilatory threshold. *Journal of Applied Physiology, 90,* 321-328.

Avontuur, J.A., Bruining, H.A., & Ince, C. (1995.) Inhibition of nitric oxide synthesis causes myocardial ischemia in endotoxemic rats. *Circulation Research, 76,* 418-425.

Ayalon, J., Simkin, A., Leichter, I., & Raifmann, S. (1987.) Dynamic bone loading exercised for postmenopausal women: effect on the density of the distal radius. *Archives of Physical Medicine and Rehabilitation, 68,* 280-283.

Baar, K., & Esser, K. (1999.) Phosphorylation of p70 (S6K) correlates with increased skeletal muscle mass following resistance exercise. *American Journal of Physiology, 276,* C120-C127.

Baar, K., Blough, E., Dineen, B., & Esser, K. (1999.) Transcriptional regulation in response to exercise. *Exercise and Sports Science Reviews, 27,* 333-379.

Babu, G.J., Lalli, M.J., Sussman, M.A., Sadoshima, J., & Periasamy, M. (2000.) Phosphorylation of elk-1 by MEK/ERK pathway is necessary for c-fos gene activation during cardiac myocyte hypertrophy. *Journal of Molecular and Cellular Cardiology, 32,* 1444-1457.

Badoer, E., McKinley, M.J., Oldfield, B.J., & McAllen, R.M. (1994.) Localization of barosensitive neurons in the caudal ventrolateral medulla which project to the rostral ventrolateral medulla. *Brain Research, 657,* 258-268.

Baehr, E.K., Fogg, L.F., & Eastman, C.I. (1999.) Intermittent bright light and exercise to entrain human circadian rhythms for night work. *American Journal of Physiology, 277,* R1598-R1604.

Bagdy, G., Calogero, A.E., Aulakh, C.S., Szemeredi, K., & Murphy, D.L. (1989.) Long-term cortisol treatment impairs behavioral and neuroendocrine responses to 5-HT1 agonists in the rat. *Neuroendocrinology, 50,* 241-247.

Bahouth, S.W. (1991.) Thyroid hormones transcriptionally regulate the beta 1-adrenergic receptor gene in cultured ventricular myocytes. *Journal of Biological Chemistry, 266,* 15863-15869.

Bahr, R., & Sejersted, O.M. (1991.) Effect of intensity exercise on excess postexercise oxygen consumption. *Metabolism, 40,* 836-841.

Bahrke, M.S., Wright, J.E., Strauss, R.H., & Catlin, D.H. (1992.) Psychological moods and subjectively perceived behavioral and somatic changes accompanying anabolic-androgenic steroid use. *American Journal of Sports Medicine, 20,* 717-724.

Bahrke, M.S., Yesalis, C.E., & Brower, K.J. (1994.) Anabolic-androgenic steroid abuse and performance-enhancing drugs among adolescents. *Child and Adolescent Psychiatric Clinics of North America, 7,* 821-838.

Bahrke, M.S., Yesalis, C.E. 3rd, and Wright, J.E. (1996.) Psychological and behavioural effects of endogenous testosterone and anabolic-androgenic steroids: an update. *Sports Medicine, 22,* 367-390.

Bailey, D.M., Davies, B., Castell, L.M., Newsholme, E.A., & Calam, J. (2001.) Physical exercise and normobaric hypoxia: Independent modulators of peripheral cholescystokinin metabolism in man. *Journal of Applied Physiology, 90,* 105-113.

Baird, D.T., Horton, R., Longcope, C., & Tait, J.F. (1969.) Steroid dynamics under steady state conditions. *Recent Progress in Hormone Research, 25,* 611-624.

Baker, E.R. (1985.) Body weight and the initiation of puberty. *Clinical Obstetrics and Gynecology, 28,* 573-579.

Baker, E.R., Mathur, R.S., Kirk, R.F., & Williamson, H.O. (1981.) Female runners and secondary amenorrhea: correlation with age, parity, mileage, and plasma hormonal and sex-hormone-binding globulin concentrations. *Fertility and Sterility, 36,* 183-187.

Baker, E.R., Mathur, R.S., Kirk, R.F., Landgrebe, S.C., Moody, L.O., & Williamson, H.O. (1982.) Plasma gonadotropins, prolactin, and steroid hormone concentrations in female runners immediately after a long-distance run. *Fertility and Sterility, 38,* 38-41.

Baker, J. (1999.) A report on alterations to the speaking and singing voices of four women following hormonal therapy with virilizing agents. *Journal of Voice, 13,* 496-507.

Baldwin, K.M., Klinkerfuss, G.H., Terjung, R.L., Mole, P.A., & Holloszy, J.O. (1972.) Respiratory capacity of white, red, and interemediate muscle: adaptive response to exercise. *American Journal of Physiology, 222,* 373-378.

Baldwin, K.M., Winder, W.W., Terjung, R.L., & Holloszy, J.O. (1973.) Glycolytic enzymes in different types of skeletal muscle: adaptation to exercise. *American Journal of Physiology, 225,* 962-966.

Ballard, H.J. (1995.) The role of intracellular pH in the control of adenosine output from red skeletal muscle. *Biological Signals, 4,* 168-173.

Ballor, D.L., & Keesey, R.E. (1991.) A meta-analysis of the factors affecting exercise-induced changes in body mass, fat mass and fat-free mass in males and females. *International Journal of Obesity, 15,* 717-726.

Balthazart, J., Foidart, A., & Harada, N. (1990.) Immunocytochemical localization of aromatase in the brain. *Brain Research, 514,* 327-333.

Banaszkiewicz, A., & Wojtczak-Jaroszowa, J. (1980.) Some problems of night and shift work: Is the physiological cost of long-lasting standard work the same at night as in daytime? *Acta Physiologica Polonica, 31,* 107-114.

Bang, P., Brandt, J., Degerblad, M., Enberg, G., Kaijser, L., Thoren, M., &. Hall, K. (1990.) Exercise-induced changes in insulin-like growth factors and their low molecular weight binding protein in healthy subjects and patients with growth hormone deficiency. *European Journal of Clinical Investigation, 20,* 285-292.

Barbe, P., Millet, L., Galitzky, J., Lafontan, M., & Berlan, M. (1996.) In situ assessment of the role of beta-1, beta-2, and beta-3 adrenoceptors in the control of lipolysis and nutritive blood flow in human subcutaneous adipose tissue. *British Journal of Pharmacology, 117,* 907-913.

Barletta, G., Stefani, L., Del Bene, R., Fronzaroli, C., Vacchiarino, S., Lazzeri, C., Fantini, F., & La Villa, G. (1998.) Effects of exercise on natriuretic peptides and cardiac function in man. *International Journal of Cardiology, 65,* 217-225.

Barnes, P.J. (1995.) Beta-adrenergic receptors and their regulation. *American Journal of Respiratory and Critical Care Medicine, 152,* 838-860.

Barnes, P.J. (1985.) Circadian variation in airway function. *American Journal of Medicine, 79,* 5-9.

Baron, A. (1994.) Hemodynamic actions of insulin. *American Journal of Physiology, 276,* E187-E202.

Baron, A., & Brechtel, G. (1993.) Insulin differentially regulates systemic and skeletal muscle vascular resistance. *American Journal of Physiology, 265,* E61-E67.

Baron, R.L., Neff, L., Tran Van, P., Nefussi, J.-R., & Vignery, A. (1986.) Kinetic and cytochemical identification of osteoclast precursors and their differentiation into multinucleated osteoclasts. *American Journal of Pathology, 122,* 363-378.

Barria, A., Leyton, V., Ojeda, S.R., & Lara, H.E. (1993.) Ovarian steroidal response to gonadotrophins and beta-adrenergic stimulation is enhanced in polycystic ovary syndrome: role of sympathetic innervation. *Endocrinology 133,* 2696-2703.

Barron, B.A., Laughlin, M.H., & Gwirtz, P.A. (1997.) Exercise effect of canine and miniswine cardiac catecholamines and enkephalins. *Medicine and Science in Sports and Exercise, 29,* 1338-1343.

Barron, W.M., Mujais, S.K., Zinaman, M., Bravo, E.L., & Lindheimer, M.D. (1986.) Plasma catecholamine responses to physiologic stimuli in normal human pregnancy. *American Journal of Obstetrics and Gynecology, 154,* 80-84.

Barry, J., Selwyn, A.P., Nabel, E.G., Rocco, M.B., Mead, K., Campbell, S., & Rebecca, G. (1988.) Frequency of ST-segment depression produced by mental stress in stable angina pectoris from coronary artery disease. *American Journal of Cardiology, 61,* 989-983.

Bartkova, J., Lukas, J., Strauss, M., & Bartek, J. (1998.) Cyclin D3: requirement for G1/S transition and high abundance in quiescent tissues suggest a dual role in proliferation and differentiation. *Oncogene, 17,* 1027-1037.

Bartness, T.J., Song, C.K., & Demas, G.E. (2001.) SCN efferents to peripheral tissues: implications for biological rhythms. Journal of Biological Rhythms, 16,196-204.

Bartsch, P., Haberli, A., & Straub, P.W. (1990.) Blood coagulation after long distance running: antithrombin III prevents fibrin formation. *Thrombosis and Haemostasis, 63,* 430-434.

Bartsch, P., Welsch, B., Albert, M. Friedmann, B., Levi, M., & Kruithoff, E.K. (1995.) Balanced activation of coagulation and fibrinolysis after a 2-h triathlon. *Medicine and Science in Sports and Exercise, 27,* 1465-1470.

Bass, J., Oldham, J., Sharma, M., & Kambadur, R. (1999.) Growth factors controlling muscle development. *Domestic Animal Endocrinology, 17,* 191-197.

Bassenge, E. (1994.) Coronary vasomotor responses: role of endothelium andnitrovasodilators. *Cardiovascular Drugs and Therapy, 8,* 601-610.

Bassett, C. (1971.) Biologic significance of piezoelectricity. *Calcified Tissue International, 1,* 252-272.

Bassett, C. (1993.) Beneficial effects of electromagnetic fields. *Journal of Cellular Biochemistry, 51,* 389-393.

Batterham, R.L., Cowley, M.A., Small, C.J., Herzog, H., Cohen, M.A., Dakin, C. et al. (2002.) Gut hormone PYY(3-36) physiologically inhibits food intake. *Nature, 418,* 595-597.

Baulieu, E.-E., & Robel, P. (1995.) Non-genomic mechanisms of action of steroid hormones. *CIBA Foundation Symposium, 191,* 24-42.

Baulieu, E.-E., & Kelly, P.A. (1990.) *Hormones: From molecules to disease.* New York: Chapman and Hall.

Baumann, G. (1991.) Growth hormone heterogeneity: Genes, isohormones, variants, and binding proteins. *Endocrine Reviews, 12,* 424-449.

Bavenholm, P.N., Pigon, J., Saha, A.K., Ruderman, N.B., & Efendic, S. (2000.) Fatty acid oxidation and the regulation of malonyl-CoA in human muscle. *Diabetes, 49,* 1078-1083.

Baxter, C., & Reilly, T. (1983.) Influence of time of day on all-out swimming. *British Journal of Sports Medicine, 17,* 122-127.

Beaupre, G.S., Orr, T.E., & Carter, D.R. (1990.) An approach for time-dependent bone modeling and remodeling-application: a preliminary remodelling simulation. *Journal of Orthopaedic Research, 8,* 662-670.

Beck, B., Stricker-Krongrad, A., Burlet, A., Max, J.P., Musse, N., Nicolas, J.P., & Burlet, C. (1994.) Macronutrient type independently of energy intake modulates hypothalamic neuropeptide Y in Long-Evans rats. *Brain Research Bulletin, 34,* 85-91.

Beck, U., & Marquetand, D. (1976.) Effects of selective sleep deprivation on sleep-linked prolactin and growth hormone secretion. *Archiv fur Psychiatrie und Nervenkrankheiten, 223,* 35-44.

Behm, D.G., & Sale, D.G. (1994.) Voluntary and evoked muscle contractile characteristics in active men and women. *Canadian Journal of Applied Physiology, 19,* 253-265.

Behncken, S.N., & Waters, M.J. (1999.) Molecular recognition events involved in the activation of the growth hormone receptor by growth hormone. *Journal of Molecular Recognition, 12,* 355-362.

Behnke, A.R., & Wilmore, J.H. (1974.) *Evaluation and regulation of body build and composition.* Englewood Cliffs, NJ: Prentice-Hall.

Behnke, A.R. (1961.) Comment on the determination of whole body density and a resume of body composition data. In J. Brozek and A. Henschel (Eds.), *Techniques for measuring body composition* (pp.118-133). Washington, DC: National Academy of Sciences National Research Council.

Behnke, A.R. (1969.) New concepts of height-weight relationships. In H.L. Wilson (Ed.), *Obesity* (pp 25-53). Philadelphia: F.A. Davis.

Behrman, H.R., Hall, A.K., & Preston, S.L. (1982.) Antagonistic interactions of adenosine and prostaglanding F2alpha modulate acute responses of luteal cells to luteinizing hormone. *Endocrinology, 110,* 38-46.

Beitins, I.Z., McArthur, J.W., Turnbull, B.A., Skrinar, G.S., & Bullen, B.A. (1991.) Exercise induces two types of human luteal dysfunction: confirmation by urinary free progesterone. *Journal of Clinical Endocrinology and Metabolism, 72,* 1350-1358.

Belchetz, P.E., Plant, T.M., Nakai, Y., Keogh, E.J., & Knobil, E. (1978.) Hypophysial responses to continuous and intermittent delivery of hypothalamic gonadotropin-releasing hormone. *Science, 202,* 631-633.

Belfrage, E., Fredholm, B.B., & Rosell, S. (1977.) Effect of catechol-O-methyltransferase (COMT) inhibition on the vascular and metabolic responses to noradrenaline, isoprenaline and sympathetic nerve stimulation in canine subcutaneous adipose tissue. *Naunyn-Schmiedeberg's Arhives of Pharmacology, 300,* 11-17.

Bellisle, F., Louis-Sylvestre, J., Demozay, F., Blazy, D., & Le Magnen, J. (1985.) Cephalic phase of insulin secretion and food stimulation in humans: New perspective. *American Journal of Physiology, 249,* E639-E645.

Bell, G.M., Atlas, S.A., Pecker, M., Sealey, J.E., James, G., & Laragh, J.H. (1990.) Diurnal and postural variations in plasma atrial natriuretic factor, plasma guanosine 3',5'-cyclic monophosphate and sodium excretion. *Clinical Science, 79,* 371-376.

Bell, N.H., Godsen, R.N., Henry, D.P., Sharp, J., & Epstein, S. (1988.) The effect of muscle-building exercise on vitamin D and mineral metabolism. *Journal of Bone and Mineral Research, 3,* 369-373.

Ben Sasson, S.A., Finestone, A., Moskowitz, M., Weinner, M., Leichter, I., Margulies, J., Stein, P., Popovtzer, M., & Rubinger, D. (1994.) Extended duration of vertical position might impair bone metabolism. *European Journal of Clinical Investigation, 24,* 421-425.

Benovic, J.L., Bouvier, M., Caron, M.G., & Lefkowitz, R.J. (1988.) Regulation of adenylyl cyclase-coupled beta-adrenergic receptors. *Annual Reviews of Cellular Biology, 4,* 405-428.

Benschop, R.J., Rodriguez-Feuerhahn, M., & Schedlowski, M. (1996.) Catecholamine-induced leukocytosis: early observations, current research and future directions. *Brain, Behavior and Immunity, 10,* 77-91.

Berga, S.L., Mortola, J.F., & Yen, S.S.C. (1988.) Amplification of nocturnal melatonin secretion in women with functional hypothalamic amenorrhea. *Journal of Clinical Endocrinology and Metabolism, 66,* 242-244.

Berger, D., Floyd, Jr., J.C., Lampman, R.M., & Fajans, S.S. (1980.) The effect of adrenergic receptor blockade on exercise-induced rise in pancreatic polypeptide in man. *Journal of Clinical Endocrinology and Metabolism, 50,* 33-39.

Berger, R.J., & Phillips, N.H. (1988.) Comparative aspects of energy metabolism, body temperature and sleep. *Acta Physiologica Scandinavica. Supplementum, 574,* 21-27.

Bergman, E.A., & Boyungs, J. C. (1991.) Indoor walking program increases lean body composition in older women. *Journal of the American Dietetic Association, 91,* 1433-1435.

Bergman, R.N., & Mittelman, S.D. (1998.) Central role of the adipocyte in insulin resistance. *Journal of Basic and Clinical Physiology and Pharmacology, 9,* 205-221.

Bergstrom, J., Hermansen, L., Hultman, E., & Saltin, B. (1967.) Diet, muscle glycogen, and physical performance. *Acta Physiologica Scandinavica, 71,* 172-179.

Berk, L.S., Nieman, D.C., Youngberg, W.S., Arabatzis, K., Simpson-Westerberg, M., Lee, J.W., Tan, S.A., & Eby, W.C. (1990.) The effect of long endurance running on natural killer cells in marathoners. *Medicine and Science in Sports and Exercise, 22,* 207-212.

Berkenbosch, F., van Oers, J., del Ray, A., Tilders, F., & Besedovsky, H. (1987.) Corticotropin-releasing factor-producing neurons in the rat activated by interleukin-1. *Science, 238,* 524-526.

Berman, D.M., Nicklas, B.J., Rogus, E.M., Dennis, K.E., & Goldberg, A.P. (1998.) Regional differences in adrenoceptor binding and fat cell lipolysis in obese, postmenopausal women. *Metabolism: Clinical and Experimental, 47,* 467-473.

Bernard, C. (1859.) *Leçons sur les propriétés physiologiques et les alterations pathologiques des liquides de l'organisme.* Paris: Bailleres.

Berne, C., & Fagius, J. (1986.) Skin nerve sympathetic activity during insulin-induced hypoglycaemia. *Diabetologia, 29,* 855-860.

Bernet, F., Bernard, J., Laborie, C., Montel, V., Maubert, E., & Dupouy, J.P. (1994.) Neuropeptide Y (NPY)- and vasoactive intestinal peptide (VIP)- induced aldosterone secretion by rat capsule/glomerular zone could be mediated by catecholamines via beta1 adrenergic receptors. *Neuroscience Letters*, *166*,109-112.

Bernstein, I.L., Zimmerman, J.C., Czeisler, A.C., & Weitzman, E.D. (1981.) Meal patterning in "free-running" humans. *Physiology and Behavior*, *27*, 621-623.

Bernstein, L., Ross, R.K., Lobo, R.A., Hanisch, R., Krailo, M.D., & Henderson, B.E. (1987.) The effects of moderate physical activity on menstrual cycle patterns in adolescents: implications for breast cancer prevention. *British Journal of Cancer*, *55*, 681-685.

Bernstein, M.S., Hunter, R.L., & Yachnin, S.(1971.) Hepatoma and peliosis hepatis developing in a patient with Fanconi's anemia. *New England Journal of Medicine*, *284*,1135-1136.

Berridge, M.J. (1993.) Inositol triphosphate and calcium signalling. *Nature*, *361*, 315-325.

Berthoud, H.R., & Powley, T.L. (1990.) Identification of vagal preganglionics that mediate cephalic phase insulin response. *American Journal of Physiology*, *258*, R523-R530.

Berthoud, H.R., & Powley, T.L. (1992.) Vagal afferent innervation of the rat fundic stomach: morphological characterization of the gastric tension receptor. *Journal of Comparative Neurology*, *319*, 261-276.

Berthoud, H.R., & Powley, T.L. (1993.) Characterization of vagal innervation of the rat celiac, suprarenal and mesenteric ganglia. *Journal of the Autonomic Nervous System*, *42*, 153-169.

Berthoud, H.R., Fox, E.A., & Powley, T.L. (1991.) Abdominal pathways and central origin of rat vagal fibers that stimulate gastric acid. *Gastroenterology*, *100*, 627-637.

Bertolini, D.R., Nedwin, G.E., Bringman, T.S., Smith, D.D., & Mundy, G.R. (1986.) Stimulation of bone resorption and inhibition of bone formation in vitro by human tumor necrosis factors. *Nature*, *319*, 516-518.

Bertorello, A.M., & Katz, A.I. (1995.) Regulation of Na$^+$-K$^+$ pump activity: pathways between receptors and effectors. *News in Physiological Sciences*, *10*, 253-259.

Bertram, R., Smolen, P., Sherman, A., Mears, D., Atwater, I., Martin, F., & Soria, B. (1995.) A role for calcium release-activated current (CRAC) in cholinergic modulation of electrical activity in pancreatic beta-cells. *Biophysical Journal*, *68*, 2323-2332.

Besharse, J.C., & Iuvone, P.M. (1983.) Circadian clock in Xenopus eye controlling retinal serotonin N-acetyltransferase. *Nature*, *305*, 133-135.

Beta-blocker Heart Attack Trial Research Group. (1982.) A randomized trial of propranolol in patients with acute myocardial infarction: mortality results. *Journal of the American Medical Association*, *247*, 1707-1714.

Bevier, W.C., Bunnell, D.E., & Horvath, S.M. (1987.) Cardiovascular function during sleep of active older adults and the effects of exercise. *Experimental Gerontology*, *22*, 329-337.

Bevier, W.C., Wiswell, R.A., Pyka, G., Kozak, K.C., Newhall, K.M., & Marcus, A. (1989.) Relationship of body composition, muscle strength, and aerobic capacity to bone mineral density in older men and women. *Journal of Bone and Mineral Research*, *4*, 421-432.

Beyer, M.E., Nerz, S., Kazmaier, S., & Hoffmeister, H.M. (1995.) Effect of endothelin-1 and its combination with adenosine on myocardial contractility and myocardial energy metabolism in vivo. *Journal of Molecular and Cellular Cardiology*, *27*, 1989-1997.

Bhasin, S., Storer, T.W., Berman, N., Callegari, C., Clevenger, B., Phillips, J., Bunnell, T.J., Tricker, R., Shirazi, A., & Casaburi, R. (1996.) The effects of supraphysiologic doses of testosterone on muscle size and strength in normal men. *New England Journal of Medicine*, *335*, 1-7.

Bianda, T.L., Hussain, M.A., Keller, A., Glatz, Y., Schmitz, O., Christiansen, J.S., Alberti, K.G., & Froesch, E.R. (1996.) Insulin-like growth factor-I in man enhances lipid mobilization and oxidation induced by a growth hormone pulse. *Diabetologia*, *39*, 961-969.

Bielinski, R., Schutz, Y., & Jequier, E. (1985.) Energy metabolism during the postexercise recovery in man. *American Journal of Clinical Nutrition*, *42*, 69-82.

Biller, B.M.K., Federoff, H.J., Koenig, J.I., & Klibanski, A. (1990.) Abnormal cortisol secretion and response to corticotropin-releasing hormone in women with hypothalamic amenorrhea. *Journal of Clinical Endocrinology and Metabolism*, *70*, 311-317.

Billington, C.J., Briggs, J.E., Harker, S., Grace, M., & Levine, A.S. (1994.) Neuropeptide Y in hypothalamic paraventricular nucleus: a center coordinating energy metabolism. *American Journal of Physiology*, *266*, R1765-R1770.

Birtwistle, J. & Martin, N. (1999.) Seasonal affective disorder: Its recognition and treatment. *British Journal of Nursing*, *8*, 1004-1009.

Bishop, V.S., & Hay, M. (1993.) Involvement of the area postrema in the regulation of sympathetic outflow to the cardiovascular system. *Frontiers of Neuroendocrinology*, *14*, 57-75.

Bishop, V.S., Ryuzaki, M., Cai, Y., Nishida, Y., & Cox, B.F. (1995.) Angiotensin II-dependent hypertension and the arterial baroreflex. *Clinical and Experimental Hypertension*, *17*, 29-38.

Bittencourt, J.C., Presse, F., Arias, C., Peto, C., Vaughan, J., Nahon, J.L., Vale, W., & Sawchenko, P.E. (1992.) The melanin concentrating hormone system of the rat brain: An immuno- and hybridization histochemical characterization. *Journal of Comparative Neurology*, *319*, 218-245.

Bjorkman, O., & Eriksson, L.S. (1983.) Splanchnic glucose metabolism during leg exercise in 60-hour fasted human subjects. *American Journal of Physiology*, *245*, E443-E448..

Bjorntorp, P. (1990.) "Portal" adipose tissue as a generator of risk factors for cardiovascular disease and diabetes. *Arteriosclerosis*, *10*, 493-496.

Bjorntorp, P. (1991.) Metabolic implications of body fat distribution. *Diabetes Care*, *14*, 1132-1143.

Bjorntorp, P. (1996a.) Growth hormone, insulin-like growth factor-I and lipid metabolism: interactions with sex steroids. *Hormone Research*, *46*, 188-191.

Bjorntorp, P. (1996b.) The regulation of adipose tissue distribution in humans. *International Journal of Obesity*, *20*, 291-302.

Bjorntorp, P. (1997a.) Body fat distribution, insulin resistance, and metabolic diseases. *Nutrition*, *13*, 793-803.

Bjorntorp, P. (1997b.) Hormonal control of regional fat distribution. *Human Reproduction*, *12* (Suppl. 1), 21-25.

Blaak, E.E., van Baak, M.A., Kempen, K.P., & Saris, W.H. (1993.) Role of alpha- and beta-adrenoceptors in sympathetically mediated thermogenesis. *American Journal of Physiology*, *264*, E11-E17.

Blair, S.N., Goodyear, N.N., Gibbons, L.W., et al. (1984.) Physical fitness and incidence of hypertension in healthy normotensive men and women. *Journal of the American Medical Association*, *252*, 487-490.

Blair, S.N., Kampert, J.B., Kohl,III, H.W., et al. (1996.) Influence of cardiorespiratory fitness and other precursors on cardiovascular disease and all-cause mortality in men and women. *Journal of the American Medical Association*, *276*, 205.

Blake, M.J., Martin, A., Manktelow, B.N., Armstrong, C., Halligan, A.W., Panerai, R.B., &. Potter, J.F. (2000.) Changes in baroreceptor sensitivity for heart rate during normotensive pregnancy and the puerperium. *Clinical Science 98*, 259-268.

Blake, M.J.F. (1971.) Temperament and time of day. In W.P.Colquhoun (Ed.), *Biological rhythms and human performance* (pp.109-148). New York: Academic Press.

Blakesley, V.A., Scrimgeour, A., Esposito, D., & Le Roith, D. (1996.) Signaling via the insulin-like growth factor receptor: does it differ from insulin receptor signaling? *Cytokine Growth Factor Reviews*, *7*,153-157.

Bloch, B., Popovici, D., LeGuellec, E., Normand, S., Chouham, S., Guitteny, A.F., & Bohlen, P. (1986.) In situ hybridization histochemistry for the analysis of gene expression in the endocrine and central nervous system tissues: a 3-year experience. *Journal of Neuroscience Research*, *16*, 183-200.

Block, J.E., Friedlander, A.L., Brooks, G.A., Steiger, P., Stubbs, H.A., & Genant, H.K. (1989.) Determinants of bone density among athletes in weight-bearing and nonweight-bearing activity. *Journal of Applied Physiology*, *67*, 1100-1105.

Blomquist, C.H., & Holt, J.P. (1994.) Chronobiology of the hypothalamic-pituitary-gonadal axis in men and women. In Y. Touitou & E. Haus (Eds.), *Biologic rhythms in clinical and laboratory medicine*(pp.315-329.). Berlin:Springer-Verlag.

Blomstrand, E., & Saltin, B. (1999.) Effect of muscle glycogen on glucose, lactate, and amino acid metabolism during exercise and recovery in human subjects. *Journal of Physiology*, *514*, 293-302.

Blum, J.W., Bianca, W., Naf, F., Kunz, P., Fischer, J.A., & Da-Prada, M. (1979.) Plasma cathecolamine and parathyroid hormone responses in cattle during treadmill exercise at simulated high altitude. *Hormone and Metabolic Research*, *11*, 246-251.

Blumenfeld, J.D., Sealey, J.E., Mann, S.J., Bragat, A., Marion, R., Pecker, M.S., Sotelo, J., August, P., Pickering, T.G., & Laragh, J.H. (1999.) Beta-adrenergic receptor blockade as a therapeutic approach for suppressing the renin-angiotensin-aldosterone system in normotensive and hypertensive subjects. *American Journal of Hypertension, 12,* 451-459.

Blundell, J.E., & King, N.A. (1999.) Physical activity and regulation of food intake: current evidence. *Medicine and Science in Sports and Exercise, 31,* S573-S583.

Boden, G. (1997.) Role of fatty acids in the pathogenesis of insulin resistance and NIDDM. *Diabetes, 46,* 3-10.

Boden, G., Chen, X., DeSantis, R., Kolaczynski, J., & Morris, M. (1993.) Evidence that suppression of insulin secretion by insulin itself is neurally mediated. *Metabolism, 42,* 786-789.

Boden, G., Chen, X., Mozzoli, M., & Ryan, I. (1996.) Effect of fasting on serum leptin in normal human subjects. *Journal of Clinical Endocrinology and Metabolism, 81,* 3419-3423.

Boden, G., Cheung, P., Stein, T.P., Kresge, K., & Mozzoli, M. (2002.) FFA cause hepatic insulin resistance by inhibiting insulin suppression of glycogenolysis. *American Journal of Physiology, 283,* E12-E19.

Boden, G., Tappy, L., Jadali, F., Hoeldtke, R.D., Rezvani, I., & Owen, O.S. (1990.) Role of glucagon in disposal of an amino acid load. *American Journal of Physiology, 259,* E225- E232.

Boeckastaens, G.E., De Man, J.G., Pelckmans, P.A., Herman, A.G., & Van Maercke, Y.M. (1993.) Alpha₂-adrenoceptor mediated modulation of the nitrergic innervation of the canine isolated ileocolonic junction. *British Journal of Pharmacology, 109,* 1079-1084.

Boesch, C., Slotboom, J., Hoppeler, H., & Kreis, R. (1997.) In vivo determination of intra-myocellular lipids in human muscle by means of localized 1H-MR-spectroscopy. *Magnetic Resonance Medicine, 37,* 484-493.

Boguszewski, C.L., Jansson, C., Boguszewski, M.C., Rosberg, S., Carlsson, B., Albertsson-Wikland, K., & Carlsson, L.M. (1997a.) Increased proportion of circulating non-22-kilodalton growth hormone isoforms in short children: a possible mechanism for growth failure. *Journal of Clinical Endocrinology and Metabolism, 82,* 2944-2949.

Boguszewski, C.L., Jansson, C., Boguszewski, M.C., Rosberg, S., Wikland, K.A., Carlsson, B., &. Carlsson, L.M. (1997b.) Circulating non-22 kDa growth hormone isoforms in healthy children of normal stature: relation to height, body mass and pubertal development. *European Journal of Endocrinology, 137,* 246-253.

Boguszewski, C.L., Svensson, P.A., Jansson, T., Clark, R., Carlsson, L.M., & Carlsson, B. (1998.) Cloning of two novel growth hormone transcripts expressed in human placenta. *Journal of Clinical Endocrinology and Metabolism, 83,* 2878-2885.

Boisvert, P., Brisson, G.R., & Peronnet, F. (1993.) Effect of plasma prolactin on sweat rate and sweat composition during exercise in men. *American Journalof Physiology, 264,* F816-F820.

Boileau, R.A., Buskirk, E.R., Horstman, D.H., Mendez, J., & Nicholas, W.C. (1971.) Body compositional changes in obese and lean men during physical conditioning. *Medicine in Science and Sports, 3,* 183-189.

Boivin, D.B., Duffy, J.F., Kronauer, R.E., & Czeisler, C.A. (1996.) Dose-response relationships for resetting of human circadian clock by light. *Nature, 379,* 540-542.

Boluyt, M.O., Zheng, J.S., Younes, A., Long, X., O'Neill, L., Silverman, H., Lakatta, E.G., & Crow, M.T. (1997.) Rapamycin inhibits alpha 1-adrenergic receptor-stimulated cardiac myocyte hypertrophy but not activation of hypertrophy-associated genes: evidence for involvement of p70 S6 kinase. *Circulation Research, 81,* 176-186.

Bommer, M., & Herz, A. (1989.) Neuropeptides and other secretagogues in bovine chromaffin cells: their effect on opiod peptide metabolism. *Neuropeptides, 13,* 245-251.

Bonaterra, M., De Paul, A., Aoki, A., & Torres A. (1998.) Residual effects of thyroid hormone on secretory activity of somatotroph population. *Experimental and Clinical Endocrinology and Diabetes, 106,* 494-499.

Bonen, A., Campagna, P., Gilchrist, L., Young, D.C., & Beresford, P. (1992.) Substrate and endocrine responses during exercise at selected stages of pregnancy. *Journal of Applied Physiology, 73,* 134-142.

Bonen, A., Haynes, F.W., & Graham, T.E. (1991.) Substrate and hormonal responses to exercise in women using oral contraceptives. *Journal of Applied Physiology, 70,* 1917-1927.

Bonen, A., Ling, W.Y., MacIntyre, K.P., Neil, R., McGrail, J.C., & Belcastro, A.N. (1979.) Effects of exercise on serum concentrations of FSH, LH, progesterone, and estradiol. *European Journal of Applied Physiology, 42,* 15-23.

Bonham, A.C. (1995.) Neurotransmitters in the CNS control of breathing. *Respiratory Physiology, 101,* 219-230.

Boone, J.B., Jr., Sherraden, T., Pierzchala, K., Berger, R., & VanLoon, G.R. (1992.) Plasma Met-enkephalin and catecholamine responses to intense exercise in humans. *Journal of Applied Physiology, 73,* 388-392.

Boppart, M.D., Hirschman, M.F., Sakamoto, K., Fielding, R.A., & Goodyear, L.J. (2001.) Static stretch increases c-Jun NH₂-terminal kinase and p38 phosphorylation in rat skeletal muscle. *American Journal of Physiology, 280,* C352-C358.

Borensztajn, J., Keig, P., & Rubenstein, A.H. (1973.) The role of glucagon in the regulation of myocardial lipoprotein lipase activity. *Biochemical and Biophysical Research Communications, 53,* 603-608.

Borensztajn, J., Samols, D.R., & Rubenstein, A.H. (1972.) Effects of insulin on lipoprotein lipase activity in the rat heart and adipose tissue. *American Journal of Physiology, 223,* 1271-1275 .

Borer, K.T. (1982.) The nonhomeostatic motivation to run in the golden hamster. In A.R. Morrison & P.L. Strick (Eds), *Changing concepts of the nervous system* (pp. 539-567), New York: Academic Press.

Borer, K.T. (1987a.) How running accelerates growth. In A.N. Epstein & A.R. Morrison (Eds.) *Progress in psychobiology and physiological psychology* (pp. 47-115). New York: Academic Press.

Borer, K.T. (1987b.) Rostromedial septal area controls pulsatile growth hormone release in the golden hamster. *Brain Research Bulletin, 18,* 485-490.

Borer, K.T., Bonna, R., & Kielb, M. (1989.) Hippocampal serotonin mediates hypoactivity in dietarily obese hamsters: a possible manifestation of aging? *Pharmacology, Biochemistry and Behavior, 31,* 885-892.

Borer, K.T., & Clover, K. (1994.) Control by light of the temperature rhythm in food-restricted hamsters. *Physiology and Behavior, 56,* 385-391.

Borer, K.T., Cornelissen, G., Halberg, F., Brook, R., Rajagopalan, S., & Fay, W. (2002.) Circadian blood-pressure overswinging in a physically-fit normotensive African American woman. *American Journal of Hypertension, 15,* 827-830.

Borer, K.T., & Kaplan, L.R. (1977.) Exercise-induced growth in golden hamsters: effects of age, weight and activity level. *Physiology and Behavior, 18,* 29-34.

Borer, K.T., & Kelch, R.P. (1978.) Increased serum growth hormone and somatic growth in exercising adult hamsters. *American Journal of Physiology, 234,* E611-E616.

Borer, K.T., & Kooi, A.A. (1975.) Regulatory defense of the exercise-induced weight elevation in hamsters. *Behavioral Biology, 13,* 301-310.

Borer, K.T., & Kuhns, L.R. (1977.) Radiographic evidence for acceleration of skeletal growth in adult hamsters by exercise. *Growth, 41,* 1-13.

Borer, K.T., Campbell, C.S., Tabor, J., Jorgenson, K., Kandarian, S., & Gordon, L. (1983a.) Exercise reverses photoperiodic anestrus in golden hamsters. *Biology of Reproduction, 29,* 38-47.

Borer, K.T., Kelch, R.P., Peugh, J. & Huseman, C. (1979a.) Increased serum growth hormone somatic growth in adult hamsters with hippocampal transections. *Neuroendocrinology, 29,* 22-23.

Borer, K.T., Kelch, R.P., White, M.P., Dolson, L., & Kuhns, L.R. (1977.) The role of septal area in the neuroendocrine control of growth in the adult golden hamster. *Neuroendocrinology, 23,* 133-150.

Borer, K.T., Nicoski, D., & Owens, V. (1986.) Alteration of the pulsatile growth hormone secretion by growth-inducing exercise: involvement of endogenous opiates and somatostatin. *Endocrinology, 118,* 844-850.

Borer, K.T., Peters, N.L., Kelch, R.P., Tsai, A.N., & Holder, S. (1979b.) Contributions of growth, fatness, and activity to weight disturbance following septo-hypothalamic cuts in adult hamsters. *Journal of Comparative Physiological Psychology, 93,* 907-918.

Borer, K.T., Potter, C.D., & Fileccia, N. (1983b.) Basis for the hypoactivity that accompanies rapid weight gain in hamsters. *Physiology and Behavior, 30,* 389-397.

Borer, K.T., Trulson, M.E., & Kuhns, L.R. (1979c.) Role of limbic system in the control of hamster growth. *Brain Research Bulletin, 4,* 239-247.

Bornstein, A., & Volker, H. (1926.) Uber die Schwankungen des Grundumsatzes. *Zeitschrift der Gesaimene und Experimentale Medicin, 53,* 439-450.

Borsheim, E., Knardahl, S., Hostmark, A.T., & Bahr, R. (1998.) Adrenergic control of post-exercise metabolism. *Acta Physiologica Scandinavica, 162,* 313-323.

Bouchard, C.(Ed.) (1994.) *Genetic determinants of obesity*. Boca Raton: CRC Press.

Boulos, Z., Campbell, S.S., Lewy, A.J., Terman, M., & Dijk, D.J. (1995.) Light treatment for sleep disorders: consensus report. VII. Jet lag. *Journal of Biological Rhythms, 10*, 167-176.

Bourguinon, J.P. (1988.) Linear growth as a function of age at the onset of puberty and sex steroid dosage: therapeutical implications. *Endocrine Reviews, 9*, 467-488.

Boyar, R.M., Katz, J., Finkelstein, J.W., Kapen, S., Weiner, H., Weitzman, E.D., & Hellman, L. (1974a.) Anorexia nervosa: immaturity of the 24-hour luteinizing hormone secretory pattern. *New England Journal of Medicine, 291*, 861-865.

Boyar, R.M., Rosenfeld, R.S., Kapen, S. Finkelstein, J.W., Roffwarg, H.P., Weitzman, E.D., & Hellman, L. (1974b.) Human puberty: simultaneous augmented secretion of luteinizing hormone and testosterone during sleep. *Journal of Clinical Investigation, 54*, 609-618.

Boyce, B.F., Aufdemorte, T.B., Garrett, I.R., Yates, A.J.P., & Mundy, G.R. (1989.) Effects of interleukin-1 on bone turnover in normal mice. *Endocrinology, 125*, 1142-1150.

Boyd, A.E., Giamber, S.R., Mager, M., & Lebowitz, H.E. (1974.) Lactate inhibition of lipolysis in exercising man. *Metabolism, 23*, 531-542.

Boyle, P.J., Avogaro, A., Smith, L., Bier, D.M., Pappu, A.S., Illingworth, D.R., & Cryer, P.E. (1992.) Role of GH in regulating nocturnal rates of lipolysis and plasma mevalonate levels in normal and diabetic humans. *American Journal of Physiology, 263*, E168- E172.

Brady, M.J., Kartha, P.M., Aysola, A.A., & Saltiel, A.R. (1999.) The role of glucose metabolites in the activation and translocation of glycogen synthase by insulin in 3T3-L1 adipocytes. *Journal of Biological Chemistry, 274*, 27497-27504.

Brain, S.D., & Williams, T.J. (1988.) Substance P regulates the vasodilator activity of calcitonin gene-related polypeptide. *Nature, 335*, 73-75.

Brandenberger, G., & Follenius, M. (1975.) Influence of timing and intensity of muscular exercise on temporal patterns of plasma cortisol levels. *Journal of Clinical Endocrinology and Metabolism, 40*, 845-849.

Brandenberger, G., Charloux, A., Gronfier, C., & Otzenberger, H. (1998.) Ultradian rhythms in hydromineral hormones. *Hormone Research, 49*, 131-135.

Brandenberger, G., Follenius, M., & Hietter, B. (1982.) Feedback from meal-related peaks determines diurnal changes in cortisol response to exercise. *Journal of Clinical Endocrinology and Metabolism, 54*, 592-596.

Brandenberger, G., Krauth, M.O., Ehrhart, J., Libert, J.P., Simon, C., & Follenius, M. (1990.) Modulation of episodic renin release during sleep in humans. *Hypertension, 15*, 370-375.

Brandenberger, G., Luthringer, R., Muller, G., Gronfier, C., Schaltenbrand, N., Macher, J.P., Muzet, A., & Follenius, M. (1996.) 5-HT2 receptors are partially involved in the relationship between renin release and delta relative power. *Journal of Endocrinological Investigation, 19*, 556-562.

Bray, G.A. (1993.) The nutrient balance hypothesis: peptides, sympathetic activity, and food intake. *Annals of the New York Academy of Sciences, 676*, 223-241.

Bray, G.A. (2000a.) Afferent signals regulating food intake. *Proceedings of the Nutrition Society, 59*, 373-384.

Bray, G.A. (2000b.) Reciprocal relation of food intake and sympathetic activity: Experimental observations and clinical implications. *International Journal of Obesity and Related Metabolic Disorders, 24* (Suppl. 2), S8-S17.

Brenner, I.K., Zamecnik, J., Shek, P.N., & Shephard, R.J. (1997.) The impact of heat exposure and repeated exercise on circulating stress hormones. *European Journal of Applied Physiology and Occupational Physiology, 76*, 445-454.

Breslau, N.A. (1996.) Calcium homeostasis. In J.E. Griffin & S.R.Ojeda (Eds.), *Textbook of endocrine physiology*, 3rd ed. (pp. 314-348). New York: Oxford University Press.

Briend, A. (1980.) Maternal physical activity, birth weight, and neonatal mortality. *Medical Hypotheses, 6*, 1157-1170.

Brindle, N.P., & Ontko, J.A. (1988.) Alpha-adrenergic suppression of very-low-density lipoprotein triacylglycerol secretion by isolated rat hepatocytes. *Biochemical Journal, 250*, 363-368.

Brisson, G.R., Peronnet, F., Ledoux, M., Pellerin-Massicotte, J., Matton, P., Garceau, R., & Boisvert, P.Jr. (1986.) Temperature-induced hyperprolactinemia during exercise. *Hormone and Metabolic Research, 18*, 283-184.

Brisson, G.R., Quirion, A., Ledoux, M., Rajotte, D., & Pellerin-Massicotte, J. (1984.) Influence of long-distance swimming on serum androgens in males. *Hormone and Metabolic Research, 16*, 160.

Bristow, J.D., Honour, A.J., Pickering, G.W., Sleight, P., & Smyth, H.S. (1969.) Diminished baroreflex sensitivity in high blood pressure. *Circulation, 39*, 48-54.

Brook, C.G., & Hindmarsh, P.C. (1992.) The somatotropic axis in puberty. *Endocrinology and Metabolism Clinics of North America, 21*, 767-782.

Brooks, D.P., De Palma, P.D., Pullen, M., & Nambi, P. (1994.) Characterization of canine renal endothelin receptor subtypes and their function. *Journal of Pharmacology and Experimental Therapeutics, 268*, 1091-1097.

Brooks, V.L. (1997.) Interactions between angiotensin II and the sympathetic nervous system in the long-term control of arterial pressure. *Clinical and Experimental Pharmacology and Physiology, 24*, 83-90.

Brooks, V.L., Eli, K.R., & Wright, R.M. (1993.) Pressure-independent baroreflex resetting produced by chronic infusion of angiotensin II in rabbits. *American Journal of Physiology, 265*, H1275-H1282.

Bross, R., Casaburi, R., Storer, T.W., & Bhasin, S. (1998.) Androgen effects on body composition and muscle function: implications for the use of androgens as anabolic agents in sarcopenic states. *Bailleres Clinical Endocrinology and Metabolism, 12*, 365-378.

Broughton Pipkin, F., Morrison, R., & O'Brien, P.M. (1989.) Prostacyclin attenuates both the pressor and adrenocortical response to angiotensin II in human pregnancy. *Clinical Science, 76*, 529-534.

Broun, J.A., & Severson, D.L. (1992.) Regulation of synthesis, processing and translocation of lipoprotein lipase. *Biochemical Journal, 287*, 337-347.

Brower, K.J., Eliopulos, G.A., Blow, F.C., Catlin, D.H., & Beresford, T.P. (1990.) Evidence for physical and psychological dependence on anabolic androgenic steroids in eight weight lifters. *American Journal of Psychiatry, 147*, 510-512.

Brown, J.H., & Wilmore, J. (1974.) The effects of maximum resistance training on the strength and body composition of women athletes. *Medicine and Science in Sports, 6*, 174-177.

Brown, M.R., & Fisher, L.A. (1985.) Corticotropin releasing factor: effects on autonomic nervous system and visceral systems. *Federation Proceedings, 44*, 243-248.

Bruneau, B.G., Piazza, L.A., & de Bold, A.J. (1997.) BNP gene expression is specifically modulated by stretch and ET-1 in a new model of isolated rat atria. *American Journal of Physiology, 273*, H2678-H2686.

Bruunsgaard, H., Galbo, H., Halkjaer-Kristensen, J., Johansen, T.L., MacLean, D.A., & Pedersen, B.K. (1997.) Exercise-induced increase in serum interleukin-6 in humans is related to muscle damage. *Journal of Physiology, 499*, 833-841.

Brutsaert, D.L. (1993.) Endocardial & coronary enothelial control of cardiac performance. *News in Physiological Sciences, 8*, 82-86.

Brzezinski, A., Lynch, H.J., Seibel, M.M., Deng, M.H., Nader, T.M., & Wurtman, R.J. (1988a.) The circadian rhythm of plasma melatonin during the normal menstrual cycle and in amenorrheic women. *Journal of Clinical Endocrinology and Metabolism, 66*, 891-895.

Buckler, J.M. (1971.) The relationship between exercise, body temperature and plasma growth hormone levels in a human subject. *Journal of Physiology (London), 214*, 25P-26P.

Buckler, J.M. (1973.) The relationship between changes in plasma growth hormone levels and body temperature occuring with exercise in man. *Biomedicine, 19*, 193-197.

Buckley, N.J., & Caulfield, M. (1992.) Transmission: acetylcholine. In G. Burnstock & C.H.V. Hoyle (Eds.), *Autonomic neuroeffector mechanisms* (pp. 257-322). Philadelphia: Harwood Academic Publishers.

Budohoski, L., Challiss, R.A., Dubaniewicz, A., Kaciuba-Uscilko, H., Leighton, B., Lozeman, F.J., Nazar, K., Newsholme, E.A., & Porta, S. (1987.) Effects of prolonged elevation of plasma adrenaline concentration in vivo on insulin-sensitivity in soleus muscle of the rat. *Biochemical Journal, 244*, 655-660.

Bugge, J.F., Opstad, P.K., & Magnus, P.M. (1979.) Changes in the circadian rhythm of performance and mood in healthy young men exposed to prolonged, heavy physical work, sleep deprivation, and caloric deficit. *Aviation Space and Environmental Medicine, 50*, 663-668.

Buguet, A., Cespuglio, R., & Radomski, M.W. (1998.) Sleep and stress in man: An approach through exercise and exposure to extreme environments. *Canadian Journal of Physiology and Pharmacology, 76*, 553-561.

Buijs, R.M., Chun, S.J., Niijima, A., Romijn, H.J., & Nagai, K. (2001.) Parasympathetic and sympathetic control of the pancreas: a role for the suprachiasmatic nucleus and other hypothalamic centers that are involved in regulation of food intake. *Journal of Comparative Neurology, 431*, 405-423.

Buijs, R.M., Hermes, M.H., & Kalsbeek, A. (1998.) The suprachiasmatic nucleus-paraventricular nucleus interactions: a bridge to the neuroendocrine and autonomic nervous system. Progress in Brain Research, 119, 365-382.

Buijsen, J.G., van Acker, B.A., Koomen, G.C., Koopman, M.G., & Arisz, L. (1994.) Circadian rhythm of glomerular filtration rate in patients after kidney transplantation. *Nephrology, Dialysis, Transplantation, 9*, 1330-1333.

Bukowiecki, L., Lupien, J., Follea, N., Paradis, A., Richard, D., & LeBlanc. J. (1980.) Mechanism of enhanced lipolysis in adipose tissue of exercise-trained rats. *American Journal of Physiology, 239*, E422-429.

Bulbring, E., & Burn, J.H. (1935.) The sympathetic dilator fibres in the muscles of the cat and dog. *Journal of Physiology 83*, 483-501.

Bulbring, E., & Tomita, T. (1987.) Chatecholamine action on smooth muscle. *Pharmacological Reviews, 39*, 49-96.

Bulbulian, R., & Darabos, B.L. (1986.) Motor nerve excitability: the Hoffmann reflex following exercise of low and high intensity. *Medicine and Science in Sports and Exercise, 18*, 697-702.

Bullen, B.A., Skrinar, G.S., & Beitins, I. Z. (1985.) Induction of menstrual cycle disorders by strenuous exercise in untrained women. *New England Journal of Medicine, 312*, 1349-1353.

Bulow, J. (1982.) Subcutaneous adipose tissue blood flow & triacylglycerol mobilization during prolonged exercise in dogs. *Pflugers Archiv, 392*, 235-238.

Bulow, J., & Madsen, J. (1976.) Adipose tissue blood flow during prolonged, heavy exercise. *Pflugers Archiv, 363*, 231-234.

Bulow, J., Madsen, J., Astrup, A., & Christiansen, N.J. (1985.) The effect of high free fatty acid/albumin molar ratios on the perfusion of adipose tissue in vivo. *Acta Physiologica Scandinavica, 125*, 661-667.

Bunnell, D.E., Bevier, W.C., & Horvath, S.M. (1983.) Nocturnal sleep, cardiovascular function, and adrenal activity following maximum capacity exercise. *Electroencephalography and Clinical Neurophysiolgy, 56*, 186-189.

Bunnell, D.E., Bevier, W.C., & Horvath, S.M. (1984.) Sleep interruption and exercise. *Sleep, 7*, 261-274.

Bunnell, D.E., Bevier, W.C., & Horvath, S.M. (1985.) Effects of exhaustive submaximal exercise on cardiovascular function during sleep. *Journal of Applied Physiology, 58*, 1909-1913.

Bunt, J.C., Boileau, R.A., Bahr, J.M., & Nelson, R.A. (1986.) Sex and training differences in human growth hormone levels during prolonged exercise. *Journal of Applied Physiology, 61*,1796-1801.

Buono, M.J., & Sjoholm, N.T. (1988.) Effect of physical training on peripheral sweat production. *Journal of Applied Physiology, 65*, 811-814.

Burckhardt, B., Delco, F., Ensinck, J.W., Meier, R., Bauerfeind, P., Aufderhaar, U., Ketterer, S., Gyr, K., & Beglinger, C. (1994.) Cholecystokinin is a physiological regulator of gastric acid secretion in man. *European Journal of Clinical Investigation, 24*, 370-376.

Burden, H.W. (1985.) The adrenergic innervation of mammalian ovaries. In N. Jonathan, J.M. Bahr & R.I. Weiner (Eds.), *Catecholamines as hormone regulators* (pp. 261-278), New York: Raven Press.

Burkhart, F., & Oswald, S. (1992.) Mechanisms of myocardial ischemia and circadian fluctuations of ischemic episodes. *Schweizerische Rundschau fur Medizin Praxis, 81*, 171-175.

Burns, T.W., Langley, P.E., Terry, B.E., & Robinson, G.A. (1978.) The role of free fatty acids in the regulation of lipolysis by human adipose tissue cells. *Metabolism, 27*, 1755-1762.

Burnstock, G. (1995.) Noradrenaline and ATP: Cotransmitters and neuromodulators. *Journal of Physiology and Pharmacology, 46*, 365-384.

Burnstock, G., & Hoyle, C.H. (1992.) *Autonomic neuroeffector mechanisms.* Philadelphia: Harwood Academic Publishers.

Burnstock, G., & Wood, J.N. (1996.) Purinergic receptors: their role in nociception and primary afferent neurotransmission. *Current Opinion in Neurobiology, 6*, 526-532.

Burr, I.M., Sizonenko, P.C., Kaplan, S.L., & Grumbach, M.M. (1970.) Hormonal changes in puberty. I. Correlation of serum luteinizing hormone and follicle stimulating hormone with stages of puberty, testicular size, and bone age in normal boys. *Pediatric Research, 4*, 25-35.

Burr, I.M., Slonim, A.E., Burke, V., & Fletcher, T. (1976.) Extracellular calcium and adrenergic and cholinergic effects on islet beta-cell function. *American Journal of Physiology, 231*, 1246-1249.

Buwalda, B., DeBoer, S.F., VanKalkeren, A.A., & Koolhaas, J.M. (1997.) Physiological and behavioral effects of chronic intracerebralventricular infusion of corticotropin-releasing factor in the rat. *Psychoneuroendocrinology, 22*, 297-309.

Buxton, O.M., L'Hermite-Baleriaux, M., Turek, F.W., & Van Cauter, E. (2000.) Daytime naps in darkness phase shift the human circadian rhythms of melatonin and thyrotropin secretion. *American Journal of Physiology, 278*, R373-R382.

Buxton, O.M., Frank, S.A., L'Hermite-Baleriaux, M., Leproult, R., Turek, F.W., & Van Cauter, E. (1997a.) Roles of intensity and duration of nocturnal exercise in causing phase delays of human circadian rhythms. *American Journal of Physiology, 273*, E536-E542.

Buxton, O.M., L'Hermite-Baleriaux, M., Hirschfeld, U., & Van Cauter, E. (1997b.) Acute and delayed effects of exercise on human melatonin secretion. *Journal of Biological Rhythms, 12*, 568-574.

Cabral, A.M., Vasquez, E.C., Moyses, M.R., & Antonio, A. (1988.) Sex hormone modulation of ventricular hypertrophy in sinoaortic denervated rats. *Hypertension, 11*, 193-197.

Cadoux-Hudson, T.A., Few, J.D., & Imms, F.J. (1985.) The effects of exercise on the production and clearance of testosterone in well trained young men. *European Journal of Applied Physiology, 54*, 321-325.

Caffrey, J.L., Gaugl, J.F., & Jones, C.E. (1985.) Local endogenous opiate activity in dog myocardium: Receptor blockade with naloxone. *American Journal ofPhysiology, 248*, H382-H388.

Caffrey, J.L., Mateo, Z., Napier, L.D., Gaugl, J.F. & Barron, B.A. (1995.) Intrinsic cardiac enkephalins inhibit vagal bradycardia in the dog. *American Journal of Physiology, 268*, H848-H455.

Cairns, S.P., & Dulhunty, A.F. (1993.) The effects of β-adrenoceptor activation on contraction in isolated fast- and slow-twitch skeletal muscle fibers of the rat. *British Journal of Pharmacology, 110*, 1133-1141.

Cairns, S.P., Westerblad, H., & Allen, D.G. (1993.) Changes of tension and [Ca2+]i during beta-adrenoceptor activation of single, intact fibres from mouse skeletal muscle. *Pflugers Archiv-European Journal of Physiology, 425*, 150-155.

Calderone, G., Leglise, M., Giampietro, M., & Berlutti, G. (1986a.) Anthropometric measurements, body composition, biological maturation and growth predictions in young female gymnasts of high agonistic level. *Journal of Sports Medicine, 26*, 263-273.

Caldarone, G., Leglise, M., Giampietro, M., & Berlutti, G. (1986b.) Anthropometric measurements, body composition, biological maturation and growth predictions in young male gymnasts of high agonistic level. *Journal of Sports Medicine & Physical Fitness, 26*, 406-415

Calingasan, N.Y., & Ritter, S. (1992a.). Hypothalamic paraventricular nucleus lesions do not abolish glucoprivic or lipoprivic feeding. *Brain Research, 595*, 25-31.

Calingasan, N.Y., & Ritter, S. (1992b.) Presence of galanin in rat vagal sensory neurons: evidence from immunohistochemistry and in situ hybridization. *Journal of the Autonomic Nervous System, 40*, 229-238.

Campbell, P.J., Carlson, M.G., Hill, J.O., & Nurjhan, N. (1992.) Regulation of free fatty acid metabolism by insulin in humans: role of lipolysis and reesterification. *American Journal of Physiology, 263*, E1063-E1069.

Campfield, L.A. (1997.) Metabolic and hormonal controls of food intake: highlights of the last 25 years—1972-1997. *Appetite, 29*, 135-152.

Campfield, L.A., Smith, F.J., Guisez, Y., Devos, R., & Burn, P. (1995.) Recombinant mouse OB protein: evidence for a peripheral signal linking adiposity and central neural networks. *Science, 269*, 546-549.

Campos, M.B., Chiocchio, S.R., Calandra, R.S., & Ritta, M.N. (1993.) Effect of bilateral denervation of the immature rat testis on testicular gonadotropin receptors and in vitro androgen production. *Neuroendocrinology, 57*, 189-194.

Camus, G., Deby-Dupont, G., Duchateau, J., Deby, C., Pincemail, J., & Lamy, M. (1994.) Are similar inflammatory factors involved in strenuous exercise and sepsis? *Internal Care Medicine, 20*, 602-610.

Canalis, E., McCarthy, T.L., & Centrella, M. (1989.) The role of growth factors in skeletal remodeling. *Endocrinology and Metabolism Clinics of North America, 18*, 903-918.

Cannon, B., & Nedergaard, J. (1996.) Adrenergic regulation of brown adipocyte differentiation. *Biological Society Transactions, 24*, 407-412.

Cannon, J.G., Meydani, S.N., Fielding, R.A., Fiatarione, M.A., Meydani, M., Farhangmehr, M., Orencole, S.F., Blumberg, J.B., & Evans, W.J. (1991.) Acute phase response in exercise: II. Association between vitamin E, cytokines, and muscle proteolysis. *American Journal of Physiology, 260*, R1235-R1240.

Cannon, J.G., Orencole, S.F., Fielding, R.A., Meydani, M., Meydani, S.N., Fiatarione, M.A., Blumberg, J.B., & Evans, W.J. (1990.) Acute phase response in exercise: interaction of age and vitamin E on neutrophils and muscle enzyme release. *American Journal of Physiology, 259*, R1214-R1219.

Cannon, J.G., & Pierre, B.A. (1998.) Cytokines in exertion-induced skeletal muscle injury. *Molecular and Cellular Biochemistry, 179*, 159-167.

Cannon, W.B. (1922.) *Bodily changes in pain, hunger, fear, and rage.* New York: Appleton.

Cannon, W.B. (1932.) *Wisdom of the body.* New York: W.W.Norton.

Cannon, W.B., McIver, M.A., & Bliss, S.W. (1924.) Studies on the conditions of activity in endocrine glands. XIII: A sympathetic and adrenal mechanism for mobilizing sugar in hypoglycemia. *American Journal of Physiology, 69*, 46-66.

Cappon, J., Brasel, J.A., Mohan, S., & Cooper, D.M. (1994.) Effect of brief exercise on circulating insulin-like growth factor I. *Journal of Applied Physiology, 76*, 2490-2496.

Carandente, F., Angeli, A., Candiani, G.B., Crosignani, P.G., Dammacco, F., de Cecco, L., Marrama, P., Massobrio, M., & Martini, L. (1989.) Rhythms in the ovulatory cycle. II: LH, FSH, estradiol and progesterone. *Chronobiologia, 16*, 353-363.

Cardosso, S.S., Feuers, R.J., Tsai, T.H., Hunter, J.D., & Scheving, L.E. (1990.) Chronobiology of exercise: the influence of scheduling upon glycemic responses of control and of subjects with diabetes mellitus. *Progress in Clinical and Biological Research, 341B*, 345-353.

Carey, D.G. (1998.) Abdominal obesity. *Current Opinion in Lipidology, 9*, 35-40.

Carlberg, K.A., Alvin, B.L., & Gwosdow, A.R. (1996.) Exercise during pregnancy and maternal and fetal plasma corticosterone and androstenedione in rats. *American Journal of Physiology, 271*, E896-E902.

Carlson, C.J., Booth, F.W., & Gordon, S.E. (1999.) Skeletal muscle myostatin mRNA expression is fiber-type specific and increases during hindlimb unloading. American Journal of Physiology, 277, R601-R606.

Carlson, K.I., Yang, H.T., Bradshaw, W.S., Conlee, R.K., & Winder, W.W. (1986.) Effect of maternal exercise on fetal liver glycogen late in gestation in the rat. *Journal of Applied Physiology*, 60, 1254-1258.

Carlson, L.A., Ekelund, L.-G., & Fröberg, S.O. (1971.) Concentration of triglycerides, phospholipids and glycogen in skeletal muscle and of free fatty acids and gamma- hydroxybutyric acid in blood in man in response to exercise. *European Journal of Clinical Investigation, 1*, 248-254.

Carpene, C., Galitzky, J., Collon, P., Esclapez, F., Dauzats, M., & Lafontan, M. (1993.) Desensitization of beta-1 and beta-2, but not beta-3, adrenoceptor-mediated lipolytic responses of adipocytes after long-term norepinephrine infusion. *Journal of Pharmacology and Experimental Therapeutics, 265*, 237-247.

Carpenter, A.J., & Nunneley, S.A. (1988.) Endogenous hormones subtly alter women's response to heat stress. *Journal of Applied Physiology, 65*, 2313-2317.

Carr, D.B., Reppert, S.M., Bullen, B., Skrinar, G., Beitins, I., Arnold, M., Rosenblatt, M., Martin, J.B., & McArthur, J.W. (1981.) Plasma melatonin increases during exercise in women. *Journal of Clinical Endocrinology and Metabolism, 53*, 224-225.

Carr, D.J., & Serou, M. (1995.) Exogenous and endogenous opioids as biological response modifiers. *Immunopharmacology, 31*, 59-71.

Carretero, O.A., & Oparil, S. (2000a.) Essential hypertension: Part I: Definition and etiology. *Circulation, 101*, 329-335.

Carretero, O.A., & Oparil, S. (2000b.) Essential hypertension: Part II: Treatment. *Circulation, 101*, 446-453.

Cartee, G.D., & Bohn, E.E. (1995.) Growth hormone reduces glucose transport but not GLUT-1 or GLUT-4 in adult and old rats. *American Journal of Physiology, 268*, E902-E909.

Carter, A.L., Lennon, D.L.F., & Stratman, F.W. (1981.) Increased acetyl carnitine in rat skeletal muscle as a result of high-intensity short-duration exercise. *FEBS Letters, 126*, 21-24.

Carter, D.R., Hayes, W.C., & Schurman, D.J. (1976.) Fatigue life of compact bone. II. Effects of microstructure and density. *Journal of Biomechanics, 9*, 211-218.

Carter, D.R., Orr,T.E., & Fyhrie, D.P. (1989.) Relationships between loading history and femoral cancellous bone architecture. *Journal of Biomechanics, 22*, 231-244.

Carter, D.R., van der Meulen, M.C., & Beaupre, G.S. (1996.) Mechanical factors in bone growth and development. *Bone, 18* (Suppl. 1), 5S-10S.

Carter-Su, C., & Okamoto, K. (1985a.) Effect of glucocorticoids on hexose transport in rat adipocytes: evidence for decreased transporters in plasma membrane. *Journal of Biological Chemistry, 260*, 11091-11098.

Carter-Su, C., & Okamoto, K. (1985b.) Inhibition of hexose transport in adipocytes by dexamethasone: role of protein synthesis. *American Journal of Physiology, 248*, E215-E223.

Carter-Su, C., Rui, L., & Herrington, J. (2000.) Role of the tyrosine kinases JAK2 in signal transduction by growth hormone. *Pediatric Nephrology, 14*, 550-557.

Caruso-Nicoletti, M., Cassorla, F., Skerda, M., Ross, J.L., Loriaux, D.L., & Cutler, G.B., Jr. (1985.) Short-term, low-dose estradiol accelerates ulnar growth in boys. *Journal of Clinical Endocrinology and Metabolism, 61*, 896-898.

Casanueva, F.F., Villanueva, L., Cabranes, J.A., Cabezas-Carrato, J., & Fernandez-Cruz, A. (1984.) Cholinergic mediation of growth hormone secretion elicited by arginine, clonidine, and physical exercise in man. *Journal of Clinical Endocrinology and Metabolism, 59*, 526-530.

Casanueva, F.F., Villanueva, L., Dieguez, C., Diaz, Y., Cabranes, J.A., Szoke, B., Scanlon, M.F., Schally, A.V., & Fernandez-Cruz, A. (1987.) Free fatty acids block growth hormone (GH) releasing hormone-stimulated GH secretion in man directly at the pituitary. *Journal of Clinical Endocrinology and Metabolism, 64*,634-642.

Casanueva, F.F., Villanueva, L., Penalva, A., Vila, T., & Cabezas-Carrato, J. (1981.) Free fatty acid inhibition of exercise-induced growth hormone secretion. *Hormone and Metabolic Research, 13*, 348-350.

Casey, A., Short, A.H., Hultman, E., & Greenhaff, P.L. (1995.) Glycogen resynthesis in human muscle fibre types following exercise-induced glycogen depletion. *Journal of Physiology, 483*, 265-271.

Cashmore, A.R., Jarillo, J.A., Wu, Y.-J., & Liu, D. (1999.) Cryptochromes: blue light receptors for plants and animals. *Science, 284*, 760-765.

Casteleijn, E., Kuiper, J., Van Rooij, H.C., Kamps, J.A., Koster, J.F., & Van Berkel, T.J. (1988.) Prostaglandin D2 mediates the stimulation of glycogenolysis in the liver by phorbol ester. *Biochemical Journal, 250*, 77-80.

Castellani, J.N., Maresh, C.M., Armstrong, L.E., Kenefick, R.W., Riebe, D., Echegarm, M., Kavouras, S., & Castracane, V.D. (1998.) Endocrine responses during exercise heat stress: effects of prior isotonic and hypotonic intravenous rehydration. *European Journal of Applied Physiology and Occupational Physiology, 77*, 242-248.

Castellani, J.W., Young, A.J., Kain, J.E., Rouse, A., & Sawka, M.N. (1999.) Thermoregulation during cold exposure: effects of prior exercise. *Journal of Applied Physiology, 87*, 247-252.

Castex, C., & Sutter, B.C.J. (1979.) Seasonal variations of insulin sensitivity in edible dormouse (Glis glis) adipocytes. *General and Comparative Endocrinology, 38*, 365-369.

Caufriez, A., Frankenne, F., Hennen, G., & Copinschi, G. (1994.) Regulation of maternal insulin-like growth factor I by placental growth hormone in pregnancy: possible action of maternal IGF-I on fetal growth. *Hormone Research, 42*, 62-65.

Caufriez, A., Frankenne, F., Hennen, G., & Copinschi, G. (1993.) Regulation of maternal IGF-I by placental GH in normal and abnormal human pregnancies. *American Journal of Physiology, 265*, E572-E577.

Caverson, M.M., & Ciriello, J. (1987.) Effect of stimulation of afferent renal nerves on plasma levels of vasopressin. *American Journal of Physiology, 252*, R801-R807.

Cawthorne, M.A., Sennitt, M.V., Arch, J.R., & Smith, S.A. (1992.) BRL 35135, a potent and selective atypical beta-adrenoceptor agonist. *American Journal of Clinical Nutrition, 55*, (1 Suppl.), 252S-257S.

Celotti F., & Negri Cesi P. (1992.) Anabolic steroids: a review of their effects on the muscles, of their possible mechanisms of action and of their use in athletics. *Journal of Steroid Biochemistry and Molecular Biology, 43*, 469-477.

Celotti, F., & Bignamini, A. (1999.) Dietary calcium and mineral/vitamin supplementation: A controversial problem. *Journal of International Medical Research, 27*,1-14.

Cersosimo, E., Danou, F., Persson, M., & Miles, J.M. (1996.) Effects of pulsatile delivery of basal growth hormone on lipolysis in humans. *American Journal of Physiology, 271*,E123-E126.

Chalew, S.A., Udoff, L.C., Hanukoglu, A., Bistritzer, T., Armour, K.M., & Kowarski, A.A. (1988.) The effects of testosterone therapy on spontaneous growth hormone secretion in boys with constitutional delay. *American Journal of Diseases of Children, 142*, 1345-1348.

Challis, J.R.G., Bassett, N., Berdusco, E.T.M., Han, V.K.M., Lu, F., Riley, S.C., & Yang K. (1993.) Foetal endocrine maturation. *Equine Veterinary Journal Supplement, 14*, 35-40.

Challis, R.A., Richards, S.J., & Budohoski, L. (1992.) Characterization of the adenosine receptor modulating insulin action in skeletal muscle. *European Journal of Pharmacology, 226*, 121-128.

Chandler, K.D., Leury, B.J., Bird, A.R., & Bell, A.W. (1985.) Effects of undernutrition and exercise during late pregnancy on uterine, fetal and uteroplacental metabolism in the ewe. *British Journal of Nutrition, 53*, 625-635.

Chandler, M.P., Rodenbaugh, D.W., & DiCarlo, S.E. (1998.) Arterial baroreflex resetting mediates postexercise reductions in arterial pressure and heart rate. *American Journal of Physiology, 275*, H1627-H1634.

Chandler, W.L., Schwartz, R.S., Stratton, J.R., & Vitiello, M.V. (1996.) Effects of endurance training on the circadian rhythm of fibrinolysis in men and women. *Medicine and Science in Sports and Exercise, 28*, 647-655.

Chapman, I.M., Hartman, M.L., Straume, M., Johnson, M.L., Veldhuis, J.D., & Thorner, M.O. (1994.) Enhanced sensitivity growth hormone (GH) chemiluminescent assay reveals lower postglucose nadir GH concentrations in men than in women. *Journal of Clinical Endocrinology and Metabolism, 78*, 1312-1319.

Charloux, A., Otzenberger, H., Gronfier, C., Lonsdorfer-Wolf, E., Piquard, F., & Brandenberger, G. (1998.) Oscillations in sympatho-vagal balance opose variations in delta-wave activity and the associated renin release. *Journal of Clinical Endocrinology and Metabolism, 83*, 1523-1528.

Charlton, B.G. (1990.) Adrenal cortical innervation and glucocorticoid secretion. *Journal of Endocrinology, 126*, 5-8.

Chatterjee, D. Chakraborty, M., Leit, M., Neff, S., Jamsa-Kellokumpu, S., Fuchs, R., & Baron R. (1992.) Sensitivity to vanadate and isoforms of subunits A and B distinguish the osteoclast proton pump from other vacuolar H+ ATPases. *Proceedings of the National Academy of Sciences, 89*, 6257-6261.

Chatterjee, S., & Laudato, M. (1995.) Gender and performance in athletics. *Social Biology, 42*, 124-132.

Cheatham, B., & Kahn, C.R. (1995.) Insulin action and insulin signaling network. *Endocrine Reviews, 16*, 117-142.

Chen, C.H., Stephens, R.L., Jr., & Rogers, R.C. (1997.) PYY and NPY: Control of gastric motility via action on Y1 and Y2 receptors in the DVC. *Neurogastroenterology and Motility, 9*, 109-116.

Chen, G.Y., Kuo, C.D., Yang, M.J., Lo, H.M., & Tsai, Y.S. (1999.) Comparison of supine and upright positions on autonomic nervous activity in late pregnancy: the role of aortocaval compression. *Anaesthesia, 54*, 215-219.

Chen, H.H., & Burnett, J.C., Jr. (1998.) C-type natriuretic peptide: the endothelial component of the natriuretic peptide system. *Journal of Cardiovascular Pharmacology, 32* (Suppl. 3), S22-S28.

Chernik, S.S., Spooner, P.M., Garrison, M.M., & Scow, R.O. (1986.) Effect of epinephrine and other lipolytic agents on intracellular lipolysis and lipoprotein lipase activity in 3T3 adipocytes. *Journal of Lipid Research, 27*, 286-294.

Chesley, A., Hultman, E., & Spriet, L. (1995.) Effects of epinephrine infusion on muscle glycogenolysis during intense aerobic exercise. *American Journal of Physiology, 268*, E127-E134.

Chhibber, V.L., Soriano, C., & Tayek, J.A. (2000.) Effects of low-dose and high-dose glucagon on glucose production and gluconeogenesis in humans. *Metabolism: Clinical and Experimental, 49*, 39-46.

Chiang, F.T., Tseng, C.D., Hsu, K.L., Lo, H.M., Tseng, Y.Z., Hsieh, P.S., & Wu, T.L. (1994.) Circadian variations of atrial natriuretic peptide in normal people and its relationship to arterial blood pressure, plasma renin activity and aldosterone level. *International Journal of Cardiology, 46*, 229-233.

Chiappelli, F., Yamashita, N., Faisal, M., Kemeny, M., Bullington, R., Nguyen, L., Clament, L.T., & Fahey, J.L. (1991.) Differential effect of beta-endorphin on three human cytotoxic cell populations. *International Journal of Immunopharmacology, 13*, 291-297.

Chiasson, J.L., Liljenquist, J.E., Finger, J.E., & Lacy, W.W. (1976.) Differential sensitivity of glycogenolysis and gluconeogenesis to insulin infusions in dogs. *Diabetes, 25*, 283.

Chick, C.L., Li, B., Ogiwara, T., Ho, A.K., & Karpinski, E. (1996.) PACAP modulates L-type Ca++ channel currents in vascular smooth muscle cells: involvement of PKC and PKA. *FASEB Journal, 10*, 1310-1317.

Chihara, K., Kashio, Y., Kita, T., Okimura, J., Kaji, H., Abe, H., &. Fushita, T. (1986.) L-dopa stimulates release of hypothalamic growth hormone releasing hormone in humans. *Journal of Clinical Endocrinology and Metabolism, 62*, 466-473.

Chirpaz-Oddou, M.F., Favre-Juvin, A., Flore, P., Eterradossi, J., Delaire, M., Grimbert, F., & Therminarias, A. (1997.) Nitric oxide response in exhaled air during an incremental exhaustive exercise. *Journal of Applied Physiology, 82*, 1311-1318.

Chou, C.L., Knepper, M.A., Hoek, A.N., Brown, D., Yang, B., Ma, T., & Verkman, A.S. (1999.) Reduced water permeability and altered ultrastructure in thin descending limb of Henle in aquaporin-1 null mice. *Journal of Clinical Investigation, 103*, 491-496.

Chowen, J.A., Evain-Brion, D., Pozo, J., Alsat, E., Garcia-Segura, L.M., & Argente, J. (1996.) Decreased expression of placental growth hormone in intrauterine growth retardation. Pediatric Research, 39, 736-739.

Christensen, N.J., & Galbo, H. (1983.) Sympathetic nervous activity during exercise. *Annual Reviews of Physiology, 45*, 135-153.

Christensen, S.E., Jorgensen, O.L., Moller, N., & Orskov, H. (1984.) Characterization of growth hormone release in response to external heating. Comparison to exercise-induced release. *Acta Endocrinologica, 107*, 295-301.

Christopoulos, A., & El-Fakahany, E.E. (1998.) The generation of nitric oxide by G protein-coupled receptors. *Life Sciences, 64*, 1-15.

Claessens, A.L., Malina, R.M., Lefevre, J., Beunen, G., Stijnen, V., Maes, H., & Veer, F.M. (1992.) Growth and menarcheal status of elite female gymnasts. *Medicine and Science in Sports and Exercise, 24*, 755-763.

Clapp, J.F., III. (1998.) *Exercising through your pregnancy.* Champaign, IL: Human Kinetics Publishers.

Clapp, J.F., III. (1989.) Oxygen consumption during treadmill exercise before, during, and after pregnancy. *American Journal of Obstetrics and Gynecology, 161*, 1458-1464.

Clapp, J.F., III. (1990.) The course of labor after endurance exercise during pregnancy. *American Journal of Obstetrics and Gynecology, 163*, 1799-1805.

Clapp, J.F., III. (1991a.) Exercise and fetal health. *Journal of Developmental Physiology, 15*, 9-14.

Clapp, J.F., III. (1991b.) The changing thermal response to endurance exercise during pregnancy. *American Journal of Obstetrics and Gynecology, 165*,1684-1689.

Clapp, J.F., III, & Capeless, E.L. (1990.) Neonatal morphometrics after endurance exercise during pregnancy. *American Journal of Obstetrics and Gynecology, 163*, 1805-1811.

Clapp, J.F., III, & Dickstein, S. (1984.) Endurance exercise and pregnancy outcome. *Medicine and Science in Sports and Exercise, 16*,556-562.

Clapp, J.F., III, & Kiess W. (2000.) Effects of pregnancy and exercise on concentrations of the metabolic markers tumor necrosis factor alpha and leptin. *American Journal of Obstetrics and Gynecology, 182*, 300-306.

Clapp, J.F., III, & Little K.D. (1995.) Effect of recreational exercise on pregnancy weight gain and subcutaneous fat deposition. *Medicine and Science in Sports and Exercise, 27*, 170-177.

Clark, A.F., & Vignos, P.J., Jr. (1979.) Experimental corticosteroid myopathy: Effect on myofibrillar ATPase activity and protein degradation. *Muscle and Nerve, 2*, 265-273.

Clark, J.T., Kalra, P.S., Crowley, W.R., & Kalra, S.P. (1984.) Neuropeptide Y and human pancreatic polypeptide stimulate feeding behavior in rats. *Endocrinology, 115*, 427-429.

Clark, J.H., & Conlee, R.K. (1979.) Muscle and liver glycogen content: diurnal variation and endurance. *Journal of Applied Physiology, 47*, 425-428.

Clark, M.G., Colquhoun, E.Q., Rattigan, S., Dora, K.A., Eldershaw, T.P., Hall, J.L., & Ye, J. (1995.) Vascular and endocrine control of muscle metabolism. *American Journal of Physiology, 268*, E797-E812.

Clark, P., & Rogol, A.D. (1996.) Growth hormone and sex steroid interactions at puberty. *Endocrinology and Metabolism Clinics of North America, 25*, 665-681.

Clark, R.G., Jansson, J.O., Isaksson, O., & Robinson I. C. (1985.) Intravenous growth hormone: growth responses to patterned infusions in hypophysectomized rats. *Journal of Endocrinology, 104*,53-61.

Clarkson, P., Montgomery, H.E., Mullen, M.J., Donald, A.E., Powe, A.J., Bull, T., Jubb, M., World, M., & Deanfield, J.E. (1999.) Exercise training enhances endothelial function in young men. *Journal of the American College of Cardiology, 33*, 1379-1385.

Claybaugh, J.R., Freund, B.J., Luther, G., Muller, K., & Bennet, P.B. (1997.) Renal and hormonal responses to exercise in man at 46 and 37 atm absolute pressure. *Aviation Space and Environmental Medicine, 68*, 1038-1045.

Cleroux, J., Van Nguyen, P., Taylor, A.W., and Leenen, F.H.H. (1989.) Effects of beta$_1$ vs beta$_1$+beta$_2$ blockade on exercise endurance and muscle metabolism in humans. *Journal of Applied Physiology, 66*,548-554.

Clutter, W.E., Bier, D.M., Shah, S.D., and Cryer, P.E. (1980.) Epinephrine plasma metabolic clearance rates and physiologic thresholds for metabolic and hemodynamic action in man. *Journal of Clinical Investigation, 66*, 94-101.

Coderre, L., Srivastava, A.K., & Chiasson, J.L. (1992.) Effect of hypercorticism on regulation of skeletal muscle glycogen metabolism by epinephrine. *American Journal of Physiology, 262*, E434-E439.

Coffer, P.J., & Woodgett, J.R. (1998.) Protein kinase B (c-Akt): a multifunctional mediator of phosphatidylinositol 3-kinase activation. *Biochemical Journal, 335*, 1-13.

Coggan, A.R., & Coyle, E.F. (1987.) Reversal of fatigue during prolonged exercise by carbohydrate infusion or ingestion. *Journal of Applied Physiology, 63*, 2388-2395.

Coggan, A.R., & Coyle, E.F. (1988.) Effect of carbohydrate feedings during high-intensity exercise. *Journal of Applied Physiology, 65*, 1703-1709.

Coggan, A.R., & Coyle, E.F. (1989.) Metabolism and performance following carbohydrate ingestion late in exercise. *Medicine and Science in Sports and Exercise, 21*, 59-65.

Cohen, J.C., & Hickman, R. (1987.) Insulin resistance and diminished glucose tolerance in powerlifters ingesting anabolic steroids. *Journal of Clinical Endocrinology and Metabolism, 64*, 960-963.

Cohen, S., Tyrell, D.A., & Smith, A.P. (1991.) Psychological stress and susceptibility to the common cold. *New England Journal of Medicine, 325*, 606-612.

Cohn, P.F. (1988.) Silent myocardial ischemia. *Annals of Internal Medicine, 109*, 312-317.

Coiro, V., Volpi, R., Bertoni, P., Finzi, G., Marcato, A., Caiazza, A., Colla, R.,Giacalone, G., Rossi, G., & Chiodera, P. (1992.) Effect of potentiation of cholinergic tone by pyridostigmine on the GH response to GHRH in elderly men. *Gerontology, 38*, 217-222.

Coker, R.H., Koyama, Y., Lacy, D.B., Williams, P.E., Rheaume, N., & Wasserman, D.H. (1999.) Pancreatic innervation is not essential for exercise-induced changes in glucagon and insulin or glucose kinetics. *American Journal of Physiology, 277*, E1122-E1129.

Coker, R.H., Krishna, M.G., Lacy, D.B., Bracy, D.P., & Wasserman, D.H. (1997.) Role of hepatic alpha- and beta-adrenergic receptor stimulation on hepatic glucose production during heavy exercise. *American Journal of Physiology, 273*, E831-E838.

Col, N.F., Pauker,S.G., Goldberg, R.J., Eckman, M.H., Orr, R.K., Ross, E.M., & Wong, J.B. (1999.) Individualizing therapy to prevent long-term consequences of estrogen deficiency in postmenopausal women. *Archives of Internal Medicine, 159*, 1458-1466.

Colberg, S.R., Simoneau, J.A., Thaete, F.L., & Kelley, D.E. (1995.) Skeletal muscle utilization of free fatty acids in women with visceral obesity. *Journal of Clinical Investigation, 95*, 1846-1853.

Coleman, D.L. (1973.) Effects of parabiosis of obese with diabetes and normal mice. *Diabetologia, 9*, 294-298.

Coleman, D.L., & Hummel, K.P. (1969.) Effects of parabiosis of normal with genetically diabetic mice. *American Journal of Physiology, 217*, 1298-1304.

Collings, C.A., Curet, A.L., & Mullin, J.P. (1983.) Maternal and fetal responses to maternal aerobic exercise program. *American Journal of Obstetrics and Gynecology, 145*, 702-707.

Colquhoun, W.P.(Ed.) (1971.) *Biological rhythms and human performance.* London: Academic Press.

Compan, I., & Touati, D. (1994.) Anaerobic activation of arcA transcription in E*scherichia coli*: roles for Fnr and ArcA. *Molecular Microbiology, 11*, 955-964.

Compston, J.E. (1993.) Thyroid hormone therapy and the skeleton. *Clinical Endocrinology, 39*, 519-520.

Conley, R.K., Hickson, R.C., Winder, W.W., Hagberg, J.M., & Holloszy, J.O. (1978.) Regulation of glycogen resynthesis in muscles of rats following exercise. *American Journal of Physiology, 235*, R145-R150.

Connolly, M.J., Crowley, J.J., Nielson, C.P., Charan, N.B., & Vestal, R.E. (1994.) Peripheral mononuclear leukocyte beta adrenoceptors and non-specific bronchial responsiveness to metacholine in young and elderly normal subjects and asthmatic patients. *Thorax, 49*, 26-32.

Considine, R.V., Sinha, M.K., Heiman, M.L., Kriauciunas, A., Stephens, T.W., Nyce, M.R., Ohannesian, J.P., Marco, C.C., McKee, L.J., Bauer, T.L., & Caro, J.F. (1996.) Serum immunoreactive leptin concentrations in normal-weight and obese humans. *New England Journal of Medicine, 334*, 292-295.

Constantin-Teodosiu, D., Carlin, J.I., Cederblad, G., Harris, R.C., & Hultman, E. (1991.) Acetyl group accumulation and pyruvate dehydrogenase activity in human muscle during incremental exercise. *Acta Physiologica Scandinavica, 143*, 367-372.

Contreras, R.J., & Kosten, T. (1981.) Changes in salt intake after abdominal vagotomy: Evidence for hepatic sodium receptors. *Physiology and Behavior, 26*, 575-582.

Convertino, V.A., Keil, L.C., Bernauer, E.M., & Greenleaf, J.E. (1981.) Plasma volume, osmolality, vasopressin, and renin activity during graded exercise in man. *Journal of Applied Physiology, 50*, 123-128.

Conway, J., Boon, N., Davies, C., Jones, J.V., & Sleight, P. (1984.) Neural and humoral mechanisms involved in blood pressure variability. *Journal of Hypertension, 2*, 203-208.

Cooke, P., Rossitch, E., Jr., Andon, N.A., Loscalzo, J., & Dzau, V.J. (1991.) Flow activates an endothelial potassium channel to release an endogenous nitrovasodilator. *Journal of Clinical Investigation, 88*, 1663-1671.

Cooper, C.J., Noakes, T.D., Dunne, T., Lambert, M.I., & Rochford K. (1996.) A high prevalence of abnormal personality traits in chronic users of anabolic androgenic steroids. *British Journal of Sports Medicine, 30*, 246-250.

Copinschi, G., Akseki, E., Moreno-Reyes, R., Leproult, R., L'Hermite-Baleriaux, M., Caufriez, A., Vertongen, F., & Van Cauter, E. (1995.) Effects of bedtime administration of zolpidem on circadian and sleep-related hormonal profiles in normal women. *Sleep, 18*, 417-424.

Coppack, S.W., Jensen, M.D., & Miles, J.M. (1994.) In vivo regulation of lipolysis in humans. *Journal of Lipid Research 35*, 177-193.

Coppes, R.P., Smit, J., Benthem, L., Van der Leest, J., & Zaagsma, J. (1995.) Co-released adrenaline markedly facilitates noradrenaline overflow through prejunctional beta 2-adrenoceptors during swimming exercise. *European Journal of Pharmacology, 274*, 33-40.

Corder, R., Mason, D.F., Perrett, D., Lowry, P.J., Clement-Jones, V., Linton, E.A., Besser, G.M., & Rees, L.H. (1982.) Simultaneous release of neurotensin, somatostatin, enkephalins and catecholamines from perfused cat adrenal glands. *Neuropeptides, 3*, 9-17.

Cori, C.F. (1925.) The fate of sugar in the animal body. *Journal of Biological Chemistry, 66*, 691-715.

Cori, C.F. (1981.) The glucose-lactic acid cycle and gluconeogenesis. *Current Topics in Cellular Regulation, 18*, 377-387.

Cornelissen, G., Halberg, F., Prikryl, P., Dankova, E., Siegelova, J., & Dusek, J. (1991.) Prophylactic aspirin treatment: the merits of timing. International Womb-to-Tomb Chronome Study Group. *Journal of the American Medical Association, 266*, 3128-3129.

Cornelissen, G., Halberg, F., Schwartzkopff, O., Delmore, P., Katinas, G., Hunter, D., Tarquini, B., Tarquini, R., Perfetto, F., Watanabe, Y., & Otsuka, K. (1999.) Chronomes, time structures, for chronobioengineering for a "full life." *Biomedical Instrumentation and Technology, 33*, 152-187.

Corp, E.S., McQuade, J., Moran, T.H., & Smith, G.P. (1993.)Characterization of type A and type B CCK receptor binding sites in rat vagus nerve. *Brain Research, 623*, 161-166.

Corpas, E., Harman, S.M., & Blackman, M.R. (1993.) Human growth hormone and human aging. *Endocrine Reviews, 14,* 20-39.

Corpas, E., Harman, S.M., Pineyro, M.A., Roberson, R., & Blackman, M.R. (1992.) Growth hormone (GH)-releasing hormone (1-29) twice daily reverses the decreased GH and insulin-like growth factor-I levels in old men. *Journal of Clinical Endocrinology and Metabolism, 75,* 530-535.

Corssmit, E.P., Van Lanschot, J.J., Romijn, J.A., Endert, E., & Sauerwein, H.P. (1995.) Truncal vagotomy does not affect postabsorptive glucose metabolism in humans. *Journal of Applied Physiology, 79,* 97-101.

Cortright, R.N., & Dohm, G.L. (1997.) Mechanisms by which insulin and muscle contraction stimulate glucose transport. *Canadian Journal of Applied Physiology, 22,* 519-530.

Cosentino, M.J., Schoen, S.R., & Cockett, A.T. (1984.) Effect of sympathetic denervation of rat internal genitalia on daily sperm output. *Urology, 24,* 587-590.

Costa, M., Furness, J.J., & Gibbons, I.L. (1986.) Chemical coding of enteric neurons. *Progress in Brain Research, 68,* 217-239.

Costelli, P., Garcia-Martinez, C., Llovera, M., Carbo, N., Lopez-Soriano, F.J., Agell, N., Tessitore, N., Baccina, F.M., & Argiles, J.M. (1995.) Muscle protein waste in tumor-bearing rats is effectively antagonized by a beta2-adrenergic agonist (clenbuterol): role of the ATP-ubiquitin-dependent proteolytic pathway. *Journal of Clinical Investigation, 95,* 2367-2372.

Costill, D.L., & Saltin B. (1974.) Factors limiting gastric emptying during rest and exercise. *Journal of Applied Physiology, 37,* 679-83.

Costill, D.L., Daniels, J., Evans, W., Fink, W., Krahenbuhl, G., & Saltin, B. (1976.) Skeletal muscle enzymes and fiber composition in male and female track athletes. *Journal of Applied Physiology, 40,* 149-159.

Costill, D.L. & Fox, E.L. (1969.) Energetics of marathon running. *Medicine and Science in Sports, 1,* 81-86.

Costill, D.L., Gollnick, P.D., Jansson, E.D., Saltin, B., & Stein, E.M. (1973.) Glycogen depletion pattern in human muscle fibres during distance running. *Acta Physiologica Scandinavica, 89,* 374-383.

Costill, D.L., Coyle, E., Dalsky, G., Evans, W., Fink, W., & Hoopes, D. (1977.) Effects of elevated plasma FFA and insulin on muscle glycogen usage during exercise. *Journal of Applied Physiology, 43,* 695-699.

Courtney, A.C., Wachtel, E.F., Myers, E.R., & Hayes, W.C. (1994.) Effects of loading rate on strength of the proximal femur. *Calcified Tissue International, 55,* 53-58.

Covasa, M., & Ritter, R.C. (1999.) Satiation in response to macronutrient signals from the intestine: mechanisms and implications for macronutrient selection. In H.-R. Berthoud & R.J. Seeley (Eds.) *Neural and metabolic control of macronutrient intake* (pp. 263- 277). London: CRC Press.

Cowan, M.M., & Gregory, L.W. (1985.) Responses of pre- and postmenopausal females to aerobic conditioning. *Medicine and Science in Sports and Exercise, 17,* 138-143.

Cowley, A.W., Jr., Merrill, D., Osborn, J., & Barber, B.J. (1984.) Influence of vasopressin and angiotensin on baroreflexes in the dog. *Circulation Research, 54,* 163-172.

Cowley, M.A., Pronchuk, N., Fan, W., Dinulescu, D.M., Colmers, W.F., & Cone, R.D. (1999.) Intergration of NPY, AGRP, and melanocortin signals in the hypothalamic paraventricular nucleus: Evidence of a cellular basis for the adipostat. *Neuron, 24,* 155-163.

Cox, B.F., Hay, M., & Bishop, V.S. (1990.) Neurons in area postrema mediate vasopressin-induced enhancement of the baroreflex. *American Journal of Physiology, 258,* H1943-H1946.

Coyle, E.F. (1992a.) Carbohydrate supplementation during exercise. *Journal of Nutrition, 122,* (3 Suppl.), 788-795.

Coyle, E.F. (1992b.) Carbohydrate feeding during exercise. *International Journal of Sports Medicine, 13* (Suppl. 1), S126-S128.

Coyle, E.F. (1998.) Cardiovascular drift during prolonged exercise and the effects of dehydration. *International Journal of Sport Medicine, 19* (Suppl. 2), S121-S124.

Coyle, E.F., Coggan, A.R., Hemmert, M.K., & Ivy, J.L. (1986.) Muscle glycogen utilization during prolonged strenuous exercise when fed carbohydrate. *Journal of Applied Physiology, 61,*165-172.

Coyle, E.F., Jeukendrup, A.E., Wagenmakers, A.J., & Saris, W.H. (1997.) Fatty acid oxidation is directly regulated by carbohydrate metabolism during exercise. *American Journal of Physiology, 273,* E268-E275.

Crampes, F., Beauville, M., Riviere, D., & Garigues, M. (1986.) The effect of physical training in humans on the response of isolated fat cells to epinephrine. *Journal of Applied Physiology, 61,* 25-29.

Crampes, F., Riviere, D., Beauville, M., Marceron, M., & Garigues, M. (1989.) Lipolytic response of adipocytes to epinephrine in sedentary and exercise trained subjects: sex-related differences. *European Journal of Applied Physiology, 59,* 249-255.

Crane, R.K., & Sols, A. (1954.) The non-competitive inhibition of brain hexokinase by glucose-6-phosphate and related compounds. *Journal of Biological Chemistry, 210,* 597-606.

Crist, D.M., & Kraner, J.C. (1990.) Supplemental growth hormone increases tumor cytotoxic activity of natural killer cells in healthy adults with normal growth hormone secretion. *Metabolism, 39,* 1320-1324.

Crist, D.M., Peake, G.T., Egan, P.A., & Waters, D.L.(1988.) Body composition response to exogenous GH during training in highly conditioned adults. *Journal of Applied Physiology, 65,* 579-584.

Crist, D.M., Peake, G.T., Loftfield, R.B., Kraner, J.C., & Egan, P.A. (1991.) Supplemental growth hormone alters body composition, muscle protein metabolism and serum lipids in fit adult: characterization of dose-dependent and dose-recovery effects. *Mechanisms of Aging and Development, 58,* 191-205.

Crist, D.M., Peake, G.T., Mackinnon, L.T., Sibbitt, W.L., Jr., & Kraner, J.C. (1987.) Exogenous growth hormone treatment alters body composition and increases natural killer cell activity in women with impaired endogenous growth hormone secretion. *Metabolism: Clinical and Experimental, 36,* 1115-1117.

Crockford, G.W., & Davies, C.T.M. (1969.) Circadian variations in responses to submaximal exercise on a bicycle ergometer. *Journal of Physiology, 201,* 94P-95P.

Cross, M.C., Radomski, M.W., VanHelder W.P. Rhind, S.G., & Shephard, R.J.(1996.) Endurance exercise with or without a thermal clamp: effects on leukocytes and leukocyte subsets. *Journal of Applied Physiology, 81,* 822-829.

Cryer, A. (1981.) Tissue liprotein lipase activity and its action in lipoprotein metabolism. *International Journal of Biochemistry, 13,* 525-541.

Cryer, P.E., Tse, T.F., Clutter, W.E., & Shah, S.D. (1984.) Roles of glucagon and epinephrine in hypoglycemic and nonhypoglycemic glucose counterregulation in humans. *American Journal of Physiology, 247,* E198-E205.

Cugini, P. Centanni, M., Murano, G., Letizia, C., Lucia, P., Scavo, D., Halberg, F., Sothern, R., & Cornelissen, G. (1984.) Toward a chronophysiology of circulating aldosterone. *Biochemical Medicine, 32,* 270-282.

Cugini, P., Lucia, P., DiPalma, L., Re, M., Canova, R., Gasbarrone, L., & Cianetti, I. (1992.) Effect of aging on circadian rhythm of atrial natriuretic peptide, plasma, renin activity, and plasma aldosterone. *Journal of Gerontology, 47,* B214-B219.

Cuif, M.H., Doiron, B., & Kahn, A. (1997.) Insulin and cyclic AMP act at different levels on transcription of the L-type pyruvate kinase gene. *FEBS Letters, 417,* 81-84.

Cumming, D.C., Brunsting, L.A.,III , Strich, G., Ries, A.L., & Rebar, R.W. (1986.) Reproductive hormone increases in response to acute exercise in men. *Medicine and Science in Sports and Exercise, 18,* 369-373.

Cumming, D.C., Vickovic, M.M., Wall, S.R., & Fluker, M.R. (1985.) Defects in pulsatile LH release in normally menstruating runners. *Journal of Clinical Endocrinology and Metabolism, 60,* 810-812.

Cumming, D.C., Wheeler, G.D., & McColl, E.M. (1989.) The effects of exercise on reproductive function in men. *Sports Medicine, 7,* 1-17.

Cumming, D.C.,Wall, S.R., Galbraith, M.A., & Belcastro, A.N. (1987a.) Reproductive hormone responses to resistance exercise. *Medicine and Science in Sports and Exercise, 19,* 234-238.

Cumming, D.C.,Wall, S.R., Quinney, H.A., & Belcastro, A.N.(1987b.) Decrease in serum testosterone levels with maximal intensity swimming exercise in trained male and female swimmers. *Endocrine Research, 13,* 31-41.

Cummings, D.E., Purnell, J.Q., Frayo, R.S., Schmidova, K., Wisse, B.E., & Weigle, D.S. (2001.) A preprandial rise in plasma ghrelin levels suggests a role in meal initiation in humans. *Diabetes, 50,* 1714-1719.

Cummings, D.E., Weigle, D.S., Frayo, R.S., Breen, P.A., Ma, M.K., Dellinger, E.P., & Purnell, J.Q. (2002.) Plasma ghrelin levels after diet-induced weight loss or gastric bypass surgery. *New England Journal of Medicine, 346,*1623-1630.

Cuneo, R., Espiner, E.A., Nicholls, M.G., Yandle, T.G., & Livesey, J.H. (1987.) Effects of physiological levels of atrial natriuretic peptide on hormone secretion: inhibition of angiotensin-induced aldosterone secretion and renin release in normal men. *Journal of Clinical Endocrinology and Metabolism, 65*, 765-771.

Cuneo, R.C., Salomon, F., Wiles, C.M., Hesp, R., & Sonksen, P.H. (1991a.)Growth hormone treatment in growth hormone-deficient adults. I. Effects on muscle mass and strength. *Journal of Applied Physiology, 70*, 688-694.

Cuneo, R.C., Salomon, F., Wiles, C.M., Hesp, R., & Sonksen, P.H. (1991b.) Growth hormone treatment in growth hormone-deficient adults. II. Effects on exercise performance. *Journal of Applied Physiology, 70*, 695-700.

Cunningham, D.J., Stolwijk, J.A.J., & Wenger, C.B. (1978.) Comparative thermoregulatory responses of resting men and women. *Journal of Applied Physiology, 45*, 908-915.

Cunningham, M.J., Clifton, D.K., & Steiner, R.A. (1999.). Leptin's actions on the reproductive axis: Perspectives and mechanisms. *Biology of Reproduction, 60*, 216-222.

Cunninghman, J., Segre, G.V., Slatopolsky, E., & Avioli, L.V. (1985.) Effect of heavy exercise on mineral metabolism and calcium regulating hormones in humans. *Calcified Tissue International, 37*, 598-601.

Cupps, T.R., & Fauci, A.S. (1982.) Corticosteroid-mediated immunoregulation in man. *Immunological Reviews, 65*, 133-155.

Cutler, R.G., & Mattson, M.P. (2001.) Sphingomyelin and ceramide as regulators of development and lifespan. *Mechanisms of Ageing and Development, 122*, 895-908.

Czech, M.P., & Corvera, S. (1999.) Signaling mechanisms that regulate glucose transport. *Journal of Biological Chemistry, 274*, 1865-1868.

Czeisler, C.A., & Klerman, E.B. (1999.) Circadian and sleep-dependent regulation of hormone release in humans. *Recent Progress in Hormone Research, 54*, 97-130.

Czeisler, C.A., Duffy, J.F., Shanahan, T.L., Brown, E.N., Mitchell, J.F., Rimmer, D.W., Ronda, J.M., Silva, E.J., Allan, J.S., Emens, J.S., Dijk, D.J., & Kronauer, R.E. (1999.) Stability, precision, and near-24-hour period of the human circadian pacemaker. *Science, 284*, 2177- 2181.

Czeisler, C.A., Kronauer, R.E., Allan, J.S., Duffy, J.F., Jewett, M.E., Brown, E.N., & Ronda JM. (1989.) Bright light induction of strong (type 0) resetting of the human circadian pacemaker. *Science, 244*, 1328-1333.

Czeisler, C.A., Weitzman, E.D., Moore-Ede, M.C., Zimmerman, J.C., & Knauer, R. S. (1980a.) Human sleep: its duration and organization depend on its circadian phase. *Science, 210*, 1264-1267.

Czeisler, C.A., Zimmerman, J.C., Ronda, J., Moore-Ede, M.C., & Weitzman, E.D. (1980b.) Timing of REM sleep is coupled to the circadian rhythm of body temperature in man. *Sleep, 2*, 329-346.

Dagenais, G.R., Tancredi, R.G., & Zierler, K.L. (1976.) Free fatty acid oxidation by forearm muscle at rest, and evidence for an intramuscular lipid pool in the human forearm. *Journal of Clinical Investigation, 58*, 421-431.

Dale, E., Gerlach, D.H., & Wilhite, A.L. (1979.) Menstrual dysfunction in distance runners. *Obstetrics and Gynecology, 54*, 47-53.

Dalsky, G.P., Stocke, K.S., Ehsain, A.A., Slatopolsky, E., Lee, W.C., & Birge, S.J. (1988.) Weight-bearing exercise training and lumbar bone mineral content in postmenopausal women. *Annals of Internal Medicine, 108*, 824-828.

Daly, M. de B. (1985.) Interactions between respiration and circulation. In *Handbook of Physiology. The Respiratory System II.* Bethesda, MD: American Physiological Society, pp. 529-594.

Damase-Michel, C., Giraud, P., Portolan, G., Montastruc, J.L., Montastruc, P., & Tran, M.A. (1994.) Effects of insulin on the release of neuropeptide Y, [Met5] enkephalin and catecholamines from dog adrenal medulla. *European Journal of Pharmacology, 258*, 277-279.

Dampney, R.A.L., & McAllen, R.M. (1988.) Differential control of sympathetic fibers supplying hind limb, skin, and muscle by subretrofacial neurones in the cat. *Journal of Physiology (London), 395*, 41-56.

Danielsen, A.G., Liu, F., Hosomi, Y., Shii, K., & Roth, R.A. (1995.) Activation of protein kinase Calpha inhibits signaling by members of the insulin receptor family. *Journal of Biological Chemistry, 270*, 21600-21605.

Danilenko, K.V., Putilov, A.A., Russkikh, G.S., Duffy, L.K. & Ebbesson, S.O. (1994.) Diurnal and seasonal variations of melatonin and serotonin in women with seasonal affective disorder. *Artic Medical Research, 53*, 137-145.

Dankowski, B.A., Imanaka-Yoshida, K., Sanger, J.M., & Sanger, J.W. (1992.) Costameres are sites of force transmission to the substratum in adult cardiomyocytes. *Journal of Cellular Biology, 118*, 1411-1420.

Darlington, T.K., Wager-Smith, K., Ceriani, M.F., Staknis, D., Gekakis, N., Steeves, T.D.L., Weitz, C.J., Takahashi, J.S., & Kay, S.A. (1998.) Closing the circadian loop: CLOCK-induced transcription of its own inhibitors per and tim. *Science, 280*, 1599-1603.

Date, Y., Ueta, Y., Yamashita, H., Yamaguchi, H., Matsukura, S., Kangawa, K., Sakurai, T., Yanagisawa, M., & Nakazato, M. (1999). Orexins, orexigenic hypothalamic peptides, interact with autonomic, neuroendocrine and neuroregulatory systems. *Proceedings of the National Academy of Sciences of the United States of America, 96*, 748-753.

Daum, G., Eisenmann-Tappe, I., Fries, H.-W., Troppmair, J., & Rapp, U.R. (1994.) The ins and outs of raf kinases. *Trends in Biochemical Science, 19*, 474-480.

Davies, C.T., & Sargeant, A.J. (1975.) Circadian variation in physiological responses to exercise on a stationary bicycle ergometer. *British Journal of Industrial Medicine, 32*, 110-114.

Davis, M.J., Wu, X., Nurkiewicz, T.R., Kawasaki, J., Davis, G.E., Hill, M.A., & Meininger, G.A. (2001.) Integrins and mechanotransduction of the vascular myogenic response. *American Journal of Physiology, 280*, H1427-H1433.

Davis, S. (1999.) Syndromes of hyperandrogenism in women. *Australian Family Physician, 28*, 447-451.

Dawson, D., & van den Heuvel, C.J. (1998.) Integrating the actions of melatonin on human physiology. *Annals of Medicine, 30*, 95-102.

Dawson-Hughes, B. (1996.) Calcium and vitamin D nutritional needs of elderly women. *Journal of Nutrition, 126* (Suppl. 4), 1165S-1167S.

de Belder, A.F., Radomski, M.W., Why, H.J., Richardson, P.J., Bucknall, C.A., Salas, E., Martin, J.F., & Moncada, S. (1993.) Nitric oxide synthase activities in human myocardium. *Lancet, 341*, 84-85.

de Boer, R.E., Brouwer, F., & Zaagsma, J. (1993.) The beta-adrenoceptors mediating relaxation of rat oesophageal muscularis mucosae are predominantly of the beta-3 but also of the beta-2 subtype. *British Journal of Pharmacology, 110*, 442-446.

Debry, G., Bleyer, R., & Reinberg, A. (1975.) Circadian, circa annual and other rhythms in spontaneous nutrient and calorie intake of healthy four-year olds. *Diabete et Metabolisme, 1*, 91-99.

Decker, M.L., Janes, D.M., Barclay, M.M., Harger, L., & Decker, R.S. (1997.) Regulation of adult cardiocyte growth: Effects of active and passive mechanical loading. *American Journal of Physiology, 272*, H2902-H2918.

Decombaz, J., Sartori,D., Arnaud, M.J., Thelin, A.L., Schurch, P., & Howald, H (1985.) Oxidation and metabolic effects of fructose or glucose ingested before exercise. *International Journal of Sports Medicine, 6*, 282-286.

De Cree, C. (1990.) The possible involvement of endogenous opioid peptides and catecholestrogens in provoking menstrual irregularities in women athletes. *International Journal of Sports Medicine, 11*, 329-348.

De Cree, C. (1998.) Sex steroid metabolism and menstrual irregularities in the exercising female. *Sports Medicine, 25*, 369-406.

De Cree, C., Ball, P., Seidlitz, B., Van Kranenberg, G., Guerten, P.. & Keiser, H.A. (1997a.) Effects of a training program on resting plasma 2-hydroxycatecholestrogen levels in eumenorrheic women. *Journal of Applied Physiology, 83*, 1551-1556.

De Cree, C., Ball, P., Seidlitz, B., Van Kranenberg, G., Guerten, P., & Keiser, H.A. (1997b.) Responses of catecholestrogen metabolism to acute graded exercise in normal menstruating women before and after training. *Journal of Clinical Endocrinology and Metabolism, 82*, 3342-3348.

De Cree, C., Van Kranenburg, G., Geurten, P., Fujimori, Y., & Keiser, H.A. (1997c.). 4-Hydroxy-catecholestrogen metabolism responses to exercise and training: possible implications for menstrual cycle irregularities and breast cancer. *Fertility and Sterility, 67*, 505-516.

DeFeo, P., Perriello, G., Torlone, E., Fanelli, C., Ventura, M.M., Santeusanio, F., Brunetti, P., Gerich, J.E., & Bolli, G.B. (1991a.) Evidence against important catecholamine compensation for absent glucagon counterregulation. *American Journal of Physiology, 260*, E203-E212.

DeFeo, P., Perriello, G., Torlone, E., Fanelli, C., Ventura, M.M., Santeusanio, F., Brunetti, P., Gerich, J.E., & Bolli, G.B. (1991b.) Contribution of adrenergic mechanisms to glucose counterregulation in humans. *American Journal of Physiology, 261*, E725-E736.

DeFronzo, R.A. (1997.) Insulin resistance: a multifaceted syndrome responsible for NIDDM, obesity, hypertension, dyslipidaemia and atherosclerosis. *Netherlands Journal of Medicine, 50*, 191-197.

DeFronzo, R.A., Ferrannini, E., Handler, R., Felig, P., & Wahren, J. (1983.) Regulation of splanchnic and peripheral glucose uptake by insulin and hyperglycemia in man. *Diabetes*, *32*,35-45.

Degen, L., Matzinger, D., Drewe, J., & Beglinger, C. (2001.) The effect of cholecystokinin in controlling appetite and food intake in humans. *Peptides*, *22*, 1265-1269.

De Glisezinski, I., Harant, I., Crampes, F., Trudeau, F., Felez, A., Cottet-Emard, J.M., Garrigues, M., & Riviere, D. (1998.) Effect of carbohydrate ingestion on adipose tissue lipolysis during long-lasting exercise in trained men. *Journal of Applied Physiology*, *84*,1627-1632.

de Jong, W., Petty, M.A., & Sitsen, J.M. (1983.) Role of opioid peptides in brain mechanisms regulating blood pressure. *Chest*, *83*, 306-308.

De Jonge, L., & Bray, G.A. (1997.) The thermic effect of food and obesity: a critical review. *Obesity Research*, *5*, 622-631.

Del Corral, P., Howley, E.T., Hartsell, M., Ashraf, M., & Younger, M.S. (1998.) Metabolic effects of low cortisol during exercise in humans. *Journal of Applied Physiology*, *84*, 939-947.

Delafontaine, P., Brink, M., & Du, J. (1996.) Angiotensin II modlulation of insulin-like growth factor I expression in the cardiovascular system. *Trends in Cardiovascular Medicine*, *6*, 187- 193.

Delday, M.I., & Maltin, C.A. (1997.) Clenbuterol increases the expression of myogenin but not myoD in immobilized rat muscles. *American Journal of Physiology*, *272*, E941-E944.

de Lecea, L., Kilduff, T.S., Peyron, C.,Gao, X., Foye, P.E., Danielson, P.E., et al. (1998.) The hypocretins: Hypothalamus-specific peptides with neuroexcitatory activity. *Proceedings of the National Academy of Sciences of the United States of America*, *95*, 322-327.

de Lignieres, B., Plas, J.N., Commandre, F., Morville, R., Viani, J.L., & Plas, F. (1976.) Secretion testiculaire d'androgenes apres effort physique prolongué chez l'homme. *Novelle Presse Medicale*, *5*, 2060-2064.

Delitala, G., Grossman, A., & Besser, G.M. (1983.) Opiate peptides control growth hormone through a cholinergic mechanism in man. *Clinical Endocrinology*, *18*, 401-405.

Delling, U., Tureckova, J., Lim, H.W., De Windt, L.J., Rotwein, P., & Molkentin, J.D. (2000.) A calcineurin-NFATc3-dependent pathway regulates skeletal muscle differentiation and slow myosin heavy-chain expression. *Molecular and Cellular Biology*, *20*, 6600-6611.

Delp, M.D. (1995.) Effects of exercise training on endothelium-dependent peripheral vascular responsiveness. *Medicine and Science in Sports and Exercise*, *27*, 1152-1157.

Delp, M.D. (1998.) Differential effects of training on the control of skeletal muscle perfusion. *Medicine and Science in Sports and Exercise*, *30*, 361-374.

Demers, L.M., Harrison, T.S., & Halbert, D.R. (1981.) Effect of prolonged exercise on prostaglandin levels. *Prostaglandin Medicine*, *6*, 413-418.

Dempsey, J.A. (1964.) Antropometrical observations on obese and non-obese young men undergoing a program of vigorous physical exercise. *Research Quarterly*, *35*, 275-287.

Deng, Y., & Kaufman, S. (1993.) The influence of reproductive hormones on ANF release by atria. *Life Sciences*, *53*, 689-696.

Derave, W., Ai, H., Ihlemann, J., Witters, L.A., Kristiansen, S., Richter, E.A., & Ploug, T. (2000.) Dissociation of AMP-activated protein kinase activation and glucose transport in contracting slow-twitch muscle. *Diabetes*, *49*, 1281-1287.

de Rijk, R., Michelson, D., Karp, B., Petrides, J., Galliven, E., Deuster, P. Paciotti, G., Gold, P.W., & Sternberg, E.M. (1997.) Exercise and circadian rhythm-induced variations in plasma cortisol differentially regulate interleukin-I beta (IL-1 beta), IL-6, and tumor necrosis factor-alpha (TNF alpha) production in humans: high sensitivity of TNF alpha and resistance of IL-6. *Journal of Clinical Endocrinology and Metabolism*, *82*, 2182-2191.

de Rijk, R., Petrides, J., Deuster, P., Gold, P.W., & Sternberg, E.M. (1996.) Changes in corticosteroid sensitivity of peripheral blood lymphocytes after strenuous exercise in humans. *Journal of Clinical Endocrinology and Metabolism*, *81*, 228-235.

Deschenes, M.R., Kraemer, W.J., Bush, J.A., Doughty, T.A., Kim, D., Mullen, K.M., & Ramsey, K. (1998.) Biorhythmic influences on functional capacity of human muscle and physiological responses. *Medicine and Science in Sports and Exercise*, *30*, 1399-1407.

Deschenes, M.R., Maresh, C.M., Armstrong, L.E., Couault, J., Kraemer, W.J., & Crivello, J.F. (1994.) Endurance and resistance exercise induce muscle fiber type specific responses in androgen binding capacity. *Journal of Steroid Biochemistry and Molecular Biology*, *50*, 175-179.

Deshaies, Y., Geloen, A., Paulin, A., Marette, A., & Bukowiecki, L.J. (1993.) Tissue-specific alterations in lipoprotein lipase activity in the rat after chronic infusion of isoproterenol. *Hormone and Metabolic Research*, *25*, 13-16.

De Souza, M.J., Luciano, A.A., Arce, J.C., Demers, L.M., & Loucks, A.B. (1994.) Clinical tests explain blunted cortisol responsiveness but not mild hypercortisolism in amenorrheic runners. *Journal of Applied Physiology*, *76*, 1302-1309.

De Souza, M.J., & Metzger, D.A. (1991.) Reproductive dysfunction in amenorrheic athletes and anorexic patients: a review. *Medicine and Science in Sports and Exercise*, *23*, 995-1007.

De Souza, M.J., & Miller, B.E. (1997.) The effect of endurance training on reproductive function in male runners: a 'volume threshold' hypothesis. *Sports Medicine*, *23*, 357-374.

De Souza, M.J., Maguire, M.S., Rubin, K.R., & Maresh, C.M. (1990.) Effects of menstrual phase and amenorrhea on exercise performance in runners. *Medicine and Science in Sports and Exercise*, *22*, 575-580.

De Souza, M.J., Maresh, C.M., Maguire, M.S., Kraemer, W.J., Flora-Ginter, G., & Goetz, K.L. (1989.) Menstrual status and plasma vasopressin, renin activity, and aldosterone exercise responses. *Journal of Applied Physiology*, *67*, 736-743.

Despres, J.P., Allard, C., Trembley, A., Talbot, J., & Bouchard, C. (1985.) Evidence for a regional component of body fatness in the association with serum lipids in men and women. *Metabolism*, *34*, 967.

Despres, J.P., Bouchard, C., Savard, R., Tremblay, A., Marcotte, M., & Theriault, G. (1984a.) The effect of a 20-week endurance training program on adipose tissue morphology and lipolysis in men and women. *Metabolism*, *33*, 235-239.

Despres, J.P., Bouchard, C., Savard, R., Tremblay, A., Marcotte, M., & Theriault, G. (1984b.). Level of physical fitness and adipocyte lipolysis in humans. *Journal of Applied Physiology*, *56*, 1157-1161.

Despres, J.P., Nadeau, A., Tremblay, A., Ferland, M., Moorjani, S., Lupien, P.J., & Theriault, G. (1989.) Role of deep abdominal fat in the association between regional adipose tissue distribution and glucose tolerance in obese women. *Diabetes*, *38*, 304-309.

Dessypris, A., Kuoppasalmi, K., & Adlercreutz, H. (1976.) Plasma cortisol, testosterone, androstenedione and luteinizing hormone (LH) in a non-competitive marathon run. *Journal of Steroid Biochemistry*, *7*, 33-37.

Deuster, P.A., Chrousos, G.P., Luger, A., DeBolt, J.E., Bernier, L.L., Trostman, U.H., Kyle, S.B., Montgomery, L.C., & Loriaux, D.L. (1989.) Hormonal and metabolic responses of untrained, moderately trained, and highly trained men to three exercise intensities. *Metabolism*, *38*, 141-148.

Deuster, P.A., Kyle, S.B., & Moser, P.B. (1986.) Nutritional survey of highly trained women runners. *Fertility and Sterility*, *46*, 636-643.

Devery, R., O'Donnell, L., & Tomkin, G.H. (1986.) Effect of catecholamines on the hepatic rate-limiting enzymes of cholesterol metabolism in normally fed and cholesterol fed rabbits. *Biochemica et Biophysica Acta*, *887*, 173-181.

Devery, R., & Tomkin, G.H. (1986.) The effect of insulin and catecholamines on the activities of 3-hydroxy-3-methyl glutaryl coenzyme A reductase and acyl-coenzyme A: cholesterol-o-acyltransferase in isolated rat hepatocytes. *Diabetologia*, *29*, 122-124.

Devlin, J.T. (1992.) Effects of exercise on insulin sensitivity in humans. *Diabetes Care*, *15*, 1690-1693.

Devlin, J.T., Hirschman, M. Horton, E.D., & Horton, E.S. (1987.) Enhanced peripheral and splanchnic insulin sensitivity in NIDDM men after single bout of exercise. *Diabetes*, *36*, 434-439.

Devlin, J.T., & Horton, E.S. (1985.) Effects of prior high-intensity exercise on glucose metabolism in normal and insulin-resistant men. *Diabetes*, *34*, 973-979.

DeVol, D.L., Rothwein, P., Sadow, J.L., Novakofski, J. & Bechtel, P.J. (1990.) Activation of insulin-like growth factor gene expression during work-induced skeletal muscle growth. *American Journal of Physiology*, *259*, E89-E95.

De Zegher, F., Devlieger, H., Eggermont, E., & Veldhuis, J.D. (1993.) Properties of growth hormone and prolactin hypersecretion by the human infant on the day of birth. *Journal of Clinical Endocrinology and Metabolism*, *76*, 1177-1181.

Diamond, M.P., Grainger, D., Diamond, M.C., Sherwin, R.S., & DeFronzo, R.A. (1998.) Effects of methyltestosterone on insulin secretion and sensitivity in women. *Journal of Clinical Endocrinology and Metabolism*, *83*, 4420-4425.

Dicker, A., Cannon, B., & Nedergaard, J. (1996.) Stimulation of nonshivering thermogenesis in the Syrian hamster by norepinephrine and beta-selective adrenergic agents: a phenomenon of refractoriness. *Comparative Biochemistry and Physiology C, 113*, 37-43.

Dickerman, R.D., Schaller, F., & McConathy, W.J. (1998.) Left ventricular wall thickening does occur in elite power athletes with or without anabolic steroid use. *Cardiology, 90*, 145-148.

Dickerman, R.D., Schaller, F., Zachariah, N.Y., & McConathy, W.J. (1997.) Left ventricular size and function in elite bodybuilders using anabolic steroids. *Clinical Journal of Sports Medicine, 7*, 90-93.

Dickson, P.R., & Vaccarino, F.J. (1994.) GRF-induced feeding: evidence for protein selectivity and opiate involvement. *Peptides, 15*, 1343-1352.

Diebert, D.C., & DeFronzo, R.A. (1980.) Epinephrine-induced insulin resistance in man. *Journal of Clinical Investigation, 65*, 717-721.

Dietz, J., & Schwartz, J. (1991.) Growth hormone alters lipolysis and hormone-sensitive lipase activity in 3T3-F442A adipocytes. *Metabolism, 40*, 800-806.

Dietz, N.M., Engelke, K.A., Samuel, T.T., Fix, R.T., & Joyner, M.J. (1997.) Evidence for nitric oxide-mediated sympathetic forearm vasodilatation in humans. *Journal of Physiology, 498*, 531-540.

Dijk, D.J., & Czeisler, C.A. (1995.) Contribution of the circadian pacemaker and the sleep homeostat to sleep propensity, sleep structure, electroencephalographic slow waves, and sleep spindle activity in humans. *Journal of Neuroscience, 15*, 3526-3538.

Dijk, D.J., Duffy, J.F., & Czeisler, C.A. (2000.) Contribution of circadian psychology and sleep homeostasis to age-related changes in human sleep. *Chronobiology International, 17*, 285-311.

Dimsdale, J.E., Hartley, H., Guiney, T., Ruskin, J.N., & Greenblatt, D. (1984.) Postexercise peril: plasma catecholamines and exercise. *Journal of the American Medical Association, 251*, 630-632.

Dimsdale, J.E., Herd, J.A., Hartley, L.H. (1983.) Epinephrine mediated increases in plasma cholesterol. *Psychosomatic Medicine, 45*, 227-232.

DiPietro, L. (1995.) Physical activity, body weight, and adiposity: an epidemiologic perspective. In J. O. Holloszy (Ed.), *Exercise and sport sciences reviews* (Vol. 23, pp. 275 – 303). Baltimore, MD: Williams and Wilkins.

Divertie, G.D., Jensen, M.D., & Miles, J.M. (1991.) Stimulation of lipolysis in humans by physiological hypercortisolemia. *Diabetes, 40*, 1228-1232.

Dix, D.J., & Eisenberg, B.R. (1991.) Myosin accumulation and myofibrillogenesis at the myotendon junction of stretched muscle fibers. *Journal of Cellular Biology, 111*,1885-1893.

Djurhuus, C.B., Gravholt, C.H., Nielsen, S., Mengel, A., Christiansen, J.S., Schmitz, O.E., & Moller. N. (2002.) Effects of cortisol on lipolysis and regional interstitial glycerol levels in humans. *American Journal of Physiology, 283*, E172-177.

Dobson, J.G., Jr., & Fenton, R.A. (1993.) Adenosine inhibition of beta-adrenergic induced responses in aged hearts. *American Journal of Physiology, 265*, H494-H503.

Dockray, G.J. (1992.) Transmission: Peptides. In G. Burnstock & C.H.V. Hoyle (Eds.), *Autonomic neuroeffector mechanisms* (pp. 409-464). Philadelphia: Harwood Academic Publishers.

Doe, R.P., Vennes, J.A., & Flink, E.B. (1960.) Diurnal variations of 17-hydroxy-corticosteroids, sodium, potassium, magnesium, and creatinine in normal subjects and in cases of treated adrenal insufficiency and Cushing's syndrome. *Journal of Clinical Endocrinology, 20*, 253-265.

Doi, H., Kugiyama, K., Ohgushi, M., Sugiyama, S., Matsumura, T., Ohta, Y., et al. (1999.) Membrane active lipids in remnant lipoproteins cause impairment of endothelium-dependent vasorelaxation. *Arteriosclerosis Thrombosis Vascular Biology, 19*, 1918-1924.

Donaldson, C.L., Hulley, S.B., Vogel, J.M., Hattner, R.S., Bayers, J.H., & McMillan, D.E. (1970.) Effect of prolonged bed rest on bone mineral. *Metabolism, 19*, 1071-1084.

Donaldson, H.H. (1919.) Quantitative studies on the growth of the skeleton of the albino rat. *American Journal of Anatomy, 26*, 237-314.

Donovan, C.M., Halter, J.B., & Bergman, R.N. (1991.) Importance of hepatic gluco-receptors in sympathoadrenal response to hypoglycemia. *Diabetes, 40*, 155-158.

Dora, E., Hines, K., Kunos, G., & McLaughlin, A.C. (1992.) Significance of an opiate mechanism in the adjustment of cerebrocortical oxygen consumption and blood flow during hypercapnic stress. *Brain Research, 573*, 293-298.

Dorflinger, L.J., & Schonbrunn, A. (1985.) Adenosine inhibits prolactin and growth hormone secretion in a clonal pituitary cell line. *Endocrinology, 111*, 2330-2338.

Doris, R.A., Thompson, G.E., Finley, E., Kilgour, E., Houslay, M.D., & Vernon, R.G. (1996.) Chronic effects of somatotropin treatment on response of subcutaneous adipose tissue lipolysis to acutely acting factors in vivo and in vitro. *Journal of Animal Science, 74*, 562-568.

Doris, R.A., Vernon, R.G., Houslay, M.D., & Kilgour, E. (1994.) Growth hormone decreases the response to anti-lipolytic agonists and decreases the levels of Gi2 in rat adipocytes. *Biochemical Journal, 297*, 41-45.

Dorlochter, M., Astrow, S.H., & Herrera, A.A. (1994.) Effects of testosterone on sexually dimorphic frog muscle: repeated in vivo observations and androgen receptor distribution. *Journal of Neurobiology, 25*, 897-916.

Dorn, G.W., II, & Brown, J.H. (1999.) Gq signaling in cardiac adaptation and maladaptation. *Trends in Cardiovascular Medicine, 9*, 26-34.

Dornonville de la Cour, C., Bjorkvist, M., Sandvik, A.K., Bakke, I., Zhao, C.M., Chen, D., & Hakanson, R. (2001.) A-like cells in the rat stomach contain ghrelin and do not operate under gastrin control. *Regulatory Peptides, 99*, 141-150.

Downes, M., Carozzi, A.J., & Muscat, G.E. (1995.) Constitutive expression of the orphan receptor, Rev-erbA alpha, inhibits muscle differentiation and abrogates the expression of the myoD gene family. Molecular Endocrinology, 9, 1666-1678.

Dreon, D.M., Frey-Hewitt, B., Ellsworth, N.,Williams, P.T., Terry, R.B., & Wood, P.D. (1988.) Dietary fat: carbohydrate ratio and obesity in middle-aged men. *American Journal of Clinical Nutrition, 47*, 995-1000.

Drewnowski, A., Kurth, C., Holden-Wiltse, L., & Saari, J. (1992.) Food preferences in human obesity: carbohydrates vs. fats. *Appetite, 18*, 207.

Drinkwater, B.L. (1984.) Women and exercise: physiological aspects. *Exercise and Sports Science Reviews, 12*, 21-51.

Driver, H.S., Meintjes, A.F., Rogers, G.G., & Shapiro, C.M. (1988.) Submaximal exercise effects on sleep patterns in young women before and after an aerobic training programme. *Acta Physiologica Scandinavica. Supplementum, 574*, 8-13.

Driver, H.S., Rogers, G.G., Mitchell, D., Borrow, S.J., Allen, M., Luus, H.G., & Shapiro, C.M. (1994.) Prolonged endurance exercise and sleep disruption. *Medicine and Science in Sports and Exercise, 26*, 903-907.

Droste, C. (1992.) Transient hypoalgesia under physical exercise—relation to silent ischemia and implications for cardiac rehabilitation. *Annals of Academic Medicine in Singapore, 21*, 23-33.

Droste, C., Greenlee, M.W., Schreck, M., & Roskamm, H. (1991.) Experimental pain thresholds and plasma beta-endorphin levels during exercise. *Medicine and Science in Sports and Exercise, 23*, 334-342.

Drummond, S.E., Crombie, N.E., Cursiter, M.C., & Kirk, T.R. (1998.) Evidence that eating frequency is inversely related to body weight status in male, but not female, non-obese adults reporting valid dietary intakes. *International Journal of Obesity and Related Metabolic Disorders, 22*, 105-112.

Duan, C., & Winder, W.W. (1992.) Nerve stimulation decreases malonyl-CoA in skeletal muscle. Journal of Applied Physiology, 72, 901-904.

Duan, C., & Winder, W.W. (1993.) Control of malonyl-CoA by glucose and insulin in perfused skeletal muscle. *Journal of Applied Physiology, 74*, 2543-2547.

Duclos, M., Corcuff, J.B., Rashedi, M., Fuguere, V., & Manier, G. (1997.) Trained vs untrained men: different immediate postexercise responses of pituitary adrenal axis. A preliminary study. *European Journal of Applied Physiology and Occupational Physiology, 75*, 343-350.

Duddleston, A.K., & Benniou, M. (1970.) Effect of diet and/or exercise on obese and non-obese young men undergoing a program of vigorous physical exercise. *Journal of the American Dietetic Association, 56*, 126-129.

Dufaux, B., Order, U., & Liesen, H. (1991.) Effect of a short maximal physical exercise on coagulation, fibrinolysis, and complement system. *International Journal of Sports Medicine, 12* (Suppl. 1), S38-S42.

Duffy, J.F., Dijk, D.J., Hall, E.F., & Czeisler, C.A. (1999.) Relationship of endogenous circadian melatonin and temperature rhythms to self- reported preference for morning or evening activity in young and older people. *Journal of Investigative Medicine, 47*, 141-150.

Duffy, J.F., Zeitzer, J.M., Rimmer, D.W., Klerman, E.B., Dijk, D.J., & Czeisler, C.A. (2002.) Peak of circadian melatonin rhythm occurs later within the sleep of older subjects. *American Journal of Physiology, 282*, E297-E303.

Duncker, D.J., Van Zon, N.S., Pavek, T.J., Herrlinger, S.K., & Bache, R.J. (1995.) Endogenous adenosine mediates coronary vasodilatation during exercise after K(ATP)⁺ channel blockade. *Journal of Clinical Investigation, 95,* 285-295.

Dunn, F.L., Brennan, T.J., Nelson, A.E., & Robertson, G.L. (1973.) The role of blood osmolality and volume in regulating vasopressin secretion in the rat. *Journal of Clinical Investigation, 52,* 3212-3219.

Dunn, S.E., Burns, J.L., & Michel, R.N. (1999.) Calcineurin is required for skeletal muscle hypertrophy. *Journal of Biological Chemistry, 274,* 21908-21912.

Dunning, B.E., & Taborsky, G.J., Jr. (1989.) Galanin release during pancreatic nerve stimulation is sufficient to influence pancreatic function. *American Journal of Physiology, 256,* E191-E198.

Dunning, B.E., & Taborsky, G.J., Jr. (1991.) Neural control of islet function by norepinephrine and sympathetic neuropeptides. *Advances in Experimental Medicine and Biology, 291,* 107-127.

Durak, E.P., Jovanovic-Peterson, L., & Peterson, C.M. (1990.) Comparative evaluation of uterine response to exercise on five aerobic machines. *American Journal of Obstetrics and Gynecology, 162,* 754-756.

Durnam, D.M., & Palmiter, R.B. (1983.) A practical approach for quantitating specific mRNAs by solution hybridization. *Analytical Biochemistry 131,* 385-393.

Durnin, J.V. (1991.) Energy requirements of pregnancy. *Acta Pediatrica Scandinavica, Supplementum, 373,* 33-42.

Durnin, J.V., McKillop, F.M., Grant, S., & Fitzgerald, G.(1987.) Energy requirements of pregnancy in Scotland. *Lancet, 2,* 897-900.

Durnin, J.V. & Passmore R. (1967.) *Energy, work, and leisure.* London: Heineman.

d'Uscio, L.V., Shaw, S., Barton, M., & Luscher, T.F. (1998.) Losartan but not verapamil inhibits angiotensin II-induced tissue endothelin-1 increase: role of blood pressure and endothelial function. *Hypertension, 31,* 1305-1310.

Duvekot, J.J., & Peeters, L.L. (1994.) Maternal cardiovascular hemodynamic adaptation to pregnancy. *Obstetrical and Gynecological Survey, 49* (12 Suppl.), S1-S14.

Dyck, D.J., Peters, S.J., Wendling, P.S., & Spriet, L.L. (1996b.) Effect of high FFA on glycogenolysis in oxidative rat hindlimb muscles during twitch stimulation. *American Journal of Physiology, 270,* R766-R776.

Dyck, D.J., Peters, S.J., Wendling, P.S., Chelsey, A., Hultman, E., & Spriet, L.L. (1996a.) Regulation of muscle glycogen phosporylase activity during intense aerobic cycling with elevated FFA. *American Journal of Physiology, 270,* E116-E125.

Dyck, D.J., Putman, C.T., Heigenhauser, G.J., Hultman, E., & Spriet, L.L. (1993.). Regulation of fat-carbohydrate interaction in skeletal muscle during intense aerobic cycling. *American Journal of Physiology, 265,* E852-E859.

Dyment, P.G. (1982.) Drug misuse by adolescent athletes. *Pediatric Clinics of North America, 29,* 1363-1368.

Eakman, G., Dallas, J., Ponder, S., & Keenan, B. (1996.) The effects of testosterone and dihydrotestosterone on hypothalamic regulation of growth hormone secretion. *Journal of Clinical Endocrinology and Metabolism, 81,* 1217-1223.

Eastell, R., Calvo, M.S., Burritt, M.F., Offord, K.P., Russell, R.G., & Riggs, B.L. (1992.) Abnormalities in circadian patterns of bone resorption and renal calcium conservation in type I osteoporosis. *Journal of Clinical Endocrinology and Metabolism, 74,* 487-494.

Eastman, C.I., & Martin, S.K. (1999.) How to use light and dark to produce circadian adaptation to night shift work. *Annals of Medicine, 31,* 87-98.

Eastman, C.I., Boulos, Z., Terman, M., Campbell, S.S., Dijk, D.J., & Lewy, A.J. (1995a.). Light treatment for sleep disorders: consensus report. VI. Shift work. *Journal of Biological Rhythms, 10,* 157-164.

Eastman, C.I., Hoese, E.K., Youngstedt, S.D., & Liu, L. (1995b.) Phase-shifting human circadian rhythms with exercise during the night shift. *Physiology and Behavior, 58,* 1287-1291.

Eastman, C.I., Young, M.A., Fogg, L.F., Liu, L., & Meaden, P.M. (1998.) Bright light treatment of winter depression: A placebo-controlled trial. *Archives of General Psychiatry, 55,* 883-889.

Eaton, R.P., Bierman, M., & Steinberg, D. (1969.) Kinetic studies of plasma free fatty acid and triglyceride metabolism in man. *Journal of Clinical Investigation, 48,* 1560-1579.

Ebadi, M., & Govitrapong, P. (1986.) Neural pathways and neurotransmitters affecting melatonin synthesis. *Journal of Neural Transmission* (Suppl.), *21,* 125-155.

Eber, B., & Schumacher, N. (1993.) Fibrinogen: its role in the hemostatic regulation in atherosclerosis. *Seminars in Thrombosis and Hemostasis, 19,* 104-107.

Eberstein, A., & Goodgold, J. (1968.) Slow and fast-twitch fibres in human skeletal muscle. *American Journal of Physiology, 215,* 535-541.

Ebert, T.J., Cowley, A.W., Jr., & Skelton, M. (1986.) Vasopressin reduces cardiac function and augments cardiopulmonary baroreflex resistance increases in man. *Journal of Clinical Investigation, 77,* 1136-1142.

Eckardt, K.-U., & Kurtz, A. (1992.) The biological role, site, and regulation of erythropoietin production. *Advances in Nephrology, Necker Hospital, 21,* 203-233.

Eckardt, K.U., LeHir, M., Tan, C.C., Ratcliffe, P.J., Kaissling, B., & Kurtz, A. (1992.) Renal innervation plays no role in oxygen-dependent control of erythropoietin mRNA levels. *American Journal of Physiology, 263,* F925-F930.

Eckberg, D.L. (1979.) Carotid baroreflex function in young men with borderline blood pressure elevation. *Circulation, 59,* 632-636.

Eddy, D.O., Sparks, K.L., & Adelizi, D.A. (1977.) The effects of continuous and interval training in women and men. *European Journal of Applied Physiology, 37,* 83-92.

Edgar, D.M., & Dement, W.C. (1991.) Regularly scheduled voluntary exercise synchronizes the mouse circadian clock. *American Journal of Physiology, 261,* R928-R933.

Edgerton. V.R., Smith, J.L., & Simpson, D.R. (1975.). Muscle fiber type populations of human leg muscles. *Histochemistry Journal, 7,* 259-266.

Edholm, O.G. (1977.) Energy balance in man-studies carried out by the Division of Human Physiology, NIMR. *Journal of Human Nutrition, 31,* 413-431.

Edholm, O.G., Adam, J.M., Healy, M.J., Wolff, H.S., Goldsmith, R., & Best, T.W. (1970.) Food intake and energy expenditure of army recruits. *British Journal of Nutrition, 24,* 1091-1107.

Edinger, J.D., Morey, M.C., Sullivan, R.J., Higginbotham, M.B., Marsh, G.R., Dailey, D.S., & McCall, W.V. (1993.) Aerobic fitness, acute exercise and sleep in older men. *Sleep, 16,* 351-359.

Edwards, P.A. (1975.) The influence of catecholamines and cyclic AMP on 3-hydroxy-3 methylglutaryl coenzyme A reductase activity and lipid biosynthesis in isolated rat hepatocytes. *Archives of Biochemistry and Biophysics, 170,* 188-203.

Edwards, A.V., & Jones, C.T. (1993.) Autonomic control of adrenal function. *Journal of Anatomy, 183,* 291-307.

Edwards, C.M., Abusnana, S., Sunter, D., Murphy, K.G., Ghatei, M.A., & Bloom, S.R. (1999.) The effect of orexins on food intake: comparison with neuropeptide Y, melanin-concentrating hormone and galanin. *Journal of Endocrinology, 160,* R7-R12.

Edwards, G.L., & Ritter, R.C. (1982.) Area postrema lesions increase drinking to angiotensin and extracellular dehydration. *Physiology and Behavior, 29,* 943-947.

Egan, J.J., Greenberg, A.S., Chang, M.K., Wek, S.A., Moos, M.C., Jr., & Londos, C. (1992.) Mechanism of hormone-stimulated lipolysis in adipocytes: translocation of hormone-sensitive lipase to the lipid storage droplet. *Proceedings of the National Academy of Sciences of the United States of America, 89,* 8537-8541.

Egawa, M., Yoshimitsu, H., & Bray, G.A. (1989.) Lateral hypothalamic injection of 2-deoxy-D-glucose suppresses sympathetic activity. *American Journal of Physiology, 257,* R1386-1392.

Ehrenreich, H., Halaris, A., Ruether, E., Hufner, M., Funke, M., & Kunert, H.J. (1999.) Psychoendocrine sequelae of chronic testosterone deficiency. *Journal of Psychiatric Research, 33,* 379-387.

Ehrhart-Bornstein, M., Bornstein, S.R., Guse-Behling, H., Stromeyer, H.G., & Rasmussen, T.N. (1994.) Sympathoadrenal regulation of adrenal androstenedione release. *Neuroendocrinology, 59,* 406-412.

Eichner, E.R., & Calabrese, L.H. (1994.) Immunology and exercise. Physiology, pathophysiology, and implications for HIV infestion. *Sports Medicine, 78,* 377-388.

Eiden, L.E., Iacangelo, A., Hsu, C.M., Hotchkiss, A.J., Bader, M.F., & Aunis, D. (1987.) Chromogranin A synthesis and secretion in chromaffin cells. *Journal of Neurochemistry, 49,* 65-74.

Eik-Nes, K.B. (1964.) On the relationship between testicular blood flow and secretion of testosterone in anesthetized dogs stimulated with human chorionic gonadotrophin. *Canadian Journal of Physiology and Pharmacology, 42*, 671-677.

Eik-Nes, K.B. (1969.) An effect of isoproterenol on rates of synthesis and secretion of testosterone. *American Journal of Physiology, 217*, 1764-1770.

Einhorn, T. A. (1996.) The bone organ system: form and function. In R. Marcus, D. Feldman, and J. Kelsey (Eds.), *Osteoporosis* (pp. 3-22). New York: Academic Press.

Ekelund, U. (1996.) Effects of angiotensin-converting enzyme inhibition on arterial, venous and capillary functions in cat skeletal muscle in vivo. *Acta Physiologica Scandinavica, 158*, 29-37.

Ekholm, E.M., & Erkkola, R.U. (1996.) Autonomic cardiovascular control in pregnancy. *European Journal of Obstetrics, Gynecology, and Reproductive Biology, 64*, 29-36.

Eldridge, F.L., Millhorn, D.E., Kiley, J.P., & Waldrop, T.G. (1985.) Stimulation by central command of locomotion, respiration, and circulation during exercise. *Respiratory Physiology, 59*, 313-337.

el-Hajj Fuleihan, G., Klerman, E.B., Brown, E.N., Choe, Y., Brown, E.M., & Czeisler, C.A. (1997.) The parathyroid hormone circadian rhythm is truly endogenous—a general clinical research study. *Journal of Clinical Endocrinology and Metabolism, 82*, 281-286.

Eliakim, A., Brasel, J.A., Barstow, T.J., Mohan, S., & Cooper, D.M. (1998.) Peak oxygen uptake, muscle volume, and the growth hormone-insulin-like growth factor-I axis in adolescent males. *Medicine and Science in Sports and Exercise, 30*, 512-517.

Elias, C.F., Lee, C., Kelly, J., Aschkenasi, C., Ahima, R.S., Couceyro, P., Kuhar, M.J., Saper, C.B., & Elmquist, J.K. (1998a.) Leptin activates hypothalamic CART neurons projecting to the spinal cord. *Neuron, 21*, 1375-1385.

Elias, C.F., Saper, C.B., Maratos-Flier, E., et al. (1998b.) Chemically defined projections linking the mediobasal hypothalamus and the lateral hypothalamic area. *Journal of Comparative Neurology, 402*, 442-459.

Ellingboe, J., Veldhuis, J.D., & Mendelson, J.H. (1982.) Effect of endogenous opioid blockade on the amplitude and frequency of pulsatile luteinizing hormone secretion in man. *Journal of Clinical Endocrinology and Metabolism, 54*, 854-857.

Ellis, G.S., Lanza-Jacoby, S., Gow, A., & Kendrick, Z.V. (1994.) Effects of estradiol on lipoprotein lipase activity and lipid availability in exercised male rats. *Journal of Applied Physiology, 77*, 209-215.

Ellison, P.T. (1982.) Skeletal growth, fatness and menarcheal age: a comparison of hypotheses. *Human Biology, 54*, 269-282.

Elmquist, J.K., Ahima, R.S., Elias, C.F., Flier, J.S., & Saper, C.B. (1998a.) Leptin activates distinct projections from the dorsomedial and ventromedial hypothalamic nuclei. *Proceedings of the National Academy of Sciences of the United States of America, 95*, 741-746.

Elmquist, J.K., Maratos-Flier, E., Saper, C.B., & Flier, J.S. (1998b.) Unraveling the central nervous system pathways underlying responses to leptin. *Nature Neuroscience, 1*, 445-450.

Elmquist, J.K., Elias, C.F., & Saper, C.B. (1999.) From lesions to leptin: hypothalamic control of food intake and body weight. *Neuron, 22*, 221-232.

El-Sayed, M.S. (1996.) Effects of exercise on blood coagulation, fibrinolysis and platelet aggregation. *Sports Medicine, 22*, 282-298.

Emond, M., Schwartz, G.J., Ladenheim, E.E., & Moran, T.H. (1999.) Central leptin modulates behavioral and neural responsivity to CCK. *American Journal of Physiology, 276*, R1545-1549.

Emorine, L.J., Marullo, S., Briend-Sutren, M.-M., Patey, G., Tate, K., Delavier-Klutchko, C., & Strosberg, A.D. (1989.) Molecular characterization of the human beta 3-adrenergic receptor. *Science, 245*, 1118-1121.

Engstrom G., Hedblad, B., & Janzon, L. (1999.) Hypertensive men who exercise regularly have lower rate of cardiovascular mortality. *Journal of Hypertension, 17*, 737-742.

Enoksson, S., Degerman, E., Hagstron-Toft, E., Large, V., & Arner, P. (1998.) Various phosphodiesterase subtypes mediate the in vivo antilipolytic effect of insulin on adipose tissue and skeletal muscle in man. *Diabetologia, 41*, 560-568.

Epstein, L. H., & Wing, R. R. (1980.). Aerobic exercise and weight. *Addictive Behaviors, 5*, 371 – 388.

Erenberg, A., Omori, K., Menkes, J.H., Oh, W., & Fisher, D.A. (1974.) Growth and development of the thyroidectomized ovine fetus. *Pediatric Research, 8*, 783-789.

Eriksson, H., Ridderstrale, M., Degerman, E., Ekholm, D., Smith, C.J., Manganiello, V.C., Belfrage, P., & Tornqvist, H. (1995.) Evidence for the key role of the adipocyte cGMP-inhibited cAMP phosphodiesterase in the antilipolytic action of insulin. *Biochimica et Biophysica Acta, 1266*, 101-107.

Eriksson, L. (1989.) Growth hormone in human pregnancy. Maternal 24-hour serum profiles and experimental effects of continuous GH secretion. *Acta Obstetrica Gynecologica Scandinavica, 144*, 1-38.

Erlanson-Albertsson, C., & York, D. (1997.) Enterostatin: a peptide regulating fat intake. *Obesity Research, 5*, 360-372.

Ernst, E. (1993.) Regular exercise reduces fibrinogen levels: a review of longitudinal studies. *British Journal of Sports Medicine, 27*, 175-176.

Ernst, M., & Rodan, G.A. (1991.) Estradiol regulation of insulin-like growth factor-I expression in osteoblastic cells: evidence for transcriptional control. *Molecular Endocrinology, 5*, 1081-1089.

Ernst, M., Parker, M.G., & Rodan, G.A. (1991.) Functional estrogen receptors in osteoblastic cells demonstrated by transfection with a reporter gene containing an estrogen response element. *Molecular Endocrinology, 5*, 1597-1606.

Esler, M., Jennings, G., Lambert, G., Meredith, I., Horne, M., & Eisenhofer, G. (1990.) Overflow of catecholamine neurotransmitters to the circulation: source, fate, and functions. *Physiological Reviews, 70*, 963-985.

Esler, M., Julius S., Randall O., DeQuattro V., & Zweifler A. (1976.) High renin essential hypertension: adrenergic cardiovascular correlates. *Clinical Science and Molecular Medicine, 51*, 181S-184S.

Esler, M., Julius, S., Zweifler, A., Randall, O., Halburg, E., Gardiner, H., & De Quiattro, V. (1977.) Mild high-renin essential hypertension. Neurogenic human hypertension? *New England Journal of Medicine, 296*, 405-411.

Espersen, G.T., Elbaek, A., Ernst, E., Toft, E., Kaalund, S., Jersild, C., & Grunnet, N. (1990.) Effect of physical exercise on cytokines and lymphocyte subpopulations in human peripheral blood. *APMIS, 98*, 395-400.

Essen, B. (1977.) Intramuscular substrate utilization during prolonged exercise. *Annals of the New York Academy of Sciences, 301*, 30-44.

Essen, B. (1978.) Studies of the regulation of metabolism in human skeletal muscle using intermittent exercise as an experimental model. *Acta Physiologica Scandinavica, Supplementum, 454*, 1-32.

Essen, B., Jansson, E., Henriksson, J., Taylor, A.W., & Saltin, B. (1975.) Metabolic characteristics of fibre types in human skeletal muscle. *Acta Physiologica Scandinavica, 95*, 153-165.

Evans, D.J., Hoffman, R.G., Kalkhoff, R.K., & Kissebah, A.H. (1983.) Relationship of androgenic activity to body fat topography, fat cell morphology, metabolic aberrations in premenopausal women. *Journal of Clinical Endocrinology and Metabolism, 57*, 304.

Evans, D.J., Hoffman, R.G., Kalkoff, R.K., & Kissebah, A.H. (1984a.) Relationship of body fat topography to insulin sensitivity and metabolic profile in premenopausal women. *Metabolism, 36*, 68-75.

Evans, D.J., Murray, R., & Kissebah, A.H. (1984b.) Relationship between skeletal muscle insulin resistance, insulin-mediated glucose disposal, and insulin binding: effects of obesity and body fat topography. *Journal of Clinical Investigation, 74*, 1515-1525.

Evans, W.J., & Cannon, J.G. (1991.) The metabolic effects of exercise-induced muscle damage. *Exercise and Sports Science Reviews, 19*, 99-125.

Exton, J.H. (1988.) Role of phosphoinositides in the regulation of liver function. *Hepatology, 8*, 152-166.

Exton, J.H. (1990a.) Hormonal regulation of phosphatidylcholine breakdown. *Advances in Second Messenger and Phosphoprotein Research, 24*, 152-157.

Exton, J.H. (1990b.) Signaling through phosphatidylcholine breakdown. *Journal of Biological Chemistry, 265*, 1-4.

Exton, J.H., Miller, T.B., Harper, S.C., & Park, C.R. (1976.) Carbohydrate metabolism in perfused livers of adrenalectomized and steroid-replaced rats. *American Journal of Physiology, 230*, 163-170.

Fabiato, A. (1985.) A calcium-induced release of calcium from sarcoplasmic reticulum. *Journal of General Physiology, 85*, 189-320.

Fagard, R.H. (1993.) Physical fitness and blood pressure. *Journal of Hypertension, 11*(Suppl. 5), S47-S52.

Fagius, J., Niklasson, F., & Berne, C. (1986.) Sympathetic outflow in human muscle nerves increases during hypoglycemia. *Diabetes, 35,* 1124-1129.

Fain, J.N. (1980.) Hormonal regulation of lipid mobilization from adipose tissue. In G. Litwack (Ed.), *Biochemical Actions of Hormones* (pp. 120-204). New York: Academic Press.

Fain, J.N. ,and Garcia-Sainz, A. (1983.) Adrenergic regulation of adipocyte metabolism. *Journal of Lipid Research, 24,* 945-966.

Fain, J.N., & Saperstein, R. (1970.) The involvement of RNA synthesis and cyclic AMP in the activation of fat cell lipolysis by growth hormone and glucocorticoids. *Hormone and Metabolic Research, 2,* 20-27.

Falduto, M.T., Czerwinski, S.M., & Hickson, R.C. (1990.) Glucocorticoid-induced muscle atrophy prevention by exercise in fast-twitch fibers. *Journal of Applied Physiology, 69,* 1058-1062.

Falkenbach, A., Sedlmeyer, A., & Unkelbach, U. (1998.) UVB radiation and its role in the treatment of postmenopausal women with osteoporosis. *International Journal of Biometeorology, 41,*128-131.

Fall, K., Hindmarsh, P., Dennison, E., Kellingray, S., Barker, D., & Cooper, C. (1998.) Programming of growth hormone secretion and bone mineral density in elderly men: A hypothesis. *Journal of Clinical Endocrinology and Metabolism, 83,* 135-139.

Fallo, F. (1993.) Renin-angiotensin-aldosterone system and physical exercise. *Journal of Sports Medicine and Physical Fitness, 33,* 306-312.

Fan, W., Boston, B.A., Kesterson, R.A., Hruby, V.J., & Cone, R.D. (1997.) Role of melanocortinergic neurons in feeding and the agouti obesity syndrome. *Nature, 385,* 165-168.

FAO, Food and Agriculture Organization of the United Nations (1957.) *Calorie requirements.* Washington, DC: U.S. Government Printing Office.

Farach-Carson, M.C., & Ridall, A.L. (1998.) Dual 1,25-dihydroxyvitamin D3 signal response pathways in osteoblasts: cross-talk between genomic and membrane-initiated pathways. *American Journal of Kidney Diseases, 31,* 729-742.

Faria, I.E., & Drummond, B.J. (1982.) Circadian changes in resting heart rate and body temperature, maximal oxygen consumption and perceived exertion. *Ergonomics, 25,* 381-386.

Farrell, P.A. (1985.) Exercise and endorphins-male responses. *Medicine and Science in Sports and Exercise, 17,* 89-93.

Farrell, P.A., Gustafson, A.B., Garthwaite, J.L., Kalkhoff, R.K., Cowley, A.W., Jr., & Morgan, W. P. (1986.) Influence of endogenous opioids on the response of selected hormones to exercise in humans. *Journal of Applied Physiology, 61,* 1051-1057.

Farrell, P.A., Gustafson, A.B., Morgan, W.P., & Pert, C.B. (1987.) Enkephalins,

catecholamines, and psychological mood alterations: effects of prolonged exercise. *Medicine and Science in Sports and Exercise, 19,* 347-353.

Fauvel, J.P., Hadj-aissa, A., Laville, M., Daoud, S., Labeeuw, M. Pozet, N., & Zech, P. (1991.) Stress-induced renal function alterations in normotensives. *American Journal of Hypertension, 4,* 77A.

Feifel, D., & Vaccarino, F.J. (1994.) Growth hormone-regulatory peptides (GHRH and somatostatin) and feeding: a model for the integration of central and peripheral function. *Neuroscience and Biobehavioral Reviews, 18,* 421-433.

Felig, P. (1973.) The glucose-alanine cycle. *Metabolism, 22,* 179-207.

Felig, P., & Wahren, J. (1971.) Influence of endogenous insulin secretion on splanchnic glucose and amino acid metabolism. *Journal of Clinical Investigation, 50,* 1702-1711.

Felig, P., Wahren, J., Sherwin, R.S., & Palaiologos, G. (1977.) Protein and amino acid metabolism in diabetes mellitus. *Archives of Internal Medicine, 137,* 507-513.

Fell, R.D., McLane, J.A., Winder, W.W., & Holloszy, J.O. (1980.) Preferential resynthesis of muscle glycogen in fasting rats after exhausting exercise. *American Journal of Physiology, 238,* R328-R332.

Felsing N.E., Brasel J.A., & Cooper D.M. (1992.) Effect of low and high intensity exercise on circulating growth hormone in men. *Journal of Clinical Endocrinology and Metabolism, 75,* 157-162.

Feoktisov, I.A., Paul, S., Hollister, A.S., Robertson, D., & Biaggioni, I. (1992.) Role of cyclic AMP in adenosine inhibition of intracellular calcium rise in human platelets: comparison of adenosine effects on thrombin- and epinephrine-induced platelet stimulation. *American Journal of Hypertension, 5,* 147S-153S.

Ferin, M., Van Vugt, D., & Wardlaw, S. (1984.) The hypothalamic control of the menstrual cycle and the role of endogenous opioid peptides. *Recent Progress in Hormone Research, 40,* 441-485.

Fernandez-Pastor, V.J., Ruiz, M., Diego-Acosta, A.M., Avila, C., Garcia, J.C., Perez, F., Guirado, F., & Noguer, N. (1999.) Metabolic and hormonal changes during aerobic exercise in distance runners. *Journal of Physiology and Biochemistry, 55,* 7-16.

Ferrannini, E., Barrett, E.J., Bevilacqua, S., & DeFronzo, R.A. (1983.) Effect of fatty acids on glucose production and utilization in man. *Journal of Clinical Investigation, 72,* 1737-1747.

Ferry, R.J., Jr., Cerri, R.W., & Cohen, P. (1999.) Insulin-like growth factor binding proteins: New proteins, new functions. *Hormone Research, 51,* 53-67.

Fiatarone, M.A., Morley, J.E., Bloom, E.T., Benton, D., Makinodan, T., & Solomon, G.F. (1988.) Endogenous opioids and the exercise-induced augmentation of natural killer cell activity. *Journal of Laboratory and Clinical Endocrinology, 112,* 544-552.

Fichter, M.M., & Pirke, K.M. (1986.) Effect of experimental and pathological weight loss upon the hypothalamo-pituitary-adrenal axis. *Psychoneuroendocrinology, 11,* 295-305.

Field, C.J., Goureon, R., & Marliss, E. (1991.) Circulating mononuclear cell numbers and function during intense exercise and recovery. *Journal of Applied Physiology, 71,* 1089-1097.

Fielding, R.A., Manfredi, T.J., Ding, W., Fiatarone, M.A., Evans, W.J., & Cannon, J.G. (1993.) Acute phase response in exercise. III. Neutrophil and IL-1 beta accumulation in skeletal muscle. *American Journal of Physiology, 265,* R166-R172.

Fife, S.K., Brogan, R.S., Giustina, A., & Wehrenberg, W.B. (1996.) Immunocytochemical and molecular analysis of the effects of glucocorticoid-treatment on the hypothalamic-somatotropic axis in the rat. *Neuroendocrinology, 64,* 131-138.

Finidori J. (2000.) Regulators of growth hormone signaling. Vitamins and Hormones, 59,71-97.

Fink, G.D., & Fisher, J.W. (1976.) Erythropoietin production after renal denervation or beta-adrenergic blockade. *American Journal of Physiology, 230,* 508-513.

Fiorica, V., Burr, M.J., & Moses, R. (1968.) Contribution of activity to the circadian rhythm in excretion of magnesium and calcium. *Aerospace Medicine,* 714-717.

Fischer, J.A., Blum, J.W., Born, W., Dambacher, M.A., & Dempster, D.W. (1982.) Regulation of parathyroid hormone secretion in vitro and in vivo. *Calcified Tissue International, 34,* 313-316.

Fisher, L.A., Rivier, J., Rivier, C., Spiess, J., Vale, W., & Brown, M.V. (1982.) Corticotropin releasing factor (CRF): central effects on mean arterial pressure and heart rate in rats. *Endocrinology, 110,* 2222-2224.

Fitzsimmons, J.T. (1998.) Angiotensin, thirst, and sodium appetite. *Physiological Reviews, 78,* 583-686.

Flatt, J.P. (1988.) Importance of nutrient balance in body regulation. *Diabetes Metabolism Review, 4,* 571-581.

Flatt, J.P. (1995.) Integration of the overall response to exercise. *International Journal of Obesity and Related Metabolic Disorders, 19,* S31-S40.

Flegal, K. M., Carroll, M. D., Kuczmarski, R. J., & Johnson, C. L. (1998.) Overweight and obesity in the United States: prevalence and trends, 1960–1994. *International Journal of Obesity, 22,* 39-47.

Flier, J.S. (1998.) Clinical review 94: What's in a name? In search of leptin's physiologic role. *Journal of Clinical Endocrinology and Metabolism, 83,* 1407-1413.

Flier, J.S., Harris, M., & Hollenberg, A.N. (2000.) Leptin, nutrition, and the thyroid: the why, the wherefore, and the wiring. *Journal of Clinical Investigation, 105,* 859-861.

Flier, J.S., & Maratos-Flier, E. (1998.) Obesity and the hypothalamus: novel peptides for new pathways. *Cell, 92,* 437-440.

Florini, J.R., Ewton, D.Z., Magri, K.A., & Mangiacapra, F.J. (1993.) IGFs and muscle differentiation. Advances in Experimental Medicine and Biology, 343, 319-326.

Foldes, J., Parfitt, A.M., Shih, M.S., Rao, D.S., & Kleerekoper, M. (1991.) Structural and geometric changes in iliac bone: relationship to normal aging and osteoporosis. *Journal of Bone and Mineral Research, 6,* 759-766.

Folkard, S. (1990.) Circadian performance rhythms: some practical and theoretical implications. *Philosophical Transactions of the Royal Society, London, Series B, 327*, 543-553.

Follenius, M., Brandenberger, G., & Hietter, B. (1982.) Diurnal cortisol peaks and their relationships to meals. *Journal of Clinical Endocrinology and Metabolism, 55*, 757-761.

Foote, S.L., Berridge, C.W., Adams, L.M., & Pineda, J.A. (1991.) Electrophysiological evidence for the involvement of locus coeruleus in alerting, orienting and attending. *Progress in Brain Research, 88*, 521-532.

Forbes, G.B. (1972.) Growth of the lean body mass in man. *Growth, 36*, 325-328.

Forbes, G.B. (1985.) The effect of anabolic steroids on lean body mass: the dose response curve. *Metabolism: Clinical and Experimental, 34*, 571-573.

Forest, M.G., Sizonenko, P.C., Cathiard, A.M., & Bertrand, J. (1974.) Hypophyso gonadal function in humans during the first year of life. I. Evidence for testicular activity in early infancy. *Journal of Clinical Investigation, 53*, 819-828.

Forger, N.G., & Breedlove, S.M. (1986.) Sexual dimorphism in human and canine spinal cord: role of early androgen. *Proceedings of the National Academy of Sciences of the United States of America, 83*, 7527-7531.

Forhead, A.J., Li, J., Gilmour, R.S., & Fowden, A.L. (1998.) Control of hepatic insulin-like growth factor II gene expression by thyroid hormone in fetal sheep near term. *American Journal of Physiology, 275*, E149-E156.

Forichon, J., Jomain, M.J., Schellhorn, J., & Minaire, Y. (1977.) Effect of epinephrine upon irreversible disposal and recycling of glucose in dogs. *Experientia, 33*, 1171-1173.

Forsling, M.L. (2000.) Diurnal rhythms in neurohypophysial function. *Experimental Physiology, 85*, 179S-186S.

Forsling, M.L., Stoughton, M.P., Zhou, Y., Kelestimur, H., & Demaine, C. (1993.) The role of the pineal in the control of the daily patterns of neurohypophysial hormone secretion. *Journal of Pineal Research, 14*, 45-51.

Foster, C., Costill, D.L., & Fink, W.J. (1979.) Effects of preexercise feedings on endurance performance. *Medicine and Science in Sports, 11*, 1-5..

Fotherby, K., & Pal, S.B. (Eds.), (1985.) *Exercise endocrinology.* Berlin: Walter de Gruyter.

Foulkes, N.S., Whitmore, D., & Sassone-Corsi, P. (1997.) Rhythmic transcription: the molecular basis of circadian melatonin synthesis. *Biology of the Cell, 89*, 487-494.

Fowden, A.L. (1992.) The role of insulin in fetal growth. *Early Human Development, 29*, 177- 181.

Fowden, A.L. (1993.) Insulin deficiency: Effects on fetal growth and development. *Journal of Paediatrics and Child Health, 29*, 6-11.

Fowden, A.L. (1995.) Endocrine regulation of fetal growth. *Reproduction, Fertility and Development, 7*, 351-363.

Fowden, A.L. (1997.) Comparative aspects of fetal carbohydrate metabolism. *Equine Veterinary Journal, Supplement, 24*, 19-25.

Fowden, A.L., Szemere, J., Hughes, P., Gilmour, R.S., & Forhead, A.J. (1996.) The effects of cortisol on the growth rate of the sheep fetus during late gestation. *Journal of Endocrinology, 151*, 97-105.

Fozard, J.R., & Hannon, J.P. (2000.) Species differences in adenosine receptor-mediated bronchoconstrictor responses. *Clinical and Experimental Allergy, 30*, 1213-1220.

Frank, S.M., el-Gamal, N., Raja, S.N., & Wu, P.K. (1996.) Alpha-adrenoceptor mechanisms of thermoregulation during cold challenge in humans. *Clinical Science, 91*, 6270631.

Frankenne, F., Closset, J., Gomez, F., Scippo, M.L., Smal, J., & Hennen G. (1988.) The physiology of growth hormones (GHs) in pregnant women and partial characterization of the placental GH variant. *Journal of Clinical Endocrinology and Metabolism, 66*, 1171-1180.

Frantz, A.G., & Rabkin, M.T. (1965.) Effects of estrogen and sex difference on secretion of human growth hormone. *Journal of Clinical Endocrinology, 25*, 1470-1480.

Franz, I.W., Lohman, F.W., Koch, G., & Quabbe, H.J. (1983.) Aspects of hormonal regulation of lipolysis during exercise: effects of chronic beta-receptor blockade. *International Journal of Sport Medicine, 4*, 14-20.

Frayn, K.N. (1983.) Calculation of substrate oxidation rates in vivo from gaseous exchange. *Journal of Applied Physiology, 55*, 628-634.

Fredholm, B.B. (1971.) The effect of lactate in canine subcutaneous adipose tissue in situ. *Acta Physiologica Scandinavica, 82*, 110-123.

Fredholm, B., & Rosell, S. (1968.) Effects of adrenergic blocking agents on lipid mobilization from canine subcutaneous adipose tissue after sympathetic nerve stimulation. *Journal of Pharmacology and Experimental Therapeutics, 159*, 1-7.

Fredholm, G.B., & Rosell, S. (1967.) The effect of alpha- and beta-adrenergic blocking agents on free fatty acid release from subcutaneous adipose tissue in vivo. *Acta Pharmacologica et Toxicologica, 25*, Suppl 4,20.

Fredholm, B.B., Abbracchio, M.P., Burnstock, G., Dubyak, G.R., Harden, T.K., Jacobson, K.A., Schwabe, U., & Williams, M. (1997.) Towards a revised nomenclature for P1 and P2 receptors. *Trends in Pharmacology, 18*, 79-82.

Freedman-Akabas, S., Colt, E., Kisseleff, H.R., & Pi-Sunyer, F.X. (1985.) Lack of sustained increase in VO2 following exercise in fit and unfit subjects. *American Journal of Clinical Nutrition, 41*, 545-549.

Freedman, M.S., Lucas, R.J., Soni, B., von Schantz, M., Munoz, M., David-Gray, Z., & Foster, R. (1999.) Regulation of mammalian circadian behavior by non-rod, non-cone, ocular photoreceptors. *Science, 284*, 502-504.

Freedman, R.R., Sabherwal, S.C., Moten, M., & Migaly, P. (1992.) Local temperature modulates alpha$_1$- and alpha$_2$-adrenergic vasoconstriction in men. *American Journal of Physiology, 263*, H1197-H1200.

Freeman, R.H., Davis, J.O., & Seymour, A.A. (1982.) Volume and vasoconstriction in experimental renovascular hypertension. *Federation Proceedings, 41*, 2409-2414.

Freund, B.J., Shizuru, E.M., Hashiro, G.M., & Claybaugh, J.R. (1991.) Hormonal, electrolyte, and renal responses to exercise are intensity dependent. *Journal of Applied Physiology, 70*, 900-906.

Freund, B.J., Wade, C.E., & Claybaugh, J.R. (1988.) Effects of exercise on atrial natriuretic factor: release mechanisms and implications for fluid homeostasis. *Sports Medicine, 6*, 364-377.

Frewin, D.B., Frantz, A.G., & Downey, J.A. (1976.) The effect of ambient temperature on the growth hormone and prolactin response to exercise. *Australian Journal of Experimental Biology and Medical Science, 54*, 97-101.

Frey, M.A.B., Doerr, B.M., Srivastava, L.M., & Glueck, C.J. (1983.) Exercise training, sex hormones and lipoprotein relationships in men. *Journal of Applied Physiology, 54*, 757-762.

Friday, B.B., Horsley, V., & Pavlath, G.K. (2000.) Calcineurin activity is required for the initiation of skeletal muscle differentiation. *Journal of Cell Biology, 149*, 657-666.

Friedewald, V.E., Jr., & Spence, D.W. (1990.) Sudden cardiac death associated with exercise: The risk-benefit issue. *American Journal of Cardiology, 66*,183-188.

Friedman, B., & Kinderman, W. (1989.) Energy metabolism and regulatory hormones in women and men during endurance exercise. *European Journal of Applied Physiology and Occupational Physiology, 59*, 1-9.

Friedman, E.M., & Irwin, M.R. (1997.) Modulation of immune cell function by the autonomic nervous system. *Pharmacology and Therapeutics, 74*, 27-38.

Friedman, G., Chajek-Shaul, T., Stein, O., Noe, L., Etienne, J., & Stein, Y. (1986.) Beta-adrenergic stimulation ehances translocation, processing and synthesis of lipoprotein lipase in rat heart cells. *Biochimica et Biophysica Acta, 877*, 112-120.

Friedman, J.M., & Halaas, J.L. (1998.) Leptin and the regulation of body weight in mammals. *Nature, 395*, 763-770.

Friedman, L., Bergmann, B.M., & Rechtschaffen, A. (1979.) Effects of sleep deprivation on sleepiness, sleep intensity, and subsequent sleep in the rat. *Sleep, 1*, 369-391.

Friend, K., Iranmanesh, A., Login, I.S., & Veldhuis, J.D. (1997.) Pyridostigmine treatment selectively amplifies the mass of GH secreted per burst without altering GH burst frequency, half-life, basal GH secretion or the orderliness of GH release. *European Journal of Endocrinology, 137*, 377-386.

Frisch, R.E. (1984.) Body fat, puberty and fertility. *Biological Reviews, 59*, 161-188.

Frisch, R.E., Canick, J.A., & Tulchinsky, D. (1980a.) Human fatty marrow aromatizes androgen to estrogen. *Journal of Clinical Endocrinology and Metabolism, 51*, 394-396.

Frisch, R.E., & McArthur, J.W. (1974.) Menstrual cycles: Fatness as a determinant of minimum weight for height necessary for their maintenance or onset. *Science, 185*, 949-951.

Frisch, R.E., & Revelle, R. (1969.) The height and weight of adolescent boys and girls at the time of peak velocity of growth in height and weight: longitudinal data. *Human Biology, 41*, 536-559.

Frisch, R.E., & Revelle, R. (1970.) The height and weight at menarche and a hypothesis of critical body weights and adolescent events. *Science, 169*, 397-399.

Frisch, R.E., & Revelle, R.(1971a.) The height and weight of girls and boys at the time of initiation of the adolescent growth spurt in height and weight and the relationship to menarche. *Human Biology, 43*, 140-159.

Frisch, R.E., & Revelle, R. (1971b.) Height and weight at menarche and a hypothesis of menarche. *Archives of Diseases of Children, 46*, 695-701.

Frisch, R.E., Canick, J.A., & Tulchinsky, D. (1980a.) Human fatty marrow aromatizes androgen to estrogen. *Journal of Clinical Endocrinology and Metabolism, 51*, 394-396.

Frisch, R.E., Gotz-Welbergen, A.V., McArthur, J.W., Albright, T., Witschi, J., Bullen, B., Birnholtz, J., Reed, R.B., & Hermann, H. (1981.) Delayed menarche and amenorrhea of college athletes in relation to age of onset of training. *Journal of the American Medical Association, 246*, 1559-1563.

Frisch, R.E., Wyshak, G., & Vincent, L. (1980b.) Delayed menarche and amenorrhea in ballet dancers. *New England Journal of Medicine, 303*, 17-19.

Frisch, R.E., Wyshak, G. , Albright, N.L., Albright, T.E., Schiff, I., Jones, K.P., Witschi, J., Shiang, E., Koff E. & Marguglio M. (1985.) Lower prevalence of breast cancer and cancers of the reproductive system among former college athletes compared to non-athletes. *British Journal of Cancer, 52*, 885-891.

Frizzell, R.T., Lang, G.H., Lowance, D.C., & Lathan, S.R. (1986.) Hyponatremia and ultramarathon running. *Journal of the American Medical Association, 255*, 772-774.

Froberg, J., Karlsson, C.-G., Levi, L., & Lidberg, L. (1972.) Circadian variations in performance, psychological ratings, catecholamine excretion, and diuresis during prolonged sleep deprivation. *International Journal of Psychobiology, 2*, 23-36.

Froberg, K., & Pedersen, P.K. (1984.) Sex differences in endurance capacity and metabolic response to prolonged heavy exercise. *European Journal of Applied Physiology, 52*, 446-450.

Froberg, S.O., & Mossfeldt, F. (1971.) Effect of prolonged strenuous exercise on the concentration of triglycerides, phospholipids and glycogen in muscle of man. *Acta Physiologica Scandinavica, 82*,167-171.

Froberg, S.O., Carlson, L.A., & Ekelund, L.-G. (1971.) Local lipid stores and exercise. In B. Pernow & B. Saltin (Eds.), *Muscle metabolism during exercise, Vol. II* (pp. 307-313). New York: Plenum.

Froberg, S.O., Hultman, E., & Nilsson, L.H. (1975.) Effect of noradrenaline on triglyceride and glycogen concentrations in liver and muscle from man. *Metabolism: Clinical and Experimental, 24*,119-126.

Froesch, E.R., Schmid, C., Schwandes, J., & Zapf, V. (1985.) Actions of insulin-like growth factors. *American Review of Physiology, 47*, 443-467.

Frokiaer, J., Marples, D., Knepper, M.A., & Nielsen, S. (1998.) Pathophysiology of aquaporin-2 water balance disorders. *American Journal of the Medical Sciences, 316*, 291-299.

Frost, H.M. (1987.) The mechanostat: a proposed pathogenic mechanism of osteoporosis and the bone mass effects of mechanical and nonmechanical agents. *Bone, 2*, 73-85.

Fry, A.C., Kraemer, W.J., Stone, M.H., Warren, B.J., Kearney, J.T., Maresh, C.M., Weseman, C.A., & Fleck, S.J. (1993.) Endocrine and performance responses to high volume training and amino acid supplementation in elite junior weightlifters. *International Journal of Sport Nutrition, 3*, 306-322.

Fry, R.W., Morton, A.R., Crawford, G.P., & Keast, D. (1992.) Cell numbers and in vitro responses of leukocytes and lymphocyte subpopulations following maximal exercise and interval training sessions of different intensities. *European Journal of Applied Physiology and Occupational Physiology, 64*, 218-227.

Fryburg, D.A., Gelfand, R.A., Jahn, L.A., Oliveras, D., Sherwin, R.S., Sacca, L., & Barrett, E.J. (1995.) Effects of epinephrine on human muscle glucose and protein metabolism. *American Journal of Physiology, 268*, E55-E59.

Frye, A.J., & Kamon, E. (1981.) Responses to dry heat of men and women with similar aerobic capacities. *Journal of Applied Physiology, 50*, 65-70.

Frye, A.J., & Kamon, E. (1983.) Sweating efficiency in acclimated men and women exercising in humid and dry heat. *Journal of Applied Physiology, 54*, 972-977.

Fujioka, S., Matsuzawa, Y, Tokunaga, K., & Tarui, S. (1987.) Contribution of intra-abdominal fat accumulation to the impirment of glucose and lipid metabolism in human obesity. *Metabolism, 36*, 54-59.

Fujiwara, S., Shinkai, S., Kurokawa, Y., & Watanabe, T. (1992.) The acute effects of experimental short-term evening and night shifts on human circadian rhythms: the oral temperature, heart rate, serum cortisol and urinary catecholamine levels. *International Archives of Occupational Health, 63*, 409-418.

Fukatsu, A., Sato, N., & Shimizu, H. (1996.) 50-mile walking race suppresses neutrophil bactericidal function by inducing increases in cortisol and ketone bodies. *Life Science 58*, 2337-2343.

Fuller, S.J., Gaitanaki, C.J., & Sugden, P.H. (1990.) Effects of catecholamines on protein synthesis in cardiac myocytes and perfused hearts isolated from adult rats: stimulation of translation is mediated through the alpha 1-adrenoceptor. *Biochemical Journal, 266*, 727-736.

Furchgott, R.F., & Zawadski, J.V. (1980.) The obligatory role of endothelial cells in the relaxation of arterial smooth muscle by acetylcholine. *Nature, 286*, 373-376.

Furlan, R., Piazza, S., Dell'Orto, S., Gentile, E., Cerutti, S., Pagani, M., & Malliani A., (1993.) Early and late effects of exercise and athletic training on neural mechanisms controlling heart rate. *Cardiovascular Research, 27*, 482-488.

Furness, J.B., & Costa, M. (1980.) Types of nerves in the enteric nervous system. *Neuroscience, 5*, 1-20.

Furukawa, K., & Nakamura, H. (1987.) Cyclic GMP regulation of the plasma membrane (Ca^{2+}- Mg^{2+})ATPase in vascular smooth muscle. *Journal of Biochemistry, 101*, 287-290.

Fyhrquist, F., Dessypris, A., & Immonen, I. (1983.) Marathon run: Effects on plasma renin activity, renin substrate, angiotensin converting enzyme, and cortisol. *Hormone and Metabolic Research, 15*, 96-99.

Gabbay, R.A., & Lardy, H.A. (1984.) Site of insulin inhibition of cAMP-stimulated glycogenolysis: cAMPdependent protein kinase is affected independent of cAMP changes. *Journal of Biological Chemistry, 259*, 6052-6055.

Gabriel, H., & Kindermann, W. (1997.) The acute immune response to exercise: What does it mean? *International Journal of Sports Medicine, 18* (Suppl. 1), S28-S45.

Gabriel, H., Schwarz, L., Steffens, G., & Kindermann, W. (1992.) Immunoregulatory hormones, circulating leucocyte and lymphocyte subpopulations before and after endurance exercise of different intensities. *International Journal of Sports Medicine, 13*, 359-366.

Gaesser, G.A., & Brooks, G.A. (1984.) Metabolic bases of excess post-exercise oxygen consumption: a review. *Medicine and Science in Sports and Exercise, 16*, 29-43.

Gaesser, G. A., & Rich, R. G. (1984.) Effects of high- and low-intensity exercise training on aerobic capacity and blood lipids. *Medicine and Science in Sports and Exercise, 16*, 269-274.

Gage, J.E., Hess, O.M., Murakami, T., Ritter, M., Grimm, J., & Krayenbuehl, H.P. (1986.) Vasoconstriction of stenotic coronary arteries during dynamic exercise in patients with classic angina pectoris: reversal by nitroglycerin. *Circulation, 73*, 865-876.

Galbo, H. (1983.) *Hormonal and metabolic adaptation to exercise*. Stuttgart: Georg Thieme Ferlag.

Galbo, H., Christensen, N.J., & Holst, J.J. (1977a.) Glucose-induced decrease in glucagon and epinephrine responses to exercise in man. *Journal of Applied Physiology, 42*, 525-530.

Galbo, H., Christensen, N.J., Mikines, K.J., Sonne, B., Hilsted, J., Hagen, C., & Fahrenkrug, J. (1981.) The effect of fasting on the hormonal response to graded exercise. *Journal of Clinical Endocrinology and Metabolism, 52*, 1106-1112.

Galbo, H., Holst, J.J., Christensen, N.J., & Hilsted, J. (1976.) Glucagon and plasma catecholamines during beta-receptor blockade in exercising man. *Journal of Applied Physiology, 40*, 855-863.

Galbo, H., Holst, J.J., & Christensen, N.J. (1975.) Glucagon and plasma catecholamine responses to graded and prolonged exercise in man. *Journal of Applied Physiology, 38*, 70-76.

Galbo, H., Houston, M.E., Christensen, N.J., Holst, J.J., Nielsen, B., Nygaard, E., and Suzuki, J. (1979.) The effect of water temperature on the hormonal response to prolonged swimming. *Acta Physiologica Scandinavica, 105*, 326-337.

Galbo, H., Hummer, L., Petersen, B., Christensen, N.J., & Bie, N. (1977c.) Thyroid and testicular hormone responses to graded and prolonged exercise in man. *European Journal of Applied Physiology, 36,*101-106.

Galbo, H., Richter, E.A., Holst, J.J., & Christensen, N.J. (1977b.) Diminished hormonal response to exercise in trained rats. *Journal of Applied Physiology, 43,* 953-958.

Galitzky, J., Lafontan, M., Nordenstrom, J., & Arner, P. (1993a.) Role of vascular alpha-2 adrenoceptors in regulating lipid mobilization from human adipose tissue. *Journal of Clinical Investigation, 91,* 1997-2003.

Galitzky, J., Taouis, M., Berlan, M., Riviere, D., Garrigues, M., & Lafontan, M. (1988.) Alpha-2 antagonist compounds and lipid mobilization: evidence for lipid mobilizing effect of oral yohimbine in healthy male volunteers. *European Journal of Clinical Investigation, 18,* 587-594.

Galitzky, J., Reverte, M., Portillo, M., Carpene C., Lafontan M., & Berlan M. (1993b.) Coexistence of beta 1-, beta 2-, and beta 3-adrenoceptors in dog fat cells and their differential activation by catecholamines. *American Journal of Physiology, 264,* E403-E412.

Galliven, E.A., Singh, A., Michelson, D., Bina, S., Gold, P.W., & Deuster, P.A. (1997.) Hormonal and metabolic responses to exercise across time of day and menstrual cycle phase. *Journal of Applied Physiology, 83,* 1822-1831.

Galster, A.D., Clutter, W.E., Cryer, P.E., Collins, J.A. & Bier, D.M. (1981.) Epinephrine plasma thresholds for lipolytic effects in man: measurements of fatty acid transport with $[1-^{13}C]$ palmitic acid. *Journal of Clinical Investigation, 67,* 1729-1738.

Galvez, R., Mesches, M.H., & McGaugh, J.L. (1996.) Norepinephrine release in the amygdala in response to footshock stimulation. *Neurobiology of Learning and Memory, 66,* 253-257.

Garber, A.J., Karl, I.E., & Kipnis, D.M. (1976.) Alanine and glutamine synthesis and release from skeletal muscle. IV: Beta-adrenergic inhibition of amino acid release. *Journal of Biological Chemistry, 84,* 388-393.

Garbers, D.L., & Lowe, D.G. (1994.) Guanylyl cyclase receptors. *Journal of Biological Chemistry, 269,* 30741-30744.

Garceau, D., Yamaguchi, N., Goyer, R., & Guitard, F. (1984.) Correlation between endogenous noradrenaline and glucose released from the liver upon hepatic sympathetic nerve stimulation in anesthetized dogs. *Canadian Journal of Physiology and Pharmacology, 62,* 1086-1091.

Garcia Alonzo, L., Garcia Penalta, X., Cornelissen, G., Siegelova, J., & Halberg, J. (2000.) About-yearly and about-monthly variation in neonatal height and weight. *Scripta Medica, 73,* 125-133.

Garcia-Garcia, F., & Drucker-Colin, R. (1999.) Endogenous and exogenous factors on sleep-wake cycle. *Progress in Neurobiology, 58,* 297-314.

Garcia-Mayor, R.V., Andrade, M.A., Rios, M., Lage, M., Dieguez, C., & Casanueva, F.F. (1997.) Serum leptin levels in normal children: relationship to age, gender, body mass index, pituitary-gonadal hormones, and pubertal stage. *Journal of Clinical Endocrinology and Metabolism, 82,* 2849-2855.

Gardemann, A., Puschel, G.P., & Jungermann, K. (1992.) Nervous control of liver metabolism and hemodynamics. *European Journal of Biochemistry, 207,* 399-411.

Gardner, D.F., Kaplan, M.M., & Stanley, C.A. (1979.) Effect of triiodothyronine replacement on the metabolic and pituitary response to starvation. *New England Journal of Medicine, 300,* 579-584.

Garetto, L.P., & Armstrong, R.B. (1983.) Influence of circadian rhythms on rat muscle glycogen metabolism during and after exercise. *Journal of Experimental Biology, 102,* 211-222.

Gaumann, D.M., Yaksh, T.L., Dousa, M.K., Tyce, G.M., Lucas, D.L., & Hench, V.S. (1987.) Effects of hemorrhage and naloxone on adrenal release of methionine-enkephalin and catecholamines in halothane anesthetized dogs. *Journal of the Autonomic Nervous System, 21,* 29-41.

Gauquelin, G., Maillet, A., Allevard, A.M., Sigaudo, D., & Gharib, C. (1996.) Diurnal rhythms of plasma renin activity, atrial natriuretic peptide and arterial pressure during head-down bed rest in humans. *European Journal of Applied Physiology and Occupational Physiology, 73,* 536-543.

Geary, N., Trace, D., McEwen, B., & Smith, G.P. (1994.) Cyclic estradiol replacement increases the satiety effect of CCK-8 in ovariectomized rats. *Physiology and Behavior, 56,* 281-289.

Gebert, C.A., Park, S.H., & Waxman, D.J. (1997.) Regulation of signal transducer and activator of transcription (STAT) 5b activation by the temporal pattern of growth hormone stimulation. *Molecular Endocrinology, 11,* 400-414.

Geelen, G., Keil, L.C., Kravik, S.E., Wade, C.E., Thrasher, T.N., Barnes, P.R., Pyka, G., Nesvig, C., & Greenleaf, J.E. (1984.) Inhibition of plasma vasopressin after drinking in dehydrated humans. *American Journal of Physiology, 247,* R968-R971.

Geisterfer, A.A., Peach, M.J., & Owens, G.K. (1988.) Angiotensin II induces hypertrophy, not hyperplasia, of cultured rat aortic smooth muscle cells. *Circulation Research, 62,* 749-756.

Gekakis, N., Staknis, D., Nguyen, H.B., Davis, F.C., Wilsbacher, L.D., King, D.P., Takahashi, J.S., & Weitz, C.J. (1998.) Role of the CLOCK protein in the mammalian circadian mechanism. *Science, 280,* 1564-1569.

Geloen, A., Collet, A.J., & Bukowiecki, L.J. (1992.) Role of sympathetic innervation in brown adipocyte proliferation. *American Journal of Physiology, 263,* R1176-R1181.

Georgopoulos, N., Markou, K., Theodoropoulou, A., Paraskevopoulou, P., Varaki, L., Kazantzi, Z., Leglise, M., & Vagenakis, A.G. (1999.) Growth and pubertal development in elite female rhythmic gymnasts. *Journal of Clinical Endocrinology and Metabolism, 84,* 4525-4530.

Gerritsen, W., Heijnen, C.J., Wiegant, V.M., Bermond, B., & Frijda, N.H. (1996.) Experimental social fear: immunological, hormonal and autonomic concomitants. *Psychosomatic Medicine, 58,* 273-286.

Geva, T., Mauer, M.B., Striker, L., Kirshon, B., & Pivarnik, J.M. (1997.) Effects of physiologic load of pregnancy on left ventricular contractility and remodeling. *American Heart Journal, 133,* 53-59.

Ghaleh, B., Bea, M.-L., DuBois-Rande, J.-L., Giudicelli, J.F., Hittinger, L., & Berdeaux, A. (1995.) Endothelial modulation of beta-adrenergic dilation of large coronary arteries in conscious dogs. *Circulation, 92,* 2627-2635.

Gibbons, L.W., Cooper, K.H., Meyer, B.M., & Ellison, R.C. (1980.) The acute cardiac risk of strenuous exercise, *Journal of the American Medical Association, 244,* 1799-1801.

Gibbs, F.P., & Peterborg, L.J. (1986.) Exercise reduces gonadal atrophy caused by short photoperiod or blinding of hamsters. *Physiology and Behavior, 37,* 159-162.

Gibbs, J., Young, R.C., & Smith, G.P. (1973.) Cholecystokinin decreases food intake in rats. *Journal of Comparative Physiology and Psychology, 84,* 488-495.

Gierz, M., Campbell, S.S., & Gillin, J.C. (1987.) Sleep disturbances in various affective psychiatric disorders. *Psychiatry Clinics of North America, 10,* 565-581.

Gillette, M.U., & Tischkau, S.A. (1999.) Suprachiasmatic nucleus: the brain's circadian clock. *Recent Progress in Hormone Research, 54,* 33-58.

Giovambattista, A., Chisari, A.N., Gaillard, R.C., & Spinedi, E. (2000.) Food intake-induced leptin secretion modulates hypothalamo-pituitary-adrenal axis response and hypothalamic Ob-Rb expression to insulin administration. *Neuroendocrinology, 72,* 341-349.

Girandola, R. N. (1976.) Body composition changes in women: effect of high and low exercise intensity. *Archives of Physical Medicine and Rehabilitation, 57,* 297-300.

Girard, J. (1997.) Is leptin the link between obesity and insulin resistance? *Diabetes and Metabolism, 23* (Suppl. 3), 16-24.

Giraudo, S.Q., Grace, M.K., Welch, C.C., Billington, C.J., & Levine, A.S. (1993.) Naloxone's anorectic effect is dependent upon the relative palatability of food. *Pharmacology and Biochemical Behavior, 46,* 917.

Giustina, A., & Veldhuis, J.D. (1998.) Pathophysiology of the neuroregulation of growth hormone secretion in experimental animals and the human. *Endocrine Reviews, 19,* 717-797.

Giustina, A., Malerba, M., Bresciani, E., Desenzani, P., Licini, M., Zaltieri, G., & Grassi, V. (1995.) Effect of two beta 2-agonist drugs, salbutamol and broxaterol, on the growth hormone response to exercise in adult patients with asthmatic bronchitis. *Journal of Endocrinological Investigation, 18,* 847-852.

Giustina, A., Bossoni, S., Bodini, C., Cimino, A., Pizzocolo, G., Schettino, M., & Wehrenberg, W.B. (1991.) Effects of exogenous growth hormone pretreatment on the pituitary growth hormone response to growth hormone-releasing hormone alone or in combination with pyridostigmine in type 1 diabetic patients. *Acta Endocrinologica, 125,* 510-517.

Glass, C.K. (1994.) Differential recognition of target genes by nuclear receptor monomers, dimers and heterodimers. *Endocrine Reviews, 15,* 391-407.

Glatz, J.F., & van der Vusse, G.J. (1996.) Cellular fatty acid-binding proteins: their function and physiological significance. *Progress in Lipid Research, 35,* 243-282.

Glatzle, J., Kreis, M.E., Kawano, K., Raybould, H.E., & Zittel, T.T. (2001.) Postprandial neuronal activation in the nucleus of the solitary tract is partially mediated by CCK-A receptors. *American Journal of Physiology, 281,* R222-R229.

Glod, C.A., Teicher, M.H., Polcari, A., McGreenery, C.E., & Ito, Y. (1997.) Circadian rest-activity disturbances in children with seasonal affective disorder. *Journal of the American Academy of Child and Adolescent Psychiatry, 36,* 188-195.

Gluckman, P.D. (1986.) The role of pituitary hormones, growth factors and insulin in the regulation of fetal growth. *Oxford Reviews in Reproductive Biology, 8,* 1-60.

Gogia, P.P., Brown, M., & al-Obaidi, S. (1993.) Hydrocortisone and exercise effects on articular cartilage. *Archives of Physical Medicine and Rehabilitation, 74,* 463-467.

Gold, R.M., & Kapatos, G. (1975.) Delayed hyperphagia and increased body length after hypothalamic knife cuts in weanling rats. *Journal of Comparative and Physiological Psychology, 88,* 202-209.

Goldberg, A.L. (1967.) Work-induced growth of skeletal muscle in normal and hypophysectomized rats. *American Journal of Physiology, 312,* 1193-1198.

Goldberg, A.L. (1968.) Role of insulin in work-induced growth of skeletal muscle. *Endocrinology, 83,* 1071-1073.

Goldberg, A.L. (1979.) Influence of insulin and contractile activity on muscle size and protein balance. *Diabetes, 28* (Suppl. 1), 18-24.

Goldberg, A.L., & Goodman, H.M. (1969.) Relationship between cortisone and muscle work in determining muscle size. *Journal of Physiology, 200,* 667-675.

Goldberg, A.L., Etlinger, J.D., Goldspink, D.F., & Jeblecki, C. (1975.) Mechanism of work- induced hypertrophy of skeletal muscle. *Medicine and Science in Sports, 7,* 248-261.

Goldsmith, R.L., Bigger, J.T, Jr., Steinman, R.C., & Fleiss, J.L. (1992.) Comparison of 24-hour parasympathetic activity in endurance-trained and untrained young men. *Journal of the American College of Cardiology, 20,* 552-558.

Goldspink, G. (1998.) Selective gene expression during adaptation of muscle in response to different physiological demands. *Comparative Biochemistry and Physiology, Part B, 120,* 5-15.

Goldspink, G., Scutt, A., Loughna, P.T., Wells, D.J., Jaenicke, T., & Gerlach, G.F. (1992.) Gene expression in skeletal muscle in response to stretch and force generation. *American Journal of Physiology, 262,* R356-R363.

Goldspink, G., Scutt, A., Martindale, J., Jaenicke, T., Turay, L., & Gerlach, G.F. (1991.) Stretch and force generation induce rapid hypertrophy and isoform gene switching in adult skeletal muscle. *Biochemical Transactions, 19,* 368-373.

Goldstein, D.S. (1983.) Plasma catecholamines and essential hypertension: an analytical review. *Hypertension, 5,* 86-99.

Gollnick, P., Armstrong, R.B., Saubert, C.W., IV, Sembrowich, W.L., Shepherd, R.E., & Saltin, B. (1973a.) Glycogen depletion pattern in human skeletal muscle fibers during prolonged work. *Pflugers Archiv, 244,* 1-12.

Gollnick, P., Armstrong, R.B, Sembrowich, W.L. , Shepherd, R.E., & Saltin, B.(1973b.) Glycogen depletion pattern in human skeletal muscle fibers after heavy exercise. *Journal of Applied Physiology, 34,* 615-618.

Gollnick, P., Karlsson, J., Piehl, K., & Saltin, B. (1974a.) Selective glycogen depletion in skeletal muscle fibers of man following sustained contractions. *Journal of Physiology, 241,* 59-68.

Gollnick, P., Piehl, K., & Saltin, B. (1974b.) Selective glycogen depletion pattern in human muscle fibres after exercise of varying intensity and at varying intensity and at varying pedalling rates. *Journal of Physiology, 241,* 45-57.

Gollnick, P., Soule, R.G., Taylor, A.W., Williams, C., & Ianuzzo, C.D. (1970.) Exercise-induced glycogenolysis and lipolysis in rat: hormonal influence. *American Journal of Physiology, 219,* 729-733.

Gonzalez-Cadavid, N.F., Taylor, W.E., Yarasheski, K., et al. (1998.) Organization of the human myostatin gene and expression in healthy men and HIV-infected men with muscle wasting. *Proceedings of the National Academy of Sciences, 95,*14938-14943.

Goodner, C.J., Walike, B.C., Koerker, D.J., Ensinck, J.W., Brown, A.C., Chideckel, E.W., Palmer, J., & Kalnasy L. (1977.) Insulin, glucagon, and glucose exhibit synchronous, sustained oscillations in fasting monkeys. *Science, 195,*177-179.

Goodpaster, B.H., & Kelley, D.E. (1998.) Role of muscle in triglyceride metabolism. *Current Opinion in Lipidology, 9,* 231-236.

Goodpaster, B.H., Thaete, F.L., Simoneau, J.-A., & Kelley, D.E. (1997.) Subcutaneous abdominal fat and thigh muscle composition predict insulin sensitivity independently of visceral fat. *Diabetes, 46,* 1579-1585.

Goodyear, L.J. (2000.) AMP-activated protein kinase: A critical signaling intermediary for exercise-stimulated glucose transport. *Exercise and Sport Sciences Reviews, 28,* 113-116.

Gooren, L. (1984.) Effect of androgens on brain function in males and females. *European Journal of Endocrinology, 130* (Suppl. 2), 77-78.

Gordeladze, J.O., Reseland, J.E., & Drevon, C.A. (2001.) Pharmacological interference with transcriptional control of osteoblasts: a possible role for leptin and fatty acids in maintaining bone strength and body lean mass. *Current Pharmaceutical Design, 7,* 275-290.

Gordon, C.R., & Lavie, P. (1985.) Day-night variations in urine excretion and hormones in dogs: role of autonomic innervation. *Physiology and Behavior, 35,* 175-181.

Gore, D.C., Jahoor, F., Wolfe, R.R., & Herndon, D.N. (1993.) Acute response of human muscle protein to catabolic hormones. *Annals of Surgery, 218,* 679-684.

Gorski, J. (1977.) Effect of lactate on FFA release and cyclic 3',5'-AMP accumulation in fat cells at different pH. *Acta Physiologica Polonica, 28,* 505-510.

Gorski J. (1992.) Muscle triglyceride metabolism during exercise. *Canadian Journal of Physiology and Pharmacology, 70,* 123-131.

Gorski, J., & Stankiewicz-Chorouszucha, B. (1982.) The effect of hormones on lipoprotein lipase activity in skeletal muscles of the rat. *Hormone and Metabolic Research, 14,* 189-191.

Gorski, J., Oscai, L.B., & Palmer, W.K. (1990.) Hepatic lipid metabolism in exercise and training. *Medicine and Science in Sports and Exercise, 22,* 213-221.

Gould, E., Wooley, C.S., Frankfurt, M., & McEwen, B.S. (1990.) Gonadal steroids regulate dendritic spine density in hippocampal pyramidal cells in adulthood. *Journal of Neuroscience, 10,* 1286-1291.

Gould, K.L., Ornish, D., Scherwitz, L., Brown, S., Edens, R.P., Hess, M.J., Mullani, N., Bolomey, L. Dobbs, F., & Armstrong, W.T. (1995.) Changes in myocardial performance abnormalities by positron emission tomography after long-term, intense risk factor modification. *Journal of the American Medical Association, 274,* 894-901.

Graham, N.M., Douglas, R.M., & Ryan, P. (1986.) Stress and acute respiratory infection. *American Journal of Epidemiology, 124,* 389-401.

Graham, T.E. (1983.) Alcohol ingestion and sex differences on the thermal responses to mild exercise in a cold environment. *Human Biology, 55,* 463-476.

Granneman, J.G. (1995.) Why do adipocytes make the beta 3 adrenergic receptor? *Cellular Signalling, 7,* 9-15.

Grant, P.J. (1990.) Hormonal regulation of the acute haemostatic response to stress. *Blood, Coagulation and Fibrinolysis, 1,* 299-306.

Grassi, G., Cattaneo, B.M., Seravalle, G., Lanfranchi, A., & Mancia, G. (1998.) Baroreflex control of sympathetic nerve activity in essential and secondary hypertension. *Hypertension, 31,* 68-72.

Grau, M., Soley, M., & Ramirez, I. (1997.) Interaction between adrenaline and epidermal growth factor in the control of liver glycogenolysis in the mouse. *Endocrinology, 138,* 2601-2609.

Gray, A.B., Telford, R.D., Collins, M., & Weidemann, M.J. (1993.) The response of leukocyte subsets and plasma hormones to intense exercise. *Medicine and Science in Sports and Exercise, 25,* 1252-1258.

Grediagin, M.A., Cody, M., Rupp, J., Benardot, D., & Shern, R. (1995.) Exercise intensity does not effect body composition change in untrained, moderately overfat women. *Journal of the American Dietetic Association, 95,* 661-665.

Green, A.L., Hultman, E., Macdonald, I.A., Sewell, D.A., & Greenhaff, P.L. (1996a.) Carbohydrate ingestion augments skeletal muscle creatine accumulation during creatine supplementation in humans. American Journal of Physiology. 271, E821-E826.

Green, A.L., Simpson, E.J., Littlewood, J.J., Macdonald, I.A., & Greenhaff, P.L. (1996b.) Carbohydrate ingestion augments creatine retention during creatine feeding in humans. *Acta Physiologica Scandinavica, 158*, 195-202.

Green, H., Morikawa, M., & Nixon, T. (1985.) A dual effector theory of growth-hormone action. *Differentiation, 29*, 195-198.

Green, H.J., Jones, S., Ball-Burnett, M., Farrance, B., & Renney, D. (1995.) Adaptations in muscle metabolism to prolonged exercise and training. *Journal of Applied Physiology, 78*, 138-145.

Green, M.L., Green, R.G., & Santoro, W. (1988.) Daily relaxation modifies serum and salivary immunoglobulins and psychophysiologic symptom severity. *Biofeedback and Self-Regulation, 13*, 187-199.

Greenhaff, P.L., Casey, A., Short, A.H., Harris, R., Soderlund, K., & Hultman, E. (1993.) Influence of oral creatine supplementation on muscle torque during repeated bouts of maximal voluntary exercise in man. *Clinical Science, 84*, 565-571.

Greenhaff, P.L., Ren, J.-M., Soderlund, K., & Hultman, E. (1991.) Energy metabolism in single human muscle fibers during contraction without and with epinephrine infusion. *American Journal of Physiology, 260*, E718-E718.

Greenhaff, P.L., Nevill, M.E., Soderlund, K., Bodin, K., Boobis, L.H., Williams, C., & Hultman, E. (1994.) The metabolic responses of human type I and II muscle fibres during maximal treadmill sprinting. *Journal of Physiology, 478*, 149-155.

Greenleaf, J.E. (1992.) Problem: thirst, drinking behavior, and involuntary dehydration. *Medicine and Science in Sports and Exercise, 24*, 645-656.

Greenway, C.V., & Stark, R.D. (1971.) Hepatic vascular bed. *Physiological Reviews, 51*, 23-65.

Gregoire, J., Tuck, S., Yamamoto, Y., & Hughson, R. (1996.) Heart rate variability at rest and exercise: influence of age, gender, and physical training. *Canadian Journal of Applied Physiology, 21*, 455-470.

Gregory, P., Low, R., & Stirewalt, W.S. (1986.) Changes in skeletal muscle myosin isoenzymes with hypertrophy and exercise. *Biochemical Journal, 238*, 55-63.

Greiwe, J.S., Holloszy, J.O., & Semenkovich, C.F. (2000.) Exercise induces lipoprotein lipase and GLUT-4 protein in muscle independent of adrenergic-receptor signaling. *Journal of Applied Physiology, 89*, 176-181.

Grekin, R.J., Dumont, C.J., Vollmer, A.P., Watts, S.W., & Webb, R.C. (1997.) Mechanism in the pressor effects of hepatic portal venous fatty acid infusion. *American Journal of Physiology, 273*, R324-R330.

Griffin, J.E. (1982.) *Manual of Clinical Endocrinology and Metabolism.* New York: McGraw-Hill, 75-96.

Griffin, J.E., and Ojeda, S.R. (Eds.), (1996.) *Textbook of endocrine physiology,* 3rd ed., (pp. 86-100). Oxford: Oxford University Press.

Grimston, S.K., Tanguay, K.E., Gundberg, C.M., & Hanley, D.A. (1993.) The calciotropic hormone response to changes in serum calcium during exercise in female long distance runners. *Journal of Clinical Endocrinology and Metabolism, 76*, 867-872.

Grobet, L., Martin, L.J., Poncelet, D., et al. (1997.) A deletion in the bovine myostatin gene causes the double-muscled phenotype in cattle. *Nature Genetics, 17*, 71-74.

Grobet, L., Poncelet, D., Royo, L.J., Brouwers, B., Pirottin, D., Michaux, C., Menissier, F., Zanotti, M., Dunner, S., & Georges, M. (1998.) Molecular definition of an allelic series of mutations disrupting the myostatin function and causing double-muscling in cattle. *Mammalian Genome, 9*, 210-213.

Grohe, C., Kahlert, S., Lobbert, K., Meyer, R., Linz, K.W., Karas, R.H., & Vetter, H. (1996.) Modulation of hypertensive heart disease by estrogen. *Steroids, 61*, 201-204.

Gromada, J., Bokvist, K., Ding, W.G., Barg, S., Buschard,K., Renstrom, E., & Rorsman, P. (1997.) Adrenaline stimulates glucagon secretion in pancreatic A-cells by increasing the Ca^{2+} current and the number of granules close to the L-type Ca^{2+} channels. *Journal of General Physiology, 110*, 217-228.

Groop, L.C., Bonadonna, R.C., Del Prato, S., Ratheiser, K., Zyck, K., Ferannini, E., & DeFronzo, R.A. (1989.) Glucose and free fatty acid metabolism in non-insulin-dependent diabetes mellitus: evidence for multiple sites of control. *Journal of Clinical Investigation, 84*, 205-213.

Groop, L.C., Bonadonna, R.C., Simonson, D.C., Petrides, A.S., Shank, M., & DeFronzo, R.A. (1992.) Effect of insulin on oxidative and nonoxidative pathways of free fatty acid metabolism in human obesity. *American Journal of Physiology, 263*, E79-E84.

Grossman, A. (1985.) Endorphins: "Opiates for the masses." *Medicine and Science in Sports and Exercise, 17*, 101-105.

Grossman, A., Bouloux, P., Price, P., Drury, P.L., Lam, K.S., Turner, T., Thomas, J., Besser, G.M., & Sutton, J. (1984.) The role of opioid peptides in the hormonal responses to acute exercise in man. *Clinical Science, 67*, 483-491.

Grossman, A., & Sutton, J.R. (1985.) Endorphins: What are they? How are they measured? What is their role in exercise? *Medicine and Science in Sports and Exercise, 17*, 74-81.

Grossman, J.D., & Morgan, J.P. (1997.) Cardiovascular effects of endothelin. *News in Physiological Sciences, 12*, 113-117.

Gruber, A.J., & Pope, H.G., Jr. (2000.) Psychiatric and medical effects of anabolic-androgenic steroid use in women. *Psychotherapy and Psychosomatics, 69*, 19-26.

Grundy, S.M. (1999.) Hypertriglyceridemia, insulin resistance, and the metabolic syndrome. *American Journal of Cardiology, 83*, 25F-29F.

Guillot, E., Coste, A., Eon, M.T., & Angel, I. (1998.) Mechanisms of the hypoglycemic effects of the alpha2-adrenoceptor antagonists SL84.0418 and deriglidole. *Life Sciences, 62*, 839-852.

Gullestad,L., Hallen, J., & Sejersted, O.M. (1993.) Variable effects of beta-adrenoceptor blockade on muscle blood flow during exercise. *Acta Physiologica Scandinavica. 149*, 257-271.

Guo, K., Wang, J., Andres, V., Smith, R.C., & Walsh, K. (1995.) MyoD-induced expression of p21 inhibits cyclin-dependent kinase activity upon myocyte terminal differentiation. *Molecular and Cellular Biology, 15*, 3823-3829.

Guo, X., & Wakade, A.R. (1994.) Differential secretion of catecholamines in response to peptidergic and cholinergic transmitters in rat adrenals. *Journal of Physiology, 745*, 539-545.

Guo, Z., Johnson, C.M., & Jensen, M.D. (1997.) Regional lipolytic responses to isoproterenol in women. *American Journal of Physiology, 273*, E108-112.

Guse-Behling, H., Ehrhart-Bornstein, M., Bornstein, S.R., Waterman, M.R., Scherbaum, W.A., & Adler, G. (1992.) Regulation of adrenal steroidogenesis by adrenaline: Expression of cytochrome P450 genes. *Journal of Endocrinology, 135*, 229-237.

Guy, J.T., & Auslander, M.O.(1973.) Androgenic steroids and hepatocellular carcinoma. *Lancet, 1*, 148.

Guyton, A.C. (1991.) *Textbook of medical physiology,* 8th ed. Philadelphia: W.B. Saunders Company.

Haarbo, J., Marslew, U., Gotfredson, A., & Christiansen, C. (1991.) Postmenopausal hormone replacement therapy prevents central distribution of body fat after menopause. *Metabolism, 40*, 1323-1326.

Hackenthal, E., Paul, M., Ganten, D., & Taugner, R. (1990.) Morphology, physiology, and molecular biology of renin secretion. *Physiological Reviews, 70*, 1067-1116.

Hackney, A.C., McCracken-Compton, M.A., & Ainsworth, B. (1994.) Substrate responses to submaximal exercise in the midfollicular and midluteal phases of the menstrual cycle. *International Journal of Sport Nutrition, 4*, 299-306.

Hackney, A.C., Sinning, W.E., & Bruot, B.C. (1988.) Reproductive hormonal profiles of endurance-trained and untrained males. *Medicine and Science in Sports and Exercise, 2*, 60-65.

Haddad, G.G., Jeng, H.J., & Lai, T.L. (1986.) Effect of endorphins on heart rate and blood pressure in adult dogs. *American Journal of Physiology, 250*, H796-H805.

Haffner, S.M. (1996.) Low levels of sex hormone-binding globulin and testosterone predict the development of non-insulin-dependent diabetes mellitus in men. MRFIT Research Group. Multiple Risk Factor Intervention Trial. *American Journal of Epidemiology, 143*, 889-897.

Hagberg, J.M., Hickson, R.C., McLane, J.A., Ehsani, A.A., & Winder, W.W. (1979.) Disappearance of norepinephrine from circulation following strenuous exercise. *Journal of Applied Physiology, 47*, 1311-1314.

Hagenfeldt, L. (1979.) Metabolism of free fatty acids and ketone bodies during exercise in normal and diabetic man. *Journal of the American Dietetic Association, 28*, 66-70.

Hagenfeldt, L., & Wahren, J. (1971a.) Human forearm muscle metabolism during exercise. VI. Substrate utilization in prolonged fasting. *Scandinavian Journal of Laboratory and Clinical Investigation, 27*, 299-306.

Hagenfeldt, L., & Wahren, J. (1971b.) Metabolism of free fatty acids and ketone bodies in skeletal muscle. In B. Pernow & B. Saltin (Eds.), *Muscle metabolism during exercise*, (pp. 153-163). New York: Plenum.

Hagenfeldt, L., & Wahren, J. (1972.) Human forearm muscle metabolism during exercise. VII. FFA uptake and oxidation at different work intensities. *Scandinavian Journal of Clinical and Laboratory Investigation, 30*, 429-436.

Hagstrom-Toft, E., Bolinder, J., Eriksson, S., & Arner, P. (1995.) Role of phosphodiesterase III in the antilipolytic effect of insulin in vivo. *Diabetes, 44*, 1170-1175.

Hakansson, M., deLecea, L. Sutliffe, J.G., Yanagisawa, M., & Meister, B. (1999.) Leptin receptor- and STAT3-immunoreactivities in hypocretin/orexin neurones of the lateral hypothalamus. *Journal of Neuroendocrinology, 11*, 653-663.

Halaas, J.L., Gajiwala, K.S., Maffei, M., Cohen, S.L., & Chait, B.T. (1995.) Weight-reducing effects of the plasma protein encoded by the obese gene. *Science, 269*, 543-546.

Halberg, F. (1959.) Physiologic 24-h periodicity: general procedural considerations with reference to the adrenal cycle. *Zeitschrift fur Vitamin- Hormone- und Fermentforschung, 10*, 225-296

Halberg, F. (1963.) Circadian (about twenty-four-hours) rhythms in experimental medicine. *Proceedings of the Royal Society of Medicine, 56*, 253-260.

Halberg, F., Cornelissen, G., Halberg, J., et al. (1998.) Circadian hyper-amplitude-tension (CHAT): a disease risk syndrome of anti-aging medicine. *Journal of Anti-Aging Medicine, 1*, 239-259.

Halberg, F., Halberg, E., Barnum, C.P., & Bittner, J.J. (1959.) Physiologic 24-hour periodicity in human beings and mice, the lighting regimen, and daily routine. In *Photoperiodism and related phenomena in plants and animals*. Washington, DC: American Association for Advancement of Science, Publication 55: 803-878.

Halberg, F., Kabat, H.F., & Klein, P.(1980a.) Chronopharmacology: a therapeutic frontier. *American Journal of Hospital Pharmacy, 37*, 101-106.

Halberg, F., Nelson, W., Levi, F., Culley, D., Bogden, A., & Taylor, D.J. (1980b.) Chronotherapy of mammary cancer in rats. *International Journal of Chronobiology, 7*, 85-99.

Halberg, F., Reinberg, A., & Haus, E. (1970.) Human biological rhythms during and after several months of isolation underground in natural caves. *Bulletin of the National Speleological Society, 32*, 89-115.

Halberg, F., Siffre, M., Engeli, M., Hillman, D., & Reinberg, A. (1965.) Etude en libre-cours des rhythmes circadiens du pouls, de l'alternance veille-someil et de l'estimation du temps pendant les deux mois de sejour souterrain d'un homme adulte jeune. *Comptes Rendues de Academie de Sciences, Paris, 260*, 1259-1262.

Halberg, F., Tong, Y.L., & Johnson, E.A. (1967.) Circadian system phase- an aspect of temporal morphology; procedures and illustrative examples (pp 20-48). In H. Von Mayersbach (Ed.) *The cellular aspects of biorhythms: Symposium on Biorhythms*, Berlin: Springer-Verlag.

Halbert, J.A., Silagy, C.A., Finucane, P., Withers, R.T., Hamdorf, P.A., & Andrews, G.R. (1997.) The effectiveness of exercise training in lowering blood pressure: a meta-analysis of randomized controlled trials of 4 weeks or longer. *Journal of Human Hypertension, 11*, 641-649.

Hales, C.N., Luzio, J.P., & Siddle, K. (1978.) Hormonal control of adipose-tissue lipolysis. *Biochemical Society Symposia, 43*, 97-135.

Halford, J.C., & Blundell, J.E. (2000.) Separate systems for serotonin and leptin in appetite control. *Annals of Medicine, 32*, 222-232.

Hall, D.C., & Kaufmann, D.A. (1987.) Effect of aerobic and strength conditioning on pregnancy outcomes. *American Journal of Obstetrics and Gynecology, 157*, 1199-1203.

Halle, M., Berg, A., Northoff, H., & Keul, J. (1998.) Importance of TNF-alpha and leptin in obesity and insulin resistance: a hypothesis on the impact of physical exercise. *Exercise Immunology Review, 4*, 77-94.

Halliwill, J.R., Taylor, J.A., & Eckberg, D.L. Investigator: Eckberg, D.L. (1996.) Impaired sympathetic vascular regulation in humans after acute dynamic exercise. *Journal of Physiology, 495*, 279-288.

Hamdy, R.C., Moore, S.W., Whalen, K.E., & Landy, C. (1998.) Nandrolone decanoate for men with osteoporosis. *American Journal of Therapeutics, 5*, 89-95.

Hamilton, K.S., Gibbons, F.K., Bracy, D.P., Lacy, D.B., Cherrington, A.D., & Wasserman, D.H. (1996.) Effect of prior exercise on the partitioning of an intestinal glucose load between splanchnic bed and skeletal muscle. *Journal of Clinical Investigation, 98*, 125-135.

Hamilton, S.G., & McMahon, S.B. (2000.) ATP as a peripheral mediator of pain. *Journal of Autonomic Nervous System, 81*, 187-194.

Hampson, E. (1990.) Oestrogen-related variations in human spatial and articulatory motor skills. *Psychoneuroendocrinology, 15*, 97-111.

Han, C., Abel, P.W., & Minneman, K.P. (1987.) Alpha1-adrenoceptor subtypes linked to different mechanisms for increasing intracellular Ca^{2+} in smooth muscle. *Nature, 329*, 333-335.

Hanafy, K.A., Krumenacker, J.S., & Murad, F. (2001.) NO, nitrotyrosine, and cyclic GMP in signal transduction. *Medical Science Monitor, 7*, 801-819.

Haneda, T., & McDermott, P.J. (1991.) Stimulation of ribosomal RNA synthesis during hypertrophic growth of cultured heart cells by phorbol ester. *Molecular and Cellular Biochemistry, 104*, 169-177.

Hanley, P.C., Zinsmeister, A.R., Clements, I.P., Bove, A.A., Brown, M.L., & Gibbons, R.J. (1989.) Gender-related differences in cardiac response to supine exercise assessed by radionuclide angiography. *Journal of the American College of Cardiology, 13*, 624-629.

Hansen, J., Sander, M., & Thomas, G.D. (2000.) Metabolic modulation of sympathetic vasoconstriction in exercising skeletal muscle. *Acta Physiologica Scandinavica, 168*, 489-503.

Haq, A., al-Hussein, K., Lee, J. & al-Sedairy, S. (1993.) Changes in peripheral blood lymphocyte subsets associated with marathon running. *Medicine and Science in Sports and Exercise, 25*, 186-190.

Harber, V.J., Petersen, S.R., & Chilibeck, P.D. (1997a.) Thyroid hormone concentrations and skeletal muscle metabolism during exercise in anorexic females. *Canadian Journal of Physiology and Pharmacology, 75*, 1197-1202.

Harber, V.J., & Sutton, J.R. (1984.) Endorphins and exercise. *Sports Medicine, 1*, 154-171.

Harber, V.J., Sutton, J.R., MacDougall, J.D., Woolever, C.A., & Bhavnani, B.R. (1997b.) Plasma concentrations of beta-endorphin in trained eumenorrheic and amenorrheic women. *Fertility and Sterility, 67*, 648-653.

Hardie, D.G., & Hawley, S.A. (2001.) AMP-activated protein kinase: the energy charge hypothesis revisited. Bioessays, 23, 1112-1119.

Hardman, A.E. (1998.) The influence of exercise on postprandial triacylglycerol metabolism. *Atherosclerosis, 141* (Suppl. 1), S93-S100.

Hardman, A.E. (1999.) Interaction of physical activity and diet: implications for lipoprotein metabolism. *Public Health Nutrition, 2*, 369-376.

Hardman, A. E., Jones, P. R. M., Norgan, N. G., & Hudson, A. (1992.) Brisk walking improves endurance fitness without changing body fatness in previously sedentary women. *European Journal of Applied Physiology and Occupational Physiology, 65*, 354-359.

Hargreave, F.E., Ryan, G., & Thompson, N.C. (1981.) Bronchial responsiveness to histamine and metacholine in asthma: measurement and clinical significance. *Journal of Allergy and Clinical Immunology, 68*, 347-355.

Hargreaves, M., Costill, D.L., Fink, W.J., King, D.S., & Fielding, R.A. (1987.) Effect of pre-exercise carbohydrate feedings on endurance cycling performance. *Medicine and Science in Sports and Exercise, 19*, 33-36.

Harma, M. (1993.) Individual differences in tolerance to shiftwork: a review. *Ergonomics, 36*, 101-109.

Haro, L.S., Lewis, U.J., Garcia, M., Bustamante, J., Martinez, A.O., & Ling, N.C. (1996.) Glycosylated human growth hormone (hGH): a novel 24 kDa hGH-N variant. *Biochemical and Biophysical Research Communications, 228*, 549-556.

Harris, S.E., Sabatini, M., Harris, M.A., Feng, J.Q., Wozney, J., & Mundy, G.R. (1994.) Expression of bone morphogenetic protein messenger RNA in prolonged cultures of fetal rat calvarial cells. *Journal of Bone and Mineral Research, 9*, 389-394.

Harshfield, G.A., Pulliam, D.A., & Alpert, B.S. (1991.) Patterns of sodium excretion during sympathetic nervous system arousal. *Hypertension, 17*, 1156-1170.

Hart, D.A., Archambault, J.M., Kydd, A., Reno, C., Frank, C.B., & Herzog, W. (1998.) Gender and neurogenic variables in tendon biology and repetitive motion disorders. *Clinical Orthopaedics and Related Research, 351*, 44-56.

Hart, P.D'A., & Verney, E.B. (1934.) Observations on the rate of water loss by man at rest. *Clinical Science, 1*, 367-396.

Harte, J.L., Eifert, G.H., & Smith, R. (1995.) The effects of running and mediation on beta-endorphin, corticotropin-releasing hormone and cortisol in plasma, and on mood. *Biological Psychology, 40*, 251-265.

Hartley, L.H., Mason, J.W., & Hogan, R.P. (1972a.) Multiple hormonal responses to graded exercise in relation to physical training. *Journal of Applied Physiology, 33*, 602-606.

Hartley, L.H., Mason, J.W., & Hogan, R.P. (1972b.) Multiple hormonal responses to prolonged exercise in relation to physical training. *Journal of Applied Physiology, 33,* 607-610.

Hartman, M.L., Clayton, P.E., Johnson, M.L., Celniker, A., Perlman, A.J., Alberti, K.K., & Thorner, M.O. (1993.) A low dose euglycemic infusion of recombinant human insulin-like growth factor I rapidly suppresses fasting-enhanced pulsatile growth hormone secretion in humans. *Journal of Clinical Investigation, 91,* 2453-2462.

Hartman, M.L., Veldhuis, J.D., Johnson, M.L., Lee, M.M., Alberti, K.K., Samojlik, E., & Thorner, M.O. (1992.) Augmented growth hormone (GH) secretory burst frequency and amplitude mediate enhanced GH secretion during two day fast in normal men. *Journal of Clinical Endocrinology and Metabolism, 74,* 757-765.

Hartmann, H., Beckh, K., & Jungermann, K. (1982.) Direct control of glycogen metabolism in the perfused rat liver by the sympathetic innervation. *European Journal of Biochemistry, 123,* 521-526.

Hasser, E.M., Bishop, V.S., & Hay, M. (1997.) Interactions between vasopressin and baroreflex control of the sympathetic nervous system. *Clinical and Experimental Pharmacology and Physiology, 24,* 102-108.

Hasson, S.M., & Barnes, W.S. (1987.) Blood glucose levels during rest and exercise: Influence of fructose and glucose ingestion. *Journal of Sports Medicine and Physical Fitness, 27,* 326-332.

Hataya, Y., Akamizu, T., Takaya, K., et al. (2001.) A low dose of ghrelin stimulates growth hormone (GH) release synergistically with GH-releasing hormone in humans. *Journal of Clinical Endocrinology and Metabolism, 86,* 4552.

Hatch, M.C., Shu, X.-O., McLean, D.E., Levin, B., Begg, M., Reuss, L., & Susser, M. (1993.) Maternal exercise during pregnancy, physical fitness and fetal growth. *American Journal of Epidemiology, 137,* 1105-1114.

Hatfield, B.D., Spalding, T.W., Santa Maria, D.L., Porges, S.W., Potts, J.T., et al. (1998.) Respiratory sinus arrhythmia during exercise in aerobically trained and untrained men. *Medicine and Science in Sports and Exercise, 30,* 206-214.

Haupt, H.A., & Rovere, G.D. (1984.) Anabolic steroids: A review of the literature. *American Journal of Sports Medicine, 12,* 469-485.

Haus, E., & Tuitou, Y. (1994.) Principles of chronobiology. In Y. Tuitou & E. Haus (Eds.), *Biologic rhythms in clinical and laboratory medicine* (pp. 6-34). Berlin: Springer-Verlag.

Hauschka, P.V., Mavrakos, A.E., Iafrati, M.D., Doleman, S.E., & Klagsburn, M. (1986.) Growth factors in bone matrix: isolation of multiple types by affinity chromatography on heparin sepharose. *Journal of Biological Chemistry, 261,* 12665-12674.

Hauser, K.F., & Toran-Allerand, C.D. (1989.) Androgen increases the number of cells in fetal mouse spinal cord cultures: implications for motoneuron survival. *Brain Research, 485,* 157-164.

Hausmann, R., Hammer, S., & Betz, P. (1998.) Performance enhancing drugs (doping agents) and sudden death—a case report and review of the literature. *International Journal of Legal Medicine, 111,* 261-264.

Havel, P.J., Mundinger, T.O., Veith, R.C., Dunning, B.E., & Taborsky, G.J., Jr. (1992.) Co-release of galanin and NE from pancreatic sympathetic nerves during severe hypoglycemia in dogs. *American Journal of Physiology, 263,* E8-E16.

Havel, P.J., Veith, R.C., Dunning, B.E., & Taborsky, G.J., Jr. (1988.) Pancreatic noradrenergic nerves are activated by neuroglucopenia but not by hypotension or hypoxia in the dog: evidence for stress-specific and regionally selective activation of the sympathetic nervous system. *Journal of Clinical Investigation, 82,* 1538-1545.

Havel, P.J., Veith, R.C., Dunning, B.E., & Taborsky, G.J., Jr. (1991.) Role for the autonomic nervous system to increase pancreatic glucagon secretion during marked insulin-induced hypoglycemia in dogs. *Diabetes, 40,* 1107-1114.

Havel, R.J., & Goldfien, A. (1959.) The role of sympathetic nervous system in the metabolism of free fatty acids. *Journal of Lipid Research, 205,* 102-108.

Havlik, R.J., Hubert, H.B., Fabsitz, R.R., & Feinleib, M. (1983.) Weight and hypertension. *Annals of Internal Medicine, 98,* 855-859.

Hay, W.W. (1991.) Glucose metabolism in fetal-placental unit. In R.M. Cowett (Ed.), *Principles of fetal and neonatal metabolism* (pp. 250-275). Berlin: Springer-Verlag.

Hayashi, T., Hirshman, M.F., Fujii, N., Habinowski, S.A., Witters, L.A., & Goodyear, L.J. (2000.) Metabolic stress and altered glucose transport: activation of AMP-activated protein kinase as a unifying coupling mechanism. *Diabetes, 49,* 527-531.

Hayashi, T., Hirshman, M.F., Kurth, E.J., Winder, W.W., & Goodyear, L.J. (1998.) Evidence for 5' AMP-activated protein kinase mediation of the effect of muscle contraction on glucose transport. *Diabetes, 47,* 1369-1373.

Hayashi, T., & Nakamura, K. (1984.) Localized effects of naloxone on local cerebral glucose utilization in rat cerebral nuclei with Met-enkephalinergic neurons. *Japanese Journal of Pharmacology, 36,* 339-348.

Hayashi, T., Wojtaszewski, J.F., & Goodyear, L.J. (1997.) Exercise regulation of glucose transport in skeletal muscle. *American Journal of Physiology, 273,* E1039-E1051.

Hayes, V.Y., Urban, R.J., Jiang, J., Marcell, T.J., Helgeson, K., & Mauras, N. (2001.) Recombinant human growth hormone and recombinant human insulin-like growth factor I diminish the catabolic effects of hypogonadism in man: metabolic and molecular effects. *Journal of Clinical Endocrinology and Metabolism, 86,* 2211-2219.

Heath, G.W., Gavin III, J.R., Hinderliter, J.M., Hagberg, J.M., Bloomfield, S.A., & Holloszy, J.O. (1983.) Effects of exercise and lack of exercise on glucose tolerance and insulin inputs. In V.B. Brooks (Ed.), *Handbook of physiology, section I. The nervous system. Vol. II.* (pp. 423-507). Bethesda, MD: The American Physiological Society.

Heath, H., III. (1980.) Biogenic amines and the secretion of parathyroid hormone and calcitonin. *Endocrine Reviews, 1,* 319-338.

Heath, J.K., Atkinson, S.J., Meikle, M.C., & Reynolds, J.J. (1984.) Mouse osteoblasts synthesize collagenase in response to bone resorbing agents. *Biochemica et Biophysica Acta, 802,* 151-154.

Hefti, M.A., Harder, B., Eppenberger, H.M., & Schaub, M.C. (1997.) Signalling pathways in cardiac myocyte hypertrophy. *Journal of Molecular and Cellular Cardiology, 29,* 2871-2892.

Heini, A., Schutz, Y., Diaz, E., Prentice, A.M., Whitehead, R.G., & Jequier, E. (1991.) Free-living energy expenditure measured by two independent techniques in pregnant and nonpregnant Gambian women. *American Journal of Physiology, 261,* E9-17.

Heino, J. (2000.) The collagen receptor integrins have distinct ligand recognition and signaling functions. *Matrix Biology, 19,* 319-323.

Hellmer, J., Marcus, C., Sonnenfeld, T., & Arner, P. (1992.) Mechanisms for differences in lipolysis between human subcutaneous and omental fat cells. *Journal of Clinical Endocrinology and Metabolism, 75,* 15-20.

Hellsten, Y., Maclean, D., Radegran, G., Saltin, B., & Bangsbo, J. (1998.) Adenosine concentrations in the interstitium of resting and contracting human skeletal muscle. *Circulation, 98,* 6-8.

Hellstrom, L., Blaak, E., & Hagstrom-Toft. E. (1996.) Gender differences in adrenergic regulation of lipid mobilization during exercise. *International Journal of Sports Medicine, 17,* 439-447.

Hellstrom, L., Rossner, S., Hagstrom-Toft, E., & Reynisdottir, S. (1997.) Lipolytic catecholamine resistance linked to alpha 2-adrenoceptor sensitivity—a metabolic predictor of weight loss in obese subjects. *International Journal of Obesity and Related Metabolic Disorders, 21,* 314-320.

Henderson, B.E., Ross, R.K., Judd, H.L., Krailo, M.D., & Pike, M.C. (1985.) Do regular ovulatory cycles increase breast cancer risk? *Cancer, 56,* 1206-1208.

Henderson, B.E., Ross., & Bernstein, L. (1988.) Estrogens as a cause of human cancer: The Richard and Hinda Rosenthal Foundation award lecture. *Cancer Research, 48,* 246-253.

Henderson, S.A., Graham, H.K., Mollan, R.A.B., Riddoch, C., Sheridan, B., & Johnston, H. (1989.) Calcium homeostasis and exercise. *International Journal of Orthopedics, 13,* 69-73.

Henneman, E., & Mendell, L.M. (1981.) Functional organization of motoneuron pool and its inputs. In V.B. Brooks (Ed.), *Handbook of physiology, section I, The nervous system, Vol. II.* (pp 423-507). Bethesda, MD: American Physiological Society Press.

Henriksson, J. (1995.) Muscle fuel selection: Effect of exercise training. *Proceedings of the Nutrition Society, 54,* 125-138.

Henriksson, J. (1977.) Training induced adaptation of skeletal muscle and metabolism during submaximal exercise. *Journal of Physiology, 270,* 661-675.

Henriksson, J., & Reitman, J.S. (1976.) Quantitative measures of enzyme activities in type I and type II muscle fibres of man after training. *Acta Physiologica Scandinavica, 97,* 392-397.

Herman-Giddens, M.E., Slora, E.J., Wasserman, R.C., Bourdony,C.J., Bhapkar, M.V., Koch, G.G., & Hasemeier, C.M. (1997.) Secondary sexual characteristics and menses in young girls seen in office practice: a study from the Pediatric Research in Office Settings Network. *Pediatrics, 99*, 505-512.

Herold, M., Cornelissen,G., Loeckinger, A., Koeberle, D., Koenig, P., & Halberg, F. (1998.) About 8-hour variation of circulating human endothelin-1. *Peptides, 19*, 821-825.

Herold, M., Cornelissen, G., Rawson, M.J., Katinas, G.S., Alinder, C., Bratteli, C., et al. (2000.) About-daily (circadian) and about-weekly (circaseptan) patterns of human salivary melatonin. *Journal of Anti-Aging Medicine, 3*, 263-267.

Herring, J.L., Mole, P.A., Meredith, C.N., & Stern, J.S. (1992.) Effect of suspending exercise training on resting metabolic rate in women. *Medicine and Science in Sports and Exercise, 24*, 59-64.

Herrington, J., & Carter-Su, C. (2001.) Signaling pathways activated by the growth hormone receptor. *Trends in Endocrinology and Metabolism, 12*, 252-257.

Herrington, J., Smit, L.S., Schwartz, J., & Carter-Su, C. (2000.) The role of STAT proteins in growth hormone signaling. *Oncogene, 19*, 2585-2597.

Herskovits, M.S., & Singh, I.J. (1984.) Effect of guanethidine-induced sympathectomy on osteoblastic activity in the rat femur evaluated by 3_H-proline autoradiography. *Acta Anatomica, 120*, 151-155.

Herz, A. (1995.) Role of immune processes in peripheral opiod analgesia. *Advances in Experimental Medicine and Biology, 373*, 193-199.

Hespel, P., Lijnen, P., Van Hoof, R., Fagard, R., Goosens, W., Lissens, W., Moerman, E., & Amery, A. (1988.) Effects of physical endurance training on the plasma renin-angiotensin-aldosterone system in normal man. *Journal of Endocrinology, 116*, 443-449.

Hetherington, A.W., & Ranson, S.W. (1940.) Hypothalamic lesions and adiposity in the rat. *Anatomical Record, 78*, 149-172.

Hew, F.L., O'Neal, D., Kamarudin, N., Alford, F.P., & Best, J.D. (1998.) Growth hormone deficiency and cardiovascular risk. *Baillieres Clinical Endocrinology and Metabolism, 12*, 199-216.

Hexum, T.D., Majane, E.A., Russett, L.R., & Yang, H.Y. (1987.) Neuropeptide Y release from the adrenal medulla after cholinergic receptor stimulation. *Journal of Pharmacology and Experimental Therapeutics, 243*, 927-930

Hexum, T.D., & Russett, L.R. (1987.) Plasma enkephalin-like peptide response to chronic nicotine infusion in guinea pig. *Brain Research, 406*, 370-372.

Hexum, T.D., & Russett, L.R. (1989.) Stimulation of cholinergic receptor mediated secretion from the bovine adrenal medulla by neuropeptide Y. *Neuropeptides, 13*, 35-41.

Hickey, M.S., Carey, J.O., Azevedo, J.L., Houmard, J.A., Pories, W.J., Israel, R.G., & Dohm, G.L. (1995a.) Skeletal muscle fiber composition is related to adiposity and in vitro glucose transport rate in humans. *American Journal of Physiology, 268*, E453-E557.

Hickey, M.S., Costill, D.L., Vukovich, M.D., Kryzmenski, K., & Widrick, J.J. (1993.) Time of day effects on sympathoadrenal and pressor reactivity to exercise in healthy men. *European Journal of Applied Physiology and Occupational Physiology, 67*, 159-163.

Hickey, M.S., Tanner, C.J., O'Neill, D.S., Morgan, L.J., Dohm, G.L., & Houmard JA. (1997.) Insulin activation of phosphatidylinositol 3-kinase in human skeletal muscle in vivo. *Journal of Applied Physiology, 83*, 718-722.

Hickey, M.S., Weidner, M.D., Gavigan, K.E., Zheng, D., Tyndall, G.L., & Houmard, J.A. (1995b.) The insulin action-fiber type relationship in humans is muscle group specific. *American Journal of Physiology, 269*, E150-E154.

Hickner, R.C., Fisher, J.S., Ehsani, A.A., & Kohrt, W.M. (1997.) Role of nitric oxide in skeletal muscle blood flow during dynamic exercise in humans. *American Journal of Physiology, 273*, H405-H410.

Hickson, R.C., Czerwinski, S.M., Falduto, M.T., & Young, A.P. (1990.) Glucocorticoid antagonism by exercise and androgenic-anabolic steroids. *Medicine and Science in Sports and Exercise, 22*, 331-340.

Hickson, R.C., Rennie, M.J., Conlee, R.K., Winder, W.W., & Holloszy, J.O. (1977.) Effects of increased plasma fatty acids on glycogen utilization and endurance. *Journal of Applied Physiology, 43*, 829-833.

Higaki, Y., Hirschman, M.F., Fujii, N., & Goodyear, L.J. (2001.) Nitric oxide increases glucose uptake through a mechanism that is distinct from the insulin and contraction pathways in rat skeletal muscle. *Diabetes, 50*, 241-247.

Higashi, Y., Oshima, T., Ozono, R., Matsura, H., & Kajiyama, G. (1997.) Aging and severity of hypertension attenuate endothelium-dependent renal vascular relaxation in humans. *Hypertension, 30*, 252-258.

Higashi, Y., Sasaki, S., Kurisu, S., Yoshimuzu, A., Sasaki, N., Matsuura, H. Kajiyama, G., & Oshima, T. (1999.) Regular aerobic exercise augments endothelium-dependent vascular relaxation in normotensive as well as hypertensive subjects. *Circulation, 100*, 1194-1202.

Highman, T.J., Friedman , J.E., Huston, L.P., Wong, W.W., & Catalano, P.M. (1998.) Longitudinal changes in maternal serum leptin concentrations, body composition, and resting metabolic rate in pregnancy. *American Journal of Obstetrics and Gynecology, 178*, 1010-1015.

Hill, D.J., Strain, A.J., & Milner, R.D.G. (1987.) Growth factors in embryogenesis. *Oxford Review of Physiology, 9*, 389-455.

Hill, D.W. (1996.) Effect of time of day on aerobic power in exhaustive high-intensity exercise. *Journal of Sports Medicine and Physical Fitness, 36*, 155-60.

Hill, D.W., & Smith, J.C. (1991.) Circadian rhythm in anaerobic power and capacity. *Canadian Journal of Sport Sciences, 16*, 30-32.

Hill, D.W., Borden, D.O., Darnaby, K.M., Hendricks, D.N., & Hill, C.M. (1992.) Effect of time of day on aerobic and anaerobic responses to high-intensity exercise. *Canadian Journal of Sport Sciences, 17*, 316-319.

Hill, D.W., Cureton, K.J., & Collins, M.A. (1989a.) Effect of time of day on perceived exertion at work rates above and below the ventilatory threshold. *Research Quarterly for Exercise and Sport, 60*, 127-133.

Hill, D.W., Cureton, K.J., & Collins, M.A. (1989b.) Circadian specificity in exercise training. *Ergonomics, 32*, 79-92

Hill, D.W., Leiferman, J.A., Lynch, N.A., Dangelmaier, B.S., & Burt, S.E. (1998.) Temporal specificity in adaptations to high-intensity exercise training. *Medicine and Science in Sports and Exercise, 30*, 450-455.

Hilsted, J., Galbo, H., Sonne, B., Schwartz, Fahrenkrug, J., Schaffalitzky, D., Muckadell, O., Lauritsen, K.B., & Tronier B. (1980.) Gastroenteropancreatic hormonal changes during exercise. *American Journal of Physiology, 239*, G136-G140.

Hilton, L.K., & Loucks, A.B. (2000.) Low energy availability, not exercise stress, suppresses the diurnal rhythm of leptin in healthy young women. *American Journal of Physiology, 278*, E43-E49.

Hinkleman, L. L., & Nieman, D. C. (1993.) The effects of a walking program on body composition and serum lipids and lipoproteins in overweight women. *Journal of Sports Medicine and Physical Fitness, 33*, 49-58.

Hinson, J.P., Kapas, S., Orford, C.D., & Vinson, G.P. (1992.) Vasoactive intestinal peptide stimulation of aldosterone secretion by the rat adrenal cortex may be mediated by the local release of catecholamines. *Journal of Endocrinology, 133*, 253-258.

Hirata, K., Nagasaka, T., Hirai, A., Hirashita, M., Takahata, T., & Nunomura, T. (1986.) Effects of human menstrual cycle on thermoregulatory vasodilation during exercise. *European Journal of Applied Physiology and Occupational Physiology, 54*, 559-565.

Hirose, H., Maruyama, H., Ito, K., Kido, K., Koyama, K., & Saruta T. (1993a.) Effects of a alpha 2- and beta-adrenergic agonism on glucagon secretion from perfused pancreata of normal and streptozotocin-induced diabetic rats. *Metabolism: Clinical and Experimental, 42*, 1072-1076.

Hirose, H., Maruyama, H., Ito, K., Koyama, K., Kido, K., & Saruta, T. (1993b.) Glucose-induced insulin secretion and alpha 2-adrenergic receptor subtypes. *Journal of Laboratory and Clinical Medicine, 121*, 32-37.

Hirsch, I.B., Marker, J.C., Smith, L.J., Spina, R.J., Parvin, C.A., Holloszy, J.O., & Cryer, P.E. (1991.) Insulin and glucagon in prevention of hypoglycemia during exercise in humans. *American Journal of Physiology, 260*, E695-E704.

Hirschberg, A.L., Lindholm, C., Carlstrom, K., & Van Schoultz, B. (1994.) Reduced serum cholecystokinin response to food intake in female athletes. *Metabolism, 43*, 217-222.

Hirshman, M.F., Wardzala, L.J., Goodyear, L.J., Fuller, S.P., Horton, E. D., & Horton, E.S. (1989.) Exercise training increases the number of glucose transporters in rat adipose tissue. *American Journal of Physiology, 257*, E520-E530.

Hjalmarson, A., Gilpin, E., Nicod, P., Henning, H., Ross, J. Jr., & the SCOR Database. (1988.) Circadian pattern of onset of symptoms in acute myocardial infarction differs among clinical subsets. *Circulation, 78* (Suppl. II), II-437.

Hjemdahl, P., Chronos, N.A., Wilson, D.J., Bouloux, P., & Goodall, A.H. (1994.) Epinephrine sensitizes human platelets in vivo and in vitro as studied by fibrinogen binding and P-selectin expression. *Arteriosclerosis and Thrombosis, 14*, 77-84.

Ho, K.Y., Evans, W.S., Blizzard, R.M., Veldhuis, J.D., Merriam, G.R., Samojlik, E., Furlanetto, R., Rogol, A.D., Kaiser, D.L., & Thorner, M.O. (1987.) Effects of sex and age on the 24-hour profile of growth hormone secretion in man: Importance of endogenous estradiol concentrations. *Journal of Clinical Endocrinology and Metabolism, 64*, 51-58.

Ho, K.Y., Veldhuis, J.D., Johnson, M.L., Furlanetto, R., Evans, W.S., Alberti, K.G., & Thorner, M.O. (1988.) Fasting enhances growth hormone secretion and amplifies the complex rhythms of growth hormone secretion in man. *Journal of Clinical Investigation, 81*, 968-975.

Hodel, A. (2001.) Effects of glucocorticoids on adrenal chromaffin cells. *Journal of Neuroendocrinology, 13*, 217-221.

Hodgetts, V., Coppack, S.W., Frayn, K.N., & Hockaday, D.R. (1991.) Factors controlling fat mobilization from human subcutaneous adipose tissue during exercise. *Journal of Applied Physiology, 71*, 445-451.

Hoeldtke, R.D., & Streeten, D.H.P. (1993.) Treatment of orthostatic hypotension with erythropoietin. *New England Journal of Medicine, 329*, 611-615.

Hoelzer, D.R., Dalsky, G.P., Clutter, W.E., Shah, S.D., Holloszy, J.O., & Cryer, P.E. (1986.) Glucoregulation during exercise: hypoglycemia is prevented by redundant glucoregulatory systems, sympathochromaffin activation, and changes in islet hormone secretion. *Journal of Clinical Investigation, 77*, 212-221.

Hoffer, L.J., Beitins, I.Z., Kyung, N.-H., & Bistrian, B.R. (1986.) Effects of severe dietary restriction on male reproductive hormones. *Journal of Clinical Endocrinology and Metabolism, 62*, 288-292.

Hoffman, D.M., Crampton, L., Sernia, C., Nguyen, T.V., & Ho, K.K. (1996.) Short-term growth hormone (GH) treatment of GH-deficient adults increases body sodium and extracellular water, but not blood pressure. *Journal of Clinical Endocrinology and Metabolism, 81*, 1123-1128.

Hoffman, K. (1972.) Stride length and frequency of female sprinters. *Track Technique, 48*, 1522-1524.

Hoffstedt, J., Arner, P., Hellers, G., & Lonnqvist, F. (1997.) Variation in adrenergic regulation of lipolysis between omental and subcutaneous adipocytes from obese and non-obese men. *Journal of Lipid Research, 38*, 795-804.

Hoffstedt, J., Shimizu, M., Sjostedt, S., & Lonnqvist, F. (1995.) Determination of beta 3-adrenoceptor mediated lipolysis in human fat cells. *Obesity Research, 3*, 447-457.

Hofman, M.A. (2000.) The human circadian clock and aging. *Chronobiology International, 17*, 245-259.

Hofman, M.A., & Swaab, D.F. (1993.) Diurnal and seasonal rhythms of neuronalactivity in the suprachiasmatic nucleus of humans. *Journal of Biological-Rhythms, 8*, 283-295.

Hofman, M.A., & Swaab, D.F. (1995.) Influence of aging on the seasonal rhythm of the vasopressin-expressing neurons in the human suprachiasmatic nucleus. *Neurobiology of Aging, 16*, 965-971.

Hofmann, S., & Pette, D. (1994.) Low-frequency stimulation of rat fast-twitch muscle enhances the expression of hexokinase II and both the translocation and expression of glucose transporter 4 (GLUT-4). *European Journal of Biochemistry, 219*, 307-315.

Hohtari, H., Salminen-Lappalainen, K., & Laatikainen, T. (1991.) Response of plasma endorphins, corticotropin, cortisol, and luteinizing hormone in the corticotropin-releasing hormone stimulation tests in eumenorrheic and amenorrheic athletes. *Fertility and Sterility, 55*, 276-280.

Holder, L.E., David, J.G., Lampkin, B.C., Nishiyama, H., & Perkins, P. (1975.) Hepatoma associated with anabolic steroid therapy. *American Journal of Roentgenology, 124*, 638-642.

Hollenberg, N.K., Williams, G.H., & Adams, D.F. (1981.) Essential hypertension, abnormal renal vascular and endocrine responses to a mild psychological stimulus. *Hypertension, 3*, 3-11.

Holloszy, J.O. (1967.) Effects of exercise on mitochondrial oxygen uptake and respiratory enzyme activity in skeletal muscle. *Journal of Biological Chemistry, 242*, 2278-2282.

Holloszy, J.O., & Booth, F.W. (1976.) Biochemical adaptations to endurance exercise in muscle. *Annual Review of Physiology, 38*, 273-291.

Holloszy, J.O., & Coyle, E.F. (1984.) Adaptations of skeletal muscle to endurance exercise and their metabolic consequences. *Journal of Applied Physiology, 56*, 831-838.

Holloszy, J.O., Oscai, L.B., Don, I.J., & Mole, P.A. (1970.) Mitochondrial citric acid cycle and related enzymes: adaptive response to exercise. *Biochemical and Biophysical Research Communications, 40*, 1368-1373.

Holma, P.K. (1977.) Effects of anabolic steroid (metandienone) on spermatogenesis. *Contraception, 15*,151-162.

Holmer, S., Rinne, B., Eckardt, K.U., Le Hir, M., Schricker, K., Kaissling, B., Rieger, G., & Kurtz, A. (1994.) Role of renal nerves for the expression of renin in adult rat kidney. *American Journal of Physiology, 266*, F738-F745.

Holmes, S.J., & Shalet, S.M. (1996.) Role of growth hormone and sex steroids in achieving and maintaining normal bone mass. *Hormone Research, 45*, 86-93.

Holsboer, F., Graser A., Fries, E., & Wiedeman, K. (1994.) Steroid effects on central neurons and implications for psychiatric and neurologic disorders. *Annals of New York Academy of Science, 746*, 345-359.

Holst, M.C., Kelly, J.B., & Powley, T.L. (1997.) Vagal preganglionic projections to the enteric nervous system characterized with phaseolus vulgaris-leucoagglutinin. *Journal of Comparative Neurology, 381*, 81-100.

Honma, K.-I., Honma, S., Nakamura, K., Sasaki, M., Endo, T., & Takahashi, T. (1995.) Differential effects of bright light and social cues on reentrainment of human circadian rhythms. *American Journal of Physiology, 268*, R528-R535.

Horber, F.F., Kohler, S.A., Lippuner, K., & Jaeger, P. (1996.) Effect of regular physical training on age-associated alteration of body composition in men. *European Journal of Clinical Investigation, 26*, 279-285.

Hori, T., Katafuchi, T., Take, S., Shimizu, N., & Niijima, A. (1995.) The autonomic nervous system as a communication channel between the brain and the immune system. *Neuroimmunomodulation, 2*, 203-215.

Horne, J.A. (1981.) The effects of exercise upon sleep: a critical review. *Biological Psychology, 12*, 241-290.

Horne, J.A. (1983.) Human sleep and tissue restitution: some qualifications and doubts. *Clinical Science, 65*, 569-578.

Horne, J. A., & Ostberg, O. (1976.) A self-assessment questionnaire to determine morningness-eveningness in human circadian rhythms. *International Journal of Chronobiology, 4*, 97-110.

Horne, J.A., & Porter, J.M. (1975.) Exercise and human sleep. *Nature, 256*, 573-575.

Horne, J.A., & Porter, J.M. (1976.) Time of day effects with standardized exercise upon subsequent sleep. *Electroencephalography and Clinical Neurophysiology, 40*, 178-184.

Horne, J.A., & Reid, A.J. (1985.) Night-time sleep EEG changes following body heating in a warm bath. *Electroencephalography and Clinical Neurophysiology, 60*, 154-157.

Horne, J.A., & Staff, L.H. (1983.) Exercise and sleep: Body-heating effects. *Sleep, 6*, 36-46.

Hornum, M., Cooper, D.M., Brasel, J.A., Bueno, A., & Sietsema, K.E. (1997.) Exercise-induced changes in circulating growth factors with cyclic variation in plasma estradiol in women. *Journal of Applied Physiology, 82*, 1946-1951.

Horowitz, J.F., & Coyle, E.F. (1993.) Metabolic responses to preexercise meals containing various carbohydrates and fat. *American Journal of Clinical Nutrition, 58*, 235-241.

Horowitz, J.F., & Klein, S. (2000a.) Lipid metabolism during endurance exercise. *American Journal of Clinical Nutrition, 72* (2 Suppl.), 558S-563S.

Horowitz, J.F., & Klein, S. (2000b.) Whole body and abdominal lipolytic sensitivity to epinephrine is suppressed in upper body obese women. *American Journal of Physiology, 278*, E1144-E1152.

Horowitz, J.F., Braudy, R.J., Martin III, W.H., & Klein, S. (1999a.) Endurance exercise training does not alter lipolytic adipose tissue blood flow sensitivity to epinephrine. *American Journal of Physiology, 277*, E325-E331.

Horowitz, J.F., Coppack, S.W., Paramore, D., Cryer, P.E., Zhao, G., & Klein, S. (1999b.) Effect of short-term fasting on lipid kinetics in lean and obese women. *American Journal of Physiology, 276*, E278-E284.

Horowitz, J.F., Leone, T.C., Feng, W., Kelly, D.P., & Klein, S. (2000c.) Effect of endurance training on lipid metabolism in women: a potential role for PPARalpha in the metabolic response to training. *American Journal of Physiology, 279*, E348-E355.

Horowitz, J.F., Mora-Rodriguez, R., Byerley, L.O., & Coyle, E.F. (1997.) Lipolytic suppression following carbohydrate ingestion limits fat oxidation during exercise. *American Journal of Physiology, 273,* E768-E775.

Horton, E.S. (1992.) Exercise and physical training: effects on insulin sensitivity and glucose metabolism. *Diabetes and Metabolism Reviews, 2,* 1-17.

Horton, E.S., & Terjung, R.L. (1988.) *Exercise, nutrition and energy metabolism.* New York: Macmillan.

Horton, M.G., & Hall, T.L. (1989.) Quadriceps femoris muscle angle: normal values and relationships with gender and selected skeletal measures. *Physical Therapy, 69,* 897-901.

Horton, T.J., Pagliassotti, M.J., Hobbs, K., & Hill, J.O. (1998.) Fuel metabolism in men and women during and after long-duration exercise. *Journal of Applied Physiology, 85,* 1823-1832.

Horvath, I., Sandor, N.T., Ruttner, Z., & McLaughlin, A.C. (1994.) Role of nitric oxide in regulating cerebrocortical blood flow during hypercapnia. *Journal of Cerebral Blood Flow and Metabolism, 14,* 503-509.

Houmard, J.A., Hickey, M.S., Tyndall, G.L., Gavigan, K.E., & Dohm, G.L. (1995.) Seven days of exercise increase GLUT-4 protein content in human skeletal muscle. *Journal of Applied Physiology, 79,* 1936-1938.

Houmard, J.A., Shaw, C.D., Hickey, M.S., & Tanner, C.J. (1999.) Effect of short-term exercise training on insulin-stimulated PI 3-kinase activity in human skeletal muscle. *American Journal of Physiology, 277,* E1055-E1060.

Houseknecht, K.L., & Bauman, D.E. (1997.) Regulation of lipolysis by somatotropin: Functional alteration of adrenergic and adenosine signaling in bovine adipose tissue. *Journal of Endocrinology, 152,* 465-475.

Houssay, B.A., Molinelli, E.A., & Lewis, J.T. (1924.) Accion de la insulina sobre la secrecion de adrenalina. *Revia Asociacion di Medicina de Argentina, 37,* 486-499.

Howard, B.V. (1999.) Insulin resistance and lipid metabolism. *American Journal of Cardiology, 84,* 28J-32J.

Howlett, K., Febbraio, M., & Hargreaves, M. (1999a.) Glucose production during strenuous exercise in humans: role of epinephrine. *American Journal of Physiology, 276,* E1130-E1135.

Howlett, K., Galbo, H., Lorentsen, J., Bergeron, R., Zimmerman-Belsing, T., Bulow, J., Feldt-Rasmussen, U., & Kjaer, M. (1999b.) Effect of adrenaline on glucose kinetics during exercise in adrenolectomised humans. *Journal of Physiology, 519,* 911-921.

Hsueh, A.J., Adashi, E.Y., Jones, P.B., & Welsh, T.H. (1984.) Hormonal regulation of the differentiation of cultured ovarian granulosa cells. *Endocrine Reviews, 5,* 76-127.

Huang, Y., Li , J., Zhang, Y., & Wu, C. (2000.) The roles of integrin-linked kinase in the regulation of myogenic differentiation. *Journal of Cell Biology, 150,* 861-872.

Huang, Z., Willett, W.C., Manson, J.E., Rosner, B., Stampfer, M.J., Speizer, F.E., & Colditz, G.A. (1998.) Body weight, weight change, and risk for hypertension in women. *Annals of Internal Medicine, 128,* 81-88.

Hubert, P., King, N.A., & Blundell, J.E. (1998.) Uncoupling the effects of energy expenditure and energy intake: appetite response to short-term energy deficit induced by meal omission and physical activity. *Appetite, 31,* 9-19.

Huddleston, A.L., Rockwell, D., Kulund, D.N., & Harrison, R.B. (1980.) Bone mass in lifetime tennis players. *Journal of the American Medical Association, 244,* 1107-1109.

Hughes, S.M., Chi, M.M., Lowry, O.H., & Gundersen, K.(1999.) Myogenin induces a shift of enzyme activity from glycolytic to oxidative metabolism in muscles of transgenic mice. *Journal of Cell Biology, 145,* 633-642.

Hughes, S.M., Koishi, K., Rudnicki, M., & Maggs, A.M. (1997.) MyoD protein is differentially accumulated in fast and slow skeletal muscle fibres and required for normal fibre type balance in rodents. *Mechanisms of Development, 61,* 151-163.

Hughes, S.M., Taylor, J.M., Tapscott, S.J., Gurley, C.M., Carter, W.J., & Peterson, C.A. (1993.) Selective accumulation of MyoD and myogenin mRNAs in fast and slow adult skeletal muscle is controlled by innervation and hormones. *Development Supplement, 118,* 1137-1147,

Hurel, S.J., Koppiker, N., Newkirk, J., Close, P.R., Miller, M., Mardell, R,. Wood, P.J., & Kendall-Taylor, P. (1999.) Relationship of physical exercise and ageing to growth hormone production. *Clinical Endocrinology, 51,* 687-691.

Hurley, B.F., Nemeth, P.M., Martin, W.H., Hagberg, J.M., Dalsky, G.P., & Holloszy, J.O. (1986.) Muscle triglyceride utilization during exercise: effect of training. *Journal of Applied Physiology, 60,* 562-567.

Huszar, D., Lynch, C.A., Fairchild-Huntress, V., Dunmore, J.H., Fang, Q., Berkemeier, L.R., et al. (1997.) Targeted disruption of the melanocortin-4 receptor results in obesity in mice. *Cell, 88,* 131-141.

Hutchinson, P.L., Cureton, K.J., Outz, H., & Wilson, G. (1991.) Relationship of cardiac size to maximal oxygen uptake and body size in men and women. *International Journal of Sports Medicine, 12,* 369-373.

Hutchinson, M.R., & Ireland, M.L. (1995.) Knee injuries in female athletes. *Sports Medicine, 19,* 288-302.

Hynes, R. (1992.) Versatility, modulation, and signaling in cell adhesion. *Cell, 69,* 11-25.

Iaccarino, G., Barbato, E., Cipoletta, E., Fiorillo, A., & Trimarco, B. (2001.) Role of the sympathetic nervous system in cardiac remodeling in hypertension. *Clinical and Experimental Hypertension, 23,* 35-43.

Ianuzzo, C.D., & Chen, V. (1979.) Metabolic character of hypertrophied rat muscle. *Journal of Applied Physiology, 46,* 738-742.

Ichiki, T., Labosky, P.A., Shiota, C., Okuyama, S., Imagawa, Y., Fogo, A., Niimura, F., Ichikawa, I., Hogan, B.L., & Inagami T. (1995.) Effects on blood pressure and exploratory behaviour of mice lacking angiotensin II type-2 receptor. *Nature. 377,* 748-750.

Iellamo, F., Legramante, J.M., Raimondi, G., & Peruzzi, G. (1997.) Baroreflex control of sinus node during dynamic exercise in humans: effects of central command and muscle reflexes. *American Journal of Physiology, 272,* H1157-H1164.

Imai, Y., Abe, K., Munakata, M., Sakuma, H., Hashimoto, J., Imai, K., Sekino, K., & Yoshinaga, K. (1990.) Circadian blood pressure variations under different pathophysiological conditions. *Journal of Hypertension – Supplement, 8,* S125-S132.

Imaki, T., Shibasaki, T., Shizume, K., Masuda, A., Hotta, M., Kiyosawa, Y., Jibiki, K., Demura, H., Tsushima, T. & Ling, N. (1985.) The effect of free fatty acids on growth hormone (GH)- releasing hormone-mediated GH secretion in man. *Journal of Clinical Endocrinology and Metabolism, 60,* 290-293.

Ingber, D.E. (1993.) Cellular tensegrity: defining new rules of biological design that govern the cytoskeleton. *Journal of Cell Science, 104,* 613-627.

Ingram, C.D., Ciobanu, R., Coculescu, I.L., Tanasescu, R., Coculescu, M., Mihai, R. (1998.) Vasopressin neurotransmission and the control of circadian rhythms in the suprachiasmatic nucleus. *Progress in Brain Research, 119,* 351-364.

Innui, M., Wang, S., Saito, A., & Fleischer, S. (1991.) Purification of the ryanodine receptor and identity with feet structures of functional terminal cisternae of sarcoplasmic reticulum from fast skeletal muscles. *Journal of Biological Chemistry, 262,* 1740-1747.

Inoue, M., Kimura, T., Matsui, K., Ota, K., Shoji, M., Iitake, K., & Yoshinaga, K. (1987.) Responses of vasopressin and enkephalins to hemorrhage in adrenalectomized dogs. *American Journal of Physiology, 253,* R467-R474.

Iranmesh, A., Lizarralde, G., & Veldhuis, J.D. (1991.) Age and relative adiposity are specific negative determinants of the frequency and amplitude of growth hormone (GH) secretory bursts and the half-life of endogenous GH in healthy men. *Journal of Clinical Endocrinology and Metabolism, 73,* 1081-1088.

Iranmanesh, A., & Veldhuis, J.D. (1992.) Clinical pathophysiology of the somatotropic (GH) axis in adults. *Endocrinology and Metabolism Clinics of North America, 21,* 783-816.

Isgaard, J., Carlsson, L., Isaksson, O.G.P., & Jansson, J.-O. (1988a.) Pulsatile intravenous growth hormone (GH) infusion to hypophysectomized rat increases insulin-like growth factor I messenger ribonucleic acid in skeletal tissues more effectively than continuous GH infusion. *Endocrinology, 123,* 2605-2610.

Isgaard, J., Moller, C., Isaksson, O.G., Nilsson, A., Mathews, C.S., & Norsted, T.C. (1988b.) Regulation of insulin-like growth factor ribonucleic acid in rat growth plate by growth hormone. *Endocrinology, 122,* 1515-1520.

Issekutz Jr., B., Shaw, W.A.S., & Issekutz, T.B. (1975.) Effect of lactate on FFA and glycerol turnover in resting and exercising dogs. *Journal of Applied Physiology, 39,* 349-353.

Ito, S., & Sved, A.F. (1996.) Blockade of angiotensin receptors in rat rostral ventrolateral medulla removes excitatory vasomotor tone. *American Journal of Physiology, 270,* R1317-R1323.

Ito, H., Hirata, Y., Adachi, S., Tanaka, M., Tsujino, M., Kioke, A., Nogami, A., Marumo, & Hiroe, M. (1993.) Endothelin-1 is an autocrine/paracrine factor in the mechanism of angiotensin II-induced hypertrophy in cultured rat cardiomyocytes. *Journal of Clinical Investigation, 92*, 398-403.

Ivarsen, P., Jensen, L.W., & Pedersen, E.B. (1995.) Circadian blood pressure rhythm and atrial natriuretic peptide in prednisolone-induced blood pressure elevation. *Scandinavian Journal of Clinical and Laboratory Investigation, 55*, 655-662.

Iwamoto, J., Pendergast, D.R., Suzuki, H., & Krasney, J.A. (1994.) Effect of graded exercise on nitric oxide in expired air in humans. *Respiration Physiology, 97*, 333-345.

Jacobs, I., Lithell, H., & Karlsson, J. (1982.) Dietary effects on glycogen and lipoprotein lipase activity in skeletal muscle in man. *Acta Physiologica Scandinavica, 115*, 85-90.

Jacobs-El, J., Zhou, M.Y., & Russell, B. (1995.) MRF4, Myf-5, and myogenin mRNAs in the adaptive responses of mature rat muscle. *American Journal of Physiology, 268*, C1045-C1052.

Jacobson, K.A., Van Gallen, P.J.M., & Williams, M. (1992.) Adenosine receptors: pharmacology, structure-activity relationships, and therapeutic potential. *Journal of Medical Chemistry, 35*, 407-422.

Jahreis, G., Kauf, E., Frohner, G., & Schmidt, H.E. (1991.) Influence of intensive exercise on insulin-like growth factor I, thyroid and steroid hormones in female gymnasts. *Growth Regulation, 1*, 95-99.

James, D.E., Burleigh, K.M., Storlien, L.H., Bennett, S.P., & Kraegen, E.W. (1986.) Heterogeneity of insulin action in muscle: influence of blood flow. *American Journal of Physiology, 251*, E422-E430.

James, D.E., Jenkins, A.B., & Kraegen, E.W. (1985b.) Heterogeneity of insulin action in individual muscles in vivo: euglycemic clamp studies in rats. *American Journal of Physiology, 248*, E567-E574.

James, D.E., Kraegen, E.W., & Chisholm, D.J. (1985a.) Muscle glucose metabolism in exercising rats: comparison with insulin stimulation. *American Journal of Physiology, 248*, E575-E580.

Jandrain, B.J., Pallikarakis, N., Normand, S., Pirnay, F., Lacroix, M., Mosora, F., Pachiaudi, C., Gauthier, J.F., Scheen, A.J., & Riou, J.P. (1993.) Fructose utilization during exercise in men: Rapid conversion of ingested fructose to circulating glucose. *Journal of Applied Physiology, 74*, 2146-2154.

Jansen, R.W., & Lipsitz, L.A. (1995.) Postprandial hypotension: epidemiology, pathophysiology, and clinical management. *Annals of Internal Medicine, 122*, 286-295.

Janssen, P.J., Brinkmann, A.O., Boersma, W.J., & Van der Kwast, T.H. (1994.) Immunohistochemical detection of the androgen receptor with monoclonal antibody F394 in routinely processed, paraffin-embedded human tissues after microwave pre-treatment. *Journal of Cytochemistry and Histochemistry, 42*, 1169-1175.

Janssen, W.M., de Zeeuw, D., van del Hem, G.K., & de Jong, P.E. (1992.) Atrialnatriuretic factor influences renal diurnal rhythm in essential hypertension. *Hypertension, 20*, 80-84.

Janssen, L.J., & Sims, S.M. (1992.) Acetylcholine activates non-selective cation and chloride conductances in canine and guinea-pig tracheal smooth muscle cells. *Journal of Physiology, 453*, 197-218.

Jansson, E., & Kaijser, L. (1982.) Effect of diet on the utilization of blood-borne and intramuscular substrates during exercise in man. *Acta Physiologica Scandinavica, 115*, 19-30.

Jansson, E., & Kaijser, L. (1987.) Substrate utilization and enzymes in skeletal muscle of extremely endurance-trained men. *Journal of Applied Physiology, 62*, 999-1005.

Jansson, E., Hjemdahl, P., & Kaijser, L. (1986.) Epinephrine-induced changes in muscle carbohydrate metabolism during exercise in male subjects. *Journal of Applied Physiology, 60*, 1466-1470.

Jansson, J.-O. , Ekberg, S., Isaksson, O.G., & Eden, S. (1984.) Influence of gonadal steroids on age- and sex-related secretory patterns of growth hormone in the rat. *Endocrinology, 114*, 1287-1294.

Jansson, J.-O. Eden, S., & Isaksson, O. (1985.) Sexual dimorphism in the control of growth hormone secretion. *Endocrine Reviews, 6*, 128-150.

Jarow, J.P., & Lipshultz, L.I. (1990.) Anabolic steroid-induced hypogonadotropic hypogonadism. *American Journal of Sports Medicine, 18*, 429-431.

Jasnoski, M.L., & Kugler, J. (1987.) Relaxation, imagery, and neuroimmunomodulation. *Annals of the New York Academy of Sciences 496*, 722-730.

Javierre, C., Ventura, J.L., Segura, R., Calvo, M., & Garrido, E. (1997.) Is physical training a good synchronizer of the performance? *Revista Espanola de Fisiologia, 53*, 239-245.

Jee, W.S.S., Ke, H.Z., & Li, X.J. (1991.) Long-term anabolic effects of prostaglandin-E2 in tibial diaphyseal bone in male rats. *Bone Mineral, 15*, 33-55.

Jehue, R., Street, D., & Huizenga, R. (1993.) Effect of time zone and game time changes on team performance: National Football League. *Medicine and Science in Sports and Exercise, 25*, 127-131.

Jemmott, J.B., III, Boryzenko, J.S., Boryzenko, M., McClelland, D.C., Chapman, R., Meyer, D., & Benson, H. (1983.) Academic stress, power motivation, and decrease in secretion of salivary immunoglobulin A. *Lancet, 1*, 1400-1402.

Jenner, M.R., Kelch, R.P., Kaplan, S.L., & Grumbach, M.M. (1972.) Hormonal changes in puberty. IV. Plasma estradiol, LH and FSH in prepubertal children, pubertal females and in precocious puberty, premature thelarche, hypogonadism, and in a child with a feminizing ovarian tumor. *Journal of Clinical Endocrinology, 34*, 521-531.

Jensen, J., Brors, O., & Dahl, H.A. (1995.) Different beta-adrenergic receptor density in different skeletal muscle fibre types. *Pharmacology and Toxicology, 76*, 382-385.

Jensen, J., Oftebro, H., Breigan, B., Johnsson, A., Ohlin, K., Meen, H.D., Stromme, S.B., & Dahl, H.A. (1991.) Comparison of changes in testosterone concentrations after strength and endurance exercise in well trained men. *European Journal of Applied Physiology and Occupational Physiology, 63*, 467-471.

Jensen, L.L., Harding, J.W., & Wright, J.W. (1992.) Role of paraventricular nucleus in control of blood pressure and drinking in rats. *American Journal of Physiology, 262*, F1068-F1075.

Jensen, M.D. (1997.) Lipolysis: contribution from regional fat. *Annual Review of Nutrition, 17*, 127-139.

Jensen, M.D., Cryer, P.E., Johnson, C.M., & Murray, M.J. (1996.) Effects of epinephrine on regional free fatty acid and energy metabolism in men and women. *American Journal of Physiology, 270*, E259-E264.

Jensen, M.D., Haymond, M.W., Rizza, R.A., Cryer, P.E., & Miles, J.M. (1989.) Influence of body fat distribution on free fatty acid metabolism in obesity. *Journal of Clinical Investigation, 83*, 1168-1173.

Jern, C., Eriksson, E., Tengborn, L., Risberg, B., Wadenvik, H., & Jern, S. (1989.) Changes in plasma coaagulation and fibrinolysis in response to mental stress. *Thrombosis and Haemostasis, 62*, 767-771.

Jewett, M.E., Wyatt, J.K., Ritz-De Cecco, A., Khalsa, S.B., Dijk, D.J., & Czeisler, C.A. (1999.) Time course of sleep inertia dissipation in human performance and alertness. *Journal of Sleep Research, 8*, 1-8.

Ježova, D., & Vigaš M. (1981.) Testosterone response to exercise during blockade and stimulation of adrenergic receptors in man. *Hormone Research, 15*, 141-147.

Ježova, D., Vigaš, M., Tatár, P., Kvetnanský, R., Nazar, K., Kaciuba-Ucilko, H., & Kozlowski, S. (1985.) Plasma testosterone and catecholamine responses to physical exercise of different intensities in men. *European Journal of Applied Physiology, 54*, 62-66.

Jilka, R.L., Hangoc, G., Girasole, G., Passeri, G., Williams, D.C., Abrams, J.S., Boyce, B., Broxmeyer, H., & Manolagas, S.C. (1992.) Increased osteoclast development after estrogen loss: mediation by interleukin-6. *Science, 257*, 88-91.

Jin, P. (1992.) Efficacy of Tai Chi, brisk walking, meditation, and reading in reducing mental and emotional stress. *Journal of Psychosomatic Research, 36*, 361-370.

Johannsson, G., & Bengtsson, B.A. (1997.) Growth hormone and the acquisition of bone mass. *Hormone Research, 48* (Suppl. 5), 72-77.

Johansson, M., Rundqvist, B., Eisenhofer, G., & Friberg, P. (1997.) Cardiorenal epinephrine kinetics: evidence for neuronal release in the human heart. *American Journal of Physiology, 273*, H2178-H2185.

Johnson, A.K., & Gross, P.M. (1993.) Sensory circumventricular organs and brain homeostatic pathways. *FASEB Journal, 7*, 678-686.

Johnson, A.K., & Thunhorst, R.L. (1997.) The neuroendocrinology of thirst and salt appetite: visceral sensory signals and mechanisms of central integration. *Frontiers of Neuroendocrinology 18*, 292-353.

Johnson, M.A., Polgar, J., Weightman, D., & Appleton, D. (1973.) Data on the distribution of fibre types in thirty-six human muscles: an autopsy study. *Journal of Neurological Sciences, 18,* 111-129.

Johnson,L.C., Fisher, G., Silvester, L.J., & Hofheins, C.C. (1972.) Anabolic steroid: effects on strength, body weight, oxygen uptake and spermatogenesis upon mature males. *Medicine and Science in Sports, 4,* 43-45.

Johnston, H.M., Payne, A.P., & Gilmore, D.P. (1992.) Perinatal exposure to morphine affects adult sexual behavior of the male golden hamster. *Pharmacology, Biochemistry and Behavior, 42,* 41-44.

Jones, J.P., & Dohm, G.L. (1997.) Regulation of glucose transporter GLUT-4 and hexokinase II gene transcription by insulin and epinephrine. *American Journal of Physiology, 273,* E682-E687.

Jones, J.P., Tapscott, E.B., Olson, A.L., Pessin, J.E., & Dohm, G.L. (1998.) Regulation of glucose transporters GLUT-4 and GLUT-1 gene transcription in denervated skeletal muscle. *Journal of Applied Physiology, 84,* 1661-1666.

Jones, K.J. (1994.) Androgenic enhancement of motor neuron regeneration. *Annals of the New York Academy of Sciences, 743,* 141-164.

Jones, N.L., Heigenhauser, G.J.F., Kuksis, A., Matsos, C.G., Sutton, J.R., & Toews, C.J. (1980.) Fat metabolism in heavy exercise. *Clinical Science, 59,* 469-478.

Jones, T.M., Fang, V.S., Landau, R.L., & Rosenfield, R.L. (1977.) The effect of fluoxymesterone administration on testicular function. *Journal of Clinical Endocrinology and Metabolism, 44,* 121-129.

Joost, H.G., Weber, T.M., Cushman, S.W., & Simpson, I.A. (1986.) Insulin-stimulated glucose transport in rat adipose cells: modulation of transporter intrinsic activity by isoproterenol and adenosine. *Journal of Biological Chemistry, 261,* 10033-10036.

Jorgensen, J.O.L., Pedersen, S.A., Thuesen, L., Jorgensen, J., Moller, N., Muller, J., Skakkebaek, N.E., & Christiansen, J.S. (1991.) Long-term growth hormone treatment in growth hormone deficient adults. *Acta Endocrinologica, 125,* 449-453.

Jorgensen, J.O.L., Thuesen, L., Ingemann-Hansen, T., Pedersen, S.A., Jorgensen, J., Skakkebaek, N.E., & Christiansen, J.S. (1989.) Beneficial effects of growth hormone treatment in GH-deficient adults. *Lancet, 1,* 1221-1225.

Jovičić, A., Ivaniševič, V., & Nikolajevič, R. (1991.) Circadian variations of platelet aggregability and fibrinolytic activity in patients with ischemic stroke. *Thrombosis Research, 64,* 487-491.

Joy, M., Pollard, C.M., & Nunan, T.O. (1982.) Diurnal variation in exercise responses in angina pectoris. *British Heart Journal, 48,* 156-160.

Judd, S.J., Stranks, & Michailov, L. (1989.) Gonadotropin-releasing hormone pacemaker sensitivity to negative feedback inhibition by estradiol in women with hypothalamic amenorrhea. *Fertility and Sterility, 51,* 257-262.

Judd, S.J., Wong, J., Saloniklis, S., Maiden, M., Yeap, B., Filmer, S., & Michailov, L. (1995.) The effect of alprazolam on serum cortisol and luteinizing hormone pulsatility in normal women and in women with stress-related anovulation. *Journal of Clinical Endocrinology and Metabolism, 80,* 818-823.

Julius, S., Krause L., Schork NJ., et al. (1991.) Hyperkinetic borderline hypertension in Tecumseh, Michigan. *Journal of Hypertension, 9,* 77-84.

Jurkowski, J.E., Jones, N.L., Walker, W.C., Younglai, E.V., & Sutton, J.R. (1978.)Ovarian hormonal responses to exercise. *Journal of Applied Physiology, 44,* 109-114.

Jurzak, M., & Schmidt, H.A. (1998.) Vasopressin and sensory circumventricular organs. *Progress in Brain Research, 119,* 221-245.

Juul, A., Hjortskov, N., Jepsen, L.T., Nielsen, B., Halkjaer-Kristensen, J., Vahl, N., Jorgensen, J.O., Christiansen, J.S., & Skakkebaek, N.E. (1995.) Growth hormone deficiency and hyperthermia during exercise: a controlled study of sixteen GH-deficient patients. *Journal of Clinical Endocrinology and Metabolism, 80,* 3335-3340.

Kadi, F., Bonnerud, P., Eriksson, A., & Thornell, L.E. (2000.) The expression of androgen receptors in human neck and limb muscles: effects of training and self-administration of androgenic steroids. *Histochemistry and Cellular Biology, 113,* 25-29.

Kadi, F., Eriksson, A., Holmner, S., & Thornell, L.E. (1999.) Effects of anabolic steroids on the muscle cells of strength-trained athletes. *Medicine and Science in Sports and Exercise, 31,* 1528-1534.

Kahlert, S., Grohe, C., Karas, R.H., Lobbert, K., Neyses, L. & Vetter, H. (1997.) Effects of estrogen on skeletal myoblast growth. *Biochemical and Biophysical Research Communications, 232,* 373-378.

Kahn, B.B. (1992.) Facilitative glucose transporters: regulatory mechanisms and dysregulation in diabetes. *Journal of Clinical Investigation, 89,* 1367-1374.

Kaijser, L., Pernow, J., Berglund, B., Grubbstrom, J., & Lundberg, J.M. (1994.) Neuropeptide Y release from human heart is enhanced during prolonged exercise in hypoxia. *Journal of Applied Physiology, 76,* 1346-1349.

Kalogeras, K.T., Nieman, L.K., Friedman, T.C., Doppman, J.L., Cutler, G.B. Jr., Chrousos, G.P., Wilder, R.L., Gold, P.W., & Yanovski, J.A. (1996.) Inferior petrosal sinus sampling in healthy subjects reveals a unilateral corticotropin-releasing hormone-induced arginine vasopressin release associated with ipsilateral adrenocorticotropin secretion. *Journal of Clinical Investigation, 97,* 2045-2050.

Kalra, S.P., Dube, M.G., Pu, S., Xu, B., Horvath, T.L., & Kalra, P.S. (1999.) Interacting appetite-regulating pathways in the hypothalamic regulation of body weight. *Endocrine Reviews, 20,* 68-100.

Kalsbeek, A., Rikkers, M., Vivien-Roels, B., & Pevet, P. (1993.) Vasopressin and vasoactive intestinal peptide infused in the paraventricular nucleus of the hypothalamus elevate plasma melatonin levels. *Journal of PinealResearch, 15,* 46-52.

Kalsbeek, A., van der Vliet, J., & Buijs, R.M. (1996.) Decrease of endogenous vasopressin release necessary for expression of the circadian rise in plasma corticosterone: a reverse microdialysis study. *Journal of Neuroendocrinology, 8,* 299-307.

Kalsbeek, A., van Heerikhuize, J.J., Wortel, J., & Buijs, R.M. (1998.) Restricted daytime feeding modifies suprachiasmatic nucleus vasopressin release in rats. *Journal of Biological Rhythms, 13,* 18-29.

Kambadur, R., Sharma, M., Smith, T.P., & Bass, J.J. (1997.) Mutations in myostatin (GDF8) in double-muscled Belgian Blue and Piedmontese cattle. *Genome Research, 7,* 910-916.

Kamegai, J., Tamura, H., Shimizu, T., Ishii, S., Sugihara, H., & Wakabayashi, I. (2000.) Central effect of ghrelin, an endogenous growth hormone secretagogue, on hypothalamic peptide gene expression. *Endocrinology, 141,* 4797-800.

Kanaley, J.A., Boileau, R.A., Bahr, J.A., Misner, J.E., & Nelson, R.A. (1992a.) Substrate oxidation and GH responses to exercise are independent of menstrual phase and status. *Medicine and Science in Sports and Exercise, 24,* 873-880.

Kanaley, J.A., Boileau, R.A., Bahr, J.A., Misner, J.E., & Nelson, R.A. (1992b.) Cortisol levels during prolonged exercise: the influence of menstrual phase and menstrual status. *International Journal of Sports Medicine, 13,* 332-336.

Kannus, P., Haapasalo, H., Sievanen, H., Oja, P., & Vuori, I. (1994.) The site-specific effects of long-term unilateral activity on bone mineral density and content. *Bone, 15,* 279-284.

Kaplowitz, P.BH., & Oberfield, S.E. (1999.) Reexamination of the age limit for defining when puberty is precocious in girls in the United States: implications for evaluation and treatment. *Pediatrics, 104,* 936-941.

Kappagoda, C.T., Linden, R.J., Snow, H.M., & Whitaker, E.M. (1974.) Left atrial receptors and the antidiuretic hormone. *Journal of Physiology, 237,* 663-683.

Kappel, M., Barington, T., Gyhrs, A., & Pederson, B.K. (1994a.) Influence of elevated body temperature on circulating immunoglobulin-secreting cells. *International Journal of Hyperthermia 10,* 653-658.

Kappel, M., Kharazmi, A., Nielsen, H., Gyhrs, A., & Pederson, B.K. (1994b.) Modulation of the counts and functions of neutrophils and monocytes under in vivo hyperthermia conditions. *International Journal of Hyperthermia 10,* 165-173.

Kappel, M., Hansen, M.B., Diamant,M., Jorgensen, J.O., Gyhrs, A., & Pedersen, B.K. (1993.) Effects of an acute bolus growth hormone infusion on the human immune system. *Hormone and Metabolic Research, 25,* 579-585.

Kappel, M., Hansen, M.B., Diamant, M., & Pedersen, B.K. (1994c.) The immune system during exposure to extreme physiologic conditions. *International Journal of Sports Medicine 15* (Suppl. 3), S116-S121.

Kappel, M., Poulsen, T.D., Galbo, H., & Pedersen, B.K. (1998a.) Influence of minor increases in plasma catecholamines on natural killer cell activity. *Hormone Research, 49,* 22-26.

Kappel, M., Poulsen, T.D., Hansen, M.B., Galbo, H., & Pedersen, B.K. (1998b.) Somatostatin attenuates the hyperthermia induced increase in neutrophil concentration. *European Journal of Applied Physiology and Occupational Physiology, 77,* 149-156.

Kappel, M., Stadeager, C., Tvede, N., Galbo, H.. & Pedersen, B.K. (1991a.) Effects of in vivo hyperthermia on natural killer cell activity, in vitro proliferative responses and blood mononuclear cell subpopulations. *Clinical and Experimental Immunology, 84,* 175-180.

Kappel, M., Tvede, N., Galbo, H., Haahr, P.M., Kjaer, M., Linstow, M., & Pedersen, B.K. (1991b.) Evidence that the effect of physical exercise on NK cell activity is mediated by epinephrine. *Journal of Applied Physiology, 70,* 2530-2534.

Kappel, M., Tvede, N., Hansen, M.B., Stadeager, C., & Pedersen, B.K. (1995.) Cytokine production ex vivo: effect of raised body temperature. *International Journal of Hypertension, 11,* 329-335.

Kapur, S., Bedard, S., Marcotte, B., Cote, C.H., & Marette, A. (1997.) Expression of nitric oxide synthetase in skeletal muscle: a novel role for nitric oxide as a modulator of insulin action. *Diabetes, 46,* 1691-1700.

Karaviti, L., Schoonmaker, J., Shryne, J.E., & Gorski, R.A. (1982.) Corticoids and sexual differentiation of brain structure. *Physiological Abstracts, 25,* 315.

Kario, K., Matsuo, T., Kobayashi, H., Imaya, M., Matsuo, M., & Shimada, K. (1996.) Nocturnal fall of blood pressure and silent cerebrovascular damage in elderly hypertensive patients: advanced silent cerebrovascular damage in extreme dippers. *Hypertension, 27,* 130-135.

Karlsson, J., & Jacobs, I. (1980.) Is the significance of muscle fiber types to muscle metabolism different in females than in males?. In J. Borms, M. Hebbelinck & A. Venerando (Eds.), *Women and sport. An historical, biological, physiological and sportsmedical approach* (pp. 97-101). Basel, Switzerland: S. Karger.

Katafuchi, T., Oomura, Y., & Kurosawa, M. (1988.) Effects of chemical stimulation of paraventricular nucleus on adrenal and renal nerve activity in rats. *Neuroscience Letters, 86,* 195-299.

Katagiri, T., Akiyama, S., Namiki, M., Komaki, M., Yamaguchi, A., Rosen, V., Wozney, J.M., Fujisawa-Sehara, A., & Suda, T. (1997.) Bone morphogenetic protein-2 inhibits terminal differentiation of myogenic cells by suppressing the transcriptional activity of MyoD and myogenin. *Experimental Cell Research, 230,* 342-351.

Katch, V.L., Martin, R., & Martin, J. (1979.) Effects of exercise intensity on food consumption in the male rat. *American Journal of Clinical Nutrition, 32,* 1401-1407.

Kather, H., & Simon, B. (1981.) Adrenoceptor of the alpha2-subtype mediating inhibition of the human fat cell adenylate cyclase. *European Journal of Clinical Investigation, 11,* 111- 114.

Katusic, Z.S., & Cosentino, F. (1994.) Nitric oxide synthetase: from molecular biology to cerebrovascular physiology. *News in Physiological Sciences, 9,* 64-67.

Katz, M.S., Dax, E.M., & Gregerman, R.I. (1993.) Beta adrenergic regulation of rat liver glycogenolysis during aging. *Experimental Gerontology, 28,* 329-340.

Katz, V.L. (1991.) Physiologic changes during normal pregnancy. *Current Opinion in Obstetrics and Gynecology, 3,* 750-758.

Katz, V.L. (1996.) Water exercise in pregnancy. *Seminars in Perinatology, 20,* 285-291.

Katz, V.L., McMurray, R., Berry, M.J., & Cefalo, R.C. (1988.) Fetal and uterine responses to immersion and exercise. *Obstetrics and Gynecology, 72,* 225-230.

Katz, V.L., McMurray, R., Goodwin, W.E., & Cefalo, R.C. (1990.) Nonweight-bearing exercise during pregnancy on land and during immersion: a comparative study. *American Journal of Perinatology, 7,* 281-284.

Kaufman, J.M., Kesner, J.S., Wilson, R.C., & Knobil, E. (1985.) Electrophysiological manifestation of luteinizing hormone-releasing hormone pulse generator activity in the rhesus monkey: Influence of alpha-adrenergic and dopaminergic blocking agents. *Endocrinology, 116,* 1327-1333.

Kaufman, M.P., Longhurst, J.C., Rybicki, K.J., Wallach, J.H., & Mitchell, J.H. (1983.) Effects of static muscular contraction in impulse activity of groups III and IV afferents in cats. *Journal of Applied Physiology, 55,* 105-112.

Kawaguchi, H., Kurokawa, T., Hanada, K., Hiyama, Y., Tamura, M., Ogata, E., & Matsumoto, T. (1994.) Stimulation of fracture repair by recombinant human basic fibroblast growth factor in normal and streptozotocin-diabetic rats. *Endocrinology, 135,* 774-781.

Kawakami, Y., Natelson, B.H., & Du Bois, A.B. (1967.) Cardiovascular effects of face immersion and factors affecting diving reflex in man. *Journal of Applied Physiology, 23,* 964-970.

Kayar, S.R., Hoppeler, H., Howald, H., Claassen, H., & Oberholzer, F. (1986.) Acute effects of endurance exercise on mitochondrial distribution and skeletal muscle morphology. *European Journal of Applied Physiology, 54,* 578-584.

Kayashima, S., Ohno, H., Fujioka, T., Taniguchi N., & Nagata, N. (1995.) Leukocytosis induced by chronic strenuous physical exercise. *European Journal of Applied Physiology and Occupational Physiology, 72,* 187-188.

Keenan, B.S., Richards, G.E., Ponder, S.W., Dallas, J.S. , Nagamani, M., & Smith, E.R. (1993.) Androgen-stimulated pubertal growth: the effects of testosterone and dihydrotestosterone on growth hormone and insulin-like growth factor-I in the treatment of short stature and delayed puberty. *Journal of Clinical Endocrinology and Metabolism, 76,* 996-1001.

Keizer, H.A., Kuipers, H., de Haan, J., Beckers, E., & Habets, L. (1987.) Multiple hormonal responses to physical exercise in amenorrheic trained and untrained women. *International Journal of Sports Medicine, 8* (Suppl. 3), 139-150.

Keizer, H.A., Poortman, J., & Bunnik, G.S.J. (1980.) Influence of physical exercise in sex-hormone metabolism. *Journal of Applied Physiology, 48,* 765-769.

Kelch, R.P., Hopwood, N.J., Sauder, S., & Marshall, J.C. (1985.) Evidence for decreases secretion of gonadotropin-releasing hormone in pubertal boys during short-term testosterone treatment. *Pediatric Research, 19,* 112-117.

Keller, U., Schnell, H., Girard, J., & Stauffacher, W. (1984.) Effect of physiological elevation of plasma growth hormone levels on ketone body kinetics and lipolysis in normal and acutely insulin-deficient man. *Diabetologia, 26,* 103-108.

Kelley, D.E., & Mandarino, L.J. (1990.) Hyperglycemia normalizes insulin-stimulated skeletal muscle glucose oxidation and storage in noninsulin-dependent diabetes mellitus. *Journal of Clinical Investigation, 86,* 1999-2007.

Kelley, D.E., & Simoneau, J.A. (1994.) Impaired free fatty acid utilization in non-insulin- dependent diabetes mellitus. *Journal of Clinical Investigation, 94,* 2349-2356.

Kelley, D.E., Mokan, M., Simoneau, J.A., & Mandarino, L.J. (1993.) Interaction between glucose and free fatty acid metabolism in human skeletal muscle. *Journal of Clinical Investigation, 92,* 91-98.

Kelley, D.E., Reilly, J.P., Veneman, T., & Mandarino, L.J. (1990.) Effects of insulin on skeletal muscle glucose storage, oxidation, and glycolysis in humans. *American Journal of Physiology, 258,* E923-E929.

Kelley, G.A. (1999.) Aerobic exercise and resting blood pressure among women: a meta-analysis. *Preventive Medicine, 28,* 264-275.

Kelley, G.A., & Kelley, K.S. (1999.) Aerobic exercise and resting blood pressure in women: a meta-analytic review of controlled clinical trials. *Journal of Womens Health and Gender-Based Medicine, 8,* 787-803.

Kelley, G., & McClellan, P. (1994.) Antihypertensive effects of aerobic exercise. A brief meta-analytic review of randomized controlled trials. *American Journal of Hypertension, 7,* 115-119.

Kelly, A.B., & Watts, A.G. (1996.) Mediation of dehydration-induced peptidergic gene expression in the rat lateral hypothalamic area by forebrain afferent projections. *Journal of Comparative Neurology, 370,* 231-246.

Kelly, A.B., & Watts, A.G. (1998.) The region of the pontine parabrachial nucleus is a major target of dehydration-sensitive CRH neurons in the rat lateral hypothalamic area. *Journal of Comparative Neurology, 394,* 48-63.

Kelly, E.W., & Bradway, L.F. (1997.) A team approach to the treatment of musculoskeletal injuries suffered by navy recruits: a method to decrease attrition and improve quality of care. *Military Medicine, 162,* 354-359.

Kelsey, J.L., Gammon, M.D., & John E.M. (1993.) Reproductive factors and breast cancer. *Epidemiological Reviews, 15,* 6-47.

Kemppainen, P., Paalasmaa, P., Pertovaara, A., & Johansson, G. (1990.) Dexamethasone attenuates exercise-induced dental analgesia in man. *Brain Research, 519,* 329-332.

Kendall, A., Levitsky, D., Strupp, B., & Lissner, L. (1991.) Weight loss on a low-fat diet: Consequences of the imprecision of the control of food intake in humans. *American Journal of Nutrition, 53,* 1124-1129.

Kennedy, A., Gettys, T.W., Watson, P., Wallace, P., Ganaway, E., Pan, Q., & Garvey, W.T. (1997.) The metabolic significance of leptin in humans: gender-based differences in relationship to adiposity, insulin sensitivity, and energy expenditure. *Journal of Clinical Endocrinology and Metabolism, 82,* 1293-1300.

Kennedy, G.C. (1953.) The role of depot fat in the hypothalamic control of food intake in the rat. *Proceedings of the Royal Society of London, B. Biological Sciences, 140,* 579-592.

Kennedy G.C. (1957.) The development with age of hypothalamic restraint upon the appetite of the rat. *Journal of Endocrinology, 16,* 9-17.

Kennedy, G.C., & Mitra, J. (1963.) Body weight and food intake as initiating factors for puberty in the rat. *Journal of Physiology, 166,* 408-418.

Kenney, W.L. (1985.) A review of comparative responses of men and women to heat stress. *Environmental Research, 37,* 1-11.

Keppens, S. (1993.) The complex interaction of ATP and UTP with isolated hepatocytes. How many receptors? *General Pharmacology*, *24*, 283-289.

Kerckhoffs, D.A., Arner, P., & Bolinder, J. (1998.) Lipolysis and lactate production in human skeletal muscle and adipose tissue following glucose ingestion. *Clinical Science*, *94*, 71-77.

Kern, P.A. (1997.) Potential role of TNF alpha and lipoprotein lipase as candidate genes for obesity. *Journal of Nutrition*, *127*, 1917S-1922S.

Kern, W., Perras, B., Wodick, R., Fehm, H.L., & Born, J. (1995.) Hormonal secretion during nighttime sleep indicating stress of daytime exercise. *Journal of Applied Physiology*, *79*, 1461-1468.

Kerrigan, J., & Rogol, A.D. (1992.) The impact of gonadal steroid hormone action on growth hormone secretion during childhood and adolescence. *Endocrine Reviews*, *13*, 281-298.

Kerrigan, D.C., Todd, M.K., & Della Croce, U. (1998.) Gender differences in joint biomechanics during walking: normative study in young adults. *American Journal of Physical Medicine and Rehabilitation*, *77*, 2-7.

Kessler-Icekson, G. (1988.) Effect of triiodothyronine on cultured neonatal rat heart cells: Beating rate, myosin subunits and CK-isozymes. *Journal of Molecular Cardiology*, *20*, 649-655.

Ketelhut, R.G., Franz, I.W., & Scholze, J. (1997.) Efficacy and position of endurance therapy in the treatment of arterial hypertension. *Journal of Human Hypertension*, *11*, 651-655.

Ketelslegers, J.M., Maiter, D., Maes, M., Underwood, L.E., & Thiessen, J.P. (1996.) Nutritional regulation of the growth hormone and insulin-like growth factor-binding proteins. *Hormone Research*, *45*, 252-257.

Khoury, S.A., Reame, N.E., Kelch, R.P., & Marshall, J.C. (1987.) Diurnal patterns of pulsatile luteinizing hormone secretion in hypothalamic amenorrhea: reproducibility and responses to opiate blockade and an alpha 2-adrenergic agonist. *Journal of Clinical Endocrinology and Metabolism*, *64*, 755-762.

Kiecolt-Glaser, J.K., Fisher, L.D., Ogrocki, P., Stout, J.C., Speicher, C.E., & Glaser, P. (1987.) Marital quality, marital disruption, and immune function. *Psychosomatic Medicine*, *49*, 13-34.

Kiecolt-Glaser, J.K., Garner, W., Speicher, C., Penn, G.M., Holliday, J., & Glaser, R. (1984.) Psychosocial modifiers of immunocompetence in medical students. *Psychosomatic Medicine*, *46*, 7-14.

Kiefel, J.M., Paul, D., & Bodnar, R.J. (1989.) Reduction in opioid and non-opioid forms of swim analgesia by 5-HT2 receptor antagonists. *Brain Research*, *500*, 231-240.

Kiens, B., Essen-Gustavsson, B., Gad, P., & Lithell, H. (1987.) Lipoprotein lipase activity and intramuscular triglyceride stores after long-term high-fat and high carbohydrate diets in physically trained men. *Clinical Physiology*, *7*, 1-9.

Kiens, B., Lithell, H. Mikines, K.J., & Richter, E.A. (1989.) Effects of insulin and exercise on muscle lipoprotein lipase activity in man and its relation to insulin action. *Journal of Clinical Investigation*, *84*, 1124-1129.

Kiess, W., & Butenand, O. (1985.) Specific growth hormone receptors on human peripheral mononuclear cells: reexpression, identification, and characterization. *Journal of Clinical Endocrinology and Metabolism*, *60*, 740-746.

Kietzmann, T., Porwol, T., Zierold, K., Jungermann, K., & Acker, H. (1998.) Involvement of a local fenton reaction in the reciprocal modulation by O_2 of the glucagon-dependent activation of the phosphoenolpyruvate carboxykinase gene and the insulin-dependent activation of the glucokinase gene in rat hepatocytes. *Biochemical Journal*, *335*, 425- 432.

Kilbourn, R.G., Traber, D.L., & Szabo, C. (1997.) Nitric oxide and shock. *Disease-A-Month*, *43*, 277-348.

Kim, Y.D., Chen, B., Beauregard, J., Kouretas, P., Thomas, G., Farhat, M.Y., Myers, A.K., & Lees, D.E. (1996.) 17 Beta-estradiol prevents dysfunction of canine cononary endothelium and myocardium and reperfusion arrhythmias after brief ischemia/reperfusion. *Circulation*, *94*, 2901-2908.

Kim, Y.S., & Sainz, R.D. (1992.) Beta-adrenergic agonists and hypertrophy of skeletal muscles. *Life Sciences*, *50*, 397-407.

Kimball, T.J., Childs, M.T., Applebaum-Bowden, D., & Sembrowich, W.L. (1983.) The effect of training and diet on lipoprotein cholesterol, tissue lipoprotein lipase and hepatic triglyceride lipase in rats. *Metabolism*, *32*, 497-503.

Kimura, Y., & Okuda H. (1994.) Effects of alpha and beta adrenergic antagonists on epinephrine-induced aggregation and intracellular free calcium concentration in human platelets. *Biochemical Biophysical Research Communications* *202*, 1069-1075.

King, N.A. (1999.) What processes are involved in the appetite response to moderate increases in exercise-induced energy expenditure? *Proceedings of the Nutrition Society*, *58*, 107-113.

King, N.A., & Blundell, J.E. (1995.) High-fat foods overcome the energy expenditure induced by high-intensity cycling or running. *European Journal of Clinical Nutrition*, *49*,114-123.

King, N.A., Burley, V.J., & Blundell, J.E. (1994.) Exercise-induced suppression of appetite: Effects on food intake and implications for energy balance. *European Journal of Clinical Nutrition*, *48*, 715-724.

King, N.A., Lluch, A., Stubbs, R.J., & Blundell, J.E. (1997a.) High dose exercise does not increase hunger or energy intake in free living males. *European Journal of Clinical Nutrition*, *51*, 478-483.

King, N.A., Snell, L., Smith, R.D., & Blundell, J.E.(1996.) Effects of short-term exercise on appetite responses in unrestrained females. *European Journal of Clinical Nutrition*, *50*, 663-667.

King, N.A., Tremblay, A., & Blundell, J.E. (1997b.) Effects of exercise on appetite control: Implications for energy balance. *Medicine and Science in Sports and Exercise*, *29*, 1076-1089.

Kingwell, B.A., Dart, A.M., Jennings, G.L., & Korner, D. I. (1992.) Exercise training reduces the sympathetic component of the blood pressure-heart rate baroreflex in man. *Clinical Science*, *82*, 357-362.

Kinlaw, W.B., Schwartz, H.L., & Oppenheimer, J.H. (1985.) Decreased serum triiodothyronine in starving rats is due primarily to diminished thyroidal secretion of thyroxine. *Journal of Clinical Investigation*, *75*, 1238-1241.

Kissebah, A.H., & Peiris, A.N. (1989.) Biology of regional fat distribution: Relationship to non- insulin-dependent diabetes mellitus. *Diabetes/Metabolism Reviews*, *5*, 83-109.

Kissebah, A.H. (1987.) Low density lipoprotein metabolism in non-insulin-dependent diabetes mellitus. *Diabetes/Metabolism Reviews*, *3*, 619-651.

Kissileff, H.R., Pi-Sunyer, F.X., Segal, K., Meltzer, S., & Foelsch, P.A. (1990.) Acute effects of exercise on food intake in obese and nonobese women. *American Journal of Clinical Nutrition*, *52*, 240-245.

Kitamura, T., Ogorochi, T., & Miyajima, A. (1994.) Multimeric cytokine receptors. *Trends in Endocrinology and Metabolism*, *5*, 8-14.

Kitayama, I., Yaga, T., Kayahara, T., Nakano, K., Murase, S., Otani, M., & Nomura, J. (1997.) Long-term stress degenerates, but imipramine regenerates, noradrenergic axons in the rat cerebral cortex. *Biological Psychiatry*, *42*, 687-696.

Kitchen, A.M., Scislo, T.J., & O'Leary, D.S. (2000.) NTS A (2a) purinoceptor activation elicits hindlimb vasodilation primarily via a beta-adrenergic mechanism. *American Journal of Physiology*, *278*, H1775-H1782.

Kjaer, M., Howlett, K., Langfort, J., Zimmerman-Belsing, T., Lorentsen, J., Bulow, J., Ihlemann, J., Feldt-Rasmussen, U., & Galbo, H. (2000.) Adrenaline and glycogenolysis in skeletal muscle during exercise: a study in adrenolectomised humans. *Journal of Physiology*, *528*, 371-378.

Kjaer, M., Kiens, B., Hargreaves, M., & Richter, E.A. (1991.) Influence of active muscle mass on glucose homeostasis during exercise in humans. *Journal of Applied Physiology*, *71*, 552-557.

Kjeldsen, S.E., Weder, A.B., Egan, B., Neubig, R., Zweifler, A.J., & Julius, S. (1995.) Effects of circulating epinephrine on platelet function and hematocrit. *Hypertension*, *25*, 1096-1105.

Klaus, S., Casteilla, L., Bouillaud, F., & Ricquier, D (1991.) The uncoupling protein UCP: a membranous mitochondrial ion carrier exclusively expressed in brown adipose tissue. *International Journal of Biochemistry*, *23*, 791-801.

Klausen, T., Breum, L., Sorenson, H.A., Schifter, S., & Sonne, B. (1993.) Plasma levels of parathyroid hormone, vitamin D, calcitonin, and calcium in association with endurance exercise. *Calcified Tissue International*, *52*, 205-208.

Klausen,T., Olsen, N.V., Poulsen, T.D., Richalet, J.P., & Pedersen, B.K. (1997.) Hypoxemia increases serum interleukin-6 in humans. *European Journal of Applied Physiology and Occupational Physiology*, *76*, 480-482.

Klebanoff, M.A., Shiono, P., & Carey, J.C. (1990.) The effect of physical activity during pregnancy on preterm delivery and birth weight. *American Journal of Obstetrics and Gynecology*, *163*, 1450-1460.

Kleiger, R.E., Miller J.P., Bigger, J.T. Jr., & Moss, A.J. (1987.) Multicenter Post-Infarction Research Group. Decreased heart rate variability and its association with increased mortality after acute myocardial infarction. *American Journal of Cardiology*, *59*, 256-262.

Klein, D.C., Coon, S.L., Roseboom, P.H., Weller, J.L., Bernard, M., Gastel, J.A., et al. (1997.) The melatonin rhythm-generating enzyme: molecular regulation of serotonin N-acetyltransferase in the pineal gland. *Recent Progress in Hormone Research, 52,* 307-357.

Klein, D.C., & Moore, R.Y. (1979.) Pineal N-acetyltransferase and hydroxyindole-O-methyltransferase: control by the retinohypothalamic tract and the suprachiasmatic nucleus. *Brain Research, 174,* 245-262.

Klein, D.C., Moore, R.Y., & Reppert, S.M. (Eds.) (1991.) *Suprachiasmatic nucleus: The mind's clock.* New York: Oxford University Press.

Klein, K.O., Baron, J., Colli, M.J., McDonnell, D.P., & Cutler, G.B., Jr. (1994a.) Estrogen level in children determined by an ultrasensitive recombinant cell bioassay. *Journal of Clinical Investigation, 94,* 2475-2480.

Klein, S., Coyle, F. F., & Wolfe, R.R. (1995.) Effect of exercise on lipolytic sensitivity in endurance-trained athletes. *Journal of Applied Physiology, 78,* 2201-2206.

Klein, S., Coyle, E.F., & Wolfe, R.R. (1994b.) Fat metabolism during low-intensity exercise in endurance trained and untrained men. *American Journal of Physiology, 267,* E934-E939.

Kleinpeter, G., Schatzer, R., & Bock, F. (1995.) Is blood pressure really a trigger for the circadian rhythm of subarachnoid hemorrage? *Stroke, 26,* 1805-1810.

Kleitman, N. (1963.) *Sleep and wakefulness,* 2nd ed. Chicago: University of Chicago Press.

Klesges, R.C., Klesges, L.M., Haddock, C.K., & Peters, J.C. (1992.) A longitudinal analysis of the impact of dietary intake and physical activity on weight change in adults. *American Journal of Clinical Nutrition, 55,* 818.

Klimaschewski, L., Kummer, W., Mayer, B., Couraud, J.Y., Preissler, U., Philippin, B., & Heym C. (1992.) Nitric oxide synthase in cardiac nerve fibers and neurons of rat and guinea pig heart. *Circulation Research, 71,* 1533-1537.

Klokker, M., Secher, N.H., Matzen, S., & Pedersen, B.K. (1993a.) Natural killer cell activity during head-up tilt-induced central hypovolemia in humans. *Aviation Space and Environmental Medicine, 64,* 1128-1132.

Klokker, M., Kharazmi, A., Galbo, H., Bygbjerg, I., & Pedersen, B.K. (1993b.) Influence of in vivo hyperbaric hypoxia on function of lymphocytes, neutrocytes, natural killer cells, and cytokines. *Journal of Applied Physiology, 74,* 1100-1106.

Kluft, C., Jie, A.F., Rijken, D.C., & Verheijen, J.H. (1988.) Daytime fluctuations in blood of tissue-type plasminogen activator (t-PA) and its fast-acting inhibitor (PAI-1. *Thrombosis and Haemostasis, 59,* 329-332.

Knepper, M.A., & Inoue, T. (1997.) Regulation of aquaporin-2 water channel trafficking by vasopressin. *Current Opinion in Cell Biology, 9,* 560-564.

Kniffki, K.-D., Mense, S.L., & Schmidt, R.F. (1981.) Muscle receptors with fine afferent fibers which may evoke circulatory reflexes. *Circulation Research, 48,* 25-31.

Kobzik, L., Reid, M.B., Bredt, D.S., & Stamler, J.S. (1994.) Nitric oxide in skeletal muscle. *Nature, 372,* 546-548.

Kobzik, L., Stringer, B., Balligand, J.L., Reid, M.B., & Stamler, J.S. (1995.) Endothelial type nitric oxide synthase in skeletal muscle fibers: mitochondrial relationships. *Biochemical and Biophysical Research Communications, 211,* 375-381.

Kocamis, H., Kirkpatrick-Keller, D.C., Richter, J., & Killefer, J. (1999.) The ontogeny of myostatin, follistatin and activin-B mRNA expression during chicken embryonic development. *Growth, Development, and Aging, 63,* 143-150.

Koeslag, J.H. (1982.) Post-exercise ketosis and the hormone response to exercise: a review. *Medicine and Science in Sports and Exercise, 14,* 327-334.

Koeslag, J.H., Noakes, T.D., & Sloan, A.W. (1980.) Post-exercise ketosis. *Journal of Physiology, 301,* 79-90.

Koeslag, J.H., Noakes, T.D., & Sloan, A.W. (1982.) The effects of alanine, glucose and starch ingestion on the ketosis produced by exercise and starvation. *Journal of Physiology, 325,* 363-376.

Kohno, K., Matsuoka, H., Takenaka, K., Miyake, Y., Nomura, G., & Imaizumi, T.(1997.) Renal depressor mechanisms of physical training in patients with essential hypertension. *American Journal of Hypertension, 10,* 859-868.

Kohrt, W.M., Landt, M., & Birge, Jr., S.J. (1996.) Serum leptin levels are reduced in response to exercise training, but not hormone replacement therapy, in older women. *Journal of Clinical Endocrinology and Metabolism, 81,* 3980-3985.

Kohrt, W. M., Obert, K. A., & Holloszy, J.O. (1992.) Exercise training improves fat distribution patterns in 60-70-year-old men and women. *Journal of Gerontology, 47,* M99- M105.

Kohut, M.L., Davis, J.M., Jackson, D.A., Colbert, L.H., Strasner, A., Essig, D.A., Pate, P.R., Ghaffar, A., & Mayer, E.P. (1998.) The role of stress hormones in exercise-induced suppression of alveolar macrophage antiviral function. *Journal of Neuroimmunology, 81,* 193-200.

Koivisto, V.A., Harkonen, M., Karonen, S.-L., Groop, P.H., Elovainio, R., Ferrannini, E., Sacca, L., & DeFronzo, R.A. (1985.) Glycogen depletion during prolonged exercise: influence of glucose, fructose, or placebo. *Journal of Applied Physiology, 58,* 731-737.

Koivisto, V.A., Karvonen, S.L., & Nikkila, E.A. (1981.) Carbohydrate ingestion before exercise: Comparison of glucose, fructose, and sweet placebo. *Journal of Applied Physiology, 51,* 783-787.

Kojima, M., Hosoda, H., Date, Y., Nakazato, M., Matsuo, H., & Kangawa, K. (1999.) Ghrelin is a growth-hormone-releasing acylated peptide from stomach. *Nature, 402,* 656-660.

Kolka, M.A., & Stephenson, L.A. (1989.) Control of sweating during the human menstrual cycle. *European Journal of Applied Physiology and Occupational Physiology, 58,* 890-895.

Komi, P.V. (1980.) Fundamental performance characteristics of females and males. In J. Borms, M. Hebbelinck, & A. Venerando (Eds.), *Women and sport. An historical, biological, physiological and sportsmedical approach* (pp. 102-108). Basel, Switzerland: S. Karger.

Konagaya, M., & Max, S.R. (1986.) A possible role for endogenous glucocorticoid in orchiectomy-induced atrophy of the rat levator ani muscle: studies with RU38486, a potent and selective antiglucocorticoid. *Journal of Steroid Biochemistry, 25,* 305-308.

Konradsen, L., & Nexo, E. (1988.) Epidermal growth factors in plasma, serum and urine before and after prolonged exercise. *Regulatory Peptides, 21,*197-203.

Koopmans, S.J., de Boer, S.F., Sips, H.C., Radder, J.K., Frolich, M., & Krans, H.M. (1991.) Whole body and hepatic insulin action in normal, starved, and diabetic rats. *American Journal of Physiology, 260,* E825-E832.

Kopp, U.C., & DiBona, G.F. (1993.) Neural regulation of renin secretion. *Seminars in Nephrology, 13,* 543-551.

Kopp, U.C., & Smith, L.A. (1991.) Inhibitory renorenal reflexes: a role for renal prostaglandins in activation of renal sensory receptors. *American Journal of Physiology, 261,* R1513-R1521.

Korbonitz, M., Trainer, P.J., Franciulli, G.,Oliva, O., Pala, A., Dettori, A., Besser, G.M., Delitala, G., & Grossman, A.B. (1996.) L-arginine is unlikely to exert neuroendocrine effects in humans via the generation of nitric oxide. *European Journal of Endocrinology, 135,* 543-547.

Kotz, C.M., Briggs, J.E., Grace, M.K., Levine, A.S., & Billington, C.J. (1998a.) Divergence of feeding and thermogenic pathways influenced by NPY in the hypothalamic PVN of the rat. *American Journal of Physiology, 275,* R471-R477.

Kotz, C.M., Briggs, J.E., Pomonis, J.D., Grace, M.K., Levine, A.S., & Billington, C.J. (1998b.) Neural site of leptin influence on neuropeptide Y signalling pathways altering feeding and uncoupling protein. *American Journal of Physiology, 275,* R478-R484.

Kouame, N., Nadeau, A., Lacourciere, Y., & Cleroux, J. (1995.) Effects of different training intensities on the cardiopulmonary baroreflex control of forearm vascular resistance in hypertensive subjects. *Hypertension, 25,* 391-398.

Kouri, E.M., Lukas, S.E., Pope, H.G. Jr., & Oliva, P.S. (1995.) Increased aggressive responding in male volunteers following the administration of gradually increasing doses of testosterone cypionate. *Drug and Alcohol Dependence, 40,* 73-79.

Koylu, E.O., Couceyro, P.R., Lambert, P.D., Ling, N.C., DeSouza, E.B., & Kuhar, M.J. (1997.) Immunohistochemical localization of novel CART peptides in rat hypothalamus, pituitary and adrenal gland. *Journal of Neuroendocrinology, 9,* 823-833.

Koylu, E.O., Couceyro, P.R., Lambert, P.D., & Kuhar M.J. (1998.) Cocaine- and amphetamine-regulated transcript peptide immunohistochemical localization in the rat brain. *Journal of Comparative Neurology, 391,*115-132.

Kozlowski, S., Chwalbinska-Moneta, J., Vigas, M., Kaciuba-Uscilko, H., & Nazar, K. (1983.) Greater serum GH response to arm than to leg exercise performed at equivalent oxygen uptake. *European Journal of Applied Physiology, 52,* 131-135.

Kraemer, R.R., Kilgore, J.L., Kraemer, G.R., & Castracane, V.D.(1992a.) Growth hormone, IGF-I, and testosterone responses to resistive exercise. *Medicine and Science in Sports and Exercise, 24*, 1346-1352.

Kraemer, W.J., Anguilera, B.A., Terada, M., Newton, R.U., Lynch, J.M., Rosendaal, G.,McBride, J.M., Gordon, S.E., & Hakkinen, K. (1995.) Responses of IGF-I to endogenous increases in growth hormone after heavy-resistance exercise. *Journal of Applied Physiology, 79*, 1310-1315.

Kraemer, W.J., Dziados, J.E., Marchitelli, L.J., Gordon, S.E., Harman, E., Mello, R., Fleck, S.J., Frykman, P.N., & Triplett, N.T. (1993a.) Effects of different heavy-resistance exercise protocols on plasma beta-endorphin concentrations. *Journal of Applied Physiology, 74*, 450-459.

Kraemer, W.J., Fleck, S.J., Dziados, J.E., Harman, E., Marchitelli, L.J., Gordon, S.E., Mello, R., Frykman, P.N., Koziris, L.P., & Triplett, N.T. (1993b.) Changes in hormonal concentrations after different heavy-resistance exercise protocols in women. *Journal of Applied Physiology, 75*, 594-604.

Kraemer, W.J., Fleck, S.J.,Callister, R., Shealy,M., Dudley, G.A., Maresh, C.M., Marchitelli, L.J., Cruthirds, C., Murray, T., & Falkel, J.E. (1989.) Training responses of plasma beta- endorphin, adrenocorticotropin, and cortisol. *Medicine and Science in Sports and Exercise, 21*,146-153.

Kraemer, W.J., Fry, A.C., Warren, B.J., Stone, M.H., Fleck, S.J., Kearney, J.T., Conroy, B.P., Maresh, C.M., Weseman, C.A., & Triplett, N.T. (1992b.) Acute hormonal responses in elite junior weightlifters. *International Journal of Sport Medicine, 13*, 103-109.

Kraemer, W.J., Gordon, S.E., Fleck, S.J., Marchitelli, L.J., Mello, R., Dziados, J.E., Friedl, K., Harman, E., Maresh, C., & Fry, A.C. (1991a.) Endogenous anabolic hormonal and growth factor responses to heavy resistance exercise in males and females. *International Journal of Sports Medicine, 12*, 228-235.

Kraemer, W.J., Patton, J.F., Knuttgen, H.G., Hannan, C.J., Kettler, T., Gordon, S.E., Dziados, J.E., Fry, A.C., Frykman, P.N., & Harman, E.A. (1991b.) Effects of high-intensity cycle exercise on sympathoadrenal-medullary response patterns. *Journal of Applied Psychology, 70*, 8-14.

Kraenzlin, M.E., Keller, U., Keller, A., Thelin, A., Arnaud, M.J., & Stauffacher, W. (1989.) Elevation of plasma epinephrine concentrations inhibits proteolysis and leucine oxidation in man via beta-adrenergic mechanisms. *Journal of Clinical Investigation, 84*, 388-393.

Kral, E.A., & Dawson-Hughes, B. (1994.) Walking is related to bone density and rates of bone loss. *American Journal of Medicine, 96*, 20-26.

Krauchi, K., Keller, U., Leonhardt, G., Brunner, D.P., van der Helde, P., Haug, H.J., & Wirz-Justice, A. (1999.) Accelerated post-glucose glycaemia and altered alliesthesia-test in seasonal affective disorder. *Journal of Affective Disorders, 53*, 23-26.

Kraus, B., & Pette, D.(1997.) Quantification of MyoD, myogenin, MRF4 and Id-1 by reverse-transcriptase polymerase chain reaction in rat muscles--effects of hypothyroidism and chronic low-frequency stimulation. *European Journal of Biochemistry, 247*, 98-106.

Krauth, M.O., Saini, J., Follenius, M., & Brandenberger, G. (1990.) Nocturnal oscillations of plasma aldosterone in relation to sleep stages. *Journal of Endocrinological Investigation, 13*, 727-735.

Krieger, D.T. (1979.) *Endocrine rhythms.* New York: Raven Press.

Kristensen, P., Judge, M.E., Thim, L., et al. (1998.) Hypothalamic CART is a new anorectic peptide regulated by leptin. *Nature, 393*, 72-76.

Krotkiewski, M. (1994.) Role of muscle capillarization and morphology in the development of insulin resistance and metabolic syndrome. *Presse Medicale, 23*, 1353-1356.

Krotkiewski, M., & Bjorntorp, P. (1986.) Muscle tissue in obesity with different distribution of adipose tissue: effects of physical training. *International Journal of Obesity, 10*, 331-341.

Kubitz, K.A., Landers, D.M., Petruzello, S.J., & Han, M. (1996.) The effects of acute and chronic exercise on sleep: a meta-analytic review. *Sports Medicine, 21*, 277-291.

Kubota, E., Sata, T., Soas, A.H., Paul, S., & Said, S.I. (1985.) Vasoactive intestinal peptide as a possible transmitter of nonadrenergic, noncholinergic relaxation of pulmonary artery. *Transactions of the Association of American Physicians, 98*, 233-242.

Kubova, K., Dakurai, T., Tamura, J., & Shirakura, T. (1987.) Is the circadian change in hematocrit and blood viscosity a factor in triggering cerebral and myocardial infarction?. *Stroke, 18*, 812-813.

Kujala, U.M., Alen, M., & Huhtaniemi, I.T. (1990.) Gonadotrophin-releasing hormone and human chorionic gonadotrophin tests reveal that both hypothalamic and testicular endocrine functions are suppressed during acute prolonged physical exercise. *Clinical Endocrinology, 33*, 219-225.

Kulin, H.E., Grumbach, M.M., & Kaplan, S.L. (1969.) Changing sensitivity of the pubertal gonadal hypothalamic feedback mechanism in man. *Science, 166*, 1012- 1013.

Kulpa, P.J., White, B.M., & Visscher, R. (1987.) Aerobic exercise in pregnancy. *American Journal of Obstetrics and Gynecology, 156*, 1395-1403.

Kunihara, M., & Oshima, T. (1983.) Effects of epinephrine on plasma cholesterol levels in rats. *Journal of Lipid Research, 24*, 639-644.

Kuntz, A. (1919.) The innervation of the gonads in the dog. *Anatomical Record, 17*, 203-220.

Kuoppasalmi, K., Naveri, H., Harkonen, M., & Adlercreutz, H. (1980.) Plasma cortisol, androstenedione, testosterone and luteinizing hormone in running exercise of different intensities. *Scandinavian Journal of Clinical and Laboratory Investigation, 40*, 403-409.

Kuoppasalmi, K., Naveri, H., Rehunen, S., Harkonen, M., & Adlercreutz, H. (1976.) Effect of strenuous anaerobic running exercise on plasma growth hormone, cortisol, luteinizing hormone, testosterone, androstenedione, estrone, and estradiol. *Journal of Steroid Biochemistry, 7*, 823-829.

Kupari, M., Koskinen, P., & Leinonen, H. (1990.) Double-peaking circadian variation in the occurrence of sustained supraventricular tachyarrhythmias. *American Heart Journal, 120*, 1364-1369.

Kupfer, J.M., & Rubin, S.A. (1992.) Differential regulation of insulin-like growth factor I by growth hormone and thyroid hormone in the heart of juvenile hypophysectomized rats. *Journal of Molecular Cardiology, 24*, 631-639.

Kusaka, M., & Ui, M. (1977.) Activation of the Cori cycle by epinephrine. *American Journal of Physiology, 232*, E145-E155.

Kuwaki, T., Cao, W.-H., & Kumada, M. (1994.) Endothelin in the brain and its effects on central control of the circulation and other functions. *Japanese Journal of Physiology, 44*, 1-18.

Kuwaki, T., Cao, W.-H., Unekawa, M., Terni, N., & Kumada, M. (1991.) Endothelin-sensitive areas in the ventral surface of the rat medulla. *Journal of the Autonomic Nervous System, 36*, 149-158.

Kyrkouli, S.E., Stanley, B.G., Seirafi, R.D., & Leibowitz, S.F. (1990.) Stimulation of feeding by galanin: anatomical localization and behavioral specificity of this peptide's effect in the brain. *Peptides, 11*, 995-1001.

Kyung, N.H., Barkan, A., Klibanski, A., Badger, T.M., McArthur, J.W., Axelrod, L., & Beitins, I.Z. (1985.) Effect of carbohydrate supplementation on reproductive hormones during fasting in men. *Journal of Clinical Endocrinology and Metabolism, 60*, 827-835.

Laartz, B., Losee-Olson, S., Ge, Y.R., & Turek, F.W. (1994.) Diurnal, photoperiodic, and age- related changes in plasma growth hormone levels in the golden hamster. *Journal of Biological Rhythms, 9*, 111-123.

Laatikainen, T., Virtanen, T., & Apter, D. (1986.) Plasma immunoreactive beta-endorphin in exercise-associated amenorrhea. *American Journal of Obstetrics and Gynecology, 154*, 94-97.

Labrie, F. (1990.) Glycoprotein hormones: gonadotropins and thyrotropin. In E.-E. Baulieu & P.A. Kelly (Eds.). *Hormones: From molecules to disease* (pp.256-275). New York: Chapman and Hall.

Lacey, R.J., Cable, H.C., James, R.F., London, N.J., Scarpello, J.H., & Morgan, N.G. (1993.) Concentration-dependent effects of adrenaline on the profile of insulin secretion from isolated human islets of Langerhans. *Journal of Endocrinology, 138*, 555-563.

LaCroix, A.Z., Leveille, S.G., Hecht, J.A., Grothaus, L.C., & Wagner, E.H. (1996.) Does walking decrease the risk of cardiovascular disease hospitalizations and death in older adults?. *Journal of the American Geriatric Society, 44*:113-120.

Ladu, M.J., Kapsas, H., & Palmer, W.K. (1991.) Regulation of lipoprotein lipase in adipose and muscle tissues during fasting. *American Journal of Physiology, 260*, R953-959.

Lafontan, M., & Berlan, M. (1993.) Fat cell adrenergic receptors and the control of white and brown fat cell function. *Journal of Lipid Research, 34*, 1057-1091.

Lafontan, M., Barbe, P., Galitzky, J., Tavernier, G., Langin, D., Carpene, C., Bousquet-Melou, A., & Berlan, M. (1997.) Adrenergic regulation of adipocyte metabolism. *Human Reproduction, 12* (Suppl. 1), 6-20.

Lafontan, M., Berlan, M., Galitzky, J., & Montastruc, J.L. (1992.) Alpha-2 adrenoceptors in lipolysis: alpha-2 antagonists in lipid-mobilizing strategies. *American Journal of Clinical Nutrition, 55* (Suppl. 1), 219S-227S.

Lafontan, M., Bousquet-Melou, A., Galitzky, J., Barbe, P., Carpene, C., Langin, D., Berlan, M., Valet, P., Castan, I., and Bouloumie, A. (1995.) Adrenergic receptors and fat cells: Differential recruitment by physiological amines and homologous regulation. *Obesity Research, 3* (Suppl. 4), 507S-514S.

Lafontan, M., Dang-Tran, L., & Berlan, M. (1979.) Alpha-adrenergic antilipolytic effect of adrenaline in human fat cells of the thigh: comparison with adrenaline responsiveness of different fat deposits. *European Journal of Clinical Investigation, 9,* 261-266.

Laker, M.E., & Mayes, P.A. (1984.) Investigations into the direct effects of insulin on hepatic ketogenesis, lipoprotein secretion and pyruvate dehydrogenase activity. *Biochimica et Biophysica Acta, 795,* 427-430.

Lalani, R., Bhasin, S., Byhower, F., Tarnuzzer, R., Grant, M., Shen, R., Asa, S., Ezzat, S., & Gonzalez-Cadavid, N.F.(2000.) Myostatin and insulin-like growth factor-I and –II expression in the muscle of rats exposed to the microgravity environment of the NeuroLab space shuttle flight. *Journal of Endocrinology, 167,* 417-428.

Lam, R.W., Goldner, E.M., & Grewal, A. (1996.) Seasonality of symptoms in anorexia and bulimia nervosa. *International Journal of Eating Disorders,10,* 35-44.

Lamb, D.R. (1999.) Benefits and limitations of prehydration. *Sports Science Exchange, 12,* 1-6.

Lamont, L.S., Romito, R.A., Finkelhor, R.S., & Kalhan, S.C. (1997.) Beta 1-adrenoceptors regulate resting metabolic rate. *Medicine and Science in Sports and Exercise, 29,* 769-774.

Lampl, M., Veldhuis, J.D., & Johnson, M.L. (1990.) Saltation or stasis: a model of human growth. *Science, 258,* 801-803.

Landis, S.C., & Fredieu, J.R. (1986.) Coexistence of calcitonin gene-related peptide and vasoactive intestinal polypeptide in cholinergic sympathetic innervation of rat sweat glands. *Brain Research, 377,* 177-181.

Landmann, R. (1992.) Beta-adrenergic receptors in human leukocyte subpopulations. European Journal of Clinical Investigation, 22, Suppl 1, 30-36.

Landsberg, L. (1996.) Insulin and the sympathetic nervous system in the pathophysiology of hypertension. *Blood Pressure* (Suppl. 1), 25-29.

Landsberg, L., & Krieger, D.R. (1989.) Obesity, metabolism, and the sympathetic nervous system. *American Journal of Hypertension, 2,*125S-132S.

Landsberg, L., & Young, J.B. (1983.) Autonomic regulation of thermogenesis. In L. Girardier & M.J. Stock (Eds.), *Mammalian Thermogenesis* (pp. 99-140). London: Chapman & Hall.

Lange, A.J., Espinet, C., Hall, R., el-Maghrabi, M.R., Vargas, A.M., Miksicek, R.J., Granner, D.K., & Pilkis, S.J. (1992.) Regulation of gene expression of rat skeletal muscle/liver 6-phosphofructo-2-kinase/fructose-2,6-biphosphatase: isolation and characterization of a glucocorticoid response element in the first intron of the gene. *Journal of Biological Chemistry, 267,* 15673-15680.

Langer, G.A. (1997.) Chasing myocardial calcium: a 35-year perspective. *News in Physiological Sciences, 12,* 238-244.

Langfort, J., Ploug, T., Ihlemann, J., Holm,C., & Galbo, H. (2000.) Stimulation of hormone-sensitive lipase activity by contractions in rat skeletal muscle. *Biochemical Journal, 351,* 207-214.

Langfort, J., Ploug, T., Ihlemann, J., Saldo, M., Holm,C., & Galbo, H. (1999.) Expression of hormone-sensitive lipase and its regulation by adrenaline in skeletal muscle. *Biochemical Journal, 340,* 459-465.

Langin, D., Tavernier, G., & Lafontan, M. (1995.) Regulation of beta 3-adrenoceptor expression in white fat cells. *Fundamental and Clinical Pharmacology, 9,* 97-106.

La Noue, K., & Martin, L.F. (1994.) Abnormal A1 adenosine receptor function in genetic obesity. *FASEB Journal, 8,* 72-80.

Lanyon, L.E. (1996.) Using functional loading to influence bone mass and architecture: Objectives, mechanisms, and relationship with estrogen of the mechanically adaptive process in bone. *Bone, 18* (Suppl. 1), 37S-43S.

Lanyon, L.E., & Rubin, C.T. (1984.) Static vs dynamic loads as an influence on bone remodelling. *Journal of Biomechanics, 17,* 897-905.

Lanza, F., Beretz, A., Stierle, A., Hanau, D., Kubina, M., & Cazenave, J.P. (1988.) Epinephrine potentiates human platelet activation but is not an aggregating agent. *American Journal of Physiology, 255,* H1276-H1288.

Lara, H.E., Ferruz, J.L., Luza, S., Bustamante, D.A., Borges, Y., & Ojeda, S.R. (1993.) Activation of ovarian sympathetic nerves in polycystic ovary syndrome. *Endocrinology, 133,* 2690-2695.

Lariviere, G., & Lafond, A. (1986.) Physical maturity in young elite ice hockey players. *Canadian Journal of Applied Sports Science, 11,* 24P-27P.

Laron, Z., & Rogol, A.D. (Eds.), (1989.) Hormones and sport. *Serono Symposia Publications from Raven Press, vol. 55,* New York: Raven Press.

Larrabee, R.C. (1902.) Leukocytosis after violent exercise. *Journal of Medical Research, 2,* 76-82.

Larrain, J., Carey, D.J., & Brandan, E. (1998.) Syndecan-1 expression inhibits myoblast differentiation through a basic fibroblast growth factor-dependent mechanism. *Journal of Biological Chemistry, 273,* 32288-32296.

Larrouy, D., Remaury, A., Daviaud, D., & Lafontan, M. (1994.) Coupling of inhibitory receptors with Gi-proteins in hamster adipocytes: comparison between adenosine A1 receptor and alpha2-adrenoceptor. *European Journal of Pharmacology, 267,* 225-232.

Larsson, H., & Ahren, B. (1996.) Androgen activity as a risk factor for impaired glucose tolerance in postmenopausal women. *Diabetes Care, 19,* 1399-1403.

Larsson, P.T., Wallen, N.H., Martinsson, A., Egberg, N., & Hjemdahl, P. (1992.) Significance of platelet beta-adrenoceptors for platelet responses in vivo and in vitro. *Thrombosis and Haemostasis, 68,* 687-693.

Larsson, P.T., Wallen, N.H., & Hjemdahl, P. (1994.) Norepinephrine-induced human platelet activation in vivo is only partly counteracted by aspirin. *Circulation, 89,* 1951-1957.

Larsson, P.T., Wiman, B., Olsson, G., Angelin, B., & Hjemdahl, P. (1990.) Influence of metoprolol treatment on sympatho-adrenal activation of fibrinolysis. *Thrombosis and Haemostasis, 63,* 482-487.

Latour, M.G., Cardin, S., Helie, R., Yamaguchi, N., & Lavoie, J.M. (1995.) Effect of hepatic vagotomy on plasma catecholamines during exercise-induced hypoglycemia. *Journal of Applied Physiology, 78,* 1629-1634.

Latzka, W.A., Sawka, M.N., Montain, S.J., Skrinar,G.S, Fielding, R.A., Mattot, R.P., & Pandolph, K.B. (1997.) Hyperhydration: thermoregulatory effects during compensable exercise-heat stress. *Journal of Applied Physiology, 83,* 860-866.

Latzka, W.A., Sawka, M.N., Montain, S.J., Skrinar,G.S, Fielding, R.A., Mattot, R.P., & Pandolph, K.B. (1998.) Hyperhydration: tolerance and cardiovascular effects during uncompensable exercise-heat stress. *Journal of Applied Physiology, 84,* 1858-1864.

Laubie, M., & Schmidt, H. (1981.) Indication for central vagal endorphinergic control of heart rate in dogs. *European Journal of Pharmacology, 71,* 401-409.

Laughlin, G.A., & Yen, S.S. (1997.) Hypoleptinemia in women athletes: absence of a diurnal rhythm with amenorrhea. *Journal of Clinical Endocrinology and Metabolism, 82,* 318-321.

Laughlin, G.A., Loucks, A.B., & Yen, S.S. (1991.) Marked augmentation of nocturnal melatonin secretion in amenorrheic athletes, but not in cycling athletes: unaltered by opioidergic or dopaminergic blockade. *Journal of Clinical Endocrinology and Metabolism, 73,* 1321-1326.

Laughlin, M.H., & Armstrong, R.B. (1982.) Muscular blood flow patterns during exercise in the rat. *American Journal of Physiology, 243,* H296-H306.

Laughlin, M.H., & Armstrong, R.B. (1985.) Muscle blood flow during exercise. *Exercise and Sports Science Reviews, 13,* 95-136.

Lautt, WW. (1979a.) Autonomic neural control of liver glycogen metabolism. *Medical Hypotheses, 5,* 1287-1296.

Lautt, W.W. (1979b.) Neural activation of alpha-adrenoreceptors in glucose mobilization from liver. *Canadian Journal of Physiology and Pharmacology, 57,* 1037-1039.

Lautt, W.W., & Wong, C. (1978.) Hepatic glucose balance in response to direct stimulation of sympathetic nerves in the intact liver of cats. *Canadian Journal of Physiology and Pharmacology, 56,* 1022-1028.

Lavoie, C., Ducros, F., Bourque, J., Langelier, H., & Chiasson, J.L. (1997a.) Glucose metabolism during exercise in man: the role of insulin and glucagon in the regulation of hepatic glucose production and gluconeogenesis. *Canadian Journal of Physiology and Pharmacology, 75,* 26-35.

Lavoie, C., Ducros, F., Bourque, J., Langelier, H., & Chiasson, J.L. (1997b.) Glucose metabolism during exercise in man: the role of insulin in the regulation of glucose utilization. *Canadian Journal of Physiology and Pharmacology, 75,* 36-43.

Lawlor, M.A., & Rotwein, P. (2000.) Coordinate control of muscle cell survival by distinct insulin-like growth factor activated signaling pathways. *Journal of Cellular Biology, 151,* 1131-1140.

Lawrence, J.C. Jr., & Roach, P.J. (1997.) New insights into the role and mechanism of glycogen synthase activation by insulin. *Diabetes, 46,* 541-547.

Lax, P., Zamora, S., & Madrid, J.A. (1998.) Coupling effect of locomotor activity on the rat's circadian system. *American Journal of Physiology, 275,* R580-R587.

Layne, M.D., & Farmer, S.R. (1999.) Tumor necrosis factor-alpha and basic fibroblast growth factor differentially inhibit the insulin-like growth factor-I induced expression of myogenin in C2C12 myoblasts. *Experimental Cell Research, 249,*177-187.

Leary, S.C., Battersby, B.J., Hansford, R.G., & Moyes, C.D. (1998.) Interactions between bioenergetics and mitochondrial biogenesis. *Biochimica et Biophysica Acta, 1365,* 522-530.

Lebrun, C.M. (1994.) The effect of the phase of the menstrual cycle and the birth control pill on athletic performance. *Clinics in Sports Medicine, 13,* 419-441.

Leduc, L., Wasserstrum, N., Spillman, T., & Cotton, D.B. (1991.) Baroreflex function in normal pregnancy. *American Journal of Obstetrics and Gynecology, 165,* 886-890.

Lee, A.D., Hansen, P.A., Schluter, J., Gulve, E.A., Gao, J., & Holloszy, J.O. (1997.) Effects of epinephrine on insulin-stimulated glucose uptake and GLUT-4 phosphorylation in muscle. *American Journal of Physiology, 273,* C1082-C1087.

Lee, H.C., Galione, A., & Walseth, T.F. (1994.) Cyclic ADP-ribose: metabolism and calcium mobilizing function. In G. Litwack (Ed.), *Vitamins and hormones* (pp. 199-254), Orlando: Academic Press.

Lee, R.M., Owens, G.K., Scott-Burden, T., Head, R.J., Mulvany, M.J., & Schiffrin, E.L. (1995.) Pathophysiology of smooth muscle in hypertension. *Canadian Journal of Physiology and Pharmacology, 73,* 574-584.

Leech, C.J., & Faber, J.E. (1996.) Different alpha-adrenoceptor subtypes mediate constriction of arterioles and venules. *American Journal of Physiology, 270,* H710-H722.

Leenen, F.H., Davies, R.A., & Fourney, A. (1995.) Role of cardiac beta 2-receptors in cardiac responses to exercise in cardiac transplant patients. *Circulation, 91,* 685-690.

Legget, M.E.,Kuusisto, J., Healy, N.L., Fujioka, M., Schwaegler, R.G., & Otto, C.M. (1996.) Gender differences in left ventricular function at rest and with exercise in asymptomatic aortic stenosis. *American Heart Journal, 131,* 94-100.

Lehmann, M., Keul, J., & Huber, G. (1981.) Plasma catecholamines in trained and untrained volunteers during graded exercise. *International Journal of Sport Medicine, 2,* 143-147.

Leibowitz, S.F. (1998.) Differential functions of hypothalamic galanin cell grows in the regulation of eating and body weight. *Annals of the New York Academy of Sciences, 863,* 206-220.

Leibowitz, S.F., Alexander, J.T., Cheung, W.K., & Weiss, G.F. (1993.) Effects of serotonin and the serotonin blocker metergoline on meal patterns and macronutrient selection. *Pharmacology, Biochemistry and Behavior, 45,* 185-194.

Le Magnen, J. (1999a.) Increased food intake induced in rats by changes in the satiating sensory input from food (first published in French in 1956). *Appetite, 33,* 33-35.

Le Magnen, J. (1999b.) Effects of the duration of pre- and postprandial fasting on the acquisition of appetite (first published in French in 1957). *Appetite, 33,* 21-26.

Le Magnen, J., & Julien, N. (1999.) Efficacy of olfactory, tactile and other food stimuli in the acquisition and manifestation of appetite in rats (first published in French in 1959). *Appetite, 33,* 43-51.

Lemon, P.W.R., & Proctor, D.N. (1991.) Protein intake and athletic performance. *Sports Medicine, 12,* 313-325.

Lemon, P.W.R., Tarnopolsky, M.A., MacDougall, J.D., & Atkinson, S.A. (1992.) Protein requirements and muscle mass/strength changes during intensive training in novice bodybuilders. *Journal of Applied Physiology, 73,* 767-775.

Leon, A. S., Conrad, J., Hunninghake, D. B., & Serfass, R. (1979.) Effects of a vigorous walking program on body composition, and carbohydrate and lipid metabolism of obese young men. *American Journal of Clinical Nutrition, 32,* 1776-1787.

Leproult, R., Copinschi, G., Buxton, O., & Van Cauter, E. (1997a.) Sleep loss results in an elevation of cortisol levels the next evening. *Sleep, 20,* 865-870.

Leproult, R., Van Reeth, O., Byrne, M.M., Sturis, J., & Van Cauter, E. (1997b.) Sleepiness, performance, and neuroendocrine function during sleep deprivation: effects of exposure to bright light or exercise. *Journal of Biological Rhythms, 12,* 245-248.

Lerario, D.D.G., Ferreira, S.R.G., Miranda, W.L., & Chacra, A.R. (2001.) Influence of dexamethasone and weight loss on the regulation of serum leptin levels in obese individuals. *Brazilian Journal of Medical and Biological Research, 34,* 479-487.

Leuenberger, U., Sinoway, L., Gubin, S., Gaul, L., Davis, D., & Zelis, R. (1993.) Effects of exercise intensity and duration on norepinephrine spillover and clearance in humans. *Journal of Applied Physiology, 75,* 668-674.

Leung, D.W., Spencer, S.A., Cachianes, G., Hammonds, G.R., Collins, C., Henzel, W.J., Ross, J.B., Waters, M.J., & Wood, W.I. (1987.) Growth hormone receptor and serum binding protein: purification, cloning and expression. *Nature, 330,* 537-543.

Leveille, G.A. (1970.) Adipose tissue metabolism: influence of periodicity of eating and diet composition. *Federation Proceedings, 29,* 1294-1301.

Levine, L., Evans, W.J., Cadarette, B.S., Fisher, E.C., & Bullen, B.A. (1983.) Fructose and glucose ingestion and muscle glycogen use during submaximal exercise. *Journal of Applied Physiology, 55,* 1767-1771.

Levine-Ross, J., Myerson, L.L., & Skerda, M. (1986.) Effect of low doses of estradiol on six month growth rates and predicted height in patients with Turner's syndrome. *Journal of Pediatrics, 109,* 950-953.

Levinsky, N., Davidson, D.G., & Berlinger, R.W. (1959) Effects of reduced glomerular filtration on urine concentration in the presence of antidiuretic hormone. American Journal of Physiology, 196, 451-456.

Lewis, D.A., Kamon, E., & Hodgson, J.L. (1986.) Physiological differences between genders: implications for sports conditioning. *Sports Medicine, 3,* 357-369.

Lewis, G.F., Uffelman, K.D., Szeto, L.W., Weller, B., & Steiner, G. (1995.) Interaction between free fatty acids and insulin in the acute control of very low density liporotein production in humans. *Journal of Clinical Investigation, 95,* 158-166.

Lewis, J.W., Tordoff, M.G., Sherman J.E., & Liebeskind, J.C. (1982.) Adrenal medullary enkephalin-like peptides may mediate opioid stress analgesia. *Science, 217,* 557-559.

Lewis, U.J., Sinha, Y.N., & Lewis, G.P. (2000.) Structure and properties of members of the hGH family: a review. *Endocrine Journal, 47* (Suppl.), S1-8,

Lewy, A., Ahmed, S., Jackson, J., & Sack, R. (1992.) Melatonin shifts human circadian rhythms according to a phase-response curve. *Chronobiology International, 9,* 380-392.

Li, J., Gilmour, R.S., Saunders, J.C., Dauncey, M.J., & Fowden, A.L. (1999.) Activation of the adult mode of ovine growth hormone receptor gene expression by cortisol during late fetal development. *FASEB Journal, 13,* 545-552.

Li, J.B., & Goldberg, A.L. (1976.) Effects of food deprivation on protein synthesis and degradation in rat skeletal muscles. *American Journal of Physiology, 231,* 441-448.

Li, P., Chang, T.M., & Chey, W.Y. (1998.) Secretin inhibits gastric acid secretion via a vagal afferent pathway in rats. *American Journal of Physiology, 275,* G2-G28.

Licinio, J., Negrao, A.B., Mantzoros, C., Kaklamani, V., Wong, M.L., Bongiorno, P.B., et al. (1998.) Synchronicity of frequently sampled 24-h concentrations of circulating leptin, luteinizing hormone, and estradiol in healthy women. *Proceedings of the National Academy of Sciences of the United States of America, 95,* 2541-2546.

Liggett, S.B., Shah, S.D., & Cryer, P.E. (1988.) Characterization of beta-adrenergic receptors of human skeletal muscle obtained by needle biopsy. *American Journal of Physiology, 254,* E795-E798.

Liggins, G.C., & Schellenberg, J.C. (1988.) Endocrine control of lung development. In W. Kunzel & A. Jensen (Eds.), *The endocrine control of the fetus* (pp. 236-245). Berlin: Springer- Verlag.

Liggins, G.C., & Thorburn, G.D. (1993.) Initiation of parturition. In G.E. Lamming (Ed.), *Marshall's physiology and reproduction* (pp. 863-1002). London: Chapman and Hall.

Light, K.C., & Turner, J.R. (1992.) Stress-induced changes in the rate of sodium-excretion in healthy black and white men. *Journal of Psychosomatic Research, 36,* 497-507.

Lightly, E.R., Walker, S.W., Bird, I.M., & Williams, B.C. (1990.) Subclassification of beta-adrenoceptors responsible for steroidogenesis in primary cultures of bovine adrenocortical zona fasciculata/reticularis cells. *British Journal of Pharmacology*, 99, 709-712.

Lin, L., Bray, G., & York, D.A. (2000.) Enterostatin suppresses food intake in rats after near-celiac and intracarotid arterial injection. *American Journal of Physiology*, 278, R1346-R1351.

Lin, L., & York, D.A. (1998a.) Changes in the microstructure of feeding after administration of enterostatin into the paraventricular nucleus and the amygdala. *Peptides*, 19, 557-562.

Lin, L. & York, D.A. (1998b.) Chronic ingestion of dietary fat is a prerequisite for inhibition of feeding by enterostatin. *American Journal of Physiology*, 275, R619-R623.

Linkowski, P., Mendlewicz, J., Leclercq, R., Brasseur, M., Hubain, P., Goldstein, J., Copinschi, G., & Van Cauter, E. (1985.) The 24-hour profile of adrenocorticotropin and cortisol in major depressive illness. *Journal of Clinical Endocrinology and Metabolism*, 61, 429-438.

Linkowski, P., Spiegel, K., Kerkhofs, M., L'Hermite-Baleriaux, M., Van Onderbergen, A., Leproult, R., Mendlewicz, J., & Van Cauter, E. (1998.) Genetic and environmental influences on prolactin secretion during wake and during sleep. *American Journal of Physiology*, 274, E909-E919.

Lins, P., Wajngot, A., Adamson, U., Vranic, M., & Efendic, S. (1983.) Minimal increases in glucagon levels enhance glucose production in man with partial insulin deficiency. *Diabetes*, 32, 633-636.

Lissner, L., Levitsky, D.A., Strupp, B.J., Kalkwarf, H.J., & Roe, D.A. (1987.) Dietary fat and the regulation of energy intake in human subjects. *American Journal of Clinical Nutrition*, 46, 886-892.

Lithell, H., Cedermark, M., Froberg, J., Tesch, P., & Karlsson, J. (1981.) Increase in lipoprotein-lipase activity in skeletal muscle during heavy exercise: relation to epinephrine excretion. *Metabolism: Clinical and Experimental*, 30, 1130-1134.

Lithell, H., Hellsing, K., Lunqvist, G., & Malmberg, P. (1979a.) Lipoprotein-lipase activity of human skeletal muscle and adipose tissue after intensive physical exercise. *Acta Physiologica Scandinavica*, 105, 312-315.

Lithell, H., Krotkiewski, M., Kiens, B., Wroblewski, Z., & Holm, G. (1985a.) Non-response of muscle capillary density and lipoprotein-lipase activity to regular training in diabetic patients. *Diabetes Research*, 2, 17-21.

Lithell, H., Orlander, J., Schele, R., Sjodin, B., & Karlsson, J. (1979b.) Changes in lipoprotein-lipase activity and lipid stores in human skeletal muscle with prolonged heavy exercise. *Acta Physiologica Scandinavica*, 107, 257-261.

Lithell, H., Schele, R., Vessby, B., & Jacobs, I. (1984.) Lipoproteins, lipoprotein lipase, and glycogen after prolonged physical activity. *Journal of Applied Physiology*, 57, 698-702.

Lithell, H., Vessby, B., Selinus, I., & Dahlberg, P.A. (1985b.) High muscle lipoprotein lipase activity in thyrotoxic patients. *Acta Endocrinologica*, 109, 227-231.

Liu, M., Shen, L., & Tso, P. (1999.) The role of enterostatin and apolipoprotein AIV on the control of food intake. *Neuropeptides*, 33, 425-433.

Livett, B.G., Marley, P.D., Wan, D.C., & Zhou, X.F. (1990.) Peptide regulation of adrenal medullary function. Journal of Neural Transmission, Supplementum, 29, 77-89.

Ljung, T., Andersson, B., & Bengtsson, B-A. (1996.) Inhibition of cortisol secretion by dexamethasone in relation to body fat distribution: a dose-response study. *Obesity Research*, 4, 277-282.

Lobban, M.C. (1960.) The entrainment of circadian rhythms in man. *Symposia In Quantitative Biology*, 25, 325-332.

Lobo, M.J., Remesar, X., & Alemany, M. (1993.) Effect of chronic intravenous injection of steroid hormones on body weight and composition of female rats. *Biochemistry and Molecular Biology International*, 29, 349-358.

Locatelli, V., Torsello, A., Redaelli, M., Ghigo, E., Massara, F., Camanni, F., & Mueller, E.E. (1986.) Cholinergic agonist and antagonist drugs modulate the growth hormone response to growth-hormone releasing hormone in the rat: evidence for mediation by somatostatin. *Journal of Endocrinology*, 111, 271-278.

Loewy, A.D., & K.M. Spyer (1990.) *Central regulation of autonomic functions*. New York: Oxford University Press.

Londos, C., Brasaemle, D.L., Gruia-Gray, J., Servetnick, D.A., Schultz, C.J., Levin, D.M., & Kimmel, A.R. (1995.) Perilipin: unique proteins associated with intracellular neutral lipid droplets in adipocytes and steroidogenic cells. *Biochemical Society Transactions*, 23, 611-615.

Londos, C., Honnor, R.C., & Dhillon, G.S. (1985.) cAMP dependent protein kinase and lipolysis in rat adipocytes: III multiple modes of insulin regulation of lipolysis and regulation of insulin responses by adenylate cyclase regulators. *Journal of Biological Chemistry*, 260, 15139-15145.

Longcope, C., Kato, T., & Horton, R. (1969.) Conversion of blood androgens to estrogens in normal adult men and women. *Journal of Clinical Investigation*, 48, 2191-2201.

Longcope, C., Pratt, J.H., Schneider, S.H., & Fineberg, S.E. (1978.) Aromatization of androgens by muscle and adipose tissue in vivo. *Journal of Clinical Endocrinology and Metabolism*, 46, 146-152.

Lonnqvist, F., Arner, P., Nordfors, L., & Schalling, M. (1995.) Overexpression of the obese (ob) gene in adipose tissue of human obese subjects. *National Medicine*, 1, 950-993.

Lonnqvist, F., Thorne, A., Large, V., & Arner, P. (1997.) Sex differences in visceral fat lipolysis and metabolic complications of obesity. *Arteriosclerosis, Thrombosis, and Vascular Biology*, 17, 1472-1480.

Lonroth, P., & Smith, U. (1986.) The antilipolytic effect of insulin in human adipocytes requires activation of the phosphodiesterase. *Biochemical and Biophysical Research Communications*, 141, 1157-1161.

Lordick, F., Hauck, R.W., Senekowitsch, R., & Emslander, H.P. (1995.) Atrial natriuretic peptide in acute hypoxia-exposed healthy subjects and in hypoxaemic patients. *European Respiratory Journal*, 8, 216-221.

Lotric, M.B. StefanovskA, a., Stajer, D., & Urbancic-Rowan, V. (2000.) Spectral components of heart rate variability determined by wavelet analysis. *Physiological Measurement*, 21, 441-457.

Loucks, A.B., & Heath, E.M. (1994a.) Dietary restriction reduces luteinizing hormone (LH) pulse frequency during waking hours and increases LH pulse amplitude during sleep in young menstruating women. *Journal of Clinical Endocrinology and Metabolism*, 78, 910-915.

Loucks, A.B., & Heath, E.M. (1994b.) Induction of low-T3 syndrome in exercising women occurs at a threshold of energy availability. *American Journal of Physiology*, 266, R817-R823.

Loucks, A.B., & Horvath SM. (1985.) Athletic amenorrhea: a review. *Medicine and Science in Sports and Exercise*, 17, 56-72.

Loucks, A.B., Mortola, J.F., Girton, L., & Yen, S.S.C. (1989.) Alterations in the hypothalamic-pituitary-ovarian and the hypothalamic-pituitary-adrenal axes in athletic women. *Journal of Clinical Endocrinology and Metabolism*, 68, 402-411.

Loucks, A.B., & Verdun, M. (1998.) Slow restoration of LH pulsatility by refeeding in energetically disrupted women. *American Journal of Physiology*, 275, R1218-R1226.

Loucks, A.B., Verdun, M., & Heath, E.M. (1998.) Low energy availability, not stress of exercise, alters LH pulsatility in exercising women. *Journal of Applied Physiology*, 84, 37-46.

Loughna, P.T., & Brownson C. (1996.) Two myogenic regulatory factor transcripts exhibit muscle-specific responses to disuse and passive stretch in adult rats. *FEBS Letters*, 390, 304-306.

Loughna, P.T., Izumo, S., Goldspink, G., & Nadal-Ginard, B. (1990.) Disuse and passive stretch cause rapid alterations in expression of developmental and adult contractile protein genes in skeletal muscle. *Development*, 109, 217-223.

Louis-Sylvestre , J., & Le Magnen, J. (1980.) Fall in blood glucose level precedes meal onset in free-feeding rats. *Neuroscience and Biobehavioral Reviews*, 4 (Suppl. 1),13-15.

Louisy, F., Guezennec, C.Y., Lartigue, M., Aldigier, J.C., & Galen, F.X. (1989.) Influence of endogenous opioids on atrial natriuretic factor release during exercise in man. European Journal of Applied Physiology and Occupational Physiology, 59, 34-38.

Lowe, D.A., & Always, S.E. (1999.) Stretch-induced myogenin, MyoD, and MRF4 expression and acute hypertrophy in quail slow-tonic muscle are not dependent upon satellite cell proliferation. *Cell and Tissue Research*, 296, 531-539.

Lowell, B.B., & Flier, J.S. (1997.) Brown adipose tissue, beta-3 adrenergic receptors, and obesity. *Annual Review of Medicine*, 48, 307-316.

Lu, J., Webb, R., Richardson, J.A., & Olson, E.N. (1999.) MyoR: a muscle-restricted basic helix-loop-helix transcription factor that antagonizes the actions of MyoD. *Proceedings of the National Academy of Sciences*, 96, 552-557.

Lubin, A., Hord, D.J., Tracy, M.L., & Johnson, L.C. (1976.) Effects of exercise, bedrest and napping on performance decrement during 40 hours. *Psychophysiology, 13,* 334-339.

Lubkin, M., & Stricker-Krongrad, A. (1998.) Independent feeding and metabolic actions of orexins in mice. *Biochemical and Biophysical Research Communications, 253,* 241-245.

Luboshitzky, R. (2000.) Endocrine activity during sleep. *Journal of Pediatric Endocrinology and Metabolism, 13,* 13-20.

Lucas, K.A., Pitari, G.M., Kazerounian, S., Ruiz-Stewart, I., Park, J., Schulz, S., Chepenik, K.P., & Waldman, S.A. (2000.) Guanylyl cyclases and signaling by cyclic GMP. *Pharmacological Reviews, 52,* 375-414.

Lucas, R.J., Freedman, M.S., Munoz, M., Garcia-Fernandez, J.-M., & Foster, R.G. (1999.) Regulation of the mammalian pineal by non-rod, non-cone, ocular photoreceptors. *Science, 284,* 505-507.

Lucini, D., Strappazzon, P., Vecchia, L.D., Maggioni, C., & Pagani, M. (1999.) Cardiac autonomic adjustments to normal human pregnancy: insight from spectral analysis of R-R interval and systolic arterial pressure variability. *Journal of Hypertension, 17,* 1899-1904.

Ludwig, D.S., Mountjoy, K.G., Tatro, J.B., Gillette, J.A., & Frederich, R.C. (1998.) Melanin-concentrating hormone: a functional melanocortin antagonist in the hypothalamus. *American Journal of Physiology, 274,* E627-E633.

Luft, F.C., Rankin, L.I., Bloch, R., Weyman, A.E., Willis, L.R., Murray, R.H., Grim, C.E., & Weinberger, M.H. (1979.) Cardiovascular and humoral responses to extremes of sodium intake in normal black and white men. *Circulation, 60,* 697-706.

Luger, A., Deuster, P.A., Gold, P.W., Loriaux, D.L., & Chrousos, G.P. (1988.) Hormonal responses to the stress of exercise. *Advances in Experimental Medicine and Biology, 245,* 273-280.

Luger, A., Watschinger, B., Deuster, P., Svoboda, T., Clodi, M., & Chrousos, G.P. (1992.) Plasma growth hormone and prolactin responses to graded levels of acute exercise and to a lactate infusion. *Neuroendocrinology, 56,* 112-117.

Luiten, P.G.M., ter Horst, G.J., Karst, H., & Steffens, A.B. (1985.) The course of paraventricular hypothalamic efferents to autonomic structures in medulla and spinal cord. *Brain Research, 329,* 374-378.

Lundberg, J., Norgren, L., Ribbe, E., Rosen, I, Steen, S., Thorne, J., & Wallin, B.G. (1989.) Direct evidence of active sympathetic vasodilation in the skin of the human foot. *Journal of Physiology, 417,* 437-446.

Lundborg, P. (1983.) The effect of adrenergic blockade on potassium concentrations in different conditions. *Acta Medica Scandinavica, 672,* 121-125.

Lusardi, P., Zoppi, A., Preti, P., Pesce. R.M., Piazza, E., & Fogari, R. (1999.) Effects of insufficient sleep on blood pressure in hypertensive patients: a 24-h study. *American Journal of Hypertension, 12,* 63-68.

Luscher, T.F., & Barton, M. (1997.) Biology of the endothelium. *Clinical Cardiology, 20* (Suppl. II), II-3-II-10.

Luscher, T.F., & Dohi, Y. (1992.) Endothelium-derived relaxing factor and endothelin in hypertension. *News in Physiological Sciences, 7,* 120-123.

Lynch, F.A., & Kirov, S.M. (1986.) Changes in blood lymphocyte population following surgery. *Journal of Clinical and Loaboratory Immunology, 20,* 75-79.

Mabuchi, K., Szvetko, D., Pinter, K., & Sreter, F.A. (1982.) Type IIB to IIA fiber transformation in intermittently stimulated rabbit muscles. *American Journal of Physiology, 242,* C373- C381.

MacConnie, S.E., Barkan, A., Lampman, R.M., Schork, M.A., & Beitins, I.Z. (1986.) Decreased hypothalamic gonadotropin-releasing hormone secretion in male marathon runners. *New England Journal of Medicine, 315,* 411-417.

MacDonald, A., Forbes, I.J., Gallagher, D., Heeps, G., & McLaughlin, D.P. (1994.) Adrenoceptors mediating relaxation to catecholamines in rat isolated jejunum. *British Journal of Pharmacology, 112,* 576-578.

Macdonald, I., Nicholson, S., Melsom, R.D., & Perry, J.D. (1991.) Brief exercise induces an immediate and a delayed leucocytosis. *British Journal of Sports Medicine, 25,* 191-195.

Mackinnon, L.T., Chick, T.W., van As, A., & Tomasi, T.B. (1987.) The effect of exercise on secretory and natural immunity. *Advances in Experimental Medicine and Biology, 216A,* 869-876.

MacLean, D.A., Imadojemu, V.A., & Sinoway, L.I. (2000.) Intersitial pH, K(+), lactate, and phosphate determined with MSNA during exercise in humans. *American Journal of Physiology, 278,* R563-R571.

MacMahon, B. (1973.) Age at menarche. *National health survey* (p.1). Washington, DC: US Government Printing Office, US Department of Health, Education and Welfare publication 74-1615.

MacMahon, B., Trichopoulos, D., Brown, J., Andersen, A.P., Aoki, K., Cole, P., et al. (1982.) Age at menarche, probability of ovulation and breast cancer risk. *International Journal of Cancer, 29,* 13-16.

Madden, K.S., Sanders, V.M., & Felten, D.L. (1995.) Catecholamine influences and sympathetic neural modulation of immune responsiveness. *Annual Reviews of Pharmacology and Toxicology, 35,* 417-448.

Madden, P.A., Heath, A.C., Rosenthal, N.E., & Martin, N.G. (1996.) Seasonal changes in mood and behavior. *Archives of General Psychiatry, 53,* 47-55.

Maehlum, S., Felig, P., & Wahren, J. (1978.) Splanchnic glucose and muscle glycogen metabolism after glucose feeding during post-exercise recovery. *American Journal of Physiology, 235,* E255-E260.

Maggi, C.A., & Meli, A. (1988.) The sensori-efferent function of capsaicin-sensitive sensory neurons. *General Pharmacology, 19,* 1-43.

Maggs, D.G., Jacob, R., Rife, F., Lange, R., Leone, P., During, M.J., Tamborlane, W.V., & Sherwin, R.S. (1995.) Interstitial fluid concentrations of glycerol, glucose, and amino acids in human quadriceps muscle and adipose tissue: evidence for significant lipolysis in skeletal muscle. *Journal of Clinical Investigation, 96,* 370-377.

Maiter, D., Underwoord, L.E., Maes, M., Davenport, M.L., & Ketelslegers, J.M. (1988.) Direct effect of intermittent and continuous growth hormone (GH) administration on serum somatomedin-C/insulin-like growth factor I and liver GH receptors in hypophysectomized rats. *Endocrinology, 123,* 1053-1059.

Malarkey, W.B., Strauss, R.H., Leizman, D.J., Liggett, M., & Demers, L.M. (1991.) Endocrine effects in female weight lifters who self-administer testosterone and anabolic steroids. *American Journal of Obstetrics and Gynecology, 165,* 1385-1390.

Malchoff, C.D., Hughes, J.M., & Carey, R.M. (1987.) Effect of upright posture on the aldosterone responses to dopamine, metoclopramide, angiotensin II, and adrenocorticotropin. *Journal of Clinical Endocrinology and Metabolism, 65,* 203-207.

Malik, M., & Camm, A.J. (Eds.) (1995.) *Heart rate variability,* Armonk, NY: Futura Publishing Company.

Malina, R. (1994.) Physical growth and biological maturation of young athletes. *Exercise and Sport Sciences Reviews, 22,* 389-433.

Malina, R., Meleski, B.W., & Shoup, R.F. (1982.) Anthropometric, body composition, and maturity characteristics of selected school-age athletes. *Pediatric Clinics of North America, 29,* 1305-1323.

Malina, R.M. (1969.) Quantification of fat, muscle and bone in man. *Clinical Orthopedics and Related Disabilities, 65,* 9-38.

Malina, R.M. (1974.) Adolescent changes in size, build, composition, and performance. *Human Biology, 46,* 117-131.

Malina, R.M. (1983.) Menarche in athletes: a synthesis and hypothesis. *Annals of Human Biology, 10,* 1-24.

Malina, R.M. (1986.) Growth of muscle tissue and muscle mass. In F. Falkner & J.M. Tanner (Eds.), *Human growth: a comprehensive treatise.* (pp. 77-99). New York: Plenum Press.

Malina, R.M., Bouchard, C., Shoup, R.F., Demirjian, A., & Lariviere, G. (1979.) Age at menarche, family size, and birth order in athletes at the Montreal Olympic Games, 1976. *Medicine and Science in Sports, 11,* 354-358.

Malina, R.M., Harper, A.B., Avent, H.H., & Campbell, D.E. (1973.) Age at menarche in athletes and non-athletes. *Medicine and Science in Sports, 5,* 11-13.

Malliani, A. (1997.) The autonomic nervous system: a Sherringtonian revision of its integrated properties in the control of circulation. *Journal of the Autonomic Nervous System, 64,* 158-161.

Malliani, A., Pagani, M., Lombardi, F., & Cerutti, S. (1991.) Cardiovascular neural regulation explored in the frequency domain. *Circulation, 84,* 482-492.

Mandarino, L.J., Consoli, A., Jain, A., & Kelley, D.E. (1996.) Interaction of carbohydrate and fat fuels in human skeletal muscle: impact of obesity and NIDDM. *American Journal of Physiology, 270,* E463-E470.

Mandarino, L.J., Wright, K.S., Verity, L.S., Nichols, J., Bell, J.M., Kolterman, O.G., & Beck-Nielsen, H. (1987.) Effects of insulin infusion on human skeletal muscle pyruvate dehydrogenase, phosphofructokinase, and glycogen synthase: evidence for their role in oxidative and nonoxidative glucose metabolism. *Journal of Clinical Investigation, 80,* 655-663.

Manganiello, V.C., Degerman, E., Taira, M., Kono, T., & Belfrage P. (1996.) Type III cyclic nucleotide phosphodiesterases and insulin action. *Current Topics in Cellular Regulation, 34*, 63-100.

Mankad, P., & Yacoub, M. (1997.) Influence of basal release of nitric oxide on systolic and diastolic function of both ventricles. *Journal of Thoracic and Cardiovascular Surgery 113*, 770-776.

Mann, S., Millar, M.W., Melville, C.D., Balasubramanian, V., & Raftery, E.B. (1979.) Physical activity and the circadian rhythm of blood pressure. *Clinical Science, 57*, 291s-294s.

Manolagas, S.C., & Jilka, R.L. (1992.) Cytokines, hematopoiesis, osteoclastogenesis, and estrogens. *Calcified Tissue International, 50*, 199-202.

Manolagas, S.C., & Jilka, R.L. (1995.) Bone marrow, cytokines, and bone remodeling. *New England Journal of Medicine, 332*, 305-311.

Mannelli, M., Maggi, M., DeFeo, M.L., Cuomo, S., Delitala, G. Giusti, G., & Serio, M. (1984.) Effects of naloxone on catecholamine plasma levels in adult men: a dose response study. *Acta Endocrinologica, 106*, 357-361.

Mantzoros, C.S. (1999.) The role of leptin in human obesity and disease: a review of current evidence. *Annals of Internal Medicine, 130*, 671-680.

Mantzoros, C.S., Flier, J.S., & Rogol, A.D. (1997.) A longitudinal assessment of hormonal and physical alterations during normal puberty in boys. V. Rising leptin levels may signal the onset of puberty. *Journal of Clinical Endocrinology and Metabolism, 82*, 1066-1070.

Marette, A., & Bukowiecki, L.J. (1989.) Stimulation of glucose transport by insulin and norepinephrine in isolated rat brown adipocytes. *American Journal of Physiology, 257*, C714-C721.

Margaria, R., Cerretelli, P., & Aghemo, P. (1963.) Energy cost of running. *Journal of Applied Physiology, 18*, 367-370.

Marin, P., & Arver, S. (1998.) Androgens and abdominal obesity. *Baillieres Clinical Endocrinology and Metabolism, 12*, 441-451.

Marin-Grez, M., Fleming, J.T., & Steinhausen, M. (1986.) Atrial natriuretic peptide causes pre-glomerular vasodilatation and post-glomerular vasoconstriction in rat kidney. *Nature, 324*, 473-476.

Marinissen, M.J., & Gutkind, J.S. (2001.) G-protein-coupled receptors and signaling networks: emerging paradigms. *Trends in Pharmacological Sciences, 22*, 368-376.

Marker, J.C., Hirsch, I.B., Smith, L.J., Parvin, C.A., Holloszy, J.O., & Cryer, P.E. (1991.) Catecholamines in prevention of hypoglycemia during exercise in humans. *American Journal of Physiology, 251*, R552-R559.

Marliss, E.B., Simantirakis, E., Purdon, C., Gougeon, R., Field, C.J., Halter, J.B., & Vranic, M. (1991.) Glucoregulatory and hormonal responses to repeated bouts of intense exercise in normal male subjects. *Journal of Applied Physiology, 71*, 924-933.

Marrero, M.B., Schieffer, B., Paxton, W.G., Heerdt, L., Berk, B.C., Delafontaine, P., & Bernstein, K.E. (1995.) Direct stimulation of Jak/STAT pathway by the angiotensin II AT1 receptor. *Nature, 375*, 247-250.

Marshall, J.C., & Griffin, M.L. (1993.) The role of changing pulse frequency in the regulation of ovulation. *Human Reproduction, 8* (Suppl. 2), 57-61.

Marshall, J.C., & Kelch, R.P. (1986.) Gonadotropin-releasing hormone: role of pulsatile secretion in the regulation of reproduction. *New England Journal of Medicine, 315*, 1459-1468.

Marshall, J.M. (2000.) Adenosine and muscle vasodilatation in acute systemic hypoxia. *Acta Physiologica Scandinavica, 168*, 561-573.

Marshall, J.M., & Timms, R.J. (1980.) Experiments on the role of the subthalamus in the generation of the cardiovascular changes during locomotion in the cat. *Journal of Physiology (London), 301*, 92P-93P.

Marshall, W.A., & Tanner, J.M. (1969.) Variations in pattern of pubertal changes in girls. *Archives of Diseases of Children, 44*, 291-303.

Marshall, W.A., & Tanner, J.M. (1970.) Variations in pattern of pubertal changes in boys. *Archives of Diseases of Children, 45*, 13-23.

Marshall, W.A., & Tanner, J.M. (1986.) Puberty. In F. Falkner & J.M. Tanner (Eds.), *Human growth: a comprehensive treatise.* (pp. 171-209). New York: Plenum Press.

Marshall, J.M., Lloyd, J., & Mian, R. (1993.) The influence of vasopressin on the arterioles and venules of the skeletal muscle of the rat during systemic hypoxia. *Journal of Physiology, 470*, 473-484.

Martel, P.J., Sharp, G.W.G., Slorach, S.A., & Vipond, H.J. (1962.) A study of the roles of adrenocortical steroids and glomerular filtration rate in the mechanism of the diurnal rhythm of water and electrolyte excretion. *Journal of Endocrinology, 24*, 159-169.

Marth, P.D., Woods, R.R., & Hill, D.W. (1998.) Influence of time of day on anaerobic capacity. *Perceptual and Motor Skills, 86*, 592-594. .

Martha, P.M., Jr., & Reiter, E.D. (1991.) Pubertal growth and growth hormone secretion. *Endocrinology and Metabolism Clinics of North America, 20*, 165-182.

Martha, P.M., Jr., Blizzard, R.M., & Rogol, A.D.(1988.) Atenolol enhances growth hormone release to exogenous growth hormone-releasing hormone but fails to alter spontaneous nocturnal growth hormone secretion in boys with constitutional delay of growth. *Pediatric Research, 23*, 393-397.

Martha, P.M., Jr., Rogol, A.D., Veldhuis, J.D., Kerrigan, J.R., Goodman, D.N., & Blizzard, R.M. (1989.) Alterations in the pulsatile properties of circulating growth hormone concentrations during puberty in boys. *Journal of Clinical Endocrinology and Metabolism, 69*, 563-570.

Martha, P.M., Jr., Gorman, K.M., Jr., Blizzard, R.M., Rogol, A.D., & Veldhuis, J.D. (1992.) Endogenous growth hormone secretion and clearance rates in normal boys, as determined by deconvolution analysis: relationship to age, pubertal status, and body mass. *Journal of Clinical Endocrinology and Metabolism, 74*, 336-344.

Marti, B., Goerre, S., Spuhler, T., Schaffner, T., & Gutzwiller, F. (1989.) Sudden death during mass running events in Switzerland 1978-1987: an epidemiologico-pathologic study. *Schweizerische Medizinische Wochenschrift Journal Suisse de Medecine, 119*, 473-482.

Martin, L.G., Clark, J.W., & Connor, T.B. (1968.) Growth hormone secretion enhanced by androgens. *Journal of Clinical Endocrinology and Metabolism, 28*, 425-428.

Martin, M.L., & Jensen, M.D. (1991.) Effects of body fat distribution on regional lipolysis in obesity. *Journal of Clinical Investigation, 88*, 609-613.

Martin, W.H., III, Dalsky, G.P., Hurley, B.F., Matthews, D.E., Bier, D.M., Hagberg, J.M., Rogers, M.A., King, D.S., & Holloszy, J.O. (1993.) Effect of endurance training on plasma free fatty acid turnover and oxidation during exercise. *American Journal of Physiology, 265*, E708-E714.

Mascaro, C., Acosta E., Ortiz, J.A., Marrero, P.F., Hegardt, F.G., & Haro, D. (1998.) Control of human muscle-type carnitine palimoyltransferase I gene transcription by peroxisome proliferator-activated receptor. *Journal of Biological Chemistry, 273*, 8560-8563.

Mason, J.W., Hartley, L.H., Kotchen, T.A., Mougey, E.H., Ricketts, P.T., & Jones, L.G. (1973.) Plasma cortisol and norepinephrine responses in anticipation of muscular exercise. *Psychosomatic Medicine, 35*, 406-414.

Massicotte, D., Peronnet, F., Allah, C., Hillaire-Marcel, C., Ledoux, M., & Brisson, G. (1986.) Metabolic response to [^{13}C]glucose and [^{13}C]fructose ingestion during exercise. *Journal of Applied Physiology, 61*, 1180-1184.

Massicotte, D., Peronnet, F., Brisson, G., Boivin, L. & Hillaire-Marcel, C. (1990.) Oxidation and exogenous carbohydrate during prolonged exercise in fed and fasted conditions. *International Journal of Sports Medicine, 11*, 253-258.

Masuda, Y., Tanaka, T., Inomata, N., Ohnuma, N., Tanaka, S., Itoh, Z., Hosoda, H., Kojima, M., & Kangawa, K. (2000.) Ghrelin stimulates gastric acid secretion and motility in rats. *Biochemical and Biophysical Research Communications, 276*, 905-908.

Mathias, P.C., Salvato, E.M., Curi, R., Malaisse, W.J., & Capinelli, A.R. (1993.) Effect of epinephrine on 86Rb efflux, 45Ca outflow and insulin release from pancreatic islets perifused in the presence of propranolol. *Hormone and Metabolic Research, 25*, 138-141.

Matsuda, H., Koyama, H., Oikawa, M., Yoshihara, T., & Kaneko, M. (1991.) Nerve growth factor I like activity detected in equine peripheral blood after running exercise. *Zentralblatt fur Veterinarmedizin, 38*, 557-559.

Matsui, T., Matsufiji, H., Tamaya, K., Kawasaki, T., & Osajima, Y. (1997.) Metabolic behavior of angiotensins in normotensive human plasma in the supine and upright postures. *Bioscience, Biotechnology and Biochemistry, 61*, 1814-1818.

Matsumine, H., Hirato, K., Yanaihara, T., Tamada, T., & Yoshida, M. (1986.) Aromatization by skeletal muscle. *Journal of Clinical Endocrinology and Metabolism, 63*, 717-720.

Matsunaga, T., Okumara, K., Tsunoda, R., Tayama, S., Tabuchi, T., & Yasue, H. (1996.) Role of adenosine in regulation of coronary flow in dogs with inhibited synthesis of endothelium-derived nitric oxide. *American Journal of Physiology, 270*, H427-H434.

Matthews, K.A., & Rodin, J. (1992.) Pregnancy alters blood pressure responses to psychological and physical challenge. *Psychophysiology, 29,* 232-240.

Maughan, R.J., & Leiper, J.B. (1993.) Post-exercise rehydration in man: effects of voluntary intake of four different beverages. *Medicine and Science in Sports and Exercise, 25* (Suppl.), S2.

Maughan, R.J., & Leiper, J.B. (1995.) Sodium intake and post-exercise rehydration in man. *European Journal of Applied Physiology, 71,* 311-319.

Maughan, R.J., Leiper, J.B., & Shirreffs, S.M. (1996.) Restoration of fluid balance after exercise-induced dehydration: effects of food and fluid intake. *European Journal of Applied Physiology and Occupational Physiology, 73,* 317-25.

Maughan, R.J., Leiper, J.B., & Shirreffs, S.M. (1997.) Factors influencing the restoration of fluid and electrolyte balance after exercise in the heat. *British Journal of Sports Medicine, 31,* 175-182.

Maughan, R.J., Watson, J.S., & Weir, J. (1984.) The relative proportions of fat, muscle and bone in the normal human forearm as determined by computed tomography. *Clinical Science, 66,* 683-689.

Mauras, N. (1995.) Estrogens do not affect whole-body protein metabolism in the pubertal female. *Journal of Clinical Endocrinology and Metabolism, 80,* 2842-2845.

Mauras, N., Blizzard, R.M., Thorner, M.O., & Rogol, A.D. (1987.) Selective beta1-adrenergic- receptor blockade with atenolol enhances basal and growth hormone releasing hormone mediated growth hormone release in man. *Metabolism, 36,* 369-372.

Mauras, N., Haymond, M.W., Darmaun, D., Vieira, N.E., Abrams, S.A., & Yergey, A.L. (1994.) Calcium and protein kinetics in prepubertal boys. *Journal of Clinical Investigation, 93,* 1014-1019.

Mauras, N., Rogol, A.D., & Veldhuis, J.D. (1989.) Specific, time-dependent actions of low-dose ethinyl estradiol administration on the episodic release of growth hormone, follicle-stimulating hormone, and luteinizing hormone in prepubertal girls with Turner's syndrome. *Journal of Clinical Endocrinology and Metabolism, 69,* 1053-1058.

Mauras, N., Rogol, A.D., Haymond, M.W., & Veldhuis, J.D. (1996.) Sex steroids, growth hormone, insulin-like growth factor-1: neuroendocrine and metabolic regulation of puberty. *Hormone Research, 45,* 74-80.

Mauriege, P., Despres, J.P., & Prud'homme, D. (1991.) Regional variation in adipose tissue lipolysis in lean and obese men. *Journal of Lipid Research, 32,* 1625-1633.

Mauriege, P., Galitzky, J., Berlan, M., & Lafontan, M. (1987.) Heterogeneous distribution of beta and alpha-2 adrenoceptor binding sites in human fat cells from various fat deposits: functional consequences. *European Journal of Clinical Investigation, 17,*156-165.

Maw, G.J., Mackenzie, I.L., & Taylor, N.A. (1998.) Human body-fluid distribution during exercise in hot, temperate and cool environments. *Acta Physiologica Scandinavica, 163,* 297-304.

Mayer, B., John, M., Heinzel, B., Werner, E.R., Wachter, H., Schultz, G., & Bohme, E. (1991.) Brain nitric oxide synthase is a biopterin- and flavin-containing multifunctional oxido-reductase. *FEBS Letters, 288,* 187-191.

Mayer, J., Marshall, N.B., Vitale, J., Christensen, J.H., Masayekhi, M.S., & Stare, F.J. (1954.) Exercise, food intake and body weight in normal rats and genetically obese adult mice. *American Journal of Physiology, 177,* 544-548.

Mazzeo, R.S., Podolin, D.A., & Henry, V. (1995.) Effects of age and endurance training on beta-adrenergic receptor characteristics in Fischer 344 rats. *Mechanisms of Ageing and Development, 84,* 157-169.

Mazzeo, R.S., Rajkumar, C., Jennings, G., & Esler, M. (1997.) Norepinephrine spillover at rest and during submaximal exercise in young and old subjects. *Journal of Applied Physiology, 82,* 1869-1874.

McAllister, R.M. (1998.) Adaptations in control of blood flow with training: splanchnic and renal blood flows. *Medicine and Science in Sports and Exercise, 30,* 375-381.

McArdle, W.D., Katch, F.I., & Katch, V.L. (2001.) *Exercise physiology: energy, nutrition, and human performance,* 5th ed. Baltimore: Lippincott, Williams & Wilkins.

McArthur, J.W. (1985.) Endorphins and exercise in females: possible connection with respiratory dysfunction. *Medicine and Science in Sports and Exercise, 17,* 82-88.

McArthur, J.W., Bullen, B.A., Beitins, I.Z., Pagano, M., Badger, T.M., & Klibanski, A. (1980.) Hypothalamic amenorrhea in runners of normal body composition. *Endocrine Research Communications, 7,* 13-25.

McCance, D.R., Roberts, G., Sheridan, B., McKnight, J.A., Leslie, H., Merrett, J.D., & Atkinson, A.B. (1989.) Variations in plasma concentration of atrial natriuretic factor across 24 hours. *Acta Endocrinologica, 120,* 266-270.

McCarthy, D.A., Grant, M., Marbut, M., Watling, M., Wade, A.J., Macdonald, I., Nicholson, S., Melsom, R.D., & Perry, J.D. (1991.) Platelet peripheral benzodiazepine receptors in repeated stress. *Life Sciences, 48,* 341-346.

McCarthy, D.A., Macdonald, I., Grant, M., Marbut, M., Watling, M., Nicholson, S., Deeks, J.J., Wade, A.J., & Perry, J.D. (1992.) Studies on the immediate and delayed leucocytosis elicited by brief (30 min.) strenuous exercise. *European Journal of Applied Physiology and Occupational Physiology, 64,* 513-517.

McConnell, G.K., Burge, C.M., Skinner, S.L., & Hargreaves, M. (1997.) Influence of ingested fluid volume on physiological response during prolonged exercise. *Acta Physiologica Scandinavica, 160,* 149-156.

McGarry, J.D. (1995.) The mitochondrial carnitine palmitoyl transferase system: its broadening role in fuel homeostasis and new insights into its molecular features. *Biochemical Society Transactions, 23,* 321-324.

McGuinness, O.P., Fugiwara, T., Murrell, S., Bracy, D., New, D., O'Connor, D., & Cherrington, A.D. (1993.) Impact of chronic stress hormone infusion on hepatic carbohydrate metabolism in the conscious dog. *American Journal of Physiology, 265,* E314-E322.

McKay, M.K., & Hester, R.L. (1996.) Role of nitric oxide, adenosine, and ATP-sensitive potassium channels in insulin-induced vasodilatation. *Hypertension, 28,* 202-208.

McKee, M.D., & Nanci, A. (1995.) Osteopontin and the bone remodelling sequence. Colloidal-gold immunocytochemistry of an interfacial extracellular matrix protein. *Annals of the New York Academy of Sciences, 760,* 177-189.

McKinley, M.J., Congiu, M., Denton, D.A., Park, R.G., Penschow, J., Simpson, J.B., Tarjan, E., Weisinger, R.S., and Wright, R.D. (1984.) The anterior wall of the third cerebral ventricle and homeostatic responses to hydration. *Journal of Physiology, 79,*421-427.

McLeod, K., Donahue, H., Levin, P., Fontaine, M., & Rubin, C. (1993.) Electric fields modulate bone cell function in a density dependent manner. *Journal of Bone and Mineral Research, 8,* 977-984.

McMurray, R.G., Berry, M.J., & Katz, V.L. (1990a.) The beta-endorphin responses of pregnant women during aerobic exercise in the water. *Medicine and Science in Sports and Exercise, 22,* 298-303.

McMurray, R.G., Berry, M.J., Katz, V.L., Graetzer, D.G., & Cefalo, R.C. (1990b.) The thermoregulation of pregnant women during aerobic exercise in the water: a longitudinal approach. *European Journal of Applied Physiology and Occupational Physiology, 61,*119-123.

McMurray, R.G., Eubank, T.K., & Hackney, A.C. (1995.) Nocturnal hormonal responses to resistance exercise. *European Journal of Applied Physiology and Occupational Physiology, 72,* 121-126.

McMurray, R.G., Forsythe, W.A., Mar, M.H., & Hardy, C.J. (1987.) Exercise intensity-related responses of beta-endorphin and catecholamines. *Medicine and Science in Sports and Exercise, 19,* 570-574.

McMurray, R.G., Katz, V.L., Poe, M.P., & Hackney, A.C. (1995.) Maternal and fetal responses to low-impact aerobic dance. *American Journal of Perinatology, 12,* 282-285.

Meglasson, M.D., & Matschinsky, F.M. (1986.) Pancreatic glucose metabolism and regulation of insulin secretion. *Diabetes-Metabolism Reviews, 2,* 163-214.

Meisner, H., & Carter, J.R Jr. (1977.) Regulation of lipolysis in adipose tissue. *Horizons in Biochemistry & Biophysics, 4,* 91-129.

Mejean, L., Kolopp, M., & Drouin, P. (1994.) Chronobiology, nutrition and diabetes mellitus. In Y. Touitou & E. Haus (Eds.), *Biologic rhythms in clinical and laboratory medicine* (pp. 375-385). Berlin: Springer Verlag.

Melby, C.L., Commerford, S.R. & Hill, J.O. (1998.) Exercise, macronutrient balance,and weight control. In D.R. Lamb & R. Murray (Eds.), *Perspectives in exercise science and sports medicine, Vol. 11: Exercise, nutrition, and weight control* (pp. 1-55), Traverse City, MI: Cooper Publishing Group.

Melby, C.L., Scholl, C., Edwards, G., & Bullough, R. (1993.) Effect of acute resistance exercise on postexercise energy expenditure and resting metabolic rate. *Journal of Applied Physiology, 75,* 1847-1853.

Melhim, A.F. (1993.) Investigation of circadian rhythms in peak power and mean power of female physical education students. *International Journal of Sports Medicine, 14*, 303-306.

Mellon, S.H. (1994.) Neurosteroids: biochemistry, modes of action, and clinical relevance. *Journal of Clinical Endocrinology and Metabolism, 78*, 1003-1008.

Mendelson, C.R. (1996.) Mechanisms of hormone action. In J.E. Griffin & S.R. Ojeda (Eds.), *Textbook of endocrine physiology* (pp. 29-65). Oxford: Oxford University Press.

Mendler, L., Zador, E., Dux, L., & Wuytack F. (1998.) mRNA levels of myogenic regulatory factors in rat slow and fast muscles regenerating from notexin-induced-necrosis. *Neuromuscular Disorders, 8*, 533-541.

Menetrey, D., & Basbaum, A.I. (1987.) Spinal and trigeminal projections to the nucleus of the solitary tract: a possible substrate for somatovisceral and viscerovisceral reflex activation. *Journal of Comparative Neurology, 255*, 439-450.

Meney, I., Waterhouse, J., Atkinson, G., Reilly, T., & Davenne, D. (1998.) The effect of one night's sleep deprivation on temperature, mood, and physical performance in subjects with different amounts of habitual physical activity. *Chronobiology International, 15*, 349-363.

Meredith, I.T., Friberg, P., Jennings, G.L., Dewar, E.M., Fazio, V.A., Lambert, G.W., & Esler, M.D. (1991.) Exercise training lowers resting renal but not cardiac sympathetic activity in humans. *Hypertension, 18*,575-582.

Merlie, J.P., Mudd, J., Cheng, T.C., Olson, E.N. (1994.) Myogenin and acetylcholine receptor alpha gene promoters mediate transcriptional regulation in response to motor innervation. *Journal of Biological Chemistry, 269*, 2461-2467.

Méry, P.F., Lohmann, S.M., Walker, V., & Fischmeister, R. (1991.) Ca²⁺ current is regulated by cyclic GMP-dependent protein kinase in mammalian cardiac myocytes. *Proceedings of the National Academy of Science, USA 88*, 1197-1201.

Metz, J.A., Anderson, J.J.B., Gallagher, P.N., Jr., (1993.) Intakes of calcium, phosphorus, and protein, and physical-activity level are related to radial bone mass in young adult women. *American Journal of Clinical Nutrition, 58*, 537-542.

Metzger, D., & Kerrigan, J. (1993.) Androgen receptor blockade with flutamide enhances growth hormone secretion in late pubertal males: evidence for independent actions of estrogen and androgen. *Journal of Clinical Endocrinology and Metabolism, 76*, 1147-1152.

Metzger, D., & Kerrigan, J. (1994.) Estrogen receptor blockade with tamoxifen diminishes growth hormone secretion in boys: evidence for a stimulatory role of endogenous estrogens during male adolescence. *Journal of Clinical Endocrinology and Metabolism, 79*, 513-518.

Metzner, H.L., Lamphiear, D.E., Wheeler, N.C., & Larkin, F.A. (1977.) The relationship between frequency of eating and adiposity in adult men and women in the Tecumseh Community Health Study. *The American Journal of Clinical Nutrition, 30*, 712-715.

Meyer, R., Linz, K.W., Surges, R., Meinardus, S., Vees, J., Hoffmann, A., Windholz, O., & Grohe, (1998.) Rapid modulation of L-type calcium current by acutely applied oestrogens in isolated cardiac myocytes from human, guinea-pig and rat. *Experimental Physiology, 83*, 305-321.

Meyer, T.E., & Habener, J.F. (1993.) Cyclic adenosine 3',5'-monophosphate response element binding protein (CREB) and related transcription-activating deoxyribonucleic acid-binding proteins. *Endocrine Reviews, 14*, 269-290.

Mian, R., Marshall, J.M., & Kumar, P. (1990.) Interactions between K⁺ and β₂-adrenoreceptors in determining muscle vasodilatation induced in the rat by systemic hypoxia. *Experimental Physiology, 75*, 407-410.

Michel, M.C., Hanft, G., & Gross, G. (1994.) Radioligand binding studies of 1-adrenoceptor subtypes in rat heart. *British Journal of Pharmacology, 111*, 533-538.

Miki, N., Ono, M., & Shizume, K. (1984.) Evidence that opiatergic and alpha-adrenergic mechanisms stimulate rat growth hormone release via growth hormone-releasing factor (GRF). *Endocrinology, 114*. 1950-1952.

Mikines, K.J., Sonne, B., Farrell, P.A., Tronier, B., & Galbo, H. (1988.) Effect of physical exercise on sensitivity and responsiveness to insulin in humans. *American Journal of Physiology, 254*, E248-E259.

Milasincic, D.J., Dhawan, J., & Farmer, S.R.(1996.) Anchorage-dependent control of muscle-specific gene expression in C2C12 mouse myoblasts: in Vitro Cellular and Developmental Biology. *Animal, 32*(2), 90-9,.

Millar-Craig, M.W., Bishop, C.N., & Raftery, E.B. (1978.) Circadian variations of blood pressure. *Lancet, 1*, 795-797.

Miller, A.E., MacDougall, J.D., Tarnopolsky, M.A., & Sale, D.G. (1993.) Gender differences in strength and muscle fiber characteristics. *European Journal of Applied Physiology and Occupational Physiology, 66*, 254-262.

Miller, J.C., & Hellander, M. (1979.) The 24-hour cycle and nocturnal depression of cardiac output. *Aviation, Space and Environmental Medicine, 50*, 1139-1144.

Miller, T.B. Jr., & Larner, J. (1973.) Mechanism of control of hepatic glycogenesis by insulin. *Journal of Biological Chemistry, 248*, 3483-3488.

Miller, W.C., Gorski, J., Oscai, L.B., & Palmer, W.K. (1989.) Epinephrine-activation of heparin-nonreleasable lipoprotein lipase in 3 skeletal muscle fiber types of the rat. *Biochemical and Biophysical Research Communications, 164*, 615-619.

Miller, W.C., Bryce, G.R., & Conlee, R.K. (1984.) Adaptations to a high-fat diet that increase exercise endurance in male rats. *Journal of Applied Physiology, 56*, 78-83.

Miller, W.C., Gorski, J., Oscai, L.B., & Palmer, W.E. (1989.) Epinephrine activation of heparin-nonreleasable lipoprotein lipase in skeletal muscle types of the rat. *Biochemical and Biophysical Research Communications, 164*, 615-619.

Millet, L., Barbe, P., Lafontan, M., Berlan, M., & Galitzky, J. (1998.) Catecholamine effects on lipolysis and blood flow in human abdominal and femoral adipose tissue. *Journal of Applied Physiology, 85*, 181-188.

Mills, J.N. (1964.) Circadian rhythms during and after three months in solitude underground. *Journal of Physiology, 174*, 217-231.

Minors, D., & Waterhouse, J. (1981.) *Circadian rhythms in the human.* Bristol: Wright.

Minors, D., Waterhouse, J., Folkard, S., & Atkinson, G. (1996.) The difference between activity when in bed and out of bed. III. Nurses on night work. *Chronobiology International, 13*, 273-282.

Minotti, S., Scicchitano, B.M., Nervi, C., Scarpa, S., Lucarelli, M., Molinaro, M., & Adam, S. (1998.) Vasopressin and insulin-like growth factors synergistically induce myogenesis in serum-free medium. *Cell Growth and Differentiation, 9*,155-163.

Mistlberger, R., Bergmann, B., & Rechtschaffen, A. (1987.) Period-amplitude analysis of rat electroencephalogram: Effects of sleep deprivation and exercise. *Sleep, 10*, 508-522.

Mitchel, J.S., & Keesey, R.E. (1977.) Defense of a lowered weight maintenance level by lateral hypothamically lesioned rats: evidence from a restriction-refeeding regimen. *Physiology and Behavior, 18*, 1121-1125.

Mitchell, J.H. (1990.) Neural control of circulation during exercise. *Medicine and Science in Sports and Exercise, 22*, 141-154.

Mitchell, T.A. Smyrl, R., Hutchins, M. Schindler, W.T., & Critchlow, V. (1972.) Plasma growth hormone levels in rats with increased naso-anal length due to hypothalamic surgery. *Neuroendocrinology, 20*, 31-45.

Mitchell, T.A., Hutchins, M., Schindler, W.T., & Critchlow, V. (1973.) Increase in plasma growth hormone concentration and naso-anal length in rats following isolation of medial basal hypothalamus. *Neuroendocrinology, 21*, 161-173.

Mittendorfer, B., Horowitz, J.F., & Klein, S. (2002.) Effect of gender on lipid kinetics during endurance exercise of moderate intensity in untrained subjects. *American Journal of Physiology , 283*, E58-65.

Miura, S., Ideishi, M., Sakai, T., Motoyama, M., Kinoshita, A., Sasaguri, M., Tanaka, H., Shindo, M., & Arakawa, K. (1994.) Angiotensin formation by an alternative pathway during exercise in humans. *Journal of Hypertension, 12*, 1177-1181.

Mizelle, H.L., Hall, J.E., Woods, L.L., Montani, J.P., Dzielak, D.J., & Pan, Y.J. (1987.) Role of renal nerves in compensatory adaptation to chronic reductions in sodium intake. *American Journal of Physiology, 252*, F291-F298.

Mohan, S., & Baylink, D.J. (1991.) The role of IGF-II in the coupling of bone formation to resorption. In Spencer, E.M. (Ed), *Modern Concepts of Insulin-like Growth Factors* (pp. 169-184), New York: Elsevier.

Mole, P.A., Oscai, L.B., & Holloszy, J.O. (1971.) Adaptation of muscle to exercise. Increase in levels of palmityl CoA synthetase, carnitine palmityl transferase, and palmityl CoA dehydrogenase and in the capacity to oxidize fatty acids. *Journal of Clinical Investigation, 50*, 2323-2330.

Molle, M., Albrecht, C., Marshall, L., Fehm, H.L., & Born, J. (1997.) Adrenocorticotropin widens the focus of attention in humans. a nonlinear electroencephalographic analysis. *Psychosomatic Medicine, 59*, 497-502.

Moller, J., Moller, N., Frandsen, E., Wolthers, T., Jorgensen, J.O., & Christiansen, J.S. (1997.) Blockade of the renin-angiotensin-aldosterone system prevents growth hormone-induced fluid retention in humans. *American Journal of Physiology, 272*, E803-E808.

Moller, N., Jorgensen, J.O., Abilgard, N., Orskov,,L., Schmitz, O., & Christiansen, J.S. (1991.) Effects of growth hormone on glucose metabolism. *Hormone Research*, *36* (Suppl. 1), 32-35.

Moller, N., Jorgensen, J.O.L., Alberti, K.G.M.M., Flyvbjerg, A., & Schmitz, O. (1990a.) Short-term effects of growth hormone on fuel oxidation and regional substrate metabolism in normal man. *Journal of Clinical Endocrinology and Metabolism*, *70*, 1179-1186.

Moller, N., Jorgensen, J.O.L., Schmitz, O., Moller, J., Christiansen, J.S., Alberti, K.G.M., & Orskov, H. (1990b.) Effects of a growth hormone pulse on total and forearm substrate fluxes in humans. *American Journal of Physiology*, *258*, E86-E91.

Momomura. S., Hashimoto. Y., Shimazaki. Y., & Irie, M. (2000.) Detection of exogenous growth hormone (GH) administration by monitoring ratio of 20kDa- and 22kDa-GH in serum and urine. *Endocrine Journal*, *47*, 97-101.

Moncada, S., Palmer, R.M.J., & Higgs. E.A. (1991.) Nitric oxide: physiology, pathophysiology, and pharmacology. *Pharmacological Reviews*, *43*, 109-143.

Montague, C.T., Farooqi, I.S., Whitehead, J.P., Soos, M.A., Rau, H., & Wareham, N.J. (1997.) Congenital leptin deficiency is associated with severe early-onset obesity in humans. *Nature*, *387*, 903-908.

Montain, S.J., Laird, J.E., Latzka, W.A., & Sawka, M.N. (1997.) Aldosterone and vasopressin responses in the heat: hydration level and exercise intensity effects. *Medicine and Science in Sports and Exercise*, *29*, 661-668.

Montano, N., Gnecchi-Ruscone, T., Porta, A., Lombardi, F., Pagani, M., & Malliani, A. (1994.) Power spectrum analysis of heart rate variability to assess the changes in sympathovagal balance during graded orthostatic tilt. *Circulation*, *90*, 1826-1831.

Monteleone, P., Maj, M., Franza, F., Fusco, R., & Kemali, D. (1993.)The human pineal gland responds to stress-induced sympathetic activation in the second half of the dark phase: preliminary evidence. *Journal of Neural Transmission - General Section*, *92*, 25-32.

Monteleone, P., Maj, M., Fusco, M., Orazzo, C., & Kemali, D. (1990.) Physical exercise at night blunts the nocturnal increase of plasma melatonin levels in healthy humans. *Life Sciences*, *47*, 1989-1995.

Montelpare, W.J., Plyley, M.J., & Shephard, R.J. (1992.) Evaluating the influence of sleep deprivation upon circadian rhythms of exercise metabolism. *Canadian Journal of Sport Sciences*, *17*, 94-97.

Montessuit, C., & Thorburn, A. (1999.) Transcriptional activation of the glucose transporter GLUT1 in ventricular cardiac myocytes by hypertrophic agonists. *Journal of Biological Chemistry*, *274*, 9006-9012.

Montgomery, I., Trinder, J., & Paxton, S. (1982.) Energy expenditure and total sleep time: effect of physical exercise. *Sleep*, *5*, 159-168.

Montgomery, I., Trinder, J., Paxton, S., Harris, D., Fraser, G., & Colrain, I. (1988.) Physical exercise and sleep: the effect of the age and sex of the subjects and type of exercise. *Acta Physiologica Scandinavica. Supplementum*, *574*, 36-40.

Monti, L.D., Brambilla, P., Caumo, A., et al. (1997.) Glucose turnover and insulin clearance after growth hormone treatment in girls with Turner's syndrome. *Metabolism: Clinical and Experimental*, *46*, 1482-1488.

Moody, D. L., Kollias, J., & Buskirk, E. R. (1969.) The effect of a moderate exercise program on body weight and skinfold thickness in overweight college women. *Medicine and Science in Sports*, *1*, 75-80.

Moore, R.Y. (1996.) Neural control of the pineal gland. *Behavioural Brain Research*, *73*, 125-130.

Moore-Ede, M. (1986.) Physiology of the circadian timing system: predictive versus reactive homeostasis. *American Journal of Physiology*, *250*, R737-R752.

Moore-Ede, M., Sulzman, F.M., & Fuller, C.A. (1982.) *The clocks that time us.* Cambridge: Harvard University Press.

Moran, T.H. (2000.) Cholecystokinin and satiety: Current perspectives. *Nutrition*, *16*, 858-865.

Moretti, C., Fabbri, A., Gnessi, L., Cappa, M., Calzolari, A., Fraioli, F., Grossman, A., & Besser, G.M. (1983.) Naloxone inhibits exercise-induced release of PRL and GH in athletes. *Clinical Endocrinology*, *18*, 135-138.

Morgan, K.G., & Suematsu, E. (1990.) Effects of calcium on vascular smooth muscle tone. *American Journal of Hypertension*, *3*, 291S-298S.

Morgan, M.J., & Loughna, P.T. (1989.) Work overload induced changes in fast and slow skeletal muscle myosin heavy chain gene expression. *FEBS Letters*, *255*, 427-430.

Morita, H., Manders, W.T., Skelton, M.M., Cowley, A.W. Jr., & Vatner, S.F. (1986.) Vagal regulation of arginine vasopressin in conscious dogs. *American Journal of Physiology*, *251*, H19-H23.

Moritz, A.R., & Zamcheck, N. (1946.) Sudden and unexpected deaths of young soldiers: Diseases responsible for such deaths during World War II. *Archives of Pathology*, *42*, 459-494.

Moriyama, M., Nakanishi, Y., Tsuyama, S., Kannan, Y., Ohta, M., & Sugano, T. (1997.) Change from beta to alpha-adrenergic glycogenolysis induced by corticosteroids in female rat liver. *American Journal of Physiology*, *273*, R153-R160.

Morris, J.L., Gibbins, I.L., Kadowitz, P.J., Herzog, H., Kreulen, D.L., Toda, N., & Claing, A. (1995.) Roles of peptides and other substances in cotransmission from vascular autonomic and sensory neurons. *Canadian Journal of Physiology and Pharmacology*, *73*, 521-532.

Morrow, N.G., Kraus, W.E., Moore, J.W., Williams, R.S., & Swain, J.L. (1990.) Increased expression of fibroblast growth factors in a rabbit skeletal muscle model of exercise conditioning. *Journal of Clinical Investigation*, *85*, 1816-1820.

Morville, R., Pequies, P.C., Guezzenec, C.Y., Serruruer, B.D., & Guignard, M. (1979.) Plasma variations in testicular and adrenal androgens during prolonged physical exercise in man. *Annales d'Endocrinologie*, *40*, 501-510.

Moser, D.K., Stevenson, W.G., Woo, M.A., & Stevenson, L.W. (1994.) Timing of sudden death in patients with heart failure. *Journal of the American College of Cardiology*, *24*, 963-967.

Moskowitz, M.S. (1977.) Diseases of the autonomic nervous system. *Clinical Endocrinology and Metabolism*, *6*, 745-768.

Mosso, U. (1887.) Recherches sur l'inversion des oscillations diurnes de la temperature chez l'homme normal. *Archivi Italiani di Biologia*, *8*, 177-185.

Mrosovsky, N., & Salmon, P.A. (1987.) A behavioural method for accelerating reentrainment of rhythms to new light-dark cycles. *Nature*, *330*, 372-373.

Mrosovsky, N. (1993.) Tau changes after single nonphotic events. *Chronobiology International*, *10*, 271-276.

Mrosovsky, N., Salmon, P.A., Menaker, M., & Ralph, M.R.(1992.) Nonphotic phase shifting in hamster clock mutants. *Journal of Biological Rhythms*, *7*, 41-49.

Mu, J., Brozinick, J.T., Jr., Valladares, O., Bucan, M., and Birnbaum, M.J. (2001.) A role for AMP-activated protein kinase in contraction- and hypoxia-regulated glucose transport in skeletal muscle. *Molecular Cell*, *7*, 1085-1094.

Muller, A., Thelen, M.H., Zuidwijk, M.J., Simonides, W.S., & Van Hardeveld, C. (1996.) Expression of MyoD in cultured primary myotubes is dependent on contractile activity: Correlation with phenotype-specific expression of a sarcoplasmic reticulum Ca$^{(2+)}$-ATPase isoform. *Biochemical and Biophysical Research Communications*, *229*, 198-204.

Muller, E.E., Locatelli, V., & Cocchi, D. (1999.) Neuroendocrine control of growth hormone secretion. *Physiological Reviews*, *79*,511-607.

Muller, J.E., Ludmer, P.L., Willich, S.N., Tofler, G.H., Aylmer, G., Klangos, I., et al. (1987.) Circadian variation in the frequency of sudden cardiac death. *Circulation*, *75*, 131-138.

Muller, J.E., Tofler, G.H., & Stone, P.H. (1989.) Circadian variation and triggers of onset of acute cardiovascular disease. *Circulation*, *79*, 733-743.

Muller, J.E., Stone, P.H., Turi, Z.G., Rutherford, J.D., Czeisler, C.A., Parker, C., Poole, W.K., Passamani, E., Roberts, R., Robertson, T., Sobel, B.T., Wilkerson, J.T., Braunwald, E., & the MILIS Study Group (1985.) Circadian variability in the frequency of onset of acute myocardial infarction. *New England Journal of Medicine*, *313*, 1315-1322.

Mulligan, K., & Butterfield, G.E. (1990.) Discrepancies between energy expenditure in physically active women. *British Journal of Nutrition*, *64*, 23-36.

Munakata, M., Kameyana, J., Kanazawa, M., Nunokawa, T., Morai, N., & Yoshinaga, K. (1997.) Circadian blood pressure rhythm in patients with higher and lower spinal cord injury: simultaneous evaluation of autonomic nervous activity and physical activity. *Journal of Hypertension*, *15*, 1745-1749.

Mundinger, T.O., Boyle, M.R., & Taborsky, G.J., Jr. (1997.) Activation of hepatic sympathetic nerves during hypoxic hypotensive and glucopenic stress. *Journal of Autonomic Nervous System*, *63*, 153-160.

Mundy, C.R. (1992.) Cytokines and local factors which affect osteoclast function. *International Journal of Cell Cloning*, *10*, 215-222.

Mundy, G.R., Boyce, B.F., Yoneda, T., Bonewald, L.F., & Roodman, G.D. (1996.) Cytokines and bone remodeling. In R. Marcus, D. Feldman & J.Kelsey (Eds.), *Osteoporosis* (pp. 301-313), New York: Academic Press.

Murphy, B.L., Arntsen, A.F., Goldman-Rakic, P.S., & Roth, R.H. (1996a.) Increased dopamine turnover in the prefrontal cortex impairs spatial working memory performance in rats and monkeys. *Proceedings of the National Academy of Science, 93,* 1325-1329.

Murphy, H.M., Wideman, C.H., & Nadzam, G.R. (1996b.) The interaction of vasopressin and the photic oscillator in circadian rhythms. *Peptides, 17,* 465-475.

Murray, P.M., Herrington, D.M., Pettus, C.W., Miller, H.S., Cantwell, J.D., & Little, W.C. (1993.) Should patients with heart disease exercise in the morning or afternoon? *Archives of Internal Medicine, 153,* 833-836.

Musaro, A., McCullagh, K.J.A., Naya, F.J., Olson, E.N., & Rosenthal, N. (1999.) IGF-I induces skeletal muscle hypertrophy through calcineurin in association with GATA-2 and NF-ATc1. *Nature, 400,* 581-585.

Musha, T., Satoh, E., Koyanagawa, H., Kimura, T., & Satoh. S. (1989.) Effect of opioid agonists on sympathetic or parasympathetic transmission to the dog heart. *Journal of Pharmacology and Experimental Therapeutics, 250,* 1087-1091.

Musi, N., Fujii, N., Hirshman, M.F., Ekberg, I., Froberg, S., Ljungqvist, O., Thorell, A., & Goodyear, L.J. (2001.) AMP-activated protein kinase (AMPK) is activated in muscle of subjects with type 2 diabetes during exercise. *Diabetes, 50,* 921-927.

Mutvei, A., & Nelson, B.D. (1989.) The response of individual polypeptides of the mammalian respiratory chain to thyroid hormone. *Archives of Biochemistry and Biophysics, 268,* 215-220.

Mutvei, A., Husman, B., Anderson, G., & Nelson, B.D. (1989a.) Thyroid hormone and of growth hormone is the priciple regulator of mammalian mitochondrial biogenesis. *Acta Endocrinologica, 121,* 223-228.

Mutvei, A., Kuzela, S., & Nelson, B.D. (1989b.) Control of mitochondrial transcription by thyroid hormone. *European Journal of Biochemistry, 180,* 235-240.

Myerson, M., Gutin, B., & Warren, M.P. (1991.) Resting metabolic rate and energy balance in amenorrheic and eumenorrheic runners. *Medicine and Science in Sports and Exercise, 23,* 15-22.

Mystkowski, P., & Schwartz, M.W. (2000.) Gonadal steroids and energy homeostasis in the leptin era. *Nutrition, 16,* 937-946.

Nadal, M. (1996.) Secretory rhythm of vasopressin in healthy subjects with inversed sleep-wake cycle: evidence for the existence of an intrinsic regulation. *European Journal of Endocrinology, 134,* 174-176.

Naesh, O., Hindberg, I, Trap-Jensen, J., & Lund, J.O. (1990.) Post-exercise platelet activation: aggregation and release in relation to dynamic exercise. *Clinical Physiology, 10,* 221-230.

Naftolin, F., Ryan, K.J., & Petro, Z. (1971.) Aromatization of androstenedione by the diencephalon. *Journal of Clinical Endocrinology and Metabolism, 33,* 368-370.

Nagase, H., Inoue, S., Tanaka, K., Takamura, Y., & Niijima, A. (1993.) Hepatic glucose-sensitive unit regulation of glucose-induced insulin secretion in rats. *Physiology and Behavior, 53,* 139-143.

Nagase, I., Yoshida, T., Kumamoto, K., Umekawa, T., Sakane, N., Nikami, H., Kawada, T., and Saito, M. (1996.) Expression of uncoupling protein in skeletal muscle and white fat of obese mice treated with thermogenic beta 3-agonist. *Journal of Clinical Investigation, 97,* 2898-2904.

Nakamura, M., Katsuura, G., Nakao, K., & Imura, H. (1985.) Antidipsogenic action of alpha-atrial natriuretic peptide administered intracerebroventricularly in rats. *Neuroscience Letters, 58,* 1-6.

Nakamura, Y., Yamamoto, Y, & Muraoka, I. (1993.) Autonomic control of heart rate during physical exercise and fractal dimension of heart rate variability. *Journal of Applied Physiology, 74,* 875-881.

Nakazato, M., Murakami, N., Date, Y., Kojima, M., Matsuo, H., Kangawa, K., & Matsukura, S. (2001.) A role for ghrelin in the central regulation of feeding. *Nature, 409,* 194-198.

Nance, D.M., Bromley, B., Bernard, J.R., & Gorski, R.A. (1977.) Sexually dimorphic effects of forced exercise on food intake and body weight in the rat. *Physiology and Behavior, 19,* 155-158.

Napoli, R., Cittadini, A., Chow, J.C., Hirshman, M.F., Smith, R.J., Douglas, P.S. & Horton, E.S. (1996.) Chronic growth hormone treatment in normal rats reduces post-prandial skeletal muscle plasma membrane GLUT1 content, but not glucose transport or GLUT4 expression and localization. *Biochemical Journal, 315,* 959-963.

Nass, R. & Baker, S. (1991.) Androgen effect on cognition: congenital adrenal hyperplasia. *Psychoneuroendocrinology, 16,* 115-120.

Natali, A., Gastaldelli, A., Galvan, A.Q., Sironi, A.M., Ciociaro, D., Sanna, G., Rosenzweig, P., & Ferrannini, E. (1998.) Effects of acute alpha 2-blockade on insulin action and secretion in humans. *American Journal of Physiology, 274,* E57-E64.

Nathan, D.M., & Cagliero, E. (2001.) Diabetes mellitus. In P. Felig, & L.A.Frohman (Eds.), *Endocrinology and metabolism, 4th ed.* (pp. 827-926). New York, McGraw-Hill.

Navarro, J.F., Mora-Fernandez, C., Rivero, A., Macia, M., Gallego, E., Chahin, J., Mendez, M.L., & Garcia, J. (1998.) Androgens for the treatment of anemia in peritoneal dialysis patients. *Advances in Peritoneal Dialysis, 14,* 232-235.

Naylor, E., Penev, P.D., Orbeta, L., Janssen, I., Ortiz, R., Colecchia, E.F., Keng, M., Finkel, S., & Zee, P.C. (2000.) Daily social and physical activity increases slow-wave sleep and daytime neuropsychological performance in the elderly. *Sleep, 23,* 87-95.

Nehlsen-Cannarella, S.L., Nieman, D.C., Jessen, J., Chang, L., Gusewitch, G., Blix, G.G., & Ashley, E. (1991.) The effects of acute moderate exercise on lymphocyte function and serum immunoglobulin levels. *International Journal of Sports Medicine, 12,* 391-398.

Nervi, C., Benedetti, L., Minasi, A., Molinaro, M., & Adamo, S. (1995.) Arginine-vasopressin induces differentiation of skeletal myogenic cells and up-regulation of myogenin and Myf-5. *Cell Growth and Differentiation, 6,* 81-89.

Neufer, P.D., Costill, D.L., Flynn, M.G., Kirwan, J.P., Mitchell, J.B., & Houmard, J. (1987.) Improvements in exercise performance: effects of carbohydrate feedings and diet. *Journal of Applied Physiology, 62,* 983-988.

Newsholme, E.A., & Calder, P.C. (1997.) The proposed role of glutamine in some cells of the immune system and speculative consequences for the whole animal. *Nutrition, 13,* 728-730.

Newsholme, E.A., & Leech, A.R. (1983.) *Biochemistry for the medical sciences.* New York: John Wiley and Sons.

Newsholme, E., Leech, T., & Duester, G. (1994.) *Keep on running* (pp. 164-165).New York: John Wiley and Sons.

Newsholme, E.A., & Parry-Billings, M. (1990.) Properties of glutamine release from muscle and its importance for the immune system. *Journal of Parenteral and Enteral Nutrition, 14* (Suppl. 4), 63S-67S.

Nexo, E., Hansen, M.G., & Konradsen, L. (1988.) Human salivary epidermal growth factor, haptocrin and amylase before and after prologed exercise. *Scandinavian Journal of Clinical and Laboratory Investigation, 48,* 269-273.

Neylon, C.B. (1999.) Vascular biology of endothelin signal transduction. *Clinical and Experimental Pharmacology and Physiology, 26,* 149-154.

Ngai, S.H., Rosell, S., & Wallenberg, L.R. (1966.) Nervous regulation of blood flow in the subcutaneous adipose tissue in dogs. *Acta Physiologica Scandinavica, 68,* 397-403.

Nielsen, H.B., Secher, N.H., Kristensen, J.H., Espersen, K. & Pedersen, B.K. (1997.) Splenectomy impairs lymphocytosis during maximal exercise. *American Journal of Physiology, 272,* R1847-R1852.

Nieman, D.C. (1997a.) Immune response to heavy exertion. *Journal of Applied Physiology, 82,* 1385-1394.

Nieman, D.C. (1997b.) Exercise immunology: practical applications. *International Journal of Sports Medicine, 18* (Suppl. 1), S91-S100.

Nieman, D.C., Miller, A.R., Henson, D.A., Warren, B.J., Gusewitch, G., Johnson, R.L., Davis, J.M., Butterworth, D.E., Herring, J.L., & Nehlsen-Cannarella, S.L. (1994.) Effect of high- versus moderate-intensity exercise on lymphocyte subpopulations and proliferative response. *International Journal of Sports Medicine, 15,* 199-206.

Nieman, A.C., Henson, D.A., Sampson, C.S., Herring, J.L., Suttles, J., Conley, M., Stone, M.H., Butterworth, D.B., & Davis, J.M. (1995.)The auto immune response to exhaustive resistance exercise. *International Journal of Sports Medicine, 16,* 322-338.

Nieman, D.C., Nehlsen-Cannarella, S.L., Donohue, K.M., Chritton, D.B., Haddock, B.L., Stout, R.W., & Lee, J.W. (1991.) The effects of acute moderate exercise on leukocyte and lymphocyte populations. *Medicine and Science in Sports and Exercise, 23,* 578-585.

Niijima, A., & Meguid, M.M. (1995.) An electrophysiological study on amino acid sensors in the hepato-portal system in the rat. *Obesity Research, 3* (Suppl. 5), 741S-745S.

Niijima, A. (1969.) Afferent discharges from osmoreceptors in the liver of the guinea pig. *Annals of the New York Academy of Sciences, 157,* 690-700.

Niijima, A. (1989.) Neural mechanisms in the control of blood glucose concentration. *Journal of Nutrition, 119,* 833-840.

Nikkila, E.A., Taskinen, M.R., Rehunen, S., & Harkonen, M.(1978.) Lipoprotein lipase activity in adipose tissue and skeletal muscle of runners: relation to serum lipoproteins. *Metabolism: Clinical and Experimental, 27,*1661-1667.

Nilsson, A., Ohlsson, C., Isaksson, O.G.P., Lindahl, A., & Isgaard, J. (1994.) Hormonal regulation of longitudinal bone growth. *European Journal of Clinical Nutrition, 48* (Suppl. 1), S150- S160.

Nilsson, L.H., & Hultman, E. (1973.) Liver glycogen in man: the effect of total starvation or a carbohydrate-poor diet followed by carbohydrate refeeding. *Scandinavian Journal of Laboratory and Clinical Investigation, 32,* 325-330.

Nimrod, A., & Ryan, K.J. (1975.) Aromatization of androgens by human abdominal and breast fat tissue. *Journal of Clinical Endocrinology and Metabolism, 40,* 367-372.

Nishiyama, S., Tomoeda, S., Ohta, T., Higuchi, A., & Matsuda, I. (1988.) Differences in basal and post-exercise osteocalcin levels in athletic and nonathletic humans. *Calcified Tissue International, 43,* 150-154.

Nishiyasu, T., Nagashima, K., Nadel, E.R., & Mack, G.W. (2000.) Human cardiovascular and humoral responses to moderate muscle activation during dynamic exercise. *Journal of Applied Physiology, 88,* 300-307.

Nishiyasu, T., Tan, N., Morimoto, K., Sone, R., & Murakami, N. (1998.) Cardiovascular and humoral responses to sustained muscle metaboreflex activation in humans. *Journal of Applied Physiology, 84,* 116-122.

Nishizuka, Y. (1992.) Intracellular signalling by hydrolysis of phospholipids and activation of protein kinase C. *Science, 258,* 607-614.

Noda, M., & Camilliere, J.J. (1989.) In vivo stimulation of bone formation by transforming growth factor. *Endocrinology, 124,* 2991-2994.

Norberg, K.-A., Persson, B., & Granberg, P.-O. (1975.) Adrenergic innervation of the human parathyroid glands. *Acta Chirurgica Scandinavica, 141,* 319-322.

Norman, A.W., & Litwack, G. (1997.) *Hormones,* 2nd ed. New York: Academic Press.

Norton, K.H., Gallagher, K.M., Smith, S.A., Querry, R.G., Welch-O'Connor, R.M., & Raven, P.B. (1999.) Carotid baroreflex function during prolonged exercise. *Journal of Applied Physiology, 87,* 339-347.

Noshiro, T., Shimizu, K., Way, D., Miura, Y., & McGrath, B.P. (1994.) Angiotensin II enhances norepinephrine spillover during sympathetic activation in conscious rabbits. *American Journal of Physiology, 265,* H1864-H1871.

Nouet, S., & Nahmias, C. (2000.) Signal transduction from the angiotensin II AT$_2$ receptor. *Trends in Endocrinology and Metabolism, 11,* 1-6.

Nurjhan, N., Campbell, P.J., Kennedy, F.P., Miles, J.M., & Gerich, J.E. (1986.) Insulin dose-response characteristics for suppression of glycerol release and conversion to glucose in humans. *Diabetes, 35,* 1326-1331.

Nuutila, P., Raitakari, M., Laine, H., Kirvela, O., Takala, T., Utriainen, T., et al. (1996.) Role of blood flow in regulating insulin-stimulated glucose uptake in humans. Studies using bradykinin, [¹⁵O]water, and [¹⁸F]fluoro-deoxy-glucose and positron emission tomography. *Journal of Clinical Investigation, 97,* 1741-1747.

Oakes, N.D., Bell, K.S., Furler, S.M., Camilleri, S., Saha, A.K., Ruderman, N.B., Chisholm, D.J., & Kraegen, E.W. (1997a.) Diet-induced muscle insulin resistance in rats is ameliorated by acute dietary lipid withdrawal or a single bout of exercise: parallel relationship between insulin stimulation of glucose uptake and suppression of long-chain fatty acyl-CoA. *Diabetes, 46,* 2022-2028.

Oakes, N.D., Camilleri, S., Furler, S.M., Chisholm, D.J., & Kraegen, E.W. (1997b.) The insulin sensitizer, BRL 49653, reduces systemic fatty acid supply and utilization and tissue lipid availability in the rat. *Metabolism, 46,* 935-942.

Obal, F., Jr., & Krueger, J.M. (2001.) The somatotropic axis and sleep. *Revue Neurologique, 157,* S12-S15.

Obarzanek, E., Schreiber, G.B., Crawford, P.B., Goldman, S.R., Barrier, P.M., Frederick, M.M., & Lakatos, E. (1994.) Energy intake and physical activity in relation to indexes of body fat: The National Heart, Lung, and Blood Institute Growth and Health Study. *American Journal of Clinical Nutrition, 60,* 15-22.

O'Connor, P.J., & Davis, J.C. (1992.) Psychobiologic responses to exercise at different times of day. *Medicine and Science in Sports and Exercise, 24,*714-719.

O'Connor, P.J., Breus, M.J., & Youngstedt, S.D. (1998.) Exercise-induced increase in core temperature does not disrupt a behavioral measure of sleep. *Physiology and Behavior, 64,* 213-217.

O'Connor, P.J., Crowley, M.A., Gardner, A.W., & Skinner, J.S. (1993.) Influence of training on sleeping heart rate following daytime exercise. *European Journal of Applied Physiology and Occupational Physiology, 67,* 39-42.

Odland, L.M., Heigenhauser, G.J., Lopaschuk, G.D., & Spriet, L.L. (1996.) Human skeletal muscle malonyl-CoA at rest and during prolonged submaximal exercise. *American Journal of Physiology, 270,* E541-E544.

O'Doherty, R.M., Bracy, D.P., Osawa, H., Wasserman, D.H., & Granner, D.K. (1994.) Rat skeletal muscle hexokinase II mRNA and activity are increased by a single bout of acute exercise. *American Journal of Physiology, 266,* E171-E178.

Ogle, W. (1866.) On the diurnal variations in the temperature of the human body in health. *Saint George's Hospital Reports, 1,* 221-245.

O'Gorman, D.J., Del Aguila, L.F., Williamson, D.L., Krishnan, R.K., & Kirwan, J.P. (2000.) Insulin and exercise differentially regulate PI3-kinase and glycogen synthase in human skeletal muscle. *Journal of Applied Physiology, 89,*1412-1419.

Ohtake, P.J., & Wolfe, L.A. (1998.) Physical conditioning attenuates respiratory responses to steady-state exercise in late gestation. *Medicine and Science in Sports and Exercise, 30,* 17-27.

Ojeda, S.R. (1996.) Female reproductive function. In J.E. Griffin & S.R. Ojeda (Eds.), *Textbook of endocrine physiology* (pp. 164-200). Oxford: Oxford University Press.

Ojuka, E.O., Jones, T.E., Nolte, L.A., Chen, M., Wamhoff, B.R., Sturek, M., & Holloszy, J.O. (2002.) Regulation of GLUT4 biogenesis in muscle: evidence for involvement of AMPK and Ca(2+). *American Journal of Physiology, 282,* E1008-1013.

Okuda, Y., Pena, J., Chou, J., & Field, J.B. (1994.) Effect of growth hormone on hepatic glucose and insulin metabolism after oral glucose in conscious dogs. *American Journal of Physiology, 267,* E454-E460.

Okuda, Y., Pena, J., Chou, J., & Field, J.B. (2001.) Acute effects of growth hormone on metabolism of pancreatic hormones, glucose and ketone bodies. *Diabetes Research and Clinical Practice, 53,* 1-8.

Okuguchi, T., Osani, T., Kamada, T., Kimura, M., Takahashi, K.. & Okumura, K. (1999.) Significance of sympathetic nervous system in sodium-induced nocturnal hypertension. *Journal of Hypertension, 17,* 947-957.

Oldfield, B.J., Badoer, E.,Hards, D.K., & McKinley, M.J. (1994.) Fos production in retrogradely labelled neurons of the lamina terminalis following intravenous infusion of either hypertonic saline or angiotensin. *Neuroscience, 60,* 255-262.

Oldham, J.M., Martyn, J.A., Sharma, M., Jeanplong, F., Kambadur, R., & Bass, J.J. (2001.) Molecular expression of myostatin and MyoD is greater in double-muscled than normal-muscled cattle fetuses. *American Journal of Physiology, 280,* R1488-R1493.

O'Leary, D.S., & Augustyniak, R.A. (1998.) Muscle metaboreflex increases ventricular performance in conscious dogs. *American Journal of Physiology, 275,* H220-H224.

O'Leary, D.S., Rossi, N.F., & Churchill, P.C. (1993.) Muscle metaboreflex control of vasopressin and renin release. *American Journal of Physiology, 264,* H1422-H1427.

Olness, K., Culbert, T., & Uden, D. (1989.) Self-regulation of salivary immunoglobulin A by children. *Pediatrics, 83,* 66-71.

Oliveria, S.A., & Christos, P.J. (1997.) The epidemiology of physical activity and cancer. *Annals of the New York Academy of Sciences, 833,* 79-90.

Olsen, N.V. (1995.) Effect of hypoxemia on water and sodium homeostatic hormones and renal function. *Acta Anaesthesiologica Scandinavica, 107,* 165-170.

O'Malley, B.W. (1990.) The steroid receptor superfamily: More excitement predicted for the future. *Molecular Endocrinology, 4,* 363-369.

O'Neill, R.G. (1990.) Aldosterone regulation of sodium and potassium transport in the cortical collecting duct. *Seminars in Nephrology, 10,* 365-374.

Onuoha, G.N., Nicholls, D.P., Patterson, A., & Beringer, T. (1998.) Neuropeptide secretion in exercise. *Neuropeptides, 32,* 319-325.

Ookuma, M., & York, D.A. (1998.) Inhibition of insulin release by enterostatin. *International Journal of Obesity and Related Metabolic Disorders, 22,* 800-805.

Opp, M.R. (1997.) Rat strain differences suggest a role for corticotropin releasing hormone in modulating sleep. *Physiology and Behavior, 63,* 67-74.

Orentreich, N., Brind, J.L., Rizer, R.L., & Vogelman, J.H. (1984.) Age changes and sex differences in serum dehydroepiandrosterone sulfate concentrations throughout adulthood. *Journal of Clinical Endocrinology and Metabolism, 59,* 551-555.

Ornato, J.P., Siegel, L., Craren, E.J., & Nelson, N. (1990.) Increased incidence of cardiac death attributed to acute myocardial infarction during winter. *Coronary Artery Disease, 1,* 199-203.

Oro, L., Wallenberg, L., & Rosell, S. (1965.) Circulatory and metabolic process in adipose tissue in vivo. *Nature, 205,* 178-179.

Ortega, E., Collazos, M.E., Maynar, M., Barriga, C., & De La Fuente, M. (1993.) Stimulation of the phagocytic function of neutrophils in sedentary men after acute moderate exercise. *European Journal of Applied Physiology and Occupational Physiology, 66,* 60-64.

Ortega, E., Rodriguez, M.J., Barriga, C., & Forner, M.A.(1996.) Corticosterone, prolactin and thyroid hormones as hormonal mediators of the stimulated phagocytic capacity of peritoneal macrophages after high-intensity exercise. *International Journal of Sports Medicine, 17,* 149-155.

Osawa, H., Printz, R.L., Whitesell, R.R., & Granner, D.K. (1995.) Regulation of hexokinase II gene transcription and glucose phosphorylation by catecholamines, cyclic AMP and insulin. *Diabetes, 44,* 1426-1432.

Osborn, J.L., DiBona, G.F., & Thames, M.D. (1982.) Role of renal alpha-adrenoceptors mediating renin secretion. *American Journal of Physiology, 242,* F620-F626.

Oscai, L.B., Essig, D.A., & Palmer, W.A. (1990.) Lipase regulation of muscle triglyceride hydrolysis. *Journal of Applied Physiology, 69,* 1571-1577.

Oscai, L.B., Gorski, J., Miller, W.C., & Palmer, W.K. (1988.) Role of the alkaline TG lipase in regulating intramuscular TG content. *Medicine and Science in Sports and Exercise, 20,* 539-544.

Oscai, L.B., & Williams, B.T. (1968.) Effect of exercise on overweight middle-aged males. *Journal of the American Geriatrics Society, 16,* 794-797.

Oscai,L.B., Babirak, S.P., Dubach, F.B., McGarr, J.A., & Spirakis, C.N. (1974.) Exercise or food restriction: effect on adipose tissue cellularity. *American Journal of Physiology, 227,* 901-904.

Oscarsson, J., Ottoson, M., & Eden, S. (1999a.) Effects of growth hormone on lipoprotein lipase and hepatic lipase. *Journal of Endocrinological Investigation, 22* (5 Suppl.), 2-9.

Oscarsson, J., Ottosson, M., Vikman-Adolfsson, K., Frick, F., Enerback, S., Lithell, H., & Eden, S. (1999b.) GH but not IGF-I or insulin increases lipoprotein lipase activity in muscle tissues of hypophysectomized rats. *Journal of Endocrinology, 160,* 247-255.

Ostrowski, K., Rohde, T., Zacho, M., Asp, S., & Pedersen, B.K. (1998.) Evidence that interleukin-6 is produced in human skeletal muscle during prolonged running. *Journal of Physiology, 508,* 949-953.

Ota, K., Share, L., Crofton, J.T., & Brooks, D.P. (1986.) Methionine-enkephalin and vasopressin in SHR: effects of dehydration. *American Journal of Physiology, 250,* R1007-R1013.

Otsuka, K., Cornelissen, G., & Halberg F. (1996.) Predictive value of blood pressure dipping and swinging with regard to vascular disease risk. *Clinical Drug Investigation, 11,* 20-31.

Otsuka, K., Cornelissen, G., & Halberg F. (1997a.) Age, gender and fractal scaling in heart rate variability. *Clinical Science, 93,* 299-308.

Otsuka, K., Cornelissen, G., Halberg, F. & Oehlerts, G. (1997b.) Excessive circadian amplitude of blood pressure increases risk of ischaemic stroke and nephropathy. *Journal of Medical Engineering and Technology, 21,* 23-30.

Ottosson, M., Lonnroth, P., Bjorntorp, P., & Eden, S. (2000.) Effects of cortisol and growth hormone on lipolysis in human adipose tissue. *Journal of Clinical Endocrinology and Metabolism, 85,* 799-803.

Ottosson, M., Vikman-Adolfsson, K., Enerback, S., Olivecrona, G., & Bjorntorp, P. (1994.)The effects of cortisol on the regulation of lipoprotein lipase activity in human adipose tissue. *Journal of Clinical Endocrinology and Metabolism, 79,* 820-825.

Owens, D.S., Macdonald, I., Benton, D., Sytnik, N., Tucker, P., & Folkard, S. (1996.) A preliminary investigation into individual differences in the circadian variation of meal tolerance: effects on mood and hunger. *Chronobiology International, 13,* 435-447.

Owens, J.A. (1991.) Endocrine and substrate control of fetal growth: placental and maternal influences and insulin-like growth factors. *Reproduction, Fertility and Development, 3,* 501-517.

Pacifici, R., Rifas, L., McCracken, R., Vered, I., McCurtry, C., Avioli, L.V., & Peck, W.A. (1989.) Ovarian steroid treatment blocks a postmenopausal increase in blood monocyte interleukin 1 release. *Proceedings of the National Academy of Sciences of the United States of America, 86,* 2398-2402.

Paffenberger, R.S., Wing, A.L., Hyde, R.T., & Jung, D.L. (1983.) Physical activity and incidence of hypertension in college alumni. *American Journal of Epidemiology, 117,* 245-257.

Pagani, M., Lombardi, F., Guzzetti, S., et al. (1986.) Power spectral analysis of heart rate and arterial pressure variabilities as a marker of sympatho-vagal interaction in man and conscious dog. *Circulation Research, 59,* 178-193.

Pagani, M., Montano, N., Porta, A., Malliani, A., Abboud, F.M., Birkett, C., & Somers, V.K. (1997.) Relationship between spectral components of cardiovascular variabilities and direct measures of muscle sympathetic nerve activity in humans. *Circulation, 95,* 1441-1448.

Painson, J.-C., Thorner, M.O., Krieg, R.J., & Tannenbaum, G.S. (1992.) Short-term adult exposure to estradiol feminizes the male pattern of spontaneous and growth hormone-releasing factor-stimulated growth hormone secretion in the rat. *Endocrinology, 130,* 511-519.

Palatini, P., Giada, F., Garavelli, G., Sinisi, F., Mario, L., Michieletto, M., & Baldo-Enzi, G. (1996.) Cardiovascular effects of anabolic steroids. *Journal of Clinical Pharmacology, 36,* 1132-1140.

Palka, Y., Liebelt, R.A., & Critchlow, V. (1971.) Obesity and increased growth following partial or complete isolation of ventromedial hypothalamus. *Physiology and Behavior, 7,* 187- 194.

Pallikarakis, N., Jandrain, B., Pirnay, F., Mosora, F., Lacroix, M., Luyckx, A.S., & Lefebvre, P.J. (1986.) Remarkable metabolic availability of oral glucose during long-duration exercise in humans. *Journal of Applied Physiology, 60,* 1035-1042.

Palmblad, J., Levi, L., Burger, A., Melander, A., Westgren, U., von Schenk, H., & Skude, G. (1977.) Effects of total energy withdrawal (fasting) on the levels of growth hormone, thyrotropin, cortisol, adrenaline, noradrenaline, T4, T3, and rT3 in healthy males. *Acta Medica Scandinavica, 201,* 15-22.

Palou A., Pico C., Bonet M.L., & Oliver P. (1998.) The uncoupling protein, thermogenin. *International Journal of Biochemistry and Cell Biology, 30,* 7-11.

Pan, D.A., Lillioja, S., Kriketos, A.D., Milner, M.R., Baur, L.A., Bogardus, C., Jenkins, A.B., & Storlien, L.H. (1997.) Skeletal muscle triglyceride levels are inversely related to insulin action. *Diabetes, 46,* 983-988.

Panza, J.A., Epstein, S.E., & Quyyumi, A.A. (1991.) Circadian variation in vascular tone and its relation to alpha-sympathetic vasoconstrictor activity. *New England Journal of Medicine, 325,* 986-990.

Parfitt, A.M. (1984.) The cellular basis of bone remodeling: The quantum concept reexamined in light of recent advances in the cell biology of bone. *CalcifiedTissue International, 36* (Suppl. 1), S37-S45.

Parfitt, A.M. (1994.) The two faces of growth: Benefits and risks to bone integrity. *Osteoporosis International, 4,* 382-398.

Parfitt, A.M. (1996.) Skeletal heterogeneity and the purposes of bone remodeling. In R. Marcus, D. Feldman & J. Kelsey (Eds.), *Osteoporosis* (pp. 315-329). New York: Academic Press.

Parfitt, D.B., Church, K.R., & Cameron, J.L. (1991.) Restoration of pulsatile luteinizing hormone secretion after fasting in rhesus monkeys (Macaca mulatta): dependence on the size of refeed meal. *Endocrinology, 129,* 749-756.

Parizkova, J. (1961.) Age trends in fat in normal and obese children. *Journal of Applied Physiology, 16,* 173-174.

Partridge, N.C., Jeffrey, J.J., Ehlich, L.S., Teitelbaum, S.L., Fliszar, C., Welgus, H.G., & Kahn, A.J. (1987.) Hormonal regulation of the production of collagenase and collagenase inhibitor activity by rat osteogenic sarcoma cells. *Endocrinology, 120,*1956-1962.

Pasumarthi, K.B., Doble, B.W., Kardami, E., & Cattini, P.A. (1994.) Over-expression of CUG- or AUG-initiated forms of basic fibroblast growth factor in cardiac myocytes results in similar effects on mitosis and protein synthesis but distinct nuclear morphologies. *Journal of Molecular and Cellular Cardiology, 26,* 1045-1060.

Pate, R.R., Barnes, C., & Mivler, W. (1985.) A physiological comparison of performance-matched male and female distance runners. *Research Quarterly for Exercise and Sport, 56,* 245-250.

Pate, R.R., Sparling, P.B., Wilson, G.E., Cureton, K.J., & Miller, B.J. (1997.) Cardiorespiratory and metabolic responses to submaximal and maximal exercise in elite women distance runners. *International Journal of Sports Medicine, 8,* S91-S95.

Patel, Y.C., & Srikant, C. (1986.) Somatostatin mediation of adenohypophysial secretion. *Annual Reviews of Physiology, 48,* 551-567.

Patten, M., Hartogensis, W.E., & Long, C.S. (1996.) Interleukin-1beta is a negative transcriptional regulator of alpha1 adrenergic induced gene expression in cultured cardiac myocytes. *Journal of Biological Chemistry, 271,* 21134-21141.

Patti, M.E., & Kahn, C.R. (1998.) The insulin receptor—a critical link in glucose homeostasis and insulin action. *Journal of Basic and Clinical Physiology and Pharmacology, 9,* 89-109.

Paul, M., & Ganten, D. (1992.) The molecular basis of cardiovascular hypertrophy: the role of the renin-angiotensin system. *Journal of Cardiovascular Pharmacology, 19*(Suppl. 5), S51-S58.

Paulev, P.E., Thorboll, J.E., Nielsen, V., Kruse, P., Jordal, R., Bach, F.W., Fenger, M., & Pokorski, M. (1989.) Opioid involvement in the perception of pain due to endurance exercise in trained men. *Japanese Journal of Physiology, 39,* 67-74.

Pead, M.J., & Lanyon, L.E. (1989.) Indomethacin modulation of load-related stimulation of new bone formation in vitro. *Calcified Tissue International, 45,* 34-40.

Pearce, F.J., & Connett, R.J. (1980.) Effect of lactate and palmitate on substrate utilization of isolated rat soleus. *American Journal of Physiology, 238,* 149-159.

Pearcey, S.M., & de Castro, J.M. (2002.) Food intake and meal patterns of weight-stable and weight-gaining persons. *American Journal of Clinical Nutrition, 76, 107-112.*

Pearse, A.G. (1969.) The cytochemistry and ultrastructure of polypeptide hormone-producing cells of the APUD series and the embryologic, physiologic and pathologic implications of the concept. *Journal of Histochemistry and Cytochemistry, 17,* 303-313.

Pearson, R., MacKinnon, M.J., Meek, A.P., Myers, D.B., & Palmer, D.G. (1982.) Diurnal and sequential grip functions in normal subjects and effects of temperature change and exercise of the forearm on grip function in patients with rheumatoid arthritis and in normal controls. *Scandinavian Journal of Rheumatology, 11,*113-118.

Pedersen, B.K. (1998.) Recovery of the immune system after exercise. *Acta Physiologica Scandinavica, 162,* 325-332.

Pedersen, B.K., & Beyer, J.M. (1986.) Characterization of the in vitro effects of glucocorticoids on NK cells activity. *Allergy, 41,* 220-224.

Pedersen, B.K., Bruunsgaard, H., Klokker, M., Kappel, M., MacLean, D.A., Nielsen, H.B., Rohde, T., Ulum, H., & Zacho, M. (1997.) Exercise-induced immunomodulation—possible roles of neuroendocrine and metabolic factors. *International Journal of Sports Medicine, 18* (Suppl. 1), S1-S7.

Pedersen, B.K., Kappel, M., Klokker, M., Nielsen, H.B., & Secher, N.H. (1994.) The immune system during exposure to extreme physiologic conditions. *International Journal of Sports Medicine, 15* (Suppl. 3), S116-S121.

Pedersen, B.K., Rohde, T., & Ostrowski, K. (1998.) Recovery of the immune system after exercise. *Acta Physiologica Scandinavica, 162,* 325-332.

Pedersen, B.K., Rohde, T., & Zacho, M. (1996.) Immunity in athletes. *Journal of Sports Medicine and Physical Fitness, 36,* 236-245.

Pedersen, B.K., Tvede, N., Klarlund, K., Christensen, L.D., Hansen, F.R., Galbo, H., Kharazmi, A., & Halkjaer-Kristensen, J. (1990.) Indomethacin in vitro and in vivo abolished post-exercise suppression of natural killer cell activity in peripheral blood. *International Journal of Sports Medicine, 11,* 127-131..

Pelleymounter, M.A., Cullen, M.J., Baker, M.B., Hecht, R., Winters, D. Boone, T., & Collins, F. (1995.) Effects of the obese gene product on body weight regulation in ob/ob mice. *Science, 269,* 540-543.

Peltenburg, A.L., Erich, W.B.M., Bernink, M.J., Zonderland, M.L., & Huisveld, I.A. (1984.) Biological maturation, body composition and growth of female gymnasts and control groups of schoolgirls and girl swimmers, aged 8 to 14 years: a cross-sectional survey of 1064 girls. *International Journal of Sports Medicine, 5,* 36-42.

Penn, B.H., Berkes, C.A., Bergstrom, D.A., & Tapscott, S.J. (2001.) How to MEK muscle. *Molecular Cell, 8,* 245-246.

Pennington, J.A.T., Young, B.E., & Wilson, D.B. (1989.) Nutritional elements in U.S. diets: results from the Total Diet Study, 1982-1986. *Journal of American Dietetic Association, 89,* 659-664.

Penttila, A., & Kormano, M. (1968.) Monoamine oxidase in the testis and epididymis of the rat. *Annales Medicine Experimentalis et Biologiae Fenniae, 46,* 557-563.

Pernow, J., & Lundberg, J.M. (1988.) Neuropeptide Y induces potent contraction of arterial vascular smooth muscle via an endothelium-independent mechanism. *Acta Physiologica Scandinavica, 134,* 157-158.

Peronnet, F., Beliveau, L., Boudreau, G., Trudeau, F., Brisson, G, & Nadeau, R. (1981a.) Regional plasma catecholamine removal and release at rest and exercise in dogs. *American Journal of Physiology, 254,* R663-R672.

Peronnet, F., Beliveau, L., Boudreau, G., Trudeau, F., Brisson, G., & Nadeau R. (1988.) Regional plasma catecholamine removal and release at rest and exercise in dogs. *American Journal of Physiology, 254,* R663-R672.

Peronnet, F., Cleroux, J., Perrault, H., Cousineau, D., de Champlain, J., & Nadeau, R. (1981b.) Plasma norepinephrine response before and after training in humans. *Journal of Applied Physiology, 51,* 812-815.

Persson, P.B., Gimpl, G., & Lang, R.E. (1990.) Importance of neuropeptide Y in the regulation of kidney function. *Annals of the New York Academy of Sciences, 611,* 156-165.

Pertovaara, A., Huopaniemi, T., Virtanen, A., & Johansson, G. (1984.) The influence of exercise on dental pain threshold and the release of stress hormones. *Physiology and Behavior, 33,* 923-926.

Perusse, L., Collier, G., Gagnon, J., Leon, A.S., Rao, D.C., Skinner, J.S., Wilmore, J.S., Nadeau, A., Zimmet, P.Z., & Bouchard, C. (1997.) Acute and chronic effects of exercise on leptin levels in humans. *American Journal of Physiology, 271,* 5-9.

Pescatello, L.S., Miller, B., Danias, P.G., Werner, M., Hess, M., Baker, C., & DeSouza, J.M. (1999.) Dynamic exercise normalizes resting blood pressure in mildly hypertensive premenopausal women. *American Heart Journal, 138,* 916-921.

Peter, M.A., Winterhalter, K.H., Boni-Schnetzler, M., Froesch, E.R., & Zapf, J. (1993.) Regulation of insulin-like growth factor-I (IGF-I) and IGF-binding proteins by growth hormone in rat white adipose tissue. *Endocrinology, 133,* 2624-2631.

Peters, J.R., Evans, P.J., Page, M.D., Hall, R.,Gibbs, J.T., Dieguez, C., & Scanlon, M.F. (1986.) Cholinergic muscarinic receptor blockade with pirenzepine abolishes slow-wave sleep-related growth hormone release in normal adult males. *Clinical Endocrinology, 25,* 213-217.

Peters, R.W., Muller, J.E., Goldstein, S., Byington, R., & Friedman, L.M. for the BHAT Study Group (1989.) Propranolol and the morning increase in the frequency of sudden cardiac death (BHAT Study). *American Journal ofCardiology, 63,* 1518-1520.

Peters, S., & Kreulen, D.L. (1985.) Vasopressin-mediated slow EPSPs in mammalian sympathetic ganglion. *Brain Research, 339,* 126-129.

Petit-Jacques, J., Bescond, J., Bois, P., & Lenfant, J. (1994.) Particular sensitivity of the mammalian heart sinus node cells. *News in Physiological Sciences, 9,* 77-79.

Petrides, J.S., Mueller, G.P., Kalogeras, K.T., Chrousos, G.P., Gold, P.W., & Deuster, P.A. (1994.) Exercise-induced activation of the hypothalamo-pituitary-adrenal axis: marked differences in the sensitivity to glucocorticoid supplementation. *Journal of Clinical Endocrinology and Metabolism, 77,* 377-383.

Pette, D., & Staron, R.S. (1990.) Cellular and molecular diversities of mammalian skeletal muscle fibers. *Reviews of Physiology, Biochemistry and Pharmacology, 116,* 1-76.

Pevet P. (2000.) Melatonin and biological rhythms. *Biological Signals and Receptors, 9,* 203-212.

Pham, C.G., Harpf, A.E., Keller, R.S., Vu, H.T., Shai, S.Y., Loftus, J.C. & Ross, R.S. (2000.) Striated muscle-specific beta(1D)-integrin and FAK are involved in cardiac myocyte hypertrophic response pathway. *American Journal of Physiology, 279,* H2916-H2926.

Phillips, D.I.W., Caddy, S., Ilic, V., Fielding, B.A., Frayn, K.N., Borthwick, A.C., & Taylor, R. (1996a.) Intramuscular triglyceride and muscle insulin sensitivity: Evidence for a relationship in nondiabetic subjects. *Metabolism, 45,* 947-950.

Phillips, M.I., Heininger, F., & Toffolo, S. (1996b.) The role of brain angiotensin in thirst and AVP release induced by hemorrhage. *Regulatory Peptides, 66,* 3-11.

Phillips, S.M., Green, H.J., Tarnopolsky, M.A., Heigenhauser, G.J.F., Hill, R.E., & Grant, S.M. (1996c.) Effects of training duration on substrate turnover and oxidation during exercise. *Journal of Applied Physiology, 81,* 2182-2191.

Physical activity and health: A report of the Surgeon General. (1996.) Atlanta, GA: U.S.Department of Health and Human Services, Centers for Disease Control and Prevention, National Center for Chronic Disease Prevention and Health Promotion.

Piccirillo, G., Bucca, C., Bauco, C., Cinti, A.M., Michele, D., Fimognari, F.L., Cacciafesta, M. & Marigliano, V. (1998.) Power spectral analysis of heart rate in subjects over a hundred years old. *International Journal of Cardiology, 63*, 53-61.

Pickard, G.E., & Rea, M.A. (1997.) Serotonergic innervation of the hypothalamic suprachiasmatic nucleus and photic regulation of circadian rhythms. Biology of the Cell, 89, 513-523.

Pickering, T.G. (1988.) The influence of daily activity on ambulatory blood pressure. *American Heart Journal, 116*, 1141-1145.

Pieper, D.R., Ali, H.Y., Benson, L.L., Shows, M.D., Lobocki, C.A., & Subramian, M.G. (1995.) Voluntary exercise increases gonadotropin secretion in male Golden hamsters. *American Journal of Physiology, 269*, R179-R185.

Pieper, D.R., Borer, K.T., Lobocki, C.A., & Samuel, D. (1988.) Exercise inhibits reproductive quiescence induced by exogenous melatonin in hamsters. *American Journal of Physiology, 255*, R718-R723.

Pierce, E.F., McGowan, R.W., Barkett, E., & Fry, R.W. (1993.) The effects of an acute bout of sleep on running economy and VO$_2$ max. *Journal of Sports Sciences, 11*, 109-112.

Pierce, E.F., & Pate, D.N. (1994.) Mood alterations in older adults following acute exercise. *Perceptual and Motor Skills, 79*, 191-194.

Pieris, A.N., Mueller, R.A., Smith, G.A., Struve, M.F., & Kissebah, A.H. (1986.) Splanchnic insulin metabolism in obesity. *Journal of Clinical Investigation, 78*, 1648-1657.

Pieris, A.N., Struve, M.F., & Kissebah, A.H. (1987a.) Relationship of body fat distribution to the metabolic clearance of insulin in premenopausal women. *International Journal of Obesity, 11*, 581-589.

Pieris, A.N., Mueller, R.A., Struve, M.F., Smith, G.A., & Kissebah, A.H. (1987b.) Relationship of androgenic activity to splanchnic insulin metabolism and peripheral glucose utilization in premenopausal women. *Journal of Clinical Endocrinology and Metabolism, 64*, 162-169.

Pieris, A.N., Struve, M.F., Mueller, R.A., Lee, M.B., & Kissebah, A.H. (1988.) Glucose metabolism in obesity: influence of body fat distribution. *Journal of Clinical Endocrinology and Metabolism, 67*, 760-767.

Pilkis, S.J., & Claus, T.H. (1991.) Hepatic gluconeogenesis/glycogenolysis: regulation and structure/ function relationships of substrate cycle enzymes. *Annual Review of Nutrition, 11*, 465-515.

Pilkis, S.J., & Granner, D.K. (1992.) Molecular physiology of the regulation of hepatic gluconeogenesis and glycolysis. *Annual Review of Physiology, 54*, 885-909.

Pinchasov, B.B., Shurgaja, A.M., Grischin, O.V., and Putilov, A.A. (2000.) Mood and energy regulation in seasonal and non-seasonal depression before and after midday treatment with physical exercise or bright light. *Psychiatry Research, 94*, 29-42.

Pincus, S.M., Gevers, E.F., Robinson, I.C., van den Berg, G., Roelfsema, F., Hartman, M.L., & Veldhuis, J.D. (1996.) Females secrete growth hormone with more process irregularity than males in both humans and rats. *American Journal of Physiology, 270*, E107-E115.

Pines, A., Fisman, E.Z., Drpry, Y., Shapira, I., Averbuch, M, Eckstein, N., Motro, M., Levo, Y., & Ayalon, D. (1998.) The effects of sublingual estradiol on left ventricular function at rest and exercise in postmenopausal women: An echocardiographic assessment. *Menopause, 5*, 79-85.

Ping, P., & Faber, JE., (1993.) Characterization of α-adrenoceptor gene expression in arterial and venous smooth muscle. *American Journal of Physiology, 265*, H1501-H1509.

Pirke K.M., Schweiger, U., Lemmel, W. Krieg, JC., & Berger M. (1985.) The influence of dieting on the menstrual cycle of healthy young women. *Journal of Clinical Endocrinology and Metabolism, 60*, 1174-1179.

Pirnay, F., Bodeux, M., Crielaard, J.M., & Franchimon, P. (1987.) Bone mineral content and physical activity. *International Journal of Sport Medicine, 8*, 331-335.

Pi-Sunyer, X., Kissileff, H.R., Thornton, J., & Smith, G.P. (1982.) C-terminal octapeptide of cholecystokinin decreases food intake in obese men. *Physiology and Behavior, 29*, 627-630.

Pittendrigh, C.S. (1993.) Temporal organization: reflections of a Darwinian clock-watcher. *Annual Review of Physiology, 55*, 16-54.

Pittendrigh, C.S., & Daan, S. (1976.) A functional analysis of circadian pacemakers in nocturnal rodents. IV. Entrainment: pacemaker as a clock. *Journal of Comparative Physiology, 106*, 291-331.

Pivarnik, J.M., Marichal, C.J., Spillman, T., & Morrow, J.R., Jr. (1992.) Menstrual cycle phase affects temperature regulation during endurance exercise. *Journal of Applied Physiology, 72*, 543-548.

Plat, L., Leproult, R., L'Hermite-Baleriaux, M., & Fery, F. (1999.) Metabolic effects of short-term elevations of plasma cortisol are more pronounced in the evening than in the morning. *Journal of Clinical Endocrinology and Metabolism, 84*, 3082-3092.

Plautz, J.D., Kaneko, M., Hall, J.C., & Kay, S.A. (1997.) Independent photoreceptive circadian clocks throughout Drosophila. *Science, 278*, 1632-1635.

Plotnick, G., Corretti, M., & Vogel, R. (1997.) Effect of antioxidant vitamins on the transient impairment of endothelium-dependent brachial artery vasoactivity following a single high-fat meal. *Journal of the American Medical Association, 278*, 1682-1686.

Pocock, N.A., Eisman, J.A., Yeates, M.G., Sambrook, P.N., & Eberl, S. (1986.) Physical fitness is a major determinant of femoral neck and lumbar spine bone mineral density. *Journal of Clinical Investigation, 78*, 618-621.

Poehlman, E.T., Arciero, P.J., & Goran, M.I. (1994.) Endurance exercise in aging humans: effects on energy metabolism. *Exercise and Sports Science Reviews, 22*, 251-284.

Poehlman, E.T., McAuliffe, T.L., Van Houten, D.R., & Danforth, Jr., E. (1990.) Influence of age and endurance training on metabolic rate and hormones in healthy men. *American Journal of Physiology, 259*, E66-E72.

Poehlman, E.T., Melby, C.L., & Goran, M.I. (1991.) The impact of exercise and diet restriction on daily energy expenditure. *Sports Medicine, 11*, 78-101.Poitout, V., Rouault, C., Guerre-Millo, M., & Reach, G. (1998.) Does leptin regulate insulin secretion? *Diabetes and Metabolism, 24*, 321-326.

Poitout, V., Rouault, C., Guerre-Millo, M., & Reach, G. (1998.) Does leptin regulate insulin secretion? *Diabetes and Metabolism, 24*, 321-326.

Pollare, T., Vessby, B., & Lithell, H. (1991.) Lipoprotein lipase activity in skeletal muscle is related to insulin sensitivity. *Arteriosclerosis and Thrombosis, 11*, 1192-1203.

Pollock, M. L., Boida, J., Kendrick, Z., Miller, H. S., Janeway, R., & Linnerud, A. C. (1972.) Effects of training two days per week at different intensities on middle-aged men. *Medicine and Science in Sports, 4*, 192-197.

Pollock, M. L., Dimmick, J., Miller, H. S., Jr., Kendrick, Z., & Linnerud, A. C. (1975.) Effects of mode of training on cardiovascular function and body composition of adult men. *Medicine and Science in Sports, 7*, 139 - 145.

Ponchon, P., & Elghozi, J.L. (1997.) Contribution of humoral systems to the recovery of blood pressure following severe hemorrhage. *Journal of Autonomic Pharmacology, 17*, 319-329.

Pontiroli, A.E., Manzoni, M.F., Malighetti, M.E., & Lanzi, R. (1996.) Restoration of growth hormone (GH) response to GH-releasing hormone in elderly and obese subjects by acute pharmacological reduction of plasma free fatty acids. *Journal of Clinical Endocrinology and Metabolism, 81*, 3990-4001.

Pope, H.G., & Katz, D.L. (1990.) Homicide and near-homicide by anabolic steroid users. *Journal of Clinical Psychiatry, 51*, 28-31.

Pope, H.G., & Katz, D.L. (1994.) Psychiatric and medical effects of anabolic-androgenic steroid use. *Archives of General Psychiatry, 51*, 375-382.

Pope, H.G., Kouri, E.M., & Hudson, J.L. (2000.) Effects of supraphysiological doses of testosterone on mood and aggression in normal men: A randomized controlled trial. *Archives of General Psychiatry, 57*, 133-140.

Poppitt S.D., Prentice A.M., Jequier E., Schutz Y., & Whitehead R.G. (1993.) Evidence of energy sparing in Gambian women during pregnancy: a longitudinal study using whole-body calorimetry. *American Journal of Clinical Nutrition, 57*, 353-364.

Porcerelli, J.H., & Sandler, B.A. (1998.) Anabolic-androgenic steroid abuse and psychopathology. *Psychiatric Clinics of North America, 21*, 829-833.

Porksen, N. (2002.) The in vivo regulation of pulsatile insulin secretion. *Diabetologia, 45*, 3-20.

Portaluppi, F., Vergnani, L., & degli Uberti, E.C. (1993.) Atrial natriuretic peptide and circadian blood pressure regulation: clues from a chronobiological approach. *Chronobiology International, 10,* 176-89.

Porte Jr, D., Smith, P.H., & Ensinck, J.W. (1976.) Neurohumoral regulation of the pancreatic islet A and B cells. *Metabolism, 25* (11 Suppl. 1), 1453-1456.

Potter, C.D., Borer, K.T., & Katz, R.V. (1983.) Opiate-receptor blockade reduces voluntary running but not self-stimulation in hamsters. *Pharmacology, Physiology and Behavior, 18,* 217-223.

Potter, E.K. (1988.) Neuropeptide Y as an autonomic neurotransmitter. *Pharmacology and Therapeutics, 37,* 251-273.

Potter, E.K., & Ulman, L.G. (1994.) Neuropeptides in sympathetic nerves affect vagal regulation of the heart. *News in Physiological Sciences, 9,* 174-177.

Powers, J.B., Jetton, A.E., & Wade, G.N. (1994.) Interactive effects of food deprivation and exercise on reproductive function in female hamsters. *American Journal of Physiology, 267,* R185-R190.

Powers, S.K., & Howley, E.T. (2001.) *Exercise physiology: Theory and application to fitness and performance.* New York: McGraw-Hill.

Powers, S.K., Wade, M., Criswell, D., Herb, R.A., Dodd, S., Hussain, R., & Martin, D. (1995.) Role of beta-adrenergic mechanism in exercise training-induced metabolic changes in respiratory and locomotor muscle. *International Journal of Sports Medicine, 16,* 13-18.

Pratley, R.E., Hagberg, J.M., Rogus, E.M., & Goldberg, A.P. (1995.) Enhanced insulin sensitivity and lower waist-to-hip ratio in master athletes. *American Journal of Physiology, 268,* E484-E490.

Pratt, J.H., Turner, D.A., McAteer, J.A., & Henry, D.P. (1985.) Beta-adrenergic stimulation of aldosterone production by rat adrenal capsular explants. *Endocrinology, 117,* 1189-1194.

Pratt, W.B. (1990.) Interaction of hsp90 with steroid receptors: organizing some diverse observations and presenting the newest concepts. *Molecular and Cellular Endocrinology, 74,* C69-C76.

Pratt, W.B. (1993.) The role of heat shock proteins in regulating the function, folding, and trafficking of the glucocorticoid receptor. *Journal of Biological Chemistry, 268*(29): 21455-21458.

Prechtl, J.C., & Powley, T.L. (1990.) The fiber composition of the abdominal vagus of the rat. *Anatomy and Embryology, 181,* 101-115.

Prentice, A.M., Black, A.E., Coward, W.A., & Cole, T.J. (1996.) Energy expenditure in overweight and obese adults in affluent societies: an analysis of 319 doubly-labelled water measurements. *European Journal of Clinical Nutrition, 50,* 93-97.

Prentki, M., & Corkey, B.E. (1996.) Are the beta-cell signaling molecules malonyl CoA and cytosolic long-chain acyl-CoA implicated in multiple tissue defects of obesity and NIDDM?. *Diabetes, 45,* 273-278.

Price, T.B., Rothman, D.L., Taylor, R., Avison, M.J., Shulman, G.I., & Shulman, R.G. (1994.) Human muscle glycogen resynthesis after exercise: insulin dependent and -independent phases. *Journal of Applied Physiology, 76,* 104-111.

Prigeon, R.L., Kahn, S.E., & Porte, D., Jr. (1995.) Changes in insulin sensitivity, glucose effectiveness, and B-cell in regularly exercising subjects. *Metabolism: Clinical and Experimental, 44,* 1259-1263.

Prince, F.P. (1992.) Ultrastructural evidence of indirect and direct autonomic innervation of human Leydig cells: comparison of neonatal, childhood and pubertal ages. *Cell and Tissue Research, 269,* 383-390.

Pritzlaff, C.J., Wideman, L., Blumer, J., Jensen, M., Abbott, R.D., Gaesser, G.A., Veldhuis, J.D., & Weltman, A. (2000.) Catecholamine release, growth hormone secretion, and energy expenditure during exercise vs. recovery in men. *Journal of Applied Physiology, 89,* 937-946.

Pruitt, L.A., Jackson, R.D., Bartells, R.L., & Lehnhard, H.J. (1992.) Weight-training effects on bone mineral density in early post-menopausal women. *Journal of Bone and Mineral Research, 7,* 179-185.

Puddey, I.B., & Cox, K. (1995.) Exercise lowers blood pressure-sometimes? Or did Pheidippides have hypertension? *Journal of Hypertension, 13,* 1229-1233.

Pugh, L.G.C., Edholm, O.G., Fox, R.H., Wolff, H.S., Harvey, G.R., Hammond, W.H., Tanner, J.M., & Whitehouse, R.H. (1960.) A physiological study of channel swimming. *Clinical Science, 19,* 257-273.

Pullan, P.T., Johnston, C.I. , Anderson, W.P., & Korner, P.I. (1980.) Plasma vasopressin in in blood pressure homeostasis and in experimental renal hypertension. *American Journal of Physiology, 239,* H81-H87.

Puri, P.L., Avantaggiati, M.L., Balsano, C., Sang, N., Graessmann, A., Giordano, A., & Levrero M. (1997.) p300 is required for MyoD-dependent cell cycle arrest and muscle-specific gene transcription. *EMBO Journal, 16,* 369-383.

Putman, C.T., Dusterhoft, S., & Pette, D.(2000.) Satellite cell proliferation in low frequency-stimulated fast muscle of hypothyroid rat. *American Journal of Physiology, 279,* C682-C690.

Puvi-Rajasingham, S., Smith, G.D., Akinola, A., & Mathias, C.J. (1998.) Hypotensive and regional haemodynamic effects of exercise, fasted and after food, in sympathetic denervation. *Clinical Science, 94,* 49-55.

Pyne, D.B. (1994.) Regulation of neutrophil function during exercise. *Sports Medicine, 17,* 245-258.

Qu, D., Ludwig, D.S., Gammeltoft, S., Piper, M., Pelleymounter, M.A., Cullen, M.J., Mathes, W.F., Przypek, R., Kanarek, R., & Maratos-Flier, E. (1996.) A role for melanin-concentrating hormone in the central regulation of feeding behaviour. *Nature, 380,* 243-247.

Quabbe, H.-J., Bumke-Vogt, C., Iglesias-Rogas, J.R., Freitag, S., & Breitinger, N. (1991.) Hypothalamic modulation of growth hormone secretion in the Rhesus monkey: evidence from intracerebroventricular infusions of glucose, free fatty acids, and ketone bodies. *Journal of Clinical Endocrinology and Metabolism, 73,* 765-770.

Quyyumi, A.A., Panza, J.A., Diodati, J.G., Lakatos, E., & Epstein, S.E. (1992.) Circadian variation in ischemic threshold. A mechanism underlying the circadian variation in ischemic events. *Circulation, 86,* 22-28.

Quyyumi, A.A., Panza, J.A., Lakatos, E., & Epstein, S.E. (1988.) Circadian variation in ischemic events: Causal role of variation in ischemic threshold due to changes in vascular resistance. *Circulation, 78 (Suppl.II),* 325-331.

Rabkin, J.G., Wagner, G.J., & Rabkin, R. (2000.) A double-blind, placebo-controlled trial of testosterone therapy for HIV-positive men with hypogonadal symptoms. *Archives of General Psychiatry, 57,* 141-147.

Radegran, G., & Hellsten, Y. (2000.) Adenosine and nitric oxide in exercise-induced human skeletal muscle vasodilatation. *Acta Physiologica Scandinavica, 168,* 575-591.

Rahkila, P., Hakala, E., Alen. M., Salminen, K., & Laatikainen, T. (1988.) Beta-endorphin and corticotropin release is dependent on a threshold intensity of running exercise in male endurance athletes. *Life Sciences, 43,* 551-558.

Rahn, K.H., Barenbrock, M., & Hausberg, M. (1999.) The sympathetic nervous system in the pathogenesis of hypertension. *Journal of Hypertension, 17* (Suppl. 3), S11-S14.

Rahn, T., Ridderstrale, M., Tornqvist, H., Manganiello, V., Fredrikson, G., Belfrage, P., & Degerman, E. (1994.) Essential role of phosphatidylinositol 3-kinase in insulin-induced activation and phosphorylation of the cGMP-inhibited cAMP phosphodiesterase in rat adipocytes: studies using the selective inhibitor wortmannin. *FEBS Letters, 350,* 314-318.

Raisz, L.G. (1996.) Interaction of local and systemic factors in the pathogenesis of osteoporosis. In R.Marcus, D. Feldman & J.Kelsey (Eds.), *Osteoporosis* (pp. 661-670). New York: Academic Press.

Raisz, L.G. (1999.) Physiology and pathophysiology of bone remodelling. *Clinical Chemistry, 45,* 1353-1358.

Raisz, L.G., Trummel, C.L., Holick, M.F., & DeLuca, H.F. (1972.) 1,25-dihydroxycholecalciferol: a potent stimulator of bone resorption in tissue culture. *Science, 175,* 768-769.

Raitakari, M., Knuuti, M.J., Ruotsalainen, U., Laine, H., Makea, P., Teras, M., et al. (1995.) Insulin increases blood volume in human skeletal muscle: studies using [^{15}O]CO and positron emission tomography. *American Journal of Physiology, 269,* E1000-E1005.

Raitakari, M., Nuutila, P., Ruotsalainen, U., Laine, H., Teras, M., Iida, H., et al. (1996.) Evidence for dissociation of insulin stimulation of blood flow and glucose uptake in human skeletal muscle. *Diabetes, 45,*1471-1477.

Rakusan, K. (1984.) Cardiac growth, maturation, and aging. In R. Zak (Ed.), *Growth of the heart in health and disease* (pp.133-137). New York: Raven Press.

Ramsay, M., Broughton Pipkin, F., & Rubin, P. (1992.) Comparative study of pressor and heart rate responses to angiotensin II and noradrenaline in pregnant and non-pregnant women. *Clinical Science, 82,* 157-162.

Ramsay, M., Broughton Pipkin, F., & Rubin, P. (1993.) Pressor, heart rate and plasma catecholamine responses to noradrenaline in pregnant and nonpregnant women. *British Journal of Obstetrics and Gynaecology, 100,* 170-176.

Rand, M.J., & Li, C.G. (1995.) Nitric oxide in the autonomic and enteric nervous systems. In S. Vincent (Ed.), *Nitric oxide in the nervous system* (pp. 227-279), New York: Academic Press.

Randle, P.J., Garland, P.B., Newsholme, E.A., & Hales, C.N. (1965.) The glucose fatty acid cycle in obesity and maturity onset diabetes mellitus. *Annals of the New York Academy of Sciences, 131*, 324-333.

Randle, P.J., Garland, P.B., Hales, C.N., & Newsholme, E.A. (1963.) The glucose-fatty acid cycle: its role in insulin sensitivity and the metabolic disturbances of diabetes mellitus. *Lancet, i*, 785-789.

Randle, P.J., Newsholme, E.A., & Garland, P.B. (1964.) Regulation of glucose uptake by muscle. *Biochemical Journal, 93*, 652-665.

Rani, C.S., Nordenstrom, K., Norjavaara, E., & Ahren, K. (1983.) Development of catecholamine responsiveness in granulosa cells from preovulatory rat follicles: dependence on preovulatory luteinizing hormone surge. *Biology of Reproduction, 28*, 1021-1031.

Rapoport, R.M., Draznin, M.B., & Murad, F. (1983.) Endothelium-dependent vasodilator- and nitrovasodilator-induced relaxation may be mediated through cyclic GMP formation and cyclic GMP-dependent protein phosphorylation. *Nature, 306*, 174-176.

Rattigan, S., Appleby, G.J., & Clark, M.G. (1991.) Insulin-like action of catecholamines and Ca^{2+} to stimulate glucose transport and GLUT4 translocation in perfused rat heart. *Biochimica et Biophysica Acta, 1094*, 217-223.

Rauramo, I., Salminen, K., & Laatikainen, T. (1986.) Release of beta-endorphine in response to physical exercise in non-pregnant and pregnant women. *Acta Obstetrica et Gynecologica Scandinavica, 65*, 609-612.

Ravussin, E., & Gautier, J.F. (1999.) Metabolic predictors of weight gain. *International Journal of Obesity and Related Metabolic Disorders, 23* (suppl 1), 37-41.

Rawson, M.J., Cornelissen, G., Holte, J., Katinas, G., Eckert, E., Siegelova, J., Fiser, B., & Halberg, F. (2000.) Circadian and circaseptan components of blood pressure and heart rate during depression. *Scripta medica, 73*, 117-124.

Rea, M.A. (1998.) Photic entrainment of circadian rhythms in rodents. Chronobiology International, 15, 395-423.

Ready, A. E., Naimark, B., Ducas, J., Sawatzky, J. V., Boreskie, S. L., Drinkwater, D. T., & Oosterveen, S. (1996.) Influence of walking volume on health benefits in women post-menopause. *Medicine and Science in Sports and Exercise, 28*, 1097-1105.

Reame, N.E., Sauder, S.E., Case, G.D., Kelch, R.P., & Marshall, J.C. (1985.) Pulsatile gonadotropin secretion in women with hypothalamic amenorrhea: evidence that reduced frequency of gonadotropin-releasing hormone secretion is the mechanism of persistent anovulation. *Journal of Clinical Endocrinology and Metabolism, 61*, 851-858.

Reardon, K., Galea, M., Dennett, X., Choong, P., & Byrne, E. (2001.) Quadriceps muscle wasting persists 5 months after total hip arthroplasty for osteoarthritis of the hip: a pilot study. *Internal Medicine Journal, 31*, 7-14.

Reaven, G.M. (1993.) Role of insulin resistance in human disease (syndrome X): an expanded definition. *Annual Review of Medicine, 44*, 121-131, 1993.

Rebar, R.W., & Yen, S.S.C. (1979.) Endocrine rhythms in gonadotrophins and ovarian steroids. In D.T. Krieger (Ed.), *Endocrine rhythms* (pp.259-298). New York: Raven Press.

Rebuffe-Scrive, M. (1991.) Neuroregulation of adipose tissue: Molecular and hormonal mechanisms. *International Journal of Obesity, 15* (Suppl. 2), 83-86.

Rebuffe-Scrive, M., Lonnroth, P., Marin, P., Wesslau, C., Bjorntorp, P., & Smith, U. (1987.) Regional adipose tissue metabolism in men and postmenopausal women. *International Journal of Obesity, 11*, 347-355.

Rebuffe-Scrive, M., Bronnegard, M., Nilsson, A., Eldh, J., Gustafsson, J.A., & Bjorntorp, P. (1990.) Steroid hormone receptors in human adipose tissues. *Journal of Clinical Endocrinology and Metabolism, 71*, 1215-1219.

Redfield, A., Nieman, M.T., & Knudsen, K.A. (1997.) Cadherins promote skeletal muscle differentiation in three-dimensional cultures. *Journal of Cell Biology, 138*, 1323-1331.

Reebs, S.G., & Mrosovsky, N. (1989.) Effects of induced wheel running on the circadian activity rhythms of Syrian hamsters: entrainment and phase-response curves. *Journal of Biological Rhythms, 4*, 39-48.

Refinetti, R., Kaufman, C.M., & Menaker, M. (1994.) Complete suprachiasmatic lesions eliminate circadian rhythmicity of body temperature and locomotor activity in golden hamsters. *Journal of Comparative Physiology, 175*, 223-232.

Refsum, H.E., & Stromme, S.B. (1975.) Urine flow, glomerular filtration, and urine solute during prolonged heavy exercise. *Scandinavian Journal of Clinical and Laboratory Investigation, 35*, 775-780.

Refsum, H.E., & Stromme, S.B. (1977.) Renal osmol clearance during prolonged heavy exercise. *Scandinavian Journal of Clinical and Laboratory Investigation, 38*, 19-22.

Regnier, M., & Herrera, A.A. (1993a.) Sensitivity to androgens within a sexually dimorphic muscle of male frogs (*Xenopus laevis*). *Journal of Physiology, 461*, 565-581.

Regnier, M., & Herrera, A.A. (1993b.) Changes in contractile properties of androgen hormones in sexually dimorphic muscles of male frogs (*Xenopus laevis*). *Journal of Physiology, 461*, 565-581.

Reid, I.A. (1996a.) Angiotensin II and baroreflex control of heart rate. *News in Physiological Sciences, 11*, 270-274.

Reid, M.B. (1996b.) Reactive oxygen and nitric oxide in skeletal muscle. *News in Physiological Sciences, 11*, 114-119.

Reid, M.B. (1998.) Role of nitric oxide in skeletal muscle: Synthesis, distribution and functional importance. *Acta Physiologica Scandinavica, 162*, 401-409.

Reilly, T. (1990.) Human circadian rhythms and exercise. *Critical Reviews in Biomedical Engineering, 18*, 165-180.

Reilly, T. (1994.) Circadian rhythms. In M. Harries, C. Williams, W.D. Stanish, and L.J. Miceli (Eds.), *Oxford textbook of sports medicine* (pp. 238-253). New York: Oxford University Press.

Reilly, T., Atkinson, G., & Waterhouse, J. (1997a.) *Biological rhythms and exercise*. New York: Oxford University Press.

Reilly, T., Atkinson, G., & Waterhouse, J. (1997b.) Travel fatigue and jet-lag. *Journal of Sports Sciences, 15*, 365-369.

Reilly, T., & Baxter, C. (1983.) Influence of time of day on reactions to cycling at a fixed high intensity. *British Journal of Sports Medicine, 17*, 128-130.

Reilly, T., & Brooks, G.A. (1982.) Investigation of circadian rhythms in metabolic responses to exercise. *Ergonomics, 25*, 1093-1097.

Reilly, T., & Brooks, G.A. (1990.) Selective persistence of circadian rhythms in physiological responses to exercise. *Chronobiology International, 7*, 59-67.

Reilly, T., & Garrett, R. (1998.) Investigation of diurnal variation in sustained exercise performance. *Ergonomics, 41*, 1085-1094.

Reilly, T., Robinson, G., & Minors, D.S. (1984.) Some circulatory responses to exercise at different times of day. *Medicine and Science in Sports and Exercise, 16*, 477-482.

Reis, D.J. (1960.) Potentiation of the vasoconstrictor action of topical norepinephrine on the human bulbar conjunctival vessels after topical application of certain adrenocortical hormones. *Journal of Clinical Endocrinology and Metabolism, 20*, 446-456.

Reisner, P.J., Moss, R.L., Giulian, G.C. Greaser, M.L. (1985.) Shortening velocity and myosin heavy chains of developing rabbit muscle fibres. *Journal of Biological Chemistry, 206*, 14403-14405.

Reiter, R.J. (1991.) Pineal melatonin: cell biology of its synthesis and of its physiological interactions. *Endocrine Reviews, 12*, 151-180.

Reiter, E.O., Fuldauer, V.G., & Root, A.W. (1977.) Secretion of adrenal androgen dehydroepiandrosterone sulfate, during normal infancy, childhood, and adolescence, in sick infants, and in children with endocrinologic abnormalities. *Journal of Pediatrics, 90*, 766-770.

Reiter, R.J., & Richardson, B.A. (1992.) Some perturbations that disturb the circadian melatonin rhythm. *Chronobiology International, 9*, 314-321.

Reitman, J., Baldwin, K.M., & Holloszy, J.O. (1973.) Intramuscular triglyceride utilization by red, white, and intermediate skeletal muscle and heart during exhausting exercise. *Proceedings of the Society for Experimental Biology and Medicine, 142*, 628-631.

Reynisdottir, S., Wahrenberg, H., Carlstrom, K., Rossner, S., & Arner, P. (1994.) Catecholamine resistance in fat cells of women with upper-body obesity due to decreased expression of beta$_2$-adrenoceptors. *Diabetologia, 37*, 428-435.

Rhee, S.G., & Choi, K.D. (1992.) Regulation of inositol phospholipid-specific phospholipase C enzymes. *Journal of Biological Chemistry, 267*, 12393-12396.

Richards, A.M., Nicholls, M.G., Espiner, E.A., Ikram, H., Cullens, M., & Hinton, D. (1986.) Diurnal patterns of blood pressure, heart rate and vasoactive hormones in normal man. *Clinical and Experimental Hypertension, 8*, 153-156.

Richardson, P.D., & Withrington, P.G. (1982.) Physiological regulation of the hepatic circulation. *Annual Reviews of Physiology, 44*, 57-59.

Richelsen, B. (1997.) Action of growth hormone in adipose tissue. *Hormone Research, 48*(Suppl. 5), 105-110.

Richelsen, B., Pedersen, S.B., Borglum, J.D., Jorgensen, J., & Jorgensen, J.O. (1994.) Growth hormone treatment of obese women for 5 wk: effect on body composition and adipose tissue LPL activity. *American Journal of Physiology, 266*, E211-E216.

Richelsen, B., Pedersen, S.B., Moller-Pedersen, T., & Bak, J.F. (1991.) Regional differences in triglyceride breakdown in human adipose tissue: effects of catecholamines, insulin, and prostaglandin E2. *Metabolism, 40*, 990-996.

Richelsen, B., Pedersen, S.B., Moller-Pedersen, T., Schnitz, O., Moller, N., & Borglum, J.D. (1993.) Lipoprotein lipase in muscle tissue influenced by fatness, fat distribution, and insulin in obese females. *European Journal of Clinical Investigation, 23*, 226-233.

Richter, E.A., Derave, W., & Wojtaszewski, J.F. (2001.) Glucose, exercise and insulin: emerging concepts. *Journal of Physiology, 535*, 313-322.

Richter, S.D., Schurmeyer, T.H., Schedlowski, M., Hadicke, A., Tewes, V., Schmidt, R.E., & Wagner, J.D. (1996.) Time kinetics of the endocrine response to acute psychological stress. *Journal of Clinical Endocrinology and Metabolism, 81*, 1956-1960.

Rios, R., Carneiro, I., Arce, V.M., & Devesa, J. (2001.) Myostatin regulates cell survival during C2C12 myogenesis. *Biochemical and Biophysical Research Communications, 280*, 561-566.

Rios, R., Carneiro, I., Arce, V.M., & Devesa, J. (2002.) Myostatin is an inhibitor of myogenic differentiation. *American Journal of Physiology, 282*, C993-C999.

Rippe, C., Berger, K., Boiers, C., Ricquier, D., & Erlanson-Albertsson, C. (2000.) Effect of high-fat diet, surrounding temperature and enterostatin on uncoupling protein gene expression. *American Journal of Physiology, 279*, E293-E300.

Ritter, R.C., Ritter, S., Ewart, W.R., & Wingate, D.L. (1989.) Capsaicin attenuates hindbrain neuron responses to circulating cholecystokinin. *American Journal of Physiology, 257*, R1162-R1168.

Ritter, S., & Dinh T.T. (1994.) 2-Mercaptoacetate and 2-deoxy-D-glucose induce Fos-like immunoreactivity in rat brain. *Brain Research, 641*, 111-120.

Ritter, S., Scheurink, A., & Singer, L.K. (1995.) 2-Deoxy-D-glucose but not 2-mercaptoacetate increases Fos-like immunoreactivity in adrenal medulla and sympathetic preganglionic neurons. *Obesity Research, 3* (Suppl. 5), 729S-734S.

Rivier, C., & Rivest, S. (1991.) Effect of stress on the activity of the hypothalamic-pituitary-gonadal axis: Peripheral and central mechanisms. *Biology of Reproduction, 45*, 523-532.

Rivier, J., Rivier, C., & Vale, W. (1986.) Stress-induced inhibition of reproductive function: role of endogenous corticotropin-releasing factor. *Science 231*, 607-609.

Rizza, R.A., Cryer, P.E., Haymond, M.W., & Gerich, J.E. (1980.) Adrenergic mechanisms for the effects of epinephrine on glucose production and clearance in man. *Journal of Clinical Investigation, 65*, 682-689.

Rizza, R.A., Mandarino, L.J., & Gerich, J.E. (1982a.) Effects of growth hormone on insulin action in man: mechanisms of insulin resistance, impaired suppression of glucose production, and impaired stimulation of glucose utilization. *Diabetes, 31*, 663-669.

Rizza, R.A., Mandarino, L.J., & Gerich, J.E. (1982b.) Cortisol-induced insulin resistance in man: impaired suppression of glucose production and stimulation of glucose utilization due to postreceptor defect of insulin action. *Journal of Clinical Endocrinology and Metabolism, 54*, 131-138.

Rizzo, V., Villatico Campbell, S., Di Maio, F., Tallarico, D., Lorido, A., Petretto, F., Bianchi, A., & Carmenini, G. (1999.) Spectral analysis of heart rate variability in elderly non-dipper hypertensive patients. *Journal of Human Hypertension, 13*, 393-398.

Roberts, C.K., Barnard, R.J., Scheck, S.H., & Balon, T.W. (1997.) Exercise-stimulated glucose transport in skeletal muscle is nitric oxide dependent. *American Journal of Physiology, 273*, E220-E225.

Roberts, W. (1860.) Observations on some of the daily changes of the urine. *Edinburg Medical Journal, 5*, 817-825.

Robertson, D., Convertino, V.A., & Vernikos, J. (1994.) The sympathetic nervous system and the physiologic consequences of spaceflight: a hypothesis. *American Journal of Medical Science, 308*, 126-132.

Robidoux, J., Pirouzi, P., Lafond, J., & Savard, R. (1995.) Site-specific effects of sympathectomy on the adrenergic control of lipolysis in hamster fat cells. *Canadian Journal of Physiology and Pharmacology, 73*, 450-458.

Robinson, S., Viira, J., Learner, J., Chan, S.P., Anyaoku, V., Beard, R.W., & Johnston, D.G. (1993.) Insulin insensitivity is associated with a decrease in postprandial thermogenesis in normal pregnancy. *Diabetic Medicine, 10*, 139-145.

Robinson, T.L., Snow-Harter, C., Taaffe, D.R., Gillis, D., Shaw, J., & Marcus, R. (1995.) Gymnasts exhibit higher bone mass than runners despite similar prevalence of amenorrhea and oligomenorrhea. *Journal of Bone and Mineral Research, 10*, 26-35.

Rochard, P., Rodier, A., Casas, F., Cassar-Malek, I., Marchal-Victorion, S., Daury, L., Wrutniak, C., & Cabello, G. (2000.) Mitochondrial activity is involved in the regulation of myoblast differentiation through myogenin expression and activity of myogenic factors. Journal of Biological Chemistry, 275, 2733-2744.

Rohde, T., MacLean, D.A., Hartkopp, A., & Pedersen, B.K. (1996.) The immune system and serum glutamine during a triathlon. *European Journal of Applied Physiology and Occupational Physiology, 74*, 428-434.

Rocco, M.B., Barry, J., Campbell, S., Nabel, E., Cook, E.F., Goldman, L., & Selwyn, A.P. (1987.) Circadian variation of transient myocardial ischemia in patients with coronary artery disease. *Circulation, 75*, 395-400.

Roemmich, J.N., Clark, P.A., Mai, V., Berr, S.S., Weltman, A., Veldhuis, J.D., & Rogol, A.D. (1998.) Alterations in growth and body composition during puberty: III. Influence of maturation, gender, body composition, fat distribution, aerobic fitness, and energy expenditure on nocturnal growth hormone release. *Journal of Clinical Endocrinology and Metabolism, 83*, 1440-1447.

Roger, S.D., Baker, L.R., & Raine, A.E. (1993.) Autonomic dysfunction and the development of hypertension in patients treated with recombinant human erythropoietin (r-HuEPO). *Clinical Nephrology, 39*, 103-110.

Rolandi, E., Franceschini, R., Cataldi, A., & Barreca, T. (1992.) Endocrine effects of sumatripan. *Lancet, 339*, 1365-1368.

Rolls, B.J., Wood, R.J., Rolls, E.T., Lind, H., Lind, W., & Ledingham, J.G.G. (1980.) Thirst following water deprivation in humans. *American Journal of Physiology, 239*, R476-R482.

Rolls, E.T. (1999.) Taste, olfactory, visual, and somatosensory representations of the sensory properties of foods in the brain, and their relation to the control of food intake. In H.-R. Berthoud & R.J. Seeley (Eds.), *Neural and metabolic control of macronutrient intake* (pp. 247-262). London: CRC Press.

Romijn, J.A., Coyle, E.F., Sidossis, L.S., Gastaldelli, A., Horowitz, J.F., Endert, E., & Wolfe, R.R. (1993.) Regulation of endogenous fat and carbohydrate metabolism in relation to exercise intensity and duration. *American Journal of Physiology, 265*, E380-E391.

Rommel, C., Clarke, B.A., Zimmermann, S., Nunez, L., Rossman, R., Reid, K., Moelling, K., Yancopoulos, G.D., & Glass, D.J. (1999.) Differentiation stage-specific inhibition of the Raf-MEK-ERK pathway by Akt. *Science, 286*, 1738-1741.

Romo, M. (1972.) Factors related to sudden death in acute ischaemic heart disease. *Acta Medica Scandinavica, 547* (Suppl.), 7-92.

Rong, H., Berg, U., Torring, O., Sundberg, C.J., Granberg, B., & Bucht, E. (1997.) Effect of acute endurance and strength exercise on circulating calcium-regulating hormones and bone markers in young healthy males. *Scandinavian Journal of Medicine and Science in Sports, 7*, 152-159.

Rosansky, S.J., Menachery, S.J., Wagner, C.M., & Jackson, K. (1995.) Circadian blood pressure variation versus renal function. *American Journal of Kidney Diseases, 26*, 716-721.

Rose, S.R., Municchi, G., Barnes, K.M., & Cutler, G.B., Jr. (1991.) Spontaneous growth hormone secretion increases during puberty in normal girls and boys. *Journal of Clinical Endocrinology and Metabolism, 73*, 428-435.

Rosell, S. (1966.) Release of free fatty acids from subcutaneous adipose tissue in dogs following sympathetic nerve stimulation. *Acta Physiologica Scandinavica, 67*, 343-351.

Rosen, E.D., Beninghof, E.G., & Koenig, R.J. (1993.) Dimerization interfaces of thyroid hormone, retinoic acid, vitamin D, and retinoid X receptors. *Journal of Biological Chemistry, 268*, 11534-11541.

Rosen, R.C., Kostis, J.B., Seltzer, L.G., Taska, L.S., & Holzer, B.C. (1991.) Beta blocker effects on heart rate during sleep: a placebo-controlled polysomnographic study with normotensive males. *Sleep, 14*, 43-47.

Rosen, S.G., Clutter, W.E., Berk, M.A., Shah, S.D., & Cryer, P.E. (1984.) Epinephrine supports the postabsorptive plasma glucose concentration and prevents hypoglycemia when glucagon secretion is deficient in man. *Journal of Clinical Investigation, 73,* 405-411.

Rosen, S.G., Clutter, W.E., Shah, S.D., Miller, J.P., Bier, D.M., & Cryer, P.E. (1983.) Direct alpha-adrenergic stimulation of hepatic glucose production in human subjects. *American Journal of Physiology, 245,* E616-E626.

Rosenbaum, M., Gertner, J.M., & Leibel, R.L. (1989.) Effects of systemic growth hormone administration on regional adipose tissue distribution and metabolism in GH-deficient children. *Journal of Clinical Endocrinology and Metabolism, 69,* 1274-1281.

Rosenbaum, M., Hirsch, J., Murphy, E., & Leibel, R.L. (2000.) Effects of changes in body weight on carbohydrate metabolism, catecholamine excretion, and thyroid function. *American Journal of Clinical Nutrition, 71,* 1421-1432.

Rosenbaum, M., Presta, E., Hirsch, J., & Leibel R.L. (1991.) Regional differences in adrenoceptor status of adipose tissue in adults and prepubertal children. *Journal of Clinical Endocrinology and Metabolism, 73,* 341-347.

Rosendorff, C. (1997.) Endothelin, vasular hypertrophy, and hypertension. *Cardiovascular Drugs and Therapy, 10,* 795-802.

Rosenfeld, B.A., Faraday, N., Campbell, D., Dise, K., Bell, W., & Goldschmidt, P. (1994.) Homeostatic effects of stress hormone infusion. *Anesthesiology, 81,* 1116-1126.

Rosenwasser, A.M., & Adler, T. (1986.) Structure and function in circadian timing systems: evidence for multiple coupled circadian oscillators. *Neuroscience and Biobehavioral Reviews, 10,* 431-448.

Rosing, D.R., Brakman, P., Redwood, D.R., Goldstein, R.E. Beiser, G.D., Astrup, T., & Epstein, S.E.(1970.) Blood fibrinolytic activity in man: diurnal variation and the response to varying intensities of exercise. *Circulation Research, 27,* 171-184.

Rosmond, R., Lapidus, L., Marin, P., & Bjorntorp, P. (1996.) Mental distress, obesity and body fat distribution in middle-aged men. *Obesity Research, 4,* 245-252.

Ross, R., Fortier, L., & Hudson, R. (1996.) Separate associations between visceral and subcutaneous adipose tissue distribution, insulin, and glucose levels in obese women. *Diabetes Care, 19,* 1404-1411.

Ross, R.S., Pham, C., Shai, S.Y., Goldhaber, J.L., Fenczik, C., Glembotski, C.C., Ginsberg, M.H., & Loftus, J.C. (1998.) Beta 1 integrins participate in the hypertrophic response of rat ventricular myocytes. *Circulation Research, 82,* 1160-1172.

Rossner S. (1999.) Physical activity and prevention and treatment of weight gain associated with pregnancy: current evidence and research issues. *Medicine and Science in Sports and Exercise, 31* (Suppl.), S560-S563.

Roth, A., Cligg, S.M., Yallow, C.P., & Berson, S.A. (1963.) Secretion of human growth hormone: physiological and experimental modification. *Metabolism, 12,* 557-559.

Roth, D.L., Bachtler, S.D., & Fillingim, R.B. (1990.) Acute emotional and cardiovascular effects of stressful mental work during aerobic exercise. *Psychophysiology, 28,* 689-700.

Roubenoff, R., Rall, L.C., Veldhuis, J.D., Kehayias, J.J., Rosen, C., Nicolson, M., Lundgren, N., & Reichlin, S. (1998.) The relationship between growth hormone kinetics and sarcopenia in postmenopausal women: The role of fat mass and leptin. *Journal of Clinical Endocrinology and Metabolism, 83,* 1502-1506.

Roussel, B., & Buguet, A. (1982.) Changes in human heart rate during sleep following daily physical exercise. *European Journal of Applied Physiology and Occupational Physiology, 49,* 409-416.

Routtenberg, A., & Kuznesof, A.W. (1967.) Self-starvation of rats living in activity wheels on a restricted feeding schedule. *Journal of Comparative and Physiological Psychology, 64,* 414-421.

Rowell, L.B., & O'Leary, D.S. (1990.) Reflex control of the circulation during exercise: chemoreflexes and mechanoreflexes. *Journal of Applied Physiology, 69,* 407- 418.

Roy, A.K., Sarkar, R., Bhadra, R., & Datta, A.G. (1985.) Effect of amines on fibrinogen synthesis. *Archives of Biochemistry and Biophysics, 239,* 364-367.

Roy, B.D., Green, H.J., Grant, S.M., & Tarnopolsky, M.A. (2000.) Acute plasma volume expansion alters cardiovascular but not thermal function during moderate-intensity prolonged exercise. *Canadian Journal of Physiology and Pharmacology, 78,* 244-250.

Roy, M., & Steptoe, A. (1991.) The inhibition of cardiovascular responses to mental stress following aerobic exercise. *Psychophysiology, 28,* 689-700.

Rubin, C.T., & McLeod, K. (1996.) Inhibition of osteopenia by biophysical intervention. In R. Marcus, D. Feldman, & J. Kelsey (Eds.), *Osteoporosis* (pp. 351-371). New York: Academic Press.

Rubin, P.C., McLean, K., & Reid, J.L. (1983.) Endogenous opioids and baroreflex control in humans. *Hypertension, 5,* 535-538.

Ruby, B.C., & Robergs, R.A. (1994.) Gender difference in substrate utilisation during exercise. *Sports Medicine, 17,* 393-410.

Ruderman, N.B., Toews, C.J., & Shafrir, E. (1969.) Role of free fatty acids in glucose homeostasis. *Archives of Internal Medicine, 123,* 299-313.

Rudman, D., & Mattson, D.E. (1994.) Serum insulin-like growth factor I in healthy older men in relation to physical activity. *Journal of the American Geriatric Society, 42,* 71-76.

Ruff, C.B., & Hayes, W.C. (1982.) Subperiosteal expansion and cortical remodelling of the human femur and tibia with aging. *Science, 217,* 945-947.

Ruffolo, R.R., Jr., & Hieble, J.P. (1994.) Alpha-adrenoceptors. *Pharmacology and Therapeutics, 61,* 1-64.

Russel, M.P., & Moran, N.C. (1980.) Evidence for lack of innervation of beta-2 adrenoceptors in the blood vessels of the gracilis muscle of the dog. *Circulation Research, 46,* 344-352.

Russell, J.B., Mitchell, D., Musey, P.I., & Collins, D.C. (1984a.) The role of beta-endorphins and catechol estrogens on the hypothalamic-pituitary axis in female athletes. *Fertility and Sterility, 42,* 690-695.

Russell, J.B., Mitchell, D., Musey, P.I., & Collins, D.C. (1984b.) The relationship of exercise to anovulatory cycles in female athletes: hormonal and physical characteristics. *Obstetrics and Gynecology, 63,* 452-456.

Rutherford, O.M., Jones, D.A., Round, J.M., Buchanan, C.M., & Preece, M.A. (1991.) Changes in skeletal muscle and body composition after discontinuation of growth hormone treatment in growth hormone deficient adults. *Clinical Endocrinology, 34,* 469-475.

Ruvolo, P.P. (2001.) Ceramide regulates cellular homeostasis via diverse stress signaling pathways. *Leukemia, 15,* 1153-1160.

Ruzicka, B.B., & Akil, H. (1995.) Differential cellular regulation of pro-opiomelanocortin by interleukin-1-beta and corticotropin-releasing hormone. *Neuroendocrinology, 61,* 136-151.

Ryu, H., Lee, H.S., Shin, Y.S., Chung, S.M., Lee, M.S., Kim, H.M., & Chung, H.T. (1996.) Acute effect of qigong training on stress hormone levels in man. *American Journal of Clinical Medicine, 24,* 193-198.

Saad, W.A., Camargo, L.A., Silveira, J.E., Saad, R., & Camargo, G.M. (1998.) Imidazoline receptors of the paraventricular nucleus on the pressor response induced by stimulation of the subfornical organ. *Journal of Physiology, 92,* 25-30.

Sackin, H. (1995.) Mechanosensitive channels. *Annual Reviews of Physiology, 57,* 333-353.

Sadoshima, J., & Izumo, S. (1993a.) Mechanical stretch rapidly activates multiple signal transduction pathways in cardiac myocytes: Potential involvement of an autocrine/paracrine mechanism. *EMBO Journal, 12,* 1681-1692.

Sadoshima, J., & Izumo, S. (1993b.) Mechanotransduction in stretch-induced hypertrophy of cardiac myocytes. *Journal of Receptor Research, 13,* 777-794.

Sadoshima, J., & Izumo, S. (1993c.) Signal transduction pathways of angiotensin II-induced c-fos gene expression in cardiac myocytes in vitro: roles of phospholipid-derived second messengers. *Circulation Research, 73,* 424-438.

Sadoshima, J., Xu, Y., Slayter, H.S., & Izumo, S. (1993.) Autocrine release of angiotensin II mediates stretch-induced hypertrophy of cardiac myocytes in vitro. *Cell, 75,* 977-984.

Sady, M.A., Haydon, B.B., Sady, S.P., Carpenter, M.W., Thompson, P.D., & Coustan, D.R. (1990.) Cardiovascular response to maximal cycle exercise during pregnancy and at two and seven months post partum. *American Journal of Obstetrics and Gynecology, 162,* 1181-1185.

Sage, D., Maurel, D., & Bosler, O. (2002.) Corticosterone-dependent driving influence of the suprachiasmatic nucleus on adrenal sensitivity to ACTH. *American Journal of Physiology, 282,* E458-E465.

Saha, A.K., Kurowski, T.G., & Ruderman, N.B. (1995.) A malonyl-CoA fuel-sensing mechanism in muscle: effects of insulin, glucose, and denervation. *American Journal of Physiology, 269,* E283-E289.

Saha, A.K., Vavvas, D., Kurowski, T.G., Apazidis, A., Witters, L.A., Shafrir, E., & Ruderman, N.B. (1997.) Malonyl-CoA regulation in skeletal muscle: its link to cell citrate and the glucose fatty acid cycle. American Journal of Physiology, 272, E641-E648.

Saini, J., Bothorel, B., Brandenberger, G., Candas, V., & Follenius, M. (1990.) Growth hormone and prolactin response to rehydration during exercise: Effect of water and carbohydrate solutions. *European Journal of Applied Physiology, 61,* 61-67.

Saito, M. (1995.) Differences in muscle sympathetic nerve response to isometric exercise in different muscle groups. *European Journal of Applied Physiology and Occupational Physiology, 70,* 26-35.

Saito, Y., & Berk, B.C. (2001.) Transactivation: a novel signaling pathway from angiotensin II to tyrosine kinase receptors. *Journal of Molecular and Cellular Cardiology, 33,* 3-7.

Saitoh, M., Yanagawa, T., Kondoh, T., Miyakoda, H., Kotake, H., & Mashiba, H. (1995.) Neurohumoral factor responses to mental (arithmetic) stress and dynamic exercise in normal subjects. *Internal Medicine, 34,* 618-622.

Saitoh, Y., Itaya, K., & Ui, M. (1974.) Adrenergic alpha-receptor-mediated stimulation of the glucose utilization by isolated rat diaphragm. *Biochimica Biophysica Acta, 428,* 492-499.

Sakoda, H., Ogihara, T., Anai, M., et al. (2002.) Activation of AMPK is essential for AICAR-induced glucose uptake by skeletal muscle but not adipocytes. *American Journal of Physiology, 282,* E1239-1244.

Sakuma, K., Watanabe, K., Sano, M., Uramoto, I., Sakamoto, K., & Totsuka, T. (1999.) The adaptive response of MyoD family proteins in overloaded, regenerating and denervated rat muscles. *Biochimica et Biophysica Acta, 1428,* 284-292.

Sakuma, K., Watanabe, K., Sano, M., Uramoto, I., & Totsuka, T. (2000.) Differential adaptation of growth and differentiation factor 8/myostatin, fibroblast growth factor 6 and leukemia inhibitory factor in overloaded, regenerating and denervated rat muscles. *Biochimica et Biophysica Acta, 1497,* 77-88.

Sakurai, T., Amemiya, A., Matsuzaki, I., et al. (1998.) Orexins and orexin receptors: a family of hypothalamic neuropeptides and G protein-coupled receptors that regulate feeding behavior. *Cell, 92,* 573-585.

Salera, M., Ebert, R., Giacomoni, P., Pironi, L., Venturi, S., Corinaldesi, R., Miglioli, M., & Barbara, L. (1982.) Adrenergic modulation of gastric inhibitory polypeptide secretion in man. *Digestive Disease or Sciences, 27,* 794-800.

Salomon, W.D., & Daughaday, W.H. (1957.) A hormonally controlled serum factor which stimulates sulfate incorporation by cartilage in vitro. *Journal of Laboratory and Clinical Medicine, 49,* 825-836.

Saltin, B., Henriksson, J., Nygaard, E., & Andersen, P. (1977.) Fiber types and metabolic potentials of skeletal muscles in sedentary man and endurance runners. *Annals of the New York Academy of Sciences, 301,* 3-29.

Saltin, B., Radegran, G., Koskoulou, M.D., & Roach, R.C. (1998.) Skeletal muscle blood flow in humans and its regulation during exercise. *Acta Physiological Scandinavica, 162,* 421-436.

Sambrook, P., Birmingham, J., Kempler, S., Kelly, P., Eberl, S., Pocock, N., Yeates, M., & Eisman, J. (1990.) Corticosteroid effects on proximal femur bone loss. *Journal of Bone and Mineral Research, 5,* 1211-1216.

Samols E., & Weir, G.C. (1979.) Adrenergic modulation of pancreatic A, B, and D cells. *Journal of Clinical Investigation, 63,* 230-238.

Samson, W.K. (1985.) Atrial natriuretic factor inhibits dehydration and hemorrhage-induced vasopressin release. *Neuroendocrinology, 40,* 277-279.

Sanborn, C.F., Martin, B.J., & Wagner, W.W. (1982.) Is athletic amenorrhea specific to runners? *American Journal of Obstetrics and Gynecology, 143,* 859-861.

Sanchez de la Pena, S., Halberg, F., Schweiger, H.G., & Eaton, J. (1984.) Circadian temperature rhythm and circadian-circaseptan (about 7 days) aspects of murine death from malaria. *Proceedings of the Society for Experimental Biology and Medicine, 175,* 196-204.

Sander, G.E., Lowe, R.F., Given, M.B., & Giles, T.D. (1989.) Interactions between circulating peptides and the central nervous system in hemodynamic regulation. *American Journal of Cardiology, 64,* 44C-50C.

Sandor, P., deJong, W., Wiegant, V., & deWied, D. (1987.) Central opioid mechanisms and cardiovascular control in hemorrhagic hypertension. *American Journal of Physiology, 253,* H507-H511.

Sane, T., Helve, E., Pelkonen, R., & Koivisto, V.A. (1988.) The adjustment of diet and insulin dose during long-term endurance exercise in type 1 (insulin-dependent) diabetic men. *Diabetologia, 31,* 35-40.

Sanfield, J.A., Cshenker, Y., Grekin, R.J., & Rosen, S.G. (1987.) Epinephrine increases plasma immunoreactive atrial natriuretic hormone levels in humans. *American Journal of Physiology, 252,* E740-E745.

Sangild, P.T., Sjostrom, H., Noren, O., Fowden, A.L., & Silver, M. (1995.) The prenatal evelopment and glucocorticoid control of brush-border hydrolases in the pig small intestine. *Pediatric Research, 37,* 207-212.

Sannerstedt, R. (1966.) Hemodynamic response to exercise in patients with arterial hypertension. *Acta Medica Scandinavica, Supplementum 180,* 458-463.

Sant'Ana Pereira, J.A., Sargeant, A.J., Rademaker, A.C., de Haan, A., & van Mechelen, W. (1996.) Myosin heavy chain isoform expression and high energy phosphate content in human muscle fibres at rest and post-exercise. *Journal of Physiology, 496,* 583-588.

Saper, C.B., Loewy, A.D., Swanson, L.W., & Cowan, W.M. (1976.) Direct hypothalamo-autonomic connections. *Brain Research, 117,* 305-312.

Sar, M., & Stumpf, W.E. (1977.) Androgen concentration in motor neurons of cranial nerves and spinal cord. *Science, 197,* 77-79.

Sassin, J.F., Parker, D.C., Mace, J.W., Gotlin, R.W., Johnson, L.C., & Rossman, C.G. (1969.) Human growth hormone release: relation to slow wave sleep and sleep-waking cycles. *Science, 165,* 513-515.

Satoh, N., Ogawa, Y., Katsuura, G., Numata, Y., Masuzaki, H., Yoshimasa, Y., & Nakao, K. (1998.) Satiety effect and sympathetic activation of leptin are mediated by hypothalamic melanocortin system. *Neuroscience Letters, 249,* 107-110.

Satozawa, N., Takezawa, K., Miwa, T., Takahashi, S., Hayakawa, M., & Ooka, H. (2000.) Differences in the effects of 20 K- and 22 K-hGH on water retention in rats. *Growth Hormone and Igf Research, 10,* 187-192.

Savard, R., Despres, J.P., Marcotte, M., Theriault,G., Tremblay, A., & Bouchard, C. (1987a.) Acute effects of endurance exercise on human adipose tissue metabolism. *Metabolism, 36,* 480-485.

Savard, R., & Greenwood, M.R. (1988.) Site-specific adipose tissue LPL responses to endurance training in female lean Zucker rats. *Journal of Applied Physiology, 65,* 693-699.

Savard, G.K., Richter, E.A., Strange, S., Kiens, B., Christensen, N.J., & Saltin, B. (1989.) Norepinephrine spillover from skeletal muscle during exercise in humans: role of muscle mass. *American Journal of Physiology, 257,* H1812-H1818.

Savard,G., Strange, S., Kiens, B., Richter, E.A., Christensen, N.J., & Saltin, B. (1987b.) Noradrenaline spillover during exercise in active versus resting skeletal muscle in man. *Acta Physiologica Scandinavica, 131,* 507-515.

Sawka, M.N., & Pandolf, K.B. (1990.) Effects of body water loss on physiological function and exercise performance. In C.V. Gisolfi, & D.R. Lamb (Eds.), *Perspectives in exercise science and sports medicine, Volume 3, Fluid homeostasis during exercise* (pp 1-30). Carmel, IN: Benchmark Press

Sawka, M.N., Toner, M.M., Francesconi, R.P., & Pandolf, K.B. (1983.) Hypohydration and exercise: effects of heat acclimation, gender, and environment. *Journal of Applied Physiology, 55,* 1147-1153.

Sawka, M.N., & Coyle, E.F. (1999.) Influence of body water and blood volume on thermoregulation and exercise performance in the heat. *Exercise and Sport Sciences Reviews, 27,* 167-218.

Saxena, P.R. (1992.) Interaction between the renin-angiotensin-aldosterone and sympathetic nervous system. *Journal of Cardiovascular Pharmacology, 19,* (Suppl. 6), S80-S88.

Sayer, J.W., Gutteridge, C., Snydercombe-Court, D., Wilkinson, P., & Timmis, A. D. (1998.) Circadian activity of the endogenous fibrinolytic system in stable coronary artery disease: effects of beta-andrenoreceptor blockers and angiotensis-converting enzyme inhibitors. *Journal of the American College of Cardiology, 32,* 1962-1968.

Scacchi, M., Pincelli, A.I., & Cavagnini, F. (1999.) Growth hormone in obesity.*International Journal of Obesity, 23,* 260-271.

Scatchard, G. (1949.) An attraction of proteins for small molecules and ions. *Annals of the New York Academy of Sciences, 51,* 660-672.

Schadt, J.C., & Gaddis, R.R. (1985.) Endogenous opiates may limit norepinephrine release during hemorrhage. *Journal of Pharmacology and Experimental Therapeutics, 232*, 656-660.

Schaefer, E.J., Lamon-Fava, S., Spiegelman, D., Dwyer, J.T., Lichtenstein, A.H., McNamara, J.R., et al. (1995.) Changes in plasma lipoprotein concentrations and composition in response to a low-fat, high-fiber diet are associated with changes in serum estrogen concentrations in premenopausal women. *Metabolism, 44*, 749-756.

Schaeffer, J.I., & Haddad, G.G. (1985.) Regulation of ventilation and oxygen consumption by delta- and mu-opioid receptor agonists. *Journal of Applied Physiology, 59*, 959-968.

Schafer, H., Koehler, U., Ploch, T., & Peter, J.H. (1997.) Sleep-related myocardial ischemia and sleep structure in patients with obstructive sleep apnea and coronary heart disease. *Chest, 111*, 387-393.

Schaub, M.C., Hefti, M.A., Harder, B.A., & Eppenberger, H.M. (1997.) Various hypertrophic stimuli induce distinct phenotypes in cardiomyocytes. *Journal of Molecular Medicine, 75*, 901-920.

Scheen, A.J., Buxton, O.M., Jison, M., Van Reeth, O., Leproult, R., L'Hermite-Baleriaux, M., & Van Cauter, E. (1998.) Effects of exercise on neuroendocrine secretions and glucose regulation at different times of day. *American Journal of Physiology, 274*, E1040-E1049.

Scheidigger, K., Robbins, D.C., & Danforth, E., Jr. (1984.) Effects of chronic beta receptor stimulation on glucose metabolism. *Diabetes, 33*, 1144-1149.

Schell, F.J., Allolio, B., & Schonecke, O.W. (1994.) Physiological and psychological effects of Hatha Yoga exercise in healthy women. *International Journal of Psychosomatics, 41*, 46-52.

Scherrer, U., Randin,D., Vollenweider, P., Vollenweider, L., & Nicod, P. (1994.) Nitric oxide release accounts for insulin's vascular effects in humans. *Journal of Clinical Investigation, 94*, 2511-2515.

Scheving, L.E., Tsai, T.H., Powell, E.W., Pasley, J.N., Halberg, F., & Dunn, J. (1983.) Bilateral lesions of supraciasmatic nuclei affect circadian rhythms in [3H]-thymidine incorporation into deoxyribonucleic acid in mouse intestinal tract, mitotic index of corneal epithelium, and serum corticosterone. *Anatomical Record, 205*, 239-249.

Schiaffino, S. & Reggiani, C. (1994.) Myosin isoforms in mammalian muscle fibres. *Journal of Applied Physiology, 77*, 493-501.

Schluter, K.-D., Zhou, X.J., & Piper, H.M. (1995.) Induction of hypertrophic responsiveness to isoprenaline by TGF-beta in adult rat cardiomyocytes. *American Journal of Physiology, 269*, C1311-C1316.

Schluter, K.-D., & Piper, H.M. (1999.) Regulation of growth in the adult cardiomyocytes. *FASEB Journal, 13* (Suppl.), S17-S22.

Schluter, K.-D., Goldberg,Y., Taimor, G., Schafer, M., & Piper, H.M. (1998.) Role of phosphatidylinositol 3-kinase activation in the hypertrophic growth of adult ventricular cardiomyocytes. *Cardiovascular Research, 40*, 174-181.

Schmid, P., Pusch, H.H., Wolf, W., Pilger, E., Pessenhofer, H., Schwaberger, G., et al. (1982.) Serum FSH, LH and testosterone in humans after physical exercise. *International Journal of Sports Medicine, 3*, 84-89.

Schmidt, E.D., Binnekade, R., Janszen, A.W., & Tilders, F.J. (1996.) Short stressor induced long-lasting increases of vassopressin stores in hypothalamic corticotropin-releasing hormone (CRH) neurons in adult rats. *Journal of Neuroendocrinology, 8*, 703-712.

Schmidt, H.H., Pollock, J.S., Nakane, M., Forstermann, U., & Murad. F. (1992.) Ca2+/calmodulin-regulated nitric oxide synthases. *Cell Calcium, 13*, 427-434.

Schmidt, T., Wijga, A., Von Zur Muhlen, A., Brabant, G., & Wagner, T.O. (1997.) Changes in cardiovascular risk factors and hormones during a comprehensive residential three month kriya yoga training and vegetarian nutrition. *Acta Physiologica Scandinavica, Supplementum 640*, 158-162.

Schoene, R.B., Robertson, H.T., Pierson, D.J., & Peterson, A.P. (1981.) Respiratory drives and exercise in menstrual cycles of athletic and non-athletic women. *Journal of Applied Physiology, 50*, 1399-1305.

Schreihofer, D.A., Amico, J.A., & Cameron, J.L. (1993.) Reversal of fasting-induced suppression of luteinizing hormone (LH) secretion in male rhesus monkeys by intragastric nutrient infusion: evidence for rapid stimulation of LH by nutritional signal. *Endocrinology, 132*, 1890-1897.

Schrier, R.W., Berl, T., & Anderson, R.J. (1979.) Osmotic and nonosmotic control of vasopressin release. *American Journal of Physiology, 236*, F321-F332.

Schuit, A.J., van Amelsvoort, L.G., Verheij, T.C., Rijneke, R.D., Maan, A.C., Swenne, C.A., & Schouten, E.G. (1999.) Exercise training and heart rate variability in older people. *Medicine and Science in Sports and Exercise, 31*, 816-821.

Schultea, T.D., Dees, W.L., & Ojeda, S.R. (1992.) Postnatal development of sympathetic and sensory innervation of the rhesus monkey ovary. *Biology of Reproduction , 47*, 760-767.

Schulz, L.O., & Schoeller, D.A. (1994.) A compilation of total daily energy expenditures and body weights in healthy adults. *American Journal of Clinical Nutrition, 60*, 676-681.

Schurmeyer, T., Jung, K., & Nieschlag, E. (1984.) The effects of an 1100 kilometer run on testicular, adrenal, and thyroid hormones. *International Journal of Andrology, 7*, 276-282.

Schwabe, U., Ebert, R., & Erbler, H.C. (1973.) Adenosine release from isolated fat cells and its significance for the effects of hormones on cyclic 3', 5'-AMP levels and lipolysis. *Naunyn-Schmiedeberg's Archives of Pharmacology, 276*, 133-148.

Schwartz, A.J., Brasel, J.A., Hintz, R.L., Mohan, S., & Cooper, D.M. (1996.) Acute effect of brief low- and high-intensity exercise on circulating insulin-like growth factor (IGF) I, II, and IGF-binding protein-3 and its proteolysis in young healthy men. *Journal of Clinical Endocrinology and Metabolism, 81*, 3492-3497.

Schwartz, B., Cumming, D.C., Riordan, E., Selye, M., Yen, S.S.C., & Rebar, R.W. (1981.) Exercise-associated amenorrhea: A distinct entity? *American Journal of Obstetrics and Gynecology, 141*, 662-670.

Schwartz, J., & Goodman, H.M. (1976.) Comparison of delayed actions of of growth hormone and somatomedin on adipose tissue metabolism. *Endocrinology, 98*, 730-737.

Schwartz, L., & Kinderman, W. (1992.) Changes in beta-endorphin levels in response to aerobic and anaerobic exercise. *Sports Medicine, 13*, 25-36.

Schwartz, M.W., Woods, S.C., Porte, Jr., D., Seeley, R.J., & Baskin, D.G. (2000.) Central nervous system control of food intake. *Nature, 404*, 661-671.

Schwartz, P.J., Rosenthal, N.E., Turner, E.H., Drake, C.L., Liberty, V., & Wehr, T.A. (1997.) Seasonal variation in core temperature regulation during sleep in patients with winter seasonal affective disorder. *Biological Psychiatry, 42*, 122-131.

Selwyn, A.P., Shea, M., Deanfield, J.E., Wilson, R., Horlock, P., & O'Brien, H.A. (1986.) Character of transient ischemia in angina pectoris. *American Journal of Cardiology, 58*, 21B-25B.

Scrutton, M.C., & Utter, M.F. (1967.) Pyruvate carboxylase. IX. Some properties of the activation by certain acyl derivatives of coenzyme A. *Journal of Biological Chemistry, 242*, 1723-1735.

Seals, D.R., & Chase, P.B. (1989.) Influence of physical training on heart rate variability and baroflex circulatory control. *Journal of Applied Physiology, 66*, 1886-1895.

Seals, D.R., & Victor, R.G. (1991.) Regulation of muscle sympathetic nerve activity during exercise in humans. *Exercise and Sports Science Reviews, 9*, 313-349.

Segal, K.R., Albu, J., Chun, A., Edano, A., Legaspi, B., & Pi-Sunyer, F.X. (1992.) Independent effects of obesity and insulin resistance on postprandial thermogenesis in men. *Journal of Clinical Investigation, 89*, 824-833.

Seger, J.Y., & Thorstensson, A. (1994.) Muscle strength and myoelectric activity in prepubertal and adult males and females. *European Journal of Applied Physiology, 69*, 81-87.

Seger, R., & Krebs, E.G. (1995.) The MAPK signaling cascade. *FASEB Journal, 9*, 726-735.

Seidell, J.C., Bjorntorp, P., Sjostrom, L., Sonnerstedt, R., Krotkiewski, M., & Kvist, H. (1988.) Regional distribution of muscle and fat mass in men: new insight into the risk of abdominal obesity using computed tomography. *International Journal of Obesity, 13*, 289-303.

Seip, R.L., Angelopoulos, T.J., & Semenkovich, C.F. (1995.) Exercise induces human lipoprotein lipase gene expression in skeletal muscle but not adipose tissue. *American Journal of Physiology, 268*, E229-E236.

Seip, R.L., Mair, K., Cole, T.G., & Semenkovich, C.F. (1997.) Induction of human skeletal muscle lipoprotein lipase gene expression by short-term exercise is transient. *American Journal of Physiology, 272*, E255-E261.

Seip, R.L., & Semenkovich, C.F. (1998.) Skeletal muscle lipoprotein lipase: molecular regulation and physiological effects in relation to exercise. *Exercise and Sports Sciences Reviews, 26,* 191-218.

Selye, H. (1956.) *The stress of life.* New York: McGraw-Hill.

Selye, H. (1974.) *Stress without distress.* Philadelphia: Lippincott.

Semsarian, C., Wu, M.-J., Ju, Y.-K., Marciniec,T., Yeoh, T., Allen, D.G., Harvey, R.P., & Graham, R.M. (1999.) Skeletal muscle hypertrophy is mediated by a Ca^{2+}-dependent calcineurin signalling pathway. *Nature, 400,* 576-581.

Sforzo, G.A. (1989.) Opioids and exercise. *Sports Medicine, 7,* 109-124.

Shall, S. (1981.) Control of cell reproduction. *British Medical Bulletin, 37,* 209-214.

Shanahan, T.L., Zeitzer, J.M., & Czeisler, C.A. (1997.) Resetting the melatonin rhythm with light in humans. *Journal of Biological Rhythms, 12,* 556-567.

Shangold, M.M., Gatz, M.L., & Thysen, B. (1981.) Acute effects of exercise on plasma concentrations of prolactin and testosterone in recreational women runners. *Fertility and Sterility, 35,* 699-702.

Shapiro, C.M., Bortz, R., Mitchell, D., Bartell, P., & Jooste, P. (1981.) Slow-wave sleep: a recovery period after exercise. *Science, 214,* 1252-1254.

Shapiro, Y., Pandolph, K.B., & Goldman, R.F. (1980.) Sex differences in acclimation to a hot-dry environment. *Ergonomics, 23,* 635-642.

Share, L., & Levy, M.N. (1962.) Cardiovascular receptors and blood titer of antidiuretic hormone. *American Journal of Physiology, 203,* 425-428.

Share, L. (1996.) Control of vasopressin release: an old but continuing story. *News in Physiological Sciences, 11,* 7-13.

Sharkey, K.M., & Eastman, C.I. (2002.) Melatonin phase shifts human circadian rhythms in plaxcebo-controlled simulated night-work study. *American Journal of Physiology, 282,* R454-R463.

Sharma, M., Kambadur, R., Matthews, K.G., Somers, W.G., Devlin, G.P., Conaglen, J.V., Fowke, P.J., & Bass, J.J. (1999.) Myostatin, a transforming growth factor-beta superfamily member, is expressed in heart muscle and is upregulated in cardiomyocytes after infarct. *Journal of Cellular Physiology, 180,* 1-9.

Sharp, W.W., Simpson, D.G., Borg, T.K., Samarel, A.M., & Terracio, L. (1997.) Mechanical forces regulate focal adhesion and costamere assembly in cardiac myocytes. *American Journal of Physiology, 273,* H546-H556.

Shattil, S.J., Budzynski, A., & Scrutton, M.C. (1989.) Epinephrine induces platelet fibrinogen receptor expression, fibrinogen bionding, and aggregation in whole blood in the absence of other excitatory agonists. *Blood 73,* 150-158.

Shephard, R.J. (1984.) Sleep, biorhythms, and human performance. *Sports Medicine, 1,* 11-37.

Shephard, R.J. (1993.) Metabolism adaptations to exercise in the cold: an update. *Sports Medicine, 16,* 266-289.

Shephard, R.J. (2000a.) Exercise and training in women, Part I: Influence of gender on exercise and training responses. *Canadian Journal of Applied Physiology, 25,* 19-34.

Shephard, R.J. (2000b.) Exercise and training in women, Part II: Influence of menstrual cycle and pregnancy. *Canadian Journal of Applied Physiology, 25,* 35-54.

Shephard, R.J., & Shek, P.N. (1997.) Interactions between sleep, other body rhythms, immune responses, and exercise. *Canadian Journal of Applied Physiology, 22,* 95-116.

Shephard, R.J., Verde, T.J., Thomas, S,G., & Shek, P. (1991.) Physical activity and the immune system. *Canadian Journal of Sports Science, 16,* 169-185.

Sherman, A.J., Davis, C.A. 3rd, Klocke, F.J., Harris, K.R., Srinivasan, G., Yaacoub, A.S., Quinn, D.A., Ahlin, K.A., & Jang, J.J. (1997.) Blockade of nitric oxide synthesis reduces myocardial oxygen consumption in vivo. *Circulation, 95,* 1328-1334.

Sherman, B., Pfohl, B., & Winokur, G. (1984.) Circadian analysis of plasma cortisol levels before and after dexamethasone administration in depressed patients. *Archives of General Psychiatry, 41,* 271-275.

Sherman, B.E., & Chole, R.A. (1995.) A mechanism for sympathectomy-induced bone resorption in the middle ear. *Otolaryngology-Head and Neck Surgery, 113,* 569-581.

Sherwin, B.B. (1988.) Estrogen and/or androgen replacement therapy and cognitive functioning in surgically menopausal women. *Psychoneuroendocrinology, 13,* 345-357.

Shi, X., Stevens, G.H., Foresman, B.H., Stern, S.A., & Raven, P.B. (1995.) Autonomic nervous system control of the heart: endurance exercise training. *Medicine and Science in Sports and Exercise, 27,* 1406-1413.

Shibasaki, T., Hotta, M., Masuda, A., Imaki, T., Obara, N., & Demura, H. (1985.) Plasma responses to GHRH and insulin-induced hypoglycemia in man. *Journal of Clinical Endocrinology and Metabolism, 60,* 1265-1267.

Shibasaki, T., Imaki, T., Hotta, M., Ling, N., & Demura, H. (1993.) Psychological stress increases arousal through brain corticotropin-releasing hormone without significant increase in adrenocorticotropin and catecholamine secretion. *Brain Research, 618,* 71-75.

Shimazu, T., & Ogasawara, S. (1975.) Effects of hypothalamic stimulation on gluconeogenesis and glycolysis in rat liver. *American Journal of Physiology, 228,* 1787-1793.

Shimazu, T., & Usami, M. (1982.) Further studies on the mechanism of phosphorylase activation in rabbit liver in response to splanchnic nerve stimulation. *Journal of Physiology, 329,* 231-242.

Shimizu, N., Kaizuka, Y., Hori, T., & Nakane, H. (1996.) Immobilization increases norepinephrine release and reduced NK cytotoxicity in spleen of conscious rats. *American Journal of Physiology, 271,* R537-R544.

Shinar, D.M., Endo, N., Halperin, D., Rodan, G.A., & Weinreb, M. (1993.) Differential expression of insulin-like growth factor-I (IGF-I) and IGF-II messenger ribonucleic acid in growing rat bone. *Endocrinology, 132,* 1158-1167.

Shinkai, S., Shore, S., Shek, P.N., & Shephard, R.J. (1992.) Acute exercise and immune function. *International Journal of Sports Medicine, 13,* 452-461.

Shintani, M., Ogawa, Y., Ebihara, K., et al. (2001.) Ghrelin, an endogenous growth hormone secretagogue, is a novel orexigenic peptide that antagonizes leptin action through the activation of hypothalamic neuropeptide Y/Y1 receptor pathway. *Diabetes, 50,* 227-232.

Shirakawa, T., Honma, S., & Honma, K.(2001.) Multiple oscillators in the suprachiasmatic nucleus. Chronobiology International, 18, 371-387.

Shirakura, S., Furugohri, T., & Tokumitsu, Y. (1990.) Activation of glucose transport by activatory receptor agonists of adenylate cyclase in rat adipocytes. *Comparative Biochemistry and Physiology, Physiol A, 97,* 81-86.

Shirasu, K., Stumpf, W.E., & Sar, M. (1990.) Evidence for direct action of estradiol on growth-hormone releasing factor (GRF) in rat hypothalamus: localization of estradiol in GRF neurons. *Endocrinology, 127,* 344-349.

Shirreffs, S.M., & Maughan, R.J. (1997.) Whole body sweat collection in humans: an improved method with preliminary data on electrolyte content. *Journal of Applied Physiology, 82,* 336-341.

Shirreffs, S.M., & Maughan, R.J. (1998.) Volume repletion after exercise-induced volume depletion in humans: replacement of water and sodium losses. *American Journal of Physiology, 274,* F868-F875.

Shirreffs, S.M., Taylor, A.J., Leiper, J.B., & Maughan, R.J. (1996.) Post-exercise rehydration in man: effects of volume consumed and drink sodium content. *Medicine and Science in Sports and Exercise, 28,* 1260-1271.

Shoemaker, J.K., Hogeman, C.S., & Sinoway, L.I. (1999.) Contributions of MSNA and stroke volume to orthostatic intolerance following bed rest. *American Journal of Physiology, 277,* R1084-R1090.

Short, K.R., & Sedlock, D.A. (1997.) Excess postexercise oxygen consumption and recovery rate in trained and untrained subjects. *Journal of Applied Physiology, 83,* 153-159.

Shulman, G.I., Rothman, D.L., Jue, T., Stein, P., DeFronzo, R.A., & Shulman, R.G. (1990.) Quantitation of muscle glycogen synthesis in normal subjects and subjects with non-insulin-dependent diabetes by 13C nuclear magnetic resonance spectroscopy. *New England Journal of Medicine, 322,* 223-228.

Sidossis, L.S., & Wolfe, R.R. (1996.) Glucose and insulin-induced inhibition of fatty acid oxidation: the glucose-fatty acid cycle revisited. *American Journal of Physiology, 270,* E733-E738.

Sigal, R.J., Fisher, S.F., Halter, J.B., Vranic, M., & Marliss, E.B. (1996.) The roles of catecholamines in glucoregulation in intense exercise as defined by islet cell clamp technique. *Diabetes, 45,* 148-156.

Sigurdson, W., Ruknufdin, A., & Sachs, F. (1992.) Calcium imaging of mechanically induced fluxes in tissue-cultured chick heart: role of stretch-activated ion channels. *American Journal of Physiology, 262,* H1110-H1115.

Silber, D.H., Sinoway, L.I., Leuenberger, U.A., & Amassian, V.E. (2000.) Magnetic stimulation of the human motor cortex evokes skin sympathetic nerve activity. *Journal of Applied Physiology, 88,* 126-134.

Sillence, M.N., Reich, M.M., & Thomson, B.C. (1995.) Sexual dimorphism in the growth response of entire and gonadectomized rats to clenbuterol. *American Journal of Physiology, 268,* E1077-E1082.

Sillence, M.N., Moor, N.G., Pegg, G.G., & Londay, D.B. (1993.) Ligand binding properties of putative beta 3-adrenoceptors compared in brown adipose tissue and in skeletal muscle membranes. *British Journal of Pharmacology, 109,* 1157-1163.

Simon, E. (2000.) Interface properties of circumventricular organs in salt and fluid balance. *News in Physiological Sciences, 15,* 61-67.

Simon, G., Reid, L., Tanner, J.M., Goldstein, H., & Benjamin, B. (1972.) Growth of radiologically determined heart diameter, lung width, and lung length from 5-19 years, with standards for clinical use. *Archives of Disease in Childhood, 47,* 373-381.

Simoneau, J.-A., Colberg, S.R., Thaete, F.L., & Kelley, D.E. (1995.) Skeletal muscle glycolytic and oxidative enzyme capacities are determinants of insulin sensitivity and muscle composition in obese women. *FASEB Journal, 9,* 273-278.

Simonson, D.C., Koivisto, V., Sherwin, R.S., Ferrannini, E., Hendler, R., Juhlin-Dannfelt, A., & DeFronzo, R.A. (1984.) Adrenergic blockade alters glucose kinetics during exercise in insulin-dependent diabetics. *Journal of Clinical Investigation, 73,* 1648-1658.

Simpson, G.E. (1926.) The effect of sleep on urinary chlorides and pH. *Journal of Biological Chemistry, 67,* 505-516.

Sinha-Hikim, I., Artaza, J., Woodhouse, L., Gonzalez-Cadavid, N., Singh, A.B., Lee, M.I., et al. (2002.) Testosterone-induced increase in muscle size in healthy young men is associated with muscle fiber hypertrophy. *American Journal of Physiology, 283,* E154-164.

Siscovick, D.S., Laporte, R.E., & Newman, J.M. (1985.) The disease-specific benefits and risks of physical activity and exercise. *Public Health Reports, 100,* 180-188.

Siscovick, D.S., Weiss, N.S., Fletcher, R.H., & Lasky, T. (1984.) The incidence of primary cardiac arrest during vigorous exercise. *New England Journal of Medicine, 311,* 874-877.

Sizonenko, P.C., Paunier, L., & Carmignac, D. (1976.) Hormonal changes during puberty. IV. Longitudinal study of adrenal androgen secretions. *Hormone Research, 7,* 288-302.

Skinner, M.R., & Marshall, J.M. (1996.) Studies of the roles of ATP, adenosine and nitric oxide in mediating muscle vasodilatation induced in the rat by acute systemic hypoxia. *Journal of Physiology, 495,* 553-560.

Sklar, C.A., Kaplan, S.L., & Grumbach, M.M. (1980.) Evidence for dissociation between adrenarche and gonadarche: studies in patients with idiopathic precocious puberty, gonadal dysgenesis, isolated gonadotropin deficiency, and constitutionally delayed growth and adolescence. *Journal of Clinical Endocrinology and Metabolism, 51,* 548-556.

Skolnik, E.Y., & Marcusohn, J. (1996.) Inhibition of insulin receptor signaling by TNF: Potential role in obesity and non-insulin-dependent diabetes mellitus. *Cytokine and Growth Factor Reviews, 7,* 161-173.

Sloan, M.A., Price, T.R., Foulkes, M.A., Marler, J.R., Mohr, J.P., Hier, D.B., Wolf, P.A., & Caplan, L.R. (1992.) Circadian rhythmicity of stroke onset. *Stroke, 23,* 1420-1426.

Small, C.A., Garton, A.J., & Yeaman, S.J. (1989.) The presence and role of hormone-sensitive lipase in heart muscle. *Biochemical Journal, 258,* 67-72.

Smith, C.C.T., Prichard, B.N.C., & Betteridge, D.J. (1992a.) Plasma and platelet-free catecholamine concentrations in patients with familial hypercholesterolemia. *Clinical Science, 82,* 113-116.

Smith, J.A. (1997.) Exercise immunology and neutrophils. *International Journal of Sports Medicine, 18* (Suppl. 1), S46-S55.

Smith, J.A., Telford, R.D., Mason, J.B., & Weidermann, M.J. (1990.) Exercise, training, and neutrophil microbicidal activity. *International Journal of Sports Medicine, 11,* 179-187.

Smith, J.C., Stephens, D.P., Winchester, P.K., & Williamson, J.W. (1997a.) Facial cooling-induced bradycardia: attenuating effect of central command at exercise onset. *Medicine and Science in Sports and Exercise, 29,* 320-325.

Smith, R., Turek, F.W., & Takahashi, J.S. (1992b.) Two families of phase-response curves characterize the resetting of the hamster circadian clock. *American Journal of Physiology, 262,* R1149-R1153.

Smith, T.P., Lopez-Corrales, N.L., Kappes, S.M., & Sonstegard, T.S. (1997c.) Myostatin maps to the interval containing the bovine mh locus. *Mammalian Genome, 8,* 742-744.

Smith, T.R., Elmendorf, J.S., David, T.S., & Turinsky, J. (1997b.) Growth hormone-induced insulin resistance: role of insulin receptor, IRS-1, GLUT-1, and GLUT-4. *American Journal of Physiology, 272,* E1071-E1079.

Smol, E., Zernicka, E., Czarnowski, D., & Langfort, J. (2001.) Lipoprotein lipase activity in skeletal muscles of the rat: effects of denervation and tenotomy. *Journal of Applied Physiology, 90,* 954-960.

Snitker, S., Tataranni, P.A., & Ravussin, E. (1998.) Respiratory quotient is inversely associated with muscle sympathetic nerve activity. *Journal of Clinical Endocrinology and Metabolism, 83,* 3977-3979.

Snow, C.M., Shaw, J.M., & Matkin, C.C. (1996.) Physical activity and risk for osteoporosis. In R. Marcus, D. Feldman, & J.Kelsey (Eds.), *Osteoporosis* (pp. 511-528). New York: Academic Press.

Snow-Harter, C., Bouxsein, M., Lewis, B., Charette, S., Weinstein, P., & Marcus, R. (1990.)

Muscle strength as a predictor of bone mineral density in young women. *Journal of Bone and Mineral Research, 5,* 589-595.

Snow-Harter, C., Whalen, R., Myburgh, K., Arnaud, S., & Marcus, R. (1992.) Bone mineral density, muscle strength, and recreational exercise in men. *Journal of Bone and Mineral Research, 7,* 1291-1296.

Snyder, F., Hobson, J.S., Morrison, D.F., & Goldfrank, F. (1964.) Changes in respiration, heart rate, and systolic blood pressure in human sleep. *Journal of Applied Physiology, 19,* 417-422.

Solanes, G., Pedraza, N., Iglesias, R., Giralt, M., & Villaroya, F. (2000.) The human uncoupling protein-3 gene promoter requires MyoD and is induced by retinoic acid in muscle cells. *FASEB Journal, 14,* 2141-2143.

Solomon, R., Weinberg, M.S., & Dubey, A. (1991.) The diurnal rhythm of plasma potassium: relationship to diuretic therapy. *Journal of Cardiovascular Pharmacology, 17,* 854-859.

Somers, V.K., Conway, J., Johnston, J., & Sleight, P. (1991.) Effects of endurance training on baroreflex sensitivity and blood pressure in borderline hypertension. *Lancet, 337,* 1363-1368.

Somjen, D., Binderman, I., Berger, E., & Harell, A. (1980.) Bone remodeling induced by physical stress is prostaglandin E2 mediated. *Biochimica et Biophysica Acta, 627,* 91-100.

Sonksen, P.H., Cuneo, R.C., Salomon, F., et al. (1991.) Growth hormone therapy in adults with growth hormone deficiency. *Acta Paediatrica Scandinavica, Supplement, 379,* 139-46.

Sorensen, T.K., Easterling, T.R., Carlson, K.L., Brateng, D.A., & Benedetti, T.J. (1992.) The maternal hemodynamic effect of indomethacin in normal pregnancy. *Obstetrics and Gynecology, 79,* 61-66.

Sorhede, M., Erlanson-Albertsson, C., Mei, J., Nevalainen, T., Aho, A., & Sundler, F. (1996.) Enterostatin in gut endocrine cells-immunocytochemical evidence. *Peptides, 17,* 609-614.

Sorrentino, V., & Reggiani, C. (1999.) Expression of the ryanodine receptor type 3 in skeletal muscle: a new partner in excitation-contraction coupling? *Trends in Cardiovascular Medicine, 9,* 54-61.

Sosa, R.E., Volpe, M., Marion, D.N., Atlas, S.A., Laragh, J.H., Vaughan, Jr., E.D., and Maack, T. (1986.) Relationship between renal hemodynamic and natriuretic effects of atrial natriuretic factor. *American Journal of Physiology, 250,* F529-F524.

Sothern, R.B., Levi, F., Haus, E., Halberg, F., & Hrushesky, W.J. (1989.) Control of a murine plasmacytoma with doxorubicin-cisplatin: Dependence on circadian state of treatment. *Journal of the National Cancer Institute, 81,* 135-145.

Sotsky, M.J., Shilo, S., & Shamoon, H. (1989.) Regulation of counterregulatory hormone secretion in man during exercise and hypoglycemia. *Journal of Clinical Endocrinology and Metabolism 68,* 9-16.

Soule, S.G., Macfarlane, P., Levitt, N.S., & Millar, R.P. (2001.) Contribution of growth hormone-releasing hormone and somatostatin to decreased growth hormone secretion in elderly men. *South African Medical Journal, 91,* 254-260.

Soultanakis, H.N., Artal, R., & Wiswell, R.A. (1996.) Prolonged exercise in pregnancy: Glucose homeostasis, ventilatory and cardiovascular responses. *Seminars in Perinatology, 20,* 315-327.

Southern, A.L., Tochimoto, S., Carmody, N.C., & Isurugi, K. (1965.) Plasma production rates of testosterone in normal adult men and women and in patients with the syndrome of feminizing testes. *Journal of Clinical Endocrinology and Metabolism, 25,* 1441-1450.

Souza, S.C., de Vargas, L.M., Yamamoto, M.T., Lien, P., Franciosa, M.D., Moss, L.G., & Greenberg, A.S. (1998.) Overexpression of perilipid A and B blocks the ability of tumor necrosis factor alpha to increase lipolysis in 3T3-L1 adipocytes. *Journal of Biological Chemistry, 273*, 24665-24669.

Sowers, M.R., Kshirsagar, A., Crutchfield, M.M., & Updike, S. (1992.) Joint influence of fat and lean body composition compartments on femoral bone mineral density in premenopausal women. *American Journal of Epidemiology, 136*, 257-265.

Spaanderman, M.E., Meertens, M., van Bussel, M., Ekhart, T.H., & Peeters, L.L. (2000.) Cardiac output increases independently of basal metabolic rate in early human pregnancy. *American Journal of Physiology, 278*, H1585-H1588.

Spagnoli, A., & Rosenfeld, R.G. (1997.) Insulin-like growth factor binding proteins.*Current Opinion in Endocrinology and Diabetes, 4*,1-9.

Sparling, P.B., & Cureton, K.J. (1983.) Biological determinants of the sex differences in 12 minute run performance. *Medicine and Science in Sports and Exercise, 15*, 218-223.

Sparling, P.B. (1980.) A meta-analysis of studies comparing maximal oxygen uptake in men and women. *Research Quarterly, 51*, 542-552.

Speake, P.F., Pirie, S.C., Kibble, J.D., Muneer, A., Taylor, D., & Green, R. (1993.) Dose-response effects of adrenergic and cholinergic stimulation on atrial natriuretic peptide secretion from beating isolated guinea-pig atria. *Clinical Science, 85*, 5-12.

Speiser, W., Langer, W., Pschaik, A., Selmayr, E., Ibe, B., Nowacki, P.E., & Muller-Berghaus, G. (1988.) Increased blood fibrinolytic activity after physical exercise: comparative study in individuals with different sporting activities and in patients after myocardial infarction taking part in a rehabilitation program. *Thrombosis Research, 51*, 543-555.

Spengler, C.M., Czeisler, C.A., & Shea, E.A. (2000.) An endogenous circadian rhythm of respiratory control in humans. *Journal of Physiology, 526*, 683-694.

Speranza, G., Verlato, G., & Albiero, A. (1998.) Autonomic changes during pregnancy: Assessment by spectral heart rate variability analysis. *Journal of Electrocardiology, 31*, 101-109.

Speth, R.C., Dinh, T.T., & Ritter, S. (1987.) Nodose ganglionectomy reduces angiotensin II receptor binding in the rat brain stem. *Peptides, 8*, 677-685.

Spiegel, K., Leproult, R., & Van Cauter, E. (1999.). Impact of sleep debt on metabolic and endocrine function. *Lancet, 354*, 1435-1439.

Spiegel, K., Leproult, R., Colecchia, E.F., L'Hermite-Baleriaux, M., Nie, Z., Copinschi, G., & Van Cauter, E. (2000.) Adaptation of the 24-h growth hormone profile to a state of sleep debt. *American Journal of Physiology, 279*, R874-R883.

Spiegelman, B.M., & Flier, J.S. (1996.) Adipogenesis and obesity: rounding out the big picture. *Cell, 87*, 377-389.

Spiegelman, B.M., & Flier, J.S. (2001.) Obesity and the regulation of energy balance. *Cell, 2001*, 531-534.

Spina, R.J., Turner, M.J., & Ehsani, A.A. (1998.) Beta-adrenergic-mediated improvement in left ventricular function by exercise training in older men. *American Journal of Pysiology, 274*, H397-H404.

Spinnewijn, W.E., Wallenburg, H.C., Struijk, P.C., & Lotgering, F.K. (1996.) Peak ventilatory responses during cycling and swimming in pregnant and nonpregnant women. *Journal of Applied Physiology, 81*, 738-742.

Spriet, L.L., Dyck, D.J., Cederblad, G., & Hultman, E. (1992.) Effects of fat availability on acetyl-CoA and acetylcarnitine metabolism in rat skeletal muscle. *American Journal of Physiology, 263*, C653-C659.

Sreter, F.A., Pinter, K., Jolesz, F., & Mabuchi (1982.) Fast to slow transformation of fast muscles in response to long-term phasic stimulation. *Experimental Neurology, 75*, 5-102.

Srivastava, M.C., Oakley, N.W., Tompkins, C.V., Sonksen, P.H., & Wynn, V. (1975.) Insulin metabolism, insulin sensitivity and hormonal responses to insulin infusion in patients taking oral contraceptive steroids. *European Journal of Clinical Investigation, 5*, 425-433.

Staessen, J., Fiocchi, R., Bouilon, R., Fagard, R., Hespel, P., Lijunen, P., Moerman, E., & Amery, A. (1988.) Effects of opiod antagonism on the haemodynamic and hormonal responses to exercise. *Clinical Science, 75*, 293-300.

Stafford, E.M., Weir, M.R., Pearl, W., Imai, W., Schydlower, M., & Gregory, G. (1989.) Sexual maturity rating: a marker for effects of pubertal maturation on the adolescent electrocardiogram. *Pediatrics, 83*, 565-569.

Stallknecht, B., Simonsen, L., Bulow, J., Vinten, J., & Galbo, H. (1995.) Effect of training on epinephrine-stimulated lipolysis determined by micro-dialysis in human adipose tissue. *American Journal of Physiology, 269*, E1059-E1066.

Stanbury, S.W., & Thomson, A.E. (1951.) Diurnal variations in electrolyte excretion. *Clinical Science, 10*, 267-293.

Stanek, K.A., Neil, J.J., Sawyer, W.B., & Loewy, A.D. (1984.) Changes in regional blood flow and cardiac output after glutamate stimulation of the A5 cell group. *American Journal of Physiology, 246*, H44-H51.

Stanhope, R., Bommen, M., & Brook, C.G.D. (1985.) Constitutional delay of growth and puberty in boys: the effect of a short course of treatment with fluoxymestrone. *Acta Pediatrica Scandinavica, 74*, 390-393.

Stanhope, R., Buchanan, C.R., Fenn, G.C., & Preece, M.A. (1988.) Double blind placebo controlled trial of low dose oxandrolone in the treatment of boys with constitutional delay of growth and puberty. *Archives of Diseases of Children, 63*, 501-505.

Stankiewicz-Chorouszucha, B., & Gorski, J. (1978.) Effect of beta drenergic blockade on intramuscular triglyceride mobilization during exercise. *Experientia 34*, 357-358.

Stanley, B.G., Daniel, D.R., Chin, A.S., & Leibowitz, S.F. (1985.) Paraventricular nucleus injection of peptide YY and neuropeptide Y perferentially enhance carbohydrate ingestion. *Peptides, 6*, 1205.

Stanley, B.G., Magdalin, W., Seirafi, A., Thomas, W.J., & Leibowitz, S.F. (1993.) The perifornical area: the major focus of (a) patchily distributed hypothalamic neuropeptide-Y sensitive feeding system(s). *Brain Research, 604*, 304-317.

Starling, R.D., Trappe, T.A., Parcell, A.C., Kerr, C.G., Fink, W.J., & Costill, D.L. (1997.) Effects of diet on muscle triglyceride and endurance performance. *Journal of Applied Physiology, 82*, 1185-1189.

Staron, R.S., & Hikida, R.S. (1992.) Histochemical, biochemical, and ultrastructural analyses of single human muscle fibers, with special reference to the C-fiber population. *Journal of Histochemistry and Cytochemistry, 40*, 563-568.

Staron, R.S., & Johnson, P. (1993.) Myosin polymorphism and differential expression in adult human skeletal muscle. *Comparative Biochemistry and Physiology, Part B, 106*, 463-475.

Steckelings, U., Lebrun, C., Qadri, F., Veltmar, A., & Unger, T. (1992.) Role of brain angiotensin in cardiovascular regulation. *Journal of Cardiovascular Pharmacology, 19* (Suppl. 6), S72-S79.

Steen, V.M., Holmsen, H., & Aarbakke, G. (1993.) The platelet-stimulating effect of adrenaline through alpha$_2$-adrenergic receptors requires simultaneous activation by a true stimulatory platelet agonist. Evidence that adrenaline per se does not induce human platelet activation in vitro. Thrombosis and Haemostasis, 70, 506-503.

Steffens A.B. (1975.) Influence of reversible obesity on eating behavior, blood glucose, and insulin in the rat. *American Journal of Physiology, 228*, 1738-1744.

Stein, P.K., Hagley, M.T., Cole, P.L., Domitrovich, P.P., Kleiger, R.E., & Rottman, J.N. (1999.) Changes in 24-hour heart rate variability during normal pregnancy. *American Journal of Obstetrics and Gynecology, 180*, 978-985.

Stein, P.K., Ehsani, A.A., Domitrovich, P.P., Kleiger, R.E., & Rottman, J.N. (1999.) Effect of exercise training on heart rate variability in healthy older adults. *American Heart Journal, 138*, 567-576.

Steinberg, H.O., Brechtel, G., Johnson, A., & Baron, A.D. (1994.) Insulin-mediated skeletal muscle vasodilation is nitric oxide dependent: a novel action of insulin to increase nitric oxide release. *Journal of Clinical Investigation, 94*, 1172-1179.

Stella, A., & Zanchetti, A. (1991.) Functional role of renal afferents. *Physiological Reviews, 71*, 659-682.

Stellar, E. (1954.) The physiology of motivation. *Psychological Reviews, 101*, 301-311 (Reprinted in 1994).

Stensel, D. J., Brooke-Wavell, K., Hardman, A. E., Jones, P. R. M., & Norgan, N. G. (1994.) The influence of a 1-year programme of brisk walking on endurance fitness and body composition in previously sedentary men aged 42-59 years. *European Journal of Applied Physiology and Occupational Physiology, 68*, 531-537.

Stenstad, P., & Eik-Nes, K.B. (1981.) Androgen metabolism in rat skeletal muscle in vitro. *Biochemica Biophysica Acta, 63*, 169-176.

Stephan, F.K. (1986a.) The role of period and phase in the interactions between feeding- and light-entrainable circadian rhythms. *Physiology and Behavior, 36*, 151-158.

Stephan, F.K. (1986b.) Interaction between light- and feeding-entrainable circadian rhythm in the rat. *Physiology and Behavior, 38*, 127-133.

Stephan, F.K. (1986c.) Coupling between feeding- and light-entrainable circadian pacemakers in the rat. *Physiology and Behavior, 38*, 537-544.

Stephens, T., & Caspersen, C.J. (1994.) The demography of physical activity. In C. Bouchard, R.J. Shephard and T. Stephens (Eds.), *Physical activity, fitness, and health: International proceedings and consensus statement* (pp. 204-213). Champaign, IL: Human Kinetics Publishers.

Stephens, T.J., Chen, Z.P., Canny, B.J., Michell, B.J., Kemp, B.E., & McConell, G.K. (2002.) Progressive increase in human skeletal muscle AMPKalpha2 activity and ACC phosphorylation during exercise. *American Journal of Physiology, 282*, E688-E694.

Stephenson, L.A., & Kolka, M.A. (1985.) Menstrual cycle phase and time of day alter reference signal controlling arm blood flow and sweating. *American Journal of Physiology, 249*, R186-R191.

Stephenson, L.A., & Kolka, M.A. (1988.) Plasma volume during heat stress and exercise in women. *European Journal of Applied Physiology and Occupational Physiology, 57*, 373-381.

Stephenson, L.A., Kolka, M.A., & Wilkerson, J.E. (1982.) Metabolic and thermoregulatory responses to exercise during the human menstrual cycle. *Medicine and Science in Sports and Exercise, 14*, 270-275.

Stephenson, L.A., Kolka, M.A., Francesconi, R., & Gonzalez, R.R. (1989.) Circadian variations in plasma renin activity, catecholamines and aldosterone during exercise in women. *European Journal of Applied Physiology and Occupational Physiology, 58*, 756-764.

Steppan, C.M., Bailey, S.T., Bhat, S., Brown, E.J., Banerjee, R.R., Wright, C.M., et al. (2001.) The hormone resistin links obesity to diabetes. *Nature, 409*, 307-312.

Stern, S., & Tzivoni, D. (1974.) Early detection of silent ischemic heart disease by 24-hour electrocardiographic monitoring of active subjects. *British Heart Journal, 36*, 481-486.

Sternfeld, B. (1992.) Cancer and the protective effect of physical activity: The epidemiological evidence. *Medicine and Science in Sports and Exercise, 24*, 1195-1209.

Stevens, G.H., Graham, T.E., & Wilson, B.A. (1987.) Gender differences in cardiovascular and metabolic responses to cold and exercise. *Canadian Journal of Physiology and Pharmacology, 65*, 165-171.

Stevenson, L. (1997.) Exercise in pregnancy. Part 1: Update on pathophysiology. *Canadian Family Physician, 43*, 97-104.

Stolwijk, J.A.J., & Hardy, J.D. (1977.) Control of body temperature. In D.H. Lee, H.L. Falk, and S.D. Murphy (Eds.), *Reactions of environmental agents: Handbook of physiology, section 9* (pp. 45-68). Bethesda: American Physiological Society.

Storlien, L., James, D.E., Burleigh, K.M., Chisholm, D.J., & Kraegen, E.W. (1986.) Fat feeding causes widespread in vivo insulin resistance, decreased energy expenditure, and obesity in rats. *American Journal of Physiology, 251*, E567-E583.

Storlien, L., Jenkins, A.B., Chisholm, D.J., Pascoe, W.S., Khouri, S., & Kraegen, E.W. (1991.) Influence of dietary fat composition on development of insulin resistance in rats. *Diabetes, 40*, 280-289.

Strack, A.M., Sawyer, W.B., Platt, K.B., & Loewy, A.D. (1989.) CNS cell groups regulating the sympathetic nervous outflow to adrenal gland as revealed by transneuronal cell body labelling with pseudorabies virus. *Brain Research, 491*, 276-296.

Stralfors, P., Bjorgell, P., & Belfrage, P. (1984.) Hormonal regulation of hormone-sensitive lipase in intact adipocytes: identification of phosphorylated sites and effects on the phosphorylation by lipolytic hormones and insulin. *Proceedings of the National Academy of Sciences of the United States of America, 81*, 3317-3321.

Straumann, E., Keller, U., Kraenzlin, M., Girard, J., Thelin, A., Arnaud, A., Perruchoud, A., & Stauffacher, W. (1988.) Interaction of cortisol and epinephrine in the regulation of leucine kinetics in man. *Experientia, 44*, 176-178.

Strauss, R.H., & Yesalis, C.E. (1991.) Anabolic steroids in the athlete. *Annual Reviews of Medicine, 42*, 449-457.

Strauss, R.H., Liggett, M.T., & Lanese, R.R. (1985.) Anabolic steroid use and perceived effects in ten weight-trained women athletes. *Journal of American Medical Association, 253*, 2871-2873.

Stroud, R.M., & Finer-Moore, J. (1985.) Acetylcholine receptor structure, function, and evolution. *Annual Reviews of Cellular Biology, 1*, 317-351.

Stubbs, R.J., Ritz, P., Coward, W.A., & Prentice, A.M. (1995.) Covert manipulation of the ratio of dietary fat to carbohydrate and energy density: effects on food intake and energy balance in free-living men eating ad libitum. *American Journal of Clinical Nutrition, 62*, 330-337.

Stumvoll, M., Chintalapudi, U., Perriello, G., Welle, S., Gutierrez, O., & Gerich, J. (1995.) Uptake and release of glucose by the human kidney: postabsorptive rates and responses to epinephrine. *Journal of Clinical Investigation, 96*, 2528-2533.

Sturis, J., Scheen, A.J., Leproult, R., Polonsky, K.S., & Van Cauter, E. (1995.) 24-hour glucose profiles during continuous or oscillatory insulin infusion: demonstrations of the functional significance of ultradian insulin oscillations. *Journal of Clinical Investigation, 95*, 1464-1471.

Suda, T., Takahashi, N., & Martin, T.J. (1992.) Modulation of osteoclast differentiation. *Endocrine Reviews, 13*, 66-80.

Sugden, M.S., Howard, R.M., Munday, M.R., & Holness, M.J. (1993.) Mechanisms involved in the coordinate regulation of strategic enzymes of glucose metabolism. *Advances in Enzyme Research, 33*, 71-95.

Sugden, P.H., & Bogoyevitch, M.A. (1996.) Endothelin-1-dependent signaling pathways in the myocardium. *Trends in Cardiovascular Medicine, 6*, 87-94.

Suh, B.Y., Liu, J.H., Berga, S.L., Quigley, M.E., Laughlin, G.A., & Yen, S.S. (1988.) Hypercortisolism in patients with functional hypothalamic-amenorrhea. *Journal of Clinical Endocrinology and Metabolism, 66*, 733-739

Suh, H.H., Hudson, P., Fannin, R., Boulom, K., McMillian, M.K., Poisner, A.M., and Hong, J.S. (1992a.) Prolonged stimulation of bovine adrenal chromaffin cells with arachidonic acid and prostaglandin E2 increases expression of the proenkephalin gene and the secretion of [Met5]-enkephalin. *Journal of Pharmacology and Experimental Therapeutics, 263*, 527-532.

Suh, H.H., Mar, E.C., Hudson, P.M., McMillian, M.K., & Hong, J.S. (1992b.) Effects of [Sar1]angiotensin II on proenkephalin gene expression and secretion of [Met5]enkephalin in bovine adrenal medullary chromaffin cells. *Journal of Neurochemistry, 59*, 993-998.

Sul, H.S., Latasa, M.J., Moon, Y., & Kim, K.H. (2000.) Regulation of the fatty acid synthase promoter by insulin. *Journal of Nutrition, 139* (2S Suppl.), 315S-320S.

Sullivan, M.L., Martinez, C.M., & Gallagher, E.J. (1999.) Atrial fibrillation and anabolic steroids. *Journal of Emergency Medicine, 17*, 851-857.

Sullivan, M.L., Martinez, C.M., Gennis, P., & Gallagher, E.J. (1998.) The cardiac toxicity of anabolic steroids. *Progress in Cardiovascular Diseases, 41*, 1-15.

Surbey, G.D., Andrew, G.M., Cervenko, F.W., & Hamilton, P.P. (1984.) Effects of naloxone on exercise performance. *Journal of Applied Physiology, 57*, 674-679.

Sutton, J.R., Coleman, M.J., & Casey, J. (1974.) Adrenocortical contribution to serum androgens during physical exercise. *Medicine and Science in Sports and Exercise, 6*, 72.

Sutton, J.R., Coleman, M.J., Casey, J., & Lazarus, L. (1973.) Androgen responses during physical exercise. *British Medical Journal, 1*, 520-522.

Sutton, J.R., & Lazarus, L. (1976.) Growth hormone in exercise: comparison of physiological and pharmacological stimuli. *Journal of Applied Physiology, 41*, 523-527.

Suzuki, H., Cleemann, L., Abernethy, D.R., & Morad, M. (1998.) Glutathione is a cofactor for H$_2$O$_2$-mediated stimulation of Ca^{2+}-induced Ca^{2+} release in cardiac myocytes. *Free Radical Biology and Medicine, 24*, 318-325.

Suzuki, M., Ide, K., & Saitoh, S. (1983.) Diurnal changes in glycogen stores in liver and skeletal muscle of rats in relation to the feed timing of sucrose. *Journal of Nutritional Science and Vitaminology, 29*, 545-552.

Svedenhag, J., Lithell, H., Juhlin-Dannfelt, A., & Henriksson, J. (1983.) Increase in skeletal muscle lipoprotein lipase following endurance training in man. *Atherosclerosis, 49*, 203-207.

Swaab, D.F., & Fliers, E. (1985.) A sexually dimorphic nucleus in human brain. *Science, 228*, 1112-1115.

Swenson, E. J., Jr., & Conlee, R. K. (1979.) Effects of exercise intensity on body compostition in adult males. *Journal of Sports Medicine, 19*, 323-326.

Swoap, S.J., Haddad, F., Bodell, P., & Baldwin, K.M. (1994.) Effect of chronic energy deprivation on cardiac thyroid hormone receptor and myosin isoform expression. *American Journal of Physiology, 266*, E254-260.

Swynghedauw, D. (1986.) Development and functional adaptations of contractile proteins in cardiac and skeletal muscles. *Physiological Reviews, 65*, 710-771.

Symons, J.D., & Stebbins, C.L. (1996.) Effects of angiotensin II receptor blockade during exercise: comparison of losartan and saralasin. *Journal of Cardiovascular Pharmacology, 28,* 223-231.

Syms, A.J., Norris, J.S., Panko, W.B., & Smith, R.G. (1985.) Mechanism of androgen-receptor augmentation. analysis of receptor synthesis and degradation by the density-shift technique. *Journal of Biological Chemistry, 260,* 455-461.

Szabo, G., Dallmann, G., Muller, G., Patthy, L., Soller, M., & Varga, L. (1998.) A deletion in the myostatin gene causes the compact (Cmpt) hypermuscular mutation in mice. *Mammalian Genome, 9,* 671-672.

Szalay, K., Razga, Z., & Duda, E. (1997.) TNF inhibits myogenesis and downregulates the expression of myogenic regulatory factors MyoD and myogenin. *European Journal of Cell Biology, 74,* 391-398.

Szollar, S.M., Dunn, K.L., Brandt, S., & Fincher, J. (1997.) Nocturnal polyuria and antidiuretic hormone levels in spinal cord injury. *Archives of Physical Medicine and Rehabilitation, 78,* 455-458.

Szymanski, L.M., & Pate, R.R. (1994.) Effects of exercise intensity, duration, and time of day on fibrinolytic activity in physically active men. *Medicine and Science in Sports and Exercise, 26,* 1102-1108.

Taborsky, G.J. Jr., Dunning, B.E., Havel, P.J., Ahren, B., Kowalyk, S., Boyle, M.R., Verchere, C.B., Baskin, D.G., & Mundinger, T.O. (1999.) The canine sympathetic neuropeptide galanin: a neurotransmitter in pancreas, a neuromodulator in liver. Hormone and Metabolic Research, 31, 351-354.

Tafari, N., Naeye, R.L., & Gobezie, A. (1980.) Effects of maternal undernutrition and heavy physical work during pregnancy on birth weight. *British Journal of Obstetrics and Gynecology, 87,* 222-226.

Tai, M.M., Castillo, P., & Pi-Sunyer, F.X. (1991.) Meal size and frequency: effect on the thermic effect of food. *American Journal of Clinical Nutrition, 54,* 783-787.

Takahashi, Y., Ohno, H., & Misawa, M. (1996.) Beta 2- but not beta 1-adrenoceptors mediate the adrenergic component of reflex tracheal dilatation during bronchoconstriction in guinea pigs in vivo. *Research Communications in Molecular Pathology and Pharmacology, 93,* 301-318.

Takamata, A., Mack, G.W., Gillen, C.M., & Nadel, E.R. (1994.) Sodium appetite, thirst, and body fluid regulation in humans during rehydration without sodium replacement. *American Journal of Physiology, 266,* R1493-R1502.

Takashima, N., & Higashi, T. (1994.) Change in fibrinolytic activity as a parameter for assessing local mechanical stimulation during physical exercise. *European Journal of Applied Physiology and Occupational Physiology, 68,* 445-449.

Takeda, H., Chodak, G., Mutchnik, S., Nakamoto, T., & Chang, C. (1990.) Immuno-histochemical localization of androgen receptors with mono- and polyclonal antibodies to androgen receptor. *Journal of Endocrinology, 126,* 17-25.

Tam, C.S., Heersche, J.M.N., Murray, T.M., & Parsons, J.A. (1982.) Parathyroid resorptive action: differential effects of intermittent and continuous administration. *Endocrinology 110,* 506-512.

Tamaki, T., Uchiyama, S., Uchiyama, Y., Akatsuka, A., Yoshimura, S., Roy, R.R., & Edgerton, V.R. (2000.) Limited myogenic response to a single bout of weight-lifting exercise in old rats. *American Journal of Physiology, 278,* C1143-C1152.

Tamaoki, J., Chiyotani, A., Sakai, N., & Konno, K. (1993.) Stimulation of ciliary motility mediated by atypical beta-adrenoceptor in canine bronchial epithelium. *Life Sciences, 53,*1509-1515.

Tamir, Y., & Bengal, E. (2000.) Phosphoinositide 3-kinase induces the transcriptional activity of MEF2 proteins during muscle differentiation. *Journal of Biological Chemistry, 275,* 34424-34432.

Tanaka, H., Cleroux, J., de Champlain, J., Ducharme, J.R., & Collu, R. (1986.) Persistent effects of a marathon run on the pituitary-testicular axis. *Journal of Endocrinological Investigation, 9,* 97-101.

Tanaka, K., Hassall, G.J.S., & Burnstock, G. (1993.) Distribution of intracardiac neurons and nerve terminals that contain a marker for nitric oxide, NADPH-diaphorase, in the guinea-pig heart. *Cell and Tissue Research, 273,* 293-300.

Tanaka, K., Inoue, S., Nagase, H., & Takamura, Y. (1990.) Modulation of arginine-induced insulin and glucagon secretion by the hepatic vagus nerve in the rat: effects of celiac vagotomy and administration of atropine. *Endocrinology, 127,* 2017-2023.

Tannenbaum, G.S. (1998.) Hypothalamic control mechanisms of sexually dimorphic growth hormone secretory patterns in the rat. In J.D. Veldhuis & A. Giustina (Eds.), *Sex-steroid interactions with growth hormone* (pp. 133-143). New York: Springer.

Tanner, J.M. (1962.) *Growth at adolescence,* 2nd ed. Oxford: Blackwell Scientific Publications.

Tanner, J.M. (1974.) Sequence and tempo in the somatic changes in puberty. In M.M. Grumbach, G.D. Grave, & F.E. Mayer (Eds), *The control of the onset of puberty* (pp.448-470). New York: John Wiley and Sons.

Tanner, J.M. (1981.) *A history of the study of human growth.* Cambridge: Cambridge University Press.

Tanner, J.M., Hughes, P.C.R., & Whitehouse, R.H. (1981.) Radiographically determined widths of bone, muscle and fat in the upper arm and calf from age 3-18 years. *Annals of Human Biology, 8,* 495-518.

Tanner, J.M., & Whitehouse, R.H. (1975.) A note on the bone age at which patients with true isolated growth hormone deficiency enter puberty. *Journal of Clinical Endocrinology and Metabolism, 41,* 788-790.

Tanner, J.M., Whitehouse, R.H., & Takaishi, M. (1966.) Standards from birth to maturity for height, weight, height velocity, and weight velocity: British children 1965. *Archives of Diseases of Children, 41,* 613-635.

Tarkovacs, G., Blandizzi, C., & Vizi, E.S. (1994.) Functional evidence that alpha 2A-adrenoceptors are responsible for antilipolysis in human abdominal fat cells. *Naunyn-Schmiedebergs Archives of Pharmacology, 349,* 34-41.

Tarnopolsky, M.A., Atkinson, S.A., Phillips, S.M., & MacDougal, J.D. (1995.) Carbohydrate loading and metabolism during exercise in men and women. *Journal of Applied Physiology, 78,* 1360-1368.

Tarnopolsky, L.J., MacDougal, J.D., & Atkinson, S.A. (1988.) Influence of protein intake and training status on nitrogen balance and lean body mass. *Journal of Applied Physiology, 64,* 187-193.

Tarnopolsky, L.J., MacDougal, J.D., Atkinson, S.A., Tarnopolsky, M.A., & Sutton, J.R. (1990.) Gender differences in substrate for endurance exercise. *Journal of Applied Physiology, 68,* 302-308.

Tarnopolsky, M.A., Atkinson, S.A., MacDougall, J.D., Chesley, A., Phillips, S., & Schwarcz, H.P. (1992.) Evaluation of protein requirements for trained strength athletes. *Journal of Applied Physiology, 73,* 1986-1995.

Tarnopolsky, M.A., Bosman, M., Macdonald, J.R., Vandeputte, D., Martin, J., & Roy, B.D. (1997.) Post-exercise protein-carbohydrate and carbohydrate supplements increase muscle glycogen in men and women. *Journal of Applied Physiology, 83,* 1877-1883.

Taskinen, M.-R., Nikkila, E.A., Rehunen, S., & Gordin, A. (1980.) Effect of acute vigorous exercise on lipoprotein lipase activity of adipose tissue and skeletal muscle in physically active men. *Artery, 6,* 471-483.

Tatar, P., & Vigas, M. (1980.) Role of alpha-1 and alpha-2 adrenergic receptors in the GH and prolactin response to insulin-induced hypoglycemia in man, *Neuroendocrinology, 39,* 275-281.

Tate, C.A., & Holtz, R.W. (1998.) Gender and fat metabolism during exercise: a review. *Canadian Journal of Applied Physiology, 23,* 570-582.

Tateishi, J., & Faber, J.E. (1995.) Inhibition of arteriole alpha 2- but not alpha 1-adrenoceptor constriction by acidosis and hypoxia in vitro. *American Journal of Physiology, 268,* H2068-H2076.

Taubman, M.B., Berk, B.C., Izumo, S., Tsuda, T., Alexander, R.W., & Nadal-Ginard, B. (1989.) Angiotensin II induces c-fos mRNA in aortic smooth muscle: Role of Ca^{2+} mobilization and protein kinase C activation. *Journal of Biological Chemistry, 264,* 526-530.

Taylor, S.R., Rogers, G.G., & Driver, H.S. (1997.) Effects of training volume on sleep, psychological, and selected physiological profiles of elite female swimmers. *Medicine and Science in Sports and Exercise, 29,* 688-693.

Tchernof, A., Calles-Escandon, J., Sites, C.K., & Poehlman, E.T. (1998.) Menopause, central body fatness, and insulin resistance: effects of hormone replacement therapy. *Coronary Artery Disease, 9,* 503-511.

Teitelman, A.M., Welch, L.S., Hellenbrand, K.G., & Bracken, A.B. (1990.) Effect of maternal work capacity on preterm birth and low birth weight. *American Journal of Epidemiology, 131,*104-113.

Tempel, D.L., & Leibowitz, S.F. (1993.) Glucocorticoid receptors in PVN: interactions with NE, NPY, and Gal in relation to feeding. *American Journal of Physiology, 265,* E794-E800.

Tenaglia, S.A., McLellan, T.M., & Klentrou, P.P. (1999.) Influence of menstrual cycle and oral contraceptives on tolerance to uncompensable heat stress. *European Journal of Applied Physiology and Occupational Physiology, 80,* 76-83.

te Pas, M.F., de Jong, P.R., & Verburg, F.J. (2000.) Glucocorticoid inhibition of C2C12 proliferation rate and differentiation capacity in relation to mRNA levels of the MRF gene family. *Molecular Biology Reports, 27*, 87-98.

Terzic, A., Puceat, M., Vassort, G., & Vogel, S.M. (1993.) Cardiac alpha 1-adrenoceptors: an overview. *Pharmacological Reviews, 45*, 147-175.

Thaik, C.M., Calderone, A., Takahashi, N., & Colucci, W.S. (1995.) Interleukin-1 beta modulates the growth and phenotype of neonatal rat cardiac myocytes. *Journal of Clinical Investigation, 96*, 1093-1099.

Theron, J.J., Oosthuizen, J.M., & Rautenbach, M.M. (1984.) Effect of physical exercise on plasma melatonin levels in normal volunteers. *South African Medical Journal. 66*, 838-841.

Thiblin, I., Runeson, B., & Rajs, J. (1999.) Anabolic androgenic steroids and suicide. *Annuls of Clinical Psychiatry, 11*, 223-231.

Thirone, A.C., Paez-Espinosa, E.V., Carvalho, C.R., & Saad, M.J. (1998.) Regulation of insulin-stimulated tyrosine phosphorylation of Shc and IRS-1 in the muscle of rats: effect of growth hormone and epinephrine. *FEBS Letters, 421*, 191-196.

Thissen, J.P., Underwood, L.E., & Ketelslegers, J.M. (1999.) Regulation of insulin-like growth factor-I in starvation and injury. *Nutrition Reviews, 57*, 167-176.

Thomas, G.D., Hansen, J., & Victor, R.G. (1994.) Inhibition of alpha2-adrenergic vasoconstriction during contraction of glycolytic, not oxidative, rat hindlimb muscle. *American Journal of Physiology, 266*, H920-H929.

Thomas, M., Langley, B., Berry, C., Sharma, M., Kirk, S., Bass, J., & Kambadur, R. (2000.) Myostatin, a negative regulator of muscle growth, functions by inhibiting myoblast proliferation. *Journal of Biological Chemistry, 275*, 40235-40243.

Thompson, C., Childs, P.A., Martin, N.J., Rodin, I., & Smythe, P.J. (1997.) Effects of morning phototherapy on circadian markers in seasonal affective disorder. *British Journal of Psychiatry, 170*, 431-435.

Thompson, D.A., Wolfe, L.A., & Eikelboom, R. (1988.) Acute effects of exercise intensity on appetite in young men. *Medicine and Science in Sports and Exercise, 20*, 222-227.

Thompson, D.R., Blandford, D.L., Sutton, T.W., & Marchant, P.R. (1985.) Time of onset of chest pain in acute myocardial infarction. *International Journal of Cardiology, 7*, 139-146.

Thompson, G.G., Blanksby, B.A., & Doran, G. (1974.) Maturity and performance in age group competitive swimmers. *Australian Journal of Physical Education, 64*, 21-25.

Thompson, P.D., Funk, E.J., Carleton, R.A., & Sturner, W.Q. (1982.) Incidence of death during jogging in Rhode Island from 1975 through 1980. *Journal of the American Medical Association, 247*, 2535-2538.

Thong, F.S.L., & Graham, T.E. (1999.) Leptin and reproduction: Is it a critical link between adipose tissue, nutrition, and reproduction? *Canadian Journal of Applied Physiology, 24*, 317-336.

Thorburn, G.D. (1974.) The role of the thyroid gland and the kidneys in fetal growth. *CIBA Foundation Symposium, 27*, 185-214.

Thorell, A., Hirshman, M.F., Nygren, J., Jorfeldt, L., Wojtaszewski, J.F.P., Dufresne, S.D., Horton, E.S., Ljungqvist, O., & Goodyear, L.J. (1999.) Exercise and insulin cause GLUT-4 translocation in human skeletal muscle. *American Journal of Physiology, 277*, E733-E741.

Thorn, P., Gerasimenko, O., & Petersen, O.H. (1994.) Cyclic ADP-ribose regulation of ryanodine receptors involved in agonist-evoked cytosolic Ca²⁺ oscillations in pancreatic acinar cells. *EMBO Journal, 13*, 2038-2043.

Thrasher, T.N., Brown, C.J., Keil, L.C., & Ramsay, D.J. (1980.) Thirst and vasopressin release in the dog: An osmoreceptor or sodium receptor mechanism? *American Journal of Physiology, 238*, R333-R339.

Thresher, R.J., Vitaterna, M.H., Miyamoto, Y., Kazantsev, A., Hsu, D.S., Petit, C., et al. (1998.) Role of mouse cryptochrome blue-light photoreceptor in circadian photoresponses. *Science, 282*, 1490-1494.

Thuma, J.R., Gilders, R., Verdun, M., & Loucks, A.B. (1995.) Circadian rhythm of cortisol confounds cortisol responses to exercise: implications for future research. *Journal of Applied Physiology, 78*, 1657-1664.

Tice, L.W., & Creveling, C.R. (1986.) Electron microscopic identification of adrenergic nerve endings on thyroid epithelial cells. *Endocrinology, 97*, 1123-1129.

Tidgren, B., Hjemdahl, P., Theodorsson, E., & Nussberger, J. (1991.) Renal neurohormonal and vascular responses to dynamic exercise in humans. *Journal of Applied Physiology, 70*, 2279-2286.

Timio, M., Lippi, G., Gentili, S., Quintaliani, G., Verdura, C., Monarca, C., Saronio, P., & Timio, F. (1997.) Blood pressure trend and cardiovascular events in nuns in a secluded order: a 30-year follow-up study. *Blood Pressure, 6*, 81-87.

Timiras, P.S., & Nzekwe, E.U. (1989.) Thyroid hormones and nervous system development. *Biology of the Neonate, 55*, 376-385.

Timiras, P.S. (1978.) Biological perspectives of aging: in search of a master plan. *American Scientist, 66*, 605-613.

Timiras, P.S. (1983.) Neuroendocrinology of aging: retrospective, current, and prospective views. In J. Meites (Ed.), *Neuroendocrinology of aging* (pp.5-30). New York: Plenum Press.

Tipton CM. (1999.) Exercise training for the treatment of hypertension: a review. *Clinical Journal of Sport Medicine, 9*, 104.

Tofler, G.H., Brezinski, D.A., Schafer, A.I., Czeisler, C.A., Rutherford, J.D., Willich, S.N., et al. (1987.) Concurrent morning increase in platelet aggregability and the risk of myocardial infarction and sudden cardiac death. *New England Journal of Medicine, 316*, 1514-1518.

Toft, P., Helbo-Hansen, H.S., Tonnesen, E., Lillevang, S.T., Rasmussen, J.W., & Christensen, N.J. (1994.) Redistribution of granulocytes during adrenaline infusion and following administration of cortisol in healthy volunteers. *Acta Anaesthesiologica Scandinavica, 38*, 254-258.

Toni, D., Argentino, C., Gentile, M., Sacchetti, M.L., Girmenia, F., Millefiorini, E., et al. (1991.) Circadian variation in the onset of acute cerebral ischemia: Ethiopathogenetic correlates in 80 patients given angiography. *Chronobiology International, 8*, 321-326.

Toor, M., Massry, S., Katz, A.I., & Agmon, J. (1965.) Diurnal variations in the composition of blood and urine of man living in hot climate. *Nephron, 2*, 334-354.

Toppila, J., Asikainen, M., Alanko, L., Turek, F.W., Stenberg, D., & Porkka-Heiskanen, T. (1996.) The effect of REM sleep deprivation on somatostatin and growth hormone-releasing hormone gene expression in the rat hypothalamus. *Journal of Sleep Research, 5*, 115-122.

Torgan, C.E., Etgen, G.J., Jr., Kang, H.Y., & Ivy, J.L. (1995.) Fiber type-specific effects of clenbuterol and exercise training on insulin-resistant muscle. *Journal of Applied Physiology, 79*, 163-167.

Torgan, C.E., Etgen, G.J., Jr., Brozinick, J.T., Jr., Wilcox, R.E., & Ivy, J.L. (1993.) Interaction of aerobic exercise training and clenbuterol: effects on insulin resistant muscle. *Journal of Applied Physiology, 75*, 1471-1476.

Torii, M., Nayakama, H., & Sasaki, T. (1995.) Thermoregulation of exercising men in the morning rise and evening fall phases of internal temperature. *British Journal of Sports Medicine, 29*, 113-120.

Torreilles, J. (2001.) Nitric oxide: one of the more conserved and widespread signaling molecules. *Frontiers in Bioscience, 6*, D1161-D1172.

Torsvall, L., Akerstedt, T., & Lindbeck, G. (1984.) Effect on sleep stages and EEG power density of different degrees of exercise in fit subjects. *Electroencephalographic Clinical Neurophysiology, 57*, 347-353.

Tortorella, L.L., Milasincic, D.J., & Pilch, P.F. (2001.) Critical proliferation-independent window for basic fibroblast growth factor repression of myogenesis via the p42/p44 MAPK signaling pathway. *Journal of Biological Chemistry, 276*, 13709-13717.

Toshinai, K., Mondal, M.S., Nakazato, M., Date, Y., Murakami, N., Kojima, M., Kangawa, K., & Matsukura, S. (2001.) Upregulation of ghrelin expression in the stomach upon fasting, insulin-induced hypoglycemia, and leptin administration. *Biochemical and Biophysical Research Communications, 281*, 1220-1225.

Tosini, G., & Menaker, M. (1996.) Circadian rhythms in cultured mammalian retina, *Science, 272*, 419-421.

Trasforini, G., Margutti, A., Portaluppi, F., Menegatti, M., Ambrosio, M.R., Bagni, B., Pansini, R., & Degli Uberti, E.C. (1991.) Circadian profile of plasma calcitonin gene-related peptide in healthy man. *Journal of Clinical Endocrinology and Metabolism, 73*, 945-951.

Treadway, J.L., & Young, J.C. (1989.) Decreased glucose uptake in the fetus after maternal exercise. *Medicine and Science in Sports and Exercise, 21*,140-145,

Tremblay, A., Coveney, S., Despres, J.P., Nadeau, A., & Prudhomme, D. (1992.) Increased resting metabolic rate and lipid oxidation in exercise-trained individuals: evidence for a role of beta adrenergic stimulation. *Canadian Journal of Physiology and Pharmacology, 70*, 1342-1347.

Trouche, D., Masutani, H., Groisman, R., Robin, P., Lenormand, J.L., & Harel-Bellan, A. (1995.) Myogenin binds to and represses c-fos promoter. *FEBS Letters, 361*, 140-144.

Tsai, K.S., Lin, J.C., Chen, C.K., Cheng, W.C., & Yang, C.H. (1997.) Effect of epinephrine and exogenous glucocorticoid on serum level of intact parathyroid hormone. *Journal of Sports Medicine 18*, 583-587.

Tsai, M.-J., & O'Malley, B.W. (1994.) Molecular mechanisms of action of steroid/thyroid receptor superfamily members. *Annual Review of Biochemistry, 63*, 451-486.

Tsai, T.H., Scheving, L.E., Marques, N., & Sanchez de la Pena, S. (1987.) Circadian infraradian intermodulation of corneal epithelial mitoses in adult female rats. *Progress in Clinical and Biological Research, 227A*, 193-198.

Tsao, T.S., Burcelin, R., & Charron, M.J. (1996.) Regulation of hexokinase II gene expression by glucose flux in skeletal muscle. *Journal of Biological Chemistry, 271*, 14959-14963.

Tsementzis, S.A., Gill, J.S., Hitchcock, E.R., Gill, S.K., & Beevers, D.G. (1985.) Diurnal variation of and activity during the onset of stroke. *Neurosurgery, 17*, 901-904.

Tsetsonis, N.V., Hardman, A.E., & Mastana, S.S. (1997.) Acute effects of exercise on postprandial lipemia: a comparative study in trained and untrained middle-aged women. *American Journal of Clinical Nutrition, 65*, 525-533.

Tsika, R.W., Herrick, R.E., & Baldwin, K.M. (1987.) Interaction of compensatory overload and hindlimb suspension on myosin isoform expression. *Journal of Applied Physiology, 62*, 2180-2186.

Tsuchihashi, T., Abe, I., Tsukashima, A., Kobayashi, K., & Fujishima, M. (1990.) Effects of meals and physical activity on blood pressure variability in elderly patients: a preliminary study. *American Journal of Hypertension, 3*, 943-946.

Tsukazaki, K., Nikami, H., Shimizu, Y., Kawada, T., Yoshida, T., & Saito, M. (1995.) Chronic administration of beta-adrenergic agonists can mimic the stimulative effect of cold exposure on protein synthesis in rat brown adipose tissue. *Journal of Biochemistry, 117*, 96-100.

Tsunekawa, B., Wada, M., Ikeda, M., Uchida, H., Naito, N., & Honjo, M. (1999.) The 20-kilodalton (kDa) human growth hormone (hGH) differs from the 22-kDa hGH in the effect on the human prolactin receptor. *Endocrinology, 140*, 3909-3918.

Tucker, K. (1995.) Micronutrient status and aging. *Nutrition Reviews, 53*, S9-S15.

Tuggle, D.W., & Horton, J.W. (1986.) Cardiovascular effects of physiological doses of beta endorphin. *Circulatory Shock, 18*, 215-225.

Tunstall, R.J., Mehan, K.A., Wadley, G.D., Collier, G.R., Bonen, A., Hargreaves, M., & Cameron-Smith, D. (2002.) Exercise training increases lipid metabolism gene expression in human skeletal muscle. *American Journal of Physiology, 283*, E66-E72.

Tur, E., Tamir, A., & Guy, R.H. (1992.) Cutaneous blood flow in gestational hypertension and normal pregnancy. *Journal of Investigative Dermatology, 99*, 310-314.

Turcotte, L.P., Richter, E.A., & Kiens, B. (1992.) Increased plasma FFA uptake and oxidation during prolonged exercise in trained vs. untrained humans. *American Journal of Physiology, 262*, E791-E799.

Turek, F.W., & Zee, P.C. (1999.) *Regulation of sleep and circadian rhythms.* New York: Marcel Dekker, Inc.

Turton, M.B., & Deegan, T. (1974.) Circadian variations of plasma catecholamine, cortisol and immunoreactive insulin concentrations in supine subjects. *Clinica Chimica Acta, 55*, 389-397.

Tvede, N., Kappel, M., Halkjaer-Kristensen, J., Galbo, H., & Pedersen, B.K. (1993.) The effect of light, moderate and severe bicycle exercise on lymphocyte subsets, natural and lymphokine activated killer cells, lymphocyte proliferative response and interleukin 2 production. *International Journal of Sports Medicine, 14*, 275-282.

Tvede, N., Kappel, M., Klarlund, K., Duhn, S., Halkjaer-Kristensen, J., Kjaer, M., Galbo, H., & Pedersen, B.K. (1994.) Evidence that the effect of bicycle exercise on blood mononuclear cell proliferative responses and subsets is mediated by epinephrine. *International Journal of Sports Medicine, 15*, 100-104.

Tvede, N., Pedersen, B.K., Hansen, F.R., Bendix, T., Christensen, L.D., Galbo, H., and Halkjaer-Kristensen, J. (1989.) Effect of physical exercise on blood mononuclear cell subpopulations and in vitro proliferative responses. *Scandinavian Journal of Immunology, 29*, 383-389.

Twidale, N., Taylor, S., Heddle, W.F., Ayres, B.F., & Tonkin, A.M. (1989.) Morning increase in the time of onset of sustained ventricular tachycardia. *American Journal of Cardiology, 12*, 1204-1206.

Uchida, H., Banba, S., Wada, M., Matsumoto, K., Ikeda, M., Naito, N., Tanaka, E., & Honjo, M. (1999.) Analysis of binding properties between 20 kDa human growth hormone (hGH) and hGH receptor (hGHR): the binding affinity for hGHR extracellular domain and mode of receptor dimerization. *Journal of Molecular Endocrinology, 23*, 347-353.

Udagawa, N., Takahashi, N., Akatsu, T., Tanaka, H., Sasaki, T., Nishihara, T., & Koga, (1990.) Origin of osteoclasts: mature monocytes and macrophages are capable of differentiating into osteoclasts under a suitable microenvironment prepared by bone marrow-derived stromal cells. *Proceedings of the National Academy of Sciences of the United States of America, 87*, 7260-7264.

Uehara, A., Sekiya, C., Takasugi, Y., Namiki, M., & Arimura, A. (1989.) Anorexia induced by interleukin 1: involvement of corticotropin-releasing factor. *American Journal of Physiology, 257*, R613-R617.

Ullrich, A., & Schlessinger, J. (1990.) Signal transduction by receptors with tyrosine kinase activity. *Cell, 61*, 203-212.

Undesser, K.P., Hasser, E.M., Haywood, J.R., Johnson, A.K., & Bishop, V.S. (1985.) Interactions of vasopressin with the area postrema in arterial baroreflex function in conscious rabbits. *Circulation Research, 56*, 410-417.

Urban, R.J., Bodenburg, Y.H., Gilkison, C., Foxworth, J.,Coggan, A.R., Wolfe, R.R., & Ferrando, A. (1995.) Testosterone administration to elderly men increases skeletal muscle strength and protein synthesis. *American Journal of Physiology, 269*, E820-E826.

Urhausen, A., & Kindermann, W. (1987.) Behaviour of testosterone, sex hormone binding globulin (SHBG) and cortisol before and after a triathlon. *International Journal of Sports Medicine, 8*, 305-308.

Vaccarino, F.J., Sovran, P., Baird, J.P., & Ralph, M.R. (1995.) Growth hormone-releasing hormone mediates feeding-specific feedback to the suprachiasmatic circadian clock. *Peptides, 16*, 595-598.

Vahl, N., Jorgensen, J.O.L., Skjaerback, C., Veldhuis, J.D., Orskov, H., & Christiansen, J. (1997a.) Abdominal adiposity rather than age and sex predicts the mass and patterned regularity of growth hormone secretion in mid-life healthy adults. *American Journal of Physiology, 272*, E1108-E116.

Vahl, N., Moller, N., Lauritzen, T., Christiansen, J.S., & Jorgensen, J.O. (1997b.) Metabolic effect and pharmacokinetics of a growth hormone pulse in healthy adults: relation to age, sex, and body composition. *Journal of Clinical Endocrinology and Metabolism, 82*, 3612- 3618.

Valdez, M.R., Richardson, J.A., Klein, W.H., & Olson, E.N. (2000.) Failure of Myf5 to support myogenic differentiation without myogenin, MyoD, and MRF4. *Developmental Biology, 219*, 287-298.

Vale, W., Vaughan, J., Yamamoto, G., Spiess, J., & Rivier, J. (1983.) Effects of synthetic human pancreatic (tumor) GH releasing factor and somatostatin, triiodothyronine and dexamethasone on GH secretion in vitro. *Endocrinology, 112*, 1553-1555.

Van Aggel-Leijssen, D.P.C., Van Bak, M.A., Tenenbaum, R., Campfield, L.A., & Saris, W.H.M. (1999.) Regulation of average 24 h human plasma leptin level; the influence of exercise and physiological changes in energy balance. *International Journal of Obesity and Related Metabolic Disorders, 23*, 151-158.

Van Beaumont, W., Strand, J.C., Petrovsky, J.S., Hipskind, S.G., & Greenleaf, J.E. (1973.) Changes in total plasma content of electrolytes and proteins with maximal exercise. *Journal of Applied Physiology, 34*, 102-106.

Van Cauter, E., Leproult, R., & Plat, L. (2000.) Age-related changes in slow wave sleep and REM sleep and relationships with growth hormone and cortisol levels in healthy men. *Journal of the American Medical Association, 284*, 861-868.

Van Cauter, E., Plat, L., Leproult, R., & Copinschi, G. (1998.) Alterations of circadian rhythmicity and sleep in aging: endocrine consequences. *Hormone Research, 49*, 147-152.

Van Cauter, E., Plat, L., Scharf, M.B., Leproult, R., & Cespedes, S. (1997.) Simultaneous stimulation of slow-wave sleep and growth hormone secretion by gamma-hydroxybutyrate in normal young men. *Journal of Clinical Investigation, 100*, 745-753.

Van Cauter, E., Sturis, J., Bryne, M.M., Blackman, J.D., Leproult, R., Ofek, G., L'Hermite-Baleriaux, M., Refetoffl, S., Turek, F., & Van Reeth, O. (1994.) Demonstration of rapid light-induced advances and delays of the human circadian clock using hormonal phase markers. *American Journal of Physiology, 266*, E953-E963.

Van den Bergh, R., Oelofsen, W., Naude, R.J., & Terblanche, S.E. (1992.) The effect of exercise and in vivo treatment with ACTH and norepinephrine on the lipolytic responsiveness of guinea pig (Cavia porcellus) adipose tissue. *Comparative Biochemistry and Physiology – B, 101,* 553-557.

Van den Burg, P.J., Van Vliet, M., Moster, D.W.L., & Huisveld, I.A. (1995.) Unbalanced haemostatic changes following strenuous physical exercise: a study in young sedentary males. *European Heart Journal, 16,* 1995-2001.

Vandenburgh, H., & Kaufman, S. (1979.) In vitro model for stretch-induced hypertrophy of skeletal muscle. *Science, 203,* 265-268.

Vandenburgh, H.H. (1987.) Motion into mass: How does tension stimulate muscle growth?. M*edicine and Science in Sports and Exercise, 19* (Suppl.), S149-S149.

Vander, A.J., Sherman, J.H., & Luciano, D.S. (2001.) *Human physiology: The mechanisms of body function,* 7th ed. New York: McGraw-Hill.

van der Meulen, M.C., Ashford, M.W., Kiratli, B.J., Bachrach, L.K., & Carter, D.R. (1996.) Determinants of femoral geometry and structure during adolescent growth. *Journal of Orthopaedic Research, 14,* 22-29.

Van Ermen, A., & Fraeyman, N. (1994.) Influence of aging on the alpha 1-receptor mediated glycogenolysis in rat hepatocytes. *Journal of Gerontology, 49,* B12-B17.

Van Harmelen,V., Lonnqvist, F., Thorne, A., Wennlund, A., Large, V., Reynisdottir, S, & Arner, P. (1997.) Noradrenaline-induced lipolysis in isolated mesenteric, omental and subcutaneous adipocytes from obese subjects. *International Journal of Obesity and Related Metabolic Disorders, 21,* 972-979.

VanHelder, W.P., Casey, K., & Radomski, M.W. (1987.) Regulation of growth hormone during exercise by oxygen demand and availability. *European Journal of Applied Physiology, 56,* 628-632.

VanHelder, W.P., Goode, R.C., & Radomski, M.W. (1984.) Effect of anaerobic and aerobic exercise of equal duration and work expenditure on plasma growth hormone levels. *European Journal of Applied Physiology, 52,* 255-257.

VanHelder, W.P., Kofman, E., & Tremblay, M.S. (1991.) Anabolic steroids in sport. *Canadian Journal of Sport Sciences, 16,* 248-257.

Van Hook, J.W., Gill, P., Easterling, T.R., Schmucker, B., Carlson, K., & Benedetti, T.J. (1993.) The hemodynamic effects of isometric exercise during late normal pregnancy. *American Journal of Obstetrics and Gynecology, 169,* 870-873.

Vanhoutte, P.M., & Miller, V.M. (1989.) Alpha 2-adrenoceptors and endothelium-derived relaxing factor. *American Journal of Medicine, 87* (Suppl. 3C),1S-5S.

Van Nieuwenhoven, F.A., van den Vusse, G.J., & Glatz, J.F.C. (1996.) Membrane-associated and cytoplasmic fatty acid-binding proteins. *Lipids, 31,* S223-S227.

van Raaij, J.M. (1995.) Energy requirements of pregnancy for healthy Dutch women. *European Journal of Obstetrics, Gynecology, and Reproductive Biology, 61,* 7-13.

van Raaij, J.M., Vermaat-Miedema, S.H., Schonk, C.M., Peek, M.E., & Hautvast, J.G.(1987.) Energy requirements of pregnancy in the Netherlands. *Lancet, 2,* 953-955.

Van Reeth O. (1998.) Sleep and circadian disturbances in shift work: strategies for their management. *Hormone Research, 49,* 158-62.

Van Reeth, O., Olivares, E., Turek, F.W., Granjon, L., & Mocaer, E. (1998.)Resynchronisation of a diurnal rodent circadian clock accelerated by a melatonin agonist. *Neuroreport, 9,* 1901-1905.

Van Reeth, O., Olivares, E., Zhang, Y., Zee, P.C., Mocaer, E., Defrance, C., & Turek, F.W. (1997.) Comparative effects of a melatonin agonist on the circadian system in mice and Syrian hamsters. *Brain Research, 762,* 185-194.

Van Reeth, O., Sturis, J., Byrne, M.M., Blackman, J.D., L'Hermite-Baleriaux, M., Leproult, R., Oliner, C., Refetoff, S., Turek, F.W., & Van Cauter, E. (1994.) Nocturnal exercise phase delays circadian rhythms of melatonin and thyrotropin secretion in normal men. *American Journal of Physiology, 266,* E964-E974.

Van Someren, E.J. (2000.) Circadian rhythms and sleep in human aging. *Chronobiology International, 17,* 233-243.

Van Stapel, F., Waebens, M., Van Hecke, P., Decanniere, C., & Stalmans, W. (1991.) Modulation of maximal glycogenolysis in perfused rat liver by adenosine and ATP. *Biochemical Journal, 277,* 597-602.

Veglio, F., Molino, P., Cat Genova, G., Melchio R., Rabbia, F., Grosso, T., Martini, G., & Chiandussi, L. (1999.) Impaired baroreflex function and arterial compliance in primary aldosteronism. *Journal of Human Hypertension, 13,* 29-36.

Veldhuis, J.D. (1992.) Deconvolution analysis of hormone data. *Methods in Enzymology, 210,* 539-575.

Veldhuis, J.D. (1996.) Neuroendocrine mechanisms mediating awakening of the human gonadotropic axis in puberty. *Pediatric Nephrology, 10,* 304-317.

Veldhuis, J.D., Evans, W.S., Demers, L.M., Thorner, M.O., Wakat, D., & Rogol, A.D. (1985.) Altered neuroendocrine regulation of gonadotropin secretion in women distance runners. *Journal of Clinical Endocrinology and Metabolism, 61,* 557-563.

Veldhuis, J.D., Iranmanesh, A., Lizzarralde, G., & Urban, R.J. (1994.) Combined deficits in the somatotropic and gonadotropic axes in healthy older men: an appraisal of neuroendocrine mechanisms by deconvolution analysis. *Neurobiology of Aging, 15,* 509-517.

Veldhuis, J.D., & Johnson, M.L. (1988.) Cluster analysis: a simple, versatile and robust algorithm for endocrine pulse detection. *American Journal of Physiology, 250,* E486-E493.

Veldhuis, J.D., King, J.C., Urban, R.J., Rogol, A.D., Evans, W.S., Kolp, L.A., & Johnson, M.L. (1987.) Operating characteristics of the male hypothalamo-pituitary gonadal axis: pulsatile release of testosterone and follicular stimulating hormone and their temporal coupling with luteinizing hormone. *Journal of Clinical Endocrinology and Metabolism, 65,* 929-941.

Veldhuis, J.D., Metzger, D.L., Martha, P.M., Jr., Mauras, N., Kerrigan, J.R., Keenan, B., Rogol, A.D., & Pincus, S.M. (1997.) Estrogen and testosterone, but not a nonaromatizable androgen, direct network integration of the hypothalamo-somatotrope (growth hormone)-insulin-like growth factor I axis in the human: evidence from pubertal pathophysiology and sex-steroid hormone replacement. *Journal of Clinical Endocrinology and Metabolism, 82,* 3414-3420.

Vergauwen, L., Richter, E.A., & Hespel, P. (1997.) Adenosine exerts a glycogen-sparing action in contracting rat skeletal muscle. *American Journal of Physiology, 272,* E762-E768.

Verhaar. H.J.J., Damen, C.A., Duursma, S.A., & Scheven, B.A.A. (1994.) A comparison of the action of progestins and estrogen on the growth and differentiation of normal adult human osteoblast-like cells in vitro. *Bone, 15,* 307-312.

Verma, S., Arikawa, E., Yao, L. Laher, I., & McNeill, J.H. (1998.) Insulin-induced vasodilation is dependent on tetrahydrobiopterin synthesis. *Metabolism: Clinical and Experimental, 47,* 1037-1039.

Verney, E.B. (1947.) The antidiuretic hormone and the factors which determine its release. *Proceedings of the Royal Society of London, Series B, 135,* 25-106.

Vernon, R.G. (1996.) GH inhibition of lipogenesis and stimulation of lipolysis in sheep adipose tissue: involvement of protein serine phosphorylation and dephosphorylation and phospholipase C. *Journal of Endocrinology, 150,* 129-140.

Vessby, B., Selinus, I., & Lithell, H. (1985.) Serum lipoprotein and lipoprotein lipase in overweight,type II diabetics during and after supplemented fasting. *Arteriosclerosis, 5,* 93-100.

Vicini, P., Bonnadonna, R.C., Utriainen, T., Nuutila, P., Raitakari, M., Yki-Jarvinen, H., & Cobelli, C. (1997.) Estimation of blood flow heterogeneity in human skeletal muscle from positron emission tomography data. *Annals of Biomedical Engineering, 25,* 906-910.

Victor, R.G., & Seals, D.R. (1989.) Reflex stimulation of sympathetic outflow during rhythmic exercise in humans. *American Journal of Physiology. 127,* H2017-H2024.

Victor, R.G., Seals, D.R., & Mark, A.L. (1987.) Differential control of heart rate and sympathetic nerve activity during dynamic exercise: insight from intraneural recordings in humans. *Journal of Clinical Investigation, 79,* 508-516.

Vidal-Puig, A., Solanes, G., Grujic, D., Flier, J.S., & Lowell, B.B. (1997.) UCP-3: an uncoupling protein homologue expressed preferentially and abundantly in skeletal muscle and brown adipose tissue. *Biochemical and Biophysical Research Communications, 235,* 79-82.

Vihko, V.J., Apter, D.L., Pukkala, E.I., Oinonen, M.T., Hakulinen, T.R., & Vihko, R.K. (1992.) Risk of breast cancer among female teachers of physical education and languages. *Acta Oncologica, 31,* 201-204.

Vikman, H.L., Savola, J.M., Raasmaja, A., & Okisalo, J.J. (1996.) Alpha 2A-adrenergic regulation of cyclic AMP accumulation and lipolysis in human omental and subcutaneous adipocytes. *International Journal of Obesity and Related Metabolic Disorders, 20,* 185-189.

Vinals, F., Ferre, J., Fandos, C., Santalucia, T., Testar, X., Palacin, M., & Zorzano, A. (1977.) Cyclic adenosine 3',5'-monophosphate regulates GLUT4 and GLUT1 glucose transporter expression and stimulates transcriptional activity of the GLUT1 promoter in muscle cells. *Endocrinology, 138,* 2521-2529.

Vinson, G.P., Ho, M.M., Puddlefoot, J.R., Teja, R., Barker, S., Kapas, S., & Hinson, J.P. (1995.) The relationship between the adrenal tissue renin-angiotensin system, internalization of the type I angiotensin II receptor (AT1) and angiotensin II function in the rat zona glomerulosa cell. *Advances in Experimental Medicine and Biology, 377*, 319-329.

Vinten, J., & Galbo, H. (1983.) Effect of physical training on transport and metabolism of glucose in adipocytes. *American Journal of Physiology, 244*, E129-E134.

Vinten, J., Petersen, L.N., Sonne, B., & Galbo, H. (1985.) Effect of physical training on glucose transporters in fat cell fractions, *Biochimica et Biophysica Acta, 841*, 223-227.

Viru, A. (1985.) *Hormones in muscular activity*. Boca Raton: CRC Press.

Viru, A., Litvinova, L., Viru, M., & Smirnova, T. (1994.) Glucocorticoids in metabolic control during exercise: alanine metabolism. *Journal of Applied Physiology, 76*, 801-805.

Vock, R., Hoppeler, H., Claassen, H., Wu, D.X., Billeter, R., Weber, J.M., & Taylor C.R. (1996.) Design of the oxygen and substrate pathways. VI. Structural basis of intracellular substrate supply to mitochondria in muscle cells. *Journal of Experimental Biology, 199*,

Vogel, R., Corretti, M., & Plotnick, G. (1997.) Effect of a single high-fat meal on endothelial function in healthy subjects. *American Journal of Cardiology, 79*, 350-354.

Volker, H. (1927.)Uber die tagesperiodischen Schwankungen einiger Lebensvorgange des Menschen. *Pflugers Archiv fur die gesamte Physiologie des Menschen und der Tiere, 215*, 43-77.

Vollenweider, P., Tappy,L., Randin, D., Schneiter, P., Jequier, E., Nicod, P., & Scherrer, U. (1993.) Differential effects of hyperinsulinemia and carbohydrate metabolism on sympathetic nerve activity and muscle blood flow in humans. *Journal of Clinical Investigation, 92*, 147-154.

Vollmer, R.R. (1996.) Selective neural regulation of epinephrine and norepinephrine cells in the adrenal medulla—cardiovascular implications. *Clinical and Experimental Hypertension, 18*, 731-751.

Vriend, J., Borer, K.T., & Thliveris, J.A. (1987.) Melatonin: its antagonism of thyroxine's antisomatotrophic activity in male Syrian hamsters. *Growth, 51*, 35-43.

Vriend, J., Sheppard, M.S., & Borer, K.T. (1990.) Melatonin increases serum growth hormone and insulin-like growth factor I (IGF-I) levels in male Syrian hamsters via hypothalamic neurotransmitters. *Growth, Development and Aging, 54*, 165-171.

Vuolteenaho, O., Koistinen, P., Martikkala, U., Takala, T., & Leppaluoto, J. (1992.) Effect of physical exercise in hypobaric conditions on atrial natriuretic peptide secretion. *American Journal of Pysiology, 263*, R647-R652.

Vuori, I., Makarainen, M., & Jaaskelainen, A. (1978.) Sudden death and physical activity. *Cardiology, 63*, 287-304.

Vuori, I., Urponen, H., Hasan, J., & Partinen, M. (1988.) Epidemiology of exercise effects on sleep. *Acta Physiologica Scandinavica. Supplementum, 574*, 3-7.

Wada, H., Zile, M.R., Ivester, C.T., Cooper, G., IV, & McDermott, P.J. (1996.) Comparative effects of contraction and angiotensin II on growth of adult feline cardiomyocytes in primary culture. *American Journal of Physiology, 271*, H29-H37.

Wade, C.E., & Claybaugh, J.R. (1980.) Plasma renin activity, vasopressin concentration, and urinary excretory responses to exercise in men. *Journal of Applied Physiology, 49*, 930-936.

Wade, C.E., & Morey-Holton, E. (1998.) Alteration of renal function of rats following spaceflight. *American Journal of Physiology, 275*, R1058-R1065.

Wade, C.E. (1984.) Response, regulation, and actions of vasopressin during exercise: a review. *Medicine and Science in Sports and Exercise, 16*, 506-511.

Wade, C.E., Ramee, S.R., Hunt, M.M., & White, C.J. (1987.) Hormonal and renal responses to converting enzyme inhibition during maximal exercise. *Journal of Applied Physiology, 63*, 1796-1800.

Wahren, J., Felig, P., Ahlborg, G., & Jorfeldt, L. (1971.) Glucose metabolism during leg exercise in man. *Journal of Clinical Investigation, 50*, 2715.

Wahren, J., Felig, P. & Hagenfeldt, L. (1978.) Physical exercise and fuel homeostasis in diabetes. *Diabetologia, 14*, 213-222.

Wahren, J., Felig, P., Hendler, R., & Ahlborg, G. (1973.) Glucose and amino acid metabolism during recovery after exercise. *Journal of Applied Physiology, 34*, 838-845.

Wahren, J., Hagenfeldt, L., & Felig, P. (1975.) Splanchnic and leg exchange of glucose, amino acids, and free fatty acids during exercise in diabetes mellitus. *Journal of Clinical Investigation, 55*, 1303-1314.

Wahrenberg, H., Lonnqvist, F., & Arner, P. (1989.) Mechanisms underlying regional differences in lipolysis in human adipose tissue. *Journal of Clinical Investigation, 84*, 458-467.

Wahrenberg, H., Bolinder, J., & Arner, P. (1991.) Adrenergic regulation of lipolysis in human fat cells during exercise, *European Journal of Clinical Investigation, 21*, 534-541.

Wahrenberg, H., Engfeldt, P., Bolinder, J., & Arner, P.(1987.) Acute adaptation in adrenergic control of lipolysis during physical exercise in humans. *American Journal of Physiology, 253*, E383-E390.

Wahrenberg, H., Lonnqvist, F., Hellner, J., & Arner, P. (1992.) Importance of beta-adrenoceptor function in fat cells for lipid mobilization. *European Journal of Clinical Investigation, 22*, 412-419.

Wakade, A.R., Garcia, A.G., & Kirpekar, S.M. (1975.) Effect of castration on the smooth muscle cells of the internal sex organs of the rat: influence of the smooth muscle on the sympathetic neurons innervating the vas deferens, seminal vesicle and coagulating gland. *Journal of Pharmacology and Experimental Therapeutics, 193*, 424-434.

Waldrop, T.G., Bauer, R.M., & Iwamoto, G.A. (1988.) Microinjection of GABA antagonists into the posterior hypothalamus elicits locomotor activity and a cardiorespiratory activation. *Brain Research, 444*, 84-94.

Waldstreicher, J., Duffy, J.F., Brown, E.N., Rogacz, S., Allan, J.S., & Czeisler, C.A. (1996.) Gender differences in the temporal organization of prolactin (PRL) secretion: evidence for a sleep-independent circadian rhythm of circulating PRL levels—a clinical research study. *Journal of Clinical Endocrinology and Metabolism, 81*, 1483-1487.

Walker, J.M., & Berger R.J. (1980.) Sleep as an adaptation for energy conservation functionally related to hibernation and shallow torpor. *Progress in Brain Research, 53*, 255-278.

Walker, J.M., Floyd, T.C., Fein, G., Cavness, C., Lualhati, R., & Feinburg, I. (1978.) Physical fitness, exercise, and human sleep. *Journal of Applied Physiology, 44*, 945-951.

Walker, R.B., & Edwards, C.R.W. (1994.) Licorice-induced hypertension and syndromes of apparent mineralocorticoid excess. *Endocrinology and Metabolism Clinics of North America, 23*, 359-377.

Wall, P.D., & Melzack R. (Eds.) (1985.) *Textbook of pain*. Edinburg: Churchill, Livingstone.

Wallace, J.D., Cuneo, R.C., Bidlingmaier, M., et al. (2001.) The response of molecular isoforms of growth hormone to acute exercise in trained adult males. *Journal of Clinical Endocrinology and Metabolism, 86*, 200-206.

Wallace, J.P., Bogle, P.G., King, B.A., Krasnoff, J.B., & Jastremski, C.A. (1999.) The magnitude and duration of ambulatory blood pressure reduction following acute exercise. *Journal of Human Hypertension, 13*, 361-366.

Wallen, N.H., Goodall, A.H., Li, N., & Hjemdahl, P. (1999.) Activation of haemostasis by exercise, mental stress and adrenaline: Effects on platelet sensitivity to thrombin and thrombin generation. *Clinical Science, 97*, 27-35.

Wallin, B.G., Esler, M., Dorward, P., Eisenhofer, G., Ferrier, C., Westerman, R., & Jennings, G. (1992.) Simultaneous measurements of cardiac noradrenaline spillover and sympathetic outflow to skeletal muscle in humans. *Journal of Physiology, 453*, 45-58.

Walls, E.K., Phillips, R.J., Wang, F.B., Holst, M.C., & Powley, T.L. (1995.) Suppression of meal size by intestinal nutrients is eliminated by celiac vagal deafferentation. *American Journal of Physiology, 269*, R1410-R1419.

Walsh, B.T., Puig-Antich, J., Goetz, R., Gladis, M., Novacenko, H., & Glassman, A.H. (1984.) Sleep and hormone secretion in women athletes. *Electroencephalography and Clinical Neurophysiology, 57*, 528-531.

Wang, D.S., Sato, K., Demura, H., Kato, Y., Maruo, N., & Miyachi, Y. (1999.) Osteo-anabolic effects of human growth hormone with 22K- and 20K Daltons on human osteoblast-like cells. *Endocrine Journal, 46*, 125-132.

Wang, J., Akabayashi, A., Dourmashkin, J., Yu, H.J., Alexander, J.T., Chae, H.J., & Leibowitz, S.F. (1998a.) Neuropeptide Y in relation to carbohydrate intake, corticosterone and dietary obesity. *Brain Research, 802*, 75-88.

Wang, J., Akabayashi, A., Yu, H.J., Dourmashkin, J., Alexander, J.T., Silva, I., Lighter, J, & Leibowitz,S.F. (1998b.) Hypothalamic galanin: control by signals of fat metabolism. *Brain Research, 804*, 7-20.

Wang, J.S., & Cheng, L.J. (1999.) Effect of strenuous, acute exercise on alpha 2-adrenergic agonist-potentiated platelet activation. *Arteriosclerosis, Thrombosis and Vascular Biology, 19*, 1559-1565.

Wang, T.W., & Apgar, B.S. (1998.) Exercise during pregnancy. *American Family Physician, 57*, 1846-52, 1857.

Ward, Jr., J.M., & Armitrage, K.B. (1981.) Circannual rhythms of food consumption, body mass, and metabolism in yellow-bellied marmots. *Comparative Biochemical Physiology, 69A*, 621-626.

Ward, O.B., Orth, J.M., & Weisz, J.A. (1983.) A possible role of opiates in modifying sexual differentiation. *Monographs in Neural Sciences, 9*, 194-200.

Warren, M.P. (1980.) The effects of exercise on pubertal progression and reproductive function in girls. *Journal of Clinical Endocrinology and Metabolism, 51*, 1150-1157.

Warren, M.P., & Constantini, N.W. (Eds.) (2000.) *Sports endocrinology*. Totowa: Humana Press.

Wasserman, D.H., Lacy, D.B., & Bracy, D.P. (1993.) Relationship between arterial and portal vein immunoreactive glucagon during exercise. *Journal of Applied Physiology, 75*, 724-729.

Watanabe, T., Fujioka, T., Hashimoto, M., & Nakamura, S. (1998.) Stress and brain angiotensin II receptors. *Critical Reviews in Neurobiology, 12*, 305-317.

Waters, D.J., Caywood, D.D., Trachte, G.J., Turner, R.T., & Hodgson, S.F. (1991.) Immobilization increases bone prostaglandin E: effect of acetylsalicylic acid on disuse osteoporosis studied in dogs. *Acta Orthopedica Scandinavica, 62*, 238-243.

Watson, P.A. (1996.) Mechanical activation of signalling pathways in the cardiovascular system. *Trends in Cardiovascular Medicine, 6*, 73-79.

Watson, W.J., Katz, V.L., Hackney, A.C., Gall, M.M., & McMurray, R.G. (1991.) Fetal responses to maximal swimming and cycling exercise during pregnancy. *Obstetrics and Gynecology, 77*, 382-386.

Watt, P.W., Finley, E., Cork, S., Clegg, R.A., & Vernon, R.G. (1991.) Chronic control of the β- and α₂-adrenergic systems of sheep adipose tissue by growth hormone and insulin. *Biochemistry Journal, 273*, 39-42.

Watts, A.G., & Swanson, L.W. (1987.) Efferent projections of the suprachiasmatic nucleus: II. Studies using retrograde transport of fluorescent dyes and simultaneous peptide immunohistochemistry in the rat. *Journal of Comparative Neurology, 258*, 230-252.

Watts, A.G., Swanson, L.W., Sanchez-Watts, G. (1987.) Efferent projections of the suprachiasmatic nucleus: I. Studies using anterograde transport of Phaseolus vulgaris leucoagglutinin in the rat. *Journal of Comparative Neurology, 258*(2), 204-229.

Weaver, D.R. (1998.) The suprachiasmatic nucleus: a 25-year retrospective. *Journal of Biological Rhythms, 13*, 100-112.

Webb, M.L., Wallace, J.P., Hamill, C., Hodgson, J.L., & Marshaly, M. (1984.) Serum testosterone concentration during two hours of moderate intensity treadmill running in trained men and women. *Endocrine Research, 10*, 27-38.

Webb, W.B. (1971.) Sleep behavior as a biorhythm. In Colquhoun, W.P. (Ed.), *Biological rhythms and human performance* (pp.149-177). London: Academic Press.

Webber, J., & MacDonald, I.A. (2000.) Signalling in body-weight homeostasis: neuroendocrine efferent signals. *Proceedings of the Nutrition Society, 59*, 397-404.

Webster, J.M., Heseltine, L., & Taylor, R. (1996.) In vitro effect of adenosine agonist GR 79236 on the insulin sensitivity of glucose utilization in rat soleus and human rectus abdominis muscle. *Biochimica Biophysica Acta, 1316*, 109-113.

Wehling, M., Cai, B., & Tidball, J.G. (2000.) Modulation of myostatin expression during modified muscle use. *FASEB Journal, 14*, 103-110.

Wehr, T.A. (1992.) Seasonal vulnerability to depression. Implications for etiology and treatment. *Encephale, 18*, 479-483.

Weidman, P., Hasler, L., Gnadinger, M.P., Lang, R.E., Uehlinger, D.E., Shaw, S., Rascher, W., & Reubi, F.C. (1986.) Blood levels and renal effects of atrial natriuretic peptide in normal men. *Journal of Clinical Investigation, 77*, 734-742.

Weight, L.M., Alexander, D., & Jacobs, P. (1991.) Strenuous exercise: analogous to the acute-phase response? *Clinical Science, 81*, 677-683.

Weinstock, C., Konig, D., Harnischmacher, R., Keul, J., Berg, A., & Nothoff, H. (1997.) Effect of exhaustive exercise stress on the cytokine response. *Medicine and Science in Sports and Exercise, 29*, 345-354.

Weis J. (1994.) Jun, Fos, MyoD1, and myogenin proteins are increased in skeletal muscle fiber nuclei after denervation. *Acta Neuropathologica, 87*, 63-70.

Weisberger, D., Redlin, U., & Mrosovsky, N. (1997.) Lengthening of circadian period in hamsters by novelty-induced wheel running. *Physiology andBehavior, 62*, 759-765.

Weiss,C., Seitel. G., & Bartsch, P. (1998.) Coagulation and fibrinolysis after moderate and very heavy exercise in healthy male subjects. *Medicine and Science in Sports and Exercise, 30*, 246-251.

Weiss, L.W., Cureton, K.J., & Thompson, F.N. (1983.) Comparison of serum testosterone and androstenedione responses to weight lifting in men and women. *European Journal of Applied Physiology, 50*, 413-419.

Weiss, R. E., & Refetoff, S. (1996.) Effect of thyroid hormone on growth: lessons from the syndrome of resistance to thyroid hormone. *Endocrinology and Metabolism Clinics of North America, 25*, 719-730.

Weissberger, A.J., & Ho, K.K.Y. (1993.) Activation of the somatotropic axis in adult males: evidence for the role of aromatization. *Journal of Clinical Endocrinology and Metabolism, 76*, 1407-1412.

Weitzman, E.D., Czeisler, C.A., & Moore-Ede, M.C. (1979.) Sleep-wake, neuro-endocrine and body temperature circadian rhythms under entrained and non-entrained (free-running) conditions in man. In M. Suda, O. Hayaishi & H. Nakagawa (Eds.), *Biological rhythms and their central mechanism*. (pp. 199-227). New York: Elsevier-North Holland.

Wellman, P.J. (2000.) Norepinephrine and the control of food intake. *Nutrition, 16*, 837- 842.

Wells, C.L. (1985.) *Women, sports and performance: a physiological perspective*. Champaign: Human Kinetics Publishers.

Weltman, A., Pritzlaff, C.J., Wideman, L., Weltman, J.Y., Blumer, J.L., Abbott, R.D., Hartman, M.L.,and Veldhuis, J.D. (2000.) Exercise-dependent growth hormone release is linked to markers of heightened central adrenergic outflow. *Journal of Applied Physiology, 89*, 629-635.

Weltman, A., Weltman, J.Y., Schurrer, G., Evans, W.S., Veldhuis, J.D., & Rogol, A.D. (1992.) Endurance training amplifies the pulsatile release of growth hormone: effects of training intensity. *Journal of Applied Physiology, 72*, 2188-2196.

Wenger, C.B., Roberts, M.F., Stolwijk, J.A., & Nadel, E.R. (1976.) Nocturnal lowering of thresholds for sweating and vasodilation. *Journal of Applied Physiology, 41*, 15-19.

Wennink, J.M.B., Delemarre-Van de Waal, H.A., Shoemaker, R., Blaauw, G., Van de Braken, C., & Shoemaker, J. (1991.) Growth hormone secretion patterns in relation to LH and estradiol secretion throughout normal female puberty. *Acta Endocrinologica, 123*, 129-135.

Wentz, M., Berend, J.Z., Lynch, N.A., Chappell, S., & Hackney, A.C. (1997.) Substrate oxidation at rest and during exercise: effects of menstrual cycle phase and diet composition. *Journal of Physiology and Pharmacology, 48*, 851-860.

Wever, R. (1979.) The circadian system of man: results of experiments under temporal isolation. New York: Springer-Verlag.

Wheeler, G.D., Wall, S.R., Belcastro, A.N., & Cumming, D.C. (1984.) Reduced serum testosterone and prolactin levels in male distance runners. *Journal of the American Medical Association, 252*, 514-516.

Wheeler, G.D., Wall, S.R., Belcastro, A.N., Conger, P., & Cumming, D.C. (1986.) Are anorexic tendencies prevalent in the habitual runner? *British Journal of Sports Medicine, 20*, 77-81.

Wheeler, M.T., Snyder, E.C., Patterson, M.N., & Swoap, S.J. (1999.) An E-box within the MHC IIB gene is bound by MyoD and is required for gene expression in fast muscle. *American Journal of Physiology, 276*, C1069-C1078.

Whipp, B.J., & Ward, S.A. (1992.) Will women soon outrun men? *Nature, 355*, 25.

White, D.P., Douglas, N.J., Pickett, C.K., Weil, J.V., & Azilich, C.W. (1983.) Sexual influence on the control of breathing. *Journal of Applied Physiology, 54*, 874-879.

White, K.A., & Marletta, M.A. (1992.) Nitric oxide synthase is a cytochrome P-450 type hemoprotein. *Biochemistry, 31*, 6627-6631.

Widdowson, E.M., & McCance, R.A. (1960.) Some effects of accelerating growth. I. General somatic development. *Proceedings of the Royal Society of London, Series B, 152*, 188-206.

Wiebe, C.G., Gledhill, N., Warburton, D.E., Jamnik, V.K., & Ferguson, S. (1998.) Exercise cardiac function in endurance-trained males versus females. *Clinical Journal of Sports Medicine, 8*, 272-279.

Wiedenman, J.L., Rivera-Rivera, I.,Vyas, D., Tsika, G., Gao, L., Sheriff-Carter, K., Wang, X., Kwan, L.Y., & Tsika, R.W. (1996.) Beta-MHC and SMLC1 transgene induction in overloaded skeletal muscle of transgenic mice. *Journal of Applied Physiology, 270*, C1111-C1121.

Wilcox, C.S., Welch, W.J., Murad, F., Gross, S.S., Taylor, G., & Levi, R. (1992.) Nitric oxide synthase in macula densa regulates glomerular capillary pressure. *Proceedings of the National Academy of Sciences of the United States of America, 89*, 11993-11997.

Wilkerson, J.E., Horvath, S.M., & Gutin, B. (1980.) Plasma testosterone during treadmill exercise. *Journal of Applied Physiology, 49*, 249-253.

Wilkerson, J.E., Gutin, B., & Horvath, S.M. (1977.) Exercise-induced changes in blood, red cell, and plasma volumes in man. *Medicine and Science in Sports, 9*, 155-158.

Wilkins, L., & Richter, C.P. (1940.) A great craving for salt by a child with cortico-adrenal insufficiency. *Journal of the American Medical Association, 114*, 866-868.

Wilkinson, K.D. (1999.) Ubiquitin-dependent signaling: the role of ubiquitination in the response of cells to their environment. *Journal of Nutrition, 129*, 1933-1936.

Williams, G.R., Robson, H., & Shalet, S.M. (1998.) Thyroid hormone actions on cartilage and bone: interactions with other hormones at the epiphyseal plate and effects on linear growth. *Journal of Endocrinology, 157*, 391-403.

Williams, M.E., Gervino, E.V., Rosa, R.M., Landsberg, L., Young, J.B., Silva P., & Epstein, F.H. (1985.) Catecholamine modulation of rapid potassium shifts during exercise. *New England Journal of Medicine, 312*, 823-827.

Williams, N.I., Young, J.C., McArthur, J.W., Bullen, B., Skrinar, G.S., & Turnbull, B. (1995.) Strenuous exercise with caloric restriction: effect on luteinizing hormone secretion. *Medicine and Science in Sports and Exercise, 27*, 1390-1398.

Williams, R.S., Caron, M.G., & Daniel, K. (1984.) Skeletal muscle beta-adrenergic receptors: variations due to fiber type and training. *American Journal of Physiology, 246*, E160-E167.

Williams, P., & Goldspink, G. (1973.) The effect of immobilization on the longitudinal growth of striated muscle fibers. *Journal of Anatomy, 116*, 45-55.

Williams, P.E., Watt, P., Bicik, P., & Goldspink, G. (1986.) Effect of stretch combined with electrical stimulation on the type of sarcomere produced at the ends of muscle fibres. *Experimental Neurology, 93*, 500-509.

Williamson, J.W., Nobrega, A.C., McColl, R., Matthews, D., Winchester, F., Friberg, L., and Mitchell, J.H. (1997.) Activation of the insular cortex during dynamic exercise in humans. *Journal of Physiology (London), 503*, 277-283.

Willich, S.N., Goldberg, R.J., Maclure, M., Periello, L., & Muller, J.E. (1992.)Increased onset of sudden cardiac death in the first 3 hours after awakening. *American Journal of Cardiology, 70*, 65-68.

Willich, S.N., Levy, D., Rocco, M.B., Tofler, G.B., Stone, P.H., & Muller, J.E. (1987.) Circadian variation in the incidence of sudden cardiac death in the Framingham Heart Study Population. *American Journal of Cardiology, 60*, 801-806.

Willich, S.N., Linderer, T., Wegscheider, K., Leizorovicz, A., Alamercery, I., Schroder, R., and the ISAM Study Group (1980.) Increased morning incidence of myocardial infarction in the ISAM Study: absence with prior beta-adrenergic blockade. *Circulation, 80*, 853-858.

Willich, S.N., Lowel, H., Lewis, M., Arntz, R., Baur, R., Winther, K., Keil, U.,Schroder, R., & Trimm Study Group (1991.) Association of wake-time and the onset of myocardial infarction: TRIMM (Triggers and Mechanisms of Myocardial Infarction): Pilot Study. *Circulation, 84* (Suppl. VI), VI-62-VI-67.

Willich, S.N., Maclure, M., Mittelman, M., Arntz, H., & Muller, J.E. (1993.) Sudden cardiac death: support for a role of triggering in causation. *Circulation, 87*, 1442-1450.

Wilmore, J.H. (1974.) Alterations in strength, body composition and anthropometry measurements consequent to a 10-week training program. *Medicine and Science in Sports, 6*, 133-138.

Wilmore, J.H., Parr, R.B., Girandola, R.N., Ward, P., Vodak, P., Barstow, T.J., et al. (1978.) Physiological alterations consequent to circuit weight training. *Medicine and Science in Sports, 10*, 79-84.

Wilmore, J.H., Wambsgans, K.C., & Brenner, M. (1982.) Is there energy conservation in amenorrheic compared to eumenorrheic distance runners? *Journal of Applied Physiology, 72*, 15-22.

Wilson, D.E., Flowers, C.M., Carlisle, S.I., & Udall, K.S. (1976.) Estrogen treatment and gonadal function in the regulation of lipoprotein lipase. *Atherosclerosis, 24*, 491-499.

Wilson, D.W., George, D., Mansel, R.E., Simpson, H.W., & Halberg, F. (1984.)Circadian breast skin temperature rhythms: overt and occult benign and occult primary malignant breast disease. *Chronobiology International, 1*, 167-172.

Wilson, E.E., Word, R.A., Byrd, W., & Carr, B.R. (1991.) Effect of superovulation with human menopausal gonadotropins on growth hormone levels in women. *Journal of Clinical Endocrinology and Metabolism, 73*, 511-555.

Wilson, J.D. (1988.) Androgen abuse by athletes. *Endocrine Reviews, 9*, 181-199.

Wilson, J.D., Foster, D.W., Kronenberg, H.M., & Larsen, P.R. (1998.) *Williams texbook of endocrinology.* Philadelphia: W.B. Saunders.

Wimalawansa, S.M., Chapa, M.T., Wei, J.N., Westlund, K., Quast, M.J., & Wimalawansa, S.J. (1999.) Reversal of weightlessness-induced musculoskeletal losses with androgens: quantification by MRI. *Journal of Applied Physiology, 86*, 1841-1846.

Winder WW. (2001.) Energy-sensing and signaling by AMP-activated protein kinase in skeletal muscle. *Journal of Applied Physiology, 91*, 1017-1028.

Winder, W.W., Braiden, R.W., Cartmill, D.C., Hutber, C.A., & Jones, J.P. (1993.) Effect of adrenodemedullation on decline in muscle malonyl-CoA during exercise. *Journal of Applied Physiology, 74*, 2548-2551.

Winder, W.W., Arogyasami, J., Elayan, I.M., & Cartmill, D. (1990.) Time course of exercise-induced decline in malonyl-CoA in different muscle types. *American Journal of Physiology, 259*, E266-E271.

Winder, W.W., Baldwin, K.M., & Holloszy, J.O. (1974.) Enzymes involved in ketone utilization in different types of muscle: adaptation to exercise. *European Journal of Biochemistry, 47*, 461-467.

Winder, W.W., Hickson, R.C., Hagberg, J.M., Ehsani, A.A., & McLane, J.A.(1979.) Training-induced changes in hormonal and metabolic responses to submaximal exercise. *Journal of Applied Physiology, 46*, 766-771.

Windle, R.J., Forsling, M.L., & Guzek, J.W. (1992.) Daily rhythms in the hormone content of the neurohypophysial system and release of oxytocin and vasopressin in the male rat: effect of constant light. *Journal of Endocrinology, 133*, 283-290.

Winget, C.M., Roshia, C.W., & Holley, D.C. (1985.) Circadian rhythms and human performance. *Medicine and Science in Sports and Exercise, 17*, 497-516.

Winget, C.M., Soliman, M.R.I., Holley, D.C., & Meyler, J.S.(1992.) Chronobiology of physical performance and sports medicine. In Y. Touitou & E. Haus (Eds.), *Biological rhythms in clinical and laboratory medicine* (pp. 229-242). Berlin: Springer-Verlag.

Winget, C.M., Vernikos-Danellis, J., Cronin, S.E., Leach, C.S., Rambaut, P.C., & Mack, P.B. (1972.) Circadian rhythm asynchrony in man during hypokinesis. *Journal of Applied Physiology, 33*, 640-643.

Winn H.N., Hess O., Goldstein I., Wackers F., & Hobbins J.C. (1994.) Fetal responses to maternal exercise: effect on fetal breathing and body movement. *American Journal of Perinatology, 11*, 263-266.

Winter, B., & Arnold, H.H. (2000.) Activated raf kinase inhibits muscle cell differentiation through a MEF2-dependent mechanism. *Journal of Cell Science, 113* (Pt 23) 4211-4220.

Winther, K., Hillegass, W., Tofler, G.H., Jimenez, A., Brezinski, D.A., Schafer, A.I., Loscalzo, J., Williams, G.H., & Muller, J.E. (1992.) Effects on platelet aggregation and fibrinolytic activity during upright posture and exercise in healthy men. *American Journal of Cardiology, 70*, 1051-1055.

Wirthenson, G., & Guder, WG. (1986.) Renal substrate metabolism. *Physiological Reviews, 66*, 469-497.

Witcher, D.R., Kovacs, R.J., Schulman, H., Cefali, D.C., & Jones, L.R. (1991.) Unique phosphorylation site on the cardiac ryanodine receptor regulates calcium channel activity. *Journal of Biological Chemistry, 266*, 11144-11152.

Witelson, S.F. (1991.) Neural sexual mosaicism: sexual differentiation of the human temporoparietal region for functional asymmetry. *Psychoneuroendocrinology, 16*, 131-153.

Wititsuwannakul, D., & Kim, K. (1977.) Mechanism of palmitoyl coenzyme A inhibition of liver glycogen synthase. Journal of Biological Chemistry, 252, 7812-7817.

Witte, E.A., & Marrocco, R.T. (1997.) Alteration of brain noradrenergic activity in rhesus monkeys affects the alerting component of covert orienting. *Psychopharmacology, 132*, 315-323.

Witte, K., Schnecko, A., Buijs, R.M., van der Vliet, J., Scalbert, E., Delagrange, P., Guardiola-Lemaitre, B., & Lemmer, B. (1998.) Effects of SCN lesions on circadian blood pressure rhythm in normotensive and transgenic hypertensive rats. *Chronobiology International, 15*, 135-145.

Wittert, G.A., Stewart, D.E., Graves, M.P., Ellis, M.J., Evans, M.J., Wells, J.E., Donald, R.A., & Espiner, E.A. (1991.) Plasma corticotrophin releasing factor and vasopressin responses to exercise in normal man. *Clinical Endocrinology, 35*, 311-317.

Wojdemann, M., Wettergren, A., Hartmann, B., & Holst, J.J. (1998.) Glucagon-like peptide-2 inhibits centrally induced antral motility in pigs. *Scandinavian Journal of Gastroenterology, 33*, 828-832.

Wojtaszewski, J.F., Jorgensen, S.B., Hellsten, Y., Hardie, D.G., & Richter, E.A. (2002.) Glycogen-dependent effects of 5-aminoimidazole-4-carboxamide (AICA)-riboside on AMP-activated protein kinase and glycogen synthase activities in rat skeletal muscle. *Diabetes, 51*, 284-292.

Wojtaszewski, J.F.P., Hansen, B.F., Gade, J., Kiens, B., Markuns, J.F., Goodyear, L.J., & Richter, E.A. (2000.) Insulin signaling and insulin sensitivity after exercise in human skeletal muscle. *Diabetes, 49*, 325-331.

Wolf, M., Ingbar, S.H., & Mose, A. (1989.) Thyroid hormone and growth hormone interact to regulate insulin-like growth factor-I messenger ribonucleic acid and circulating levels in the rat. *Endocrinology, 125*, 2905-29

Wolfe L.A., Preston R.J., Burggraf G.W., & McGrath M.J. (1999.) Effects of pregnancy and chronic exercise on maternal cardiac structure and function. *Canadian Journal of Physiology and Pharmacology, 77*, 909-917.

Wolfe, L.A.,and Mottola, M.F. (1993.) Aerobic exercise in pregnancy: an update. *Canadian Journal of Applied Physiology, 18*, 119-147.

Wolfe, R.R., Klein, S., Carraro, F., & Weber, J.M. (1990.) Role of triglyceride-fatty acid cycling in controlling fat metabolism in humans during and after exercise. *American Journal of Physiology, 258*, E382-E389

Wolff, J. (1986.) *The law of bone remodeling.* Berlin: Springer Verlag. Original: (1892.) *Das Gesetz der Transformazion der Knochen.* Berlin: Verlag von August Hirschwald.

Wollert, K.C., Taga,T., Saito, M., et al. (1996.) Cardiotrophin-1 activates a distinct form of cardiac muscle dell hypertrophy: assembly of sarcomeric units in series via gp130/leukemia inhibitory factor receptor-dependent pathways. *Journal of Biological Chemistry, 271*, 9535-9545.

Woltman, T., & Reidelberger, R. (1996.) Role of cholecystokinin in the anorexia produced by duodenal delivery of glucose in rats. *American Journal of Physiology, 271*, R1521.

Woltman, T., Castellanos, D., & Reidelberger, R. (1995.) Role of cholecystokinin in the anorexia produced by duodenal delivery of oleic acid in rats. *American Journal of Physiology, 269*, R1420.

Wong, T.S., & Booth, F.W. (1990a.) Protein metabolism in rat tibilais anterior muscle after stimulated chronic eccentric exercise. *Journal of Applied Physiology, 69*, 1718-1724.

Wong, T.S., & Booth, F.W. (1990b.) Skeletal muscle enlargement with weight-lifting exercise by rats. *Journal of Applied Physiology, 69*, 950-954.

Woo, R., & Pi-Sunyer, F.X. (1985.) Effect of increased physical activity on voluntary intake in lean women. *Metabolism: Clinical and Experimental, 34*, 836-841.

Woo, R., Garrow, J.S., & Pi-Sunyer, F.X. (1982a.) Effect of exercise on spontaneous calorie intake in obesity. *American Journal of Clinical Nutrition, 36*, 470-477.

Woo, R., Garrow, J.S., & Pi-Sunyer, X. (1982b.) Voluntary food intake during prolonged exercise in obese women. *The American Journal of Clinical Nutrition, 36*, 478-484.

Woods, S.C., Chavez, M., & Park, C.R. (1996.) The evaluation of insulin as a metabolic signal influencing behavior via the brain. *Neuroscience and Biobehavioral Review, 20*, 139-144.

Wojdemann, M., Wettergren, A., Hatrmann, B., & Holst, J.J. (1998.) Glucagon-like peptide-2 inhibits centrally induced antral motility in pigs. *Scandinavian Journal of Gastroeneterology, 33*, 828-832.

Wright, J.E., Vogel, J.A., Sampson, J.B., Knapik, J.J., Patton, J.F., & Daniels, W.L. (1983.) Effects of travel across time zones (jet-lag) on exercise capacity and performance. *Aviation Space and Environmental Medicine, 54*, 132-137.

Wroe, S.J., Sandercock, P., Bamford, J., Dennis, M., Slattery, J., & Warlow, C. (1992.) Diurnal variation in incidence of stroke: Oxfordshire community stroke project. *British Medical Journal, 304*, 155-157.

Wu, F.C., Brown, D.C., Butler, G.E., Stirling, H.F., & Kelnar, C.J. (1993.) Early morning plasma testosterone is an accurate predictor of imminent pubertal development in prepubertal boys. *Journal of Clinical Endocrinology and Metabolism, 76*, 26-31.

Wu, Z., Woodring, P.J., Bhakta, K.S., Tamura, K., Wen, F., Feramisco, J.R., Karin, M., Wang, J.Y., & Puri, P.L. (2000.) P38 and extracellular signal-regulated kinases regulate the myogenic program at multiple steps. *Molecular and Cellular Biology, 20*, 3951-3964.

Wurtman, R.J. (1985.) Melatonin as a hormone in humans: a history. *The Yale Journal of Biology and Medicine, 58*, 547-552.

Wurtman, R.J., & Wurtman, J.J. (1996.) Brain serotonin, carbohydrate-craving, obesity and depression. *Advances in Experimental Medicine and Biology, 398*, 35-41.

Wyndham, C.H., Morrison, J.F., Williams, C.G., Bredell, G.A.G., Peter, J., von Rahden, M.J.E., Holdsworth, L.D., van Graan, C.H., van Rensburg, A.J., & Munro, A. (1964.) Physiological reactions to cold of Caucasian females. *Journal of Applied Physiology, 19*, 877-880.

Wyse, J.P., Mercer, T.H., & Gleeson, N.P. (1994.) Time-of-day dependence of isokinetic leg strength and associated interday variability. *British Journal of Sports Medicine, 28*, 167-170.

Xi, X., Han, J., & Zhang, J.Z. (2001.) Stimulation of glucose transport by AMP-activated protein kinase via activation of p38 mitogen-activated protein kinase. *Journal of Biological Chemistry, 276*, 41029-41034.

Xu, Q., & Wu, Z. (2000.) The insulin-like growth factor-phosphatidylinositol 3-kinase-Akt signaling pathway regulates myogenin expression in normal myogenic cells but not in rhabdomyosarcoma-derived RD cells. *Journal of Biological Chemistry, 275*, 36750-36757.

Yagami, T. (1995.) Differential coupling of glucagon and beta-adrenergic receptors with the small and large forms of the stimulatory G protein. *Molecular Pharmacology, 48*, 849-854.

Yagita, K., Tamanini, F., Yasuda, M., Hoeijmakers, J.H., van der Horst, G.T., & Okamura, H. (2002.) Nucleocytoplasmic shuttling and mCRY-dependent inhibition of ubiquitylation of the mPER2 clock protein. *EMBO Journal, 21*, 1301-1314.

Yakar, S., Liu, J.L., Fernandez, A.M., Wu, Y., Schally, A.V., Frystyk, J., Cheranusek, S.D., Mejia, W., & Le Roith, D. (2001.) Liver-specific igf-1 gene deletion leads to muscle insulin insensitivity. *Diabetes, 50*, 1110-1118.

Yamada, T., Nakao, K., Itoh, H., Shirakami, G., Sugawara, A., Saito, Y., Mukoyama, M., Arai, H., Hosoda, K., & Shiono, S. (1989.) Effects of naloxone on vasopressin secretion in conscious rats: evidence for inhibitory role of endogenous opioids in vasopressin secretion. *Endocrinology, 125*, 785-790.

Yamamoto, Y., & Hughson, R.L. (1991.) Coarse graining spectral analysis: new method for studying heart rate variability. *Journal of Applied Physiology, 71*, 1143-1150.

Yamamoto, Y., & Hughson, R.L. (1993.) Extracting fractal components from time series. *Physica D, 68*, 250-264.

Yamamoto, Y., Hughson, R.L., & Peterson, J.C. (1991.) Autonomic control of heart rate during exercise studied by a heart rate variability spectral analysis. *Journal of Applied Physiology, 71*, 1136-1142.

Yamane, A., Takahashi, K., Mayo, M., Vo, H., Shum, L., Zeichner-David, M., & Slavkin, H.C. (1998.) Induced expression of myoD, myogenin and desmin during myoblast differentiation in embryonic mouse tongue development. *Archives of Oral Biology, 43*, 407-416.

Yamashita, H., Kannan, H., Inenaga, K., & Koizumi, K. (1984.) The role of cardiovascular and muscle afferent systems in control of body water balance. *Journal of the Autonomic Nervous System, 10*, 305-316.

Yamazaki, T., Komuro, I., & Yazaki, Y. (1996.) Molecular aspects of mechanical stress-induced cardiac hypertrophy. *Molecular and Cellular Biochemistry, 163-164*, 197-201.

Yanagisawa, M., Kurihara, H., Kimura, S., Tomobe, Y., Kobayashi, M., Mitsui,Y., Yazaki, Y., Goto, K., & Masaki, T. (1988.) A novel potent vasoconstrictor peptide produced by vascular endothelial cells. *Nature, 332*, 411-415.

Yang, R.H., Jin, H., Wyss, J.M., Chen, Y.F., & Oparil, S. (1990.) Blockade of endogenous anterior hypothalamic atrial natriuretic peptide with monoclonal antibody lowers blood pressure in spontaneously hypertensive rats. *Journal of Clinical Investigation, 86*, 1985-1990.

Yang, R.H., Jin, H., Wyss, J.M., Chen, Y.F., & Oparil, S. (1992.) Pressor effect of blocking atrial natriuretic peptide in nucleus tracti solitarii. *Hypertension 19*, 198-205.

Yang, S., Bjorntorp, B., Liu, X., & Eden, S. (1996a.) Growth hormone treatment of hypophysectomized rats increases catecholamine-induced lipolysis and the number of beta-adrenergic receptors in adipocytes: no differences in the effects of growth hormone on different fat depots. *Obesity Research, 4*, 471-478.

Yang, S.Y., Alnaqeeb, M., Simpson, H., & Goldspink, G. (1996b.) Molecular cloning, regulation and mRNA processing of an insulin-like growth factor I which is expressed in skeletal muscle induced to undergo rapid growth. *Journal of Muscle Research and Cellular Motility, 17*, 487-497.

Yang, S.Y., Alnaqeeb, M., Simpson, H., & Goldspink, G. (1997.) Changes in muscle fiber type, muscle mass and IGF-I gene expression in rabbit skeletal muscle subject to stretch. *Journal of Anatomy, 190*, 613-622.

Yang, Y.T., & McElligott, M.A. (1989.) Multiple actions of β-adrenergic agonists on skeletal muscle and adipose tissue. *Biochemical Journal, 261*, 1-10.

Yaoita, H., Sato, E., Kawaguchi, M., Saito, T., Maehara, K., & Maruyama, Y. (1994.) Nonadrenergic, noncholinergic neurons regulate basal coronary flow via release of capsaicin-sensitive neuropeptides in the rat heart. *Circulation Research, 75*, 780-788.

Yarasheski, K.E. (1994.) Growth hormone effects on metabolism, body composition, muscle mass, and strength. *Exercise and Sports Science Reviews, 22*, 285-312.

Yarasheski, K.E., Zachwieja, J.J., Campbell, J.A., & Bier, D.M. (1995.) Effect of growth hormone and resistance exercise on muscle growth and strength in older men. *American Journal of Physiology, 268*, E268-276.

Yasuda, I. (1954.) Piezoelectric activity of bone. *Journal of Japanese Orthopedic Surgery Society, 28*, 267.

Yasuda, Y., Itoh, T., Miyamura, M., & Nishino, H. (1997.) Comparison of exhaled nitric oxide and cardiorespiratory indices between nasal and oral breathing during submaximal exercise in humans. *Japanese Journal of Physiology, 47*, 465-470.

Yasue, H., Omote, S., Takizawa, A., Nagao, M., Miwa, K., & Tanaka, S. (1979.) Circadian variation of exercise capacity in patients with Prinzmetal's variant angina: Role of exercise-induced coronary arterial spasm. *Circulation, 59*, 938-946.

Yates, A., Leehey, K., & Shisslak, C.M. (1983.) Running—an analogue of anorexia? *The New England Journal of Medicine, 308*, 251-255.

Yen, S.S.C. (1983.) Clinical applications of gonadotropin releasing hormone and gonadotropin releasing-hormone analogs. *Fertility and Sterility, 39*, 257-266.

Yen, S.S.C. (1998.) Effects of lifestyle and body composition on the ovary. *Endocrinology and Metabolism Clinics of North America, 27*, 915-926.

Yen, S.S.C. (1999a.) Neuroendocrinology of reproduction. In S.S.C. Yen, R.B. Jaffe, & R.L.Barbieri (Eds.), *Reproductive endocrinology: physiology, pathophysiology, and clinical management* (pp. 30-80). New York: W.B. Saunders.

Yen, S.S.C. (1999b.) The human menstrual cycle: neuroendocrine regulation. In: S.S.C. Yen, R.B. Jaffe, & R.L.Barbieri (Eds.), *Reproductive endocrinology: physiology, pathophysiology, and clinical management* (pp. 191-217). New York: W.B. Saunders.

Yen, S.S.C., Apoter, D., Butzow, T., & Laughlin, G.A. (1993.) Gonadotropin releasing hormone pulse generator activity before and during sexual maturation in girls: new insights. *Human Reproduction, 8* (Suppl. 2), 66-71.

Yip, R.G., & Goodman, H.M. (1999.) Growth hormone and dexamethasone stimulate lipolysis and activate adenylyl cyclase in rat adipocytes by selectively shifting Gi alpha2 to lower density membrane fractions. *Endocrinology, 140*, 1219-1227.

Young N., Formica, C., Szmuckler, G., & Seeman, E. (1994.) Bone density at weight-bearing and nonweight bearing sites in ballet dancers: the effects of exercise, hypogonadism, and body weight. *Journal of Clinical Endocrinology and Metabolism, 78*, 449-454.

Young, D.A., Wallberg-Henriksson, H., Cranshaw, J., Chen, J., & Holloszy, J.O. (1985.) Effect of catecholamines on glucose uptake and glycogenolysis in rat skeletal muscle. *American Journal of Physiology, 248*, C406-C409.

Young, J.B., & Landsberg, L. (1997.) Suppression of sympathetic nervous system during fasting. *Obesity Research, 5*, 646-649.

Young, J.B., & Landsberg, L. (1998.) Catecholamines and the adrenal medulla. In J.D. Wilson, D.W. Foster, H.M. Kronenberg, & P.R. Larsen (Eds.), *Williams textbook of endocrinology*, 9th ed. (pp. 665-725). Philadelphia: W.B. Saunders Company.

Young, J.B., & Landsberg, L. (1979a.) Effect of diet and cold exposure on norepinephrine turnover in pancreas and liver. *American Journal of Physiology, 236*, E524-E533.

Young, J.B., & Landsberg, L. (1997.) Suppression of sympathetic nervous system during fasting. *Obesity Research, 5*, 646-649.

Young, J.B., & Landsberg, L. (1979b.) Sympathoadrenal activity in fasting pregnant rats: dissociation of adrenal medullary and sympathetic nervous system responses. *Journal of Clinical Investigation, 64*, 109-116.

Young, J.B., & Cohen, W.R. (1991.) Regional stimulation of sympathetic nervous system activity during rat pregnancy: potential role of progesterone. In M. Yoshikawa, M. Uono, H. Tanabe, & S. Ishikawa (Eds.), *New trends in autonomic nervous system research: Basic and clinical integration.* Amsterdam: Elsevier Science, pp. 468-472.

Young, J.B., Rosa, R.M., & Landsberg, L. (1984.) Dissociation of sympathetic nervous system and adrenal medullary responses. *Americal Journal of Physiology, 247*, E35-E40.

Young, J.C., & Treadway, J.L. (1992.) The effect of prior exercise on oral glucose tolerance in late gestational women. *European Journal of Applied Physiology and Occupational Physiology, 64*, 430-433.

Young, M.A., Meaden, P.M., Fogg, L.F., Cherin, E.A., & Eastman, C.L. (1997.) Which environmental variables are related to the onset of seasonal affective disorder? *Journal of Abnormal Psychology, 106*, 554-562.

Young, M.E., & Leighton B. (1998.) Evidence for altered sensitivity of the nitric oxide/cGMP signalling cascade in insulin-resistant skeletal muscle. *Biochemical Journal, 329*, 73-79.

Young, V.R., Steffe, W.P., Pencharz, P.B., Winterer, J.C., & Scrimshaw, N. S. (1975.) Total human body protein synthesis in relation to protein requirements at various ages. *Nature, 253*, 192-194.

Youngstedt, S.D., O'Connor, P.J., & Dishman, R.K. (1997.) The effects of acute exercise on sleep: a quantitative synthesis. *Sleep, 20*, 203-214.

Yox, D.P., Stokesberry, H., & Ritter, R.C. (1991.) Vagotomy attenuates suppression of sham feeding induced by intestinal nutrients. *American Journal of Physiology, 260*, R503-R508.

Yu, W.-H.A. (1989.) Administration of testosterone attenuates neuronal loss following axotomy in the brainstem motor nuclei of female rat. *Annals of the New York Academy of Sciences, 9*, 3908-3914.

Zacharia, M., Rocco, S., Noventa, D., Varnier, M., & Opocher, G. (1998.) *Journal of Clinical Endocrinology and Metabolism, 83*, 570-574.

Zachman, M. (1992.) Interrelations between growth hormone and sex hormones: physiological and therapeutic consequences. *Hormone Research, 38* (Suppl. 1), 1-8.

Zachwieja, J.J., Smith, S.R., Sinha-Hikim, I., Gonzalez-Cadavid, N., & Bhasin, S. (1999.) Plasma myostatin-immunoreactive protein is increased after prolonged bed rest with low-dose T$_3$ administration. *Journal of Gravitational Physiology, 6*, 11-15.

Zador, E., Dux, L., & Wyytack, F. (1999.) Prolonged passive stretch of rat soleus muscle provokes an increase in the mRNA levels of the muscle regulatory factors distributed along the entire length of the fibres. *Journal of Muscle Research and Cell Motility, 20*, 395-402.

Zahrt, J., Taylor, J.R., Mathew, R.G., & Arnsten, A.F. (1997.) Supranormal stimulation of D$_1$ dopamine receptors in the rodent prefrontal cortex impairs spatial working memory performance. *Journal of Neuroscience, 17*, 8528-8535.

Zambraski, E.J., Tucker, M.S., Lakas, C.S., Grassl, S.M., & Scanes, C.G. (1984.) Mechanism of renin release in exercising dog. *American Journal of Physiology, 246*, E71-E76.

Zehender, M., Meinertz, T., Hohnloser, S., Geibel, A., Gerisch, U., Olschewski, M., et al. (1992). Prevalence of circadian variations and spontaneous variability of cardiac disorders and ECG changes suggestive of myocardial ischemia in systemic ventricular tachycardia. *Circulation, 85*, 1808-1815.

Zeitler, P., Argente, J., Chowen-Breed, J.A. , Clifton, D.K., & Steiner, R.A. (1990.) Growth hormone releasing hormone messenger ribonucleic acid in the hypothalamus of the adult male rat is increased by testosterone. *Endocrinology, 127*, 1362-1368.

Zeitzer, J.M., Ayas, N.T., Shea, S.A., Brown, R., & Czeisler, C.A. (2000.) Absence of detectable melatonin and preservation of cortisol and thyrotropin rhythms in tetraplegia. *Journal of Clinical Endocrinology and Metabolism, 85*, 2189-2196.

Zeman, R.J., Hirschman, A., Hirschman, M.L., Guo, G., & Etlinger, J.D. (1991.) Clenbuterol, a beta$_2$-receptor agonist, reduces net bone loss in denervated hindlimbs. *American Journal of Physiology, 261*, E285-E289.

Zeng, G., Nystrom, F.H., Ravichandran, L.V., Cong, L.-N., Kirby, M., Mostowski, H., & Quon, M.J. (2000.) Roles for insulin receptor, PI3-kinase, and Akt in insulin-signaling pathways related to production of nitric oxide in human vascular endothelial cells. *Circulation, 101*, 1539-1545.

Zhang, J.M., Wei, Q., Zhao, X., & Paterson, B.M. (1999a.) Coupling of the cell cycle and myogenesis through the cyclin D1-dependent interaction of MyoD with cdk4. *EMBO Journal, 18*, 926-933.

Zhang, X., Xie, Y.W., Nasjletti, A., Xu, X., Wolin, M.S., & Hintze, T.H. (1997.) ACE inhibitors promote nitric oxide accumulation to modulate myocardial oxygen consumption. *Circulation, 95*, 176-182.

Zhang, Y., Proenca, R., Maffei, M., Barone, M., Leopold, L., & Friedman, J.M. (1994.) Positional cloning of the mouse obese gene and its human homologue. *Nature, 372*, 425-432.

Zhang Z., Xiao, Z., & Diamond, S.L. (1999b.) Shear stress induction of C-type natriuretic peptide (CNP) in endothelial cells is independent of NO autocrine signalling. *Annals of Biomedical Engineering, 27*, 419-426.

Zhou, G., Myers, R., Li, Y., Chen, Y., Shen, X., Fenyk-Melody, J., et al. (2001.) Role of AMP-activated protein kinase in mechanism of metformin action. *Journal of Clinical Investigation, 108*, 1167-1174.

Zhou, J., & Olson, E.N. (1994.) Dimerization through the helix-loop-helix motif enhances phosphorylation of the transcription activation domains of myogenin. *Molecular and Cellular Biology, 14*, 6232-6243.

Zhou, Q., & Dohm, G.L. (1997.) Treadmill running increases phosphatidyl inositol 3-kinase activity in rat skeletal muscle. *Biochemical and Biophysical Research Communications, 236*, 647-650.

Zhu, J.L., & Leadley, R.J., Jr. (1995.) Contribution of cardiac and arterial baroreceptors to enhanced vasopressin release during hemorrhage with autonomic blockade. *Proceedings of the Society for Experimental Biology and Medicine, 208*, 361-369.

Zhu, Z., & Miller, J.B. (1997.) MRF4 can substitute for myogenin during early stages of myogenesis. *Developmental Dynamics, 209*, 233-241.

Zimmer, H.G., Kolbeck-Ruhmkorff, C., & Zierhut, W. (1995.) Cardiac hypertrophy induced by alpha- and beta-adrenergic receptor stimulation. *Cardioscience, 6*, 47-57.

Zinzen, E., Clarijs, J.P., Cabri, J., Vanderstappen, D., & Van den Berg, T.J.(1994.) The influence of triazolam and flunitrazepam on isokinetic and isometric muscle performance. *Ergonomics, 37*, 69-77.

Zir, L.M., Smith, R.H., & Parker, D.C. (1971.) Human growth hormone release in sleep: effect of daytime exercise. *Journal of Clinical Endocrinology, 32*, 662-665.

Zubiran, S., & Gomez-Mont, F. (1952.) Endocrine disturbance in chronic human malnutrition. *Vitamins and Hormones, 11*, 97-102.

Zucker, I., & Stephan, F.K. (1973.) Light-dark rhythms in hamster eating, drinking and locomotor behaviors. *Physiology and Behavior, 11*, 239-250.

Zwiren, L.D., Cureton, K.J., & Hutchinson, P. (1983.) Comparison of circulatory responses to submaximal exercise in equally trained men and women. *International Journal of Sports Medicine, 4*, 255-259.

Zylstra, S., Hopkins, A., Erk, M., Hreshchyshyn,M.M., & Anbar, M. (1989.) Effect of physical activity on lumbar spine and femoral neck densities. *International Journal of Sports Medicine, 10*, 181-186.

Index

The letters *f* and *t* after a page number indicate figure and table, respectively.

About the Author

Katarina Borer, PhD, is a professor of movement science and researcher at the University of Michigan. She studied the neuroendocrine controls of growth in an animal model from 1974 to 1994, and she developed and validated radioimmunoassays for hamster growth hormone and prolactin in the 1980s. Since 1995, Borer has been studying the effects of training intensity on hormonal and cardiovascular adaptations in postmenopausal women.

Dr. Borer is a 2002 member of the NIA Special Emphasis Panel (Older Americans Independence Centers). In 1990, she was the recipient of the Fulbright Hayes Research Scholarship to Sweden. From 1993 to 1995, she was a member of the Council for International Exchange of Scholars, Area Committee for Scandinavia. She also is on the editorial board of *Kinesiology*, the international journal of fundamental and applied kinesiology.